Physical Chemistry of Ionic Materials
Ions and Electrons in Solids

Joachim Maier
*Max-Planck-Institut für Festkörperforschung, Heisenbergstraße 1,
70569 Stuttgart, Germany*

John Wiley & Sons, Ltd

Telephone (+44) 1243 779777

Email (for orders and customer service enquiries): cs-books@wiley.co.uk
Visit our Home Page on www.wileyeurope.com or www.wiley.com

Other Wiley Editorial Offices

John Wiley & Sons Inc., 111 River Street, Hoboken, NJ 07030, USA

Jossey-Bass, 989 Market Street, San Francisco, CA 94103-1741, USA

Wiley-VCH Verlag GmbH, Boschstr. 12, D-69469 Weinheim, Germany

John Wiley & Sons Australia Ltd, 33 Park Road, Milton, Queensland 4064, Australia

John Wiley & Sons (Asia) Pte Ltd, 2 Clementi Loop #02-01, Jin Xing Distripark, Singapore 129809

John Wiley & Sons Canada Ltd, 22 Worcester Road, Etobicoke, Ontario, Canada M9W 1L1

Wiley also publishes its books in a variety of electronic formats. Some content that appears in print may not be available in electronic books.

British Library Cataloguing in Publication Data

A catalogue record for this book is available from the British Library

ISBN 0-470-99991-1 (HB)
0-470-87076-1 (PB)

Typeset in LaTeX using author files
Printed and bound in Great Britain by Antony Rowe Ltd, Chippenham, Wiltshire
This book is printed on acid-free paper responsibly manufactured from sustainable forestry in which at least two trees are planted for each one used for paper production.

Preface

The book that you are about to read is, in a broad sense, concerned with the physical chemistry of solids. More specifically, it deals with the ionic and electronic charge carriers in ionic solids. The latter species are the major players in the game when one attempts a detailed understanding or deliberate tuning of kinetic properties. The charge carriers that we refer to are not necessarily identical with the charged particles that constitute the solid, but rather with the effective particles that transport charge, i.e. in the case of ionic crystals the ionic point defects, in addition to excess electrons and holes. These ionic and electronic charge carriers constitute the redox and acid–base chemistry in the same way as is the case for aqueous solutions, they permit charge and matter transport to occur, and are also reactive centres in the sense of chemical kinetics. This explains the central role of defect chemistry in this book.
The more classical introductory chapters on chemical bonding, phonons, and thermodynamics of the perfect solid may, on one hand, be considered as preparation for the key chapters which deal with thermodynamics of the real solid, as well as with kinetics and electrochemistry — both being unthinkable without the existence of defects; on the other hand, they provide the complement necessary for the book to serve as a textbook of physical chemistry of solids. (In fact the different chapters correspond to classical fields of physical chemistry but referred to the solid state.)
The structure of the book is expected to be helpful in view of the heterogeneity of the potential readership: This addresses chemists who traditionally consider solids from a static, structural point of view and often ignore the "internal life" enabled by defect chemistry, physicists who traditionally do not take pertinent account of composition as a state parameter, and materials scientists who traditionally concentrate on materials properties and may not adequately appreciate the basic role of electrochemistry.
Of course the book cannot fully cover the materials space or the world of properties. If the reader is a chemist, he or she may miss special chapters on covalent and disordered solids (e.g. polymers); the physicist will certainly find electronic properties under–represented (e.g. metals), and the materials scientist may have expected a detailed consideration of mechanical and thermal properties. Nonetheless the author is convinced — and this is based on lectures on Physical Chemistry and Materials Science given to very different audiences in Cambridge (USA), Tübingen and Stuttgart (Germany) and Graz (Austria) — that he made a germane selection to highlight the physical chemistry of charge carriers in solids. A certain preference

1 Introduction

1.1 Motivation

It may seem odd to ask the reader in the first sentence of the book he or she has just opened to put it down for a moment (naturally with the intention of picking it up and reading it again with greater motivation). Consider, however, your environment objectively for a moment. The bulk of it is (as we ourselves are to a large degree) made up of solid matter. This does not just apply to the materials, from which the house in which you live is built or the chair in which you may be seated is made, it also applies to the many technical products which make your life easier, and in particular to the key components that are hidden from your eyes, such as the silicon chip in the television set, the electrodes in the radio battery, and the oxide ceramics in the oxygen sensors of modern automobiles. It is the rigidity of solids which endows them with characteristic, advantageous properties: The enduring structure of our world is inconceivable without solid matter, with its low diffusion coefficients at least for one component (the reader may like to consider for a moment his or her surroundings being in spatial equilibrium, i.e. with all diffusion barriers having been removed). In addition and beyond the mere mechanical functionality, solids offer the possibility of subtly and reproducibly tailoring electromagnetic, chemical and thermal functions.

The proportion of functional materials and, in particular, electrical ceramics in daily life is going to increase enormously in the future: Chemical, optical or acoustic sensors will analyse the environment for us, actuators will help us influence it. More or less autonomous systems, controlled by computers and powered by an autarchic energy supply (battery) or by an "electrochemical metabolism" (fuel cell) are by no means visions for the distant future. Wherever it is possible, attempts are being made to replace fluid systems by solid ones, for instance, liquid electrolytes by solid ion conductors. In short: The importance of (inorganic or organic) solids can hardly be overestimated (even if we ignore the crowning functionality of biopolymers, as (almost)[1] done in this book). Furthermore, solid state reactions were not only of importance for and during the creation of our planet, they also constitute a large portion of processes taking place, nowadays, in nature and in the laboratory.

Perhaps you are a chemistry student in the midst of your degree course or a chemistry graduate already with a complete overview of the syllabus. You will then certainly agree that the greater part of a chemist's education is concerned with liquids and, in particular, with water and aqueous solutions. Solids, when they are considered, are almost always considered from a naive "outer" point of view, i.e. as chemically invariant entities: Interest is chiefly concerned with the perfect structure and chemical bonding; in aqueous solutions it either precipitates or dissolves. Only the surface is considered as a site of chemical reactions. The concept of a solid having an "internal

[1]See Section 6.10 of Chapter 6 for systems far from equilibrium.

Physical Chemistry of Ionic Materials J. Maier

ISBN: 0-471-99991-1 (HB); 0-470-87076-1 (PB)

chemical life", which makes it possible for us to tailor the properties of a solid, in the same manner that we can those properties of aqueous solutions, sounds — even now — somewhat adventurous.
On the other hand, solid state physicists have influenced the properties of semiconductors such as silicon, germanium or gallium arsenide by defined doping in a very subtle way. If the reader is a physicist, I believe he or she would agree that the role of composition as a parameter is not sufficiently appreciated in physics. Even though internal chemical equilibria are sometimes considered and doping effects are generally taken into account, concentration is still too strongly focused on singular compositions and electronic carriers. In fact, a large number of functional materials are based on binary or multinary compounds, for which stoichiometric effects play an enormous role.
Lastly this text is addressed to materials scientists for whom the mechanical properties frequently and traditionally are of prime interest. Electrochemical aspects are generally not sufficiently considered with respect to their importance for the preparation and durability of the material and optimization of its function. Thus, the fields of ceramics, in general, and electroceramics, in particular, are addressed.

The chemistry and physics of defects play a key role in the following text [1,2]. After all, in the classical examples of water in chemistry and silicon in physics it is not so much the knowledge of the structure or of the chemical bonding that has made it possible to carry out subtle and controllable tuning of properties, but rather the phenomenological knowledge of the nature of relevant particles, such as H_3O^+ ions, OH^- ions or foreign ions in water that determine its acid–base and redox chemistry. In the case of silicon the relevant particles are conduction electrons and electron holes, which, on account of their properties, determine the (redox) chemistry and the electronic properties.
Focusing on such relevant particles leads to the generalized concept of defect chemistry that permits the treatment of internal chemical processes within the solid state (in this context Fig. 1.1 is illustrative). In processes, in which the structure of the

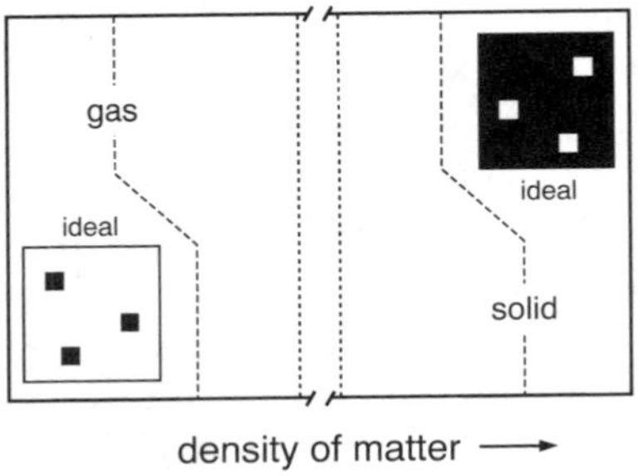

Fig. 1.1: In the same way as the treatment of ideal gases is simple — since the particles are dilute and uncorrelated (l.h.s.) —, the treatment of the solid state becomes equally simple from the viewpoint of the (dilute) defects (r.h.s.). (The portion of matter increases from the left to the right, while the portion of vacancies correspondingly decreases.)

phase does not alter, the perfect state can be regarded as invariant and all the chemical occurrences can then be reduced to the behaviour of the defects, that is, the deviations from the perfect state. The foundation stone of defect chemistry was laid by Frenkel, Schottky and Wagner [1,2] as early as the 1930s; there is an extensive technical literature covering the field [3–14], but in chemistry and physics it has not yet become an adequate and generally accepted component of our training. In this sense this text is intended to motivate the chemist to deal with the internal chemistry of solid bodies. I hope that the effort will be rewarded with a density of "aha experiences" that will be adequate to compensate for the trouble caused by the physical language which is sometimes necessary. The physicist should be stimulated by the text to examine the internal equilibria of solid materials, changes in their composition and, in particular, the properties of more complex materials. The motivation here ought to be the fact that the formalism of defect chemistry is largely material independent, at least as long as the defect concentrations are sufficiently low, and that it offers a universal phenomenological description in such cases. Finally the text is intended to help the materials scientist to optimize the functional properties of materials, but also to understand the preparation and degradation of structural materials.

If this attempt at motivation is an "attack on open doors", then the sentences I have written may at least act as a guide for the path ahead.

The text concentrates on ionic materials and on electrical and electrochemical properties in order to keep the contents within bounds. On the whole, we will refer to a "mixed conductor", for which ion and electron transport are both important and with regard to which the pure electronic conductor and the pure ionic conductor represent special cases. We will specifically address material transport with regard to its significance for electrochemistry and reaction kinetics. Whenever necessary, indications of the generality of the concepts are interspersed. In order to make the treatment reasonably complete, references are given whenever a detailed consideration is beyond the scope of the book.

We start with an extensive introduction to the perfect solid, its bonding and its vibrational properties, knowledge of which is necessary for understanding the physical chemistry of the processes involved. In order not to lose sight of the purpose of the book these sections have been kept as simple as possible (but as precise as necessary). The same applies to the general thermodynamic and kinetic sections, which also serve to introduce the formal aspects. Nevertheless, in view of the heterogeneity of the potential readership, this detailed mode of presentation has been chosen deliberately in order to be able to assume a uniform degree of knowledge when discussing defect chemistry. Some material may be repeated later in the text and this is intended to ensure that some chapters can be omitted by the advanced readers without loss of internal consistency.

The text will have fulfilled its purpose in an ideal manner, if it not only conveys to the reader the elegance and power of the defect concept, when it not only puts him or her in the position of being able to recognize the common aspects of different properties and processes such as doping and neighbouring phase effects, ionic and electronic conductivity, passivation and corrosion of metals, diffusion and reaction processes, synthesis kinetics and sintering kinetics in solids, electrode reactions and catalysis, sensor processes and battery processes; but also puts the reader in the position to optimize the solid state "strategically" at the "writing desk" in those situations in which the desired parameters are already known.

1.2 The defect concept: Point defects as the main actors

As already mentioned, phenomenological understanding of condensed phases with regard to the tunability of chemical and electrical properties implies knowledge of the defects as the relevant particles and their interactions, rather than (or at least in addition to) knowledge of the structure of the perfect state. This is known to be the key to phenomenological understanding of the aqueous phase and to control its chemical and electrical properties. In pure water these defects or "chemical excitations" are H_3O^+ and OH^- ions. Let us consider the first row in Fig. 1.2. As shown, it is advantageous, in a purely phenomenological sense, to substract from

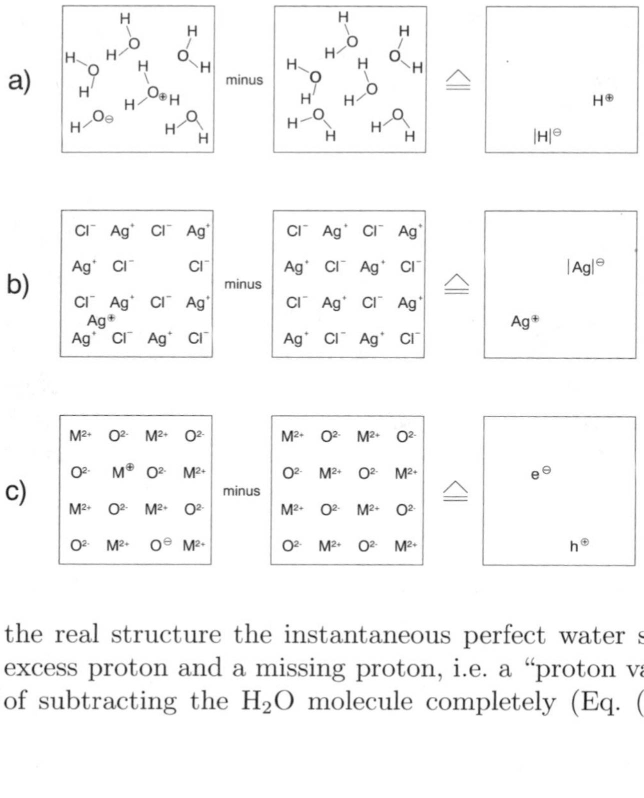

Fig. 1.2: a) If the basic compositionally unperturbed structure (chemical ground structure) is subtracted from the real structure, the point defects shown on the right remain. Naturally each is surrounded by a distorted region (effective radius of the point defect) which affects at least the immediate neighbourhood. In the case of fluid phases (see above) this procedure can only be regarded as an instantaneous picture. Owing to the absence of defined sites no distinction is made between various types of defect reactions as is done in the solid state. b) Frenkel disorder is sketched in the second row. c) The third row shows the case of purely electronic disorder whereby localized charge carriers are assumed for the sake of clarity [14].

the real structure the instantaneous perfect water structure. We are left with an excess proton and a missing proton, i.e. a "proton vacancy". This is also the result of subtracting the H_2O molecule completely (Eq. (1.1c)) from the autoprotolysis reaction[2] according to

$$2H_2O \rightleftharpoons H_3O^+ + OH^- \quad |{-H_2O} \tag{1.1a}$$
$$H_2O \rightleftharpoons H^+ + OH^- \quad |{-H_2O} \tag{1.1b}$$
$$\mathrm{Nil} \rightleftharpoons H^+ + |H|^- . \tag{1.1c}$$

Here $|H|^-$ denotes a proton vacancy[3].

Let us now consider disorder in a crystalline phase, to be specific, in solid AgCl. In this case some silver ions have left their regular sites[4], and hence, left vacancies behind. As one can see (Fig. 1.2b), there is a very close analogy to Eq. (1.1). Even the nature of the driving force for the internal dissociation, namely the gain in entropy of configuration, is identical. The "subtraction" of the perfect structure as an invariant yields an excess cation ($Ag^{\cdot}$) and a cation vacancy ($|Ag|'$) as the relevant particles. Just as in Eq. (1.1c) we can write

$$\mathrm{Nil} \rightleftharpoons Ag^{\cdot} + |Ag|' . \tag{1.2a}$$

The old–fashioned charge designations (dot and dash) denote the relative charges: The crystal segments containing the defects such as

$$\begin{bmatrix} Ag^+ & Cl^- & \\ & Ag^+ & \\ Cl^- & & Ag^+ \end{bmatrix}^+ \text{ and } \begin{bmatrix} Ag^+ & Cl^- \\ & \\ Cl^- & \end{bmatrix}^-$$

are positively or negatively charged in an absolute sense, but the local positive or negative charge at the proper defect–site — interstitial or vacant site — represents a charge relative to the perfect situation. This differentiation between absolute and relative charge was naturally not necessary for H_2O.

---

[2]It is naturally possible to formulate the whole of aqueous acid–base chemistry in water in this minimal notation. Eq. (1.1c) would also formally describe autoprotolysis in liquid ammonia. An analogous Cl^- disorder reaction would be suitable for describing the dissociation of $SOCl_2$:

$$\mathrm{Nil} \rightleftharpoons |Cl|^+ + Cl^-$$

instead of

$$SOCl_2 \rightleftharpoons SOCl^+ + Cl^- .$$

Conversely, these considerations emphasize that internal acid–base chemistry of solids involves point defects [15]. In Chapter 5 we will see that such acid–base reactions, together with redox–reactions, constitute defect chemistry.

[3]According to Eq. (1.1c) we could also refer to the OH^- ion (more precisely to the difference of OH^- and H_2O) as a proton hole or an anti excess–proton. The following joke may be instructive in this context: A mathematics teacher notices that 10 pupils left the class–room even though only 9 have been in. His comment: If one pupil enters the room, the occupation is zero and the world is in order again.

[4]Perfect AgCl is dissociated in the sense that it consists of ions; but it is not dissociated into free particles since Ag^+ and Cl^- are trapped in their deep Coulomb potentials. The latter, "superionic" dissociation is described by Eq. (1.2).

The disorder in the electron shells is to be comprehended in an analogous manner. Here the bonding electrons, more precisely the valence electrons, have left their "regular" positions and have been excited into the conduction band. This also creates excess particles and missing particles, which are conduction electrons (e') and electron holes ($h^{\cdot}$). Let us take a metal oxide with the (perfect) composition MO as our model compound and for the purpose of better visualization assume that, to a good approximation, the valence band is composed of the oxygen p–orbitals, while the conduction band is composed of the outer metal orbitals. Hence, the reaction can also be formulated as an internal redox reaction

$$O^{2-} + M^{2+} \rightleftharpoons O^{-} + M^{+}. \tag{1.3a}$$

The minimal notation (subtraction of the perfect phase MO on both sides of the equation, see Fig. 1.2c) then becomes

$$\mathrm{Nil} \rightleftharpoons h^{\cdot} + e'. \tag{1.3b}$$

In silver chloride this corresponds to the charge transfer from Cl^- to Ag^+. The advantage of the notation used in Eq. (1.3b) is that it is independent of such detailed bonding considerations. The building element formulation of Eqs. (1.2a), (1.3b) and also of Eq. (1.1c) is an adequate notation for the thermodynamic, i.e. phenomenological treatment and emphasizes the superposition of perfect and defect components in energetic questions.

Unfortunately the formulation suffers from a lack of vividness precisely on account of the high degree of abstractness. From a descriptive point of view, structure element formulation is to be preferred. Nonetheless, descriptions utilizing very detailed structural elements, as in Eq. (1.1b) or Eq. (1.1a), are not employed in the case of ionic solids, although the disorder in AgCl (Eq. (1.2a)) could certainly be analogously formulated as the "dissociation reaction of the lattice molecule" according to

$$2AgCl_{(AgCl)} \rightleftharpoons Ag_2Cl^{+}_{(AgCl)} + Cl^{-}_{(AgCl)} \tag{1.2b}$$

(the lower index refers to the perfect state), or even — by analogy to $H_9O_4^+$ — by inclusion of further regular neighbours (see Chapter 5). The author has resisted, for two reasons, the temptation to select such a "molecular" notation: Firstly, the field of the "internal chemistry" of solids is already conceptually overloaded, and secondly, such a chemical notation would be clumsy for complicated solids or in kinetic considerations. Instead the conventional Kröger–Vink notation [3] is used: It also considers structural elements, in that it refers to absolute structures, but "boiled down" to only the "atomic" particles actually reacting. That means in the case of our silver chloride example that the anion–sublattice is completely omitted from the description, while, on the other hand, vacancies (here in the cation–sublattice) are explicitly taken into account as structural elements using the symbol V. So instead of Eq. (1.2a) or (1.2b) we write

$$(Ag^{+}_{Ag^{+}}) + (V^{0}_{i}) \rightleftharpoons (Ag^{+}_{i})^{\cdot} + (V^{0}_{Ag^{+}})' \tag{1.2c}$$

or abbreviated by the omission of all absolute charges

$$\mathrm{Ag_{Ag}} + \mathsf{V_i} \rightleftharpoons \mathrm{Ag_i^{\cdot}} + \mathsf{V'_{Ag}}. \tag{1.2d}$$

The superscript in Eq. (1.2d) again represents the relative charge, i.e. the difference between the charge in the real case and that in the perfect case ($' \hat{=} -1, \cdot \hat{=} +1$). An effective charge of zero is not indicated or is sometimes indicated by means of a cross ("×"). The subscript denotes the crystallographic position in the perfect structure (i: interstitial site). Specifically Eq. (1.2d) indicates that a silver ion (Ag^+) has moved from a regular silver ion position (subscript Ag) to a vacant (V stands for vacancy) interstitial site (subscript i) where it becomes an interstitial silver ion $Ag_i^{\cdot}$ and leaves a vacancy (V'_{Ag}) in the silver ion lattice. The regular components such as Ag_{Ag}, V_i or Cl_{Cl} do not carry an effective charge, while the interstitial silver ion bears the relative charge +1 (= +1 − 0) and the silver vacancy the relative charge -1 (= 0 − (+1)). Structure element notation is not used in the case of electronic defects, rather the building element notation of Eq. (1.3b). This has the advantage that, firstly, the formulation is independent of the nuances of the bonding and that, secondly, a possible double counting of electronic states[5] is avoided. Since an ionic crystal is a rigid body with well–defined sites and the picture is only slightly complicated by vibrations (or rotations) about the equilibrium state, it is possible (and necessary) to distinguish in contrast to the fluid phase water, between several defect types, as will be discussed in Section 5.5.
In addition to defects intrinsically formed in pure substances by thermal disorder, defects are also generated by the incorporation of foreign substances. Hence, the (substitutional) incorporation of a D^{2+} cation in place of an M^+ cation leads to a point defect $(D^{2+}_{M^+})^{\cdot} \equiv D^{\cdot}_M$, while the (additive) interstitial incorporation of a cation of higher valence leads to a defect with a higher effective charge, namely $D_i^{\cdot\cdot}$. Although dissolved foreign substances are similarly important in liquids, it is necessary to point out another basic difference between liquid and solid phases: The simultaneous dissolution of cations and anions does not normally constitute a difficulty in the case of deformable fluids, and electroneutrality is automatically guaranteed. The normal case in the solid state is that either only the anion or the cation (here D^{2+}) is soluble. The dissolution of D^{2+} then takes place either by substitution of M^+ or by taking up an unoccupied interstitial position[6]. The change in charge must be compensated by creation of another defect. Thus, the introduction of D^{2+} (by substituting[7] for an M^+ or by occupation of a free interstitial position) is associated with the formation of negatively charged ionic and electronic defects

[5] The electronic shell is included in the symbol for the element.

[6] As in organic chemistry, it is possible to distinguish between addition, substitution and elimination reactions (in and on the "giant molecule solid"). Rearrangement reactions also occur (cf. phase transitions).

[7] Such substitution reactions also play a role in aqueous solution: The precipitation reaction,

$$AlCl_3 + 3H_2O \rightleftharpoons Al(OH)_3(s) + 3HCl(aq)$$

such as cation vacancies (V_M') and conduction electrons. Such electronic (redox) and ionic (acid–base) effects are not independent of each other but appear simultaneously in a coupled way. Figure 1.3 shows cation vacancies created by substitution of a

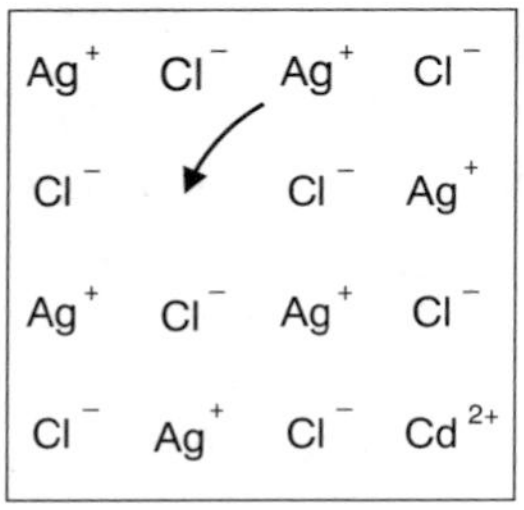

Fig. 1.3: Doping AgCl with $CdCl_2$ (only Cd^{2+} dissolves) leads to the formation of silver vacancies. The arrow indicates that silver conductivity is produced in this manner (migration of the vacancies in the opposite direction). This doping process generates electronic effects too, although to a small degree (see Section 5.6).

monovalent silver ion by a divalent cadmium ion. A further example is the creation of a high concentration of oxygen vacancies, and thus a high ionic conductivity in ZrO_2 by doping with CaO or Y_2O_3. This important ceramic material is used in automobile exhaust gas sensors and in high temperature fuel cells. The oxygen vacancies ($V_O^{\cdot\cdot}$) compensate for the charges of the Ca_{Zr}'' or Y_{Zr}' defects. Electronic effects play a very minor role in this oxide. In defect notation the incorporation can be written as

$$CaO + Zr_{Zr} + O_O \rightarrow ZrO_2 + Ca_{Zr}'' + V_O^{\cdot\cdot}. \tag{1.4}$$

A third example is the partial replacement of La^{3+} in La_2CuO_4 by Sr^{2+} (forming the defect Sr_{La}'). As in the previous example oxygen vacancies and electron holes are also produced here; but unlike in the previous case the electronic effects are significant. Hence, the substitution brings about marked oxydation of the crystal. The oxydation is necessary for the occurrence of "high temperature superconduction" in this oxide.

The procedure is similar for covalently bonded materials, such as silicon or organic polymers, although sites in the latter are not always sharply defined. If pentavalent phosphorus is introduced into silicon, this leads to the formation of some $P_{Si}^{\cdot}$ defects (that is P^+ on Si), since the fifth valence electron in the sp^3 hybridized basic silicon structure can be readily delocalized as a quasi–free electron. In the same manner trivalent aluminium takes electrons from silicon, that is from the valence band (Fig. 1.4). This defect (Al_{Si}') bears the formal charge (-1) ($(Al_{Si^0}^{-})'$). The electronic counterdefect is a delocalized electron hole ($h^{\cdot}$):

$$Al + Si_{Si} \rightarrow Si + Al_{Si}' + h^{\cdot}. \tag{1.5}$$

Ionic and electronic defects can also be created by an excess or deficit of a native component[8] instead of by the introduction of foreign species. This takes place,

e.g., corresponds to the substitution of hydroxide groups by chloride in the aqueous substrate. This process too is associated with considerable changes in chemical (cf. acidity) and electrical (cf. proton conduction) properties.

[8]Native components are understood as components occurring in the pure material (i.e. M and X in MX). In binary compounds the phase width corresponds to the toleration of redox effects.

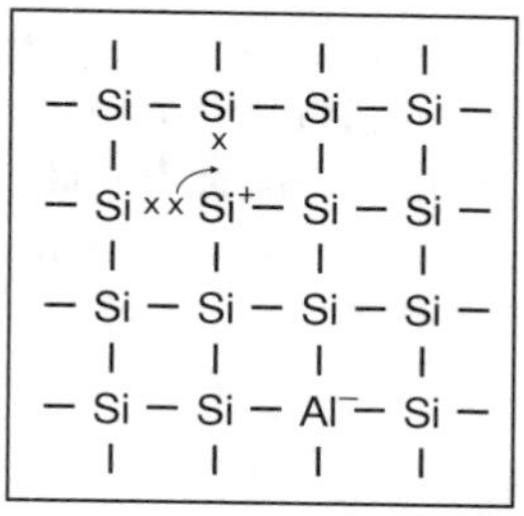

Fig. 1.4: Al doping of silicon effects the formation of a "vacancy" in the electronic shell (see arrow tip). The arrow indicates that this results in electronic (hole) conductivity. The electron hole migrates in the opposite direction.

for instance, in an oxide by interaction with the oxygen of the neighbouring phase (see Fig. 1.5). At sufficiently high temperatures, varying the oxygen content of the

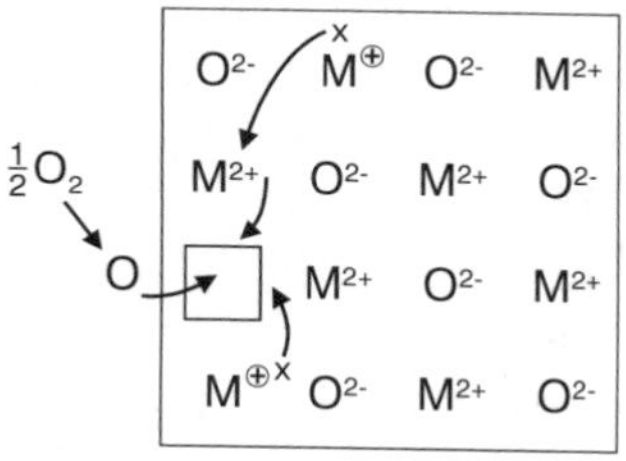

Fig. 1.5: Oxygen incorporation resulting from the jump of an adsorbed oxygen particle into a vacant oxygen site with the uptake of two electrons. In the example, the excess electron states correspond to monovalent metal ions. It is assumed, for the sake of the example, that the absorbed oxygen before passing into the vacancy, is in the neutral state, but this is generally not the case mechanistically.

gas phase makes possible a continuous tuning of the precise position in the phase diagram. Such phase widths are often tiny and the changes obtained often negligible with respect to total mass or to the energy of the phase, but the change in defect density and all the properties specifically associated with it can be immense. Thus, in n–conducting SnO_2 an increase of the oxygen partial pressure of the surroundings leads to a drastic reduction in conductivity, according to:

$$\frac{1}{2}O_2 + V_O^{\cdot\cdot} + 2e' \rightleftharpoons O_O. \tag{1.6}$$

In this case the oxygen introduced occupies oxygen vacancies in the lattice. It is incorporated in the form of O^{2-}, and electrons are required for this; these are available in SnO_2 in the form of conduction electrons. In more chemically oriented terms we may state that reduced Sn states (Sn^{m+}, m<4) are oxydized and, thus, annihilated. If, as in La_2CuO_4, there are (almost) no excess electrons in the material, the incorporation of oxygen is associated with the consumption of bonding electrons. Holes are then created in the valence band and the p–type conductivity increases. Chemically[9] speaking this corresponds to the oxydation of Cu^{2+} or O^{2-} to Cu^{3+} or O^-. The hole density in La_2CuO_4 is increased in this manner and induces superconductivity at low temperatures.

[9]This simple chemical notation in terms of defined valence changes only provides a correct picture in those cases in which valence or conduction bands can be overwhelmingly assigned to the cation or anion (see Section 5.3). However, in general, there is a hybridization, as for the Cu and O orbitals in the case of cuprates.

For simplicity of presentation individual mechanisms have been emphasized until now. However, in practice, various different defect states occur simultaneously. Their distribution, i.e. their concentration, can be obtained from solution of the overall reaction scheme. This will be treated systematically in Chapter 5. Here we just mention that oxygen vacancies cannot only be destroyed by redox reactions but also by pure acid–base reactions. Thus, H_2O can be dissolved in many oxides with the formation of internal OH groups (OH^- on O^{2-} positions). According to

$$H_2O + V_O^{\cdot\cdot} + O_O \rightleftharpoons 2OH_O^{\cdot} \tag{1.7}$$

the "OH^- part" of the water molecule occupies the vacancy, while the "H^+ part" is added to a regular O^{2-} (i.e. O_O).
In this manner defects do not just constitute the "internal chemical life" but also the (chemical) "communication with the environment". Naturally, the detailed kinetics must also be based on the defect concept. Every chemical or electrochemical process is made up of an interfacial reaction (more precisely a coupled scheme of individual elementary reactions at an interface) and the "transport reaction", that is a site exchange process within the bulk of the solid. This is also valid for true solid state reactions involving the formation of a new phase. The necessary internal mobility is also made possible by the presence of defects, as shown in Figs. 1.3, 1.4 and also 1.5. The arrows indicate that an atomic vacancy or an electronic hole migrates by neighbouring atoms or electrons occupying the vacant sites. Hence, in this case defect transport and mass transport are in opposite directions. When excess ions or excess electrons migrate, the direction of defect transport and mass transport is identical. Diffusion processes, as described in this manner, frequently constitute the rate–determining step in solid state chemical kinetics. However, elementary reactions at interfaces, which are defect reactions, too, may also dominate the kinetics in many cases[10].
Since the defects involved carry charges, they play a prominent role in the conversion of chemical signals to electrical signals (and vice versa). The example given in Fig. 1.6 can serve as leitmotif for this. As already explained, varying the oxygen content alters the internal chemistry (i.e. defect concentrations) of oxides, with immense effects on the electrical conductivity. The measurement of the electrical signal "conductivity" can be used for elegant and accurate detection or even for control of the oxygen content of the environment. Such a chemical sensor is only one of many highly interesting electrochemical applications.

This example also indicates the arrangement of this book. The thermodynamics of defects (Chapter 5) will enable us to specify the concentrations of the individual ionic and electronic defects and, thus, the ionic and electronic conductivities as a function of the relevant thermodynamic parameters (such as temperature and composition

[10] In the spirit of Chapter 5 the interface represents a (higher–dimensional) defect itself. Point defects within the interface are centres of enhanced reactivity (excitations within the higher–dimensional defect).

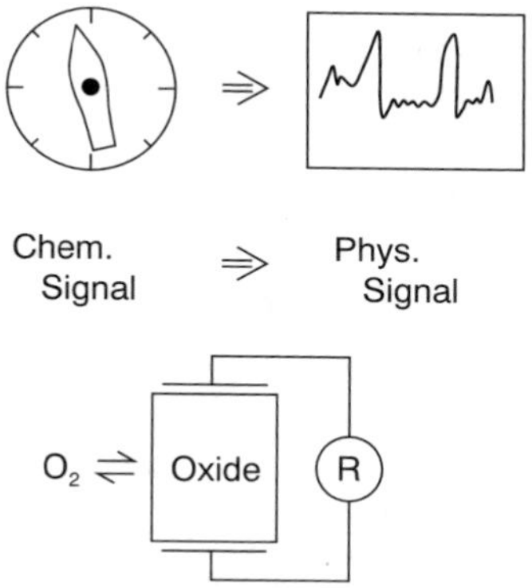

Fig. 1.6: The conversion of a chemical signal into a physical signal, using a conductivity sensor as an example. There the interaction with oxygen takes place in accordance with Fig. 1.5 and significantly influences the carrier concentrations. The implication arrow can usually be reversed and the physical signal can be applied to adjust the "chemistry".

of neighbouring phase). In this way the (thermochemical[11]) equilibrium state of the solid is precisely defined. In the defect chemical kinetics chapter (Chapter 6) we then consider mechanistically how and how rapidly changed state parameters bring about compositional changes. It is of particular importance for our example to elucidate how rapidly the defect concentrations and the conductivities change as the oxygen content changes. The charge of the defects makes electrochemistry a relevant subject *per se*. Since the thermodynamic and kinetic considerations already involve electrochemical equilibrium and nonequilibrium effects, the special electrochemical chapter (Chapter 7) at the end of the book is devoted to electrochemical systems that are of scientific importance — for the measurement, interpretation and control of interesting properties — or of technological importance for energy utilization or information conversion.

This main part of the book (Chapters 5, 6, 7) is obviously related to Materials Science; it is almost a triviality to conclude that, at least for ionic solids, now defect chemistry is fundamental to and, to a large extent, the substance of this interdisciplinary area[12]. Materials research is synonymous with the strategic exploitation of structure–property relationships, with a view to the optimization of properties. If the question posed by material research is, say, the optimization of electrical properties via selection of materials and control parameters, then this question immediately refers back to the thermodynamics and kinetics of defect chemistry. As far as materials are concerned, we concentrate on ceramics and in particular on

[11] The external shape, i.e. the surface, also belongs to the detailed description. Like the precise macrostructure (bulk plus surfaces) the "microstructure" (inclusion of internal interfaces, dislocations etc.) is almost without exception a nonequilibrium structure but highly relevant on account of its metastability (see Section 5.4).

[12] Materials research implies structure optimization on all relevant scales. The most important step is the selection of the solid ground state, i.e. synthesis of the chemical compound that is able to provide the optimizable "property world". Within this framework tuning the defects allows fine–adjustment, but this is frequently accompanied by enormous variation in properties. The adjustment of the supra–atomic architecture in the form of nano–, micro– and macrostructure (external shape) completes the procedure. The appropriate combination of different materials then leads to the design of relevant systems (as treated in Chapter 7).

electroceramics. Although implicitly discussed, classical semiconductor materials as well as polymers are only touched on in order to emphasize the generality of the concepts.

The first four chapters dealing with the perfect state are more introductory. According to Fig. 1.2 we formally construct our real solids by superposition of the perfect solid ("chemical groundstate") and the defects ("chemical excitations"). Both ensembles are not independent of each other but strictly coupled in equilibrium. For this reason we start with a concise treatment of the chemically perfect solid. Firstly, there is a discussion (Chapter 2) of the chemical bonding and then of the formation of the solid state, followed by a discussion of lattice vibrations (Chapter 3).

The purpose of the chapter dealing with equilibrium thermodynamics of the perfect solid (Chapter 4) is to elaborate, on the one hand, simple expressions for the thermodynamic functions of the "chemical ground state" and, on the other hand, to make the reader familiar with questions of internal and external equilibria, not least with the intention to provide the equipment to deal with the thermodynamics of defect formation. (The major portion of the free enthalpy at absolute zero consists of bonding energy, while the temperature dependence is largely determined by the vibration properties.)

Evidently, the structure of this book refers to the classical areas of Physical Chemistry (bonding theory, thermodynamics, kinetics, electrochemistry) related, however, to the solid state. In this sense, the monograph may partly serve as a textbook of physical chemistry of solids. In order not to lose track of our subject we will confine ourselves to the simplest cases. Nonetheless the introductory chapters are rather comprehensive in order to take account of the heterogeneity of potential readers. This, however, should not distract from the actual topic of the book. Those, who are familiar with these aspects or who are only interested operatively in the application of the defect–chemical formalism can skip these introductory chapters with an acceptable loss of continuity or they can refer to them later if relevant problems arise in understanding.

2 Bonding aspects: From atoms to solid state

Strictly speaking, a knowledge of the state variables (these are temperature, pressure, etc. and the numbers of the different particles involved) is all that is required to calculate the equilibrium composition, structure and even the external equilibrium form (i.e. the shape of the solid body) using the Schrödinger equation (or more precisely its relativistic generalization, i.e. Dirac's equation). However, in view of the many–body problem such a statement is almost without exception purely academic. This is even more true in the case of nonequilibrium, and particularly for instationary systems. Calculations of this type are limited to the simplest of examples even if we separate the electronic from the nuclear motion and treat the systems in the time–independent single–electron approximation neglecting relativistic effects. Even then the uncertainties of the numerical solution are frequently of the order of magnitude of the differences of interest, when, for example, the stability of a given crystallographic structure is considered. For this reason the procedure generally employed is a combination of a priori chemical knowledge with respect to atomic and molecular properties and a posteriori knowledge with respect to the crystallographic structure.
Since a solid body represents a giant three–dimensional molecule ("3–D polymer") with possible anisotropy in the chemical bonding and with terminal groups constituting its surfaces, the description according to bonding theory is a many–body problem — in terms of both nucleons and electrons. It is appropriate to start out from the simplest type of chemical bond, namely the two–atom problem in the single–electron approximation. This is not only didactically appropriate, but also useful in so far as the energetics of the whole solid state are frequently largely represented by the short range interactions.

2.1 Chemical bonding in simple molecules

2.1.1 Ideal covalent bonding

Let us first consider an arrangement of two (a, b) atoms of the same nature (X_a, X_b) produced according to

$$\text{Reaction B} = \qquad 2X \rightleftharpoons X_2, \qquad (2.1)$$

in which it is only necessary to take account of one electron, a condition that is only strictly met in the H_2^+ molecule; hence, there are two relevant wave functions $|\widehat{ab}\rangle$ and $|\breve{ab}\rangle$, which correspond to bonding and antibonding states and which are, to an approximation (according to the LCAO method), composed of the wave functions

Physical Chemistry of Ionic Materials J. Maier
©2004 John Wiley & Sons, Ltd ISBN: 0-471-99991-1 (HB); 0-470-87076-1 (PB)

of the single atom problems $|a\rangle$ and $|b\rangle$ as follows[1,2]:

$$|\widehat{ab}\rangle \propto |a\rangle + |b\rangle \tag{2.2a}$$
$$|\breve{ab}\rangle \propto |a\rangle - |b\rangle . \tag{2.2b}$$

If the probability densities were merely summed, then, in a quasi–classical manner, the total probability density would be proportional to the sum of the squares (integrated and taking into account complex functions: $\langle ab|ab\rangle \propto \langle a|a\rangle + \langle b|b\rangle$). Since the wave functions interfere, there is, however, a higher electron density between the nuclei in the case of the bonding state ($\langle\widehat{ab}|\widehat{ab}\rangle \propto \langle a|a\rangle + \langle b|b\rangle + 2\langle a|b\rangle$) and a greatly reduced electron density in the case of an antibonding state ($\langle\breve{ab}|\breve{ab}\rangle \propto \langle a|a\rangle + \langle b|b\rangle - 2\langle a|b\rangle$). Nonetheless, it would be wrong to make the increased potential energy due to the increased electron density between the nuclei (cf. the term $2\langle a|b\rangle$) solely responsible for the bonding. The potential energy balance is not so favorable since the accumulation or removal of electrons from the centre takes place at the cost of the density at the atoms. Attention must therefore be paid to the conservation of charge and thus to the normalization (cf. proportionality factor in Eq. (2.2)). Moreover, the kinetic energy makes a significant proportion of the chemical bonding, as will be considered again below[3]. Overall, the situation in the two–atom problem now corresponds to the modified energy states $\widehat{\epsilon}$ and $\breve{\epsilon}$ which are approximately[4] altered by $\pm\beta$ with respect to the initial energy $\epsilon_a = \epsilon_b$:

$$\widehat{\epsilon} = \epsilon_a - |\beta| \quad \text{and} \quad \breve{\epsilon} = \epsilon_a + |\beta| . \tag{2.3}$$

[1] The procedure mentioned here is that of L. Pauling [16] and corresponds to a linear combination of atomic orbitals (LCAO) [17] which is extensively employed in the molecular orbital theory [18].

[2] In agreement with literature the Dirac bra/ket notation is used here to represent functions as vectors that are formed from the infinite set of the function values (or in a more operational representation as vectors in Hilbertian space spanned by the countable infinite set of basis–functions in which the function under regard is developed): $\langle c|$ designates the complex conjugate of $|c\rangle$, the scalar product $\langle c|d\rangle$ is then the sum over the products of the individual function values that is the integral over the corresponding product of the functions. The scalar product $\langle c|c\rangle$ is a measure of the integrated electron density referring to c and may be normalized to unity. Linear combination is naturally an approximation. That linear combination, which corresponds to the lowest energy, is, hence, not necessarily identical to the "true function", but, nevertheless, the readily proved variation theorem shows that it is the nearest of all possible alternative linear combinations to this. A variation calculation leads, in the case of a linear combination, to a normal minimax problem in the coefficients and to the solutions given above, as described in all quantum chemical textbooks (e.g. [19–22]). There is a clear treatment of the physical basis in Ref. [23].

[3] With respect to the complex interaction between kinetic and potential energy as a function of the nuclear distance and the importance for chemical bonding, cf. Ref. [24].

[4] Strictly speaking the above model yields

$$\widehat{\epsilon} = \frac{H_{aa} + H_{ab}}{1 + S} = (H_{aa} + H_{ab})(1 - S_{ab} + S_{ab}^2 - S_{ab}^3 + \ldots)$$

If we now adopt for the two–electron problem (one bonding electron per X) the energy levels from this single–electron model[5] and compare with the initial states, the bonding energy of the two atoms X_a and X_b in the molecule X_2 formed, i.e. the reaction energy in Eq. (2.1), is obtained as $\Delta_B\epsilon = 2\widehat{\epsilon} - 2\epsilon_a \cong -2|\beta|$. The (negative) quantity β here represents the reduced resonance integral[4]

$$\beta = \langle a|\mathcal{H}|b\rangle - \langle a|\mathcal{H}|a\rangle\langle a|b\rangle. \tag{2.4}$$

In these integrals $\mathcal{H}$ represents the Hamiltonian operator, that is the energy operator in the Schrödinger equation

$$\mathcal{H}|ab\rangle = \epsilon|ab\rangle. \tag{2.5}$$

As is well known this is obtained from the operators for the potential energy and the kinetic energy[6]. The integrals $\langle a|\mathcal{H}|b\rangle$ (resonance integral $\equiv H_{ab} = H^*_{ba} = H_{ba}$) and $\langle a|b\rangle$ (overlap integral $\equiv S_{ab} = S^*_{ba} = S_{ba}$) are measures of the overlap of the atomic orbitals, for they only have contributions differing from zero at those positions, for which both $|a\rangle$ and $|b\rangle$ differ from zero. Conversely, for a given internuclear distance, the contribution of the Coulomb integral $\langle a|\mathcal{H}|a\rangle$ ($\equiv H_{aa} = H_{bb} \equiv \langle b|\mathcal{H}|b\rangle$, also known as α) is only appreciably different from zero in the region of the nucleus. However, there the interaction with the neighbouring nucleus may be neglected and $\mathcal{H}$ be set as equal to the Hamiltonian operator of the single atom problem; in consequence α can be represented by $\epsilon_a (= \epsilon_b)$, as already used in Eq. (2.3). Nevertheless, the presentation of the energy states by

$$\widehat{\epsilon}_{ab} = \alpha - |\beta| \quad \text{and} \quad \check{\epsilon}_{ab} = \alpha + |\beta| \tag{2.6}$$

is more generally valid than Eq. (2.3). Figure 2.1 illustrates the matrix elements and energy functions discussed as functions of the internuclear distance of H_2^+.

$$\check{\epsilon} = \frac{H_{aa} - H_{ab}}{1 - S} = (H_{aa} - H_{ab})(1 + S_{ab} + S_{ab}^2 + S_{ab}^3 + \ldots).$$

Obviously the distance of $\widehat{\epsilon}$ to H_{aa} is less than that of H_{aa} to $\check{\epsilon}$. If $S_{ab} \ll 1$, β in Eq. (2.3) and Eq. (2.6) might be identified with the resonance integral; this assumption is generally unjustified (see [20]), but is frequently employed. The better approximation (Eq. (2.3) and Eq. (2.4)) does not completely agree with the second approximation of this presentation, but is a favourable approximation for the present problem since the missing term ($S_{ab}H_{ab}$) and the second order terms partially compensate (see sign).

[5]We obviously ignore electron–electron interactions. Later on, some situations in which such interactions will be important, will be briefly considered (see in particular Mott–Hubbard criterion). In these cases the ratio of the interaction energy and β is decisive for the strength of the effect.

[6]While the first is derived from classical momentum ($\mathbf{p}$) considerations via the transformation $\mathbf{p} \to \frac{h}{2\pi i}\nabla$, the classical expression is retained in the case of the potential energy that only depends on the local coordinate. $\mathcal{H}$ is, thus, given in the end by the space functions and the second space derivatives (kinetic energy $\propto$ (momentum)^2). It can be shown that $\mathcal{H}$ is a Hermitian operator, i.e. $\langle a|\mathcal{H}|b\rangle = \langle b|\mathcal{H}|a\rangle^*$. The star denotes the complex conjugate. Such Hermitian operators have, as they must, real eigenvalues: Because $\langle a|\mathcal{H}|a\rangle = \epsilon\langle a|a\rangle$ and $\langle a|\mathcal{H}|a\rangle^* = \epsilon^*\langle a|a\rangle^* = \epsilon^*\langle a|a\rangle$, it follows that $\epsilon = \epsilon^*$.

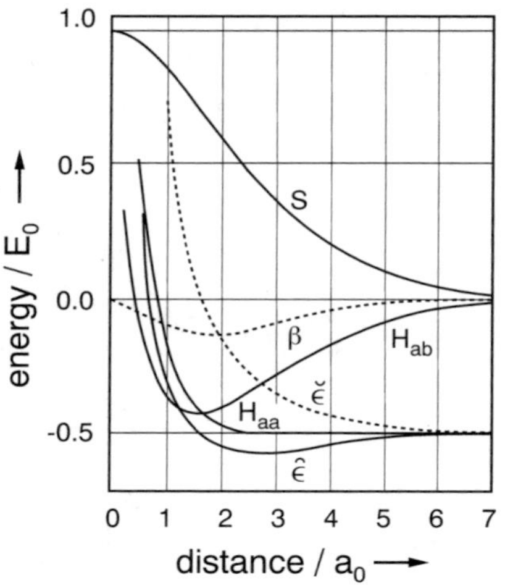

Fig. 2.1: The matrix elements $S_{ab} \equiv S, H_{ab}, H_{aa}$, the reduced resonance integral β and energy eigenvalues of H_2^+ as functions of nuclear distance. The equilibrium value corresponds to the minimum of $\tilde{\epsilon}$ in the LCAO approximation used (a_0 = Bohr unit of length = 0.529 Å; E_0 = Hartree energy unit = 27.21 eV). From Ref. [20].

The centring or symmetrical "sharing" of two bonding electrons can be generalized to cover the bonding of homonuclear atoms and, in the concrete case of the bonding of say a Cl_2 molecule, corresponds formally to the creation of two noble gas configurations (Ar–configuration$\equiv$/Ar/):

$$|\overline{Cl}^x \quad + \quad {}^x\overline{Cl}| \quad \rightleftharpoons \quad |\overline{Cl} - \overline{Cl}|$$

$$KL3s^23p^5 \quad KL3s^23p^5 \quad \underbrace{KL3s^23p^4_{y,z}}_{/Ar/} \underbrace{\left(\overset{\frown}{3p_x 3p_x} \right)^2 3p^4_{y,z}3s^2 LK.}_{/Ar/} \qquad (2.7)$$

This is naturally a very approximate description in which not even all the outer electrons are included in bond formation. A more precise procedure creates 8 orbitals from the outer s and p electrons (which are referred to as σ and π orbitals depending on symmetry: $\sigma(s)$, $\sigma^*(s)$, $\sigma(p)$, $\sigma^*(p)$ and two $\pi(p)$ and $\pi^*(p)$ orbitals[7]). The number of "real bonds" in the case of the example given above, e.g. in the case of the Cl_2 molecule is 1, since the corresponding antibonding orbitals are also filled for all orbitals apart from $\sigma(p)$ (14 outer electrons) and, thus, energetically nullify the effect of filling the respective bonding states approximately (Eq. (2.3)) (cf. "bond order" [21]). Figure 2.2 applies to the atomic dimers of the first octet in the periodic table. The energetic order of the levels shown comes from the fact that the s orbitals are energetically lower than the p orbitals and that the degree of overlap of the π orbitals is less than that of the σ orbitals and so, to an approximation, the corresponding

---

[7]The overlap of s–orbitals and of p–orbitals lying in the direction of the bonding axis (p_x see above) leads to σ orbitals, which are rotationally symmetrical about their bonding axis, while the p_y, p_z orbitals perpendicular to the axis form π bonds. For every MO orbital type there is a bonding ($\tilde{\sigma}$, $\tilde{\pi}$ or simply σ, π) and an antibonding ($\tilde{\sigma}$, $\tilde{\pi}$ or σ^*, π^*) level.

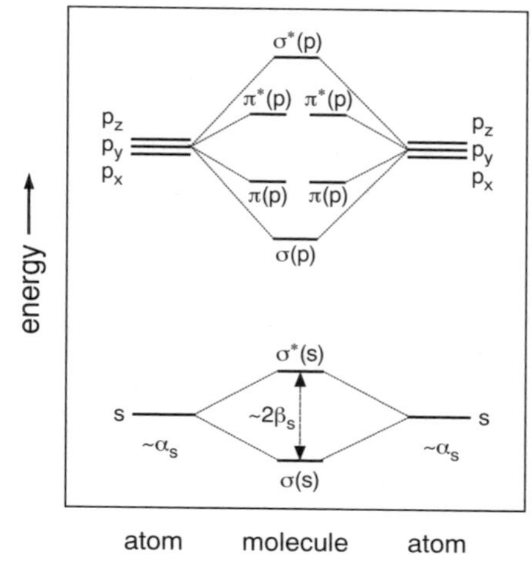

Fig. 2.2: The approximate position of the energy levels of the molecular orbitals in the homonuclear molecule X_2 formed from the atomic levels in X. Consider N_2 as an example. Since every nitrogen contributes 5 outer shell electrons, the lowest 5 MOs are doubly occupied. It is the occupation of the σ(p) and of the two π(p) orbitals, that contribute to bonding, the s interaction is nonbonding: $|N \equiv N|$. In the same manner a double bond is formed for O_2 from the p orbitals. However, here the π^*(p) orbitals are singly occupied and the ground state is a triplet state. This explains the paramagnetism of the O_2 molecule.

$|\beta|$ values are smaller and the size of the splitting too. When orbital interactions are taken into account, these energy levels are altered; the changes are perceptible, at least for the light dimers of the first octet. The approximate Eq. (2.6) and Fig. 2.2 do not take into account that, strictly speaking, the splitting of the levels is asymmetrical[4]. The distance of the nonbonding states to the antibonding levels is greater than to the bonding levels. Already this explains why Ne_2 is unstable, while Fig. 2.2 merely predicts a nonbonding situation. In such cases the electron correlation, which has also been neglected, is of additional importance.
A slightly more complex example is the methane molecule CH_4 which exhibits four single bonds. Since each C–H two–centre interaction generates a bonding and an antibonding orbital and the four bonds are identical for reasons of symmetry, it is advantageous to consider these as having been formed from hybrid sp^3 atomic orbitals. These hybrids are linear combinations of the atomic orbitals, the molecular orbitals approximately linear combinations of the hybrid orbitals and, hence, as before, linear combinations of atomic orbitals. When, in hydrocarbons, there are three bonding neighbours it is advantageous to refer to sp^2 hybridized bonds, and in the case of two neighbours to sp hybridized bonds. The remaining p orbitals can form π bonds. The extensive literature on chemical bonding should be consulted for further discussion and in particular, if d and f orbitals are involved (e.g. [19–25]).

2.1.2 Polar covalent bonding

For reasons of symmetry the X_2 molecule already treated does not exhibit any permanent dipole moment. Permanent dipole moments are always the result of the charge being distributed asymmetrically in the bonded state, and thus occur in diatomic molecules formed from different atoms. According to

$$|\widehat{ab}\rangle \propto |a\rangle + \lambda\,|b\rangle \qquad (2.8)$$

a different weighting of the atomic wave functions[8] ($\lambda \neq 1$) is necessary. The degree of the effect (in molecule XY) is reflected by the difference in the electronegativity. According to Pauling this is obtained from the square root of the difference in the bonding energies of the virtually unpolarized molecule and of the actual molecule containing the ionic contributions. The first contribution is estimated from the arithmetic or geometric mean of the bonding energies of X_2 and Y_2. If the electronegativities of the bonding partners are very different, the charge transfer is almost complete ($\lambda \to 0$ or $\lambda \to \infty$). In these cases the orbital relevant for bonding approximates to a pure atomic orbital ($|a\rangle$ or $|b\rangle$) and the corresponding resonance integral β is almost zero in such cases. This corresponds to the limiting case of ion formation described below. The α values for the two atoms naturally differ considerably and are also no longer identifiable with the energy of the single–atom problem.

If we take into account the difference in the α values by $\alpha_a = \bar{\alpha} + \Delta\alpha$ and $\alpha_b = \bar{\alpha} - \Delta\alpha$, then a Hückel calculation, which is only justifiable for weakly polar bonds (see Ref. [20]), yields instead of Eq. (2.6)

$$\widehat{\epsilon}_{ab} = \bar{\alpha} - \Delta\alpha/\gamma \quad \text{and} \quad \breve{\epsilon}_{ab} = \bar{\alpha} + \Delta\alpha/\gamma, \tag{2.9}$$

where $\gamma = \Delta\alpha/\sqrt{(\Delta\alpha)^2 + \beta^2}$ represents the charge displacement or polarity[9]. $\Delta\alpha = (\alpha_a - \alpha_b)/2$ is sometimes also referred to as the polar energy.

Let us consider once again the H_2 molecule as the extreme case of the ideal covalent bond ($\Delta\alpha \to 0$, $\Delta\alpha/\gamma \to |\beta|$). Since two electrons occupy the lowest energy state twice in the case of the H_2 molecule and since we neglect electronic correlations, the bonding energy of two hydrogen atoms, is $2\beta = -2|\beta|$. Naturally there is no dipole moment in this case. According to Eq. (2.9) a weakly asymmetrical bond results in an approximate bonding energy[10] of $2\beta(1 + (1/2)(\Delta\alpha/\beta)^2)$. Obviously there is only a second order correction on account of $|\Delta\alpha/\beta| \ll 1$, while the effect on the charge displacement is of first order and identical to $|\Delta\alpha/\beta|$ (see above).

A polar atomic bond occurs, for instance, when a hydrogen atom comes into contact

[8] A dipole moment may also occur if the charges attributed to X and Y in a molecule XY are identical. This "homopolar dipole contribution" is caused by an asymmetry of the electron density centred between the two nuclei (overlap of "orbitals of differing sizes").

[9] The analogously defined parameter $\frac{|\beta|}{(\beta^2 + (\Delta\alpha)^2)^{1/2}}$ can be considered as the covalency of the bond [26].

[10] $2\widehat{\epsilon} - \epsilon_a - \epsilon_b = 2(\widehat{\epsilon} - \bar{\alpha}) = -2\sqrt{(\Delta\alpha)^2 + \beta^2}$. Note, it follows for small x that $\sqrt{1 + x} \simeq 1 + x/2$ since $(\sqrt{1 + x} - \sqrt{1})/(1 + x - 1) \simeq d\sqrt{1 + x}/dx|_{x=0} = \frac{1}{2}$. If we define the bond order as half the change in bonding energy with β [27], the result coincides with the covalency defined in footnote 9.

with a chlorine atom

$$\begin{array}{llll} H^{x} \quad + & {}^{x}\underline{\overline{Cl}}| & \rightleftharpoons & H - \underline{\overline{Cl}}| \\ 1s^1 & KL3s^23p^5 & & \underbrace{\underbrace{\left(\widehat{1s3p_x}\right)^2}_{/He/} \quad 3p^4_{y,z}3s^2LK}_{/Ar/} . \end{array} \qquad (2.10)$$

In the language of the molecular orbital theory the orbital, that is relevant for bonding, is an energetically low, fully occupied $\sigma(1s, 3p_x)$ orbital, while the corresponding antibonding orbital remains unoccupied.

In principle, as for the X_2 molecule (see Fig. 2.2), the participation of other orbitals (e.g. $3p_{y,z}$(Cl) or 2s(H)) must be considered. However, in the case of HCl these differ so much energetically that Eq. (2.10) represents a good approximation[11].

On account of the finite charge transfer within the bond corresponding to an admixture of ionic, mesomeric structures, the stable configuration is best represented by H$\triangleleft$Cl or $H^{\delta+} - Cl^{\delta-}$.

Once the electron distribution has been derived quantum–mechanically, it is possible to use classical considerations to calculate the forces occurring in the molecule, and thus also dipole forces; this is an expression of the Hellmann–Feynman theorem [28,29]. Such dipole–dipole interactions are naturally important when considering intermolecular interactions. The energy of dipole–dipole interactions decays with the third power of the distance of separation, the energies of dipole–multipole interactions fall off with a correspondingly higher power. The energy of dispersion interaction, which is still an interaction of comparatively long range and which is decisive for the interaction between neutral particles without permanent dipole, quadrupole or octupole moments, falls off more rapidly ($\propto R^{-6}$ for large R). This is responsible for the very weak form of bonding, which occurs, for instance, amongst noble gas atoms. It emerges from a higher order solution of the Schrödinger equation and can be regarded as resulting from interactions between mutually induced dipoles. This dipole moment disappears in the time average, but the interaction does not.

The hydrogen bond is a special sort of such an interaction. The bonding energies are generally considerably greater (typically 10...100 kJ/mol) and the bonding distances correspondingly smaller. One reason is the nature of the dipole moments that occur; in addition the electronegative partners can approach each other so closely that the hydrogen bonds take on the character of three–centre bonds with an electron excess. Exchange processes involving the protons (naked elementary particles) are expected to be important in many cases. The moderate values of the bonding energies and activation energies for formation and separation, which are less than

[11] In the case of LiH it is not merely the two s orbitals but also 1s(H) and 2p(Li) that interact. However, these do not contribute appreciably to bonding.

for ionic or covalent bonds but perceptible, predestine them for a fundamental role in biochemistry.

2.1.3 The ionic bonding

In the limiting case of ionic bonding ($\beta \to 0$) the bonding state corresponds to the double occupation of the relevant atomic orbital of the electronegative partner:

$$\text{Reaction B} = \qquad \mathrm{M} + \mathrm{X} \rightleftharpoons \mathrm{M^+X^-}. \tag{2.11}$$

According to the treatment given in the previous section, a charge transfer of 1 (since $\Delta\alpha/\sqrt{(\Delta\alpha)^2 + \beta^2} \to 1$ for $\beta \to 0$) is correctly predicted. The predicted value for the energy of bonding, i.e. the energy of reaction B (2.11), $\Delta_\mathrm{B}\epsilon = -2\Delta\alpha$, does emphasize the importance of the differing α values of the reactants, but is quantitatively incorrect, since the marked polarity of the bond exceeds the range of validity of Eq. (2.9). In order to avoid a more exact consideration, which would fall outside the terms of this sketch, we shall agree to use the following semiempirical approach. For the purpose of discussing the bonding energy of our "salt molecule" (such as occurs e.g. in the gas phase), we break up Eq. (2.11) into the partial steps

$$\text{Reaction I} = \qquad \mathrm{M} \rightleftharpoons \mathrm{M^+} + \mathrm{e^-} \tag{2.12a}$$
$$\text{Reaction A} = \qquad \mathrm{X} + \mathrm{e^-} \rightleftharpoons \mathrm{X^-} \tag{2.12b}$$
$$\text{Reaction Z} = \qquad \mathrm{M^+} + \mathrm{X^-} \rightleftharpoons \mathrm{M^+X^-}, \tag{2.12c}$$

i.e. we first ionize atom M to yield the cation ($\Delta_\mathrm{I}\epsilon = \mathrm{I_M}$ = ionization potential of the metal) and transfer the released electron to X with the formation of the anion ($-\Delta_\mathrm{A}\epsilon = \mathrm{A_X}$ = electron affinity of X). Evidently, the third contribution then corresponds — assuming the nuclear distance (R) is sufficiently large — to the energy of the Coulomb interaction ($\Delta_\mathrm{Z}\epsilon \propto (+1)(-1)/\mathrm{R}$). Thus, the formation energy from the elements ($\Delta_\mathrm{B}\epsilon$) only differs from the formation energy from the ions ($\Delta_\mathrm{Z}\epsilon$) by the difference between the ionization potential of the electropositive partner and the electron affinity of the electronegative one:

$$\Delta_\mathrm{B}\epsilon = (\mathrm{I_M} - \mathrm{A_X}) + \Delta_\mathrm{Z}\epsilon = (\mathrm{I_M} - \mathrm{A_X}) - \mathrm{const/R}. \tag{2.13}$$

Electron transfer does not occur solely as the result of the difference $\mathrm{I_M} - \mathrm{A_X}$ even in the case of caesium fluoride[12]. The Coulomb energy between $\mathrm{M^+}$ and $\mathrm{X^-}$ is essential. The third term in Eq. (2.13) is the only contribution that is directly dependent on both partners; however, it would be misleading to consider this contribution as the true bonding term, because the decision concerning whether complete charge transfer occurs, is dependent on the combination M/X and, hence, on $\mathrm{I_M} - \mathrm{A_X}$. The

[12] In the case of AgCl the formation of separate, gaseous ions of $\mathrm{Ag^+}$ and $\mathrm{Cl^-}$ from the neutral gaseous components Ag and Cl (referred to 300K) requires ca. 3.8 eV ($\mathrm{I_{Ag}} \simeq 7.55$eV, $\mathrm{A_{Cl}} = 3.76$eV). It is bringing these together in the form of a gaseous "salt molecule" (gain of $|\Delta_\mathrm{Z}\epsilon| = 6.9$eV) that finally leads to the release of energy (3.1 eV).

combination of an alkali metal element (e.g. Na) as a strongly electropositive component with a halogen (e.g. Cl) as a strongly electronegative component according to

$$\begin{array}{ccccc} \mathrm{Na}^{\mathrm{x}} & + & {}^{\mathrm{x}}\overline{\underline{\mathrm{Cl}}}| & \rightleftharpoons & \mathrm{Na}^{+} \quad \mathrm{Cl}^{-} \\ \mathrm{KL3s}^{1} & & \mathrm{KL3s}^{2}\mathrm{3p}^{5} & & /\mathrm{Ne}/ \; /\mathrm{Ar}/ \end{array} \tag{2.14}$$

is a prototype example of a primarily ionic bond. The transfer of the outer electron from sodium to chlorine means that both entities achieve a noble gas configuration: Na^+Cl^- is isoelectronic with NeAr, but exhibits a different charge distribution, reflecting the different character and strength of bonding.

Considering the last paragraphs it is not surprising that the important criterion for the formation of various types of bonds, that is the electronegativity, is also given by the arithmetical mean of the ionization and electron affinity: Thus, whether a molecule takes the form M^+X^- or the form M^-X^+ depends, according to Eq. (2.13), upon the difference $(I_M - A_X) - (I_X - A_M)$, so, in fact, upon $(I_M + A_M) - (I_X + A_X)$. The contribution described by Eq. (2.12c) does not enter the consideration because it is the same in each case. The equivalence of this concept by Mulliken with that of Pauling (see above) can be demonstrated by means of a simple Hückel calculation (see, for example, Ref. [20]).

The following ought to be mentioned concerning transition metal ions before we turn to the metallic bonding: Because of the differing spatial forms of the d orbitals (which are incompletely occupied in all relevant cases) an important correction is due to the splitting of the energy levels depending on the configuration of the nearest neighbours. If the ligands are octahedrally positioned the energy levels of the d_{xy}–, d_{xz}– and d_{yz} orbitals lie below that of the $d_{x^2-y^2}$ and the d_{z^2} orbitals. This is reversed if the ligand environment is tetrahedral; the degree of splitting is however lower in this case. These effects are of particular relevance in the discussion of the stability of complex ions and elementary ions in the crystal lattice, optical transition, magnetic effects and with regard to the correction of the lattice energy in crystals (cf. Section 2.2.2 and Ref. [30]).

2.1.4 Metallic bonding

Very electropositive atoms, such as the alkali metal elements, can only achieve the noble gas configuration by joint release of electrons

$$\begin{array}{ccc} \mathrm{Na}^{\mathrm{x}} & \rightleftharpoons & \mathrm{Na}^{+} + \mathrm{e}^{-}. \\ \mathrm{KL3s}^{1} & & \mathrm{KL} \end{array} \tag{2.15}$$

However, the complete expulsion of an electron is very unfavourable energetically; let us recall that the energy of reaction for this process is the highly positive ionization potential (5.2eV for Na). Stabilization in the form of bond formation only occurs in a many–body system (N$\ggg$1):

$$\mathrm{N} \quad \mathrm{Na}^{\mathrm{x}} \rightleftharpoons (\mathrm{Na}^{+}\mathrm{e}^{-})_{\mathrm{N}}. \tag{2.16}$$

This will be discussed in more detail in Sections 2.2.1 and 2.2.5. In the molecular orbital picture many more nearest neighbours are bonded (Na: 8), than valence atom orbitals (4) and, in particular, many more than there are valence electrons (1) available. The bond formation in the metal is determined by ionization potential and the energy contributions ($\Delta\epsilon'$) required to condense[13] the isolated charged particles to a solid and to delocalize the electrons (in the form of a quasi electron gas) in this structure:

$$\Delta_B\epsilon = I_M + \Delta\epsilon'. \tag{2.17}$$

The latter effect, which is of particular importance for the solid state electronic properties, can be explained in terms of the behaviour of electrons in extended systems. This will be discussed in Section 2.2.
A bonding mesomerism similar to that in metal crystals (see Section 2.2.1) can also occur under certain conditions within molecular subunits. If the ratio of metal to nonmetal element is unusually large with respect to the normal valence, metal clusters can occur as complex cations within which metal bonding is dominant. Such bonding pecularities are discussed later (Section 2.2.6). The well–known situation in conjugated hydrocarbons is similar in a certain sense. In conjugated, unsaturated hydrocarbons the neighbouring carbons are joined by sp^2 bonds. As far as the remaining p–electrons are concerned, alternative limiting structures can be formulated, whose superposition is equivalent to partial delocalization.

2.1.5 Further intermediate forms of chemical bonding

Just as the polar atomic bonding already treated, is an intermediate form between ionic and covalent bonding (compare the series $NaCl, MgCl_2, AlCl_3, SiCl_4, PCl_3, SCl_2, ClCl$, in which the electronegativity of the more electropositive M element bonded to the constant X element is varied), there are also, according to the measuring rod of electronegativity, intermediate forms from covalent bonds and from ionic bonds in the direction of metallic bonds. While strongly electropositive elements, such as Na, form purely metallic bonds in a many–particle environment, that is the electrons play the role of (delocalized) anions, in the semimetals bonds occur that are intermediate between metallic and covalent bonding[14]. Examples include

[13] Cf. Eq. (2.12c) with $X^- \equiv e^-$.

[14] The term "localization" is not unambiguous (chemical localization vs. positional localization, electron localization vs. bond localization). Cf. Refs. [20,31,32] with regard to the difficulty and the demarcation of the terms "localized" and "delocalized", and with regard to the correspondence of local bonds and overall molecular orbitals in giant molecules. There it is shown that the local bond model can offer an appropriate description also in the many–particle system, namely for the limiting case of the covalent crystal with filled bands and significant gap ("equivalent orbitals"). A similar statement is true for the ionic crystal. However, in this case the local picture is to be preferred, as anion and cation states hardly overlap. Yet, it naturally fails in the case of the metallic bonding and for the intermediate forms discussed in Section 2.1.5.

graphite or bismuth[15,16]. The transition from predominantly metallic behaviour to an exclusively covalent bonding is made clear by the following variation within a given row of the periodic system: NaNa, MgMg, AlAl, SiSi, PP, SS, ClCl.
A variation of the more electronegative partner X bonded to a constant electropositive atom M also transforms metals into nonmetals: Na, Na_xMg, Na_xAl, Na_xSi, Na_3P, Na_2S, NaCl. Here, in contrast to the previous case, the electron density is more and more shifted towards the (electronegative) partner[14], eventually leading to ionic compounds: The intermediates in this row are the intermetallic compounds (cf. e.g. Na_xAl in the above row, or MgBi which may be compared with a Mg–metal on one side and with a Mg–halide on the other). These examples demonstrate that a deeper understanding requires the treatment of many–particle systems which are dealt with in Section 2.2.

2.1.6 Two–body potential functions

Although the various types of bonds may differ very greatly from each other they are very similar in one respect. The bond reacts to an increase in distance with a moderate restoring force, while the repulsive forces, that come into effect when the distance is reduced and which act over a small distance, react very sensitively to reduction of the equilibrium position. Hence, the potential function[17] is of the form shown for the "bonding energy eigenvalue" in Fig. 2.1, and can be approximated, for example, by a superimposition of an r^{-m} and an r^{-n} term with n>m (Mie potential [33]).
In the case of an ionic bond the first term is rather precise with m=1. Concerning the second term an exp–(const. r) law has a more fundamental basis as repulsion term (see Fig. 2.3). However, since the exponential term is also not strictly valid and the various models become equivalent for small displacements (cf. Morse potential [34], see also Lennard–Jones potential [35] and Born–Mayer potential [36] below), we will restrict ourselves to the Mie potential

$$\epsilon = Ar^{-n} - Br^{-m} = Ar^{-n}\left(1 - \frac{B}{A}r^{n-m}\right) = -Br^{-m}\left(1 - \frac{A}{B}r^{m-n}\right). \qquad (2.18)$$

The condition for the equilibrium distance ($\widehat{r}$), viz. $d\epsilon/dr = 0$, yields as a correlation between the bonding parameters A and B

$$\left(\frac{nA}{mB}\right)^{\frac{1}{n-m}} = \widehat{r} \qquad (2.19a)$$

[15] One band is almost full, the overlapped higher one almost empty (see Section 2.2).

[16] The fact that copper (a typical metal) has an electronegativity similar to bismuth (a semimetal) shows that electronegativity is not the decisive factor here.

[17] Since the motion of the electron is very rapid and therefore electronic equilibrium is usually established, with respect to r, it is possible to decouple electronic and nuclear problems. In the effective Schrödinger equation responsible for nuclear motion, the electronic energy plays the role of a potential energy. For this reason it is justifiable to call $\epsilon(r)$ the "potential function".

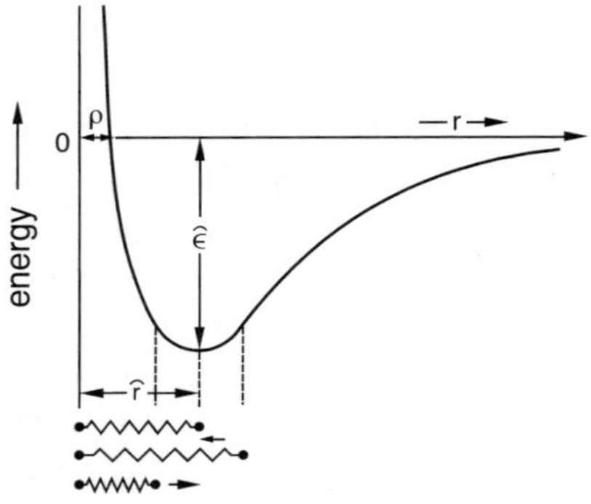

Fig. 2.3: Typical potential curve of a bond. The harmonic region of the displacement is indicated.

and hence for the equilibrium energy

$$\widehat{\epsilon} = A\widehat{r}^{-n}\left(1 - \frac{n}{m}\right) = -B\widehat{r}^{-m}\left(1 - \frac{m}{n}\right) . \tag{2.19b}$$

This makes it possible to express the potential in terms of the equilibrium parameters:

$$\frac{\epsilon}{\widehat{\epsilon}} = \frac{1}{m - n}\left[m\left(\frac{\widehat{r}}{r}\right)^n - n\left(\frac{\widehat{r}}{r}\right)^m\right] . \tag{2.20a}$$

In the following text we will denote an equilibrium value by using an arc over the relevant species. Please note the different meaning of $\widehat{\epsilon}$ compared with the previous paragraphs. There the same symbol denoted the energy of the bonding level at a given nuclear distance, while here it indicates the special value at the equilibrium distance. The Taylor expansion of the expression of Eq. (2.20a) yields a harmonic behaviour (Hooke's law) for small displacements ($r \simeq \widehat{r}$) (see Fig. 2.3):

$$\frac{\epsilon - \widehat{\epsilon}}{\widehat{\epsilon}} = -\frac{1}{2}mn\left(\frac{r - \widehat{r}}{\widehat{r}}\right)^2 . \tag{2.21}$$

A different representation refers r to $\rho \equiv r(\epsilon{=}0)$, that is to the value of r at the zero–point of the potential curve before ϵ rises rapidly (see Fig. 2.3). The parameter ρ is therefore a measure of effective particle size. Since for $r \equiv \rho$ the square brackets in Eq. (2.20a) equal zero, $\widehat{r}$ can be substituted by ρ via $\widehat{r} = \rho(n/m)^{\frac{1}{n-m}}$, yielding

$$\frac{\epsilon}{\widehat{\epsilon}} = \frac{1}{m - n}\left(\frac{n^n}{m^m}\right)^{\frac{1}{n-m}}\left[\left(\frac{\rho}{r}\right)^n - \left(\frac{\rho}{r}\right)^m\right] . \tag{2.20b}$$

For n = 2m = 12 this is the usual representation of the Lennard–Jones potential. The prefactor is then −4.

These considerations are of particular importance when treating lattice energy and lattice vibrations. In the case of ionic bonding a Mie potential, supplemented with a term proportional to r^{-6}, is used usually for the mathematical simulation of the static and also the dynamic[18] behaviour of atomic aggregates. Polarizability and covalency

[18]Molecular dynamics simulation or MD simulation for short.

effects are frequently accounted for by means of charge correction or by implementing a so–called shell model. In simple cases the conceptual difficulties with respect to the validity of the potential functions employed for such "computer experiments"[19] can be avoided or diminished by carrying out a numerical ab initio calculation for the whole problem or by combining quantum–mechanical and empirical calculations [36–38].

2.2 Many atoms in contact: The solid state as a giant molecule

The formation of solids relies on the fact that bonding forces are not saturated in individual molecules. At sufficiently low temperatures the thermal energy is not high enough to "destroy" the intermolecular forces and a many–particle ensemble spontaneously forms a three–dimensional "polymer"[20], i.e. the solid. If the bonding forces in individual molecules are far from being saturated, the many–particle ensemble forms a solid body with a relatively large energy of formation (Fig. 2.4) from the molecules of which it is constituted. If they are approximately saturated then

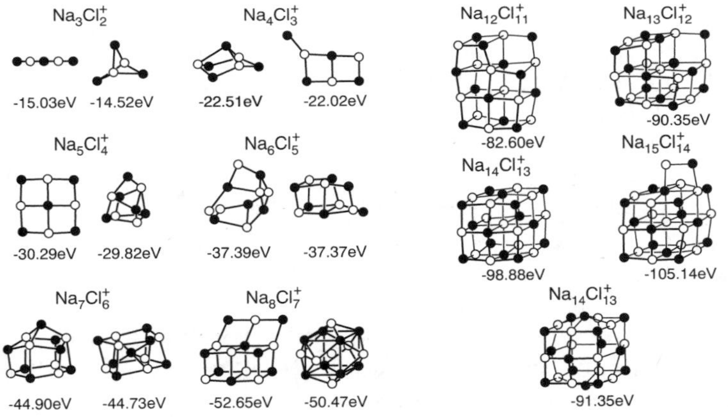

Fig. 2.4: Local stable (ionized) NaCl structures as a function of particle number (as detected in the mass spectrum) ($(NaCl)_N Na^+$). The cubic rock salt structure is already established when N>10. The formation energies (cf. numbers) refer to the reaction of the isolated ions. From Ref. [39].

the solid formed has a relatively low energy of formation and a low melting point. In this case there are large differences between intra– and intermolecular bonding forces.

Figure 2.4 shows the structural development from dimeric Na^+Cl^- to a very strongly bonded solid via (locally) stable, NaCl clusters (as a function of the particle number

[19]The expression "computer experiment" should not be taken too literally. It involves numerical mathematics with artificial input and not an experiment as such, in which, after all, "questions are put to nature". However, it is possible to study the response to varying "external" conditions just as — or sometimes easier than — in a real experiment, and just as there the response needs to be interpreted.

[20]The bonding conditions as a function of the number of particles have been the subject of intensive investigation. This is particularly of interest for oligomeric clusters.

N). In this case the rock salt structure is realized very early on (N>10)[21]. Molecular units can only be compositionally defined but not structurally or energetically identified. The interior of the "giant molecule", the bulk, is structurally "screened" by the surfaces from the ambient. The surface region is, as it were, constituted by the terminal groups of the giant molecule, and is of fundamental importance in particular for the kinetics (see Section 5.4). Phase transformations correspond to rearrangement reactions of the total giant molecule. But let us now address the behaviour of electrons in the solid.

2.2.1 The band model

2.2.1.1 An electron in a potential–free box

As already mentioned the possibility of delocalization of electrons within a macroscopic aggregate is a characteristic of the crystalline solid state. The outer electrons of metals, in particular, behave as a "confined" electron gas. This is readily recognized by considering an arrangement of 4 Na atoms[22]. Such an ensemble represents a markedly electron–deficient state (one valence electron per Na); the energetically relevant mesomeric structures are[23]

$$\begin{array}{ccccccccccc}
\mathrm{Na} - \mathrm{Na} & & \mathrm{Na} \quad \mathrm{Na} & & & & & & \\
 & \leftrightarrow & | \qquad | & \leftrightarrow & & & & & \\
\mathrm{Na} - \mathrm{Na} & & \mathrm{Na} \quad \mathrm{Na} & & & & & & \\
 & & & & & & & & \\
\mathrm{Na}^{+} \quad \mathrm{Na} & & \mathrm{Na} \quad \mathrm{Na}^{+} & & \mathrm{Na} - \mathrm{Na}^{-} & & \mathrm{Na}^{-} - \mathrm{Na} \\
\qquad | & \leftrightarrow & | \qquad\quad & \leftrightarrow & \qquad | & \leftrightarrow & | \qquad\quad \\
\mathrm{Na} - \mathrm{Na}^{-} & & \mathrm{Na}^{-} - \mathrm{Na} & & \mathrm{Na}^{+} \quad \mathrm{Na} & & \mathrm{Na} \quad \mathrm{Na}^{+}
\end{array} \qquad (2.22)$$

all of which contribute to the overall state with the result that the 3s electrons can be regarded as being distributed through the system without any perceptible localization effect and thus as being almost freely mobile.

For a further description, we use the single–electron approximation and confine a single electron for simplicity in a one–dimensional box of length L. We set the potential energy within the box to zero, while it is infinite at the walls of the box.

---

[21]If a(N) is the mean energy of particles at the surface, b(N) the corresponding mean energy of particles in the interior, then it follows approximately for the total energy of a cube that $E = 6aN^{2/3} + b\left(N - 6N^{2/3}\right)$ or $E/N \simeq 6(a - b)N^{-1/3} + b$. (However, the cluster must however be so large that it is possible to neglect edge and corner effects.) According to Ref. [40] this relationship (a, b constant) is already fulfilled for surprisingly small NaCl–clusters (N≥10).

[22]In the real Na crystal the deficit situation is more marked on account of the higher co–ordination number (8).

[23]See also Ref. [25] here. Reference has already been made to the similarity of the situation in conjugated hydrocarbons (see Section 2.1.4) with respect to the p electrons remaining after sp^2 hybridization, even though there is not an electron deficit situation in the strict sense.

As a result of this the Hamiltonian operator reduces to the operator for the kinetic energy that is $-h^2(\partial/\partial x)^2/(8\pi^2 m)$.
The Schrödinger equation for the wave function $|k\rangle$ leads to a linear homogeneous differential equation of the form $(\partial/\partial x)^2|k\rangle \propto -|k\rangle$ for which sine and cosine functions are relevant solutions. At the two ends of our box $\langle k|k\rangle = 0$ and, hence, $|k\rangle = 0$ must hold². The disappearance of the function at $x = 0$ only permits the sine function:

$$|k\rangle \propto \sin kx = \sin \frac{2\pi}{\lambda} x. \tag{2.23}$$

The parameter k designates the wave vector which is reduced here to a component in the x direction; λ is the associated wavelength ($\lambda = 2\pi/k$)[24]. The relationship between k (and hence also λ) and the energy eigenvalues ϵ is obtained by substitution in the Schrödinger equation $(-h^2/8\pi^2 m)(\partial/\partial x)^2|k\rangle = \epsilon|k\rangle$ as

$$k = \pi\sqrt{8m\epsilon/h^2}, \tag{2.24}$$

and is, thus, a square root function.
However, only those wavelengths and, hence, values of k and energy eigenvalues, of which the corresponding wave functions are also zero at the other wall, that is at x=L, are permitted. This is obviously the case if k·L is an integral multiple of π or, in less abstract terms, if the length of the box is an integral multiple of the half wavelength. In other words (Eq. (2.24)) the following simple relationship between the energy eigenvalues and the box length must be obeyed:

$$\epsilon = \frac{h^2}{8m}\frac{n^2}{L^2} \quad \text{with } n = 1, 2, 3, ... \tag{2.25}$$

As can be seen from Eq. (2.25) and Fig. 2.5 the energy increases quadratically as a function of the quantum number n and falls quadratically with the length of the box. The energies of the highest occupied and the lowest unoccupied states at T=0K, ϵ_{HO} and ϵ_{LU}, which also vary with L^{-2} are of particular interest (see below).
Fig. 2.5 then shows us how, at given n in a one–dimensional arrangement, the energy levels decrease with the number of constituent nuclei and thus with the number of electrons introduced. Doubling the size of the one–dimensional atomic box results in a decrease of the energy term by a factor of four (in this model solely as a result of "kinetic energy"). This is — as already indicated above — of great importance in chemical bonding[25]. But back to our problem: The two–centre bond showed us that two identical energy levels become two new levels displaced upwards or downwards by approximately β. In a similar manner 4 Na atoms yield four energetically different levels and N Na atoms N different levels (inset in Fig. 2.5). From this it does not

[24] This interpretation of k follows from the periodicity: $\sin kx = \sin(kx+2\pi) = \sin[k(x+2\pi/k)] = \sin[k(x+\lambda)]$.

[25] However, this must not be taken as evidence of the dominance of kinetic energy in molecule formation, since in the approximate picture the partitioning of the total energy into E_{kin} and E_{pot} is not the correct one.

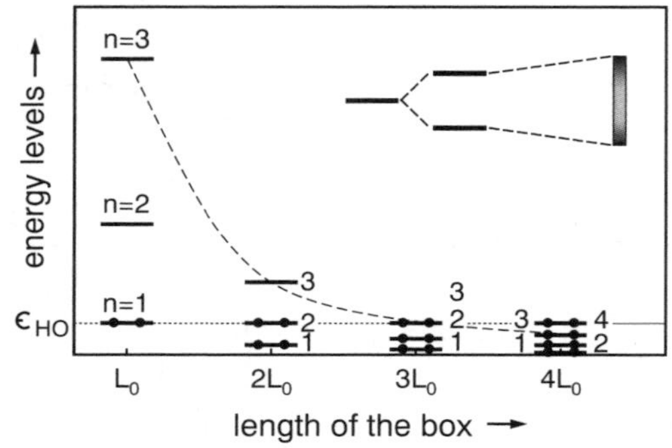

Fig. 2.5: The dependence of the energy level of an electron in a box on the quantum number and the length of the box. If we assume that a box of size L_0 contains just two particles, that of size $2L_0$ four particles etc., the dotted line shows the constancy of the width of the occupied zone (more generally: see Fermi level, Section 5.3). The broken line emphasizes for n=3 the reduction in energy with increasing box length for a given quantum number. The inset shows the splitting of a one–electron state at the start to a continuous band in the MO–LCAO model (cf. Fig. 2.2) for a chain of many atoms. Please note the difference in the density of states.

yet follow that individual sharp energy levels become bands; this is a consequence of the fact that according to Eq. (2.25), the energy difference between two levels decrease strongly (quadratically) as L increases:

$$\epsilon_{n+1} - \epsilon_n = \frac{2n+1}{L^2}\frac{h^2}{8m^2}. \tag{2.26}$$

The $\epsilon(k)$ curve then becomes virtually continuous. Equation (2.26) predicts a density of states which decays with increasing energy. Since the s–band of a sodium metal crystal (to remain with the example above) contains a limited number of states, Eq. (2.25) derived for the free electrons, cannot offer a sufficient description. The alternative treatment, which starts from atomic orbitals and yields the molecular orbitals of a one–dimensional chain via linear combination (cf. Section 2.1), leads to the nonmonotonic density of states shown in the inset of Fig. 2.5. (It follows from the fact that $\epsilon(k)$ depends on k via a cos–function (see following section, Eq. (2.32)); the density of states is great where the graph of $\epsilon(k)$ is flat.) One expects from Fig. 2.2 (and it will become obvious from Eq. (2.32) on page 43) that the width of the band will be of the same order of magnitude, namely $\sim \beta$. Accordingly the splitting of neighbouring levels is of the order of β/N. If there is 1 mol particles and $\beta = 6$eV then β/N is $\sim 10^{-23}$eV corresponding to 10^{-18}J/mol. This may be compared with the thermal molar energy RT which is still of the order of 10J/mol even at 1K.
In addition the density of states changes with dimensionality. This is already so in the case of the quasi–free electron approximation. A three–dimensional box with dimensions $L_x \times L_y \times L_z$, in which the energy eigenfunction has, on account of the factorizability of the wave function, the general form

$$\epsilon = \frac{h^2}{8m}\left(\frac{n_x^2}{L_x^2} + \frac{n_y^2}{L_y^2} + \frac{n_z^2}{L_z^2}\right); \tag{2.27}$$

there are already so many degeneracies at higher quantum numbers for each value of ϵ that, in fact, the density of the states increases with the energy ϵ. Let us consider

a cube with $L_x = L_y = L_z$: Since $n_x^2 + n_y^2 + n_z^2 = n^2 =$ const the triplets (1,1,2), (1,2,1), (2,1,1) all have the same energy value.
If the permitted energy states are set out in a three–dimensional coordinate system with the axes n_x, n_y, n_z then states of the same energy fall into that eighth of the spherical surface which is characterized by exclusively positive coordinates (see Fig. 2.6). Each point on this spherical surface is characterized by a distance

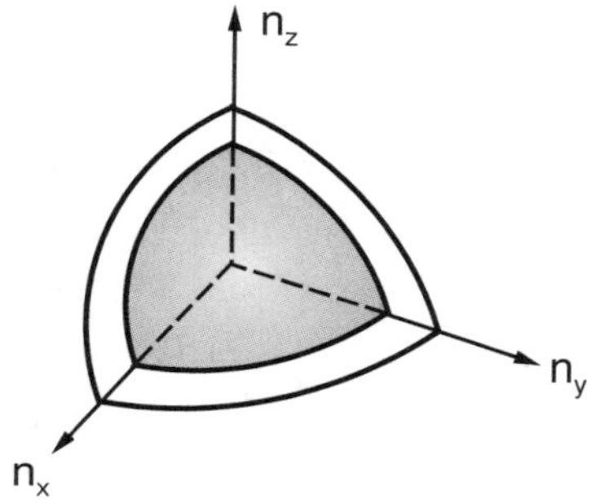

Fig. 2.6: In the space described by the three quantum numbers n_x, n_y, n_z each spherical surface (more accurately only the eighth of the sphere characterized by positive values) is a locus of constant energy values. Two surface sections are shown that have distances from the origin of $n = \sqrt{n_x^2 + n_y^2 + n_z^2}$ and n+dn.

$n = \sqrt{n_x^2 + n_y^2 + n_z^2} \propto \sqrt{\epsilon}$ from the origin of the coordinates and is, thus, associated with the same ϵ value. The same procedure yields the corresponding energy values associated with the energy $\epsilon + d\epsilon$. The radius of the sphere is now greater by $dn \propto d\sqrt{\epsilon} \propto d\epsilon/\sqrt{\epsilon}$. The number of states dZ enclosed by the sections of these two surfaces and thus belonging to energies between ϵ and $\epsilon + d\epsilon$, is simply the volume of the shell section and, hence, is proportional to $(n + dn)^3 - n^3 \propto n^2 dn$ and, hence, to $\epsilon d\sqrt{\epsilon} \propto \epsilon^{1/2} d\epsilon$. The result is that the density of the states increases proportionally with $\epsilon^{1/2}$ as the energy increases:

$$D(\epsilon) \equiv \frac{dZ}{d\epsilon} \propto \epsilon^{1/2}. \tag{2.28}$$

Again, this result does not hold for all band states owing to the finite number of levels. (Realistic densities of states are shown e.g. by Fig. 2.12 on page 45).
Let us return to the one–dimensional box problem and to Fig. 2.5. The enormous stabilization of the electron as a consequence of increased box size is reflected by the energy level ϵ_{HO}, the corresponding quantum number of which, n_{HO}, represents the half particle number and hence marks the occupation limit. In so far as the box length scales proportionally with the particle number, ϵ_{HO} remains constant.
In the case where $L \longrightarrow \infty$ the energies (at T=0K) of the highest occupied (ϵ_{HO}) and the lowest unoccupied (ϵ_{LU}) bands will be almost identical. We term the limiting value ϵ_F. The more precise significance of this Fermi energy will be apparent later. Obviously ϵ_F is independent of the size of the system (see however Section 5.8). One might initially suppose that this is only the case for ^{1}D, since for ^{3}D the number of electrons increases as L^3 and not as L^2. But one can verify from Eq. (2.27) that the invariance also applies to ^{3}D and also to the nonisotropic case ($L_z \neq L_x \neq L_y \neq L_z$). These considerations naturally only apply at constant electron concentration under

the idealized conditions of Fig. 2.5. Let us, for simplicity, consider a cube of side L. According to Eq. (2.27) the Fermi energy is $\epsilon_F = \frac{h^2}{8m}\frac{n_F^2}{L^2}$, n_F^2 being the sum of the squares of the three corresponding quantum numbers. The relation between n_F and the total number of electrons N can be obtained from Fig. 2.6: Owing to Eq. (2.27) surfaces of spheres (more accurately the eighth containing the positive numbers) in the $n_x n_y n_z$–space are loci of constant energy. At ϵ_F, i.e. for n_F there are exactly $1/8(\frac{4\pi}{3}n_F^3)$ energy levels and twice as many electrons in the eighth of the sphere. If N is the total number of electrons it follows that $n_F = \left(\frac{3}{\pi}N\right)^{\frac{1}{3}}$ and with Eq (2.27) $\epsilon_F = \frac{h^2}{8m}\left(\frac{3}{\pi}\frac{N}{L^3}\right)^{2/3} \propto (\text{electron concentration})^{2/3}$. Such a proportionality also applies to the total width of a band generated from a single level according to the inset of Fig. 2.5.

2.2.1.2 An electron in a periodic potential

Until now we have neglected the periodic potential produced by the atomic cores. We will now turn to it. In this case the probability density of the electron and, hence, — apart from a phase factor e^{ikx}, which disappears in the square of the absolute value — the wave function itself is of this periodicity. These Bloch waves are, thus, plane waves whose amplitudes are modulated according to the lattice period. Let us consider once again a one–dimensional model in which the potential thresholds are approximated by rectangular boxes, distributed with translational symmetry at a distance a from each other in the solid. Now let us allow the width to approach zero, while the height approaches infinity at the same rate, so that the area, which is a measure of the local obstacle, remains constant (delta function–shaped potentials). A lengthy, but elementary calculation[26] reveals that the solution of the Schrödinger equation now requires the fulfilment of the following relationship:

$$\Gamma\frac{\sin\kappa a}{\kappa a} + \cos\kappa a = \cos ka. \qquad (2.29)$$

Γ is proportional to the area, and hence is a measure of the strength of the potential wall; the parameter κ is proportional to the square root of the energy value. The left hand side of Eq. (2.29) consists of an amplitude–modulated, periodic function, while the amplitude of the cos–function of the right hand side always lies between +1 and −1. Hence, Eq. (2.29) is only fulfilled for those values of κa, and hence of ϵ, for which the amplitude of the left hand side varies between these limits. This is shown graphically in Fig. 2.7. One can see that the width of a permitted band increases with increasing ϵ. It also increases with a decreasing threshold. The band centres are equidistant in κa, i.e. in $\sqrt{\epsilon}$; the distances, thus, increase with energy, just as it is the case for the individual levels in a one–dimensional, potential–free box.

[26] In addition to the problem of the electron in the box, it must be taken into account that the amplitude function of the Bloch wave fulfils the periodicity and remains continuous and differentiable at the points at which the potential exhibits discontinuities [41–43].

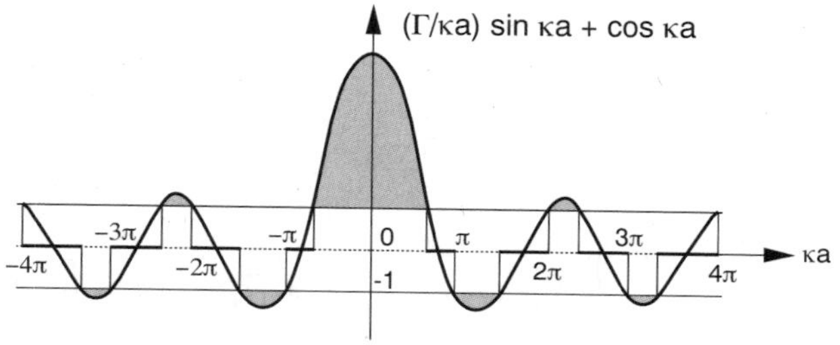

Fig. 2.7: Eq. (2.29) can only be satisfied in the ordinate range +1 to −1. The dependence of the left hand side of Eq. (2.29) on κa and, hence, on the energy and lattice constant is shown ($\Gamma = 3\pi/2$). The forbidden regions are shaded.

Now we will let the barriers in the solid body disappear by allowing Γ in Eq. (2.29) to approach zero. The result is then cosκa = coska which must be identically satisfied. From κ = k we find again the continuous eigenvalues of the electron in the macroscopic box of length L ($\epsilon \propto k^2$, Eq. (2.25)). However, if the barrier strength becomes infinite, we then break down our solid into unbonded regions, the left hand term in Eq. (2.29) predominates, so that the equation can only be satisfied if sinκa simultaneously approaches zero. It is only then that the expression Γsinκa/κa remains finite as demanded by the right hand side. Then κa must be an integral multiple of π and there is a relationship for ϵ analogous to Eq. (2.25), now, however, with a as the effective length of the box:

$$\epsilon \propto \frac{n^2}{a^2}. \tag{2.30}$$

Interpretation of these results with regard to a real solid is simple but far–reaching (Fig. 2.8): Low energy (inner) electrons experience the local box potential, i.e. the atom potential as insurmountable and are trapped in the box of atomic dimensions.

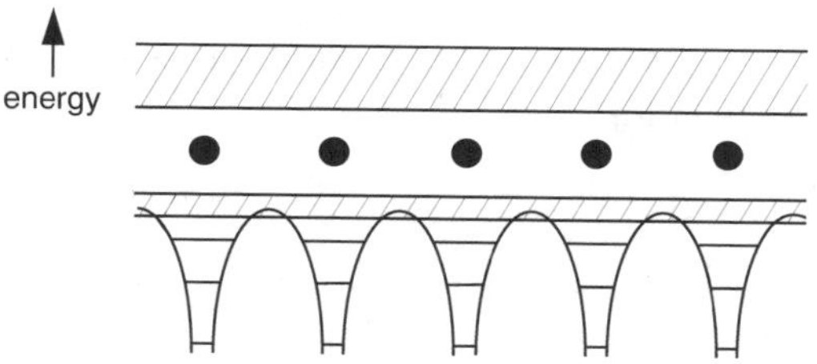

Fig. 2.8: Electrons in the periodic potential of the nuclei (•), schematic (e.g. Na). Inner electrons (in the case of Na: 1s, 2s, 2p) may be compared to electrons at an isolated atom, outer electrons (3s) are quasi–free. The sharpness of the levels of the inner electrons has been exaggerated for the purpose of illustration.

The energy levels are discrete (splitting $\propto a^{-2}$). (Note, that the unrealistic increase of the difference between the energy levels with increasing ϵ predicted by Eq. (2.30) is a consequence of the unrealistic potential.) The situation is different for the more energetic "outer" electrons, which are responsible for bonding: They "overlook" the local periodic potentials, only experience the actually insurmountable outer walls of the box, and so are delocalized over the whole crystal. The splitting of the energy levels is tiny ($\propto L^{-2}$), and bands are formed. The width of these bands increases with energy. Highly energetic electrons are naturally the most delocalized ones, that is, they have the largest $|\beta|$ (see above).

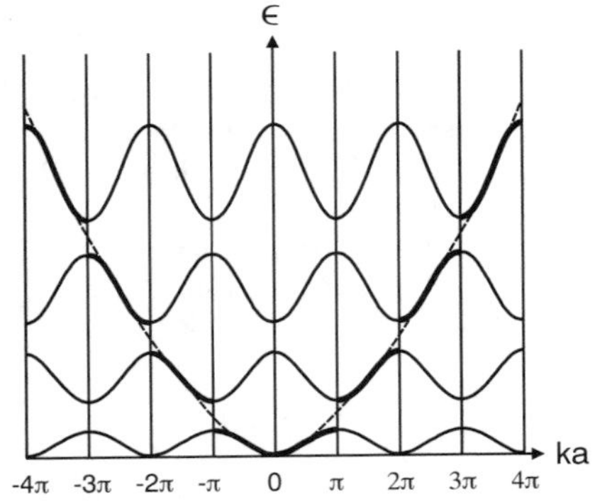

Fig. 2.9: Energy as a function of the wave vector in the model of the nearly free electron[27]. The presentation as an extended scheme is redundant. The whole information is contained in the inner zone ($[-\pi, +\pi]$), i.e. the 1st Brillouin zone. The parabola for the free electron is drawn in.

Figure 2.9 illustrates the approximate dependence of the energy on the wave vector[27]. The picture is very similar to the parabolic form of a free electron (see Eq. (2.23)); however, there are deviations (see the thick lines) as a result of the obstacles we have inserted (a,2a,3a etc.)[28]. We remember that the Schrödinger equation is a wave equation. We expect diffraction effects at the relevant positions in the reciprocal space (k space) marked in Fig. 2.9[29,30]. In the case of a small box, it is true that ϵ quadratically depends on a, but there are only a few discrete points. For a large box the function becomes continuous. Since we imagine our periodic solid as composed (cf. Fig. 2.2) of small boxes forming a large box[28], we expect a behaviour according to Fig. 2.9.
Owing to the periodicity of the thresholds reflected by the periodicity of $\cos \mathrm{ka}$ in Eq. (2.29), κ and hence, ϵ must exhibit the periodicity of the lattice. The thinner lines in Fig. 2.9 correspond to the presentation in the extended energy scheme. On account of the periodicity the presentation in the reduced scheme, which limits itself to the inner zone, the so–called 1st Brillouin zone, suffices. At small k values, i.e. sufficiently distant from the critical positions responsible for band gaps, the parabolic form is expected. In fact for small ka values it holds that

$$\cos(\mathrm{ka}) \cong 1 - (\mathrm{ka})^2/2 + \cdots \tag{2.31}$$

[27]The real solutions of the Kronig–Penney problem are slightly different to Fig. 2.9 [42].

[28]This is expression of the "bonding of the unit cells". Compare the splitting in Fig. 2.9 with Fig. 2.2.

[29]While the meaning of a reciprocal space in ^{1}D is trivial (cf. periodicity in Fig. 2.9), it is defined vectorially in ^{3}D. If $\mathbf{x}_1$, $\mathbf{x}_2$, $\mathbf{x}_3$ are vectors of real space and $\check{\mathbf{x}}_1, \check{\mathbf{x}}_2, \check{\mathbf{x}}_3$ those of reciprocal space, it then follows that $\check{\mathbf{x}}_1 = \frac{\mathbf{x}_2 \times \mathbf{x}_3}{[\mathbf{x}_1, \mathbf{x}_2, \mathbf{x}_3]}$ (1, 2, 3 cyclic). The denominator represents the scalar triple product of the vectors $\mathbf{x}_1$, $\mathbf{x}_2$, $\mathbf{x}_3$ and, hence, the volume of the unit cell of the real lattice (= reciprocal volume of the unit cell of the reciprocal lattice). It can be seen that $\mathbf{x}_\mathrm{i}\check{\mathbf{x}}_\mathrm{j} = \delta_\mathrm{ij}$ ($\delta_\mathrm{ij} = 0$ for $\mathrm{i} \neq \mathrm{j}$ and $\delta_\mathrm{ij} = 0$ for i=j). The k–space is realized by multiplying all reciprocal vectors by $2\,\pi$.

[30]Roth in Ref. [44] provides a very vivid analogy, evidently from his own experience, which therefore fails in many respects: A very badly sprung small car is travelling over a desert track that has been made rough by the passage of camels (the periodic roughness is of the order of m). At velocities of 30±2 km/h there is an uncomfortable resonance so that a velocity range of $\sim$ 4km/h is "forbidden".

On the other hand cos–functions are, as already mentioned, also obtained for $\epsilon(\mathrm{k})$ on construction of the solid state from atomic functions (cf. Section 2.1)[31]. Consideration of an infinite chain of hydrogen atoms with the period a in a Hückel approximation analogous to the two–centre model yields solutions of the form[32,33]:

$$\epsilon(\mathrm{k}) = \alpha + 2\beta \cos \mathrm{ka}. \tag{2.32}$$

Each individual solution (see Fig. 2.10) corresponds to a band which can be thought to be formed from the individual two–centre orbitals (1s, 2s, etc.). One can see

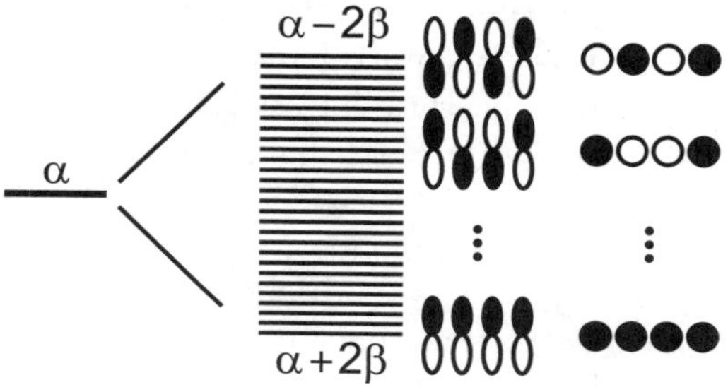

Fig. 2.10: The overlap of orbitals leads to a band. The energy increase with the number of nodes (separating white from black parts) in the wave function is shown for the formation of π–bands from p–orbitals and of σ–bands from s–orbitals (energy increases with increasing k–value)[33,34].

with the aid of these considerations that $|\beta|$ is also quantitatively a measure of the band width, as was presumed in Section 2.2.1.1: The maximum of this function is at $\alpha + 2|\beta|$, the minimum at $\alpha - 2|\beta|$. The difference between them, the band width[34], is hence $4|\beta|$. The band gap is the distance between the maximum of one band (e.g. 1s) and the minimum of the next higher (2s) band. This distance (at least) must be traversed for electron transition between the bands. The gaps between 1s band and 2s band[34] follows from Eq. (2.32) as

$$\epsilon_{\mathrm{g}} = (\alpha_{2\mathrm{s}} - \alpha_{1\mathrm{s}}) + 2(\beta_{2\mathrm{s}} + \beta_{1\mathrm{s}}). \tag{2.33}$$

As expected, ϵ_{g} is substantially influenced by the difference in the Coulomb integrals[35].

---

[31] LCAO–MO–theory in Hückel approximation ("tight binding").

[32] Here, too, it is only possible to neglect the overlap integral to a coarse approximation. If we take it into account, the antibonding effect turns out to be stronger than the bonding effect (analogously to the X_2 problem, see Fig. 2.2) (The expression on the right hand side of Eq. (2.32) must then be divided by $1 + 2\mathrm{S}\cos \mathrm{ka}$.)

[33] Considering a chain of s–states (cf. Eq. (2.32)) or the formation of π–bonds from p–orbitals (cf. conjugated double bonds in polyenes): The most bonding state is at k=0 (all atomic functions are of the same sign), while, in the state of greatest oscillation, the energetic situation is unfavourable (alternatingly opposite sign); this corresponds to the most antibonding state. When, however, p states overlap to form σ bonds (i.e. linear alignment of p–orbitals) the maximum obviously lies at k=0 (see e.g. Ref. [45]).

[34] In the three–dimensional case Eq. (2.32) has to be modified correspondingly. The treatment for the primitive cubic lattice yields an analogous relation, but with a sum of three cosine functions. Hence energies between $\alpha + 6\beta$ and $\alpha - 6\beta$ become possible within a band, and Eq. (2.33) has to be corrected accordingly.

[35] There is a detailed discussion in Ref. [31].

If the relevant maxima and minima lie at the same k value as it is fulfilled in Fig. 2.9 we speak of a direct transition. In the case of an indirect transition, that is if these extreme values do not lie at the same value of k, as in the case of the s–bands (H chain) discussed, the parameter ϵ_g in Eq. (2.33) merely represents the thermal energy gap[36]. An optical energy transition is (without phonon support) only possible at constant k value. The "optical energy gap" is then accordingly greater.

The distinction between nonmetals and metals is based on the occupation of the bands at absolute zero. If the energetically highest nonempty band is not completely occupied, we speak of a metal (as in our artificially equidistant H chain)[37,38]. Electrons within it are almost freely mobile. If this nonempty band is completely occupied (as in the case of an analogous He chain), substantial electronic excitation is required for transport — either thermally or optically. If the band gap is small, this excitation is perceptible, and one speaks of semiconductors (such as Si, Ge); if the gap is very large, of (electronic) insulators (such as diamond or NaCl). The demarcation is arbitrary. These electronic phenomena belong to a discussion of defects (see Section 5.3). Here it is sufficient to point out the sensitivity of the effect with respect to the value of the band gap ϵ_g between the topmost occupied band (valence band) and the lowest unoccupied band (at T=0K) (conduction band): The fraction of electrons overcoming this gap at finite temperature is governed by the square root of the Boltzmann factor $\exp -(\epsilon_g/RT)$; at 300K this square root term is 4×10^{-9} for the semiconductor silicon ($\epsilon_g \simeq$1eV), $\sim 10^{-42}$ for the insulator diamond (5eV) and as low as 10^{-84} for the insulator NaCl (10eV). A more accurate consideration is given in Section 5.3; there, in particular, the role of the Fermi energy is clarified in regulating the occupation.
The crystals of the alkali elements are metals, since the outer orbital is only occupied by one (s) electron, and thus, the orbital interaction of N atoms leads to a half–filled

[36]In the model involving a nearly free electron in a one–dimensional periodic box discussed on the previous pages the maxima and minima are at the same k–value, since the interrupted parabola in Fig. 2.9 must result. In the case of s–bands (i.e. transition from 1s– to 2s–band, see Eqs. (2.32,2.33)) (H(1s)–chain), however, a direct band gap opens when one proceeds from an equidistant H–chain (1s) to a H_2–chain. Note the analogous situation in Fig. 2.9 (see e.g. [46] and following footnote).

[37]It should not remain unmentioned that, under normal conditions, the assumed equidistance within the H chain is very artificial. A chain of H_2 pairs naturally has a very much lower energy. Such a perturbation of the translational symmetry, as a result of reduction in the energy by local movement together and apart, is referred to as Peierls distortion. It occurs also in those one–dimensional systems, in which such an effect is not evident from the molecular standpoint (see polyacetylene, Fig. 6.15, page 293).

[38]This is not necessarily correct for narrow bands. Owing to insufficient overlap, band conduction may be impossible. (Section 2.2.5). Care must be taken to ensure that the particle distance specified by the Mott–Hubbard criterion [47], is not exceeded, so that delocalization is possible (see Section 2.2.5).

band with N levels and a capacity to hold 2N electrons:

$$\begin{array}{ccc} \mathrm{N\,Na} & \rightleftharpoons & \mathrm{Na_N} \\ \mathrm{N\,3s^{1(2)}} & & (\mathrm{3s-band})^{\mathrm{N(2N)}}. \end{array} \tag{2.34}$$

These electrons can be excited and transported without any appreciable energy input, i.e. by smallest electrical driving forces, and the conductivity is said to be metallic. Figure 2.11 shows the energy splitting as a function of the distance between

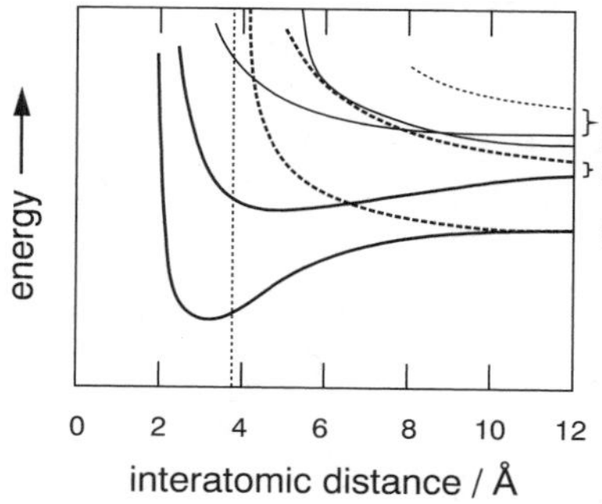

Fig. 2.11: Illustration of the broadening and energetic position of the energy levels in the sodium crystal as a function of interatomic distance. The 3s and 3p bands overlap at the equilibrium distance (at $\sim$ 4Å). From Ref. [48].

the sodium nuclei. On closer inspection it can be seen that the s and p bands overlap at the equilibrium distance and the relationship (2.34) above is not complete[39]. Figure 2.12a also shows this on the basis of the density of states. (Here it should be

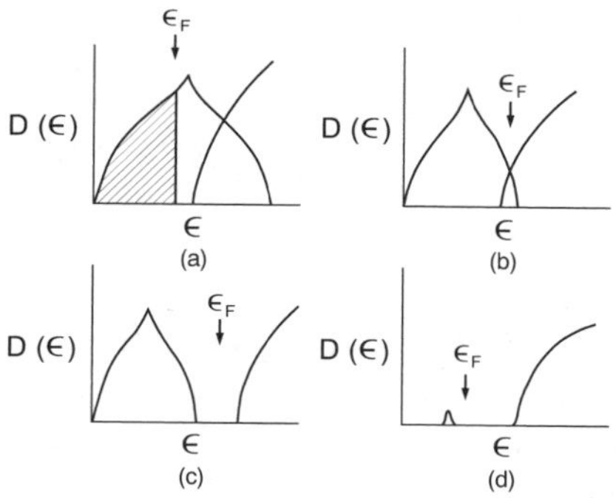

Fig. 2.12: Density of state as a function of the energy in the band model, simple examples. The levels at T=0 are occupied upto ϵ_F.
a) Normal metal
b) Semimetal
c) Semiconductor or insulator
d) Semiconductor or insulator containing impurities (see Chapter 5)

noted that this quantity cannot increase according to a $\sqrt{\epsilon}$ function for all energies, as predicted by Eq. (2.28)). Rather it must — since the number of levels in a band is fixed — return to zero again.) The s–p overlap mentioned is the reason that the crystals of the alkaline earth elements (Fig. 2.12b) are also metals. However, their metallic properties are less marked. These elements, whose outer s orbital is doubly occupied, would otherwise possess completely occupied s bands. Rather it holds that

$$\begin{array}{ccc} \mathrm{N\,Mg} & \rightleftharpoons & \mathrm{Mg_N} \\ \mathrm{N\,3s^{2(2)}3p^{0(6)}} & & (\mathrm{3s-3p-band})^{\mathrm{2N(6N)}}. \end{array} \tag{2.35}$$

[39]Such overlaps are obviously not surprising in view of the band width in the crystal and the distance between the sharp energy states in the isolated atom.

Conversely, a silicon crystal, in spite of the ground state configuration $KL3s^23p^2$ of the element, is not a metal but a semiconductor (Fig. 2.12c). As discussed in Section 2.2.1 the bonds formed are sp^3 bonds. The splitting between bonding and antibonding levels is not compensated by band broadening. Figure 2.13a illustrates this point. The sp^3 hybrid orbital energy ϵ_h is obtained from ϵ_s and ϵ_p according to

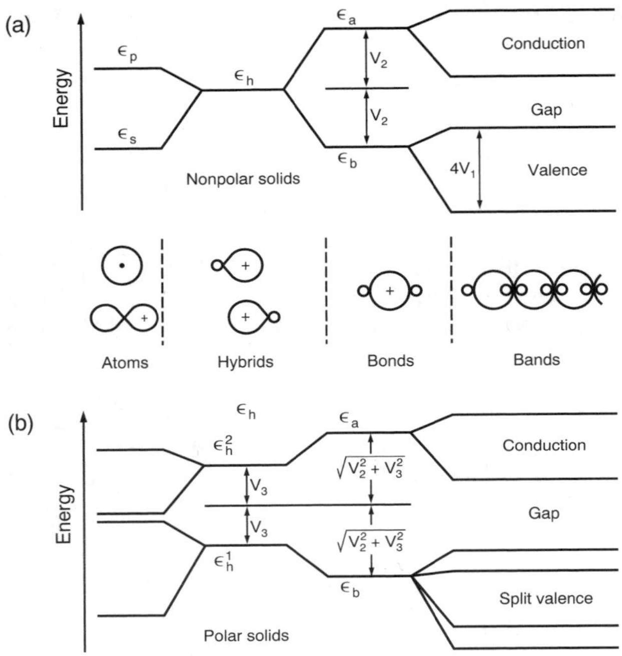

Fig. 2.13: a) The formation of a nonpolar covalent crystal made up of sp^3 bonds. b) The formation of a polar covalent crystal via sp^3 bonds. From Ref. [26].

$\epsilon_h = \frac{1}{4}\epsilon_s + \frac{3}{4}\epsilon_p$. The split in level is approximately given by the "covalent energy" V_2, which corresponds to our resonance integral (cf. Section 2.1), but referred to the hybrid functions. For the purpose of a qualitative discussion the overlap integral is neglected. Figure 2.14 displays the energy as a function of the interatomic distance[40]. At large distances Si would indeed be metallic corresponding to the lower

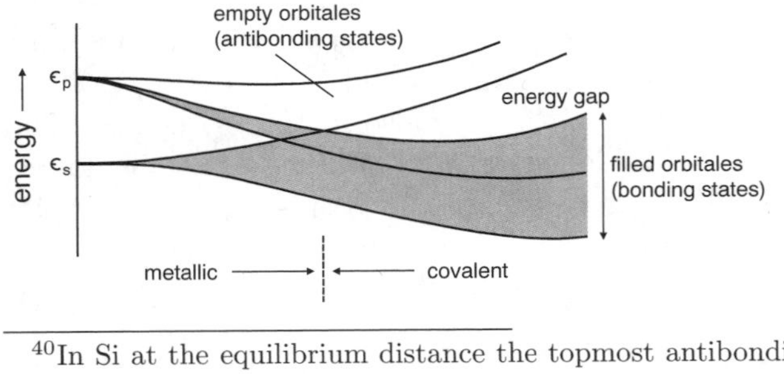

Fig. 2.14: Energy levels as a function of the interatomic distance in the case of elements of the IVth main group[40] (diamond, silicon, germanium, α–tin). The interatomic distance decreases from the left to the right. From Ref. [26].

[40]In Si at the equilibrium distance the topmost antibonding s orbital is over its p counterpart. This does not correspond to Fig. 2.14. The situation represented there corresponds rather to germanium [49]. However, such nuances are not of significance for our treatment.

overlap. The p band would be one–third occupied in accordance with the ground configuration. As the distance becomes smaller the bands do not merely broaden as a result of increased overlap, a gap also forms as a result of the pronounced splitting between bonding and antibonding hybrid levels. The lower band is completely occupied corresponding to the formation of a "closed shell" in the molecular picture. The gap is ca. 1eV at the equilibrium distance so that silicon is a semiconductor. As can be seen from Fig. 2.13 and as can be demonstrated [31], $\epsilon_p - \epsilon_s$ approximately determines the width of the sp^3 valence band ($4V_1$). Hence, the ratio of the s–p splitting and the bonding–antibonding splitting (V_2) is essential for the question of whether the solid is metallic or semiconducting. V_1 is frequently referred to as the "metallic energy" in this context[41].

At the same time Fig. 2.14 also represents the variation in the equilibrium bonding situation for the IVth main group from the (semi–) metal α–Sn via the semiconductors Ge and Si to the insulator diamond.

In ionic solids such as NaCl the electrons under study belong (almost) exclusively to the electronegative atoms. The degree of delocalization is small as reflected by the formation of narrow bands (cf. Section 2.2.2). Figures 2.13b and 2.15 illustrate the situation for polar solids. We consider the iso–electronic series Ge, GaAs, ZnSe

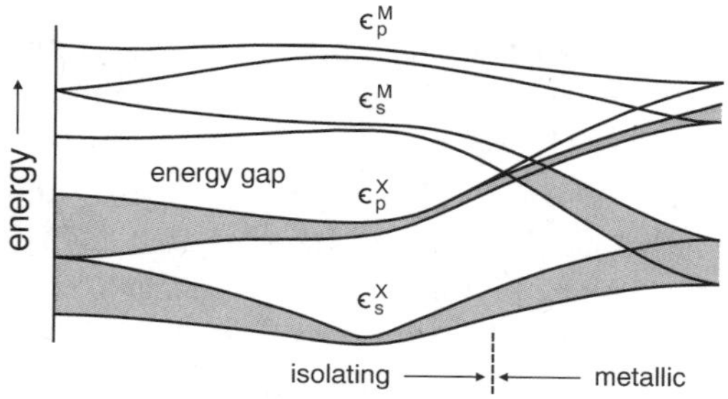

Fig. 2.15: The change in energy levels in MX crystals on variation of ionicity (at the expense of covalency) and metallicity (see text, cf. also Fig. 2.14). From Ref. [26].

and CuBr. They all possess the same crystal structure and our discussion concerns the sp^3 orbitals (see Fig. 2.13b). In analogy to $\Delta\alpha$ in Section 2.1.2 it is sensible to define a "polar energy" with respect to the two, now different hybrid levels (2.13b) which is termed V_3 [26]. The bond is then substantially characterized by $\sqrt{V_2^2 + V_3^2}$ analogously to $\sqrt{(\Delta\alpha)^2 + \beta^2}$ in Section 2.1.2. Figure 2.15 reveals how, as the polarity or ionicity ($V_3/(V_2^2 + V_3^2)^{1/2}$, see page 28) increases, individual s and p bands are formed again, and that they become narrower on account of the overlap becoming less. The ratio $V_1/(V_2^2 + V_3^2)^{1/2}$ is referred to as metallicity [26], while covalency is defined as the ratio $V_2/(V_2^2 + V_3^2)^{1/2}$ (cf. definition for the dimer, footnote 9).

If we move further to the right in Fig. 2.15 we destroy the strong polarity by making the atoms more similar while metallicity is allowed to increase (s–p splitting). In the resulting metallic state it is no longer the difference between the electronegativities

[41] A detailed treatment is given in Ref. [26].

but between ϵ_s and ϵ_p that is important. The interplay between the bonding–antibonding splitting, the s–p splitting and the polarity, as reflected in the parameters covalency, metallicity and ionicity[42], approximately defines the various bonding types, as is seen from the "phase diagram" represented by Fig. 2.16.

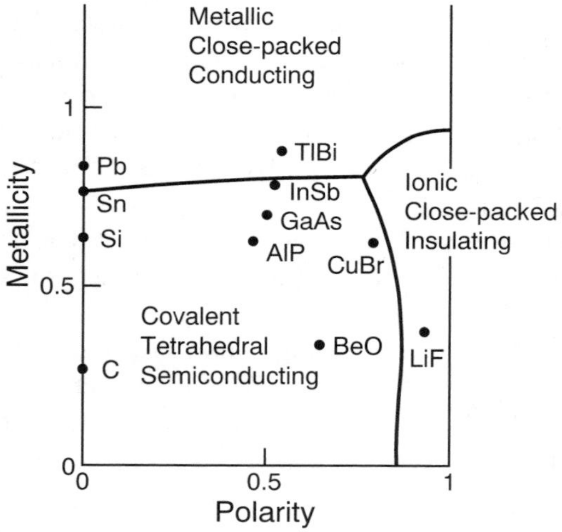

Fig. 2.16: Demarcation of the various simple bonding types by means of the parameters metallicity and polarity (ionicity)[42]. From [26].

While in the local picture[14] the bonding electrons of polar solids with completely occupied bands exhibit a probability density which is shifted toward the anion coordinate and can be approximately considered as being situated at the anion, the electrons of the intermediate forms to the metallic bonding are partially freely mobile. In the case of semimetals the upper edge of the valence band is just a little above the lower edge of the conduction band (see α–Sn), so that the former always exhibits a small number of holes and the latter a small number of electrons. In the case of bismuth[43] this concentration is $\sim 3 \times 10^{17}/\mathrm{cm}^3$, corresponding to a relatively low conductivity[44]. In contrast to the semiconductor, the conductivity does not disappear at absolute zero. However, the majority of the outer electrons is tightly bound (Fig. 2.12b).
The situation is similar in the case of the intermetallic compounds (e.g. Mg_2Pb); here, however, the majority of the outer electrons is situated at the anion. In Fig. 2.16 these intermediate forms are to be found in the neighbourhood of the (upper) right hand boundary line.
Figure 2.17 illustrates realistic band structures in k–space[29] for crystals of the zinc-blende type (see below, Fig. 2.23). The $\epsilon(\mathbf{k})$ dependence naturally corresponds to a four–dimensional presentation. In order to present this graphically, ϵ is normally

[42] Note that covalency and polarity are not independent of each other.

[43] Bismuth has 5 outer electrons, but has 2 atoms per lattice cell and, thus, pairs can be thought to be formed (Bi_2 as the smallest structural unit).

[44] For conductivity of metals an effective electron concentration must be taken into account, that is proportional to $(\mathrm{d}\epsilon/\mathrm{dk})$ at the Fermi level.

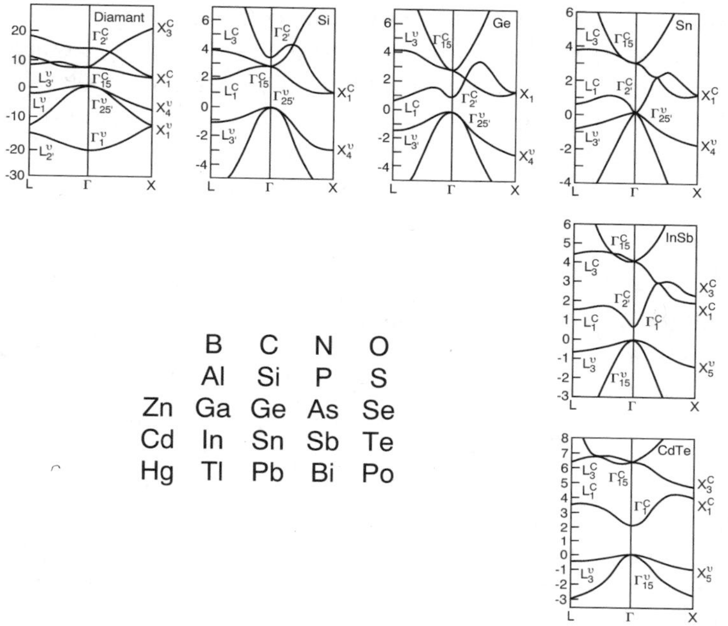

Fig. 2.17: Band structure of AB crystals of the zinc–blende type. The IVth main group — from top to bottom — is represented in the upper row (diamond structure if A=B). In the right hand column the vertical distance in the periodic table is increased iso–electronically. According to Ref. [31].

represented along a labelled path in reciprocal space[29]. The points L, Γ, X are points of high symmetry[45]. The Γ point corresponds to the origin (zone centre). The upper row in Fig. 2.17 illustrates the variation discussed from the insulator diamond, via the semiconductors Si, Ge to the semimetal Sn. The column refers to the isoelectronic increase in polarity by which the gap becomes greater and the bands narrower (Sn, InSb, CdTe) as discussed above.
At this point it is worth pointing out once again that this method of treatment, which assumes nearly free electrons[46], and that of the other extreme, namely taking electrons tightly bound to individual atoms as a starting point, both lead to similar results. Since the deeper–lying electrons do not play any significant role in many problems, it is frequently unnecessary to take into account all atomic nuances: The very successful concept of the so–called pseudopotentials[47] is based on this fact. The reader is referred to the literature for further details [51]. In general it can be stated that the nearly–free electron models are better suited for the treatment of electrons in the conduction band while the approximation starting from overlap of atomic orbitals is better suited for description of the electrons in the valence band. It is not possible to go here into the reasons why the one–electron approximation functions so surprisingly well in spite of the enormous simplifications. We also do not touch

[45]Cf. Refs. [42,49].

[46]The Kronig–Penney model (described in Section 2.2.1.2) is naturally useless for more accurate calculations. It has come into prominence once again for the treatment of semiconductor superlattices [50].

[47]The pseudo–potential acts as a , comperatively weak, effective perturbation of the free electron.

upon the numerous extensions to take account of electronic correlations[48]. Disregarding the detailed description of electronic properties it is evident that bonding descriptions of the solid state correspond well to what is expected from the molecular picture[14]. So we will now consider the types of solids corresponding to the various bond types and take the two–centre model as our starting point. According to their significance for the following chapters we start with ionic crystals.

2.2.2 Ionic crystals

What happens on the contact of very many, we shall say N, sodium and chlorine atoms? In a thought experiment let us consider Na and Cl atoms in pairs. The sodium atoms will transfer their 3s electrons to the chlorine atoms. The ions formed interact primarily via the Coulomb forces, which are now completely nondirectional in nature, i.e. all the pairs of the thought experiment will arrange themselves spontaneously into a giant, translationally symmetrical, three–dimensional Coulomb polymer, i.e. an ionic crystal. The quantum-mechanical repulsion, which is only effective at small distances, prevents penetration of the ionic shells. (This would correspond to the unfavourable penetration of two orbital systems of noble gas configuration, see Section 2.) The bonding energy of the "giant Coulomb polymer" is now, according to Eq. (2.13) primarily, apart from the ionization potential and the electron affinity, determined by the Coulomb energy of the total ensemble. (As shown in Fig. 2.4 (page 35), this crystal behaviour is established for even relatively small N.) When there is pronounced ionic bonding, the contribution of repulsion is not more than 10%. So we will ignore these effects together with polarization effects for the moment[49]. The summation over all Coulomb effects for a particular crystal structure leads in the case of the rock salt structure of NaCl (z: charge number, here $|z_i| = |z_j| = 1$) to

$$\begin{aligned} E_{Cou} &= \frac{1}{4\pi\varepsilon_0}\frac{1}{2}\sum_{i\neq j}\sum_j z_i z_j \frac{e^2}{R_{ij}} \\ &= \frac{1}{4\pi\varepsilon_0} N\frac{2e^2}{a}\left(-\frac{6}{\sqrt{1}}+\frac{12}{\sqrt{2}}-\frac{8}{\sqrt{3}}+\frac{6}{\sqrt{4}}\pm\cdots\right) = -\frac{1}{4\pi\varepsilon_0}N\frac{2e^2}{a}1.748 \\ &= -\frac{1}{4\pi\varepsilon_0}N\frac{2e^2}{a}f = -\frac{1}{4\pi\varepsilon_0}N\frac{e^2}{b}f. \end{aligned} \quad (2.36)$$

As can readily be concluded from the crystal structure in Fig. 2.18, each Na^+ ion possesses 6 Cl^- ions as nearest neighbours at a distance a/2 (a: lattice constant), as next nearest neighbours 12 Na^+ ions at a distance of $(a/2)\sqrt{2}$, after that 8 Cl^- atoms at a distance of $(a/2)\sqrt{3}$, etc. The environment for a Cl^- ion is analogous. The distances R_{ij}, over which the summations are made, are all naturally proportional

[48]The fact that the correlation energy — as well as the total energy — is a functional of electron density is very helpful in this context. The Nobel prize of the year 1998 was awarded to W. Kohn [52] not least for the proof of this significant theorem.

[49]However, we implicitly make use of such effects when we introduce Madelung number and lattice constant. Both are determined by the crystal structure, which (apart from the charge) critically depends on ionic radii (and polarizabilities) and hence on repulsion terms (and polarizabilities).

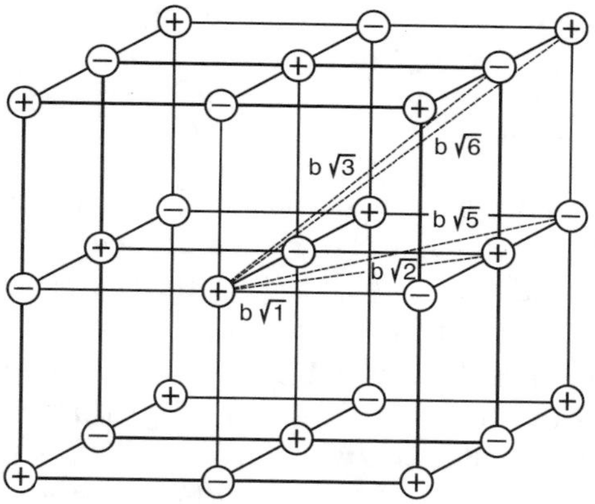

Fig. 2.18: The distances in the rock salt lattice to the nearest neighbours, next nearest neighbours, etc. The nearest neighbour distance (b) is half the lattice constant.

to the lattice constant a, and thus to the distance between nearest neighbours b (here = a/2). The resulting, poorly converging sum of these proportionality factors, multiplied by the corresponding co–ordination numbers yields the Madelung constant[50] f in Eq. (2.36). It is characteristic of the lattice type and independent of the lattice constant[51] (and in the symmetrical case of equal absolute charge, $|z_i| = |z_j|$, independent of the charge number). For the rock salt structure it is 1.748. The absolute value of the Coulomb energy per mol substance (i.e. formed from 1 mol monomer, N_m=Avogadro's number) is normally referred to as the Madelung energy:

$$E_{Mad} = -\frac{1}{4\pi\varepsilon_0} N_m \frac{z_1 z_2 e^2}{b} f. \tag{2.37a}$$

The literature contains differing definitions for Madelung constants[52]. This is particularly true for ionic crystals with differing absolute charges of the cation and anion such as for CaF_2 or Al_2O_3. In order not to be so dependent on the absolute charges in such cases it is appropriate to define reduced Madelung constants f^*. If M_mX_x is

[50] Because of this poor convergence the range of electrostatic interactions, with respect to the chemistry of ionic crystals, is frequently wrongly estimated. It can be shown that a summation over suitably collected dipolar NaCl units leads to rapid convergence (the effective Coulomb potential decays then with the fifth power of the distance) [53]. Direct evidence of the dominance of the immediate environment is also provided by the information that sublimation and evaporation energies only differ slightly, even for markedly ionic crystals [54]. Compare in this connection Fig. 2.4 and footnote 21 on page 36.

[51] For an equidistant chain of alternating positive and negative ions separated by a distance b we immediately obtain the following result

$$f = 2\left(\frac{b}{b} - \frac{b}{2b} + \frac{b}{3b} \mp \ldots\right) = 2 \lim_{x\to 1} \ln(1+x) = 2\ln 2.$$

[52] It is always necessary to check whether the Madelung constant is referred to the shortest cation–anion distance (as here), the lattice constant, to the edge length of a cube containing a formula unit, whether the largest common denominator is included in f or not etc., see [55,56,26].

the chemical formula of such a heterovalent crystal, the usual definition is

$$E_{Mad} = -\frac{1}{4\pi\varepsilon_0} N_m \frac{z_1 z_2 e^2 (m + x)}{2b} f^*. \tag{2.37b}$$

Madelung constants obtained in this manner are similar to each other even for different charges. For our purpose it suffices to note that the Madelung energy is given by the structure– and charge–defining parameters f, a, z_M, z_X, and that this Madelung energy makes up a major proportion of the total lattice energy.
The bilinear influence of the charge number is particularly worthy of mention. MgO crystallizes in the rock salt structure, possesses the same Madelung constant as NaCl, and the lattice constant is only slightly smaller (4.2Å compared to 5.3Å). The doubling of the charge number is primarily responsible for the fact that the Madelung energy is 5 times greater. The same effect causes the enormous stabilities of Al_2O_3 or ZrO_2 (see Chapter 4).
The lattice energy of ionic crystals is defined as the negative reaction energy of

$$\text{Reaction G} = \quad mM^+(g) + xX^-(g) \rightleftharpoons M_mX_x(s) \tag{2.38}$$

and can amount, for predominantly ionic bonding, to values of the order of 1MJ/mol even for the alkali halides composed of singly charged ions (see Table 2.1). As already mentioned, it is related, via the parameters I_M and A_X, in a simple manner to the bonding energy, i.e. the reaction energy of (cf. Eq. (2.11))

$$\text{Reaction B} = \quad mM(g) + xX(g) \rightleftharpoons M_mX_x(s). \tag{2.39}$$

Table 2.1: Contributions to the lattice energy of alkali halides (kJ/mol)

crystal	E_{Mad}	$E_{repulsion}$	$E_{v.d.Waals}$	E_{total}	lattice energy from Born-Haber cycle*
LiF	1194.4	184.5	16.3	1026.3	1003
NaF	1038.0	147.7	18.8	909.1	920
NaCl	854.7	98.3	21.7	778.2	787
NaBr	807.1	86.2	23.0	743.9	747
NaI	744.7	71.5	26.3	698.5	700
KCl	766.4	89.9	29.7	706.2	716
RbCl	735.5	83.2	33.0	685.3	670
CsCl	679.8	74.0	48.9	654.8	627

* The lattice energy ($E_{Lat} = -\Delta_G E$) is obtained as the sum of the energies of decomposition (i.e. negative formation energy), metal sublimation, (cat–)ionization, half nonmetal dissociation and (an–)ionization. In general, only the enthalpies of the individual reactions are experimentally accessible. The lattice enthalpy, thus obtained, does not differ very much from the lattice energy. The difference is negligible because $-\Delta_G(pV) < 2RT < 0.1$kJ/mol (see Eq. (2.38)) [57].

The thermodynamic energy of formation from the elements, normally from solid M and gaseous X_2, experimentally accessible and tabulated for standard conditions according to

$$\text{Reaction F} = \qquad mM(s) + \frac{x}{2}X_2(g) \rightleftharpoons M_mX_x(s) \tag{2.40}$$

is another significant quantity in this context. The difference between formation and bonding energies then further involves the energy of sublimation of M(s) and the dissociation energy of X_2. Since these are also experimentally accessible, the lattice energy can be obtained from purely experimental data (Born–Haber process). The values obtained in this manner are also included in Table 2.1 and take into account the necessary corrections for pure Madelung energy. These contain bonding corrections (polarization and dispersion effects) and zero–point energy.
Even though it is a good first approximation for the understanding of the bonding energy, the limitation to the Coulomb term can bring about errors of an order of magnitude that is the same as the energy difference between relevant crystal structures. Hence it is usually not possible to make a decision concerning the crystal structure from the Madelung energy[49,53].
The quantum–mechanical repulsion, which is only effective over very short distances, can be formally described using the Mie potential given in Section 2.1.6. For alkali halides the value of n in the repulsion term ($\propto r^{-n}$) is $n \sim 9$. In the summation in Eq. (2.36), a potential according to Eq. (2.20) has then to be taken into account instead of the pure Coulomb potential (m=1). As far as the repulsion term is concerned it is, owing to its small range, sufficient to consider only the nearest neighbours. The result is an expression of the form of Eq. (2.18), but now for the total energy. In the same way there is a relationship analogous to Eq. (2.19b) for the energy at the equilibrium distance $\hat{r}$. Thus, at the end the Madelung energy is to be multiplied by a factor $(1 - \frac{1}{n})$ and, hence, it is necessary to correct the preliminary lattice energy for n=9 ca. 10% (alkali halides) downwards:

$$E_{Lat} = -\Delta_G E = E_{Mad}\left(1 - \frac{1}{n}\right). \tag{2.41}$$

If the entire lattice energy is formally re–apportioned to pairwise interaction energies, an effective pair bonding energy is obtained between anions and cations. In the case of NaCl, with a co–ordination number of 6, this obviously leads to $E_{Na^+/Cl^-} = E_{Lat}/6$. Refinements of Eq. (2.41) consist of taking account of the van der Waals and polarization effects (multipolar effects particularly in structures of low symmetry), which can frequently also be represented by power law functions (typically r^{-6}, r^{-8}), and also of zero–point vibrational contributions (see the next section); on account of the increased attraction the former lead to an increase in E_{Lat} (only $\sim$ 1% for alkali halides, but, in the case of the highly polarizable silver ion, the neglect of such corrections causes significant errors), the latter lead to a slight reduction of the same

[53]Cf. Ref. [55].

order of magnitude. Further effects, that can be of importance under certain circumstances and are not independent of the above, are covalent contributions as a result of orbital overlap — of particular importance in complex ions — and additional effects as a result of energetic orbital splitting in the case of transition metal cations. This crystal field effect has been mentioned as ligand field effect in Section 2.1.3. Lattice vibrations can make a significant contribution (see next chapter) at higher temperatures, in addition the lattice constant may change as a result of anharmonicities[53].

Until now we have not taken explicit account of the effects that lead to band formation in our ionic crystal[54]. Even though the Na–Cl resonance integrals are negligible in NaCl since it is a very ionic compound, this does not strictly apply for the band widths. These are determined by the β values of the Na–Na (conduction band) and the Cl–Cl interaction (valence band). On account of the importance of oxides in our context the situation for the main group and transition metal oxides is described in somewhat more detail. Figure 2.19 shows the order of bands, occupation and densities of states ($D(\epsilon)$) for MgO. Here, as in the previous example, only s and p levels are of importance. Because of the strong ionicity the Mg orbitals lie appreciably above the O orbitals (cf. Fig. 2.15). The electron transfer is almost complete and the electrons are strongly bonded to the oxygen: At T=0K the outer Mg orbitals are empty and those of oxygen fully occupied. The transfer of an electron from the valence to the conduction band, thus, approximately represents the internal reaction $O^{2-} + Mg^{2+} \longrightarrow O^{-} + Mg^{+}$ (7eV). According to the lower band gap, this internal charge transfer is still more readily accomplished than in the more strongly ionic NaCl (10eV). The band widths of the orbitals are determined by the small resonance

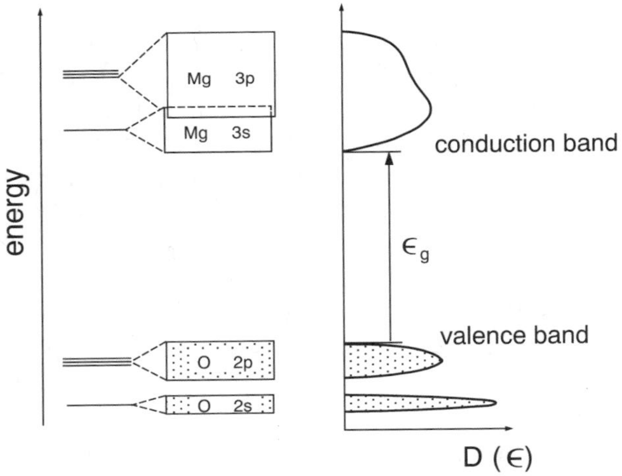

Fig. 2.19: Development of the band structure in the main group metal oxide MgO. In the isolated state (not shown) the Mg–3s orbital lies below the O–2p orbital, i.e. the bonding electrons are localized at Mg. The condensation (in particular due to the Madelung energy), reverses the order. The two 3s electrons of the Mg become 2p electrons of the oxygen ($Mg+O \rightarrow Mg^{2+}O^{2-}$). Orbital overlap produces bands, whose state ($D(\epsilon)$) and occupation densities are illustrated to the right [58]. According to Ref. [59].

[54]Ref. [32] provides a very good treatment.

integrals of the Mg–Mg and O–O interactions, while the resonance integral of the electronic Mg–O interaction is negligible as in NaCl.
The stability of the ions is primarily caused by the Madelung energy. If the particles are isolated from each other, neutral atoms are formed. These are more stable than the isolated ions (see Eq. (2.13)) according to the difference of electron affinity and ionization potential ($A_X - I_M$) (see Section 2.1.3). In contrast to Fig. 2.19 the Mg 3s orbitals then lie below the O 2p orbitals.

In the case of a transition metal oxide of type $M^{2+}O^{2-}$ the nonclosed shell points to metallic conduction. However, the ligand field effect already mentioned in Section 2.1.3 has a significant influence. The relationships in the octahedral ligand field are those depicted in Fig. 2.20. Ligand field effect (strength of the $e_g t_{2g}$ splitting)

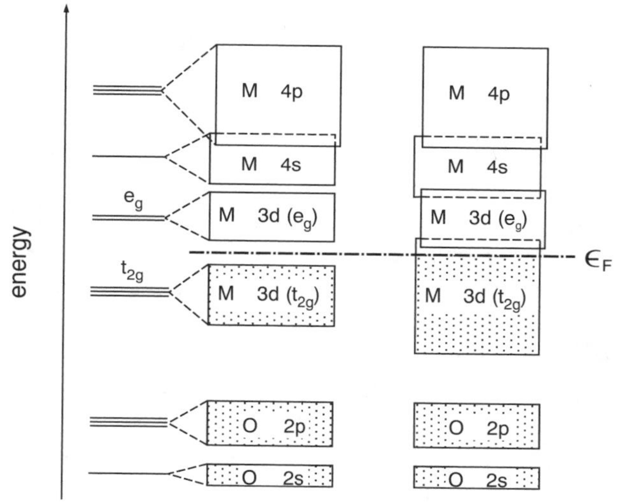

Fig. 2.20: The development of band structure in transition metal oxides of the formula $M^{2+}O^{2-}$ in an octahedral ligand field. Here, too, the relevant M orbitals (3d) lie below those of the O–2p orbitals (not shown) in the isolated state [58]. According to Ref. [59].

and interaction between atomic orbitals of the same energy (band width) determine whether partially or completely filled bands (at T=0K) are present. In the latter case the transition metal oxide is a semiconductor and the band–band transition corresponds to a redox disproportionation of M^{2+}. Attention must also be paid to the fact that partial occupation is not a sufficient condition for metallic conduction. The nearest neighbour distance must also be less than a critical value (see Section 2.2.5).
In the case of those higher oxides in which the metal atom has lost all its outer electrons such as in the case of the "insulator" TiO_2 (Ti^{4+}) the situation is similar to that for the main group metals, in that the relevant orbitals (here d orbitals) are unoccupied and the valence band is approximately provided by the O 2p orbitals (conduction band for TiO_2: Ti–3d).

2.2.3 Molecular crystals

Let us now consider an ensemble of many electronegative elements. Since covalent bonds are directional and locally saturable to a good approximation, the final picture depends strongly on the nature of the elements in the ensemble. Under standard conditions an accumulation of Cl atoms reacts to yield a collection of saturated Cl_2 molecules. No condensed phase at all is formed here at room temperature. A solid is only formed at low temperatures for which bonding forces of higher order (this reflects the fact that the two–centre bond is after all only approximately saturated), the dispersion or London forces mentioned above, which correspond to the interaction of induced dipoles, are relevant (see Section 2.1.2). These intermolecular bonding forces are very weak compared with the strong intramolecular forces. Let us mark for our purpose the covalent bonds with square brackets, and the van der Waals bonds with straight brackets; this makes it possible to denote this mixed bonding form by $|[Cl_2]|_{\overset{3}{\infty}}$[55]. The thermodynamic energies of formation (see Eq. (2.42)) of such 3–D van der Waals polymers or molecular crystals formed according to

$$\mathrm{N\,Cl_2(g) \rightleftharpoons (Cl_2)_N(s)}, \tag{2.42}$$

are typically of the order of −10kJ/mol. They are identical with the negative sublimation energies. It is also appropriate to define the lattice energy using Eq. (2.42), whereby it is necessary to include the zero–point energy of the vibration which is appreciable in this case. The intermolecular bonding energies for noble gas crystals, i.e. for atomic crystals such as $|Ar|_{\overset{3}{\infty}}$, are between 1 and 20 kJ/mol (the zero–point energy is of the order of 1 kJ/mol); however, the interaction is considerably greater in the case of polar molecular crystals such as HCl (s) owing to the interaction of the permanent dipoles. For HCl it is also necessary to take into account the special role of hydrogen bonds, as discussed in Section 2.1.2.
The Mie form of the potential function (Eqs. (2.18, 2.19)) serves well for molecular and atomic crystals too, and the lattice energy can be described by

$$\mathrm{E_{Lat} \propto \hat{r}^{-m}(1 - m/n)}. \tag{2.43}$$

Here both the attraction and repulsion terms appear with high exponents ($m \simeq 6, n \simeq 12$) making the bond of short range. The presentation in the form of the Lennard–Jones potential, as given in Eq. (2.20b), is better known in this connection.

If the ensemble of atoms is made up of hydrogen and carbon, then, depending on the conditions (temperature, pressure; type and number of particles), various cases can be realized. If only carbon–hydrogen bonds are formed then the methane molecular crystals result $|[CH_4]|_{\overset{3}{\infty}}$. Because of the multivalence of carbon, pairwise carbon–carbon bonds are not saturated. So, for example, one–dimensional polymers with

[55]The lower index $\overset{1}{\infty}$, $\overset{2}{\infty}$, $\overset{3}{\infty}$ indicates the infinite extension in 1, 2 or 3 dimensions. The formal index $\overset{0}{\infty}$ (in $|[Cl_2]_{\overset{0}{\infty}}|_{\overset{3}{\infty}}$) is omitted.

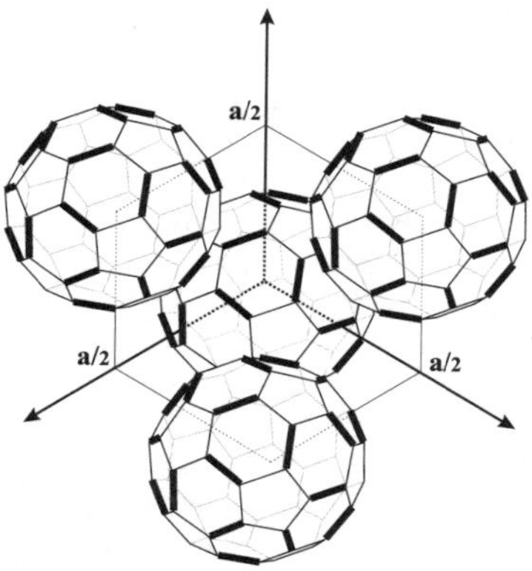

Fig. 2.21: Fullerite solids consist of C_{60} molecules, held together by van der Waals bonds. In the hexagons double and single bonds alternate nominally. In reality there is partial delocalization. The arrangement is face–centred cubic at room temperature (see Section 2.2.7). The lattice constant a is 14.17Å. According to Ref. [60].

a carbon backbone can be formed in which the remaining bonds are saturated by H, and which are themselves held together by van der Waals forces, in the limiting case $|[CH_2]_{\frac{1}{\infty}}|_{\frac{2}{\infty}}$.
Graphite crystals are examples of a two–dimensional, covalent C–C network; they are made up of covalently bonded C planes (sp^2), which are themselves kept together by means of van der Waals bonds $|[C]_{\frac{2}{\infty}}|_{\frac{1}{\infty}}$. In this mixed case (see Section 2.2.6) the lattice energy which refers to formation from isolated C atoms reflects basically the contribution of covalent bonding. Modern examples of carbon–based molecular solids are fullerene–crystals (fullerite) which consist e.g. of C_{60} molecules ($|[C_{60}]|_{\frac{3}{\infty}}$ see Fig. 2.21) held together by van der Waals bonding. The intermolecular bonding energy is 0.15eV (per single bond) [61], the intramolecular one is ~7eV on average [60] (see also below, Eq. (2.46)).

2.2.4 Covalent crystals

The extreme case of a giant 3D covalent polymer is realized in the case of diamond $[C]_{\frac{3}{\infty}}$ (see Fig. 2.22):

$$\mathrm{N\,C(g)} \rightleftharpoons \mathrm{C_N(s)}. \tag{2.44}$$

Such covalent crystals possess high energies of formation from the atoms (e.g. ~ -1MJ/mol); this value reflects the negative lattice energy which is the negative energy of sublimation, or the energy of bonding. In contrast to the crystal types referred to above, there is no difference in this case between intra– and intermolecular bonding (as in ionic crystals) and, at the same time, the bonding is directional and to a good approximation of very short range (different to ionic crystals). As far as the lattice energy is concerned, we do not, therefore, need to resort to any of the more or less empirical potential functions but can turn directly to the effective bonding energy of the two–centre problem[14]. We have seen in detail that the bonding energy in the hydrogen problem is given approximately by the reduced resonance integral. In the case of diamond (or Si, Ge, α–Sn) the situation is more complex, as discussed in Section 2.2.1, in that, in particular, the energy of excitation from

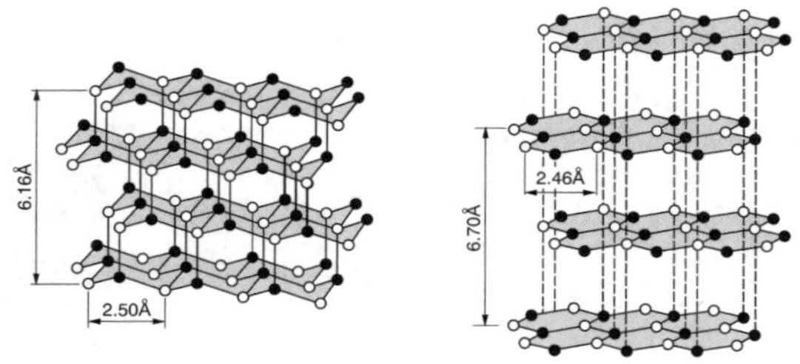

Fig. 2.22: Structures of diamond (left) and graphite, the latter being slightly more stable under standard conditions. While diamond represents the case of a true covalent crystal, graphite is characterized by a mixture of covalent bonding (within the planes) and of van der Waals bonding (between the planes). The symbols are chosen differently for the purpose of illustration. Both solids are composed of identical C–atoms. If one attributes the different symbols to different elements, the zinc–blende structure is realized (see Fig. 2.23). From Ref. [62].

the electronic ground state into the sp^3 valence state must be taken into account; nevertheless, theory[14] and practice show that the bonding energy of diamond results, to a good approximation, from the summation of the local bonding energies. Thus, the bond strengths and bond lengths of diamond deviate from those of long–chain aliphatic hydrocarbons only by a few percent. The lattice energy for diamond can be reasonably well represented in terms of the experimentally accessible bonding energies for the C–C sp^3 bond for paraffins, E_{C-C}, as

$$E_{Lat} = 4\frac{E_{C-C}}{2} \simeq 2 \times 348\text{kJ/mol} = 696\text{kJ/mol}. \quad (2.45)$$

The deviation from the experimental value of 716kJmol^{-1} can be attributed to the van der Waals attraction (between noncovalently bonded atoms)[56]. If one constructs graphite from double ($E_{C-C} = 615\text{kJmol}^{-1}$) and single bonds with the weights of 1/3 and 2/3, a mean bonding energy of 437kJmol^{-1}results. This yields, in view of the number of nearest neighbours a lattice energy of $1.5 \times 437\text{kJmol}^{-1} = 656\text{kJmol}^{-1}$ which is much lower than the experimental value (compared with diamond, graphite is stabilized by 1.9kJmol^{-1}). If we assume a van der Waals contribution similar as for diamond, a little more than 40kJmol^{-1} must be attributed to the aromatization (significant delocalization over the ^{2}D network of conjugated double bonds). In other words we ought to use a mean sp^2–bonding energy (E'_{C-C}) of ca. 465kJmol^{-1} according to

$$E_{Lat} = 3\frac{E'_{C-C}}{2} \simeq 1.5 \times 465\text{kJ/mol}. \quad (2.46)$$

Roughly speaking, such a value can be justified by the experimental bond energies of aromatic hydrocarbons.

As in the case of metals and in contrast to ionic and molecular crystals, covalent crystals yield extended bands in the band model, because of the significant overlap.

[56]The van der Waals attraction between nearest neighbours is included in E_{C-C}. Cf. Ref. [25].

Large splitting between bonding and antibonding sp^3 states in diamond or silicon with the lower sp^3 band being completely occupied is an expression of strong, saturated covalent bonds. The configuration for Si_N reads at T=0: ($3sp^3$ valence band)$^{4N(4N)}$ ($3sp^3$ conduction band)$^{0(4N)}$ (cf. Section 2.2.1.2). At finite temperatures there is, to a small degree, an internal charge transfer of the form

$$2Si^0 \rightleftharpoons Si^+ + Si^-, \tag{2.47}$$

i.e. conduction electrons and holes in the valence band are created. The cost of 1eV is reimbursed by the entropy gain (see Chapter 5).
As expected the bandgap drastically decreases from diamond to α–Sn. Diamond is a typical insulator with a band gap of 5eV; in silicon the band gap is still 1eV; in germanium only 0.7eV; α–Sn is already a semimetal (as is graphite). In the case of α–Sn the covalency is not very pronounced and the s–p splitting (metallicity) predominates (see Section 2.2.1 and Fig. 2.14).

2.2.5 Metallic crystals

The most important electronic aspects have already been dealt with in Section 2.2. From the energetic point of view it remains to be added that the energy of formation of the elementary metals ($\langle M \rangle_{\substack{3\\\infty}}$) represents the negative sublimation energy, i.e. the reaction energy of the following reaction:

$$N\,M(g) \rightleftharpoons M_N. \tag{2.48}$$

The energy of formation is of the order of magnitude of 10^2 kJ/mol for alkali metals. There is disagreement concerning the definition of the lattice energy in the case of metallic crystals: It is, corrected for the zero–point energy, usually identified with the negative energy of reaction (2.48), i.e. referring to the formation of $\langle M \rangle_{\substack{3\\\infty}}$ from neutral atoms; sometimes, however — in view of the fact that metals are composed of positive metal ions and electrons as "anionic cement" — it is identified with the formation from gaseous cations and electrons. Then the ionization potential of the element also has to be included. The values obtained for lattice energies defined in this manner are of the same order of magnitude as those for ionic and covalent crystals. It is of interest that a Madelung estimate according to $\frac{N_m f e^2}{b}\left(1 - \frac{1}{n}\right)$ yields a reasonable approximation if the electrons are treated as anions in an ionic crystal and assigned to their own lattice (see [63,64]). The value of n is small (≃3) in agreement with the high compressibility of the metal. Accurate calculations are complex. Because of the small "space requirement" of the bonding electrons, metal structures generally follow the principle of close–packed structures (see following section). Higher order co–ordination spheres do not play a large role on account of the screening due to the quasi–free electrons; accordingly the energies of melting and sublimation are very similar and metals are very ductile (ready formation of dislocations (see Section 5.4)).

For the phenomenon of metallic conduction it is important that there is a certain minimum nearest neighbour distance, corresponding to a certain minimum degree of overlap. This finds its expression in the Mott–Hubbard criterion, and is also of particular relevance for transition metal and rare earth metal compounds as well as for strongly doped semiconductors. It can be understood immediately if we carry out a thought experiment involving delocalization of the localized electrons. Let us assign to each atom in a chain an electron and then try to raise one out of its localization and to bring it in the environment of a neighbouring atom; to do this we expend the ionization energy of the atom but we get back its electron affinity. Since the latter parameter measures the willingness of an atom already possessing an electron to take up another, the difference I–A is also the expression of the electron–electron repulsion (Mott–Hubbard energy) [47,58], which was neglected in Section 2.2.1). Only if the interaction of the orbitals, which is measured via the band width (cf. β), is large enough, does delocalization and metallic conduction occur. The Mott–Hubbard criterion (band width > I–A) is qualitatively plausible if we remember that I–A refers to the splitting of the states under discussion (ground state ... MMM ... MMM ..., and excited state ... $M^{\cdot}MM$ ... $M'MM$ ...). The corresponding gap is closed at a particular band width. This refers directly to Eq. (2.22). When the bands are very narrow, it is better to speak of a high density of states of individual orbitals. Thus, partially filled narrow "bands" do not necessarily imply metallic conductivity. The solid state physics and chemistry literature should be referred to for further details and also with respect to the types of crystals which result from the intermediate forms of semimetallic and intermetallic bonding.

2.2.6 Mixed forms of bonding in solids

In contrast to the intermediate forms of bonding (discussed in particular in the Section 2.1.5) let us now turn to the simultaneous occurrence of different bond types (bonding anisotropies and inhomogeneities). This is normally associated with the occurrence of anisotropy or inhomogeneity of properties. In particular the simultaneous occurrence of strong and weak bonding leads to the possibility of distinguishing between "intramolecular" and "intermolecular bonding" in solids. Crystals, such as graphite, held together by covalent and van der Waals bonds have already been mentioned (see Fig. 2.22). Combinations of other bond types naturally occur too. Let us use the notation specific to this book, already introduced in Section 2.2.3, and label covalent bonds with square brackets, ionic bonds with braces, metallic bonds with pointed brackets, and van der Waals bonded units we will mark between two straight lines. In this sense we label[57] the methane crystal as $\|[CH_4]\|_{\overset{3}{\infty}}$, an ideal polyethylene crystal as $\|[CH_2]_{\overset{1}{\infty}}\|_{\overset{2}{\infty}}$, graphite and diamond as $\|[C]_{\overset{2}{\infty}}\|_{\overset{1}{\infty}}$ and $[C]_{\overset{3}{\infty}}$.

[57] The formal index $\overset{0}{\infty}$, e.g. $\|[CH_4]_{\overset{0}{\infty}}\|_{\overset{3}{\infty}}$ is suppressed. In this text the symbols $\overset{1}{\infty}, \overset{2}{\infty}, \overset{3}{\infty}$ are used as r.h.s. indices, since they make a statement on the number of units (cf. solid conceived as a giant molecule). In this sense the numbers 1,2,3 are to be formally understood as exponents.

Polycationic and polyanionic compounds are of special interest [65–67]. $\{Cs_2O\}_{\frac{3}{\infty}}$ is an oxide that crystallizes in the form of a pure ionic crystal. Metal–rich oxides (suboxides) such as $Cs_{11}O_3$, $\{\langle Cs_{11}^{6+}\rangle(O^{2-})_3\}_{\frac{3}{\infty}}$, contain metal clusters that are, in correspondence with the electron–deficiency, stabilized by metallic bonds and act as polycations [65]. It is remarkable that (in spite of the presence of electronegative O) the reduction of the effective room available for the quasi–free electrons (see Eq. (2.25)) in comparison with $\langle Cs\rangle_{\frac{3}{\infty}}$, by inclusion of insulating "Coulomb bubbles", leads to a further reduction of the (very small) ionization energy of Cs [65]. There are also polycationic compounds in which the cationic cluster is covalently bonded (e.g. $[S_8^{2+}]$, $[Te_6^{4+}]$). Polyanionic compounds are equally common (cf. Zintl phases), such as Na_3P_7 or KP_{15} [66]. Here the anions are covalently bonded with each other, and the whole cluster acts as an anion. The Na_3P_7 is more accurately represented as $\{(Na^+)_3[P_7^{3-}]\}_{\frac{3}{\infty}}$, four of the seven P atoms are triply bonded and the remaining three doubly bonded — corresponding to P^- — so that the cluster makes up a complex trivalent anion. A coexistence of covalent and ionic bonds in the cationic and anionic part can be found in molecules or solids as simple as NH_4NO_3: $\{[NH_4^+][NO_3^-]\}_{\frac{3}{\infty}}$. Some silicates, e.g. the chain silicate enstatite $\{(Mg^{2+})_{\frac{1}{\infty}}[SiO_3^{2-}]_{\frac{1}{\infty}}\}_{\frac{2}{\infty}}$ are more complicated examples. Metallically bonded strands occur in $RhBi_4$; these are connected by van der Waals bonds. On account of the long range of Coulomb forces mixtures of ionic bonding and pure van der Waals bonding do not exist together.
It should be noted that materials with anisotropic transport properties (e.g. many superionic conductors) frequently exhibit such bonding inhomogeneities or anisotropies (see Chapter 6).

2.2.7 Crystal structure and solid state structure

As already mentioned it is frequently difficult to predict in which crystal structure a particular compound will crystallize even in the equilibrium state at low temperatures [68]. Important decision criteria involve the electronegativity, the polarizability, preferred co–ordination spheres and in particular the ionic radius. Textbooks of crystallography and structural chemistry are recommended to the reader for the details[58]. Here attention will only be drawn to the fruitful ordering principle of the close–packing of spheres[59].
The previous sections showed that a high bonding energy demands a relatively short distance from the bonding partner. The limit at which strong quantum–mechanical repulsion begins, is expressed by the size of the ionic or atomic radius. When the arrangement is balanced, it is — in this sense — generally possible for ions of the same charge to approach each other very closely in ionic crystals. Since anions are usually larger than cations, the following concept is often very serviceable:

[58] See Refs. [25,67,69,70]

[59] The assumption of hard spheres with defined charges is not a good approximation in cases of significant polarizabilities.

The anions form close–packed spheres while the cations approximately occupy the tetrahedral (4 nearest neighbours) or the (more roomy) octahedral interstices (6 nearest neighbours). It must be remembered that in a close–packed structure there are twice as many tetrahedral interstices as octahedral ones and that the number of octahedral interstices corresponds to the number of close–packed spheres. In addition spheres can be hexagonal or cubic close–packed.
If the anions in NaCl (see Fig. 2.18) or in NiAs are approximately allocated to cubic or hexagonal close–packing respectively, then formally the cations occupy all of the octahedral interstices. Niggli formulae provide information concerning the mutual co-ordination numbers: Thus, $\{NaCl_{6/6}\}^{3}_{\infty}$ or $\{NiAs_{6/6}\}^{3}_{\infty}$ mean that both anions and cations are octahedrally coordinated by counterions. Considering ZnS as an example the zinc–blende (sphalerite) structure (Fig. 2.23) can be viewed as a cubic

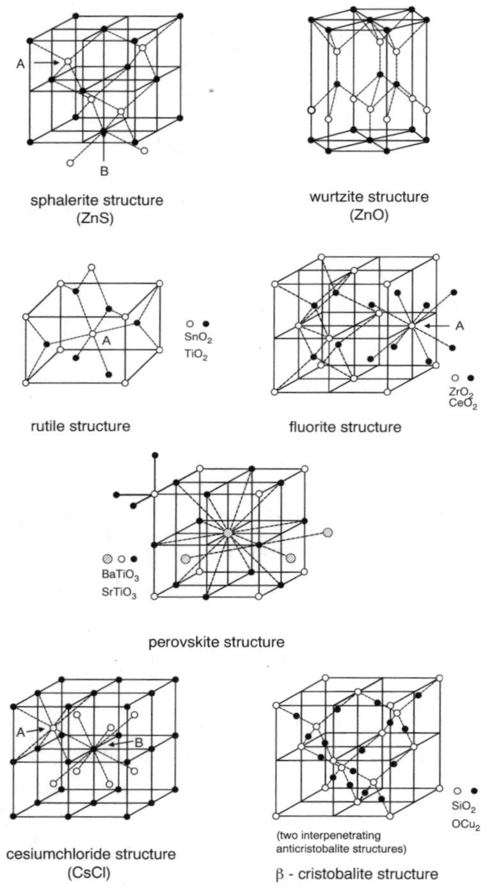

Fig. 2.23: Some relevant crystal structures. From Ref. [70].

close–packing of the sulfide ions, in which one half the tetrahedral interstices are occupied by zinc ions $\{ZnS_{4/4}\}^{3}_{\infty}$; the wurtzite structure (Fig. 2.23) exhibits the same Niggli formula, but here there is hexagonal close–packing. If A and B atoms become identical, one passes from the zinc–blende structure to the diamond structure ($[C_{4/4}]_{\infty}^{3}$), which is also the form in which the semiconductors Si, Ge and α–Sn[60] (grey tin) crystallize. In the case of the cubic fluorite structure (Fig. 2.23) all the tetrahedral sites are occupied $\{CaF_{8/4}\}_{\infty}^{3}$, here the cations are formally considered to form a cubic close–packed structure, in the case of $\{ONa_{8/4}\}_{\infty}^{3}$ the roles of anions and cations are interchanged (antifluorite structure). Further useful Niggli formulae are $\{TiO_{6/3}\}_{\infty}^{3}$ for TiO_2 in the rutile structure and $\{CsCl_{8/8}\}_{\infty}^{3}$ for the caesium chloride structure. The perovskite structure, which is illustrated together with the structures mentioned above in Fig. 2.23, is an extraordinarily important one in the context of this book. The perovskite shown here is of cubic symmetry. The Niggli formula $\{Ba(TiO_{6/3})_{8/8}\}$ for $BaTiO_3$ stresses that the structure combines structure principles of the rutile and caesium chloride structure. Noncubic perovskite phases are of even higher technological relevance. In these structures ion displacements can cause polarization effects that are responsible for the phenomena of ferroelectricity, anti–ferroelectricity, pyro– and piezoelectricity. The principle of close–packing of spheres also works for many metals: Here the cations, as it were, form a close–packed arrangement while the geometrical space required by the electron gas is formally negligible. For details, Refs. [42,71,72] are recommended. Specific crystal structures are explained in the text[61].

All these considerations referred to the "perfect structure" of a single crystal which is normally the domain of structural chemistry. It refers, as it were, to the virtually defect–free crystal structure. The "chemically excited" crystal structure resulting from a superposition of perfect structure and defect structure is of prime significance in the context of this text (cf. Chapters 5, 6 and 7)[62].

Coming to the end of this section, it must be emphasized that the solid state structure of even a single–phase crystalline substance is much more than just the crystal structure. A solid body is usually made up of many single crystalline regions (so–called "grains"), that are tilted and twisted with respect to each other and connected by grain boundaries[63]. Together with other higher–dimensional defects (see Section 5.4) such grain boundaries make up the microstructure. This microstructure plays the role of a superstructure on the micrometre scale with, however, a rather fuzzy periodicity, if we may use this term at all (see Sections 5.4.4 and 7.3.7). Finally the complete solid structure of the material is determined by the surface, which characterizes the shape, and thus the macrostructure. Since surfaces and grain boundaries can be regarded as higher–dimensional defects, we will return to them in Chapter 5. Microstructures and macrostructures are generally markedly metastable nonequilib-

[60] at the boundary to a metal

[61] These "ground structures" will be compared with the "(atomically) excited structures" in Section 5.2.

[62] The lattice vibrations as thermal but nonchemical excitations take an intermediate position (see next chapter).

[63] In the same way as weak intermolecular forces characterize molecular crystals, weak intergranular bonding leads to "soft matter".

rium structures. There are large kinetic barriers present in each case. These usually prevent both the disappearance of the former and the achievement of the equilibrium state of the latter. It is the existence of these kinetic effects that permits the material structuring and shaping that is so important in daily life.
Special cases are macroscopically aperiodic solids, materials with macroscopic composition gradients, as well as amorphous solids. The latter may not be regarded as periodic, neither are they absolutely structureless; many properties, however, can be described on a coarse grained scale as translationally invariant on account of averaging effects.
Finally those solids, which (more precisely those conditions under which the solids) do not reach a spatial equilibrium on the nanometre scale, deserve attention. Owing to their memory of the synthesis path, it is possible to create artificial materials by appropriate chemical precursors [73] or by spatially precise synthesis for instance by successive deposition of atomic layers using molecular beam epitaxy [74–76]. In such supra–atomic but not yet macroscopic regions mesoscopic effects (oligomers!) occur with respect to many properties (cf. also Section 5.8). In some cases more sophisticated treatments allow structuring to be carried out in all spatial dimensions. In this manner artificial inorganic solids may be prepared which contain a very large information content. Other techniques, such as the translation of atoms or molecules with the atomic force microscope probe, makes structuring possible in a very direct manner. A presupposition to all this is that the constituents are sufficiently immobile. In organic and, in particular, in biologically relevant solids such local metastability is the rule and the basis of biological and biomimetic structuring [76].
In addition, materials are often heterogeneous, i.e. they are composed of various separate kinetically or thermodynamically more or less stable phases. Their overall properties are then determined not only by the relative proportions of the phases but also by their distribution topology, i.e. by orientation, by the arrangement of the phase interfaces and their respective properties (see Sections 5.4, 5.8, 6.6.2).

3 Phonons

3.1 Einstein and Debye models

In this chapter we examine the energetic contributions of the lattice vibrations. These are the most important, nonchemical, thermal excitations and involve motion of the nuclei[1]. Lattice vibrations are quantized. In the same way as photons are, as the respective quasi–particles, equivalent to electromagnetic waves, there are quasi–particles allocated to these elastic waves termed phonons.

If you bring a solid from absolute zero to a finite temperature, the atomic constituents of the solid begin to vibrate around their equilibrium positions. In the simplest model, the Einstein model [77], all constituents vibrate with the same frequency (ν_E). Increasing the temperature effects an increase in the mean vibration amplitude, but not in ν_E. As shown in Section 2.1.6 the expansion of the Mie function yields for small displacements a harmonic[2] potential (see Eq. (2.21)) with spring constant $mn|\widehat{\epsilon}|/\widehat{r}^2$. Hence, the frequency of an oscillator vibrating in this potential is

$$\nu = \frac{1}{2\pi\widehat{r}}\sqrt{\frac{mn\epsilon_{dis}}{M_{red}}}; \tag{3.1}$$

ϵ_{dis} is the dissociation energy, that corresponds — if the potential at infinity is taken as zero — to the (negative) minimum of the potential curve ($-\widehat{\epsilon}$) at $\widehat{r}$ (see Fig. 2.3). The quantities ϵ_{dis} and $\widehat{r}$ are related to the bonding parameters A, B, n, m via Eq. (2.19)[2]; M_{red} is the reduced mass.

A crystal composed of N identical vibrators possesses 3N degrees of freedom; if the degrees of freedom related to internal translation and rotation of the particles are neglected and the (six) external degrees of freedom of the total crystal are subtracted, 3N-6 degrees of vibrational freedom remain. In the case of a macroscopic solid 3N-6$\simeq$ 3N, and the vibration energy of a monatomic solid is

$$E_{vib} = 3N\,\bar{\epsilon}_{vib}; \tag{3.2}$$

$\bar{\epsilon}_{vib}$ is not the individual vibrator energy of a particular vibration under consideration, but rather a mean value because of the fact that not all constituents vibrate

[1]See footnote 17 on page 33.

[2]We remind ourselves:
In a harmonic potential the energy is proportional to the square of the displacement ($\mathrm{const}(\Delta r)^2/2$) while the force is proportional to the displacement itself ($-\mathrm{const}\Delta r$). On the other hand, the force is determined by the product of mass and acceleration ($M\frac{d^2\Delta r}{dt^2}$). Hence, the solution function for Δr is a sine function with the argument $2\pi\nu t = \sqrt{\mathrm{const}/M}\,t$. A comparison with Eq. (2.21) yields const $= mn|\widehat{\epsilon}|/\widehat{r}^2$. If two coupled masses are vibrating, the analogous equation of motion applies to the coordinates of the centre of gravity [78]. In the result it is then necessary to interpret M as the reduced mass (M_{red}). M_{red} is obtained from the individual masses by harmonically averaging. In the case of an isotropic, atomic crystal, M_{red} is directly given by the atomic mass and the co–ordination number.

Physical Chemistry of Ionic Materials J. Maier

ISBN: 0-471-99991-1 (HB); 0-470-87076-1 (PB)

with the same amplitude. Quantum–mechanically the harmonic oscillator possesses a spectrum of eigenvalues (ϵ_v) which can be realized with varying probabilities. The Boltzmann distribution is to a good approximation applicable, so that:

$$\bar{\epsilon}_\mathrm{vib} = \Sigma_\mathrm{v} \left(\frac{\exp(-\epsilon_\mathrm{v}/\mathrm{k_B T})}{\Sigma_\mathrm{v} \exp(-\epsilon_\mathrm{v}/\mathrm{k_B T})} \right) \epsilon_\mathrm{v} = \frac{\Sigma_\mathrm{v} \epsilon_\mathrm{v} \exp(-\epsilon_\mathrm{v}/\mathrm{k_B T})}{\Sigma_\mathrm{v} \exp(-\epsilon_\mathrm{v}/\mathrm{k_B T})}. \tag{3.3}$$

We write the sum in the denominator of Eq. (3.3) as $\mathrm{Z_{vib}}$ which is known as the partition sum (here for vibrations), and from which the thermodynamic state functions (see following chapter) can be derived, e.g. the mean energy in Eq. (3.3). The latter quantity is given directly by

$$\bar{\epsilon}_\mathrm{vib} = \mathrm{k_B T^2} \frac{\partial \ln \mathrm{Z_{vib}}}{\partial \mathrm{T}}, \tag{3.4}$$

as readily verifiable. On account of this it is only necessary to consider a single sum, which can be evaluated in a straightforward way. As demonstrated in the relevant textbooks [23] the energy eigenvalues of a harmonically vibrating oscillator are equidistant and depend on the (Einstein) frequency ν_E, according to

$$\epsilon_\mathrm{v} = \mathrm{h}\nu_\mathrm{E}(\mathrm{v} + 1/2) \quad \text{with} \quad \mathrm{v} = 0, 1, 2 \cdots \tag{3.5}$$

Hence $\mathrm{Z_{vib}}$ becomes

$$\mathrm{Z_{vib}} \equiv \Sigma_\mathrm{v} \exp -\frac{\epsilon_\mathrm{v}}{\mathrm{k_B T}} = \exp -\frac{\mathrm{h}\nu_\mathrm{E}}{2\mathrm{k_B T}} \Sigma_\mathrm{v} \left(\exp -\frac{\mathrm{h}\nu_\mathrm{E}}{\mathrm{k_B T}} \right)^\mathrm{v}. \tag{3.6}$$

The remaining sum is a geometric series of the form $\Sigma_\mathrm{v}\mathrm{q_v} = 1 + \mathrm{q} + \mathrm{q}^2 + \cdots$. Since ν_E is positive so that q is less than unity, the sum converges to $1/(1-\mathrm{q})$; for $(1 + \mathrm{q} + \mathrm{q}^2 + \cdots)(1 - \mathrm{q}) = (1 + \mathrm{q} + \mathrm{q}^2 + \cdots) - (\mathrm{q} + \mathrm{q}^2 + \mathrm{q}^3 + \cdots) = 1 + \mathrm{F}$, that is 1 with ever reducing error F, the more members are taken into account. We obtain the energy of vibration on differentiating $\ln \mathrm{Z_{vib}}$ with respect to T and multiplying by $3\mathrm{Nk_B T^2}$ in accordance with Eq. (3.2) and Eq. (3.4):

$$\mathrm{E_{vib}} = 3\mathrm{N}\frac{\mathrm{h}\nu_\mathrm{E}}{2} + \frac{3\mathrm{Nh}\nu_\mathrm{E}}{\exp(\mathrm{h}\nu_\mathrm{E}/\mathrm{k_B T}) - 1}. \tag{3.7}$$

Differentiation with respect to T yields the specific heat of vibration, which plays a fundamental role in the temperature dependence of the thermodynamic functions of the solid (see following chapter)

$$\mathrm{C_{vib}} = 3\mathrm{Nk_B} \left(\Theta_\mathrm{E}/\mathrm{T}\right)^2 \frac{\exp\left(\Theta_\mathrm{E}/\mathrm{T}\right)}{\left(\exp\left(\Theta_\mathrm{E}/\mathrm{T}\right) - 1\right)^2}. \tag{3.8}$$

In Eq. (3.8) the Einstein temperature $\Theta_\mathrm{E} \equiv \mathrm{h}\nu_\mathrm{E}/\mathrm{k_B}$ is introduced as an abbreviation that remains as the only material parameter. The term $3\mathrm{Nh}\nu_\mathrm{E}/2 = 3\mathrm{Nk_B}\Theta_\mathrm{E}/2$ in Eq. (3.7) is the zero–point energy already mentioned above. For a typical Einstein temperature of 500 K corresponding to $\nu_E = 10^{13}s^{-1}$ it amounts to 6 kJ/mol. For temperatures greatly above Θ_E we can simplify[3] $\exp(\Theta_E/T)$ to $1 + \Theta_E/T$. Thus, at the classical limit, that is for temperatures at which the discrete form of the levels no longer plays a role, the well–known Dulong and Petit law results:

$$E_{vib} \simeq 3Nk_BT \quad \text{and} \quad C_{vib} \simeq 3Nk_B. \tag{3.9}$$

One can see that the vibrational frequency no longer contributes to Eq. (3.9). The simple Einstein theory provides a qualitative and surprisingly often a quantitative picture of the temperature behaviour of the specific heat. As required C_{vib} tends to zero as zero Kelvin is approached but exponentially, that is more steeply $\left(\rightarrow 3Nk_B \frac{(\Theta_E/T)^2}{\exp(\Theta_E/T)}\right)$ than the experimentally well–established T^3 law requires. The assumption of a single and temperature–independent frequency is obviously too great an approximation.

The Debye model provides a better description [79]. A whole spectrum of frequencies occurs in a three–dimensional system of coupled springs even if the individual vibrators are identical. If as before we restrict ourselves to harmonic vibrations alone, which is guaranteed at low temperatures, then it can be shown[4] that the frequency distribution $dN/d\nu$ is proportional to the square of the frequency as illustrated in Fig. 3.1. Hence, when calculating the mean energy (Eq. (3.2)) it is not merely necessary to average over the energy levels at a given frequency, but also to take into account the frequency distribution. (In the case of the Einstein model this distribution function corresponded to a delta function.) In all cases the frequency integral of $dN/d\nu$ must correspond to the total number of vibrators: However, the integral of the ν^2 function would diverge. In other words: As this distribution function is used at higher frequencies, for which it no longer strictly applies, it is necessary to cut off the frequency spectrum at the Debye frequency[5] ν_D. This provides the conditions for ν_D or a Debye temperature[6] $\Theta_D = h\nu_D/k_B$ defined in a similar manner to Θ_E.

[3]This corresponds to taking account of the absolute and the linear term of the Taylor series, or — what is equivalent to this — it follows with $x_E = \Theta_E/T \ll 1$ from $[\exp x_E - \exp 0]/[x_E - 0] \simeq d\exp x_E/dx_E|_{x_E=0} = 1$.

[4]As for the case involving free electrons (Section 2.2.1) the condition for a standing wave is that the cube edge L must be an integral multiple of the half wavelength, i.e. $n^2 = \frac{4L^2}{\lambda^2}$. This applies in the form $n^2 = n_x^2 + n_y^2 + n_z^2$ also to waves which are not propagating parallel to the axes, since the squares of the direction cosines add up to 1. The above relation is the equation of a sphere of radius $2L/\lambda$ (more precisely: one–eighth of a sphere). The number of standing waves in the frequency range between ν and $\nu + d\nu$ is obtained from the volume (more precisely: one–eighth of the volume) of the corresponding shell (see Fig. 2.6, page 39). Introduction of the velocity of sound, $v_s = \lambda\nu$, yields const $\left(\nu^2/v_s^3\right) L^3$ for the number of possible waves.

[5]If the distribution were correct, ν_D would actually be the real highest frequency of vibration and defined by the lattice distance.

[6]Since the number density of the standing waves is proportional to $\left(\nu^2/v_s^3\right) \times$ (volume) (see footnote 4) the parameter ν_D is found to be proportional to v_s (molar volume)$^{-1/3}$ via const $\int_0^{\nu_D}$ (volume) $\times \nu^2 v_s^{-3} d\nu = 3\times$ (particle number).

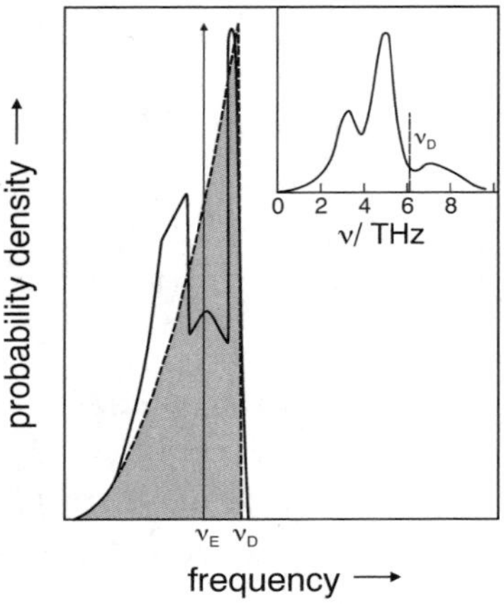

Fig. 3.1: The frequency distribution in real solids, as a function of the temperature compared with the Einstein (delta function at ν_E) and the Debye (shaded area) models. In the case of Ag the Debye approximation (dashed line) is very well fulfilled. In the case of complicated solids a quadratic increase at lower temperatures can be seen, too, but the respective ν_D value is no longer identical with ν_D as overall fit parameter. The inset illustrates the situation for NaCl [55,64,80]. According to Ref. [48].

Again the energy of vibration is defined by this parameter alone and application[7] of this additional averaging process to Eq. (3.7) with $x \equiv (h\nu/k_BT)$ yields the result:

$$E_{vib} = E_{vib}(T=0) + 9Nk_BT\left(\frac{T}{\Theta_D}\right)^3 \int_0^{\Theta_D/T} \frac{x^3}{\exp x - 1}dx \tag{3.10}$$

and on differentiation with respect to temperature

$$C_{vib} = 9Nk_B\left(\frac{T}{\Theta_D}\right)^3 \int_0^{\Theta_D/T} x^4 \frac{\exp x}{(\exp x - 1)^2}dx. \tag{3.11}$$

At high temperatures[8], $T \gg \Theta_D$, the Dulong–Petit law applies once again. For $T \ll \Theta_D$ the integrand[9] becomes $x^4 \exp -x$. Partial integration allows successive reduction of the x^4 term. At the end the exponential function remains. It disappears at the upper limit $\Theta_D/T \gg 1$ and becomes 1 at the lower limit. Hence, $C_{vib} \propto T^3$ results as a low temperature approximation in agreement with experiment[10,11]. Both theories presented[12] have C_{vib} as a universal function of the reduced temperature

[7] Multiplication with the distribution function ($\propto x^2$) and integration.

[8] In this case $0 \le x \le \Theta_D/T$ is also small and in Eq. (3.11) the integral is to be taken over $x^4\left(1 + x + \frac{1}{2}x^2 + \ldots\right) / \left(1 + x + \frac{1}{2}x^2 + \ldots - 1\right)^2 dx \simeq x^2dx$.

[9] Here we are obviously neglecting the area under the curve of the integrand for comparably small x. A more precise discussion sets Θ_D/T to ∞ and calculates the improper integral via the gamma function (Ref. [81]). The identity $I = \int \frac{\partial^n}{\partial a^n}\left(e^{ax}\right)dx = \frac{\partial^n}{\partial a^n}\left(\frac{1}{a}\right)e^{ax} + \text{const.}$ is a helpful calculation trick for integrals of type $I = \int e^{ax}x^n dx$.

[10] In the above three–dimensional Debye model the frequency density is proportional to ν^2. The exponents of the low temperature law in the one– or two–dimensional case are reduced appropriately. In fact it is found experimentally for crystals with a marked layer structure (graphite, boron nitride) that $C_{vib} \propto T^2$.

[11] The proportionality of the energy to T^4 is analogous to Stefan's radiation law.

[12] Comparing $E_{vib}(T=0)$ in both theories yields $\frac{9}{8}k_B\Theta_D \sim \frac{3}{2}k_B\Theta_E$, i.e. $\Theta_D \sim \frac{4}{3}\Theta_E$. This only applies as an estimate since both theories cannot be fulfilled simultaneously.

T/Θ_E or T/Θ_D. Figure 3.2 confirms how well the last theory is fulfilled in the case of simple solids.

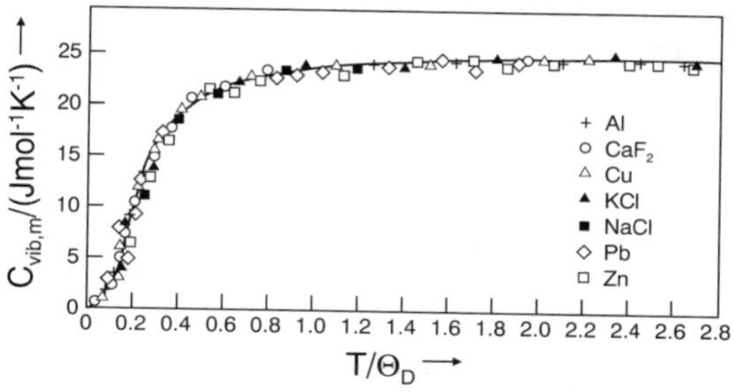

Fig. 3.2: According to the Debye theory the specific heat of many solids is, to a good approximation, a universal function of the reduced temperature T/Θ_D [82]. From Ref. [48].

3.2 Complications

Figure 3.3 illustrates the behaviour of the alkali metals K, Rb, Cs close to absolute zero. The specific heat divided by T is plotted against T^2. In agreement with the

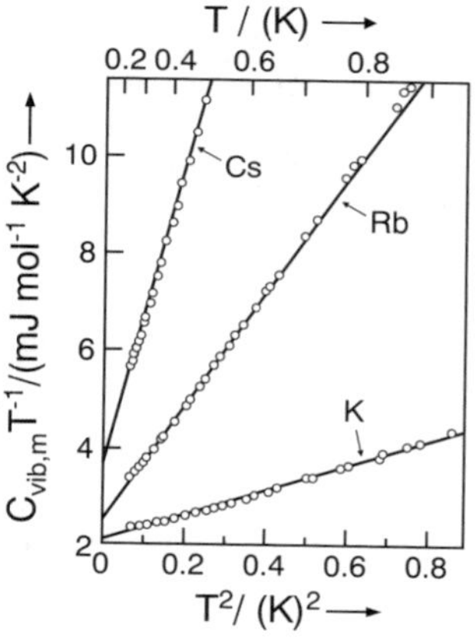

Fig. 3.3: The plot of the specific heat of metals divided by T against T^2 reveals the electronic component as the intercept with the ordinate axis [83]. From Ref. [55].

T^3 law the result is a straight line, but this does not, in contrast to what is expected from Debye's theory, go through the origin. Hence, in this case, a linear contribution in T is included in the specific heat. This is attributable to the translation of the electron gas in the box and can be calculated from the box energy (Eq. (2.25)) by means of Fermi–Dirac statistics. Since metals are not at the focus of our interests we will omit any further discussion here[13,14].

[13] Cf., for example, Refs. [42,84].

[14] We just mention that, owing to the density of states and Pauli's exclusion principle (see Sections 2.2 and 5.3), the electronic contribution to the specific heat is much smaller (by the ratio of Fermi energy and thermal energy [42]) than expected classically ($\frac{3}{2}Nk_B$), and hence is only perceived at low temperatures. Only the electrons with an energy that is comparable with the Fermi energy contribute to C_V.

In molecular crystals or in crystals composed of complex ions it is necessary to take into account intramolecular vibrations in addition to the vibrations of the molecules with respect to each other. If both modes are approximately independent, the former can be treated using the Einstein model. In the case of covalent molecules specifically, it is necessary to pay attention to internal rotations. The behaviour is especially complicated[15] in the case of the compounds discussed in Section 2.2.6.
The pure lattice vibrations are also more complex than has been described so far[13]. In addition to (transverse and longitudinal) acoustical phonons, i.e. vibrations by which the constituents are moved coherently in the same direction without charge separation, there are so–called optical phonons. The name is based on the fact that the latter lattice vibrations are — in polar compounds — now associated with a change in the dipole moment and, hence, with optical effects. The inset to Fig. 3.1 illustrates a real phonon spectrum for a very simple ionic crystal. A detailed treatment of the lattice dynamics lies outside the scope of this book. The formal treatment of phonons (cf. $\epsilon(\mathbf{k})$, $D(\epsilon)$) is very similar to that of crystal electrons. (Observe the similarity of the vibration equation to the Schrödinger equation.) However, they obey Bose rather than Fermi statistics (cf. page 119).
In general Θ_D is a pure fit parameter in complex cases. If an acceptable approximation is obtained with a reasonably temperature–independent value of Θ_D, then this usually deviates from the value obtained at low temperatures (see Fig. 3.1). In addition, the above treatment was limited to harmonic behaviour. The importance of anharmonicity is already demonstrated by the occurrence of thermal expansions. (In the next chapter the specific heat of a complicated compound is given as an example (Fig. 4.2 on page 85).)
All these complicating points lead to the fact that, in practice, the specific heat is usually represented in the form of a power law function in T with empirical constants. At high temperatures it is frequently given in tables[16] in the form $A+BT+CT^{-2}$. In such empirically derived relationships contributions arising from point defects (see Chapter 5) are naturally included.
It remains important to remember that — when the Debye model is valid — the Debye temperature represents the sole parameter. Θ_D can be estimated from measurable parameters such as the velocity of sound or the melting point and is a "measure of the softness" of a crystal. Unlike the lattice energy, the vibrational frequencies are very essentially determined by the repulsion term (see Eq. (3.1), Eq. (2.21)). In this context consider the series[17]

$$\Theta_D(\mathrm{As}) \simeq 80\mathrm{K} < \Theta_D(\mathrm{Pb}) \simeq 90\mathrm{K} < \Theta_D(\mathrm{AgI}) \simeq 130\mathrm{K} < \Theta_D(\mathrm{NaCl}) \simeq 300\mathrm{K} <$$

$$< \Theta_D(\mathrm{ZnO}) \simeq 400\mathrm{K} < \Theta_D(\mathrm{Si}) \simeq 500\mathrm{K} < \Theta_D(\mathrm{diamond}) \simeq 1400\mathrm{K}.$$

[15] $NH_4^+NO_3^-$ may serve as an example. The vibrational modes of this compound are detailed in Ref. [6].
[16] The T^{-2} term results from expanding the e–function in Eq. (3.11) for $x \ll 1$, to a higher order.
[17] Source is Ref. [85]. The data for AgI and ZnO refer to the zinc–blende structure.

(The literature values are subject to a scatter of up to 20%.) In addition to the correlation with the velocity of sound (see footnote 4 on page 67) it is the relationship with the melting point that is primarily of interest here, since it can be associated with defect formation in the compound[18]. According to Lindemann [86], melting commences very roughly when the vibration amplitudes reach the magnitude of interatomic distances. Equating vibrational energy ($\propto \nu^2$) and thermal energy ($\propto T_m$) yields the relationship[19]

$$\Theta_D = \text{const}\sqrt{\frac{T_m}{MV_m^{2/3}}} \qquad (3.12)$$

(T_m: melting point, M: molecular weight, V_m: molar volume). It is important to note that the packing density and, hence, the intermolecular distance (cf. critical amplitude) are included in Eq. (3.12) via V_m.

A further useful relation is given by the Grüneisen–equation. It postulates an inverse proportionality of Θ_D to a power of the molar volume. This characteristic exponent carriers the name Grüneisen's constant [88] and interrelates within the range of its validity thermal expansion, compressibility and specific heat (see e.g. [89]).

In nonmetallic systems phonons play a highly important role also for the phenomenon of heat conduction. (In the case of metals the mobile electrons are essential.) Phenomenologically speaking, heat transport is described by relations isomorphous to the diffusion equations (see Chapter 6). The thermal conductivity is the decisive transport coefficient and is very high for perfect, chemically simple and strongly bonded materials (Θ!). The prototypical example is diamond (which for this reason is cold to the touch). The slightest disorder (such as isotope disorder) perturbs the phonon transport perceptibly. Another example is AlN which is well suited as a substrate material for electrical circuits on account of its high thermal and low electrical conductivity. For both properties a defect–free solid is important. In particular oxygen defects ($O_N^{\cdot}$ see Chapter 5) are detrimental in this specific case. The significance of the phonons with respect to the temperature dependence of thermodynamic data will be outlined in the following chapter, their relevance for the carrier formation in Chapter 5.

Phonons are of special importance for the transport of charge carriers (Section 6.2). The lattice vibrations — determining the solid's "breathing frequency" [90] are indispensable for the phenomenon of ionic conduction, and also set an upper limit for it. As far as electron conduction is concerned, the phonon scattering of electrons limits the mobility of these carriers, on the other hand electron–phonon coupling represents the basic mechanism of superconductivity.

---

[18]As discussed in the next chapters, point defect formation does not only depend on the bonding properties but also on the vibrational properties. Point defect concentrations in many substances are comparable close to the melting point.

[19]For simple metal crystals the constant can be given as $134\mathrm{K}^{1/2}\mathrm{g}^{1/2}\mathrm{mol}^{-2/3}\mathrm{cm}$. To a better approximation different constants are used for different structural families of materials (see e.g. Ref. [87] in which a value of $200\mathrm{K}^{-1/2}\mathrm{g}^{1/2}\mathrm{mol}^{-2/3}$ is shown to lead to a good fit for various crystals).

4 Equilibrium thermodynamics of the perfect solid

4.1 Preliminary remarks

The aim of the equilibrium thermodynamic (i.e. in fact thermostatic) treatment in this book will be to define the equilibrium state of real solids, in particular to express the equilibrium concentrations of the defects and, hence, the precise composition of the solid as a function of the relevant variables of state (Chapter 5). These are, in this context, temperature and particle number[1]. Dependences on total pressure will only be dealt with marginally. External electrical fields are of special interest, and are extensively considered in Chapter 7. Internal electrical fields are of importance in general for inhomogeneous and heterogeneous systems. They are discussed briefly at the end of this chapter, but are, above all, significant for the discussion of the distribution of charge carriers in boundary layers (see Section 5.8) and for kinetics in general (Chapter 6). Consideration of surfaces and thus of morphology is shifted to Section 5.4.

According to Fig. 1.2 we decompose thermodynamic functions into contributions that arise from (chemically)[2] perfect solids and contributions that are brought in by defects. At this point we are now interested in the equilibrium thermodynamics of the (chemically) perfect state. Our aim is to sketch the free enthalpy of the perfect solid with the aid of the previous chapters on chemical bonding and phonons, as well as to consider relevant aspects of the thermodynamic formalism and its application to solids, in particular in view of interactions with the chemical environment.

Let us first build up the necessary thermodynamic apparatus. Readers familiar with solid state thermodynamics can omit this chapter.

4.2 The formalism of equilibrium thermodynamics

The aim of thermodynamics is to express functions of state such as Gibbs energy (free enthalpy) in terms of state variables and to obtain relevant information on the equilibrium state. Both the first and second laws make statements concerning the variation of a particular extensive state function of a given system with regard to

[1] Later we essentially use the intensive parameters temperature and chemical component potentials (component partial pressures) as variables. Note that fixing the particle numbers in a given equilibrium system of given pressure or of given volume also defines the component partial pressures.

[2] The addendum "chemical" is intended to emphasize that phonons are elements of the perfect solid as it is defined here. On the other hand, "chemical" at this point also includes effects that can, with some justification, be regarded as crystallographic.

Physical Chemistry of Ionic Materials J. Maier

ISBN: 0-471-99991-1 (HB); 0-470-87076-1 (PB)

internal and external effects. For the sake of simplicity we will consider a homogeneous system at this point[3]. We understand a state function as being a function, let us call it M, that is unambiguously determined by the variables of state. It does not matter how the state was reached for the value of this function in the state concerned; in other words: dM is a total differential. M shall be an extensive function in the following, the differential of which is made up of changes within the system (δ_iM) (see Fig. 4.1) and changes across the boundary of the system (δ_eM):

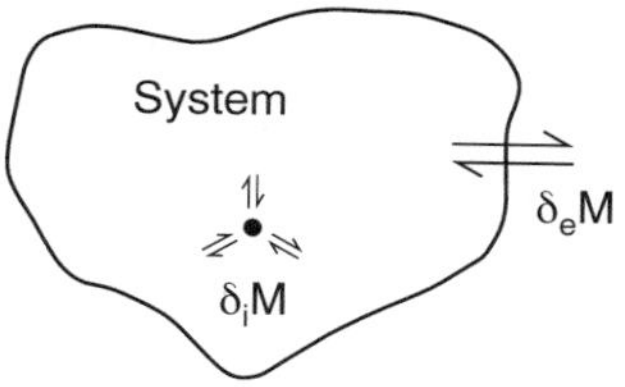

Fig. 4.1: Breakdown of the change in parameter M of the system into internal and external contributions.

$$\mathrm{dM} = \delta_e \mathrm{M} + \delta_i \mathrm{M}. \tag{4.1}$$

In contrast to the total change the two partial contributions to it are not necessarily total differentials. The total change in time ($\dot{\mathrm{M}} \equiv \mathrm{dM/dt}$) is made up of the "(net) production" of M per unit time ($\delta_i\mathrm{M}/\delta\mathrm{t}$) and the "(net) import" of M per unit time ($\delta_e\mathrm{M}/\delta\mathrm{t}$)[4].
For simplicity let us consider a system for which material exchange is excluded, which is open to heat exchange and at which mechanical work can be carried out (i.e. volume changes). The first part of the first law of thermodynamics makes the following statement with respect to the function of state U, which is the internal energy[5] ($\mathrm{M} \equiv \mathrm{U}$):

(1st law a) $$\delta_i\mathrm{U} = 0, \tag{4.2}$$

i.e. energy can neither be created nor destroyed within the system. The second part

(1st law b) $$\delta_e\mathrm{U} = \delta\mathrm{q} + \delta\mathrm{w}, \tag{4.3}$$

lays down that energy changes are possible as a result of exchange of heat (δq) and work (δw). Under the conditions given, the last contribution in Eq. (4.3) is the differential mechanical work, so that (p: pressure, V: volume)[6] it follows for the total change of the internal energy:

[3]In contrast to heterogeneous or inhomogeneous systems we need not specify where the changes of state are taking place.

[4]In Chapter 6 we will formulate such relations more locally (continuity equation); $\delta_e\mathrm{M}/\delta\mathrm{t}$ then will be replaced by the divergence of the respective flux density.

[5]Total energy minus external kinetic energy and potential energy of the system.

[6]For an ideal elastic solid, pdV must be replaced in Eq. (4.4) by the product of the pressure tensor and the differential of the deformation tensor. In the limiting case of a fluid phase, pressure anisotropies disappear. In the treatment that follows, we will neglect pressure anisotropies and assume sufficient mobilities of the components [91,92], which then allows the introduction of a scalar chemical potential for the components without any problem. We will return to this point again in Section 4.3.7 and in Section 5.4.4.

(1st law) $$\mathrm{dU} = \delta \mathrm{q} - \mathrm{pdV}. \quad (4.4)$$

Equation (4.4) neglects the work contributions resulting from changes in the surface area (A), which strictly speaking always play a role even in the perfect solid in equilibrium (γdA, γ: surface tension)[7]. Amongst other contributions left out in Eq. (4.4) are electrical work terms (ϕdQ; ϕ: electrical potential, Q: electrical charge) which will become important when we deal with charge carriers in boundary zones[7]. (In open systems we also have to take account of external material input ($\mu_k \delta_e n_k$; μ_k: chemical potential, n_k: mole number of component k).)
The second law refers to the state function entropy (M≡ S). In contrast to the internal energy, entropy can certainly be created inside the system, and this entropy creation is always positive[8]. In fact, it only disappears at equilibrium:

(2nd law a) $$\delta_i \mathrm{S} \geq 0. \quad (4.5)$$

The entropy import is only due to the heat exchange, and thus

(2nd law b) $$\delta_e \mathrm{S} = \delta \mathrm{q}/\mathrm{T}. \quad (4.6)$$

Brought together the 2nd law becomes[9]

(2nd law) $$\mathrm{TdS} = \delta \mathrm{q} + \mathrm{T}\delta_i \mathrm{S} \geq \delta \mathrm{q}. \quad (4.7)$$

Equations (4.4) and (4.7) can be combined to yield the fundamental relationship

(Fundamental equation) $$-\mathrm{T}\delta_i \mathrm{S} = \mathrm{dU} + \mathrm{pdV} - \mathrm{TdS} \leq 0. \quad (4.8)$$

(In general cases $-$pdV is replaced by a sum of isomorphic product terms consisting of intensive work coefficients and differentials of extensive work coordinates [93].) Now in order to modify Eq. (4.8) to a statement concerning the change of an appropriate function of state, it is necessary to hold certain variables constant. Thus, it can be seen immediately that the Helmholtz energy F (free energy)[10] is the proper measure for constant temperature and constant volume, since it follows (dV=dT=0) in this case

$$-\mathrm{T}\delta_i \mathrm{S} = \mathrm{dU} - \mathrm{d(TS)} = \mathrm{d(U - TS)} \equiv \mathrm{dF} \leq 0. \quad (4.9)$$

Under these conditions the function F=U-TS decreases with time until equilibrium is reached and then no longer alters. The analogous quantity — even more important

[7]Analogously there are also line tension and point tension contributions from the edges and corners of the surface. See Sections 4.3.6 and 4.3.7 for a short discussion of electric and elastic contributions to the chemical potential and Section 5.4.4 for surface thermodynamics.

[8]This determines the arrow of time corresponding to the irreversibility of history.

[9]It can be seen that, in the general case, the entropy of the system also may remain constant or can even decrease as is characteristic for structure formation (see Section 6.10).

[10]Since the recommended symbol A is used in the text to denote various other quantities, the free energy will be denoted by F.

in practice — for the case that pressure and temperature are kept constant is the Gibbs energy (free enthalpy) $G \equiv U + pV - TS = F + pV$. It follows from Eq. (4.8) that

$$-T\delta_i S = d(U + pV) - d(TS) \equiv d(H - TS) \equiv dG \leq 0 \qquad (4.10)$$

with enthalpy $H \equiv U + pV$ being the energy parameter analogous to U. In view of the importance of the function G in the following description, we will consider its differential more closely. From the definition itself it follows that

$$dG = d(U + pV - TS) = dU + pdV + Vdp - TdS - SdT. \qquad (4.11)$$

If we introduce our thermostatic knowledge by means of Eq. (4.8), pdV and TdS are removed from the balance and the expression is reduced to

$$dG = Vdp - SdT - T\delta_i S. \qquad (4.12)$$

Naturally $(dG)_{p,T} = -T\delta_i S \leq 0$, as already stated in Eq. (4.10).
Even in the simple case considered here G is not a function of pressure and temperature alone. Changes in mole number can occur in the irreversibility term $-T\delta_i S$ on account of internal chemical reactions, even when we have excluded mass transport. Hence G is a function of T, p and n_k (the vector $\mathbf{n}$ stands below as a representative for the whole sequence of mole numbers of the various components, $n_1, n_2 ...$)[11,12]

$$\begin{aligned} dG(p, T, \mathbf{n}) &= \left.\frac{\partial G}{\partial T}\right)_{p,\mathbf{n}} dT + \left.\frac{\partial G}{\partial p}\right)_{T,\mathbf{n}} dp + \left.\frac{\partial G}{\partial \mathbf{n}}\right)_{T,p} d\mathbf{n} \\ &= -SdT + Vdp + \boldsymbol{\mu} d\mathbf{n} \equiv -SdT + Vdp + \Sigma_k \mu_k dn_k. \end{aligned} \qquad (4.13)$$

In Eq. (4.13) the chemical potential of the component k, μ_k, has been introduced by definition [95]. The derivatives of G follow accordingly[11]:

$$\left.\frac{\partial G}{\partial T}\right)_{p,\mathbf{n}} = -S \qquad (4.14a)$$

$$\left.\frac{\partial G}{\partial p}\right)_{T,\mathbf{n}} = V \qquad (4.14b)$$

$$\left.\frac{\partial G}{\partial \mathbf{n}}\right)_{p,T} = \boldsymbol{\mu}, \qquad (4.14c)$$

thus specifically $\left.\partial G/\partial n_k\right)_{p,T,n_{k'\neq k}} = \mu_k$.

[11] The abbreviation $\partial G/\partial \mathbf{n}$ stands for the gradient of the parameter G in the composition space, i.e. $(\partial G/\partial n_1, \partial G/\partial n_2, \ldots)$. Hence $\boldsymbol{\mu}$ is not only a reasonable abbreviation but also has a very vivid significance.

[12] Please note that the entropy production is assumed to be caused by mole number changes only, and that the formalism presupposes local equilibrium [94].

At constant pressure and constant temperature our fundamental thermodynamic statement becomes

$$\boldsymbol{\mu}\mathrm{d}\mathbf{n} \equiv \sum_{\mathrm{k}} \mu_{\mathrm{k}} \mathrm{dn}_{\mathrm{k}} \leq 0. \tag{4.15}$$

In the case of a system permeable for matter it is necessary to distinguish between internal mole number changes ($\delta_{\mathrm{i}}\mathrm{n}_{\mathrm{k}}$) brought about by internal chemical reactions, and external mole number changes ($\delta_{\mathrm{e}}\mathrm{n}_{\mathrm{k}}$) brought about by transport over the boundaries of the system with[13] $\mathrm{dn}_{\mathrm{k}} = \delta_{\mathrm{i}}\mathrm{n}_{\mathrm{k}} + \delta_{\mathrm{e}}\mathrm{n}_{\mathrm{k}}$. Equation (4.13) then has a more general meaning. Equation (4.15), however, refers to the internal change.
Heterogeneous or inhomogeneous systems [96] can be built up from small (if necessary infinitely small) partial systems[4,14]. In particular, it is possible to prove using the fundamental Equation (4.8), that an equilibrium contact of two phases requires the equality of temperature, pressure and chemical potential of the components. (The condition referring to pressure will be refined in Sections 4.3.7 and 5.4.4, while the condition concerning the chemical potentials will be revisited in Section 4.3.6.)

On account of the importance of the chemical potential for chemical thermodynamics it is appropriate to make some comments about this quantity.
According to Eq. (4.14c) μ_{k} describes the increase in free enthalpy of a homogeneous system on infinitesimal addition of component k under conditions of constant temperature and pressure (and other work coefficients). It is a measure of how much the component k is "disliked" under these conditions[15]. For instance, if further sodium is added to metallic sodium with a proportional increase in volume, the free enthalpy increases proportionally to the quantity, since the chemical potential of Na in Na is constant ($\mu_{\mathrm{Na}} = \mu^{\circ}_{\mathrm{Na}}$). If Na is added to a system containing an excess of pure $\mathrm{Cl_2}$ gas, low energy NaCl is formed immediately. The chemical potential of sodium in this NaCl co–existing with $\mathrm{Cl_2}$ is relatively low. If traces of Na are dissolved in pure stoichiometric NaCl, the free enthalpy of the phase increases overproportionally and, hence, μ_{Na} increases, too. In accordance with the very low solubility limit in NaCl the value $\mu^{\circ}_{\mathrm{Na}}$ is soon reached, that is the value of μ_{Na} at which the phase equilibrium with Na(s) occurs[16] (see Section 4.3.5). Thus, in accordance with the steep increase in the chemical potential it is not possible to incorporate appreciable

---

[13] Even though in systems permeable for matter the concepts of heat and work need further consideration and an extension, it can, nevertheless, be concluded that the term $\boldsymbol{\mu}\mathrm{d}\mathbf{n} = \boldsymbol{\mu}(\delta_{\mathrm{i}}\mathbf{n} + \delta_{\mathrm{e}}\mathbf{n})$ occurs in the differential of the functions of state (see Eqs. (4.8), (4.11), (4.12)) since $\boldsymbol{\mu}\delta_{\mathrm{i}}\mathbf{n}$ and $\boldsymbol{\mu}\delta_{\mathrm{e}}\mathbf{n}$ are introduced via $\delta_{\mathrm{i}}\mathrm{S}$ and $\delta_{\mathrm{e}}\mathrm{U}$ or $\delta_{\mathrm{e}}\mathrm{S}$, respectively. On account of $\delta_{\mathrm{i}}\mathrm{U} = 0$ the internal mole number change does not appear in the energy balance.

[14] More precisely gradient energies have to be taken into account. They can be regarded as "smeared out" interfacial energies of the walls of the infinitesimal subsystems [97]. The occurrence of gradient effects emphasize that not only the local concentration is important but also (due to interactions) the concentrations at neighbouring positions. Only for very small gradients do the particles behave locally as being in a homogeneous environment (see also footnote 46 on page 142).

[15] In economics an analogous quantity is the "marginal profit", i.e. the increase of the profit function with the production of the goods under consideration (k).

[16] Strictly speaking the finite solubility of Cl in solid sodium must be taken into account, too.

Na excess in NaCl. In phases with greater phase width with respect to the elements the chemical potential increase is comparatively low. This is treated more precisely in Section 4.3.5. The intensive quantity μ is not a function of the absolute quantity, but merely of the concentration. In contrast the extensive quantity G is a proportional, i.e. linear homogeneous function of the mole numbers. If the solid (M) is monocomponent, then it follows that (with G_m as an abbreviation of the molar free enthalpy)

$$\mu_{M\,in\,M} = \mu_M^\circ = G_M/n_M \equiv G_{m,M}. \quad (4.16)$$

An analogous relationship also applies to multicomponent compounds in which appreciable stoichiometric changes do not occur and hence the phase width is small[17], e.g.

$$\mu_{MX\text{ in "MX"}} \simeq \text{ const } = \mu_{MX}^\circ = G_{MX}/n_{MX} \equiv G_{m,MX}. \quad (4.17)$$

It should be noted that relationships of the type of Eq. (4.17) do not apply for the elemental components of these compounds, e.g. $\mu_{M\text{ in "MX"}}$ and $\mu_{X\text{ in "MX"}}$ (let alone the chemical potentials of the ions). These alter markedly within the homogeneity range when the range of stoichiometry is slight as already discussed above for NaCl. This can be seen alone from the fact that the addition of a small quantity of M or X to "MX" changes the composition much more significantly than the addition of exactly stoichiometric MX to "MX"[18]. The detailed reasons are explored in Chapter 5. Yet, in phases with a low range of homogeneity the sum of the potentials $\mu_{M\text{ in "MX"}}$ and $\mu_{X\text{ in "MX"}}$ is over the whole phase range almost invariant, namely equal to μ_{MX}° (see also Eq. (4.17))[19]. Such standard parameters — suitably normalized — are tabulated in standard works and represent most important thermochemical material constants (in Tables 4.1, 4.2, termed G_m° there). In the normal case chemical potentials are concentration–dependent and the following notation is commonly used:

$$\mu_k(c) = \mu_k^\circ + RT \ln a_k(c_k). \quad (4.18)$$

If a(c) is not specified, nothing is gained. The concentration dependence of the chemical potential is merely included in this quantity, which is termed activity. Nevertheless the expression used in Eq. (4.18) is useful since in limiting cases the activity is identical to (the suitably normalized) concentration (see Section 4.3.5). We are especially interested in the limit of very dilute states (Henryan

---

[17] "MX" here designates the compound composed of M and X with only approximate 1:1 stoichiometry over the whole range of homogeneity.

[18]

$$M_{1+\delta}X + \epsilon M \rightarrow M_{1+\delta+\epsilon}X;$$

$$M_{1+\delta}X + \epsilon X \rightarrow M_{1+\delta}X_{1+\epsilon} \approx M_{1+\delta-\epsilon}X;$$

$$M_{1+\delta}X + \epsilon MX \rightarrow M_{1+\delta+\epsilon}X_{1+\epsilon} \approx M_{1+\delta-\epsilon\delta}X \approx M_{1+\delta}X$$

(ϵ and δ are of the same order of magnitude and assumed to be small with respect to 1)

[19] $d\mu_M + (1+\delta)d\mu_X \simeq d\mu_M + d\mu_X = 0$, see also Eq. (4.21).

Table 4.1: Standard molar thermodynamic data. Δ_f values represent the reaction values for the formation from the elements. While these, by definition, are zero at all temperatures for the elements, the H°_m–values are only set to zero for 298.15 K (and 1 bar). The values for other temperatures are obtained via $C^\circ_{p,m}$. The entropy data are absolute and enter G°_m via $H^\circ_m - TS^\circ_m$. According to Ref. [98].

TiO_2 (rutile)

Phase	T [K]	$C^\circ_{p,m}$ J/(K mol)	S°_m J/(K mol)	H°_m kJ/mol	G°_m kJ/mol	$\Delta H^\circ_{f,m}$ kJ/mol	$\Delta G^\circ_{f,m}$ kJ/mol	log K_f
SOL	298.15	55.103	50.292	-944.747	-959.741	-944.747	-889.406	155.820
	300.00	55.288	50.633	-944.645	-959.835	-944.746	-889.063	154.800
	400.00	62.836	67.675	-938.703	-965.773	-944.364	-870.544	113.681
	500.00	67.204	82.207	-932.182	-973.286	-943.603	-852.173	89.026
	600.00	69.930	94.719	-925.316	-982.147	-942.681	-833.972	72.604
	700.00	71.762	105.645	-918.226	-992.177	-941.718	-815.930	60.885
	800.00	73.074	115.317	-910.981	-1003.234	-940.781	-798.025	52.106
	900.00	74.066	123.984	-903.622	-1015.207	-939.907	-780.233	45.284
	1000.00	74.849	131.829	-896.174	-1028.004	-939.116	-762.535	39.831
	2000.00	78.872	185.163	-819.004	-1189.330	-949.283	-585.830	15.300

TiO_2 (anatase)

Phase	T [K]	$C^\circ_{p,m}$ J/(K mol)	S°_m J/(K mol)	H°_m kJ/mol	G°_m kJ/mol	$\Delta H^\circ_{f,m}$ kJ/mol	$\Delta G^\circ_{f,m}$ kJ/mol	log K_f
SOL	298.15	55.271	49.907	-938.722	-953.602	-938.722	-883.266	154.745
	300.00	55.472	50.249	-938.620	-953.694	-938.720	-882.922	153.730
	400.00	63.591	67.437	-932.626	-959.600	-938.286	-864.372	112.875
	500.00	68.144	82.162	-926.018	-967.099	-937.439	-845.986	88.380
	600.00	70.889	94.848	-919.056	-975.965	-936.421	-827.789	72.065
	700.00	72.659	105.918	-911.873	-986.015	-935.364	-809.768	60.426
	800.00	73.863	115.703	-904.543	-997.106	-934.343	-791.896	51.705
	900.00	74.718	124.455	-897.112	-1009.121	-933.397	-774.148	44.930
	1000.00	75.349	132.362	-889.607	-1021.969	-932.549	-756.500	39.515
	2000.00	77.544	185.498	-812.827	-1183.822	-943.105	-580.322	15.156

BaO

Phase	T [K]	$C^\circ_{p,m}$ J/(K mol)	S°_m J/(K mol)	H°_m kJ/mol	G°_m kJ/mol	$\Delta H^\circ_{f,m}$ kJ/mol	$\Delta G^\circ_{f,m}$ kJ/mol	log K_f
SOL	298.15	47.278	70.417	-553.543	-574.538	-553.543	-525.346	
	300.00	47.332	70.709	-553.455	-574.668	-553.535	-525.171	
	400.00	49.898	84.695	-548.588	-582.466	-553.140	-515.784	
	500.00	51.785	96.042	-543.499	-591.520	-553.386	-506.435	
	600.00	53.223	105.616	-538.246	-601.616	-554.540	-496.941	
	700.00	54.395	113.911	-532.863	-612.601	-554.602	-487.344	
	800.00	55.406	121.242	-527.372	-624.366	-554.993	-477.709	
	900.00	56.313	127.821	-521.785	-636.824	-555.201	-468.034	
	1000.00	57.153	133.798	-516.112	-649.910	-555.325	-458.342	

$BaTiO_3$

Phase	T [K]	$C^\circ_{p,m}$ J/(K mol)	S°_m J/(K mol)	H°_m kJ/mol	G°_m kJ/mol	$\Delta H^\circ_{f,m}$ kJ/mol	$\Delta G^\circ_{f,m}$ kJ/mol	log K_f
SOL-3	298.15	102.467	107.901	-1659.797	-1691.968	-1659.797	-1572.440	275.485
	300.00	102.844	108.536	-1659.607	-1692.168	-1659.787	-1571.898	273.692
	394.65*	115.797	138.665	-1649.180	-1703.904	-1658.834	-1544.292	204.398
			0.509	0.201				
SOL-2	394.65*	115.797	139.175	-1648.979	-1703.904	-1658.633	-1544.292	204.398
	400.00	116.279	140.737	-1648.358	-1704.653	-1658.571	-1542.742	201.461
	500.00	122.794	167.458	-1636.371	-1720.100	-1657.678	-1513.902	158.156
	600.00	126.585	190.208	-1623.888	-1738.012	-1657.546	-1485.163	129.295
	700.00	129.090	209.921	-1611.096	-1758.041	-1656.326	-1456.536	108.688
	800.00	130.910	227.283	-1598.092	-1779.918	-1655.513	-1428.052	93.242
	900.00	132.332	242.787	-1584.927	-1803.436	-1654.628	-1399.671	81.235
	1000.00	133.506	256.792	-1571.634	-1828.426	-1653.789	-1371.389	71.634

* Cf. footnote 31 page 86.

Table 4.2: See Table 4.1 for definition of the parameters. According to Ref. [98].

O_2(g)

Phase	T [K]	$C^\circ_{p,m}$ J/(K mol)	S°_m J/(K mol)	H°_m kJ/mol	G°_m kJ/mol	$\Delta H^\circ_{f,m}$ kJ/mol	$\Delta G^\circ_{f,m}$ kJ/mol	log K_f
SOL	298.15	29.376	205.147	0.000	-61.165	0.000	0.000	0.000
	300.00	29.385	205.329	0.054	-61.544	0.000	0.000	0.000
	400.00	30.106	213.871	3.025	-82.523	0.000	0.000	0.000
	500.00	31.091	220.693	6.084	-104.262	0.000	0.000	0.000
	600.00	32.089	226.451	9.244	-126.626	0.000	0.000	0.000
	700.00	32.981	231.466	12.499	-149.528	0.000	0.000	0.000
	800.00	33.734	235.921	15.836	-172.901	0.000	0.000	0.000
	900.00	34.354	239.931	19.241	-196.697	0.000	0.000	0.000
	1000.00	34.870	243.578	22.703	-220.875	0.000	0.000	0.000

Cu

Phase	T [K]	$C^\circ_{p,m}$ J/(K mol)	S°_m J/(K mol)	H°_m kJ/mol	G°_m kJ/mol	$\Delta H^\circ_{f,m}$ kJ/mol	$\Delta G^\circ_{f,m}$ kJ/mol	log K_f
SOL	298.15	24.443	33.164	0.000	-9.888	0.000	0.000	0.000
	300.00	24.464	33.315	0.045	-9.949	0.000	0.000	0.000
	400.00	25.318	40.481	2.538	-13.654	0.000	0.000	0.000
	500.00	25.912	46.196	5.100	-17.998	0.000	0.000	0.000
	600.00	26.477	50.971	7.720	-22.862	0.000	0.000	0.000
	700.00	26.995	55.092	10.394	-28.170	0.000	0.000	0.000
	800.00	27.494	58.731	13.120	-33.865	0.000	0.000	0.000
	900.00	28.032	61.999	15.895	-39.904	0.000	0.000	0.000
	1000.00	28.676	64.985	18.730	-46.255	0.000	0.000	0.000

Cu_2O

Phase	T [K]	$C^\circ_{p,m}$ J/(K mol)	S°_m J/(K mol)	H°_m kJ/mol	G°_m kJ/mol	$\Delta H^\circ_{f,m}$ kJ/mol	$\Delta G^\circ_{f,m}$ kJ/mol	log K_f
SOL	298.15	62.544	92.341	-170.707	-198.238	-170.707	-147.880	25.908
	300.00	62.666	92.728	-170.591	-198.410	-170.709	-147.739	25.724
	400.00	67.668	111.505	-164.052	-208.654	-170.641	-140.084	18.293
	500.00	70.939	126.976	-157.113	-220.601	-170.356	-132.475	13.840
	600.00	73.475	140.141	-149.888	-233.973	-169.950	-124.935	10.877
	700.00	75.650	151.634	-142.430	-248.574	-169.468	-117.470	8.766
	800.00	77.626	161.866	-134.765	-264.258	-168.922	-110.078	7.187
	900.00	79.484	171.117	-126.909	-280.914	-168.319	-102.758	5.964
	1000.00	81.267	179.585	-118.871	-298.455	-167.681	-95.507	4.989

CuO

Phase	T [K]	$C^\circ_{p,m}$ J/(K mol)	S°_m J/(K mol)	H°_m kJ/mol	G°_m kJ/mol	$\Delta H^\circ_{f,m}$ kJ/mol	$\Delta G^\circ_{f,m}$ kJ/mol	log K_f
SOL	298.15	42.244	42.593	-156.063	-168.762	-156.063	-128.292	22.476
	300.00	42.363	42.855	-155.985	-168.841	-156.057	-128.120	22.308
	400.00	46.808	55.727	-151.500	-173.791	-155.551	-118.875	15.523
	500.00	49.264	66.457	-146.687	-179.915	-154.829	-109.786	11.469
	600.00	50.937	75.595	-141.672	-187.029	-154.014	-100.853	8.780
	700.00	52.241	83.548	-136.511	-194.995	-153.155	-92.061	6.870
	800.00	53.348	90.598	-131.231	-203.709	-152.268	-83.393	5.445
	900.00	54.340	96.939	-125.845	-213.091	-151.361	-74.838	4.343
	1000.00	55.260	102.713	-120.365	-223.077	-150.446	-66.385	3.468

normalization)[20]. There Boltzmann distribution is generally valid (see Chapter 5), i.e.

$$\mu_k = \mu_k^\circ + RT\ln(c_k/c^\circ). \tag{4.19a}$$

Other important special cases of Eq. (4.18) are represented by

$$\mu_k = \mu_k^\circ + RT\ln(\frac{c_k}{c^\circ \pm c_k}), \tag{4.19b}$$

whereby the minus sign applies to the Fermi–Dirac type distribution, while the plus sign applies to Bose–Einstein type distributions. (How Eq. (4.19a,b) is derived from the combinatorial analysis of the problem, is shown in Section 5.2 using defects as an example.) While for Eq. (4.19a) the number of states that can be occupied is inexhaustible, it is limited in the first case (see Eq. (4.19b)). There the denominator in Eq. (4.19b) (number of effectively available states) falls on occupation (Fermi–Dirac), while it increases in the second case (Bose–Einstein)[21]. The Fermi–Dirac correction is of general relevance for Chapters 4, 5, 6 when we consider electronic or ionic carriers at high concentrations. In Section 4.3.5, more complex dependencies of $\mu(c)$ are discussed.
As long as the system is homogeneous, the free enthalpy is given by the chemical potentials of the components multiplied by the mole numbers, yielding

$$G(T, p, \mathbf{n}) = \Sigma_k n_k \mu_k \equiv \mathbf{n}\boldsymbol{\mu} \tag{4.20a}$$

or in differentiated form

$$dG(T, p, \mathbf{n}) = d\Sigma_k n_k \mu_k \equiv d(\mathbf{n}\boldsymbol{\mu}). \tag{4.20b}$$

In view of Eq. (4.13) this may appear surprising, but follows immediately from the extensivity of the G–function[22]. More vividly, Eq. (4.20) follows from the fact that it is possible to integrate Eq. (4.13) while keeping the chemical potentials constant. This corresponds to the build–up of the total homogeneous system of subunits with the same composition (at constant T, p). It becomes particularly evident if we use the vector notation. The integration over $\boldsymbol{\mu}d\mathbf{n}$ is carried out by a line integral,

---

[20] The chemical potential is strictly defined thermodynamically; however, there is the freedom of normalizing μ° (at the cost of a) or a (at the cost of μ°). The problems in this area only need to be touched upon marginally here (see Section 4.3.5). If we write $a(c) = f(c) \cdot c$ with f as the activity coefficient, then μ and c must be invariant with respect to any normalization undertaken (let us indicate it with a and b) so: ${}^a\mu^\circ - {}^b\mu^\circ = RT\ln({}^bf/{}^af)$.

[21] Compare the "Fermi pressure" and the tendency to "Bose condensation" [99].

[22] The precise derivation is as follows: At constant p and T all mole numbers are equally homogeneously multiplied by a factor λ. Thus $G(T, p, \lambda\mathbf{n}) = \lambda G(T, p, \mathbf{n})$. Differentiation of the left hand side with respect to λ yields $\frac{\partial G(T,p,\lambda\mathbf{n})}{\partial \lambda\mathbf{n}}\frac{\partial \lambda\mathbf{n}}{\partial \lambda} = \frac{\partial G(T,p,\lambda\mathbf{n})}{\partial \lambda\mathbf{n}}\mathbf{n}$. Differentiation of the right hand side yields $G(T, p, \mathbf{n})$. If we select $\lambda = 1$, $G(T, p, \mathbf{n}) = \frac{\partial G}{\partial \mathbf{n}}\mathbf{n}$ is obtained, that is Eq. (4.20). Note that $\mathbf{n}$ and $\partial/\partial\mathbf{n}$ represent vectors in composition space. For the chemical potential as an intensive function $(\mu_l(T, p, \lambda\mathbf{n}) = \mu_l(T, p, \mathbf{n}))$ it follows that $\Sigma_k n_k (\partial\mu_l/\partial n_k) = 0$.

whose integration path we can select in such a manner that these conditions remain fulfilled. The comparison with Eq. (4.13) then shows that

$$\mathbf{n}\mathrm{d}\boldsymbol{\mu} \equiv \Sigma_\mathrm{k}\mathrm{n}_\mathrm{k}\mathrm{d}\mu_\mathrm{k} = -\mathrm{SdT} + \mathrm{Vdp}, \tag{4.21}$$

i.e. in particular $\Sigma \mathrm{n_k d}\mu_\mathrm{k})_{\mathrm{p,T}} = 0$. In order to emphasize this once again, this equation, known as the Gibbs–Duhem equation, does not follow from the first or second law but is a mathematical expression of the homogeneity of the system. We can concisely summarize the major results of this section by stating:

$$\mathrm{dG})_{\mathrm{p,T}} = \mathrm{d(H - TS)}_{\mathrm{p,T}} = \boldsymbol{\mu}\mathrm{d}\mathbf{n} = \mathrm{d}\left(\boldsymbol{\mu}\mathbf{n}\right) \leq 0. \tag{4.22}$$

Let us now consider chemical reactions in the system and write them in the form

$$\mathrm{Nil} \rightleftharpoons \Sigma_\mathrm{k}\nu_\mathrm{k}\mathrm{A}_\mathrm{k}, \tag{4.23}$$

i.e. we take the stoichiometric coefficients of the products as positive, those of the educts as negative. There are obviously conditions of the form

$$\frac{\mathrm{dn_k}}{\nu_\mathrm{k}} = \mathrm{d}\xi \tag{4.24}$$

for the mole number changes of $\mathrm{A_k}$ with a substance–independent progress variable ξ.

The formation of $BaTiO_3$ from the oxides can serve as an example:

$$\mathrm{Nil} \rightleftharpoons \mathrm{BaTiO_3 - TiO_2 - BaO} \tag{4.25}$$

or the formation of Cu_2O from the elements

$$\mathrm{Nil} \rightleftharpoons \mathrm{Cu_2O} - 2\mathrm{Cu} - \frac{1}{2}\mathrm{O_2}. \tag{4.26}$$

In the first case $\mathrm{dn_{BaTiO_3}} = -\mathrm{dn_{TiO_2}} = -\mathrm{dn_{BaO}}$. If the oxides concerned are also involved in other reactions this statement naturally only applies to the changes with respect to the reaction under consideration (4.25). Analogously for reaction (4.26): $\mathrm{dn_{Cu_2O}} = -2\mathrm{dn_{Cu}} = -\frac{1}{2}\mathrm{dn_{O_2}}$. Using Eq. (4.24) the basic equation Eq. (4.15) becomes

$$\mathrm{d}\xi\left(\boldsymbol{\nu\mu}\right) \equiv \mathrm{d}\xi\Sigma_\mathrm{k}\nu_\mathrm{k}\mu_\mathrm{k} \leq 0. \tag{4.27}$$

With respect to the process from left to right ($\mathrm{d}\xi > 0$) the condition for the reaction progress (nonequality) is

$$\boldsymbol{\nu\mu} \equiv \Sigma_\mathrm{k}\nu_\mathrm{k}\mu_\mathrm{k} \equiv \Delta_\mathrm{r}\mathrm{G_m} \equiv -\mathcal{A}_\mathrm{m} < 0, \tag{4.28}$$

and for the reaction equilibrium (equality)

$$\boldsymbol{\nu\mu} \equiv \Delta_\mathrm{r}\mathrm{G_m} \equiv -\mathcal{A}_\mathrm{m} = 0. \tag{4.29}$$

The usual abbreviations $\Delta_r G_m$, the free molar enthalpy of reaction, and its negative, the molar reaction affinity $\mathcal{A}_m$, are introduced here. A comparison with Eq. (4.13) reveals that

$$\Delta_r G_m \equiv -\mathcal{A}_m \equiv \left.\frac{\partial G}{\partial \xi}\right)_{p,T} . \tag{4.30}$$

When we take into account that $d\xi/dt = (dn_k/dt)/\nu_k$ represents the reaction rate ($\mathcal{R}$) — which is not dependent on the index k — it is possible to formulate Eq. (4.27) concisely as

$$\mathcal{A}\mathcal{R} \geq 0. \tag{4.31}$$

De Donder's equation (4.31) states directly that when the affinity is positive the reaction rate $\mathcal{R}$ is also positive, that is the reaction must run from left to right until the equilibrium is reached, while $\mathcal{A} < 0$ allows the reaction only to run in the reverse direction[23] [96]. The comparison of Eq. (4.31) with Eq. (4.12) and Eq. (4.10) reveals that the product of affinity and reaction rate gives the entropy production, more precisely $+T(\delta_i S/\delta t)$. It is not only at equilibrium ($\mathcal{R} = 0 = \mathcal{A}$) that no entropy is produced, but also in frozen states ($\mathcal{R} = 0$, $\mathcal{A} \neq 0$). That the entropy production can, in general, be described as a product of generalized forces (here $\mathcal{A}$) and generalized fluxes (here $\mathcal{R}$) is a general finding of irreversible thermodynamics [94] and is considered in more detail in Chapter 6.
When expressed in terms of activities, i.e. using Eq. (4.18), our conditions for reaction progress and reaction equilibrium become

$$\Delta_r G_m = \Delta_r G_m^\circ + RT\Sigma_k \ln a_k^{\nu_k} = \Delta_r G_m^\circ + RT \ln \Pi_k a_k^{\nu_k} \leq 0, \tag{4.32}$$

whereby the free standard enthalpy of reaction is introduced as an abbreviation

$$\Delta_r G_m^\circ \equiv \Sigma_k \nu_k \mu_k^\circ . \tag{4.33}$$

According to Eq. (4.29), in equilibrium $\Delta_r G_m$ is zero and hence $RT \ln \Pi a_k^{\nu_k}$ is a constant[24], namely $(-\Delta_r G_m^\circ)$. We denote the product $\Pi a_k^{\nu_k}$ by Q and its equilibrium value $\Pi \widehat{a}_k^{\nu_k}$, i.e. $\widehat{Q}$, by K, the equilibrium constant. Hence

$$\widehat{Q} \equiv K = \exp -\frac{\Delta_r G_m^\circ}{RT} = \exp -\frac{\Sigma_k \nu_k \mu_k^\circ}{RT} = \Pi_k \widehat{a}_k^{\nu_k}, \tag{4.34}$$

and our reaction condition takes on the simple form (Guldberg–Waage law)[25]

$$Q/K \begin{cases} \leq 1 & (\rightarrow) \\ \geq 1 & (\leftarrow). \end{cases} \tag{4.35}$$

---

[23] This is only a statement about the direction. The dependence of the rate of the driving force is, on a phenomenological level, only predictable for very low driving forces ($|\mathcal{A}| \ll RT$) (see Chapter 6).

[24] Constant with respect to concentrations.

[25] The usual name, mass action law, is unfortunate. Concentration action law would be better.

The upper inequality sign applies for the process from left to right ($d\xi > 0$, i.e. $\mathcal{R} > 0$), the lower for the process from right to left ($d\xi < 0$, i.e. $\mathcal{R} < 0$). Equation (4.35) states the following: If the activity product Q is less than (greater than) calculated for equilibrium, i.e. smaller (greater) than $K \equiv \widehat{Q}$, the concentrations of the products (educts) are "too small" and of educts (products) "too great", and the reaction runs from left to right (from right to left).
As discussed in greater detail in Chapter 6, the quotient Q/K represents the ratio of the backward and forward (partial) rate in the case of an elementary reaction. Hence, as the reaction proceeds from left to right we can write:

$$\frac{\mathrm{Q}}{\mathrm{K}} = \frac{\overleftarrow{\mathcal{R}}}{\overrightarrow{\mathcal{R}}} = \exp\frac{\Delta_r G_m}{RT} \leq 1. \tag{4.36}$$

Since $\partial\Delta G/\partial T = \Delta\partial G/\partial T = -\Delta S$ (see Eq. (4.14)) and $\Delta G = \Delta H - T\Delta S$ (see Eq. (4.10)) the temperature dependence of K is obtained as

$$\frac{\partial \ln K}{\partial T} = -\frac{1}{R}\frac{\partial}{\partial T}\frac{\Delta_r G_m^\circ}{T} = -\frac{1}{R}\left(-\frac{\Delta_r S_m^\circ}{T} - \frac{1}{T^2}\left(\Delta_r H_m^\circ - T\Delta_r S_m^\circ\right)\right) = \frac{\Delta_r H_m^\circ}{RT^2} \tag{4.37a}$$

or

$$\frac{\partial \ln K}{\partial(-1/RT)} = \Delta_r H_m^\circ. \tag{4.37b}$$

For an exothermic (endothermic) reaction (i.e. $\Delta_r H_m^\circ > 0\,(< 0)$) the equilibrium constant is reduced (increased) as the temperature increases.

Before we discuss a few characteristic illustrations of equilibrium thermodynamics, let us consider the temperature dependence of thermodynamic functions somewhat more closely. We already know, at this point, the temperature dependences of the free enthalpy (Eq. (4.14a)) and of the equilibrium constant (Eq. (4.37)). However, integration over larger temperature ranges requires knowledge of the temperature dependence of the enthalpy and the entropy. For this we require the concept of specific heat. The specific heat (or better heat capacity), which has already been discussed with regard to the case of vibrations in Chapter 3, is a reciprocal measure of the increase in temperature on the supply of heat:

$$\text{specific heat} \equiv \delta q/\delta T. \tag{4.38}$$

In other words, the specific heat is high when the supply of a large quantity of heat only leads to a small increase of the temperature of the system and is, thus, a measure of the "capacity to store heat"[26]. It is obvious and readily demonstrable that

[26] In a similar manner, we can regard $\partial n_k/\partial\mu_k$ as "chemical storage capacity" (chemical capacitance); this parameter is discussed again in Section 6.5. An intimately related quantity is the so–called "thermodynamic factor" $\partial \ln a/\partial \ln c$ which plays an important role in Chapter 6. The "charge storage capacity" (electrical capacitance) is defined in an analogous manner: $\partial Q/\partial\phi$ (ϕ: electrical potential).

the specific heat must always be positive[27]. In order to make it an unambigous thermodynamic parameter we define one specific heat for constant volume C_V and one for constant pressure C_p. In both cases δq can be converted to an exact differential. On the basis of the first law it follows according to Eq. (4.4) that $\delta q)_V = dU)_V$ and $\delta q)_p = d(U + pV)_p = dH)_p$. So we obtain

$$C_V = \frac{\partial U}{\partial T}\Big)_V \quad \text{and} \quad C_p = \frac{\partial H}{\partial T}\Big)_p \tag{4.39}$$

for the temperature dependence of U and H. Let us consider a first order phase transition: Here all particles change their thermodynamic state collectively in an abrupt way. Taking the melting of H_2O as a concrete example: Increasing the temperature above 0°C, the solid ice changes its structure and converts into a liquid. The latter state possesses a higher energy. This, therefore, involves structural storage of energy. Since we can describe a phase transition as a rearrangement reaction of the giant molecule solid, this is, at the bottom line, a reactive storage of the thermal energy input. The internal energy, as a function of the temperature, behaves as a step–function and C_V becomes a delta function at the phase transition.

In the stability range of a given phase, the thermal energy is primarily taken up in the form of vibrational excitations. As we discussed in more detail in Chapter 3, this is not a two–state problem, rather lattice vibrations of ever higher states are available and U increases continuously with T. At high temperatures, for which the differences of the energy levels are small with respect to k_BT, U increases linearly with T and C_V and reaches the classical limit $3Nk_B$, whereby N is the number of vibrators. At low temperatures for which differences between levels play a great role, C_V must approach zero with positive curvature. All this is in agreement with the findings of Chapter 3. The parameters E_{vib} and C_{vib} derived statistically in Chapter 3, are strictly speaking, internal energies and C_V–values, respectively. For our purpose it is possible to neglect the difference between C_p and C_V (and between ΔH and ΔU)[28]. C_p is generally a few percent larger than C_V at room temperature. It is only at very high values of T that the difference is significant and this is very frequently the reason for the Dulong–Petit limit being exceeded. Figure 4.2 illustrates an example for the specific heat ($C_p \simeq C_V$) of a complex sodium ion conductor, termed Nasicon[29] ($Na_{1+x}Zr_2Sr_xP_{3-x}O_{12}$). In particular one can see the increase with the number of vibrators in accordance with $\Delta C_{p,m}/\Delta x \simeq 3R$, as well as a clear superimposed phase transition for x=2.

[27]Otherwise, a system, which has been virtually displaced from the state of equilibrium, would not be driven back to the initial state. In an isolated system the entropy must form a maximum with respect to all the perturbation variables T, p, **n** ($\delta(\delta S) < 0$). Analysis reveals that $C_V > 0$ is a necessary condition of thermal stability, a positive compressibility $\chi \equiv -\partial \ln V/\partial p$ for mechanical stability and $\partial\mu_k/\partial c_k > 0$ for chemical stability (see e.g. Ref. [92]). Note that $C_p \geq C_V$.

[28]Even for ideal gases $C_p - C_V = \frac{d(H-U)}{dT} = \frac{d(pV)}{dT} = \frac{d(RT)}{dT} = R$ is only for the order of 10J/molK. In condensed phases $\frac{d(pV)}{dT}$ is very small.

[29]Nasicon stands for Na–SuperIonicCONductor.

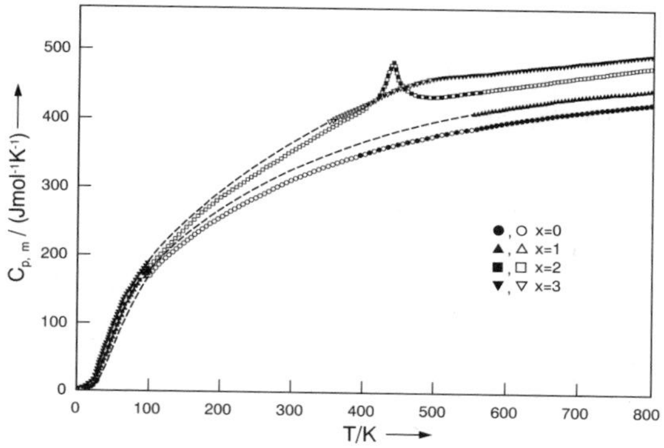

Fig. 4.2: The specific heat of $Na_{1+x}Zr_2Si_xP_{3-x}O_{12} \equiv$ NASICON(x) as a function of temperature for various x. (The Dulong–Petit limit is $(18+x)3R \simeq (450\ldots525)\,\mathrm{Jmol^{-1}K^{-1}}$) [100].

The temperature dependence of the entropy is also obtained via C_p or C_V, namely:

$$\left.\frac{\partial S}{\partial T}\right)_{p\,\mathrm{or}\,V} = \frac{C_{p\,\mathrm{or}\,V}}{T}. \tag{4.40}$$

This follows for constant p because of

$$-S = \frac{\partial G}{\partial T} = \frac{\partial}{\partial T}(H - TS) = C_p - T\frac{\partial S}{\partial T} - S. \tag{4.41}$$

The derivation for constant V is analogous. Specific heats are generally readily measured, e.g. via heats of mixing (adiabatic) or via isothermal calorimetric methods (DSC)[30]. If the functions of state are known at a temperature T_0, then the specific heat can be used to determine the whole temperature variation, e.g. at constant pressure:

$$H(T) = H(T_0) + \int_{T_0}^{T} C_p(T)dT \tag{4.42a}$$

$$S(T) = S(T_0) + \int_{T_0}^{T} C_p d\ln T \tag{4.42b}$$

$$G(T) = H(T_0) - TS(T_0) + \int_{T_0}^{T} C_p dT - T\int_{T_0}^{T} \frac{C_p}{T} dT = H(T_0) - TS(T_0) - \iint_{T_0}^{T} \frac{C_p}{T}(dT)^2. \tag{4.42c}$$

[30]While in DTA (differential thermal analysis) the difference in temperature between sample and reference is detected for a given heat input, making it possible to determine phase transition temperatures very accurately, it is appropriate for thermodynamic analysis to work isothermally and to measure the energy input (e.g. electric current), necessary to keep the temperatures between sample and reference the same. This is the measurement principle of DSC (differential scanning calorimetry).

The left hand side of Eq. (4.42c) follows from the combination of Eq. (4.42a) and Eq. (4.42b), the right hand side from double integration according to Eq. (4.14a) and Eq. (4.40). Both sides can be converted to each other by partial integration. In most cases we are interested in the temperature dependences of changes in thermodynamic properties, e.g. enthalpies of reaction: On account of the commutability of the operators $\Delta_r \equiv \frac{d}{d\xi}$ and $\frac{d}{dT}$, $\Delta_r H(T)$ follows from $\Delta_r H(T_0)$ and integration of $\Delta_r C_p$ analogously to Eq. (4.42a). $\Delta_r C_p$ is obtained according to $\Delta_r C_p = \Sigma_k \nu_k C_{p,k}$ from the C_p data of the reaction participants ($\nu_k < 0$ for educts, $\nu_k > 0$ for products). Since in solid state problems the number of vibrators frequently does not change and the specific heat of various substances per vibrator varies similarly with T (in fact in the classical limit they are identical), $\Delta_r C_{p,V}$ is frequently negligible and the reaction parameters $\Delta_r H$ and $\Delta_r S$ are thus constant, if the temperature range is not too large. $\Delta_r G$ then varies linearly with temperature. In practice C_p is often given in the form of a semiempirical series, as already described in Chapter 3: $C_{p,k} = \Sigma_j a_{jk} f_j(T)$. $\Delta_r C_p$ is, hence, determined via $\Sigma_j f_j(T)\Delta_r a_j$. Tabulated values of C_p are to be found for some substances in Tables 4.1 and 4.2 as examples.
Let us return now to the equilibrium criteria and, in particular, to "chemical reactions", whereby we will understand this term in a wide sense in order to include heterogeneous equilibria.

4.3 Examples of equilibrium thermodynamics

4.3.1 Solid–solid phase transition

The simplest (solid state) "chemical reaction" of all is probably a solid–solid phase transition which is accompanied by a change in structure but not in composition. There is some justification in regarding such a modification transition of the "giant molecule" solid as a rearrangement reaction, even though there are frequently only slight changes in bond lengths and bond angles, and usually no topological changes at all. The transition from the cubic high temperature phase of $BaTiO_3$ to the ferroelectric tetragonal phase[31] at $T_c \simeq 130°C$

$$\text{Reaction U} = \qquad BaTiO_3\,(tet) \rightleftharpoons BaTiO_3\,(cub) \tag{4.43}$$

may serve as an example.
The equilibrium condition demands equality of the chemical potentials (see Eq. (4.29)). In this case ($\mu = \mu°$)

$$\mu^{\circ}_{BaTiO_3(tet)}(T_c) = \mu^{\circ}_{BaTiO_3(cub)}(T_c). \tag{4.44}$$

Since there is no compositional variability, the transition temperature at a given pressure is defined unequivocally by the intersection of the $\mu°(T)$- (i.e. G_m (T)-)

[31] The transition point of a very pure single–crystalline sample is at 131°C, while it is up to 10°C lower for polycrystalline samples of typical purity.

curves[32] (see Fig. 4.3). More precisely: Slight deviations from the ideal stoichiometry ("Dalton composition") do not play a significant role in determining the free en-

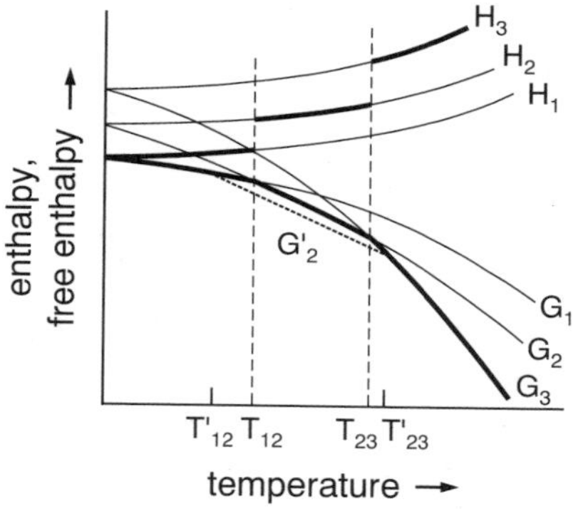

Fig. 4.3: Enthalpy and free enthalpy for a substance undergoing two first order transitions at T_{12} and T_{23}. The bold lines correspond to the thermodynamic equilibrium of the pure phases. The dotted curve shows the effect of stabilizing phase 2, by doping. If the phases 1, 2 and 3 are identified with solid, liquid and gaseous phases, the dotted line mirrors the effect of freezing point depression ($T'_{12} < T_{12}$) and boiling point elevation ($T'_{23} > T_{23}$).

thalpy of the phase (cf. Eq. (4.20)), so that, to a good approximation, the transition temperature is independent of the component activities — say the partial pressure of oxygen in the surroundings of an oxide. Only the total pressure has a slight possibility of influence via Eq. (4.14b). Table 4.1 (see page 78) shows the equivalence of the free enthalpy of the $BaTiO_3$ phases for $T_C = 122°C$[31]. In the case of such a first–order phase transition, as is realized here and illustrated schematically in Fig. 4.3, there is a discontinuity in $H_m(T)$ and, hence, in $S_m(T) = -\partial G_m/\partial T$ whereby $\Delta_U H_m = T_m \Delta_U S_m$ (r=U). At phase transitions of second or higher (nth) order there are, according to Ehrenfest's classification, discontinuities only in $\partial^2 G_m/\partial T^2$ or in the respective higher differentials ($\partial^n G_m/\partial T^n$). Table 4.1 (see page 78) also lists the G_m values for TiO_2 (rutile) and TiO_2 (anatase). It can be seen that, at least for p=1bar, rutile is the more stable modification over the entire temperature range. Since the rutile modification is also the denser one, that is, has the lower molar volume (V_m), the stability of rutile relative to anatase even increases at higher pressures[33] because of (see Eq. (4.14b))

$$\left.\frac{\partial \Delta_{U'} G_m}{\partial p}\right)_T = \Delta_{U'} V_m > 0 \tag{4.45}$$

referring to

$$\text{Reaction } U' = \qquad TiO_2(\text{rutile}) \rightleftharpoons TiO_2\,(\text{anatase}). \tag{4.46}$$

Hence, anatase is thermodynamically unstable in all these cases and is merely obtained as a metastable phase from an initial state of high free energy (free enthalpy) (e.g. as a result of the hydrolysis of $TiCl_4$).

[32]Because $\partial G/\partial T\,)_p = -S$, $\partial S/\partial T\,)_p = C_p/T$, $S > 0, C_p > 0$ curves fall with T and decrease with negative curvature (see Fig. 4.3). Enthalpies and entropies increase with T because $C_p > 0$ (cf. also Eq. (4.16)).

[33]This is true, as long as the condition $\Delta_{U'} V_m(p) > 0$ is fulfilled. Interestingly, the relevant stabilities seem to be reversed at very small crystal sizes [101], which may be traced back to the different surface free energies (see Section 5.4.4).

4.3.2 Melting and evaporation

The melting "reaction"

$$\text{Reaction S} = \qquad H_2O(s) \rightleftharpoons H_2O(l) \qquad (4.47)$$

is similarly characterized by a transition temperature that is only slightly dependent on pressure. To make such a pressure effect (on the bulk structure) responsible for the gliding effect when skating — as is frequently done in physics textbooks — is probably a "scientist's tale" (see Section 5.8). The solubility of foreign substances (doping) can have a significant impact on T_c: The solubility of NaCl in $H_2O(l)$ is large compared with that in $H_2O(s)$[34]. The stabilization associated with this brings about a one–sided reduction in the free enthalpy of the liquid phase and lowers the melting point (cf. Fig. 4.3), which is much appreciated by car drivers in winter time. (The opposite effect of melting point elevation as a result of solid–state doping is also known (see Section 5.2)).

In the case of the evaporation process

$$\text{Reaction V} = \qquad H_2O(l) \rightleftharpoons H_2O(g) \qquad (4.48)$$

the stabilization of the liquid phase by salt addition results in an increase in the boiling point. This is shown in Fig. 4.3. Even for the pure phase there is a new feature in Eq. (4.48) compared to Eq. (4.47): The molar free enthalpy of H_2O (g), i.e. the chemical potential and, hence, the activity of H_2O are no longer (even not approximately) invariant. In the case of ideal conditions (Eq. (4.19)) it holds that

$$\mu_{H_2O(g)} = \mu^{\circ}_{H_2O(g)} + RT\ln(P_{H_2O}/P^{\circ}). \qquad (4.49)$$

Now, the chemical potential can be varied greatly with respect to the (tabulated) μ–value at $P_{H_2O}/P^{\circ} = 1$. This significant dependence is the reason why mountaineers have great difficulty in boiling eggs at high altitudes. Water then boils at temperatures appreciably below 100°C. Since at the transition temperature $\mu_{H_2O(g)} = \mu_{H_2O(l)} \simeq \mu^{\circ}_{H_2O(l)}$ is fulfilled, the following relation is obtained from Eq. (4.49) or directly from the law of mass action ($\Delta_V S^{\circ}$ and $\Delta_V H^{\circ}$ are taken as being approximately T–independent)

$$\widehat{P}_{H_2O}(T)/P^{\circ} = K_V(T) \propto \exp -(\Delta_V H^{\circ}/RT), \qquad (4.50)$$

which is the well–known vapour pressure–temperature relationship and which can also be obtained by integration of the Clausius–Clapeyron equation.

[34]The corresponding effect on solid–solid transition is distinctly lower but often measurable.

4.3.3 Solid–solid reaction

Let us now consider a "proper" chemical reaction, namely

$$\text{Reaction C} = \qquad \mathrm{BaO + TiO_2 \rightleftharpoons BaTiO_3}. \tag{4.51}$$

According to our treatment it follows that $\Delta_C G_m = \mu^\circ_{BaTiO_3} - \mu^\circ_{TiO_2} - \mu^\circ_{BaO} = \Delta_C G^\circ_m$. Concentration–dependent terms do not occur here (more precisely: they can be neglected), and $\Delta_C G^\circ_m(T)$ can be calculated at any temperature from the μ° values of the phases (see Table 4.1, see page 78). $BaTiO_3$ is thermodynamically stable with respect to decomposition into the oxides at all temperatures at which $\Delta_C G^\circ_m < 0$. There is no mass action effect. If $\Delta_C G^\circ_m$ is known at a temperature T_0, it is possible to calculate the reaction parameters for all other temperatures via the specific heats using Eq. (4.42). If C_p is approximated by means of a series with coefficients a_{jk} it is technically much more economical to carry out the calculation via integration of $\Delta_C C_p$, i.e. via $\Delta_C a_j$ (see page 86), than to determine the thermodynamic functions of the reaction participants individually at temperature T and then to calculate the difference (operation Δ_C).

4.3.4 Solid–gas reaction

There is an additional degree of freedom if the chemical potential of a component varies via the concentration term, which is the case when a gaseous phase is involved. Let us consider the oxydation of Cu to Cu_2O

$$\text{Reaction G} = \qquad 2\mathrm{Cu(s)} + \frac{1}{2}\mathrm{O_2(g)} \rightleftharpoons \mathrm{Cu_2O(s)}. \tag{4.52}$$

With

$$\Delta_G G_m = \mu^\circ_{Cu_2O} - 2\mu^\circ_{Cu} - \frac{1}{2}\mu^\circ_{O_2} - \frac{1}{2}RT\ \ln(P_{O_2}/P^\circ) = \Delta_G G^\circ_m - \frac{1}{2}RT\ \ln(P_{O_2}/P^\circ) \tag{4.53}$$

i.e. in terms of the mass action law

$$(\widehat{P}_{O_2}(T)/P^\circ)^{-1/2} = K_G(T) = \exp -(\Delta_G G^\circ_m/RT) \tag{4.54}$$

or

$$Q_G/K_G = (P_{O_2}/\widehat{P}_{O_2})^{-1/2}, \tag{4.55}$$

$\Delta_G G^\circ_m$ and, hence, K_G can be obtained from Table 4.2 (see page 79).
Equation (4.54) means that at a particular temperature the coexistence of Cu and Cu_2O corresponds to a defined oxygen partial pressure ($\widehat{P}_{O_2}$). If the external partial pressure of O_2 is less than the equilibrium value $\widehat{P}_{O_2}$ given by K_G, that is if Q_G/K_G is smaller than 1, copper(I) oxide gives up O_2 in accordance with Eq. (4.35) until $\widehat{P}_{O_2}$ is re–established. If $P_{O_2}/P^\circ > K_G^{-2}$, copper is consumed accordingly. The presence

of a mixture of a metal and co–existing metal oxide[35] is a convenient manner of producing low activities of oxygen in the gas phase that are no longer controllably set up by means of mixtures of O_2 and inert gases (10^{-5}bar is generally the lower limit for the latter method). The use of a mixture[36] of redox active gases, such as CO/CO_2 or H_2/H_2O, is a related method.
If all Cu has been oxydized to Cu_2O, then the system Cu–O has the possibility of forming CuO according to

$$Cu_2O + \frac{1}{2}O_2 \rightleftharpoons 2CuO. \tag{4.56}$$

Here too a two–phase mixture Cu_2O/CuO fixes a temperature–determined (this time much higher) equilibrium partial pressure. At a given temperature, the equilibrium partial pressure of oxygen can only be varied[37] in the case of a binary oxide, if a

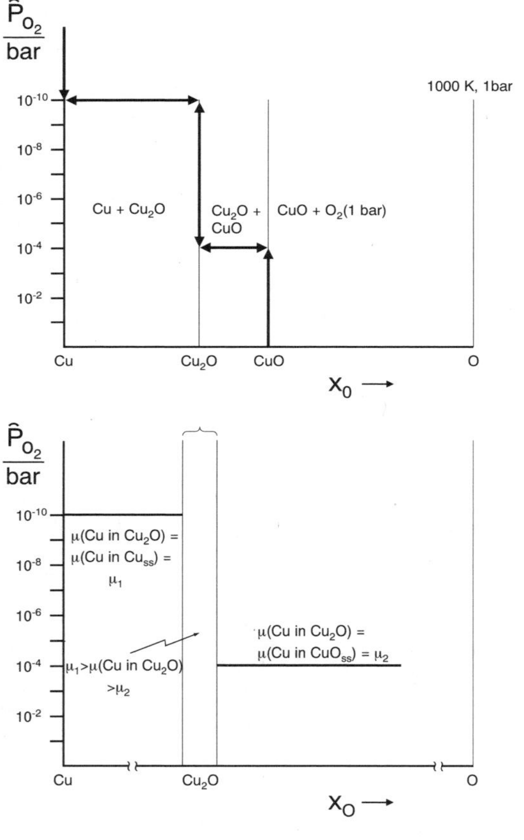

Fig. 4.4: Equilibrium partial pressure of oxygen in the system Cu–O as a function of oxygen content x_O. The "line phase" Cu_2O is resolved in the lower Figure. The region left of the Dalton composition ($Cu_{2.00...0}O_{1.00...0}$) is exaggerated (see page 174). The change of P_{O_2} within the phase is discussed in Chapter 5. Cu_{ss} means Cu saturated with O, CuO_{ss} means CuO saturated with Cu.

[35]or two co–existing binary oxides

[36]A constant partial pressure of CO_2 can be set up in an analogous manner using, say, $CaCO_3/CaO$ mixtures.

[37]In general, this is described by the phase rule [95]. Equilibrium between the phases implies that the intensive parameters (T, p, $\boldsymbol{\mu}$) cannot all be varied independently of each other. If we

single phase is present. There are three possibilities of achieving a single–phase situation in the system under consideration: 1) oxygen partial pressures so small that Cu_2O does not form (only Cu present), 2) medium pressures at which Cu no longer exists, but CuO is not yet formed (only Cu_2O present) or 3) a P_{O_2} value so high that only CuO is present. Figures 4.4 and 4.5 illustrate these states and show, in particular, how the fields of stability depend on the partial pressure of oxygen. Figure 4.4 emphasizes how drastic the change in partial pressure is when passing through a phase compared with the constant values which are obtained in the two–phase regions. The atomistic discussion of the internal changes undergone by a homogeneous solid is the subject of the later chapters (in particular of the following one)[38].

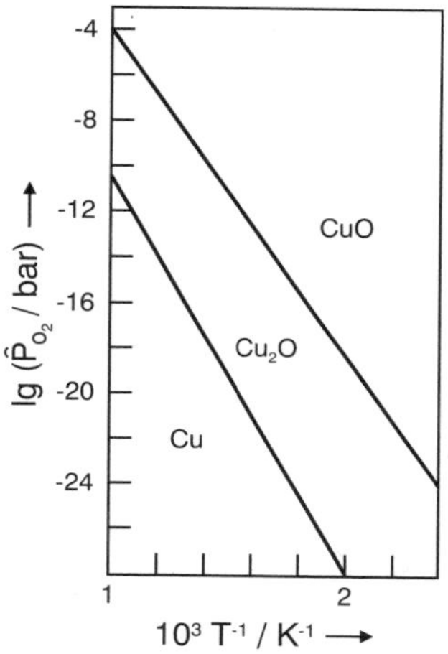

Fig. 4.5: The plot of the equilibrium partial pressure as a function of the inverse temperature separates the existence ranges from each other. The slope is proportional to the enthalpy of reaction.

call the number of intensive parameters, that can be independently varied, "degree of freedom" or variance (υ), the phase rule says:

$$\upsilon = 2 + (\chi - \rho) - \pi$$

(π: number of phases, χ: number of components, ρ: number of chemical reactions (apart from the phase reaction itself)). In partial equilibria, which occur frequently in reality, υ is correspondingly greater [102]. The above relation is simple to verify: When there are π phases ($\alpha, \beta \ldots$) and χ components (1, 2, ..., k, ...) it is possible to define $\pi\chi$ chemical potentials. In addition, there is the fixing of the two parameters p and T. On account of the phase equilibrium there exist $\chi(\pi - 1)$ relationships of the form $\mu_k^\alpha = \mu_k^\beta$ and — since the μ's depend on the concentrations but not on the total amount — π relationships of the type $\Sigma_k x_k^\alpha = 1$ (x: mole fraction). In addition to the phase equilibria, there may also be ρ reaction equilibria (see Eq. (4.29)), yielding ρ relationships amongst the μ's. This leads to the above relationship with $(\pi\chi + 2) - (\pi + \chi(\pi - 1) + \rho)$. $(\chi - \rho)$ can also be considered to be the number of chemically independent components. Thus, in the case of the equilibrium between H_2, O_2 and H_2O $(\chi - \rho) = 2$ because $\chi = 3$, $\rho = 1$, i.e. setting the partial pressures of H_2 and H_2O fixes the partial pressure of O_2. If we refer, from the start, to the chemically independent components (2 in this example), it is not necessary to take the reactions into account, i.e. $\upsilon = 2 + \chi' - \pi$ with $\chi' = \chi - \rho$.

[38]The following preview may be allowed: Within the homogeneity range we write μ_O (up to a constant) as a linear function in the logarithm of the activity a_O; equilibrium with the gas phase

It remains to be added that logarithmic representations of equilibrium pressures as a function of temperature are helpful in mapping out the stability window of the oxides or to reveal affinities of inorganic redox–solid reactions of the form $mM + M'_{m'}O \rightleftharpoons m'M + M_mO$. Figure 4.6 illustrates such an affinity diagram for a variety of oxides.

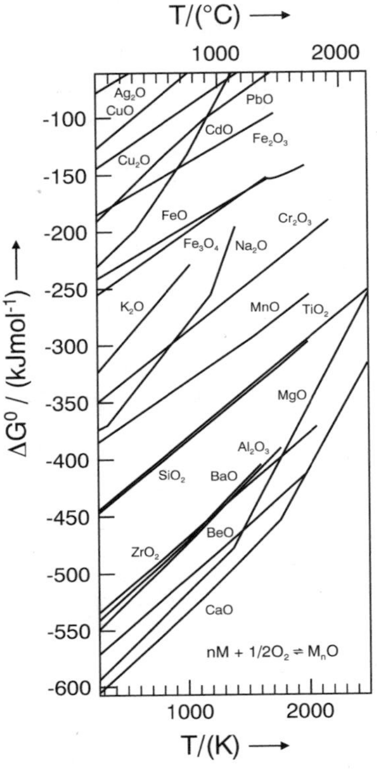

Fig. 4.6: The free standard enthalpies of reaction for $mM + 1/2O_2 \rightleftharpoons M_mO$ (e.g. $3/4Fe + 1/2O_2 \rightleftharpoons Fe_{3/4}O$) as a function of temperature. The ordinate also reflects the equilibrium partial pressure via Eq. (4.54). The kinks are the result of phase changes. The slope is given by $-\Delta S^\circ$. According to Ref. [6].

In particular, it is possible to read off that at equilibrium in air at room temperature even Ag must be present in the form of the oxide (in contrast to gold). We also recognize that Al_2O_3 and ZrO_2 exhibit extremely high stabilities with respect to the elements, even compared with Na_2O or K_2O. This is ascribable to the high lattice energy (cf. Section 2.2.2) of Al_2O_3 or ZrO_2 and is the reason why many — even very electronegative — metals can be fabricated "aluminothermically" from the oxides. At $T > 1000K$ the relative stability of ZrO_2 is greater than that of Al_2O_3.

yields $a_O \propto P_{O_2}^{1/2}$. If we consider an oxide ($MO_{1+\delta}$ with $\delta < 0$) in which oxygen is incorporated on a vacancy ($V_O^{\cdot\cdot}$) while two conduction electrons (e') are consumed, the mass action law $[V_O^{\cdot\cdot}][e']^2 \propto P_{O_2}^{-1/2}$ can be formulated for dilute defect concentrations (see Chapter 5). We recognize that the oxygen activity is determined by the defect concentrations: here $a_O \propto ([V_O^{\cdot\cdot}][e']^2)^{-1}$. If no further defects are relevant $2[V_O^{\cdot\cdot}] = [e']$ and $a_O \propto |\delta|^{-3}$. Cf. also footnote 49.

4.3.5 Phase equilibria and mixing reactions

It is sufficient here to state that every phase can be regarded as a mixture of elements (even though mostly absolutely nonideal), that is, possesses a finite phase width. (At this point we tacitly leave the perfect solid!) Cu possesses a finite solubility for oxygen before it forms Cu_2O, consequently the metal at the Cu/Cu_2O contact exhibits a maximum oxygen content $CuO_{\delta_{max}}$ with $\delta_{max} \ll 1$. (Note: It is still the copper structure that we are dealing with.) Correspondingly Cu_2O exhibits a maximum copper content at the contact, viz. $Cu_{2+\varepsilon_{max}}O$ (with $|\epsilon_{max}| \ll 1$). Again, it does not interest us here how such deviations from the idealized Dalton compositions (for which ϵ, δ are exactly zero) are achieved. This is the subject of the following chapter[38]. In an analogous manner, considering Eq. (4.51) as an example, the reaction of TiO_2 with excess BaO yields a (most highly) TiO_2–deficient $BaTiO_3$, while a titania excess leads to a TiO_2–saturated $BaTiO_3$. All these phenomena can be formally[38] treated under the concept of the thermodynamics of mixtures. Larger solubilities occur when the phases are more similar. Thus, the mutual solubilities increase from AgCl/AgI via AgBr/AgI to AgCl/AgBr. In the case of the latter system the rocksalt structure is exhibited for all compositions, and the substances are completely miscible (provided that the temperatures are not too low). Even when miscibilities are so slight that they are irrelevant regarding thermodynamic stability, the effects are of considerable importance for the kinetics of chemical reactions and for the electrical properties (Chapters 5, 6, 7). The chemical potentials of the partners are variable within the phase range. Let us consider Cu_2O once again (see Fig. 4.4) in a parameter window in which Cu_2O is thermodynamically co–existing with Cu. Now it must be true that the chemical potential of the copper in the Cu_2O is equal to the chemical potential of copper in copper (that is μ°_{Cu}) (see Section 4.2). We simply can formulate a "chemical reaction"[39]

$$\mathrm{Cu}(\text{“}\mathrm{Cu_2O}\text{”}) \rightleftharpoons \mathrm{Cu}(\text{“}\mathrm{Cu}\text{”}) \tag{4.57}$$

for which the general equilibrium condition (Eq. (4.29)) yields the condition for the phase equilibrium, namely

$$\mu_{\mathrm{Cu\ in\ “Cu_2O”}} = \mu^\circ_{\mathrm{Cu}}. \tag{4.58}$$

A continuous increase in the oxygen partial pressure causes μ_{Cu} in "Cu_2O" to fall until "CuO" forms. Now, the copper potential is the same in both oxide phases. The corresponding oxygen potential is coupled to the Cu potential. This can be demonstrated in the following way: For "Cu_2O" we may formulate

$$2\mathrm{Cu}(\text{“}\mathrm{Cu_2O}\text{”}) + \mathrm{O}(\text{“}\mathrm{Cu_2O}\text{”}) \rightleftharpoons \mathrm{Cu_2O}(\text{“}\mathrm{Cu_2O}\text{”}). \tag{4.59}$$

[39] "MX" represents a phase of about the same composition as MX, i.e. the interval of all stable compositions $[MX_{1+\delta_{min}}, MX_{1+\delta_{max}}]$ with constant "ground structure", whereby δ can also be negative.

Again it is possible to apply Eq. (4.29). Since, as already discussed (Eq. (4.17)), $\mu_{\text{Cu}_2\text{O in "Cu}_2\text{O"}} \simeq \mu^{\circ}_{\text{Cu}_2\text{O}}$, we obtain the desired coupling

$$2\mu_{\text{Cu in "Cu}_2\text{O}} + \mu_{\text{O in "Cu}_2\text{O"}} = \mu^{\circ}_{\text{Cu}_2\text{O}} \tag{4.60}$$

or differentially $2d\mu_{\text{Cu in "Cu}_2\text{O"}} \equiv -d\mu_{\text{O in "Cu}_2\text{O"}}$[40]. This could have been immediately derived via the Gibbs–Duhem equation (Eq. (4.21)), too, on account of $n_{\text{Cu}}/n_{\text{O}} \simeq 2$.
The connection with the gas phase follows from:

$$2\text{O}(\text{"Cu}_2\text{O"}) \rightleftharpoons \text{O}_2(\text{g}), \tag{4.61}$$

i.e.

$$2\mu_{\text{O in "Cu}_2\text{O"}} = \mu_{\text{O}_2} = \mu^{\circ}_{\text{O}_2} + \text{RT}\ln\left(\text{P}_{\text{O}_2}/\text{P}^0\right). \tag{4.62}$$

Equation (4.60) can be formulated in terms of activities as

$$\text{a}_{\text{Cu in "Cu}_2\text{O"}} \cdot \text{P}_{\text{O}_2}^{1/4} = \text{const}, \tag{4.63}$$

whereby const is determined by the formation mass action constant of the oxide. (μ_{Cu} can be obtained in this manner for the contact $\text{Cu}_2\text{O}/\text{CuO}$ from Eqs. (4.63) and (4.56) and for the contact $\text{Cu}/\text{Cu}_2\text{O}$ from Eqs. (4.63) and (4.52)).
It is the purpose of the next chapter, which deals with defect chemistry, to demonstrate not only that the deviation from the Dalton composition is determined by the defect concentrations, but also that the chemical potentials of the components are determined by the chemical potentials of the defects[38]. In many cases of interest, the latter quantities can be represented by a Boltzmann relation (Eq. (4.19a)) as a function of defect concentrations. Such simple functions do not in almost all relevant cases apply to the relationships between component potentials and component concentrations $\left(\mu_{\text{Cu in "Cu}_2\text{O"}} \neq \text{const} + \text{RT}\ln[\text{Cu}]\right)$.

There is, however, a Boltzmann relationship (and then over the entire solubility range), if the partners are so similar that there is no heat of mixing produced and the entropy of mixing only occurs as a result of the (ideal) configuration effect. Thus moving far away from the above example, let us consider a mixture[41] of A and B at $\text{A}_\text{x}\text{B}_{1-\text{x}} \equiv \text{C}$ formed according to

$$\text{Reaction M} = \qquad \text{xA} + (1-\text{x})\text{B} \rightleftharpoons \text{A}_\text{x}\text{B}_{1-\text{x}}, \tag{4.64}$$

thus, it is a necessary (and, under realistic conditions, a sufficient) condition for ideal mixing that $\Delta_\text{M}\text{U} = 0$. Simplified, this means that it does not matter whether A-A bonds, B-B bonds or A–B bonds occur, i.e. the energy of the reaction

[40] Equation (4.60) is also implicitly based on the Gibbs–Duhem equation.
[41] In the sense of the following chapter this is a substitutional mixture, since A and B take comparable positions [87,89].

$$\text{Reaction M}' = \begin{array}{c} \text{A} \\ | \\ \text{A} \end{array} + \begin{array}{c} \text{B} \\ | \\ \text{B} \end{array} \rightleftharpoons \begin{array}{c} \text{A} - \text{B} \\ + \\ \text{A} - \text{B} \end{array} \tag{4.65}$$

is zero. If $\Delta_M U$ and $\Delta_{M'} U$ are less than zero, mixing effects occur, that reduce the activity with respect to the concentration and, thus, additionally stabilize the mixture. Conversely $\Delta_{M'} U > 0$ means a tendency to phase separation.

In the case of an ideal mixture $\Delta_M U = 0 \simeq \Delta_M H$ and $\Delta_M G = -T\Delta_M S$. The simple site–statistical analysis[42] yields $\Delta_M G_m = RT\Sigma_k x_k \ln x_k = RT(x \ln x + (1 - x)\ln(1-x))$ and $\mu_k = \frac{\partial G}{\partial n_k} = \frac{\partial \Delta_M G}{\partial n_k}$ is thus obtained in the Boltzmann form[43] as $\mu_k = \mu_k^\circ + RT \ln x_k$, i.e. the activity coefficient is 1. $\Delta_M G_m$ is negative (i.e. the mixture is thermodynamically stable with respect to separation into A and B), describes in the binary case a symmetrical curve with a minimum at $x = 1/2$ and does not exhibit a point of inflection ($d^2\Delta_M G/dx^2 = RT\left(\frac{1}{x} + \frac{1}{1-x}\right) > 0$). This means that it is not possible to apply a double tangent[44], i.e. all compositions are thermodynamically stable with respect to any sort of demixing along the coordinate (and not just with respect to demixing in pure A and pure B).

---

[42] Similar to the treatment in the next chapter (Eqs. (5.12)–(5.15) on page 118) it turns out that $\Delta_M S = \Delta_M(k_B \ln \Omega) = k_B \ln \Omega_C = k_B \ln \frac{(N_A+N_B)!}{N_A! N_B!} \propto -(x_A \ln x_A + x_B \ln x_B)$. The Stirling formula and the definition $x_k = N_k/\Sigma N_k$ (N: number of particles; Ω: statistical weight) were used.

[43] $G_c = n_A G_{Am} + n_B G_{Bm} + \Delta_M G = n_A \mu_A^\circ + n_B \mu_B^\circ + RT\Sigma n_k \ln x_k$. Thus it follows for component A: $\mu_A \equiv \partial G_c/\partial n_A)_{n_B} = \mu_A^\circ + RT \ln x_A$.

[44] Let us recapitulate briefly, how we determine equilibrium compositions and quantities from G(x) curves:

We assume that we know the G curves of two phases α and β and calculate the co–existing *compositions* $\widehat{x}^\alpha$, $\widehat{x}^\beta$ at a given temperature. (It does not matter here whether the G curves merge with one another as assumed above (see Fig. 4.7) or not (see Fig. 4.9)). Let us assume that two phases α and β of different composition are possible with G functions $G^\alpha(x)$ and $G^\beta(x)$. On account of conservation of particles it is necessary to find a compromise between the quantities of α and β.

Phase equilibrium requires the chemical potentials of A and B in both phases α and β to be equal, that is $\mu_A^\alpha(\widehat{x}_A^\alpha) = \mu_A^\beta(\widehat{x}_A^\beta)$ and $\mu_B^\alpha(\widehat{x}_A^\alpha) = \mu_B^\beta(\widehat{x}_A^\beta)$. Here we write x_A for x explicitly for clarity. These are two equations with the two unknowns $\widehat{x}_A^\alpha$ and $\widehat{x}_A^\beta$ from which it is possible to calculate the equilibrium concentrations if the analytical relationships are known. Here, however, we will solve them graphically by applying a double tangent. For this purpose it is sufficient to show that the slopes of the tangents at the equilibrium concentrations are (i) identical and (ii) equal to the slope of the straight line connecting the equilibrium points (see Fig. 4.7). From $G_m^\alpha = (1-x_A^\alpha)\,\mu_B^\alpha(x_A^\alpha) + x_A^\alpha\,\mu_A^\alpha(x_A^\alpha)$ and the analogous equation of G_m^β it follows after differentiation that $dG_m^\alpha/dx_A = \mu_A^\alpha - \mu_B^\alpha$ and $dG_m^\beta/dx_A = \mu_A^\beta - \mu_B^\beta$. (The derivatives of μ with respect to x cancel because of the Gibbs–Duhem equation (Eq. (4.21), page 81)). On account of the phase equilibrium the tangent slopes so defined are equal in size at just the mole fractions $\widehat{x}_A^\alpha$ and $\widehat{x}_A^\beta$ which we want to calculate. On the other hand, the gradients are identical to $\left[G_m^\beta(\widehat{x}_A^\beta) - G_m^\alpha(\widehat{x}_A^\alpha)\right] / \left[\widehat{x}_A^\beta - \widehat{x}_A^\alpha\right]$, i.e. both tangents fall on the line through the solution points.

The *quantities* are given by the overall amounts (they correspond to the gross composition x_{A0}) and the distances in the phase diagram in accordance with the "lever rule": We assume that the co–existing compositions are $\widehat{x}_A^\alpha$ and $\widehat{x}_A^\beta$. Then the equilibrium quantities are determined by $N^\alpha/N^\beta = \left[\widehat{x}_A^\beta - x_{A0}\right] / \left[x_{A0} - \widehat{x}_A^\alpha\right]$, that is behave inversely to the corresponding distances in the

The "model of the regular solution" [89], which is the next most simple model after the ideal mixture, is still a symmetrical model. It takes $\Delta_{M'}U_m$ as constant (but not as zero), nevertheless neglects (actually inconsistently) nonidealities in entropy. This model already explains the occurrence of miscibility gaps [103] and provides a simple example of the relevance of activity coefficients in different reference systems. In contrast to ideal mixing, there is an additional term in the energy of mixing which is proportional to the half particle number, the co–ordination number æ, the concentrations x_A and x_B and the energy of reaction $\Delta_{M'}U$. This follows from simple probability considerations concerning the pairs AA, AB, BA, BB with pair energies ϵ_{AA}, $\epsilon_{AB} = \epsilon_{BA}$ and ϵ_{BB}: We take particles A and B of approximately the same size and similar to each other. We can then ignore elastic effects in energy and assume the same co–ordination numbers (æ) for A and B. The probability of meeting the configuration AA is simply $x_A^2 \equiv x^2$, the probability for BB is $(1-x)^2$ and finally the probability of meeting AB or BA is $x(1-x)+(1-x)x$. It follows that the internal energy of C is

$$U_C = \frac{1}{2}N æ \epsilon_{AA} x^2 + \frac{1}{2}N æ \epsilon_{BB}(1-x)^2 + \frac{1}{2}2N æ \epsilon_{AB} x(1-x). \tag{4.66}$$

Since $\Delta_M H \simeq \Delta_M U = U_C - xU_A - (1-x)U_B$, whereby $U_A = U_C(x{=}1) = \frac{1}{2}N æ \epsilon_{AA}$ and $U_B = U_C(x{=}0) = \frac{1}{2}N æ \epsilon_{BB}$, we find for the molar quantity that

$$\Delta_M U_m = x(1-x)W. \tag{4.67}$$

Hence, in what follows, we can, by neglecting the nonidealities in the entropy, also write the excess term $\Delta_M G_m - \Delta_M G_m^{id}$ as $\Delta_M G_m^{ex} = Wx_A(1-x_A)$.
W is then, apart from the co–ordination number æ, the half molar pair rearrangement energy according to Eq. (4.65), that is $W/(æN_m) = \epsilon_{AB} - (\epsilon_{AA}+\epsilon_{BB})/2 = \Delta_{M'}U_m/(2N_m)$ (N_m: Avogadro's number). Or, expressed in another manner: $W/(æN_m)$ describes the deviation of the bonding energy of the AB pair from the mean energies of the AA and BB pairs. If the temperature is sufficiently high, we always expect — assuming the solid phase to be adequately stable — complete miscibility. However, W becomes important at lower temperatures and leads to an additional contribution of $\Delta_M G_m^{ex\prime\prime} = -2W$ to the curvature $\Delta_M G_m''$ (in addition to $\Delta_M G_m^{id\prime\prime}$). Points of inflection now occur in the G–curve demanding demixing according to the double tangent rule[44]. At the critical temperature at which demixing just occurs, the points of inflection coincide[45] and $\Delta_M G_m'' = \Delta_M G_m''' = 0$. The critical temperature at which demixing starts is[46] readily computed as $T_c = W/2R$ and is,

phase diagram. This follows immediately from the solution for the mass balances $\widehat{x}_A^\alpha N^\alpha + \widehat{x}_A^\beta N^\beta = x_{A0}\left(N^\alpha + N^\beta\right)$ and $\left(1-\widehat{x}_A^\alpha\right)N^\alpha + \left(1-\widehat{x}_A^\beta\right)N^\beta = \left(1-x_{A0}\right)\left(N^\alpha+N^\beta\right)$.

[45] $\lim_{\Delta x_{infl}\longrightarrow 0}\left(\frac{\Delta G''_{infl}}{\Delta x_{infl}}\right) = \Delta G'''_{infl} = 0$.

[46] $\Delta_M G_m''(x_c, T_c) = RT_c\left(\frac{1}{x_c} + \frac{1}{1-x_c}\right) - 2W = 0$

$\Delta_M G_m'''(x_c, T_c) = RT_c\frac{2x_c - 1}{x_c^2(1-x_c)^2} = 0$.

thus, directly proportional to $\Delta_{M'}U_m$. The critical mole fraction[46] lies (as expected from the symmetry conditions with respect to A and B) at 1/2. In higher mixing models this symmetry is lost. Figure 4.7 shows the creation of an asymmetrical

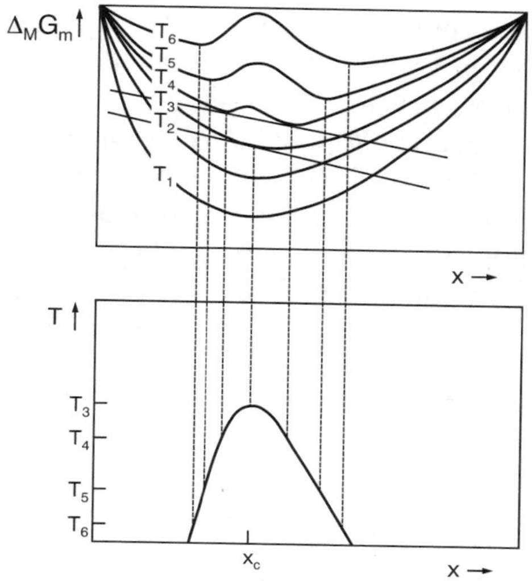

Fig. 4.7: Application of a double tangent to the $\Delta_M G_m(x)$ curves reveals co–existing compositions at the edges of the miscibility gap (see the bottom part of the diagram). The critical parameters of demixing are $T_c = T_3$ and x_c.

miscibility gap at lower temperatures for such more complicated situations.
Here it is interesting to investigate how far the shape of the curve predicts kinetic inhibition of the demixing process. Let us consider the hypothetical decomposition of the initial composition $A_{x_0}B_{1-x_0}$ (index 0) into the compositions 1 ($A_{x_1}B_{1-x_1}$) and 2 ($A_{x_2}B_{1-x_2}$) (see Fig. 4.8). If n is the mole number, the free enthalpy before

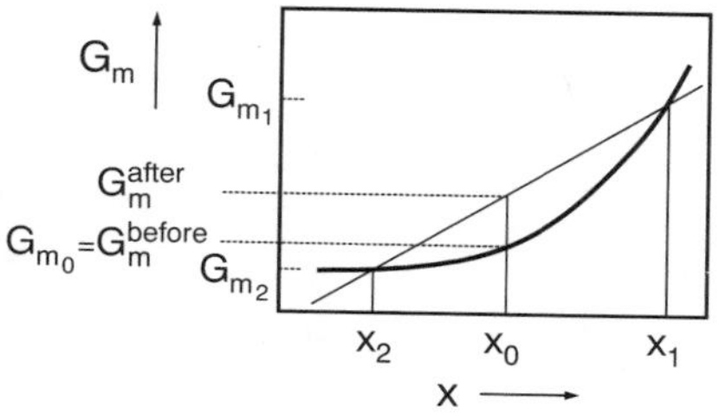

Fig. 4.8: The effect of the curve shape of the $G_m(x)$ relationship on the demixing.

the decomposition is n_0G_{m0}; after the decomposition it is $n_1G_{m1} + (n_0 - n_1)\,G_{m2}$. On account of the lever rule[44] this can be written in the form $G_m^{final} = \frac{x_2-x_0}{x_2-x_1}G_{m1} + \frac{x_0-x_1}{x_2-x_1}G_{m2}$. This, on the other hand, is just the value of the secant through the points (x_1, G_{m1}) and (x_2, G_{m2}) at mole fraction x_0 which is easily proven[47]. Accordingly the reaction value is given by the distance to the curve and is always positive when the curvature is positive. In agreement with the above considerations there is no

[47] The secant equation is $G_m(x) = \frac{G_{m2}-G_{m1}}{x_2-x_1}x + \frac{G_{m1}x_2-G_{m2}x_1}{x_2-x_1}$. It follows for $x = x_0$ that $G_m(x) = G_m^{final}$.

demixing in the case of an ideal mixture, which always exhibits a curved shape as shown in Fig. 4.8. Now let us consider the general case. In Fig. 4.7 we consider an initial composition that lies in the region of positive curvature, for example, directly to the right of the left hand minimum of the T_6 isotherm in Fig. 4.7, and investigate whether or not we can reach our final state in a continuous manner. The initially slight changes in composition refer to the G(x) curve in the direct neighbourhood of the starting position. The free enthalpy necessary here is positive, as in the example just described, and a spontaneous demixing is not possible in this manner. Therefore nucleation (see Section 5.4) is essential. The situation is different in the region of negative curvature between the points of inflection. In this so–called spinodal region[48] a fluctuation in composition leads to a reduction in G. Spontaneous demixing can occur here if no other inhibitions are important.
Figures 4.9, 4.10 reveal how the equilibrium compositions and phase diagrams can

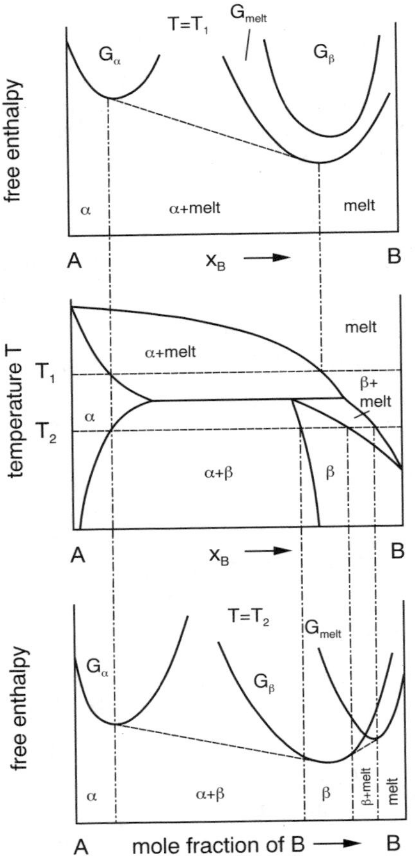

Fig. 4.9: Derivation of the phase diagram from $G_m(x)$ curves according to the double tangent method[44]. The index m is suppressed for simplicity in this figure. According to Ref. [64].

[48]The more general treatment (101) which includes elastic effects and gradient energies leads to modified criteria [104].

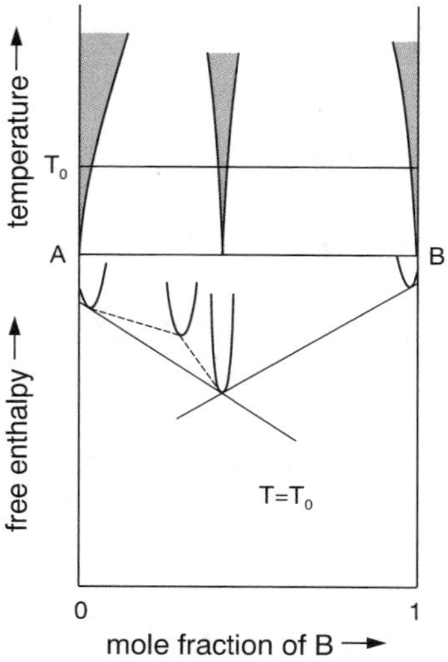

Fig. 4.10: The derivation of the phase width from $G_m(x)$ for almost stoichiometric compounds.

be obtained according to the double tangent method, also for more complicated cases. The G curves of Fig. 4.10 refer to almost stoichiometric phases, i.e. those phases which are of greatest importance in this text. The free enthalpy changes very rapidly on either side of the minimum and only permits narrow phase widths. Generally, asymmetrical miscibility gaps, as shown in Fig. 4.7, require explanations more complex than regular mixture models ("subregular mixtures"). An interesting phenomenological procedure involves treating of Eq. (4.65) in terms of the mass action formalism, i.e. by the introduction of chemical pair potentials (quasi–chemical method) [105]. While these models yield good semiquantitative approximations for alloy systems, for ionic compounds it is only of value if the partners in the mixture are themselves binary or multinary and similar to each other, i.e. when the mixing process does not involve too large an energy of reaction. For a detailed modelling of phases with small phase widths it is necessary to leave the rough mixture models and to consider the atomistic situation, as is done in the next chapter. It will turn out there that the other extreme case, namely the case of almost vanishing phase widths is statistically easily accessible.

In anticipation of the next chapters let us emphasize that it is necessary to correctly identify the particles for which a random configurational entropy is approximately valid. To give an example: In a similar way to NaCl, which when dissolved in water, splits into two components, viz. Na^+ and Cl^-, that may be approximately treated independently[49], the dissolution of H_2O in solid oxides comprises a variety of charged components (cf. page 195ff, see Eq. 5.176). This illustrates why attempts

[49] Note that the activity of NaCl in liquid water is not approaching the NaCl concentration for high dilution but its square ($\mu_{NaCl} = \mu_{Na^+} + \mu_{Cl^-}$ hence $a_{NaCl} \propto c_{Na^+} \cdot c_{Cl^-} \propto c^2_{salt}$). Cf. also footnote 38.

to circumvent defect chemistry in the thermodynamic treatment of solids lead to either clumsy or incorrect results.

Finally, let us discuss the problem of the normalization of activity coefficients and activities (cf. footnote 20), on the basis of the regular solution model.
The excess term in the chemical potential $\mu_A^{ex} = \mu_A - \mu_A^{id}$ and, hence, $RT \ln f_A$ is obtained from $\Delta_M G_m^{ex}$ and is[50] $W(1-x_A)^2$. This term goes to zero when $x_A \to 1$ (cf. Raoult's law). The activity coefficient just considered therefore, is referred to as the Raoultian activity coefficient. When $x_A \to 0$ it is constant (cf. Henry's law). The displacement of the constant W from the μ^{ex} term to the μ° term makes possible another normalization method, which is advantageous for describing dilute states. The first normalization chosen above and characterized by $f_A \to 1$ for $x_A \to 1$ is known as the Raoultian normalization, and the second possibility with $f_A \to 1$ for $x_A \to 0$ as the Henryan normalization. There are then two equivalent representations[20] of μ_A, namely as ${}^R\mu_A^\circ + RT \ln x_A + RT \ln {}^R f_A$ or as ${}^H\mu_A^\circ + RT \ln x_A + RT \ln {}^H f_A$. In the regular model ${}^R\mu_A^\circ = {}^H\mu_A^\circ - W$ and, hence, $\ln {}^H f_A = W(x_A^2 - 2x_A)/RT$ while $\ln {}^R f_A = (1 - x_A)^2 W/RT$. Figure 4.11 displays the relations.

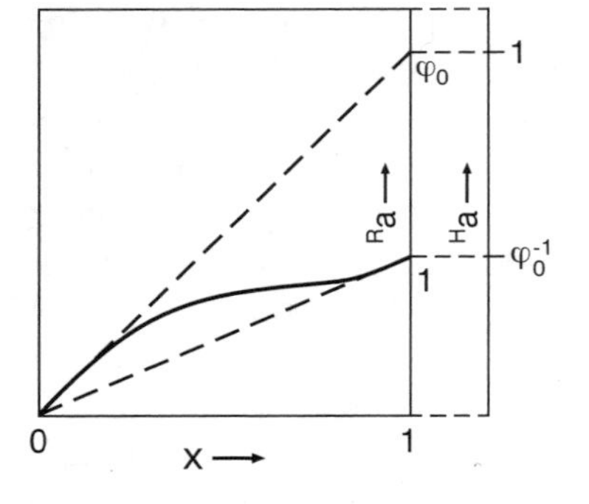

Fig. 4.11: Henryan and Raoultian activities. In the regular model (here W>0) both activity scales are proportional: $\varphi \equiv {}^R a/{}^H a = {}^R f/{}^H f = \exp(W/RT)$. Generally the scale is only qualitatively valid. The intersections denoted by φ_0 and φ_0^{-1} refer to infinite dilution ($\varphi_0 \equiv \varphi(x{=}0)$).

4.3.6 Spatial equilibria in inhomogeneous systems

The formalism of chemical reactions can be taken even further and applied to transport within one and the same phase. Let us consider two different sites x and x′ in medium MX, which is considered to be isotropic. Equilibrium conditions of the form[51]

$$\mu_M(x) = \mu_M(x') \quad \text{and} \quad \mu_X(x) = \mu_X(x') \tag{4.68}$$

[50] $\mu_A = \mu_A^\circ + RT \ln x_A + \frac{\partial \Delta_M U_m}{\partial n_A} = \mu_A^\circ + RT \ln x_A + RT \ln f_A$,
$\frac{\partial \Delta_M U}{\partial n_A} = RT \ln f_A = W\,(1 - x_A)^2$. Please pay attention to the difference between $\Delta_M U$ and $\Delta_M U_m$ in Eq. (4.67) and to the fact that the differentation has to be performed at constant n.

[51] The spatial coordinate can be converted to a reaction coordinate if the process is formulated as a rearrangement reaction.

then apply to both M and X. This means that no gradients in component potentials may occur in a state of complete thermodynamic equilibrium (this naturally also applies to μ_{MX}). Otherwise processes of the form

$$\text{Reaction T} = \qquad \mathrm{M(x)} \rightleftharpoons \mathrm{M(x')} \quad \text{or} \quad \mathrm{X(x)} \rightleftharpoons \mathrm{X(x')} \tag{4.69}$$

take place leading to the realization of Eq. (4.68). The correspondence of these transport processes to chemical reactions will be discussed further in Chapter 6. As long as we can consider the μ°–term in $\mu = \mu^{\circ} + \mathrm{RT}\ln \mathrm{a}$ as spatially invariant, Eq. (4.68) is identical with the condition for uniformity of composition under isotropic boundary conditions. (Considering the site exchange as the simplest version of a "chemical reaction", viz. a reaction with vanishing $\Delta_r G^{\circ}$ values, the G° profile taken along the reaction coordinate is symmetrical.) This uniformity of composition only strictly applies within the solid, remote from its interfaces. There can be variations in structure near interfaces (i.e. variations in μ°), as is e.g. demonstrated by the change in the band gaps of semiconductors or insulators [106].

Until now we have not addressed the following important point: In addition to the above phase equilibrium, the condition of contact equilibrium demands that charged constituents, such as electrons or ions, have to redistribute themselves individually to a limited degree in order to achieve a minimum in the global free enthalpy. On account of the occurrence of electrical fields this effect is limited in size and in spatial extent, but in boundary layers, where most of the interesting processes occur or at least start, it is of very great relevance. We have neglected electrical fields until now[52]. Electrical field effects lead to an additional term in the differential of the G function that takes the form $\phi \mathrm{dQ}$ (ϕ: electrical potential, Q: charge). The variation in charge caused by particles introduced into the system or particles brought into the system is $\mathrm{dQ_k} = \mathrm{z_k F dn_k}$ ($\mathrm{z_k}$ is the charge number, $\mathrm{z_k F}$ the molecular charge, and $\mathrm{n_k}$ the mole number of the charged species, F = Faraday constant). This electrical term can be combined with the chemical term $\mu_k \mathrm{dn_k}$ to yield an electrochemical term $\widetilde{\mu}_k \mathrm{dn_k}$. So analogous expressions are obtained for charged particles simply by replacing the chemical potential by the electrochemical potential

$$\widetilde{\mu}_k = \mu_k + \mathrm{z_k F}\phi. \tag{4.70}$$

In particular, the relevant equilibrium condition is

$$\Sigma \nu_k \widetilde{\mu}_k = 0 \tag{4.71}$$

(replacing Eq. (4.29)). More generally and instead of Eqs. (4.12, 4.13) we have to consider the relation

$$-\mathrm{T}\delta_i \mathrm{S} = \Sigma_k \widetilde{\mu}_k \mathrm{dn_k} \leq 0. \tag{4.72}$$

These intricacies are not, of course, important when the particles are neutral, but also not if processes between charged particles are considered within a region where

[52] Analogously it is necessary, in the general case, to take account of elastic fields, gravity fields etc.

potential electrical differences do not play a role, that is in the bulk or if the process is restricted to the same spatial coordinate. Then the electrical terms in Eq. (4.71) cancel on account of ϕ being constant and the conservation of charge within a chemical reaction ($\Sigma_k \nu_k z_k = 0$). In the region of interfaces, however, we are concerned with charge transfer processes of the form[53] (see Fig. 4.12)

$$\text{Reaction } Te^- = \quad e^-(x) \rightleftharpoons e^-(x') \tag{4.73a}$$
$$\text{Reaction } TM^+ = \quad M^+(x) \rightleftharpoons M^+(x') \tag{4.73b}$$
$$\text{Reaction } TX^- = \quad X^-(x) \rightleftharpoons X^-(x') \tag{4.73c}$$

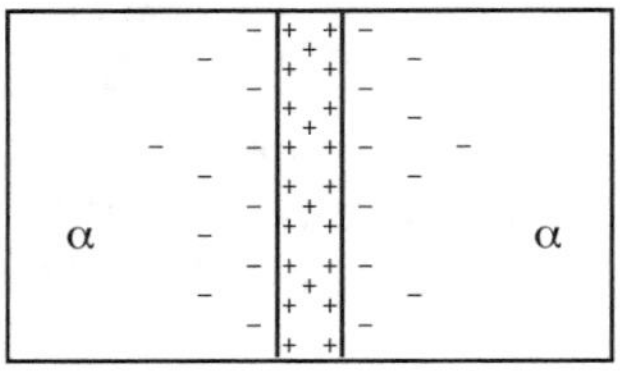

Fig. 4.12: The introduction of a heterogeneity (interface region) into a homogeneous ionic crystal or ionic solution leads to redistribution of the charges.

whereby the transition from one phase to the other or the transport within the boundary areas of a phase are meant. In equilibrium $\Delta\mu_k \propto \Delta\phi$ holds due to Eq. (4.71). At the level of the mass action law it is now found that

$$\frac{a_k(x')}{a_k(x)} = K_{Tk}\kappa_{xx'}. \tag{4.74}$$

Here $\ln K_{Tk}$ measures the pure "chemical effect" (total effect if the species were uncharged), that is $-(\mu_k^\circ(x') - \mu_k^\circ(x))/RT$, while $\ln \kappa_{xx'}$ measures the "electrical effect", $-(\phi(x') - \phi(x))zF/RT$. However, the two parameters are not independent of each other; a detailed analysis must refer κ to chemical constants and variables of state (cf. Section 5.8). In particular Eq. (4.74) shows that concentration differences now also appear within the same structure ($\Delta\mu^\circ = 0$), and that this is at the cost of an electrical field. Such contact problems are not exotic in nature, but rather the normal case when mobile, charged particles are involved:

1. On contact between two different semiconductors both interfaces become charged. Well–known consequences are, for example, the thermo–electric effect for semiconductors, p–n transitions, and many other major issues in electronics[54].

2. If a metal is dipped into a solution, metal ions can go into solution from the boundary regions; at the same time dissolved ions can be adsorbed at the surface of the metal. Phenomena of this sort dominate "liquid electrochemistry" and are the basis for galvanic effects (cf. Chapter 7 in particular). Similar phenomena determine the stability of colloids.

[53]These charge transfer processes now include the phase equilibrium described by $M(x) \rightleftharpoons M(x')$ and $X(x) \rightleftharpoons X(x')$.

[54]However, interfaces are frequently avoided in electronics by introducing step functions in the doping content within the same matrix structure.

3. Charging processes at membranes, demanded by the constancy of the electrochemical potential of ions that can pass through, are the basis of electrophysiological processes and are of paramount importance in nerve signal propagation.

These are only a few examples (see also Fig. 5.70 in Chapter 5). In general, the introduction of an intermediate phase (or interface) into phase α which contains charged constituents leads to a separation of charge. The significance of such electrochemical boundary effects for the chemistry of the (usually charged) defects can be hardly overestimated. Special emphasis will be placed, in Chapter 5, on the redistribution of mobile ions in the boundary regions of solids.

4.3.7 Thermodynamics of elastically deformed solids

So far we considered phases with sufficient atomic mobilities and vanishing pressure anisotropies such that we could use the term -pdV to describe the mechanical energy increment (see also footnote 6). Generally, for elastic deformations (i.e. usually small deformations), this increment has to be expressed in terms of the stress tensor components s_{ij} and the differential strain tensor components $d\varepsilon_{ij}$:

$$dU_{elastic} = V\Sigma_{ij}s_{ij}d\varepsilon_{ij} \tag{4.75}$$

(i, j are the positional coordinates).
For a hydrostatic compression, during which the volume changes but not the form of the stressed sample, s_{ij} is simply given by $-p\delta_{ij}$ (i.e. $-p$ for i=j and zero otherwise). Hence $\Sigma_{ij}s_{ij}d\varepsilon_{ij} = -p\Sigma_{ij}\delta_{ij}d\varepsilon_{ij} = -p\Sigma_i d\varepsilon_{ii}$. The diagonal values ε_{11}, ε_{22}, ε_{33} describe the volume change upon deformation. The relation between a differentially small volume increment before and after deformation is given by [91]

$$\frac{\delta V' - \delta V}{\delta V} = \Sigma_i\varepsilon_{ii} \simeq \frac{V' - V}{V}. \tag{4.76}$$

(V': volume in the deformed state). For constant ε_{ii}, also the integrated form on the r.h.s. applies. As a consequence of Eq. (4.76) $\Sigma_{ij}s_{ij}d\varepsilon_{ij} = -pdV/V$ and Eq. (4.4) is obtained.
Another important special case of a homogeneous deformation (i.e. ε_{ij} is positionally constant) in which now, however, pressure anisotropies are effective, is the uniaxially stressed cubic crystal. Let us assume this time that there is tensile stress in the x–direction. There it holds for small effects (i.e. Hooke's law fulfilled) that $s_{xx} = p$, $d\varepsilon_{xx}/ds_{xx} = \text{const} = \varepsilon_{xx}/s_{xx}$ and also $d\varepsilon_{yy}/ds_{xx} = d\varepsilon_{zz}/ds_{xx} = \varepsilon_{yy}/s_{xx} = \varepsilon_{zz}/s_{xx} =$ const.
The first constant is the inverse of the elastic modulus while the second is the negative ratio of Poisson's number and the elastic modulus.
The fact that homogeneous samples shrink in the y– and z–direction if elongated in x–direction leads to a Poisson's number being greater than 0. (It must be also not greater than 1/2 owing to thermodynamic criteria [91].)

In equilibrium the (total) chemical potential of mobile components (k) is positionally constant. Accordingly, it exhibits an excess term (compared to the value of the undeformed solids) which amounts to[55]

$$\mu^{ex}_{elastic,k} = -\frac{s_{xx}}{3} v_k \tag{4.77}$$

for the case of an uniaxially homogeneously and elastically deformed solid; v_k is the respective partial molar volume. In the case of triaxial, homogeneous and elastic deformations s_{xx} has to be replaced by $\Sigma_i s_{ii}$. (Note the formal similarity to Eq. (5.73) for the case of surface free energy contributions.)
Such effects lead to modified solubilities of components (or for charged particles to modified surface charges) in stressed solids. One example that has been considered in some detail in the literature is hydrogen solubility in metals or alloys: In iron, in which hydrogen is mobile in the interstitial lattice, the chemical potential of hydrogen is adequately decreased by applying tensile stress and increased by applying compressive stress [6,107].
In simple cases it is possible to consider a solid as being composed of more or less hard spheres with fixed (nominal) charges connected by springs (see Fig. 2.3). Then many chemical effects can be advantageously conceived as elastic effects, e.g. the generation of "internal stress" by substitution of an ion by a larger ion of the same charge.
If we leave the regime of elastic deformations we introduce irreversible changes, e.g. in the form of dislocations. This is considered in more detail in Section 5.4.3 when we consider higher–dimensional defects.

4.3.8 The thermodynamic functions of state of the perfect solid

To conclude this chapter, it is necessary to consider — to a rough approximation — the thermodynamic state functions with regard to the results of previous chapters. For this purpose we turn to the (unstressed) bulk again, for which electrical (and elastic) field effects play no role. As described in detail above, knowledge of a single thermodynamic function of state in terms of its characteristic variables, that is e.g. knowledge of the free enthalpy as a function of pressure, temperature and composition, is sufficient for the discussion of the thermodynamic situation. Since we wish to describe changes in composition later by means of the defect concentration in a perfect solid of given Dalton composition and also can, to a first approximation, neglect pressure dependences, it is primarily the G function as a function of temperature that interests us in any discussion of the perfect solid. For the sake

[55]Eq. (4.77) follows from the definition of the chemical potential (Eq. (4.14c)), the neglect of entropic effects and from Eq. (4.75). Note that in the example considered, the change in length of the x–coordinate with mole number of k is proportional to a third of the partial molar volume $(= \partial V/\partial n_k)$.

of generality we will only carry out this discussion in a semiquantitative manner, using coarse approximations. G(T) is made up of its value at absolute zero and the thermal contributions, which are obtained from the specific heats. The two most important contributions to the energy of the solid have already been mentioned:
1) The bonding energy at a given temperature T: This itself is made up of the bonding energy at absolute zero and temperature-dependent contributions which are mainly caused by the thermal expansion. The latter contributions are small and negligible in the case of a harmonic solid.
2) The vibrational energy: The zero–point contribution is negligible here and the temperature-dependent term is the main parameter determining the T dependence of the energy.
In the limiting case of high temperatures we can approximately formulate

$$U_{perfect} \cong U_0 + U_{Lat}(T{=}0) + 3Nk_BT. \tag{4.78}$$

for a simple solid (see Chapter 3). The constant U_0 is (as H_0 used below in Eq. (4.82)) determined by the normalization. For the two limiting cases important in this context, namely ionic and covalent solids (e.g. NaCl and Si), the bonding energy (determined by the lattice energy at T=0) is given by (see Chapter 2):

$$U_{Lat}(0) \cong \left\{ \begin{array}{ll} N\left(\frac{2e^2}{4\pi\varepsilon_0}\right) z^2 \frac{f}{a} & (\{NaCl\}_{\frac{3}{\infty}}) \\ \frac{N\text{æ}\epsilon_{x-x}}{2} & ([Si]_{\frac{3}{\infty}}) \end{array} \right\} = Ns\frac{\text{æ}}{2} \times \left\{ \begin{array}{lll} \frac{2e^2}{4\pi\varepsilon_0} z^2 \frac{f}{a\text{æ}} & (\{NaCl\}_{\frac{3}{\infty}} & s=2, \text{æ}=6) \\ \epsilon_{x-x} & ([Si]_{\frac{3}{\infty}} & s=1, \text{æ}=4) \end{array} \right. \tag{4.79}$$

As far as the entropy is concerned, the zero–point contributions are nil in the simplest cases. Nonzero values are obtained if i) ground states are exactly degenerate, ii) the energy levels are so closely neighboured that their resolution "in practice" is only relevant at temperatures below the lowest temperatures of measurement, or iii) energetically demanded ordering processes do not take place for kinetic reasons. We neglect such true or apparent zero–point entropies in our semiquantitative approach. If we restrict ourselves to phases with small homogeneity ranges, the most important contribution to the entropy stems from the vibrations. Since at high temperatures the specific heat is $3Nk_B$, the vibrational entropy (see Eq. (4.40)) is logarithmically dependent on T in this regime:

$$S_{vib} \cong \text{const} + 3Nk_B \ln T. \tag{4.80}$$

A more accurate calculation (see e.g. Ref. [11]) yields a (T–) constant of $3Nk_B - 3Nk_B \ln \Theta_E$ using the Einstein model.

In combination with the internal energy, the following simple relationship is then produced for G_{vib}

$$G_{vib} \cong +3Nk_BT \ln(\Theta_E/T). \tag{4.81}$$

If we refer to the term to the right in braces in Eq. (4.79) as the effective pair energy ($\langle\epsilon\rangle$), we obtain for the free enthalpy

$$G = H_0 + \frac{Ns\text{æ}\langle\epsilon\rangle}{2} + 3Nk_BT\ln(\Theta_E/T). \tag{4.82}$$

In the general case of a Debye solid the temperature dependence of G is determined via the parameter Θ_D (see Chapter 3). Hence, the free enthalpy of a perfect solid is determined for all temperatures by the parameters effective pair bonding energy[56], co–ordination number and Debye–temperature. While $\langle\epsilon\rangle$ is a meaningful parameter in the case of covalent crystals (for elemental crystals ϵ_{X-X}, cf. Section 2.2.4), in the case of ionic crystals charge numbers, lattice constant and Madelung number are the appropriate parameters. Hence

$$G(T) = \begin{cases} G_{z,a,f;\Theta_D} & \text{for ionic crystals} \\ G_{\epsilon_{X-X},\,\text{æ};\,\Theta_D} & \text{for covalent crystals.} \end{cases} \tag{4.83}$$

For simplicity we have, when writing ϵ_{X-X}, considered an elemental covalent crystal. One should note that also in the case of ionic crystals the quantum–mechanical repulsion forces are relevant even in this rough approximation; they enter implicitly in Θ_D, f and a (cf. Sections 2.1.6, 2.2.7, Chapter 3).
Since we referred to compounds with small phase widths (otherwise we have to consider also the composition dependence, see Section 4.3.5) Eq. 4.83 yields (after suitable normalization in view of the constant in Eq. (4.82)) the materials parameters $G^\circ_m = \mu^\circ$ (see Tables 4.1, 4.2, pages 78, 79), which, in the end, as discussed in Chapter 2, depend on the atomic parameters. Even though it is possible to calculate the Gibbs energy in simple examples from such atomic parameters, the above given parametrization is very helpful, in particular, as far as an intuitive understanding is concerned.
The total free enthalpy of a real solid now has also to take account of the contributions from the defects:

$$G(T) \simeq (G_{bond} + G_{vib})_{perfect} + \Delta G_{defect} = G_{perfect} + \Delta G_{defect}. \tag{4.84}$$

In the following chapter we are going to try to estimate the energetic and entropic contributions of point defects at equilibrium and, finally, to derive an expression for the free energy of a simple, real solid and, thus, for the chemical potential of the point defects in this system. We will see that lattice energy and dielectric constant are the fundamental parameters with respect to the defect formation energies emphasizing the trivial point that perfect structure and real structure are not independent of each other. It is not the magnitude (which is usually negligible) but the concentration dependence of the last term in Eq. (4.84) that is of particular interest, since it is fundamental for the chemical and functional variability. We will see under what conditions we can formulate mass action laws for internal and external equilibria

[56] This is determined from the resonance integral in the case of an ideal covalent crystal.

such as chemists usually do for the aqueous phase and physicists for electronic states in silicon, and what lies behind chemical parameters, such as the mass action constants for defect formation.

5 Equilibrium thermodynamics of the real solid

5.1 Preliminary remarks

Having finished our detailed introduction, we can now turn to the central topics of this book, the first of which is the thermodynamics of the real solid at equilibrium. That means that we have to study the defects which we can classify according to their dimensionality.

We distinguish between point defects[1] (zero dimensional defects) — these are atomic and electronic imperfections; line (one–dimensional) defects — these are essentially dislocations; plane (two–dimensional) defects — i.e. surfaces and basically internal interfaces; and pores or inclusions as three–dimensional defects. We will not discuss other variants of higher–dimensional disorder, which can be very complex, particularly in multiphase systems. Since we concentrate on the equilibrium state in this chapter, we are primarily interested in point defects and surfaces. Point defects exist at equilibrium on account of entropy; surfaces are a necessary consequence of the requirement that the amount of substance is finite. Defects of other types are necessarily nonequilibrium phenomena[2], which will be demonstrated in Section 5.4. Nonetheless, the higher–dimensional defects will, as metastable structure elements, be important for us later (see Sections 5.4, 5.8).

The total free enthalpy of a real solid is given by the G value of the perfect solid plus the changes that occur on the introduction of defects. This parameter $\Delta_d G$ is the free enthalpy of reaction of the process

$$\text{Reaction d} = \qquad \text{perfect solid} \rightleftharpoons \text{real solid} \tag{5.1}$$

and, hence,

$$G_{real} = G_{perfect} + \Delta_d G. \tag{5.2}$$

As for $G_{perfect}$ the quantity $\Delta_d G$ is composed of static bonding components ($\Delta_d G_{bdg}$) and vibrational components ($\Delta_d G_{vib}$). In addition, there is a configurational component ($\Delta_d G_{cfg}$), which does not occur in the perfect solid and which makes "internal chemistry" possible. It is the purpose of the next section to explain this. First we state

$$\Delta_d G = \Delta_d G_{bdg} + \Delta_d G_{vib} + \Delta_d G_{cfg}. \tag{5.3}$$

[1]For the purpose of a unified treatment we will also subsume excess electrons and holes into the concept of point defects, even though the wave functions concerned can be very extended.

[2]Those internal interfaces, such as domain walls in ferroelectrics or ferromagnetics, whose existence can be traced back to energetic origins, are best not referred to as defects but as super–structural elements. In the context of higher–dimensional defects limiting cases are naturally conceivable as regards higher–dimensional defects with an extremely low energy of formation, which may be compensated by the low configurational entropy. However, we will not concern ourselves with such exceptional cases here.

Physical Chemistry of Ionic Materials J. Maier
ISBN: 0-471-99991-1 (HB); 0-470-87076-1 (PB)

It is to be established that Eq. (5.1) has not been very precisely formulated, since it is possible to add species from outside to the perfect solid (or vice versa), so that, in addition to the "re–arrangement", it is also possible to encounter "substitution", "addition" and "elimination processes". In all cases it is possible to describe, with thermodynamic precision, the real solid as a superposition of defect building units and the constituents of a perfect solid (the monomeric units are sometimes termed "lattice molecules"), as we will now see in some detail.

5.2 Equilibrium thermodynamics of point defect formation

Let us take a simple example, namely the introduction of a vacancy into an elemental crystal, for which it is already possible to demonstrate the relevant points. Let us first consider a typical representative of an ideal covalent crystal, namely diamond, the ground structure of which was shown in Fig. 2.22.
The creation of a carbon vacancy means the transfer of an internal C atom to the surface[3]. This leads to a crystal with slightly reduced density. In order to remove an internal carbon atom from its crystal assembly it is necessary to break 4 bonds and, thus, to provide energy equal to $4\epsilon_{C-C}$. Since this C atom is not moved to infinity but to the surface of the crystal, the energy $2\epsilon_{C-C}$ is recovered[3]. The net loss of $2\epsilon_{C-C}$ is exactly the lattice energy ($4\epsilon_{C-C}/2$, see Eq. (2.45))[4]. If the energy loss is estimated in this manner we get

$$\Delta_d\epsilon \simeq 2|\epsilon_{C-C}| \simeq |\epsilon_{git}| \simeq 7\text{eV}, \tag{5.4}$$

[3]We assume that the individual surface crystallography is not important which implies sufficiently large crystals. A representative formation process is the transfer from the interior (6 contacts) to the half–crystal position or kink position (3 contacts cf. Fig. 5.24 on page 145). Figure 5.24 also shows that the formation of a vacancy within the surface is less expensive: The transfer of a cube from a regular surface site (5 contacts) to the half–crystal position involves the loss of 2 instead of 3 contacts. Thus one expects, in a rough approximation and by neglecting further structural differences, a reduction of the formation energy (without relaxation) by a factor 2/3.

[4]This can be generalized in the first approximation (point defects in a rigid lattice) for other materials. The treatment also applies to ionic solids: It is essentially the lattice energy which is required for the transfer of a cation or anion to the gas phase. (Effectively the "cation-anion" bonds are broken, cf. Eq. (4.82). We ignore the changes due to the violation of electroneutrality; they are nullified in the incorporation that follows.) One half is recovered by the transfer from the gas phase to the interface. No energy is required in this extreme approximation on transfer from the gas phase to an interstitial site: If we formulate the bonding of an interstitial particle to the environment, bonds between other partners are correspondingly weakened. Hence, the lattice energy is, on this level of approximation, the reaction energy for all elementary ionic disorder types discussed below (Section 5.5.1). However, the numbers obtained for specific examples may not fit even to an order of magnitude as described in the text (see also Eq. (5.8)). The relaxation (or polarization) processes are not only of considerable but of quite specific importance.

a value that is almost double as high as the experimental result of 4eV (see Ref. [108]). The difference of 3eV is readily explained: When the lattice assembly loses an atom, it naturally attempts, for energetic reasons, to limit the damage as much as possible by changing the local structure[5]. Considered in terms of bonding theory, the local ensemble will so arrange and relax ($\Delta_d \epsilon_{rlx}$), that the valences that have become free, are involved in partial bond formation. This is shown schematically in Fig. 5.1. The bonding contribution to the free enthalpy is thus:

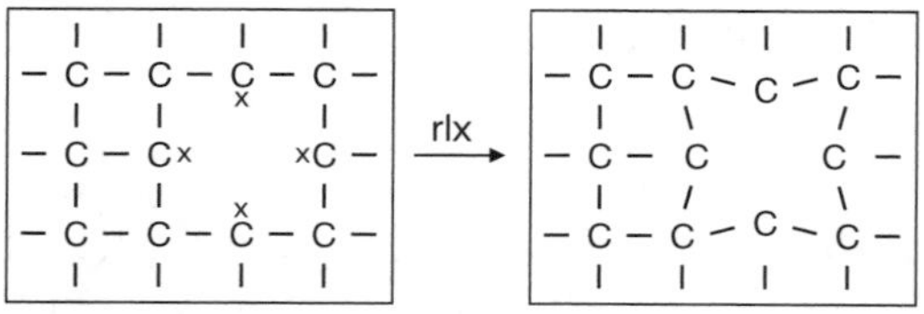

Fig. 5.1: The interaction of radicals, resulting from removal of a C atom, leads, on energetic grounds, to the relaxation of the neighbouring atoms, as indicated schematically in the right hand diagram. The relevant electrons are not shown on the right in order to emphasize the new bonding situation.

$$\Delta_d g_{bdg} \simeq \Delta_d \epsilon = |\epsilon_{git}| - |\Delta_d \epsilon_{rlx}|. \tag{5.5}$$

The alteration of the bonding component of G, on introducing N_d defects, is obtained by multiplication of the above result by N_d, as long as the defects do not interact with each other. Here, we wish to assume such dilute states (but see Section 5.7). The formation of defects in ionic crystals can be viewed in a completely analogous manner[4]. Here, too, on removal of a Na^+ ion and of a Cl^- ion, twice the lattice energy is required initially, half of which is recovered by incorporation at the surface. Here, too, ϵ_{git} (6...8eV for alkali halides) is far too high an estimate of the bonding contribution — the experimental value is about 2eV —, since the lattice also relaxes. If the ion removed is positive, the positive ions in the environment of the vacancy are attracted and the negative ones repelled, and vice versa, as shown in Fig. 5.2. In the case of an ionic crystal the relaxation as shown in Fig. 5.2. In the case of an

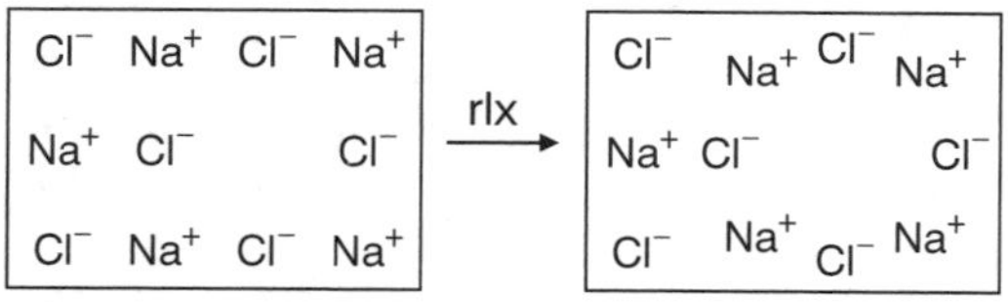

Fig. 5.2: The absent positive charge leads to an increased repulsion of the Cl^- neighbours and to an attraction of the Na^+ neighbours.

ionic crystal the relaxation thus corresponds to a local ionic polarization. Classical dielectricity theory yields, for the polarization energy $\Delta_d \epsilon_{pol}$ of charged cavities (z=1) with mean radius d in a homogeneous medium of dielectric number[6] ε_r, the

[5] The perfect crystal structure is only valid for the perfect, completely homogeneous, bulk.

[6] ε_r= dielectric number; $\varepsilon = \varepsilon_r \varepsilon_0$ = dielectric permittivity. The dielectric number is a measure of the displaceability of electronic and ionic charges (electronic and ionic polarizability) and can, in principle, be traced back to the atomic parameters discussed in Chapter 2. (The mean orientation and, hence, the temperature is of explicit importance in the case that permanent dipoles are present.)

relationship[7] [110]

$$\Delta_d \epsilon_{pol} \simeq -\frac{e^2}{4\pi\varepsilon_0 d}\left(1-\frac{1}{\varepsilon_r}\right). \tag{5.6}$$

It is obvious that it is the static dielectric constant that must be inserted (for dilute defects); a typical value for alkali halides is 5.

If we tentatively identify d with b in Eq. (2.37), it follows for the formation energy

$$\Delta_d \epsilon = \epsilon_{git}\left(1+\frac{\Delta_d \epsilon_{pol}}{\epsilon_{git}}\right) = \epsilon_{git}\left(1-\frac{1-\frac{1}{\varepsilon_r}}{f_M\left(1-\frac{1}{n}\right)}\right) \sim \epsilon_{git}\left(1-\frac{1}{f_M}\right) \tag{5.7}$$

which predicts values of the right order of magnitude: If $\varepsilon_r \sim 5$ and $n \sim 10$ and $f_M \sim 1.7$, there is a reduction of more than a half in ϵ_{git}, which is in satisfactory agreement with the data. Since $\Delta_d \epsilon_{pol}$ in Eq. (5.6) represents the sum of the cationic and anionic contributions, we should, however, expect that the effective radius d is given by one half of the harmonic mean of the ionic diameters, while b as the sum of the ionic radii is the arithmetic mean of the ionic diameters (Na^+ : 1.9Å, Cl^- : 3.6Å) and, thus, approximately twice as large (2.76Å compared to 1.24Å). The vagueness of the effective parameter d, which appears in Eq. (5.6), reflects the oversimplification of the above approach. Mott and Littleton have given a self–consistent procedure for calculation of the polarization term, which forms an important basis for defect energy computations. Here the interested reader is referred to the literature [111]. In a qualitative sense, Eq. (5.7) correctly reveals the role played by the ionic radius in the bonding energy. Hence, ca. 2/3 of the Schottky disorder energy of NaCl stems from the formation of Cl^- vacancies and only 1/3 from the formation of cation vacancies (concerning vacancy formation, cf. also Fig. 5.5, below).

A further fundamental point defect type is the interstitial particle. In this case sites are occupied that are energetically unfavorable in comparison to the "chemical ground structure". Figure 5.3 illustrates some examples of such "excited structures" (compare with the "ground structures" on page 51 and page 62). In the sodium chloride structure (see Fig. 5.3a), in which the anions (or formally also the cations) can be thought to have formed a cubic close–packed lattice, the regular counterions occupy octahedral interstices, while the tetrahedral interstices are favourable interstitial positions[8] (e.g. for Ag^+ in AgCl). In the anionically disordered CaF_2 structure

[7] A relationship analogous to Eq. (5.6) is used in liquid systems to describe solvation effects [109]. The derivation is as follows:
The electrical potential at the surface of a charged cavity (charge q, radius r) is given by $\phi = q/4\pi\varepsilon r$. The reversible work of charging such a cavity in a medium of dielectric permittivity $\varepsilon = \varepsilon_0\varepsilon_r$ is $\int_0^e \phi dq = e^2/8\pi\varepsilon r$, the reversible work for charging it in vacuum then $e^2/8\pi\varepsilon_0 r$. Summing up the contributions for cations and anions and, for simplicity, introducing a mean radius d, leads to Eq. (5.6). This polarization contribution corresponds to a free energy. The entropy contribution is negligible if we ignore the temperature dependence of ε.

[8] The local environment of the interstitial particle is altered in a similar manner, as an excess proton distorts the local water structure ($H_3O^+, H_9O_4^+$). So it is structurally more precise to

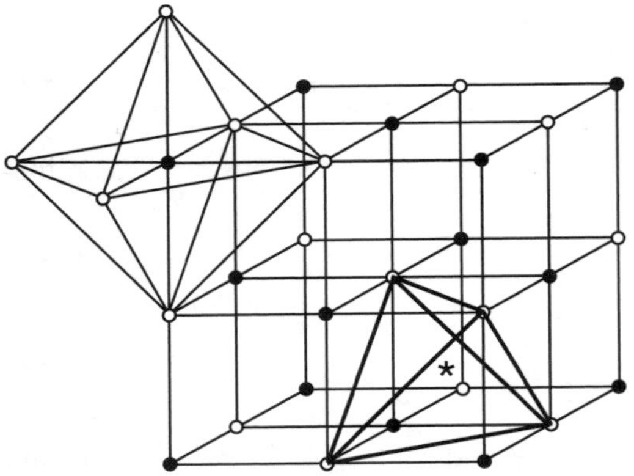

Fig. 5.3a: In the rock salt structure (e.g. white spheres: Cl, black spheres: Ag) all the octahedral interstices of the close–packed anion sublattice are occupied by cations (filled–in circles). Accordingly, the tetrahedral spaces are interstitial sites (see asterisk). The Niggli–formula (cf. Section 2.2.7) for the cluster that contains the interstitial ion is $(Ag_i(ClAg_{6/6})_{4/1})^+$. For the vacancy (remove a black sphere) we have to formulate $(V_{Ag}(ClAg_{5/6})_{6/1})^-$.

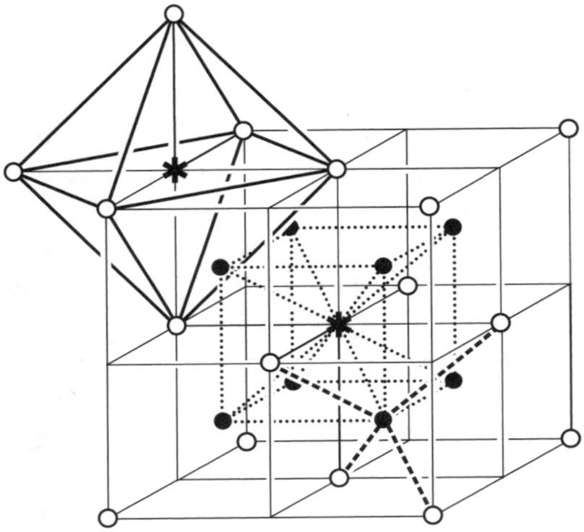

Fig. 5.3b: In the CaF_2 structure (white spheres: Ca, black spheres: F) the fluoride ions occupy all the tetrahedral interstices of a formally close–packed Ca^{2+} sublattice. The octahedral spaces are interstitial sites (see asterisk). As indicated, these are also the centres of cubes formed by regular F^- ions (filled–in circles). (Niggli–formulae would be $(F_i(CaF_{8/4})_{6/1})^-$ for the interstitial and $(V_F(CaF_{7/4})_{4/1})^+$ for the vacancy cluster.)

(anti–Frenkel disorder), which crystallizes in the fluorite structure illustrated in Fig. 5.3b, the situation is reversed: There all the tetrahedral interstices of the cubic, formally close–packed cation lattice are filled by regular F^- and the octahedral interstices can be additionally occupied by interstitial F^- ions. In the Ag^+ conductors β– or γ–AgI (wurtzite or zinc–blende) only half the tetrahedral interstices of the close–packed I–lattice are occupied by regular Ag^+ ions, so that both the octahedral and the remaining tetrahedral interstices of the close–packed iodine arrangement are available as interstitial positions. A more subtle interstitial position, identified

formulate an interstitial Ag^+, as Ag_2Cl^+ or better $Ag_5Cl_4^+$. The vacancy containing cluster would be best formulated as $Ag_5Cl_6^-$ (cf. legends of Figs. 5.3a, 5.3b).

Regular dumb–bell structures have been detected in face–centred cubic elementary crystals [113]: After the relaxation of the structure, which involves the interstitial atom at the centre of the unit cell (octahedral interstice) and a face–centring atom (corner of the octahedron), such a symmetrical situation is created. However, none of these structural details influences the phenomenological description of dilute defects.

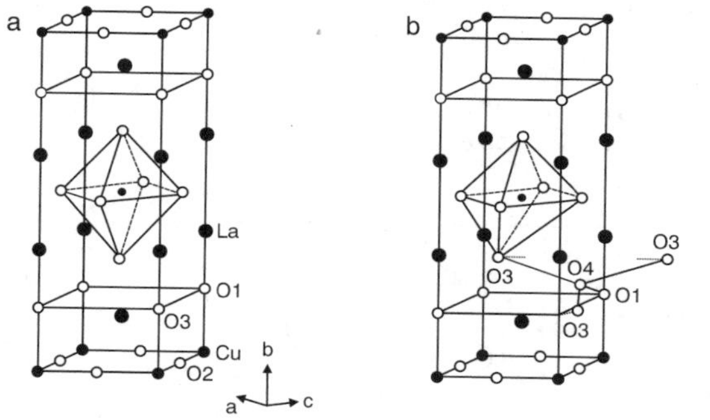

Fig. 5.3c: The position of the interstitial oxygen in La_2CuO_4 (O4) as determined by neutron diffraction. Compare the perfect structure (a) with the real structure (b). In contrast to Figs. 5.3a, 5.3b, the lattice relaxation is also indicated. From Ref. [112].

by neutron diffraction in the more complicated compound La_2CuO_4, is illustrated in Fig. 5.3c.

In a Frenkel reaction in a silver halide, a silver ion is "excited" from its regular position into an interstitial site. The exact energetic calculation is similar[4] to that described above for the Schottky reaction. It is understandable that here polarization effects are even more important for the lattice energy and, in particular, the effective ionic radius and the electronic polarizability play substantial roles. Thus, these types of disorder are very significant in cases in which either the ion under consideration is very deformable, as in the case of Ag^+; or the ion is small (in particular in the case of a polarizable counterion), as the F^-–ion in the anion–Frenkel reaction in alkaline earth fluorides (below we refer to this disorder as anti–Frenkel disorder). In AgCl the energy of formation of a (separated) Frenkel pair is of the order of 1eV (compare with the lattice energy of 10eV!). Empirically one finds, in many cases, that

$$\Delta_d \epsilon \sim \text{const}\epsilon_{\text{git}}/\varepsilon_r. \tag{5.8}$$

For Schottky–disorder in alkali halides [114], Frenkel–disorder in silver halides and anti–Frenkel–disorder in alkaline earth halides it is found that const $\sim 1.5 \ldots 2$ [115]. Anti–Schottky disorder, i.e. the interstitial incorporation of both cation and anion, is extremely rare, primarily on account of the high energy of large anions in the interstices and is only possible in loose structures. It has, for instance, been postulated for orthorhombic PbO [116,117]. A detailed description of these disorder types is given in Section 5.5. Table 5.1 lists formation data for a series of halides.

The discussion naturally becomes specific on the introduction of foreign particles. Figure 5.4 based on a computer simulation, illustrates the relaxation that occurs if an aliovalent impurity has been introduced into MgO. Here it is primarily the obvious and experimentally well–established finding, that the effective particle diameter is a significant criterion for possible substitution and addition processes. Corresponding to a tolerable expenditure of enthalpy, substitutions of native particles by foreign particles generally demand small differences in radius, while, under given conditions,

Table 5.1: Thermodynamic standard parameters for point defect formation.
(S: Schottky, F: Frenkel (also termed cation–Frenkel), $\overline{\text{F}}$: anti–Frenkel (also termed anion–Frenkel)) [14]. The last two columns refer to the migration (cf. Chapter 6) (1eV is equivalent to 96.5kJ/mol).

crystal	disorder type	formation enthalpy in MJ/mol	formation entropy in units of R	migration enthalpy in kJ/mol	migration entropy in units of R
LiF	S	0.23	9.6	70 (V'_{Li})	1 (V'_{Li})
LiCl	S	0.21		40 (V'_{Li})	
LiBr	S	0.18		40 (V'_{Li})	
LiI	S	0.11		40 (V'_{Li})	
NaCl	S	0.24	9.8	70 (V'_{Na})	1-3 (V'_{Na})
KCl	S	0.25	9.0	70 (V'_{K})	2.4 (V'_{K})
RbCl	S	0.21		50-100 (V'_{Rb})	
CsCl	S	0.18	10	60 (V'_{Cs})	19 (V'_{Cs}) (?)
AgCl	F	0.14	9.4	28 (V'_{Ag}) 1-10 ($Ag_i^{\bullet}$)	-1 (V'_{Ag}) -3 ($Ag_i^{\bullet}$)
AgBr	F	0.11	6.6	~30 (V'_{Ag}) 5-20 ($Ag_i^{\bullet}$)	
CaF_2	$\overline{\text{F}}$	0.27	5.5	40-70 ($V_F^{\bullet}$) 80-100 (F'_i)	1-2 ($V_F^{\bullet}$) 5 (F'_i)
SrF_2	$\overline{\text{F}}$	0.17		50-100 ($V_F^{\bullet}$) 80-100 (F'_i)	
BaF_2	$\overline{\text{F}}$	0.19		40-70 ($V_F^{\bullet}$) 60-80 (F'_i)	

small foreign particles find a place in the interstices more easily. Too high an effective charge is also less favourable energetically.
Let us consider the $SrTiO_3$ perovskite[9]: For both geometric (Zr^{4+} has a similar ionic radius as Ti^{4+}: 0.6Å; interstitial particles are improbable in the tightly packed $SrTiO_3$) and charge reasons (cf. $Zr_{Ti}^{\times}$ vs. $Zr_i^{\bullet\bullet\bullet\bullet}$, cf. charge stability of tetravalent Zr) Zr is incorporated at Ti positions (formally) as Zr^{4+}. Depending on the conditions, e.g. temperature and partial pressure of oxygen, Fe is incorporated[10] essentially into Ti^{4+} positions as Fe^{3+} (0.6Å) or Fe^{4+} (0.5Å) (see Section 5.6), while the larger La (La^{3+}: 1.1Å) trivalently ocupies Sr^{2+} positions[9] (1.1Å). Figure 5.5 refers to the

[9]These are naturally not all–or–nothing phenomena. More unfavourable states — but with accordingly lower statistical weight — are also realized. The precise valence and site distribution depends on the state parameters (see Section 5.6).

[10]In $SrTiO_3$ the effective charge is minimal for Fe^{4+} (this is in sharp contrast to aqueous solutions).

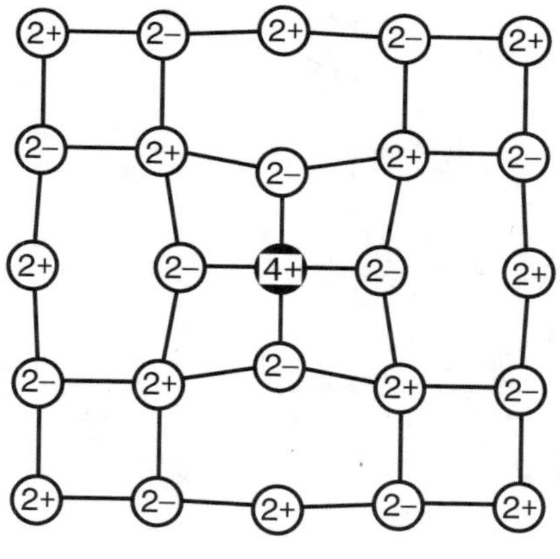

Fig. 5.4: Polarization effect as a result of the substitution of a cation in MgO by a more highly charged one of the same size. According to Ref. [118].

substitution of O^{2-} by OH^- which are of almost equal size. The defect formed gives rise to proton conduction (see Section 5.6). The l.h.s. shows the reaction of the oxygen and cerium neighbours on the removal of the O^{2-}, the r.h.s. shows the situation after introduction of OH^-.

Particularly at high temperatures, the site exchange between various types of ions becomes important. While this is naturally of very little relevance in the case of pure binary ionic crystals, such pronounced site distribution and anti–site disorder processes in cation or anion lattices require attention in doped or in multinary compounds. $YBa_2Cu_3O_{6+x}$ provides an example in which there is a considerable cation disorder with respect to Y and Ba ions at the high preparation temperatures [121]. Naturally one must not lose sight of the local effects of chemical bonding in each case. This is of particular importance for transition metal cations (see Section 2.2), where characteristically preferred co–ordinations occur (depending on the charge).

The incorporation of a defect also alters the vibrations locally. The influence of these energetic changes on the energy of formation of defects is relatively small and mostly negligible (cf. Dulong–Petit limit for a conserved number of vibrators, Chapter 3). However, the effect on entropy and, hence, on the temperature–dependence of the thermodynamic formation balance, as well as on the free enthalpy directly, is important. Even at small values of S the TS term can be of significance at high temperatures.

According to Eqs. (4.80, 4.81) ΔS_{vib} is proportional to the logarithm of the vibration frequency. If this changes from ν to ν_d during defect formation, then the resulting entropy of reaction is:

$$\Delta_d S_{vib} \cong -\Delta_d G_{vib}/T = -N_d æ k_B \ln\left(\frac{\nu_d}{\nu}\right). \tag{5.9}$$

In Eq. (5.9) it is assumed that the defect merely influences the vibrational frequency of the æ nearest neighbours. If the local vibrational frequency is halved compared with the perfect state, then this corresponds to a $\Delta_d S_{vib}$ of the order of $(10^{-4} \ldots 10^{-3})$eV/K, corresponding to a G value, that may be negligible at low

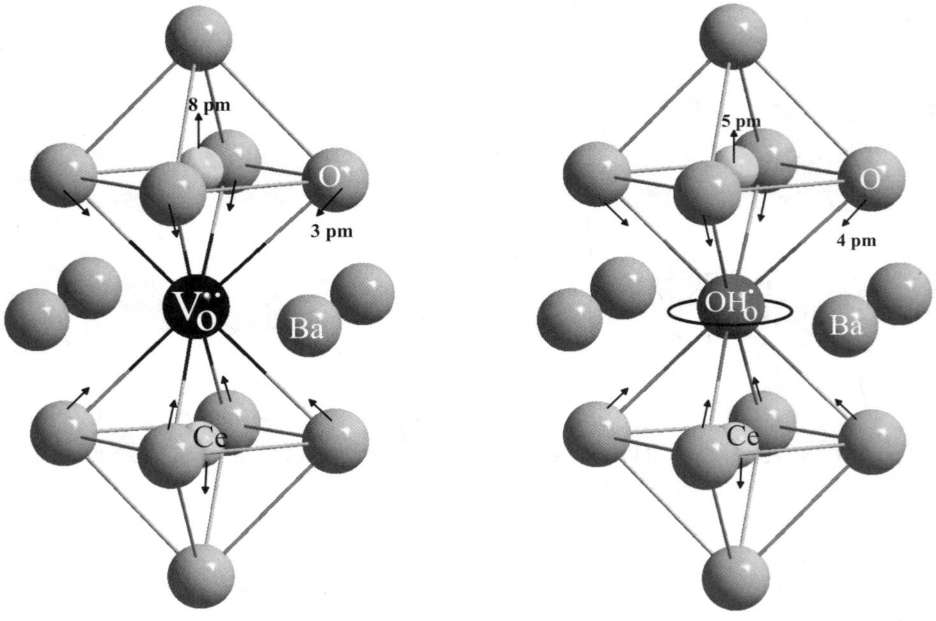

Fig. 5.5: Quantum mechanical simulations show that the formation of an oxygen vacancy in $BaCeO_3$ results in a displacement of the oxygen ions towards it and of the cerium ions away from it (l.h.s.) [119]. On ocupation of the vacancy by OH^- [120] (see Section 5.6) the Ce relaxation is substantially reduced (r.h.s.). The increased displacement of the oxygens towards the $OH_O^{\cdot}$ defect is caused by the directive hydrogen bridges (here partitioned between 8 oxygen neighbours). The time resolution under consideration is coarse compared to the fast proton rotation, which is sketched schematically. The "jump" of the proton occurs on an even coarser time scale (see Fig. 6.13, page 291). Detailed discussion is given in Section 6.2.1. According to Ref. [120].

temperatures, in comparison with the energy of formation, but is of a considerable magnitude at T=1000K.

The change in the frequency of vibration and, hence, in the vibrational entropy is not always easy to estimate. One rough rule of thumb is: When a vacancy is produced, the amplitude of the vibration increases (see Chapter 3) and the frequency is reduced, while the reverse[11] is generally the case when an interstitial defect is created. So, in the first case, $\Delta_d G_{vib}$ is less than zero and greater than zero in the second. Therefore, in the second case, defect formation is additionally opposed by the reduction of the vibrational entropy, while, in the first case, it is made easier by the increase of the vibrational entropy. Both (bonding and vibrational) contributions

[11] The mean values of kinetic and potential energy are equal for a harmonic vibration and proportional to the squares of mass (m) and maximum amplitude (x_0). This follows readily from the averaging of $m\dot{x}^2/2$ and integrating the force over the space coordinate x ($x = x_0 \sin \omega t$).

have in common that they refer locally to the individual defect. Hence, the following is obtained (we assume small concentrations)

$$\Delta_d G^* = N_d \Delta_d g^* = N_d(\Delta_d g_{bdg} + \Delta_d g_{vib}). \tag{5.10}$$

If this were already the entire G balance, Eq. (5.10) would predict that there should not be any point defects at equilibrium. For positive $\Delta_d g^*$ (as is the case in the stability range of the phase under consideration) the formation of even a single defect would be forbidden; if $\Delta_d g^*$ were negative, the whole phase would be unstable.

The third, the concentration–dependent contribution[12], is decisive for the existence of the defective (and thus also reactive) centres[13]:
The introduction of point defects causes an enormous increase in the number of possible site configurations of constant energy. A single vacancy in an elemental crystal, with N regular sites, creates N possible micro–states with regard to site distribution. The calculation of the number of micro–states possible with N_d defects can be visualized by considering a lottery game. In the case of random distribution the number of micro–states possible with six defects in a lattice of 36 particles (Fig. 5.6) is equivalent to the number of possible combinations. When filling in the lottery

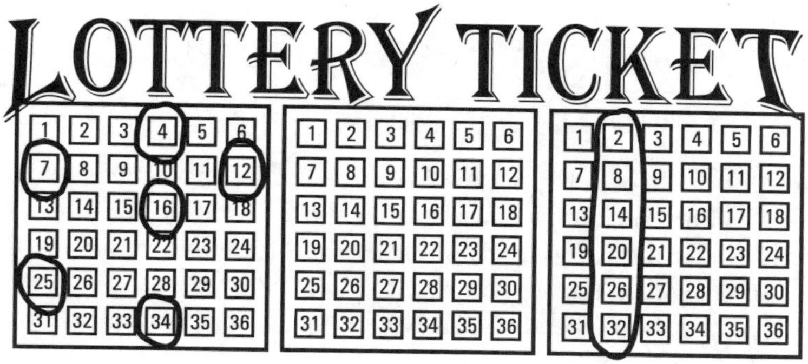

Fig. 5.6: The relevance of a lottery ticket concerning the statistics of zero–dimensional and one–dimensional defects (see text).

ticket shown in Fig. 5.6, we keep the total number N constant as we keep constant the number of lattice sites when introducing defects[14]. In order to calculate the number of combinations we leave our lottery game for a moment. Let us think of the 36 elements in a statistically predetermined order in one line one after the other

[12]At high defect concentrations interactions come into play in the form of additional further concentration–dependent effects that can lead to a decrease or increase of the concentrations calculated for the interaction–free case. In this manner, it is possible, in principle, to maintain a limited defect concentration even if there is a negative free energy of formation of an individual defect (see Section 5.7), more precisely of the first defect formed.

[13]The thermodynamic basis for the treatment of point defects was laid down by Frenkel, Schottky and Wagner [1,2].

[14]The fact that we consider the "crystal" as a distribution of the crosses (N_d) over the boxes (N lattice molecules), i.e. as a substitutional solution of defects (crosses) within the ensemble of boxes (lattice molecules), corresponds to the introduction of defects in form of building units (defect minus regular particle). In this way N remains constant (cf. Refs. [122,9]).

and agree that we will regard the first 6 elements as selected (corresponding to the vacancies). The first problem is the calculation of the possible sequences of the 36 elements. This number, more precisely the number of possible permutations of 36 elements (P_{36}), is evidently 36 times the permutation of 35 elements, since it is possible to select a differently numbered element 36 times while leaving a sequence of 35 elements to be permutated. Immediately this gives

$$P_{36} = 36 \cdot P_{35} = 36 \cdot 35 \cdot P_{34} = \ldots = 36 \cdot 35 \cdots 1 = 36! \tag{5.11}$$

and 36! is the result. Now, those states in which the vacancies or in which the regular particles are exchanged with each other are statistically identical. In other words we can exchange the first 6 elements or the remaining 30 elements amongst each other without changing the result[15]. Hence, it is necessary to divide 36! by $P_6 = 6!$ and $P_{36-6} = 30!$. In the lottery game[15] this expresses the fact that the order in which the numbers are drawn does not matter[16]. But back to our scientific problem: Obviously the number of micro–states of N_d vacancies with N regular positions and, hence, $(N - N_d)$ regular particles is given by

$$\Omega_d = \binom{N}{N_d} = \frac{N!}{N_d!(N - N_d)!} \tag{5.12}$$

with the entropy associated

$$S_{cfg} = k_B \ln \Omega = \Delta_d S_{cfg} = -\Delta_d G_{cfg}/T. \tag{5.13}$$

Since Ω=1 holds in the perfect state, this is also the reaction parameter sought, namely $\Delta_d S_{cfg}$. The pure configuration contribution to the free enthalpy refers to the same driving force that is responsible for the dispersion of a cloud of smoke in a room, the distribution of a drop of ink in water or for isotope exchange.
We use the Stirling approximation to calculate this contribution, which is applicable for large numbers[17]: $\ln N! \simeq N \ln N - N$. The result obtained is $N \ln N - N - N_d \ln N_d + N_d - (N - N_d) \ln(N - N_d) + (N - N_d)$. Since the linear terms cancel, we have

$$\ln \Omega = N \ln \frac{N}{N - N_d} - N_d \ln \frac{N_d}{N - N_d}. \tag{5.14}$$

[15]The number of possible combinations in bridge is 52!/(13!13!13!13!), since exchanges within each hand do not alter the situation.

[16]The low chances of winning in the lottery, which have caused it to be called "a special tax on the stupid" would be even worse if you also had to predict the order of drawing of the numbers (36!/30!). Another variant, to which we will return, involves returning each ball to the pool after it has been drawn so that repeats are possible. Then N of Eq. (5.12) is replaced by $(N + N_d - 1)$, that is by $(36 + 6 - 1)$. In the text we refer to this mode as the "Bose–lottery".

[17]This can be demonstrated very readily by mathematical induction. Let us assume that the relationship applies for N = M and then show that it also applies to N = M+1. This can be seen because $\ln(M+1)! = \ln(M+1) + \ln M! = \ln(M+1) + (M \ln M - M) \simeq (M+1)\ln(M+1) - (M+1)$. The error involved is reduced as M is increased. When M = 50 the Stirling formula already applies with an error of less than 2%.

At this point we will assume that the number of defects is always very small ($N_d \ll N$), i.e. we can use the Boltzmann approximation. The second term in Eq. (5.14) is predominant and we find the following contribution to the free enthalpy:

$$\Delta_d G_{cfg} = +k_B T N_d \ln \frac{N_d}{N}. \tag{5.15}$$

It is important that $\Delta_d G_{cfg}$ is always negative. It should be remembered that the ratio of the numbers, that constitute the corresponding binomial expression in Eq. (5.12), appears as the concentration in the logarithm.

We can now write the total free enthalpy of the real solid with dilute defects as

$$G = G_{perfect} + N_d \left(\Delta_d g^* + k_B T \ln \frac{N_d}{N} \right), \tag{5.16}$$

whereby $\Delta_d g^*$ is defined by Eq. (5.10).

We obtain the chemical potential of the defect by differentiation[18,19]

$$\mu_d = \frac{\partial G}{\partial n_d} = N_m \frac{\partial G}{\partial N_d} = \mu_d^* + RT \ln x_d, \tag{5.17}$$

in the well–known Boltzmann form with $\mu_d^* = N_m \Delta_d g^* = \Delta_d G_m^*$ and $x_d = N_d/N = n_d/n$ ($n \equiv N/N_m =$ mole number). If we use the more exact relation given by Eq. (5.14), differentiation leads more precisely to

$$\mu_d = \mu_d^* + RT \ln \frac{x_d}{1 - x_d}. \tag{5.18}$$

This corresponds to the chemical potential of a Fermi–Dirac–like distribution, a form that will be important again later. (It is left to the reader to demonstrate that the "Bose–lottery" described in footnote 16 leads to a Bose–Einstein–like distribution.) In equilibrium the free enthalpy of the elemental crystal must possess a minimum with respect to defect concentration. Here, this condition simply means

[18] If we form the derivation of Eq. (5.16) then Eq. (5.17) only follows approximately (there will appear an additive term of the magnitude of RT). This inaccuracy is a consequence of the approximation made in Eq. (5.15). Equation (5.17) follows more precisely and directly from Eq. (5.18) for $x_d \ll 1$.

[19] It is also interesting to differentiate the left hand side of Eq. (5.14) with respect to N at constant N_d, leading to $\partial \ln \Omega / \partial N = \ln \frac{N}{N - N_d} = \ln \frac{1}{1-x} \simeq +x$. This yields the chemical potential of the lattice molecule (monomeric unit MX, see page 109) as $\mu_{MX} = (G_{perfect}/n_{MX}) - RTx$. This configurational component (r.h.s. term) is usually negligible. Nevertheless, the treatment shows by comparison of $G = n_{MX}\mu_{MX} + n_d\mu_d$ with Eq. (5.16) (or better Eq. (5.14)) that $G_{perfect}$ is not exactly identical to $n_{MX}\mu_{MX}$ (and that $N_d \Delta_d g$ is not exactly identical to $n_d \mu_d$, cf. Eqs. (5.14 - 5.18)); note that μ_{MX} refers to the real solid. Of course $G_p = n_{MX}\mu_{MX}(x = 0)$. Rather the term $N_d \Delta_d g$ contributes weakly to the derivative with respect to N. This is primarily of importance for interactions (see Section 5.7).

that $\mu_d(x_d=\widehat{x}_d) = 0$ and it follows, in Boltzmann–approximation, for the equilibrium concentration of point defects (in an elemental crystal) that

$$\widehat{x}_d = \frac{\widehat{N}_d}{N} = \frac{\widehat{n}_d}{n} = \exp -\frac{\mu_d^*}{RT} \simeq \exp -\frac{\Delta_d \epsilon_{bdg} - T\Delta_d s_{vib}}{k_B T}. \tag{5.19}$$

It should be noted that the left hand side of this fundamental expression contains, as a measure of the concentration, the ratio of the actual number of defects to the number of sites which could be made defective (more precisely, see Eq. (5.18), to the number of regular particles, i.e. $N - N_d$), while the free energy of formation to be expended locally (mainly binding energy and vibrational entropy contributions) occurs in the exponent on the right hand side. The configuration entropy no longer appears on the right hand side, but has led — since the equilibrium condition is ultimately derived from a balance between configuration terms and local free enthalpy terms — exactly to the left hand side expression. Equation (5.19) corresponds to the trivial form of a mass action law (as discussed in Eq. (4.53) in Chapter 4), and corresponds here to the reaction[20]

$$\text{Nil} \rightleftharpoons \text{defect}. \tag{5.20}$$

It must be emphasized once again that the nonproportionality of $\Delta_d G_{cfg}$ to N_d is what permits compositional variations and thus a finite equilibrium concentration of defects. The individual contributions are illustrated in Fig. 5.7. Without such a term

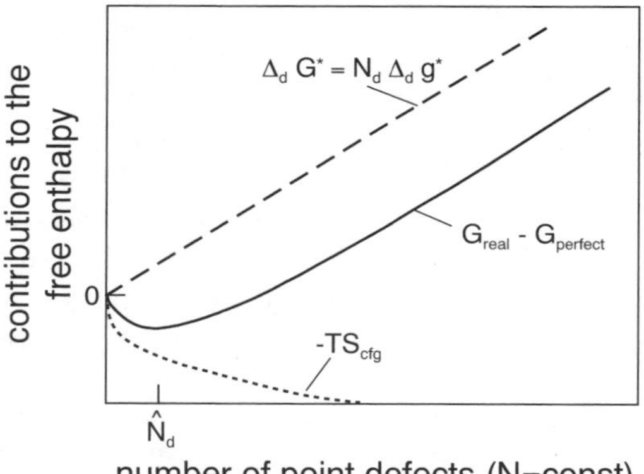

Fig. 5.7: Contributions to the free enthalpy of the solid by defect formation with a constant total number of sites.

the G dependence of N_d would be a straight line. Depending on the slope there would be no point defects at equilibrium or the whole crystal would be unstable. Now, owing to the favorable configuration contribution and the logarithmic dependence

[20] Reaction (5.20) is identical to the reaction

$$\text{perfect solid} \rightleftharpoons \text{real solid} \tag{5.1}$$

and is explicitly obtained from it by subtracting the perfect solid from both sides of the equation.

of $\Delta_d G_{cfg}(N_d)$, implying a steep decay characterized by a slope of $-\infty$ at $N_d \rightarrow 0$ followed by a flattening as N_d increases, a minimum in G always occurs, even though this usually lies at a very small defect concentration because of the usually high (free) formation enthalpy. At $\Delta_d g^* \sim 1$eV, corresponding to $\Delta_d G_m^* \sim 10^2$kJ/mol, the equilibrium concentration $\widehat{x}_d$ is calculated to be only $\sim 10^{-5}$ even at T=1000K. On account of the fact that it is only the defects that are mobile within the solid — just like H_3O^+ and OH^- particles in water — these micro–effects are of crucial macroscopic importance with respect to transport and reactivity. However, the absolute value of G (see Eq. (4.20)) is not particularly affected.

Because of the charge the situation of ionic crystals is more complex. Internal defect formation always involves the simultaneous creation of positive and negative imperfections according to

$$\text{Nil} \rightleftharpoons \text{positive defect} + \text{negative defect}, \tag{5.21}$$

for which $|z_{d_+}| = |z_{d_-}| = 1$ has been assumed for simplicity. From our general thermodynamic treatment (Section 4.2), since $dn_{d_+} = dn_{d_-} \equiv d\xi$ it follows that

$$dG = \Sigma_k \mu_k dn_k = (\mu_{d_+} + \mu_{d_-})d\xi \tag{5.22}$$

must vanish. This corresponds to the condition

$$\Sigma_k \nu_k \mu_k = 0 = \mu_{d_+} + \mu_{d_-}. \tag{5.23}$$

Strictly speaking, it is necessary to consider the electrochemical potential ($\widetilde{\mu}$) (see Section 4.3.6); however, because of the local electroneutrality, the electrical potentials cancel each other out (cf. page 101), so that $\Sigma_k \widetilde{\mu}_k dn_k = \Sigma_k \mu_k dn_k$. It is obvious that the contributions of both defects in the free enthalpy are simply additive[21], as long as they are, apart from the electroneutrality condition[22], independent of each other. Since the derivative, $\partial G/\partial n_k$ is calculated, while keeping the relevant counter–defect constant, in order to establish μ_k, it is found that, as in Eq. (5.17), there is a linear relation between μ_k and the logarithm of x_k, and hence an ideal mass action law results

$$\left(\frac{\widehat{N}_{d+}}{N_+}\right)\left(\frac{\widehat{N}_{d-}}{N_-}\right) = \left(\frac{\widehat{n}_{d+}}{n_+}\right)\left(\frac{\widehat{n}_{d-}}{n_-}\right) = K^*(T) = \exp -\frac{\mu^*_{d_+} + \mu^*_{d_-}}{RT} = \exp -\frac{\Delta_d G_m^*}{RT}. \tag{5.24}$$

with the free enthalpy of reaction $\Delta_d G_m^*$ being composed of the bonding and vibrational properties of the (separated) defect pair.

[21] This applies quite evidently to local contributions. But it also applies to S_{cfg}, since probabilities multiply in Ω_d, that is are additive in $\ln \Omega_d$.

[22] Strictly speaking it is not simply possible to introduce a charged particle into the volume, on account of the high energy that is connected with the violation of electroneutrality. Thus, we preferably speak of $(\mu_{d_+} + \mu_{d_-})$ as the chemical potential of the (separated) pair or of twice the mean chemical potential of the individual defects.

Examples are the mass action laws for Frenkel and Schottky reactions, which are discussed in detail later. Complicated defect reactions can be treated in an analogous manner[23].

The number of available sites N (cf. N in Eq. (5.19) or N_+ and N_- in Eq. (5.24)) that occurs in Ω and then in μ or in the concentration terms varies from problem to problem[23]. In the case of vacancies this is identical with the number of regular lattice sites and, in a crystal of composition MX, it is equal to the number of lattice molecules, N_{MX} (number of monomeric units). This is more complex in the case of interstitial positions. In AgCl, for example, the tetrahedral sites are the accessible interstitial sites (see Fig. 5.3a). Since there are twice as many as there are octahedral sites (these are the regular sites), we get, for this specific structure, $2N_{MX}$ as the reference parameter. In the case of the dilute state such differing measures of concentration can be readily taken account of by including the correction parameter, here a factor of 2, in the K^* value (or an excess term of $RT\ln 2$ in the μ^* value). Since in the following we assume that the $\Delta_d S_m^*$ or $\Delta_d H_m^*$ values are temperature–independent (i.e. $\Delta_d C_{pm}^* \simeq 0$), this T–independent correction of the order of $R\ln 2 \sim R$ appears formally in the $\Delta_d S_m^*$ term, which typically amounts to (5–10) R.

As an example of a more detailed consideration, let us refer to the previous example and let us face the point that the occupation of neighbouring interstitial sites is energetically very unfavourable. Such interactions are looked at more closely in Section 5.7. The more exact statistical treatment rapidly becomes complicated. Here we will briefly consider an "all or nothing" approximation, which treats defects that are distant from each other as previously, while neighbouring defects are simply excluded, i.e. their energy is set to infinity [81]. To make it simple, let us hypothetically assume that we can assign a pair of neighbouring interstitial sites to each regular position, and that the simultaneous occupancy of both sites is energetically excluded. Then there would be $2N_{MX}$ positions available just for the very first defect; the second, introduced into the perfect solid, would then only have $2N_{MX} - 2$. The two interstitial positions associated with MX are to be regarded as split positions, in the sense that both positions cannot be occupied at the same time. In this case $\binom{2N_{MX}}{N_d}$ in Eq. (5.12) is replaced by $2^{N_d} \cdot \binom{N_{MX}}{N_d}$. The prefactor multiplies the number of possibilities of selecting N_d of the N_{MX} double positions, with the number of various possibilities within the N_d pairs (that is 2 per double position)[24].

[23] Let us resume the formal transition from statistical weight to mass action law:

$$\Omega = \Pi_k \binom{A_k}{B_k} \rightarrow \mu_k = \mu_k^* + RT\ln\frac{B_k}{A_k - B_k} \simeq \mu_k^* + RT\ln\frac{B_k}{A_k} \rightarrow K^* = \Pi_k\left(\frac{B_k}{A_k}\right)^{\nu_k}.$$

[24] On our lottery ticket (Fig. 5.6) it would be necessary to double the length of our box and allow for two possibilities for making the cross, but forbid a double cross in the box. If we label the two possibilities with a and b and the boxes with 1, 2, 3, etc. then 1a, 1b, 2a, 2c, etc. represent different states. Combinations, such as 1a & 1b, are not permitted. After taking logarithms the factor 2

In reality, of course, it is not possible to distinguish a pair of interstitial positions crystallographically, that would be assigned to a regular position, rather one has to pay attention to the co–ordination number. In any event the result for small concentrations is the same: The degeneracy correction of the order of R (here $R\ln 2$) can be included in $\Delta_d S^*_m$. Hence, when not stated otherwise, we agree always to relate the defect numbers to the number of formula units in the perfect crystal (number of lattice molecules) and to denote this concentration by x_k ($\equiv N_k/N_{MX} = n_k/n_{MX}$). We will designate appropriately rescaled K^* or μ^* values, etc. with K or μ° in the following treatment. Confusion with component potentials in the pure phase ought not to occur on account of the indices. Later it will be sensible to define volume concentrations $\left(\frac{n_k}{V_{MX}} = \frac{x_k}{V_{m,MX}}, V = \text{volume}\right)$, which we will abbreviate as c_k. In order to avoid tedious indexing we will also retain the designations K, μ°, ΔS°, ΔH°, etc. in these cases. The conversion factors are then included in K, μ° and ΔS°. The meaning ought to be unequivocal from the context. Similarly, the useful designation [k] will be employed to denote site fractions or volume concentrations.

In the next section it will be shown that, in dilute systems, the same formalism also applies to electronic defects. Hence, using Eq. (4.20a), the free enthalpy of the whole crystal, which is made up of the components (k′) M and X, can be generally written in terms of the contributions of lattice molecules and (dilute ionic and electronic) defect building elements k as

$$G = \Sigma_{k'}\mu_{k'}n_{k'} \simeq n_{MX}\mu^\circ_{MX} + \Sigma_k n_k \mu^\circ_k + \Sigma_k RTn_k \ln x_k, \tag{5.25}$$

$\mu^\circ_{MX} = G_{perfect}/n_{MX}$ is hereby the chemical potential of the lattice molecule. Configuration effects with respect to the regular particles have been and can be safely ignored here in view of the low defect concentrations[18,19,25].
Effects, at which the electrical field is of importance, will be treated separately. For such cases it is necessary to explicitly introduce the electrochemical potential of the defect

$$\widetilde{\mu}_k = \mu_k + z_k e\phi = \mu^*_k + RT\ln x_k + z_k e\phi \tag{5.26}$$

(see Section 5.8).
As already mentioned the case of higher defect concentrations deserves separate treatment. Here it is necessary to introduce activity coefficients in Eq. (5.26) (see Section 5.7). In particular, it is possible for a second minimum to occur in the $G(N_d)$ curve, which gives rise to transitions to "superionic" phases. In the latter phases, as in molten phases, the distinction between defects and regular components is essentially meaningless in the sub–lattice affected (cf. Section 5.7).

---

appears in the denominator of the concentration term in Eq. (5.14). Exactly as above, 2N appears in the denominator of the concentration expression in the case of the Boltzmann approximation. In the general case, we get $2(N_{MX} - N_d)$ instead of $(2N_{MX} - N_d)$.

[25] Nonetheless, such effects can express themselves by melting point elevation (doped solids vs. pure solids). More precisely (see Footnote 19) $\mu_{MX} = \mu^\circ_{MX} - RT\Sigma_k \frac{n_k}{n_{MX}-n_k} \simeq \mu^\circ_{MX} - RT\Sigma_k x_k$.

As a preview to later sections and chapters it may be stressed that many electrical and chemical–kinetic properties are very sensitive towards point defect concentrations, so that they can, in turn, be used for measuring them. Fairly large defect concentrations can be detected very directly by means of comparably insensitive methods, such as density, mass and volume changes[26]. Scanning probe techniques have become available for investigation of surface point defects (see Fig. 5.25 on page 145).

Coming towards the end of this section, let us consider more closely the concept of the (electro–)chemical potential of a defect and the correlation with the (electro–)–chemical potential of the ionic components.

As an example, let us take Frenkel–disordered AgCl with the charge carriers $|\mathrm{Ag}|'$ (i.e. Ag^+ vacancies) and $\mathrm{Ag}^{\cdot}$ (i.e. interstitial Ag^+) as set out in Section 1.2. On account of the equilibrium conditions (cf. Eqs. (1.2a) and (5.23)) it follows that $-\mu_{|\mathrm{Ag}|'} = +\mu_{\mathrm{Ag}^{\cdot}}$ (more precisely $-\widetilde{\mu}_{|\mathrm{Ag}|'} = +\widetilde{\mu}_{\mathrm{Ag}^{\cdot}}$). Let us now formulate the incorporation of a silver ion into AgCl from the gas phase. Defect chemistry allows us to write (at the very right of Eq. (5.27) we use structure units, Section 1.2)

$$\mathrm{Ag}^+(\mathrm{g}) \rightleftharpoons \mathrm{Ag}^{\cdot} \equiv \mathrm{Ag}_\mathrm{i}^{\cdot} - \mathrm{V}_\mathrm{i}, \tag{5.27}$$

i.e. we bring in the ion as an interstitial particle. Phenomenonologically we would simply say that the silver ion is transferred from the gas phase to the solid phase:

$$\mathrm{Ag}^+(\mathrm{g}) \rightleftharpoons \mathrm{Ag}^+(\mathrm{AgCl}). \tag{5.28}$$

On account of the conditions of equilibrium we find that the (electro–)chemical potential of Ag^+, as an ionic component (not to be confused with a regular Ag^+ structure element[27] Ag_Ag), is identical to the (electro–)chemical potential of the interstitial particle (as building element) or, on account of Eq. (5.23), with the negative (electro–)chemical potential of the vacancy (as building element). This correspondence between the defects ($\mathrm{Ag}^{\cdot}$, $|\mathrm{Ag}|'$) and the components (Ag^+), which is frequently overlooked, clarifies many virtual problems encountered in the literature of defect chemistry (cf. Ref. [122]). On the basis of the analogy to electrons treated in the next chapter, $(\widetilde{\mu}_{\mathrm{Ag}^+}/\mathrm{N_m})$ may be referred to as the "ionic Fermi level". All this will become much clearer in the later parts of the chapter.

The electronic disorder problems in the case of semiconductors and electronic "insulators" at higher temperatures, at which the Boltzmann distribution can be regarded

---

[26] The anomalous expansion of a gold rod with temperature, because of the formation of vacancies [123], is a classical experiment.

[27] Please note that $\mu_{\mathrm{Ag}^+(\mathrm{s})} = -\mu_{|\mathrm{Ag}|'}$. Since $|\mathrm{Ag}|'$ can be written as the structure element combination $\mathrm{V}'_\mathrm{Ag} - \mathrm{Ag}_\mathrm{Ag}$, the component $\mathrm{Ag}^+(\mathrm{AgCl})$ is, as far as the chemical potential is concerned, to be identified with $\mathrm{Ag}_\mathrm{Ag} - \mathrm{V}'_\mathrm{Ag}$ (but not with Ag_Ag!) [7].

as valid, are largely isomorphic with the ionic disorder problem[28,29] (cf. Fig. 5.9 in the next section) and will be discussed now.
The fundamental electronic disorder reaction is equivalent to the overcoming of the gap between valence band and conduction band and, hence, represents the generation of conduction electrons e′ (in the conduction band) and holes h˙ (in the valence band). Relations between the (electro–) chemical potentials of defects (e′, h˙) and components (e^-) are analogous to the ionic case (see below).

5.3 Equilibrium thermodynamics of electronic defects

If we initially consider solids at high temperatures, that possess adequately large band gaps, low doping levels and a sufficient number of available electronic levels, Fermi–Dirac statistics can be replaced by Boltzmann statistics and the formalism is substantially that applied to ions[28]. Here, too, there is an expression of the form of Eq. (5.15) whereby the "number of positions available" has a somewhat more complex meaning and can differ for conduction electrons (excess electrons) and holes ("electron vacancies"). We will leave this discussion for later and name these reservoir quantities, the effective numbers of states for the conduction ($\bar{N}_C$) and for the valence band ($\bar{N}_V$). Conditions under which we cannot neglect correlations (i.e. energy levels, in particular, become dependent on occupation) will also be considered later. The chemical potentials are then obtained in the expected form

$$\begin{aligned} \mu_{e'} &= \mu^*_{e'} + RT\ln(N_{e'}/\bar{N}_C) &= \mu^\circ_{e'} + RT\ln[e'] \\ \mu_{h^\cdot} &= \mu^*_{h^\cdot} + RT\ln(N_{h^\cdot}/\bar{N}_V) &= \mu^\circ_{h^\cdot} + RT\ln[h^\cdot]. \end{aligned} \tag{5.29}$$

On the right hand side they have been expressed as logarithmic functions of $[e'] \equiv \frac{N_{e'}}{N_{MX}}$ or $[h^\cdot] \equiv \frac{N_{h^\cdot}}{N_{MX}}$ by rescaling the μ^*'s to μ°'s accordingly. As with ions it is necessary to use the electrochemical potentials in the presence of electric fields:

$$\begin{aligned} \widetilde{\mu}_{e'} &= \mu_{e'} - F\phi &= \mu^\circ_{e'} + RT\ln[e'] - F\phi &= \widetilde{\mu}^\circ_{e'} + RT\ln[e'] \\ \widetilde{\mu}_{h^\cdot} &= \mu_{h^\cdot} + F\phi &= \mu^\circ_{h^\cdot} + RT\ln[h^\cdot] + F\phi &= \widetilde{\mu}^\circ_{h^\cdot} + RT\ln[h^\cdot]. \end{aligned} \tag{5.30}$$

As already discussed in Section 1.1 we can describe the electronic transitions from the valence into the conduction band as

[28] In fact it is not the necessity of applying Fermi–Dirac statistics per se which distinguishes the ionic problem from the electronic one (Fermi–Dirac statistics have also been applied for the ionic defects, cf. Eq. (5.18)), it is the different state densities. In the Boltzmann approximation, this difference is reflected by a different standard state.

[29] In the case of electronic conductors, the role of the superionic state is played by the metallic state ("superelectronic" state) characterized by almost invariable chemical potentials of the charge carriers. Analogously, the number of electronic "defects" in metals is very large, and the term "defect" is no longer appropriate. As a result the relative change in concentration ($\delta \ln c$) can be neglected and the associated chemical potential (μ_{e^-} for metals) is constant ($\delta\mu \propto \delta \ln a(c) \propto \delta a/a$) (cf. also Fig. 5.68 on page 214).

$$\text{Reaction B} = \qquad \mathrm{Nil} \rightleftharpoons \mathrm{e}' + \mathrm{h}^{\cdot}. \tag{5.31}$$

This is the fundamental electronic disorder reaction for which because of

$$\widetilde{\mu}_{\mathrm{e}'} + \widetilde{\mu}_{\mathrm{h}^{\cdot}} = \mu_{\mathrm{e}'} + \mu_{\mathrm{h}^{\cdot}} = 0 \tag{5.32}$$

and because of Eq. (5.29) a mass action law can be written. The band gap, whose relationship to the bonding parameters has already been discussed in detail in Section 2.2, corresponds to the free standard enthalpy of reaction $\Delta_{\mathrm{B}}\mathrm{G}^*$, in other word to the standard formation values of the intrinsic electronic imperfections. In view of the discussion in Section 2.2 it is not unexpected that there exist, as is illustrated in Fig. 5.8, empirical relationships between $\mathrm{E_g}$ and the electronegativities of anions

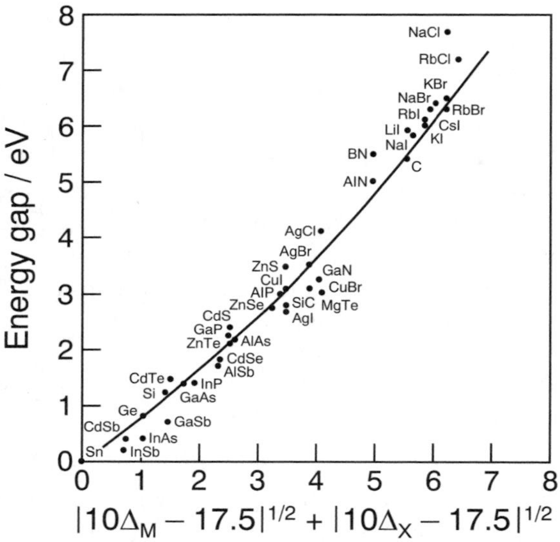

Fig. 5.8: Energy gap for a variety of elements and binary compounds (MX) as a function of electronegativities Δ_{M} and Δ_{X}. From Ref. [124].

and cations. Since we ignore — apart from the configurational components — the entropy effects of electron transitions (see however Ref. [125]) it is of no relevance here whether the individual band gaps are written as a difference in the energy or free energy (or free enthalpy; pressure dependences are weak, anyway).

On account of the equivalence of $\Delta_{\mathrm{B}}\mathrm{G}^*_{\mathrm{m}} = \widetilde{\mu}^*_{\mathrm{e}'} + \widetilde{\mu}^*_{\mathrm{h}^{\cdot}} = \mu^*_{\mathrm{e}'} + \mu^*_{\mathrm{h}^{\cdot}}$ and the band gap, it is appropriate to allocate the value $\widetilde{\mu}^*_{\mathrm{e}'}$ to the lower edge of the conduction band and the value $-\widetilde{\mu}^*_{\mathrm{h}^{\cdot}}$ to the upper edge of the valence band. According to the same arguments, as used in the previous chapter, $\widetilde{\mu}_{\mathrm{e}'} = -\widetilde{\mu}_{\mathrm{h}^{\cdot}}$ can be identified with the electrochemical potential of the electronic component e^-. The quantity $\widetilde{\mu}_{\mathrm{e}^-}/\mathrm{N_m} = \widetilde{\mu}_{\mathrm{e}'}/\mathrm{N_m} = -\widetilde{\mu}_{\mathrm{h}^{\cdot}}/\mathrm{N_m}$ is identical to the (electronic) Fermi level [122]. According to Eq. (5.29) the distances from the Fermi level to the band edges are a logarithmic measure of the inverse concentration of charge carriers. In pure intrinsic semiconductors, where $[\mathrm{e}'] = [\mathrm{h}^{\cdot}]$, the Fermi level lies approximately in the centre (not exactly because $\bar{N}_C \neq \bar{N}_V$). All these relationships will become much clearer later.

Figure 5.9 illustrates the isomorphy between electron and ion disorder, on the one hand, and between solid and liquid solutions, on the other hand, in the physical

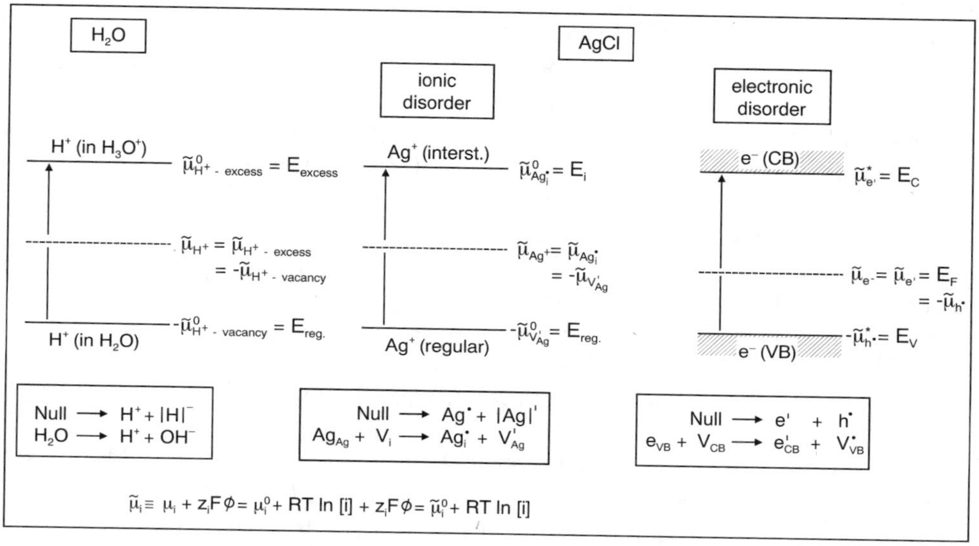

Fig. 5.9: Electronic and ionic disorder in ionic solids and in water in "physical" and "chemical" language (see text) [14]. The coupling of the ionic and electronic Fermi levels takes place via the chemical potential of the neutral components (here: $\widetilde{\mu}_{Ag^+} + \widetilde{\mu}_{e^-} = \mu_{Ag}$) and thus via the precise position in the phase diagram (see Section 5.5.2). (Note that the energy levels of the liquid fluctuate in time and space.)

energy level language. The progress of the pseudochemical reactions (see Fig. 5.9) corresponds to the surmounting of the gap: Electrons are heaved from the valence to the conduction band leading to the formation of a conduction electron with an electron hole remaining behind; Ag^+ ions can be forced on input of energy into an interstitial position leaving a vacancy behind. Protons can be removed from a regular H_2O yielding a proton vacancy (OH^-) and assume a higher energy position (H_3O^+) by associating with a neutral H_2O. In pure water (pH=7 under standard conditions) the "Fermi level of the protons in water" (($\widetilde{\mu}_{H^+}$) is (approximately) midway between the levels[30] illustrated. Taking into account the various energetic positions possible also allows the construction of higher levels[31] and, hence, a formal similarity to

[30]The distance of $\widetilde{\mu}_{H^+}$ (which may be called Brønsted level) from the two levels reflects pH and pOH, while the difference between them is a measure of $pK_W \equiv -\lg K_W$ being the autoprotolysis constant (w.r.t. the correspondence of the Brønsted picture and defect chemistry see Ref. [15]).

[31]This ought to be of special importance for ions in polymers or glasses (see page 290). In addition, the levels vary spatially here, whilst in liquids they also vary as a function of time. In

the manifold of states for electrons. The basic difference, in terms of density and occupation of the states however remains (see below and Section 2.2.1). The decisive difference between electrons and ions lies in the wave–like nature of the first. While ions behave classically to a good approximation, the electrons are, in the case of strong orbital overlap, delocalized and can tunnel through potential barriers[32]. With the exception of superconducting systems, there is always interaction with the lattice vibrations that limits the mobility of the electrons. The similarity to ionic disorder is greater when localization effects create additional barriers. This is most likely the case for pronounced ionic crystals with narrow bands (cf. Section 2.2.5).

The concrete structure of such a localized electronic defect has been demonstrated for NaCl under oxydizing conditions [126]. There, a hole which is "trapped" by a Cl^- forms together with a further regular Cl^-, a Cl_2^- dumb–bell which bears the hole, a situation which, to some extent, reflects the stability of the Cl_2 molecule. In the Kröger–Vink nomenclature this is described as $Cl^{\cdot}_{Cl}$ or $(Cl_i)'_2(V^{\cdot}_{Cl})_2$ and corresponds to a "self–trapped" hole or a "small polaron". In a simplistic conception the elastic energy of distortion expended on polarization ($\propto$(displacement)^2) and the polarization energy acquired ($\propto$ dipole moment $\propto$ displacement) are in competition, which yields a configuration of minimum energy characterized by the ratio of the two constants of proportionality. One speaks of "large polarons" if the respective electrons move in a fairly broad band with slightly enhanced masses. These effects are of particular importance for the electronic transport (see Section 6.2.2).

Conduction electrons and holes can also occur as a result of the ionization (dissociation) of donor or acceptor states (e.g. P or Al in silicon). The corresponding levels

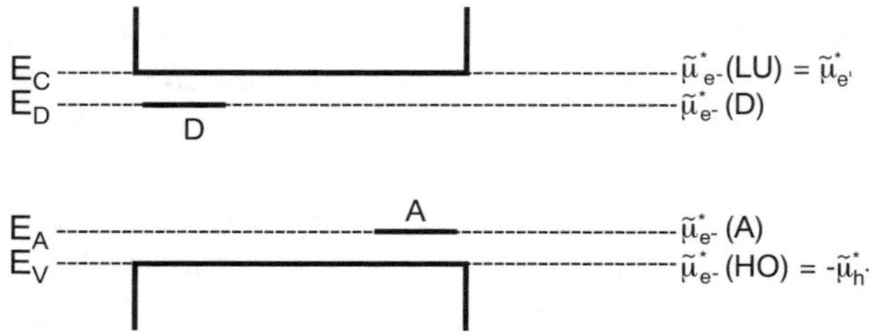

Fig. 5.10: Conduction electrons and holes can be created energetically more easily from donor or acceptor states, that occur within the band gap. If the distances to the band edges (at T=0 in a pure semiconductor, LU and HO denote the lowest unoccupied and the highest occupied orbitals respectively) are very small (shallow state) almost all donors and acceptors are ionized ($D^{\cdot}$, A') (cf. also Fig. 5.61, page 203).

are illustrated in Fig. 5.10. Thus, the transfer of an electron from the donor level to the conduction band requires the free enthalpy energy of the following reaction

$$\text{Reaction D} = \qquad D^{x} \rightleftharpoons D^{\cdot} + e' \qquad (5.33)$$

ion conducting polymers the relevant ionic excitation is the dissociation of ion pairs (see Section 6.2).

[32] In our context, the only ions, for which the tunnelling effect may not be neglected, are protons.

to be overcome. Strictly speaking (see below) the upper level (LU), i.e. the edge of the conduction band, represents the (electro–)chemical standard potential of the "free" electron, while the donor level reflects the (electro–)chemical potential of the electron at the donor atom, i.e. $\Delta_D G^*_m = E_C - E_D = \widetilde{\mu}^*_{e'} - \widetilde{\mu}^*_{e^-(D)}$. The comparison with Eq. (5.33) reveals that the level $E_D = \widetilde{\mu}^*_{e^-(D)}$ corresponds to the difference $\widetilde{\mu}^*_{D^x} - \widetilde{\mu}^*_{D^{\cdot}}$, more accurately to the $\widetilde{\mu}^*$–value of the building element $D^x - D^{\cdot}$. The transformation of Eq. (5.33) into

$$\text{Reaction D} = \qquad (D^x - D^{\cdot}) \rightleftharpoons e' \quad \text{or} \quad e^-(D) \rightleftharpoons e^-(CB) \tag{5.34}$$

makes this conclusion obvious. One should keep in mind that the levels usually considered in semiconductor physics only refer to electronic (free) energies. Whenever ionic free energy contributions are also of relevance, the chemical formalism which will be discussed later, is more powerful and hence advantageous for the description of complex defect reactions.

The transfer of an electron from the valence band to the donor level, leaving behind a hole (e.g. the consumption of a bonding electron to neutralize the ionized donor state $P^{\cdot}_{Si}$) corresponds to the reaction

$$\text{Reaction D}' = \qquad \text{Nil} \rightleftharpoons (D^x - D^{\cdot}) + h^{\cdot}, \tag{5.35}$$

which can be transformed into

$$\text{Reaction D}' = \qquad D^{\cdot} \rightleftharpoons D^x + h^{\cdot}. \tag{5.36}$$

It follows that $E_D - E_V = \Delta_{D'} G^*_m$. The free energy change necessary for this transition,

$$\Delta_{D'} G^*_m = E_D - E_V = (E_C - E_V) + E_D - E_C = E_g - \Delta_D G^\circ_m \tag{5.37}$$

is unfavourably high for strong donors (shallow donors).

Conversely, the transfer of an electron from the valence band to a typical acceptor level is comparatively easy. The acceptor ionization (e.g. of Al_{Si}) can be described as

$$\text{Reaction A} = \qquad A^x \rightleftharpoons A' + h^{\cdot}, \tag{5.38}$$

whereby $E_A = \widetilde{\mu}^*_{e^-(A)} = \widetilde{\mu}^*_{A'-A^x}$ applies. If we, as above, use chemical potentials for the individual structure elements, $E_A = \widetilde{\mu}^*_{A'} - \widetilde{\mu}^*_{A^x}$ results. The edge of the valence band (HO) represents the negative (electro–)chemical standard potential of the hole $(-\widetilde{\mu}^*_{h^{\cdot}})$.

Although the "level language" can be subsumed into our defect chemical notation (the latter is of more general validity) we will, in what follows — not least in order to overcome language difficulties — return to the level diagram whenever helpful.

In order to look more closely at both types of language and to understand the meaning of effective density of states and the applicability of the Boltzmann formulation more deeply, the equilibrium thermodynamics of the electrons in a semiconductor will be analysed in more depth for the interested reader. Let us first consider a narrow energetic section of the conduction band, labelled by ℓ. The section is so narrow that we can neglect the energetic variations within it. The number of possibilities to distribute N_ℓ electrons on the Z_ℓ levels of the section is then (see page 118)

$$\Omega_\ell = \binom{Z_\ell}{N_\ell}. \tag{5.39}$$

By analogy to the treatment of ionic defects (see page 118ff), this leads to

$$\widetilde{\mu}_\ell - \widetilde{\mu}_\ell^* = \mu_\ell - \mu_\ell^* = RT \ln \frac{N_\ell}{Z_\ell - N_\ell} = RT \ln \frac{N_\ell/Z_\ell}{1 - N_\ell/Z_\ell} \tag{5.40}$$

(the electronic index is suppressed here) or, solved for the concentration term

$$N_\ell/Z_\ell = \frac{1}{1 + \exp -\frac{\widetilde{\mu} - \widetilde{\mu}_\ell^*}{RT}} = \frac{1}{1 + \exp -\frac{\epsilon_F - \epsilon_\ell}{k_B T}}. \tag{5.41}$$

Owing to the internal equilibrium

$$e^-(\ell{=}1) \rightleftharpoons e^-(\ell{=}2) \rightleftharpoons \ldots \tag{5.42}$$

$\widetilde{\mu}_\ell$ is independent of the section selected and can, therefore, be freed from the index ℓ. When $\widetilde{\mu}$ is identified with $\epsilon_F N_m$ and $\widetilde{\mu}_\ell^*$ with $\epsilon_\ell N_m$, Eq. (5.41) is the known result of the Fermi–Dirac treatment, which leads to a Boltzmann expression when $N_\ell \ll Z_\ell$. The fact that for Boltzmann distribution $N_m \epsilon_\ell$ corresponds to the partial, molar, free enthalpy is also consistent with the following points:

Since the invariance of the Fermi energy is an expression of the thermodynamic equilibrium, ϵ_F is strictly assigned to the Gibbs energy at constant pressure and temperature. However, local changes in entropy are not, in general, relevant for electronic transitions, so that we wish to retain the expression "energy level". It is more important that this term represents a partial parameter in the sense of chemical thermodynamics; this is in agreement with the Gibbs–Duhem relationship Eq. (4.20): $E = \Sigma_\ell N_\ell \epsilon_\ell$. The fact that ϵ_ℓ is classified as a standard potential expresses the fact that it is not dependent on occupation (corresponding to a rigid band model).

Note that — apart from taking the Pauli exclusion principle into account — there has been no explicit account taken of electron interaction. Otherwise an excess term dependent on occupation appears in the free enthalpy leading to a μ^{ex} term, that would have to be formally included in ϵ_ℓ[33].

[33]Thus, the attractive interaction between e' and $h^{\cdot}$ leads to a reduction of the band gap (see Section 5.7). Note that the energy levels, more generally, correspond to the electrochemical potentials after subtraction of the configurational term, that is to the sum of the standard values and (nonconfigurational) interaction terms. Note also, that interactions affect the distribution statistics.

Equation (5.41) also applies to the distribution of electrons over acceptor and donor levels. Here, however, it is necessary to take degeneracy factors into account similarly, as is done on page 122f for ionic defects. The occupied donor term P^{x}_{Si}, e.g., exhibits an excess electron that can have two spin quantum numbers. The factor 2^{N_ℓ} by which Eq. (5.39) has to be supplemented becomes an additive term ($k_B \ln 2$) in the numerator of the exponential function (Eq. (5.41)) and can be incorporated in ϵ_ℓ (as $\epsilon_\ell - k_B T \ln 2$).
Let us now turn to the calculation of the concentration of charge carriers in the bands. Since N_ℓ and Z_ℓ represent the electron numbers and the numbers of states in the differentially narrow energy interval ϵ_ℓ with the width $d\epsilon$, it is also possible to express the distribution function $F_\ell = N_\ell / Z_\ell$ as

$$F(\epsilon) = dN(\epsilon)/dZ(\epsilon). \tag{5.43}$$

The total number of electrons in the conduction band (CB) is obtained as

$$N_{e'} = \int\limits_{CB} dN_{e'} = \int\limits_{CB} \frac{dN(\epsilon)}{dZ(\epsilon)} \frac{dZ(\epsilon)}{d\epsilon} d\epsilon \tag{5.44}$$

by integration over all energy states of this band or, introducing the density of states $D(\epsilon) = dZ(\epsilon)/d\epsilon$, as

$$N_{e'} = \int\limits_{CB} F_{e'}(\epsilon) D(\epsilon) d\epsilon. \tag{5.45}$$

The number of holes in the valence band (VB) is obtained in an analogous manner from the difference between the number of states and the number of electrons

$$N_{h^{\cdot}} = \int\limits_{VB} \frac{dZ}{d\epsilon} d\epsilon - \int\limits_{VB} \frac{dN_{e^-}}{dZ} \frac{dZ}{d\epsilon} d\epsilon = \int\limits_{VB} (1 - F_{e^-}) D(\epsilon) d\epsilon = \int\limits_{VB} F_{h^{\cdot}} D(\epsilon) d\epsilon. \tag{5.46}$$

In both cases the calculation of the density of states[34] remains. In Section 2.2 it was shown that, to a first approximation for small values of k ($k \ll \pi/a$), the $\epsilon(k)$ function can be approximated by a parabolic function ($\epsilon - \epsilon(k{=}0) \propto k^2$).
Thus, (see Section 2.2)

$$D_C = \frac{dZ_C}{d\epsilon} \propto (\epsilon - \epsilon_C)^{1/2} \tag{5.47}$$

is valid for the conduction band in the three–dimensional case[35]. Note that account must be taken of the zero–point $\epsilon(k{=}0) = \epsilon_C$. The detailed calculation reveals

[34] This refers to the major difference between the electronic and the ionic picture. In the cases of simple ionic disorder described in the previous section, the density of states corresponded to delta functions.

[35] We remember that the density of the electron levels in the one–dimensional box falls with increasing energy. Because of the occurrence of three independent quantum numbers in the three–dimensional case (with the result that very many number combinations appear with the same ϵ value), the previously mentioned reduction is overcompensated by the degeneracy effect.

that the constant of proportionality includes the effective mass (m^*) to a power of 3/2. For adequately high temperatures the evaluation of the integral yields the expressions[36]

$$\frac{N_{e'}}{\bar{N}_C} = \exp - \frac{\epsilon_C - \epsilon_F}{k_B T} \quad \text{and} \quad \frac{N_{h^{\cdot}}}{\bar{N}_V} = \exp - \frac{\epsilon_F - \epsilon_V}{k_B T}. \tag{5.48}$$

already employed. The detailed calculation for the effective state densities leads to

$$\frac{\bar{N}_{C,V}}{V} = 2\left(\frac{2\pi}{h^2} m^*_{e',h^{\cdot}} kT\right)^{3/2} = 4.2 \times 10^{-5} \frac{\text{mol}}{\text{cm}^3} \times N_m \left(\frac{m^*_{e',h^{\cdot}}}{m_{e^-}}\right)^{3/2} \left(\frac{T}{300\text{K}}\right)^{3/2}. \tag{5.49}$$

To a first approximation, $m_{e^-} \simeq m^*_{e'} \simeq m^*_{h^{\cdot}}$ is frequently used, although the differences can be considerable.

The fact that we have set the Fermi level as constant means that we have concerned ourselves with equilibrium states, although we left off the equilibrium arc for simplicity. We will now take this notation more seriously. After introducing the electronic concentrations referred to N_{MX} (e.g. $[e'] \equiv \frac{N_{e'}}{N_{MX}} = \frac{N_{e'}}{\bar{N}_C}\frac{\bar{N}_C}{V}\frac{V_m}{N_m}$), we write the mass action constant of the band–band transfer as

$$\widehat{[e']}\widehat{[h^{\cdot}]} = K_B(T) = \exp - \frac{\Delta_B G^\circ_m}{RT} = \exp \frac{\Delta_B S^\circ_m}{R} \exp - \frac{\Delta_B H^\circ_m}{RT}. \tag{5.50}$$

The comparison with

$$\left(\frac{\widehat{N}_{e'}}{\bar{N}_C}\right)\left(\frac{\widehat{N}_{h^{\cdot}}}{\bar{N}_V}\right) = \exp - \frac{E_g}{RT} \tag{5.51}$$

leads to the conclusion that $\Delta_B H^\circ_m \simeq E_g$ if $E_g, \bar{N}_{C,V}, \Delta_B H^\circ_m$ and $\Delta_B S^\circ_m$ are approximately independent of temperature. The "entropy term", that comes up formally, is primarily a correction term here and basically contains the effective state densities and the molar volume. It is found empirically that in many materials E_g does depend on the temperature — and usually in a linear manner (Fig. 5.11). The main reason is the expansion of the lattice. In this case the energy parameter in the exponential term obviously corresponds to the band gap at 0K, while the prefactor includes the corresponding temperature coefficients (see insert in Fig. 5.11). While such divisions are formally permissible, we must be cautious with the thermodynamic interpretation. If ΔH° is T–dependent, ΔS° must also be T–dependent (e.g. if ΔH° is linearly dependent on T, ΔS° must depend logarithmically on T (Eqs. (4.39), (4.40)).

[36]It is not the detailed form of the density of states which finally results in Boltzmann–relations for dilute systems. Rather, it is the fact that the distance between Fermi level and edge is sufficiently large. Thus in the Fermi distribution treatment (cf. Eq. (5.41)) the 1 in the denominator can be neglected for all relevant energy levels, and the Boltzmann factor (cf. r.h.s. of Eq. (5.48)) can be taken out of the integral. The remaining integral yields $\bar{N}$. A different distribution of states would result in a modified form of the effective state density. Similar arguments must be used to explain in more depth the validity of mass action laws in fluid and solid disordered systems.

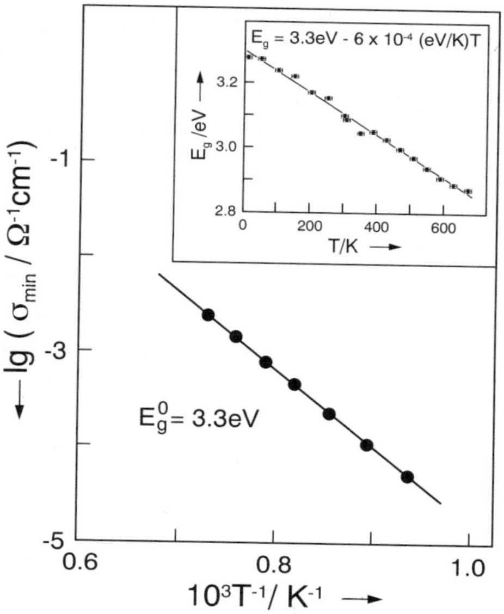

Fig. 5.11: The (thermal) band gap in $SrTiO_3$ determined from the minimum of the electronic conductivity (see Section 5.6) is displayed as a function of the inverse temperature. The inset shows the (optical) band gap obtained from optical absorbance as a function of temperature. Since E_g is a linear function of temperature, plotting $\lg \sigma_{min}$ against 1/T yields E_g^0 ($\equiv E_g(T{=}0K)$) as the slope [127]. Obviously optical bandgap and thermal bandgap are identical or very similar.

Irrespective of such complications as the thermal expansion, the above approximation fails at high concentration and the more general result for the conduction band (see Ref. [128]) follows, by taking account of the exact Fermi–Dirac relation in Eq. (5.45) (see Eqs. (5.40), (5.41)) as

$$\frac{\widehat{N}_{e'}}{\bar{N}_C} = \frac{2}{\sqrt{\pi}}\mathcal{F}_{1/2}\left(\frac{\widehat{\epsilon}_F - \epsilon_C}{k_B T}\right) \tag{5.52}$$

with the definition of the Fermi function:

$$\mathcal{F}_{1/2}(y) \equiv \int_0^\infty \frac{\tau^{1/2}d\tau}{\exp(\tau + y) + 1}. \tag{5.53}$$

This correction, which formally leads to an activity coefficient, is explained in more detail in Section 5.7. All the considerations for the valence band are analogous.

To conclude this section some remarks are included concerning the position of the Fermi level in the energy gap.
In the case of the pure semiconductor, in which no defects are present apart from e′ and h˙, $\widehat{N}_{e'} = \widehat{N}_{h^\cdot}$. From Eq. (5.48) it follows (with $E = N_m \cdot \epsilon$) that

$$\widehat{E}_F = \frac{E_C + E_V}{2} - \frac{1}{2}RT\ln\frac{\bar{N}_C}{\bar{N}_V}. \tag{5.54}$$

In the case of equal densities of states (i.e. $m^*_{e'} = m^*_{h^\cdot}$) the Fermi level then lies in the centre of the band. In the presence of acceptors or donors this no longer

applies, because $\widehat{N}_{e'} \neq \widehat{N}_{h^{\cdot}}$. The following consideration, in which we set the effective densities of state equal, illustrates this.
The equilibrium constant of the donor reaction[37] (Eq. (5.33)) is given by $K_D = \widehat{[D^{\cdot}]}\left(\widehat{N}_{e'}/\ \bar{N}_C\right)/\widehat{[D^x]} = \exp -\frac{\Delta_D G_m^*}{RT}$ where $\Delta_D G_m^* = E_C - E_D$. Taken together with the Fermi level from Eq. (5.48) it follows that $RT\ln\left(\left(\frac{\widehat{N}_{e'}}{\bar{N}_C}\right)/K_D\right) = RT\ln\frac{\widehat{[D^{\cdot}]}}{\widehat{[D^x]}} = E_D - \widehat{E}_F$. If the Fermi level is at the donor level, then this level is half filled and the degree of ionization is 50%. If $\widehat{E}_F$ lies above (below) E_D, the degree of ionization is less (greater). However, as $T \to 0$, the position of $\widehat{E}_F$ is shifted to the centre between level and band edge (in correspondence with the undoped case)[38].

5.4 Higher–dimensional defects

5.4.1 Equilibrium concentration

The fact that the entropy of configuration brings about a certain concentration of energetically expensive chemical excitations at equilibrium, quickly loses importance as the defects reach a certain size. This is readily demonstrated in the case of an internal interface, a grain boundary. Let us consider our lottery ticket once again (Fig. 5.6), but this time the example on the right hand side of the figure. The condition that all the selected numbers have to be one below the other reduces the number of configurations from 2 million to only 6. Let us consider a cube composed of N atoms of the same type and discuss the formation of an internal grain boundary, which, for simplicity's sake, is to run parallel to the surface of the cube. There are evidently $3N^{1/3}$arrangement possibilities. The enthalpy of configuration of N_k such surfaces is then

$$S = k_B \ln \binom{3N^{1/3}}{N_K} \tag{5.55}$$

and the chemical potential of the boundary[23]

$$\mu = \mu^* + RT\ln\frac{N_K}{3N^{1/3}}. \tag{5.56}$$

[37]It is worthy of note that the introduction of a separate Boltzmann estimate for $D^{\cdot}$ and D^x yields an equation that — owing to mass conservation — "simulates" Fermi–Dirac statistics which is valid over the entire concentration range for the building elements, i.e. the electrons in the acceptor term. The same would be true for the band–band transition, if we formulated Eq. (5.31) in terms of structure elements (see Eqs. (5.94c, 5.94d)). However here the correction would only be qualitatively correct (cf. Eq. (5.209) and Fig. 5.69).

[38]This is obtained, for instance, in the donor case from $E_F = E_C + RT\ln\frac{N_{e'}}{N_C}$ (Boltzmann approximation suffices, since $N_{D^{\cdot}} = N_{e'} \to 0$), $E_F = E_D + RT\ln\frac{[D^x]}{[D^{\cdot}]}$ and $N_{D^{\cdot}} = N_{e'}$. The result is $E_F = \frac{E_D+E_C}{2} + \frac{RT}{2}\ln\frac{[D^x]}{N_C}$ corresponding to Eq. (5.54). See also the semiconductor physics literature [128,129,70] and Ref. [5].

The standard value is the molar free enthalpy of grain boundary formation $\Delta_K G^*_m$, i.e., the G–change during the reaction[20]

$$\text{Reaction K} = \qquad \text{Nil} \rightleftharpoons \text{grain boundary.} \tag{5.57}$$

For the equilibrium concentration it follows that

$$\frac{\widehat{N}_K}{N^{1/3}} \simeq \exp -\frac{\Delta_K G^*_m}{RT}. \tag{5.58}$$

In Eq. (5.58) we have incorporated the factor 3 into $\Delta_K G^*$ as $RT \ln 3$. It does not play any role in the following order of magnitude consideration.
If we compare this with the concentration of point defects $\widehat{N}_d$ and make the statement that $\Delta_K G^*$ is α times the free energy of formation of a single defect, then with Eq. (5.19) we get[39]

$$\widehat{N}_K = \widehat{N}_d^\alpha / N^{\alpha - 1/3}, \quad \text{i.e.} \quad \frac{\widehat{N}_K}{\widehat{N}_d} = \frac{1}{N^{2/3}} \left(\frac{\widehat{N}_d}{N} \right)^{\alpha - 1}. \tag{5.59}$$

An absolutely unrealistic, but safe upper estimate for N_K is obtained if we set the energy of formation of the whole interface as equal to the energy of formation of a single atomistic defect ($\alpha = 1$). Even at this value, which is far too small, the equilibrium number is reduced by a factor[40] of $N^{2/3}$ with respect to $\widehat{N}_d$, and if $\widehat{N}_d \simeq 10^{15}$ and $N \simeq 10^{21}$ a value of 10 for $\widehat{N}_K$ is predicted. Already a further doubling of the energy ($\alpha = 2$) leads to a $\widehat{N}_K$ value of 10^{-5}, i.e. there is not one single grain boundary at equilibrium. A more realistic estimate is for $\Delta_K G^*$ to take a value that is obtained from the sum of the individual point defect energies, that is $\alpha \sim N^{2/3}$. The result using the above number game yields an astronomically small value

$$\widehat{N}_K = 10^7 (10^{-6})^{10^{14}} \lll 1. \tag{5.60}$$

Let us now consider the same situation from another point of view and regard the grain boundary as an aggregation of ν point defects d to a higher–dimensional defect $(d)_\nu$:

$$\text{Reaction P} = \qquad \nu d \rightleftharpoons (d)_\nu. \tag{5.61}$$

Assuming Boltzmann distribution, the following law of mass action is valid at equilibrium (N_1, N_ν are numbers for the possible states of educt and product)

$$\frac{\widehat{N}((d)_\nu)}{\widehat{N}(d)^\nu} = \frac{N_\nu}{N_1^\nu} \exp -\frac{\Delta_P G^*_m}{RT}. \tag{5.62}$$

---

[39]In the case of one–dimensional defects 1/3 is replaced by 2/3 and vice versa. In general, $\widehat{N}_D = \widehat{N}_d^\alpha / N^{\alpha - 1 + D/3}$ for defects of the dimensionality D.

[40]In the case of line defects by $N^{1/3}$ instead of $N^{2/3}$.

When $\nu = N^{2/3}$, $N_\nu = N^{1/3}$, $N_1 = N$ we refer to the above grain boundary example whose last estimate represented the assumption of a vanishingly small free enthalpy of reaction. We recognize the enormous influence of the stoichiometric factor ν on the equilibrium concentrations and can conclude that higher–dimensional defects, even one–dimensional ones, do not occur at thermodynamic equilibrium[2]. Nevertheless, they are present as metastable structure elements and play an important role as "boundary conditions" for the consideration of point defect equilibria in real systems: This is particularly true of grain boundaries, whose elimination generally requires very high temperatures; individual dislocations are more readily annealed out. In other words: The treatment of higher–dimensional defects in this chapter on equilibrium thermodynamics refers to the significance of partial equilibria (similarly to the discussion of dopants in the previous and in the following sections).
The outer surface plays a special role; it does possess a higher molar free enthalpy than the volume, but it has, nevertheless, to exist on account of the finite amount of material. However, here, too, the actual surface, as found in reality, is generally not the equilibrium surface demanded by thermodynamics[41] (see Section 5.4.4).
On account of this dependence on the sample's pre–history we meet a whole range of different cases with respect to higher–dimensional defects. Even though the absolute G minimum is not normally achievable, more or less marked local minima in the free enthalpy can still be realized. We will content ourselves with a few remarks concerning structure and energetics at this point.

5.4.2 Dislocations: Structure and energetics

Dislocations play a large role in determining mechanical properties[42]. They can be regarded as formed by displacements of parts of the crystal. Figure 5.12 illustrates the formation of the two limiting cases, an edge and a screw dislocation, by mechanical action. On account of the peculiarities of bonding discussed in Section 2.2, the atoms of metals, in particular, are relatively easily displaced. For this reason many metals can be readily worked, e.g. by hammering or bending, without the formation of cracks. The energy is invested in the creation of dislocations with the result of plastic material deformation. In addition, an important role is played by dislocations as sites of repeatable growth (sources and sinks for point defects), as rapid diffusion paths, as nucleation centres for the formation of new phases, and — in an aggregated state — as constituents of grain boundaries.

---

[41] At this stage let us tackle the problem of forming point defects within the surface of an elemental crystal which will become important later. If the formation energy is reduced by a factor of β compared with the formation value of the bulk defects (a rough approximation is $\beta \simeq 2/3$ according to footnote 3, page 109), a procedure being analogous to the one above leads to $\widehat{N}_{ds}/N_s = (\widehat{N}_{d\infty}/N_\infty)^\beta$, where N_{ds}, $N_{d\infty}$ represent the numbers of point defects in the surface and in the bulk, while N_s, N_∞ are the number of regular particles there. For an approximately equal packing density of surface layer and of bulk, the molar volume concentration is $c_{ds} = c_{d\infty}^\beta V_m^{\beta-1}$ (V_m: molar volume) [130].

[42] See [91,131–133].

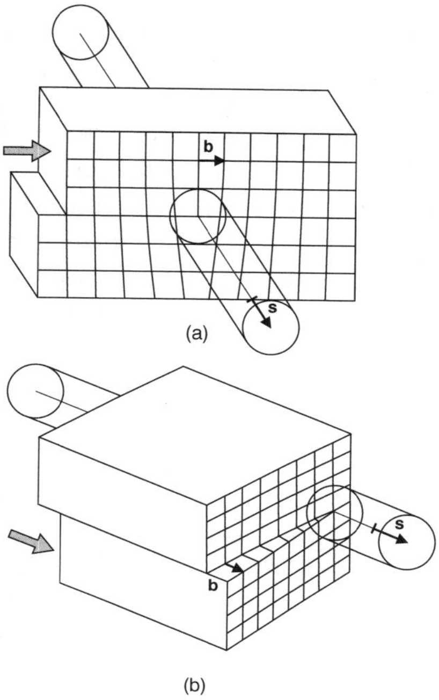

Fig. 5.12: The formation of edge (a) and screw dislocations (b) as a result of mechanical action. The Burgers vector **b** gives the displacement [133,134]. The vector **s** characterizes the direction of the dislocation line.

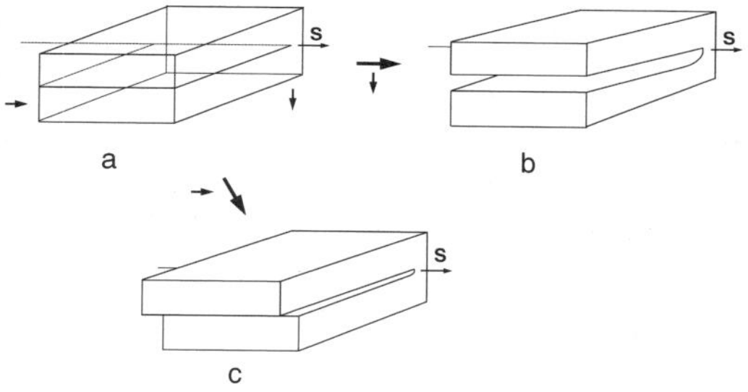

Fig. 5.13: Thought experiment creating edge (b) and screw dislocations (c) in an ideal crystal (a). The crystal is cut open as far as the indicated line (a); the lower part of the crystal is then moved downwards (b) or to the right (c).

The difference between edge and screw dislocations will be clearer systematically on consideration of Fig. 5.13. We cut a perfect crystal (a) as far as the line indicated in the figure (later termed the line of dislocation, whose direction is characterized by **s**) and displace part of the crystal by a certain amount either downwards, that is perpendicular to the line of displacement (b) or parallel to this (c). In the first case ($\mathbf{b} \perp \mathbf{s}$) we insert a further crystal plane into the slit and an edge dislocation is produced. The lattice relaxes and the structure illustrated in Fig. 5.12a is produced, in which the structural change is essentially limited to the immediate region around the line of dislocation, forming the greatly perturbed core of the dislocation. In this way, a one–dimensional defect is produced. The dislocation line must end at the surface or be closed in itself corresponding to a net plane, that is inserted internally. The displacement parallel to the dislocation line (Fig. 5.13c) ($\mathbf{b} \| \mathbf{s}$) followed by relaxation leads to the screw dislocation illustrated in Fig. 5.12b. Again the "dislocation pipe" contains the actual structural modification. Outside the dislocation pipe it is merely the atomic displacement that is of significance. This displacement is described by the Burgers vector **b** (in the case of Fig. 5.12 $|\mathbf{b}|=$ atomic distance) that can, a posteriori, be determined as follows. If one encircles the perturbed region of the crystal, as in Fig. 5.12, and compares it with an analogous path in a perfect crystal, the difference is the Burgers vector, which characterizes nature and strength of the dislocation. The major proportion of the energy is stored in the elastic deformation of the surroundings. According to Hooke's law, for small angles of shear the elastic energy is proportional to the square of the Burgers vector

$$\Delta \mathrm{U}_{\mathrm{el}} \propto \mathbf{b}^2, \tag{5.63}$$

whereby the phenomenological shear modulus as the material constant is contained in the constant of proportionality.

Equation (5.63) has various consequences: On the one hand, the dislocations with the lowest (always positive) energy are usually those with the shortest Burgers vector[43], that generally coincides with the shortest possible distance vectors. On the other hand, the quadratic dependence on Burgers vector energetically favours the splitting of a dislocation with a large **b** into two smaller ones with correspondingly smaller values.

In Eq. (5.63) we have neglected the perturbation in the dislocation core (pipe) for the energy balance. This is generally justified: Let us estimate the line energy of a dislocation core for a metal, by regarding the molecules there as being molten, which is equivalent to an energy of ~0.1eV per atom; a typical $|\mathbf{b}|$ value of 3×10^{-10}m leads then to a core energy of 5×10^{-11}J/m. However, typical dislocation energies are often on the order of 10^{-9}J/m, so that, in fact, more than 90% results from the elastic deformation of the crystal. This applies equally to edge and screw dislocations. Note that the resulting average energy of ~2eV per atom in the core is of the order of magnitude of the point defect formation.

Although properly belonging in the chapter on kinetics (Chapter 6), the mobility of dislocations being an important issue for the properties of real crystals will be briefly discussed here (Chapter 6 focusses on point defects). Figure 5.14 shows how a dislocation migrates, by gliding over macroscopic distances without the re–arrangement of atoms or material exchange between the dislocation line and the surroundings, i.e. in the form of a conservative motion. The elastic stress field migrates too. The elementary activation energy to be overcome is known as the Peierls energy. This gliding of dislocations — in the case of edge dislocations occurring in the glide plane spanned by the vector **b** and the dislocation line — is a most important elementary process in plastic deformation. In the case of screw dislocations, gliding is conceivable at any plane, but for kinetic reasons this gliding takes place on close–packed planes i.e. planes with a low index. The mechanism of climbing permits the movement of a dislocation (line) also outside such glide planes. Figure 5.15 shows that

[43] This naturally only applies for an elementary crystal in this generality.

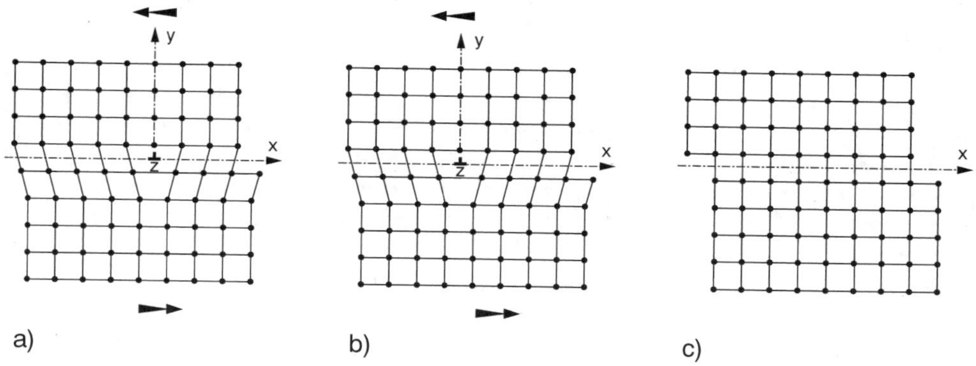

Fig. 5.14: Gliding of a step dislocation in the x direction along a glide plane spanned by the line of dislocation (**s**∥ z axis, i.e. ⊥ plane of the paper) and the Burgers vector (**b**∥ x axis) driven by mechanical stress (see arrows). From Ref. [132].

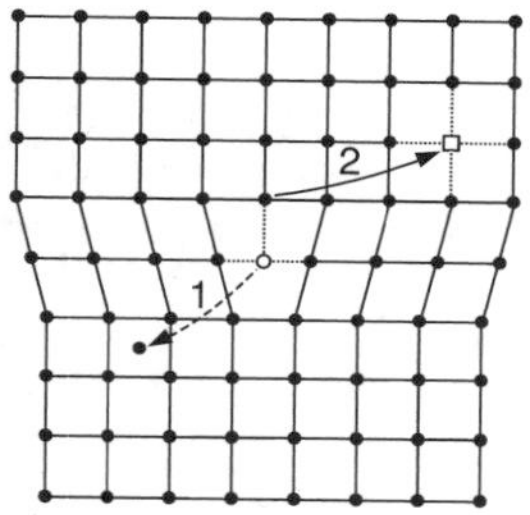

Fig. 5.15: The line of dislocation can migrate upwards by formation of interstitial atoms or occupation of vacancies. (In the converse case, the core of the dislocation migrates downwards. From Ref. [132].)

point defects play a significant role in this case (nonconservative motion). Dislocations, like interfaces, are generally internal sinks or sources of point defects. Figures 5.16 and 5.17 provide an overview of interaction mechanism of dislocations. Here, too, structure elements, such as vacancies, interstitial particles, jogs or kinks, play an important role. They are not involved when two dislocations on the same glide plane meet ($\mathcal{G}$). Then the dislocations annihilate each other as displayed in Fig. 5.16. This can be concisely written as

$$\text{dislocation}\,(\mathbf{b},\mathcal{G}) + \text{dislocation}\,(-\mathbf{b},\mathcal{G}) \rightarrow \text{Nil}. \tag{5.64}$$

It should be remembered that such reactions are not equilibrium processes and therefore a spontaneous back reaction need not be considered. On account of the metastability, the nature, number and distribution of the dislocations naturally depend on the sample's history. The number of dislocations ranges between 0 (for not too large crystals of silicon after careful preparation) and values as high as $10^{11}/\text{cm}^2$ (in the case of heavily worked copper or gold).

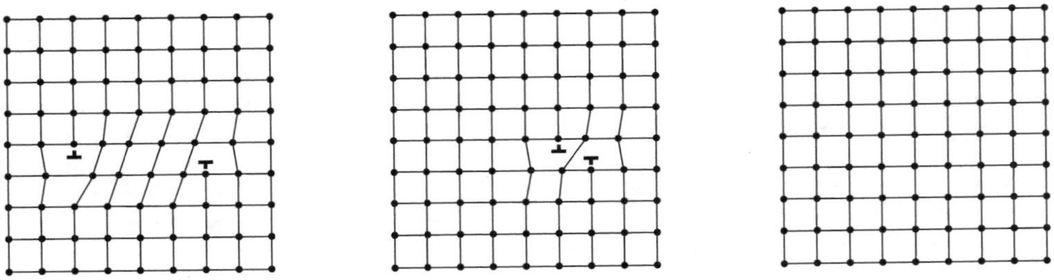

Fig. 5.16: If two dislocations with opposite Burgers vectors meet on the same glide plane, they annihilate each other. If the glide planes are nearest neighbours, the result is a chain of point defects. From Ref. [132].

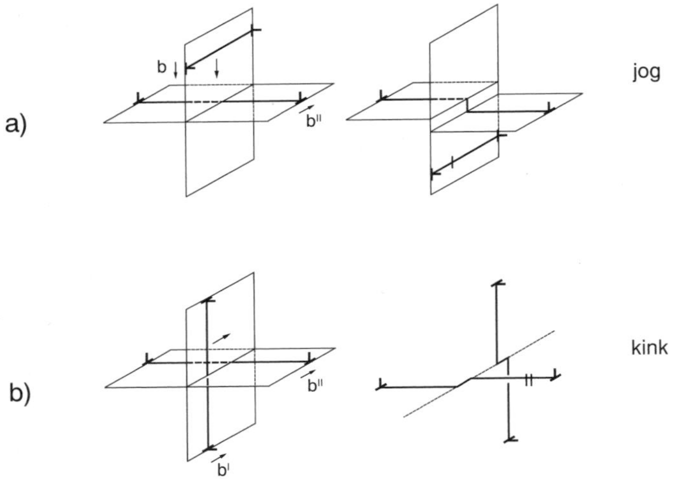

Fig. 5.17: When two dislocations on nonparallel glide planes meet, steps are produced. Depending on whether the Burgers vectors of the step dislocations I and II are perpendicular (a) or parallel (b) to each other, jog or kink sites are produced (that is, steps from glide plane to glide plane or steps in the glide plane). From Ref. [132].

5.4.3 Interfaces: Structure and energetics

The expression "interface" strictly denotes the two–dimensional transition region between three–dimensional regions that are homogeneous in the equilibrium case. However, in general, this transition region is not limited to a single contact plane, but can itself be extended (cf. Section 5.4.4).

Internal interfaces include both phase boundaries (interfaces between grains of differing structures and usually of differing compositions) and also the proper grain boundaries (boundaries between grains of the same structure and composition, but differing orientation) and can, for instance, be classified according to the degree of their coherence. If the interface is simply a common surface of the two lattices[44] — as the phase boundary illustrated in Fig. 5.18a —, then this is referred to as a

[44]Elastic distortions are possible.

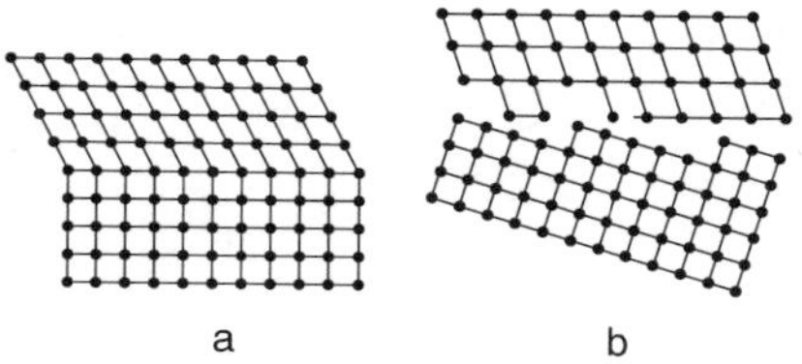

Fig. 5.18: Limiting cases of coherent (a) and incoherent (b) phase boundaries [133].

coherent boundary. If points of coincidence[45] are only present at low density or not at all, then we speak of an incoherent phase boundary (Fig. 5.18b). A more precise classification considers the differences of orientation. Limiting cases of grain boundaries are pure twist grain boundaries and pure tilt grain boundaries. The description of formal formation of grain boundaries is similar to that for dislocations; however, in Fig. 5.13 the whole crystal is cut through and a rotation rather than a shear executed (see Fig. 5.19). If the axis of rotation is normal to the grain boundary we

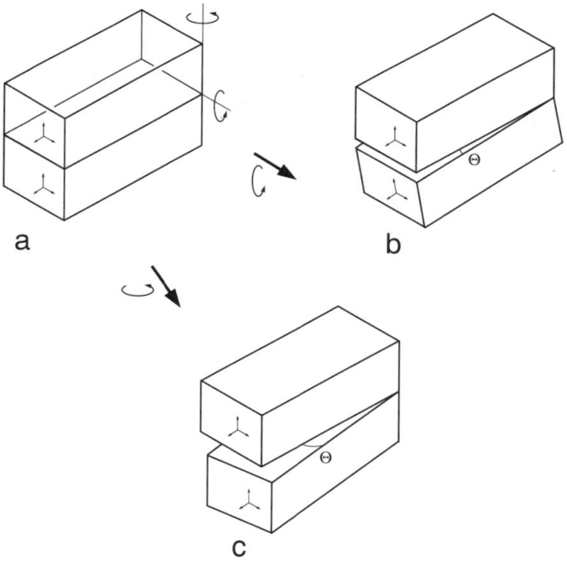

Fig. 5.19: Formation of tilt (b) and twist grain boundaries (c) by cutting a single crystal and rotating one half by Θ about an axis of rotation either in the plane of the cut (b) or normal (c) to it.

speak of a twist grain boundary; if it is in the grain boundary we speak of a tilt grain boundary. If the angle of rotation Θ is less than 5° to 10° (depending on definition), we speak of a low–angle boundary, otherwise of a high–angle boundary. Low–angle boundaries, in contrast to high–angle boundaries, can — as illustrated in Fig. 5.20 — be regarded as aggregation of dislocations: Edge dislocations constitute tilt grain boundaries and screw dislocations twist grain boundaries. The mean distance of the dislocations is given by b/Θ and the whole elastic energy of dislocation is proportional to $b^2(b/\Theta)^{-1}$ according to Eq. (5.63). Taking into account the energy of the

[45] Points being elements of both lattices. The density of sites of coincidence is denoted by the Σ–nomenclature: In the case of a Σn boundary, 1/n gives the fraction of lattice points that form the coincidence site lattice.

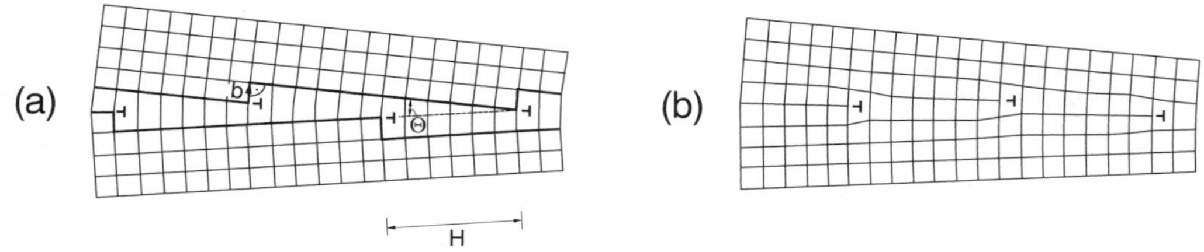

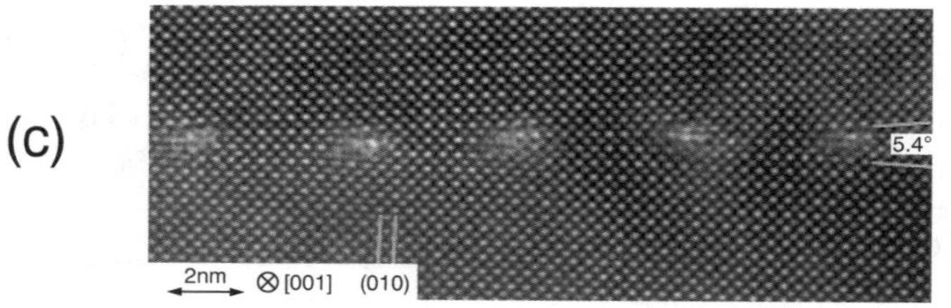

Fig. 5.20: Symmetrical tilt grain boundary a) before relaxation, b) after relaxation. Part c shows a HRTEM–image of such a boundary in $SrTiO_3$ ($\Theta = 5.4°$, HRTEM = high–resolution transmission microscopy) [135].

core including the dislocation interactions yields an expression of the form [136]

$$\Delta U \propto \Theta\,(\text{const} - \ln\Theta) \tag{5.65}$$

for the interfacial energy. Equation (5.65) predicts that the interfacial energy (cf. also Section 5.4.4) runs through a maximum as a function of the angle (l.h.s. in Fig. 5.22) as has been confirmed for many examples. In the case of atomically sharp boundaries the core contribution can be estimated by consideration of local bonding. In very simple cases the regular mixing model discussed in Section 4.3.5 is found to be helpful[46]. When the boundaries are smeared out what is termed the gradient energy is of relevance [97].

In the case of high–angle grain boundaries (see Fig. 5.21) both the extended dislocation models and coincidence site lattice models have proven worthwhile. The

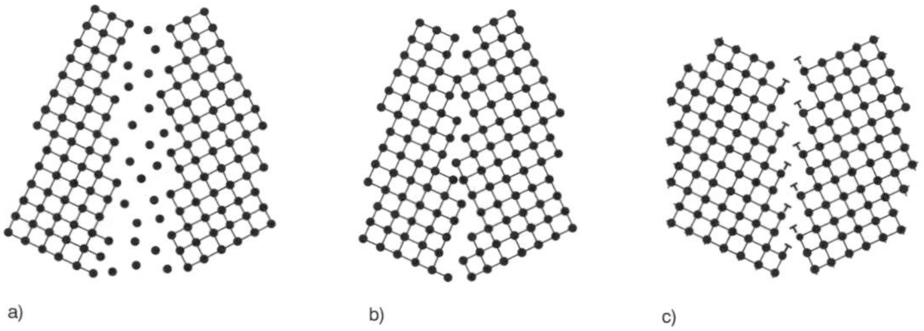

Fig. 5.21: High–angle grain boundaries with (a) and without (b) amorphous interfacial layer. Part c) illustrates a dislocation model for a high–angle grain boundary. From Ref. [132].

[46]If the regular mixing model applies, at the contact of two mixtures of A and B particles, simple local bonding considerations lead to an excess bonding energy that is proportional to W (see pages 96ff) and the square of the difference of the concentrations in both phases [137]. In continuous systems this results in the gradient energy. Since in a compositional gradient the number of neighbours are different from plane p to plane p+1, the bonding energy between particles of the two planes is obtained by replacing in Eq. (4.66) x_A^2 by $x_{A,p}x_{A,p+1}$, x_B^2 by $x_{B,p}x_{B,p+1}$ and $2x_Ax_B$ by $x_{A,p}x_{B,p+1} + x_{B,p}x_{A,p+1}$.

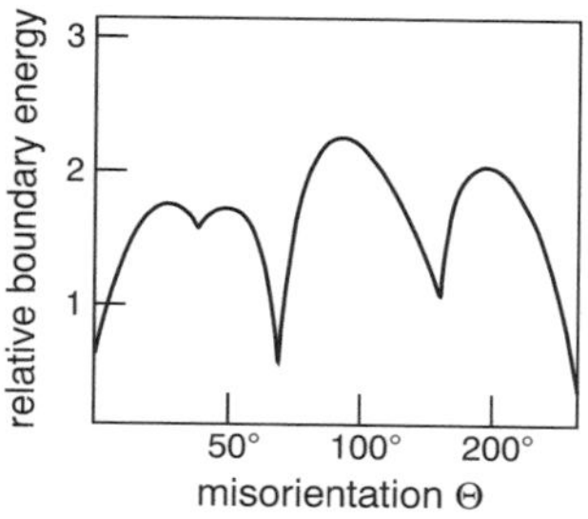

Fig. 5.22: The variation of grain boundary energy with misorientation in Al. (Tilt grain boundary, rotation axis: ⟨110⟩). According to Ref. [138].

density of coincidence points naturally varies with Θ. If γ simply scales with this density, as expected in simple models, then $\gamma(\Theta > \Theta_{\max})$ does not behave monotonically, but exhibits marked minima at the corresponding angles of coincidence (see Fig. 5.22). However, experiments and calculations have shown that such minima, as are predicted by naive lattice coincidence models, do not always occur, even in the case of elemental crystals. Occasionally, grain boundaries are discussed in terms of an alternation of more or less coherent regions with very incoherent regions (island models). Emphasizing the lateral inhomogeneity of the grain boundary is of particular relevance for the electrical properties (cf. Section 7.3.7). In many cases grain boundaries exhibit a lower density than the bulk — such void models have been discussed for nanocrystalline metals —, in others interphases or amorphous stationary interfacial layers[47] have been detected (see Fig. 5.23). In those cases in which

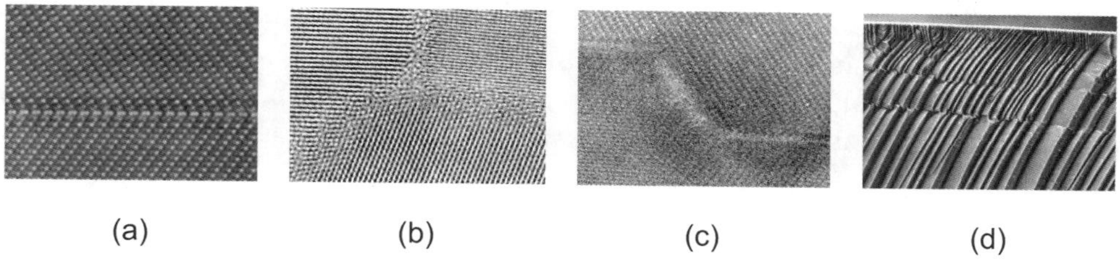

Fig. 5.23: Grain boundaries in $SrTiO_3$ ($m_{Fe} = 9.5 \cdot 10^{19} cm^{-3}$) recorded by high–resolution electron microscopy: a) atomically sharp, low–energy grain boundary (symmetrical $\Sigma3$ $(11\bar{1})$, without amorphous interfacial layer), b) triple point with amorphous intergranular phase (thickness: $\approx$ 1 nm) c) faceted grain boundary with amorphous interfacial layer (thickness $\approx$ 1 nm). Fracture surface (d) with energetically favourable, step–shaped surfaces. From Refs. [139,140].

glassy layers permit the grain–grain matching, the free enthalpy of contact will vary with the film thickness. In fact, the existence of an "equilibrium film thickness" is expected[48] and has been demonstrated experimentally many times [142].

[47] Amorphous phases are to be expected, in particular, at high–energy interfaces (cf. Fig. 5.22).

[48] This can be a mere consequence of double–layer repulsion. Beyond that, van der Waals effects [141] as well as structural effects play a great role.

Grain boundaries (more exactly: homophase boundaries) must disappear at complete thermodynamic equilibrium. This is, however, kinetically very demanding — much more than is the case for dislocations — since it is necessary for the grains to re–orient or at least a large number of atoms must be re–arranged. For details on grain boundary migration the reader is referred to the literature (e.g. [143]). In temperature ranges of usual interest, particularly in the case of multinary compounds, it is only possible to lower the free enthalpy by changing the local morphology, such as by faceting a high–energy surface into surfaces of lower energies at the cost of a larger total area (cf. Figs. 5.23c,d) or by setting up an optimal contact angle. Such processes are analysed below in terms of interfacial thermodynamics. Generally speaking the (partial) equilibrium composition and structure of a particular grain boundary[49] are unequivocally determined by the orientation and the thermodynamic state parameters (component potentials, temperature, pressure). In realistic cases this is, however, a purely academic statement.

In the context of the book higher–dimensional defects are especially significant with respect to their effects on the charge carrier concentration. In an intermediate temperature range of interest to us, the assumption of structural invariance will be a worthwhile approximation. This exploits the fact that the (mobile) point defect kinetics and the interfacial kinetics take place in general on completely different time scales (dislocations occupy an intermediate position): Thus, as is the case for the bulk, the interfacial core is regarded as a locus of constant ground structure (and composition) which is modified with respect to the bulk. The response to variations in state and control parameters consists of modifying point defect concentrations in the interfacial core and in its neighbourhood. The metastable ground structure is regarded, so–to–speak, as an ex situ parameter, that can be affected during preparation at very high temperatures (sinter and creep processes), but not in–situ during the conditions of measurement or performance.

[49]In addition to crystal structures and lattice parameters further parameters are necessary in order to "geometrically" characterize a bicrystal (see e.g. [144]). Two respective normal unit vectors (corresponding to 4 parameters) define the planes to be in contact. (By a tilt operation the vectors are made parallel corresponding to the establishment of two different internal coordinate systems.) The fifth parameter which is the angle of rotation of one bicrystal half about an axis normal to the interface completes the list of the macroscopic degrees of freedom. (In an analogous procedure, referring to Fig. 5.19, a crystal is split into two parts, which are then rotated (twisted and/or tilted) and by sintering converted into a bicrystal. Then it is necessary to select one plane (2 degrees of freedom), the position of the rotation axis (2) and the angle (1).) Beyond that there are three further, microscopic parameters (i.e. related to the atomistic inhomogeneity) which refer to possible translations parallel and normal to the boundary plane. In the case of noncentrosymmetrical crystals the handedness is a further macroscopic degree of freedom; another one comes into play in the case of crystals whose basis contains more than one atom (knowledge of exact position of the interface within the elementary cell is necessary). For mobile boundaries it is in addition useful to define the position with respect to the normal coordinate. The name "degree of freedom" which is commonly used in the literature should be restricted to those parameters (cf. five macroscopic parameters) which can be independently varied in the partial equilibrium under concern.

The same applies for interfaces between grains of different compositions. In contrast to what is generally understood under the phrase "grain boundary" (viz. homophase boundary), their very existence can be demanded as part of the global equilibrium. This is the case for the surface on account of the finite mass, i.e. the interface to the ambient (usually gaseous) phase. But here, too, equilibrium morphology is only achieved in very rare cases (cf. Fig. 5.28 on page 150). Figure 5.24 shows the Kossel model of the surface, which emphasizes the existence of various defective and reactive

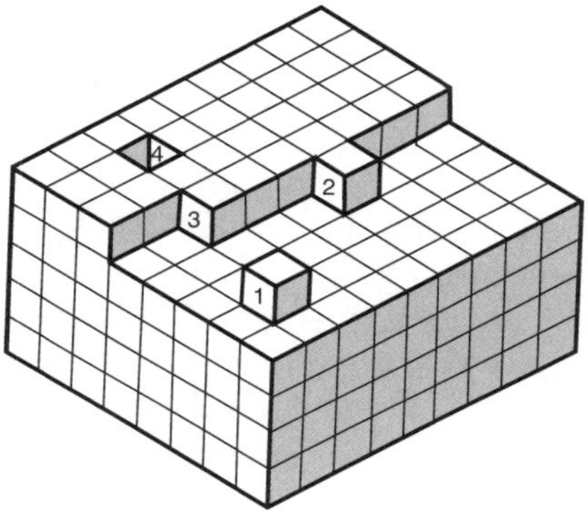

Fig. 5.24: The Kossel model of a surface: lower dimensional defects within the surface as a two dimensional imperfection. The differing energies of the structural elements with varying numbers of contacts (see numbering) are of particular importance for reactivity and growth [104]. The triple contact corresponds to the so–called half–crystal site.

centres (lower dimensional defects). The surface structure is naturally never that which would be obtained by an actual virtual cleavage of a single crystal. Relaxation always occurs. Also, on account of various bonding states (e.g. interaction with the gas phase by adsorption) a reconstruction of the surface may take place. The development of the scanning tunnelling microscope has brought about enormous experimental advances in this respect. Figure 5.25 shows the STM[50] picture of a

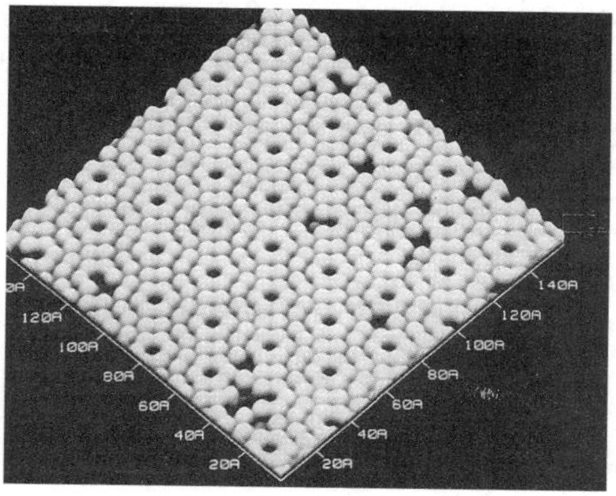

Fig. 5.25: STM picture of a reconstructed Si surface: Si(111)-(7×7). The lattice constant is enlarged 7-fold with respect to the underlying (111) surface. From Ref. [132].

[50]STM stands for scanning tunnelling microscopy.
The tunnelling current between surface atoms and atoms of the scanning tip depends sensitively on their distance. When the tunnelling current is held constant an approximate image of the surface structure may be generated in this way [145].

reconstructed Si surface. Another instructive example is ice, for the surface of which structures similar to the liquid phase have been demonstrated [146]. In many cases the surface crystallography is a sensitive function of the gas phase. In the catalytic oxydation of CO on Pt it is even possible to observe oscillations of the surface structure [147] (see Section 6.10).
Two specific two–dimensional defects are also worthy of mention. These are firstly the translation boundary at which $\Theta = 0$, such as, for instance, the stacking fault. As all higher–dimensional defects, stacking faults can be thought of as being made up of strongly correlated point defects. The energy gained by these correlations can be so high that these arrangements may be energetically favoured under certain conditions (see Fig. 5.109 in Section 5.7.1) — similarly to the shear surfaces of transition metal oxides (cf. Fig. 5.64 in Section 5.7). Here, the term "defect" looses its meaning. A special energetic role is also played by domain boundaries, that are separate regions differing in orientation with respect to specific properties, such as the boundaries between ferroelectric domains in $BaTiO_3$, across which the electrical polarization changes its vector. In this case the energetics are so favourable that domain boundaries with temperature–dependent structure and concentration (that is, separation) occur at equilibrium. Again, we ought to speak of superstructure elements rather than defects.

5.4.4 Interfacial thermodynamics and local mechanical equilibria

Some aspects of interfacial thermodynamics will now be addressed. For simplicity we will neglect anisotropy effects[51] (except for the consideration of equilibrium shape). The free interfacial enthalpy of the interface (of area a) can be characterized by defining an excess parameter G^{Σ}, which represents the excess of G[52,53] over the values obtained by extrapolating the bulk values up to the geometrical separation plane between the two phases (grains). This difference is the (quasi–static) reaction quantity for the process

$$\text{solid without interface} \rightarrow \text{solid with interface.}$$

(The reaction is equivalent to Eq. (5.57) for a grain boundary. Cf. also Fig. 5.19 in this context.) This value is (apart from exotic exceptions) positive.

[51] A more detailed treatment is given in Refs. [87,92,148,149].

[52] Such a splitting, as recommended by Gibbs, is useful, generally, if the parameters are additive and one is not interested in the local resolution.

[53] The more detailed consideration identifies the interfacial tension with $\int_{-\infty}^{+\infty} (p_N - p_T)\,dz$ where z is the coordinate normal to the layer [150]. The integrand is the difference between the normal and tangential component of the pressure tensor. From this definition of γ, we immediately obtain the differential mechanical work of deformation of a fluid phase as $-pdV + \gamma da$. We will use the Gibbs treatment in what follows.

The parameter most relevant for the formal treatment is the interfacial tension γ[51,53,54,55]. It describes the increase of free enthalpy[56] (p, T, **n** constant) on increase of the interfacial area

$$\gamma \equiv \left.\frac{\partial \mathrm{G}}{\partial \mathrm{a}}\right)_{\mathrm{p,T,\mathbf{n}}}, \tag{5.66}$$

so that the total free enthalpy of the (isotropic) solid is represented in differential form[56] as (cf. also Section 4.2)

$$\mathrm{dG} = \mathrm{Vdp} - \mathrm{SdT} + \boldsymbol{\mu}\mathrm{d}\mathbf{n} + \gamma\mathrm{da}. \tag{5.67}$$

By subtraction[57] of the G–contribution of the bulk the change of free surface enthalpy is (S^{Σ}, $\mathbf{n}^{\Sigma}$ are defined analogously to G^{Σ}) obtained as

$$\mathrm{dG}^{\Sigma} = -\mathrm{S}^{\Sigma}\mathrm{dT} + \boldsymbol{\mu}\mathrm{d}\mathbf{n}^{\Sigma} + \gamma\mathrm{da} \tag{5.68}$$

which yields

$$\gamma = \left.\frac{\partial \mathrm{G}^{\Sigma}}{\partial \mathrm{a}}\right)_{\mathrm{T},\mathbf{n}^{\Sigma}}. \tag{5.69}$$

While Eq. (5.66) defines γ as the change in free enthalpy of the whole system at constant mole numbers, Eq. (5.69) represents γ as the change in excess Gibbs energy at constant excess mole number[58] of the surface region. The possibility of integrating Eq. (5.68) at constant $\boldsymbol{\mu}$, γ and T yields

$$\mathrm{G}^{\Sigma} = \boldsymbol{\mu}\mathbf{n}^{\Sigma} + \gamma\mathrm{a}. \tag{5.70}$$

According to Eq. (5.70) γ can be only identified with the surface–related excess Gibbs energy ($\mathrm{G}^{\Sigma}/\mathrm{a}$) if we ignore effects caused by changes in the mole number at the cost of the bulk composition. The validity of Eq. (5.70) is equivalent to the

[54]In an analogous manner, it is necessary to define a line tension which is relevant for proper one–dimensional defects, crystal edges and generally boundaries of two–dimensional defects. The treatment is analogous; the area element is replaced by a line element. Analogously corners cause point tensions.

[55]A distinction is made between interfacial tension (γ) and interfacial stress. While the first (discussed here) term characterizes the work necessary to form a new interface, the latter refers to the work required to deform an interface. The difference between the two parameters involves the dependence of the interfacial tension of the components of the strain tensor.

[56]Note that — as far as the free energy effect is concerned — γ plays a similar role with respect to the area as p and $\boldsymbol{\mu}$ do with respect to V and **n**. The asymmetrical appearance of p and γ in Eq. (5.67) is a consequence of the definition of G (relevant state function for p, T). In the free energy the symmetry is present. Note also that in the definition used here dG differs from dF only by the volume work term; another, more general definition of a free enthalpy includes surface terms a priori ($\mathrm{F} + \Sigma_\alpha \mathrm{p}_\alpha \mathrm{V}_\alpha - \gamma\mathrm{a}$).

[57]Let $\mathrm{dG}^{\alpha,\beta} = \mathrm{V}^{\alpha,\beta}\mathrm{dp} - \mathrm{S}^{\alpha,\beta}\mathrm{dT} + \boldsymbol{\mu}\mathrm{d}\mathbf{n}^{\alpha,\beta}$ be the change in free enthalpy of the pure phase α or β; the bulk term cancels by definition ($\mathrm{V}^{\Sigma} = 0$). Note that $\boldsymbol{\mu}$, **n** represent the vectors in the composition space (and not in position space).

[58]In contrast to $\mathrm{n_k}$ it is possible for $\mathrm{n_k^{\Sigma}}$ to be either positive or negative.

existence of a Gibbs–Duhem equation[59]. Comparison with Eq. (5.68) leads to the Gibbs adsorption isotherm for a given component, e.g. component 1:

$$\left(\frac{\mathrm{d}\gamma}{\mathrm{d}\mu_1}\right)_{\mu_{k\neq 1}} = -\frac{n_1^{\Sigma}}{a}. \qquad (5.71)$$

Equation (5.71) directly refers to the adsorption effects at the expense of the bulk composition. Adsorption effects that merely correspond to charge separation and that are compensated by space charges, are not included in $\mathbf{n}^{\Sigma}$ in Eq. (5.71), since, by definition, the whole space charge zones are part of the interface. Therefore, it is sufficient to consider neutral components.

This changes, however, if electrical potential differences are relevant. In such cases, it is necessary to include the individual ions — and, hence, their electrochemical potentials — in the calculation [151–153]. If this electrochemical potential is split into chemical and electrical potential terms (Section 4.3.6), it can be seen that changes in the interfacial tension ($\mathrm{d}\gamma$) are not only caused by changes in chemical potential ($\mathrm{d}\mu$, cf. Eq. 5.71)) but also by changes in electrical potential ($\mathrm{d}\phi$). The dependence of the interfacial tension (morphology!) on the electrical potential is a phenomenon known as electrocapillarity. It can be demonstrated in a straightforward way [148] that the interfacial tension of the contact of an ideally polarizable[60] electrode with an electrolyte depends on the potential U (the counterelectrode is assumed to be ideally nonpolarizable[60]) according to the Lippmann relationship:

$$-(\partial\gamma/\partial U)_{P,T,\boldsymbol{\mu}} = Q_E/a_E. \qquad (5.72)$$

Q_E/a_E is the surface charge density of the electrode. Note that this relationship only applies if the chemical potentials are kept constant. It should be emphasized, in particular, that the interfacial tension reaches an extremum for a voltage at which charging vanishes[61]. The fact that the extremum constitutes a maximum is reasonable mechanistically, since, independent of the sign, charging makes enlargement of the surface easier on account of the lateral repulsion effect, i.e. γ is reduced.

Until now we have limited ourselves to planar interfaces. In fact the free enthalpy of a system is dependent on size and curvature in several ways. First of all, on account of the greater surface area, an ensemble of smaller crystals has a greater surface energy than an ensemble of larger crystals of the same total mole number, same composition and comparable shape. For this reason large crystals grow at the expense of small ones (Ostwald ripening) and for the same reason there is an activation threshold to be overcome (nucleation) in the process of phase formation

[59] Consider the difference between the differential of Eqs. (5.70) and (5.68) (cf. page 81).

[60] In the case of an ideally polarizable electrode, the transfer resistance of the charge carrier through the interface is infinitely high (the classical example from electrochemistry is the interface between Hg and an inert aqueous electrolyte such as KCl or sulfuric acid); in the case of the ideally nonpolarizable electrode, in contrast, it is zero (approximately for Ag/AgCl) (cf. Chapter 7).

[61] This does not mean that any double–layer purely on the electrolyte side must disappear (see Section 5.8).

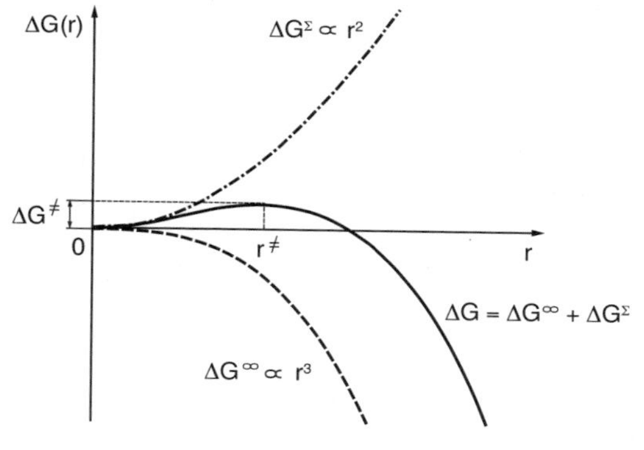

Fig. 5.26: The reduction in the free enthalpy on formation of a new bulk phase contrasts with the positive interfacial contribution, which leads to an activation threshold for a nucleus of critical radius $r^{\neq}$.

as displayed schematically in Fig. 5.26. The following simple consideration shows this: First we neglect adsorption effects and suppose the thermodynamic conditions for the precipitation of a macroscopic phase to be fulfilled, i.e. the macroscopic free energy of formation ΔG^{∞} in Fig. 5.26 to be negative. As a pure bulk term this parameter falls with the third power of the grain radius. The positive surface component ΔG^{Σ} is proportional to r^2 (see Eq. (5.70)). This leads to an activation threshold at the critical radius[62] $r^{\neq}$. In addition, γ is also a function of the curvature at very small radii and, in mechanical equilibrium, the pressures of two neighbouring phases separated by a curved interface, are different. The complications that arise with regard to the thermodynamic relationships and the phase rule[63] will not be discussed here. Ref. [149] offers a very competent presentation.

It is immediately evident that the equilibrium form of a (deformable) amorphous system is spherical; this follows from G^{Σ} being positive and from a sphere being the geometrical body with the smallest surface to volume ratio. What, however, is the equilibrium shape of an ionic crystal? This is a more complex problem, since the surface tension depends on the crystallography and, hence, on the orientation. From experience and intuition it is clear that the equilibrium form is a polyhedron. Such an equilibrium form is illustrated in Fig. 5.27. There we take the centre of the polyhedron as the origin and designate the distances to the surfaces as $h_{\mathbf{a}}$, with $\gamma_{\mathbf{a}}$ being the surface tension of the surface associated with the normal vector $\mathbf{a}$. It can then be shown that the ratio $\gamma_{\mathbf{a}}/h_{\mathbf{a}}$ is a constant. This leads to what is known as the Wulff construction[64] of the equilibrium form. Starting from the origin we draw in the vectors assigned to the crystallographic directions. The length is chosen to be

[62]Strictly speaking explicit kinetic considerations must be used in calculation of the parameters of activation.

[63]When two phases are in planar contact with χ components then, according to Chapter 4, the number of degrees of freedom (variance) is: variance $= 2 + \chi - 2 = \chi$. It holds true that $T^{\alpha} = T^{\beta}$, $\boldsymbol{\mu}^{\alpha} = \boldsymbol{\mu}^{\beta}$ and $p^{\alpha} = p^{\beta}$. When the interface is curved the latter equality no longer applies, rather variance $= 3 + \chi - 2 = \chi + 1$.

[64]The Wulff–construction follows from minimization of the free surface energy at constant volume. A straight forward derivation takes into account that volume and area are homogeneous functions of degree 3 and 2 in the position coordinates, respectively [154,155]. (Cf. footnote 22 of

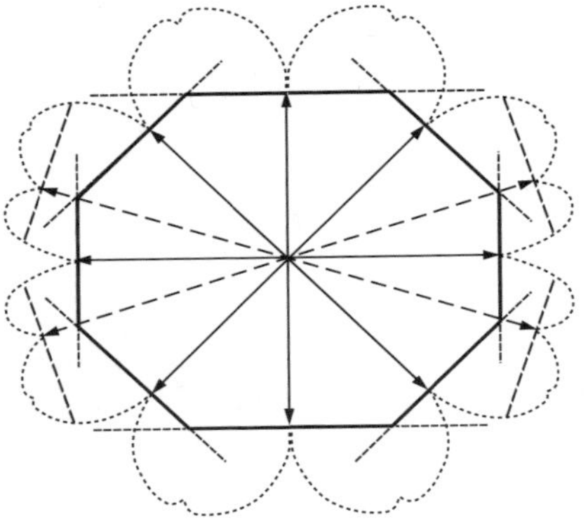

Fig. 5.27: The Wulff construction [154] for the generation of the equilibrium shape of a crystal (see text). The dotted lines represent schematically the direction–dependent surface tension, the thick lines the normals (normal planes) to the direction vector the magnitude of which is determined by the surface tension of low–energy surfaces. The inner envelopes (not dashed parts) represents the equilibrium shape.

proportional to the associated γ values. (In Fig. 5.27 the proportionality constant is set to unity, and the normal planes are set up at the end points.) Figure 5.28 shows such an equilibrium morphology for Si_3N_4 obtained in a computer simulation [157].

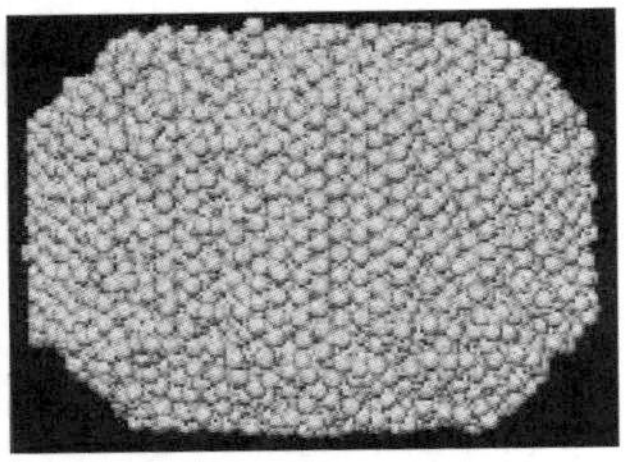

Fig. 5.28: Molecular dynamic simulation of the Wulff shape of a small Si_3N_4 crystal. According to Ref. [157].

In Refs. [148,149] it is shown that the chemical potential of small Wulff–crystals (MX) μ_{MX} differs from the value μ_{MX}^{∞} for large crystals according to

$$\mu_{MX} = \mu_{MX}^{\infty} + \frac{2\gamma_{\mathbf{a}}}{h_{\mathbf{a}}} V_m \qquad (5.73)$$

(V_m: molar volume). Equation (5.73) explains e.g. the increased vapour pressure and the decreased melting point of tiny crystals.
The ratio $\gamma_{\mathbf{a}}/h_{\mathbf{a}}$ can advantageously be expressed by averaged quantities[65], as $\bar{\gamma}/\bar{r}$, where $\bar{\gamma} = \Sigma_j a_j \gamma_j / \Sigma_j a_j$ and $\bar{r} = \Sigma_j a_j h_j / \Sigma_j a_j$ with j being a running index of the equilibrium planes. Note that $\Sigma_j a_j \gamma_j / a = (\gamma_j/h_j)\Sigma_j h_j a_j / a$ on account of Wulff's theorem. For spherical bodies (fluid phases) the well–known Kelvin–equation ($\bar{\gamma}/\bar{r} = \gamma/r$) follows from Eq. (5.73).
On account of the unfavourable kinetics such a treatment is only relevant for small crystals and high temperatures[66]. The morphology is generally under kinetic control

page 80: G treated as a homogeneous function of first degree in the mole numbers.) Not only has it to be modified for very small crystals (see footnote 65) but also if substrates are present [156].

65 Please note, that $\bar{r}$ is given by 3V/a, as $\frac{1}{3}a_j h_j$ represents the volume of the partial pyramid defined by crystal plane and origin (see Fig. 5.27). The ensemble of all partial pyramids forms the crystal volume (V).

66 In the case of small crystals, line tensions are already of importance and the equilibrium shape differs from the Wulff form. Equation (5.73) has then to be generalized accordingly [149]. This

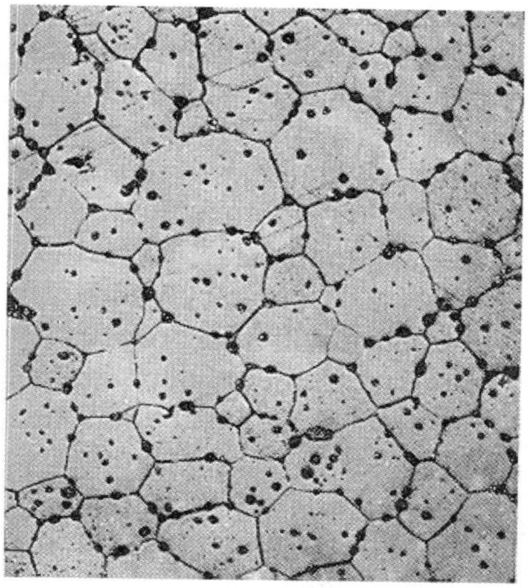

Fig. 5.29: Micrograph of a CaF_2 ceramic. The angle of contact of three grains is roughly 120°. The differing sizes of the grains lead to curvatures, which can be regarded as a driving force for further homogenization [134]. From Ref. [133].

(see Fig. 5.29). Nevertheless, as mentioned and illustrated in Figs. 5.23c,d highly energetic interfaces can re–arrange locally by faceting. The kinetic requirement is tolerable here, and because of the decreased surface tension there is an energy gain in spite of the formation of an increased surface area. Now we will concern ourselves with some examples important for questions of local morphology. For simplicity, we will ignore orientation dependences.

In Fig. 5.30 we consider the contact of three phases α, β, γ with three corresponding interfacial tensions $\gamma_{\alpha\beta}, \gamma_{\alpha\gamma}, \gamma_{\beta\gamma}$. The force equilibrium[67] requires

$$\gamma_{\alpha\beta}\mathbf{e}_{\alpha\beta} + \gamma_{\alpha\gamma}\mathbf{e}_{\alpha\gamma} + \gamma_{\beta\gamma}\mathbf{e}_{\beta\gamma} = \mathbf{0} \tag{5.75}$$

with the unit vectors $\mathbf{e}_{\alpha\beta}, \mathbf{e}_{\alpha\gamma}, \mathbf{e}_{\beta\gamma}$. The unit vectors are the tangential vectors perpendicular to the common contact line of the three phases (perpendicular to the plane of the drawing).

The Young equation is a well–known special case of Eq. (5.75) valid for the contact of an amorphous, deformable phase (l) with a rigid phase (s) and a gas phase (g)

phenomenological argument builds a bridge to the equilibrium shapes of cluster chemistry (see also Fig. 2.4).

[67]Neumann's formulation (Eq. (5.75)) [149] is equivalent to the Gibbs presentation $\Sigma_j \gamma_j dl_j = 0$ whereby a sum is also made over all interfaces and dl_j represents the propagation of the lines of intersection along the interface j [158]. It can, of course, also be derived from energy considerations [149,158]. In the case of orientation dependence the tendency of the interface to contract is supplemented by a tendency to rotate to a favourable orientation, and the general Herring relationship must be obeyed [159]

$$\sum_{j=1}^{3} \gamma_j \mathbf{e}_j + \sum_{j=1}^{3} \frac{\partial \gamma_j}{\partial \alpha_j} \mathbf{e}_j \times \mathbf{n} = \mathbf{0}; \tag{5.74}$$

here $\mathbf{n}$ is the unit vector along the line of contact ($\perp \mathbf{e}_j$, in Fig. 5.30 it stands up out of the plane of the paper), the angle α_j measures the crystallographical orientation of the boundary j.

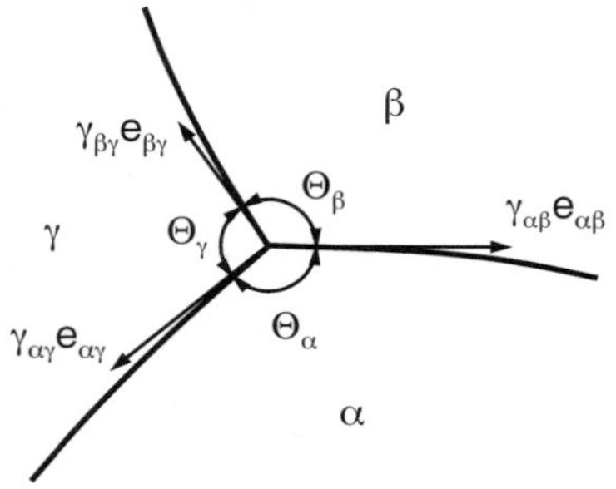

Fig. 5.30: The mechanical equilibrium of a three–phase contact.

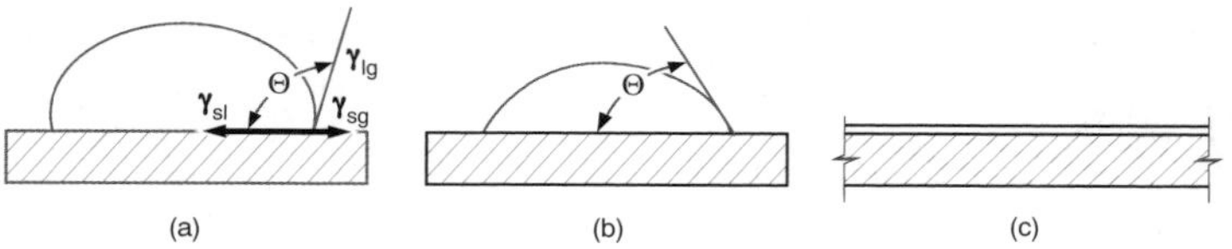

Fig. 5.31: Contact angle Θ of a liquid drop on a solid for varying wettability. For simplicity $\boldsymbol{\gamma}_{\alpha\beta}$ stands for $\gamma_{\alpha\beta}\mathbf{e}_{\alpha\beta}$.

(as illustrated in Fig. 5.31). Equation (5.75) then simplifies to

$$\gamma_{\mathrm{sg}} = \gamma_{\mathrm{sl}} + \gamma_{\mathrm{lg}}\cos\Theta, \tag{5.76}$$

which defines the situation completely by means of the contact angle Θ. If γ_{sl} is reduced — for given γ_{sg} and γ_{lg} values — then Θ decreases. Presupposing similar interfacial tensions of the two condensed phases with respect to the gas phase, the contact angle Θ eventually reaches zero (Fig. 5.31c) and wetting is optimal. If $\gamma_{\mathrm{sl}} \simeq \gamma_{\mathrm{sg}}$, then $\Theta \simeq 90°$. If γ_{sl} is much greater than γ_{sg} and γ_{lg}, Θ becomes very large, the wetting very poor and the s/l contact is avoided as much as possible (Fig. 5.31a).

In turn wetting experiments make it possible to determine interfacial tensions as a function of temperature and to separate the enthalpy and entropy contributions. (A concrete example is considered in Section 5.8.5.) When γ_{sg} values are compared with calculated values, account must be taken of the fact that adsorption processes can play an important role in real atmospheres (when the experiments are carried out in vacuum, there is the danger of the preferential loss of components).

Equation (5.76) is also able to provide a qualitative explanation for the formation of islands that are sometimes observed in the formation of thin films on substrates. Let us consider, for instance, a very thin glass film fabricated under metastable conditions and assume a pronounced γ_{sl}–value (corresponding to an appreciable contact angle Θ under local equilibrium conditions). If the viscosity is such that a single drop, with a contact angle Θ, cannot form for kinetic reasons but local re–arrangements may well be possible, then since the mass is constant separate islands (exhibiting this angle) will be formed.

Equation (5.75) is of great usefulness, above all, when discussing grain morphology within a polycrystalline structure. If the orientation dependence of γ is ignored, it says that for a contact of three grains the equilibrium angle is 120° and for a contact of four grains it is 109.5° (see Fig. 5.30). In fact, hexagon–like grain distributions are frequently observed in micro–graphs. A polyhedron, that approximately fulfils the conditions described and which fills the space in 3 dimensions [160] is the fourteen–faced solid shown in Fig. 5.32.

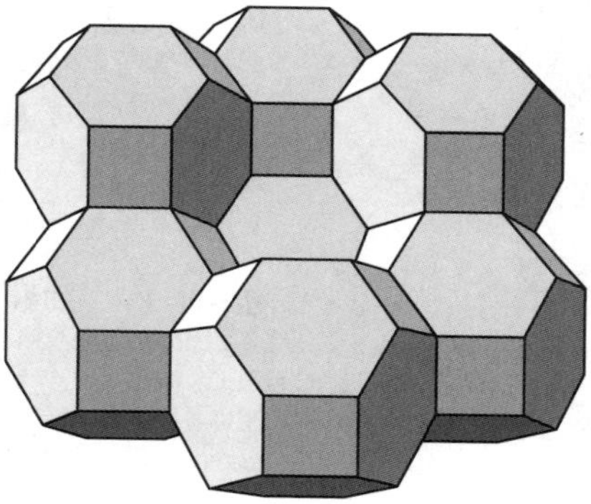

Fig. 5.32: The space–filling tetrakaidecahedron approximately fulfils the criterion of an ideal triple–grain contact [160]. From Ref. [131].

The relevant technical process which is of enormous importance and which makes use of local equilibration, is the process of sintering. Here, the individual grains are "bonded" together into a mechanically stable solid, the aim being to produce a microstructure with as low an energy as possible. The micrograph of a CaF_2 ceramic shown in Fig. 5.29 may serve as a typical example. The local triple–grain contact angle of the structure is close to 120°, even though by no means all the grains exhibit the equilibrium[68] shape. The differing grain sizes lead to the interfaces being curved; this interfacial curvature can be considered to be a driving force of progressive homogenization on further annealing. Pores and solid phase inclusions are also to be seen in Fig. 5.29.

Figure 5.33a illustrates three copper grains in contact on the way to "equilibrium" morphology, and Fig. 5.33b the molecular dynamic simulation of the analogous process for Si_3N_4. The resulting contact angles are almost exactly 120°. The sinter neck (Figs. 5.33a,b), that is forming, is a characteristic microstructural feature, as is the form of the triple–grain junctions in the centre between the three grains (see Fig. 5.33c for ZnO). The latter is (energetically) a preferred "site of retreat" for condensed impurity phases (inclusions) or gas phases (pores). This process also follows (if kinetically possible) the interfacial thermodynamics.

Figure 5.34 presents various situations, from the case of a (amorphous) second phase which wets extremely well (Figs. 5.34f and 5.33c) to its absolute opposite (Fig. 5.34a), the complete separation of the second phase into the triple–grain junction. While the surface tension is accessible from wetting experiments (see Eq. (5.76)), the grain boundary tension can be determined by measuring the grooving angle of the grain boundary trench (see Fig. 5.35). The relevant relationship follows from

[68]The reader should keep in mind that also the equilibrium shape referred to is not a state of global equilibrium which would be the single crystal.

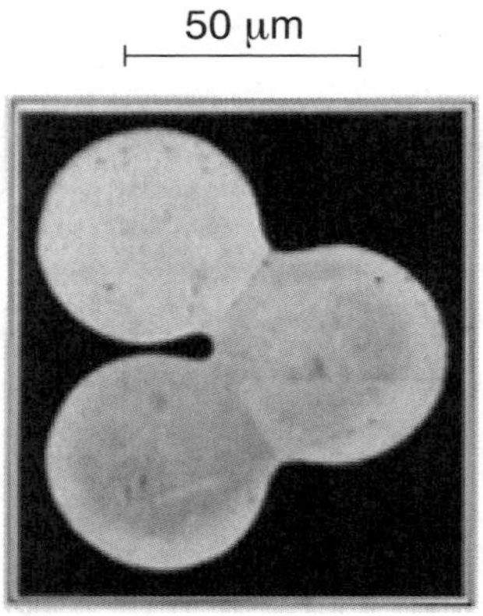

Fig. 5.33a: Three copper grains, sintered at 1300 K for 8h, on the way to an ideal triple–grain junction [161]. From Ref. [162].

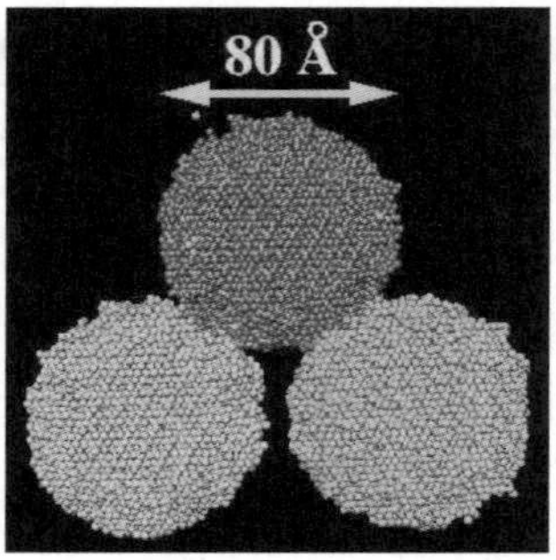

Fig. 5.33b: Molecular dynamic simulation of the sintering of three Si_3N_4 grains [157]. From Ref. [163].

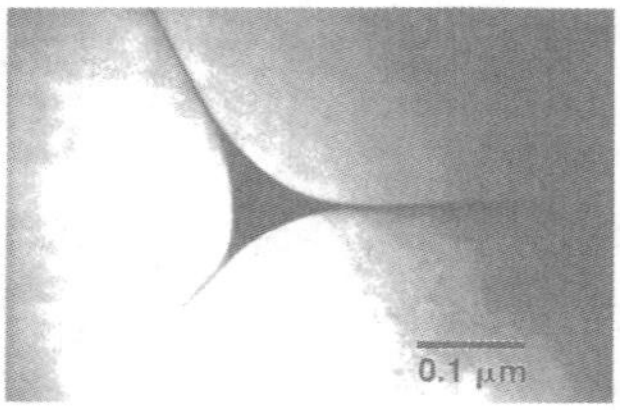

Fig. 5.33c: Triple–grain junction in ZnO with vanishing dihedral angle (cf. Fig. 5.34). The grain boundary phase is rich in bismuth oxide, wets extremely well and is essential for the varistor properties (see Section 5.8). From Ref. [164].

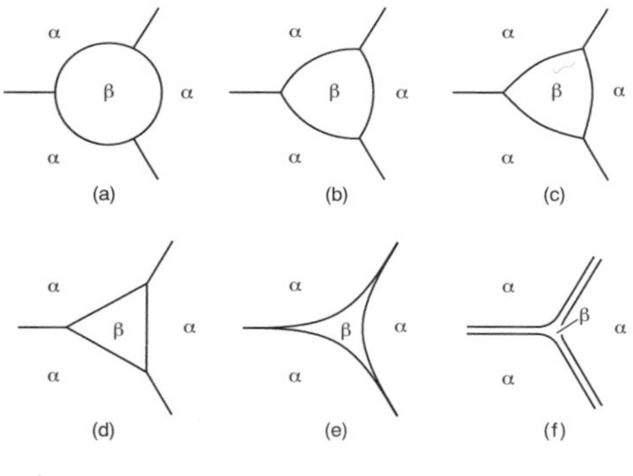

Fig. 5.34: The morphology of an inclusion as a function of the interfacial tension between the grain phase and the phase in the triple–grain junction. It can be characterized by means of the dihedral angle. This is defined by the internal angle of the tangents to the α–β boundaries at the point of intersection and varies between 180° and 0° (cf. also Fig. 5.33c). Diagrams a to f illustrate the characteristic morphologies corresponding to angles of 180° (a), 135° (b), 90° (c), 60° (d), 30 (e) and 0° (f). From Ref. [165].

Eq. (5.75) as

$$\gamma_{\text{grain,grain}} = 2\gamma_{\text{grain,air}}\cos\frac{\psi}{2} \tag{5.77}$$

(again neglecting orientation effects)[69].
The purely qualitative fact that the surface cannot, at local equilibrium, be planar at sites where the grain boundaries intersect with the surface, is important for grain size determination via micrographs (thermal etching). Special etching methods for visualization of the contact morphology, depend on the increased reactivity and solubility of grain boundary regions on account of the higher local free enthalpy.

[69]If the third medium is not the gas phase, but a grain of the same composition, then from Eq. (5.77) we obtain the result ($\cos(\psi/2) = 1/2, \psi = 120°$) for triple–grain contact already used.

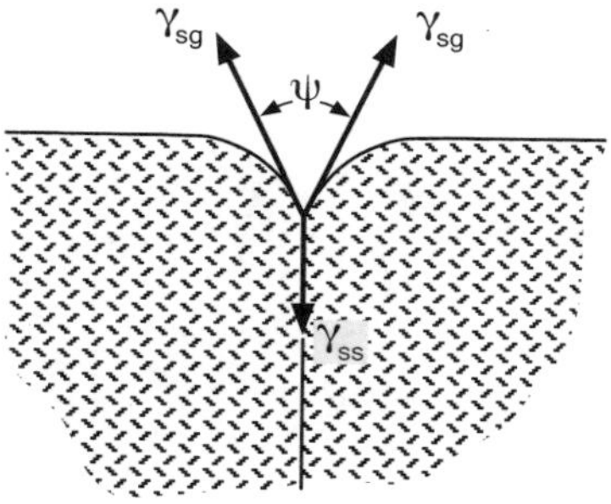

Fig. 5.35: The grooving angle of the grain boundary is a function of the grain boundary interfacial tension and of the surface tension [138].

Now, after this excursion, let us focus on point defects, which are the heart of the text. We will return in Section 5.8 to higher–dimensional defects in the form of "boundary conditions" for the distribution of point defects.

5.5 Point defect reactions

5.5.1 Simple internal defect equilibria

We have previously established (i) that, at finite temperatures, ionic and electronic point defects are required as local chemical excitations at equilibrium, (ii) that we can write ideal mass action laws in all cases for low concentrations of defects, and (iii) that we know what parameters influence our mass action constants. We can now turn to a specific consideration of defect chemistry. Let us consider first internal defect reactions and pure single crystals: By internal defect reactions in pure substances we mean processes that occur as a consequence of nonzero temperature in the otherwise perfect crystal without neighbouring phases being involved. (For two of these reaction types, however, we will need surfaces as sinks or sources of monomeric units, i.e. of lattice molecules.) In binary systems such processes leave the composition within the solid unchanged. If we refer to the Dalton composition[70], we also speak of the intrinsic case.

The first basic disorder type to be discussed is the Frenkel disorder. Here a few cations have left their regular positions, on account of the favourable configurational entropy (thermal influence), thus leaving behind vacancies, and now occupy interstitial sites. Such defects typically occur in the cationic sublattice and preferentially in cases of high polarizability, such as in the Ag^+ sublattices of the silver halides AgCl, AgBr and AgI (see Section 5.2).

[70]The composition of the perfect solid is meant here, e.g. $AgCl_{1+\epsilon}$ with ϵ strictly zero. We use this designation on account of the historic controversy between Dalton and Berthollet concerning the validity of the principle of constant proportions. Even in the case of ionic solids, Dalton, the protagonist of the principle, was only approximately correct. The composition can be changed via interaction with neighbouring phases (see Section 5.5.2).

In absolute notation we obtain for AgCl[71]:

$$\left[\begin{array}{cccc} Ag^+ & Cl^- & Ag^+ & Cl^- \\ Cl^- & Ag^+ & Cl^- & Ag^+ \\ Ag^+ & Cl^- & Ag^+ & Cl^- \\ Cl^- & Ag^+ & Cl^- & Ag^+ \end{array}\right] \rightleftharpoons \left[\begin{array}{cccc} & Cl^- & Ag^+ & Cl^- \\ Cl^- & Ag^+ & Cl^- & Ag^+ \\ Ag^+ & Cl^- & \begin{array}{cc} Ag^+ & Cl^- \\ & Ag^+ \end{array} & \\ Cl^- & Ag^+ & Cl^- & Ag^+ \end{array}\right] \qquad (5.78)$$

or, if we only consider the marked and initially neutral crystal sections:

$$\left[\begin{array}{cc} Ag^+ & Cl^- \\ Cl^- & Ag^+ \end{array}\right]^0 + \left[\begin{array}{cc} Ag^+ & Cl^- \\ Cl^- & Ag^+ \end{array}\right]^0 \rightleftharpoons \left[\begin{array}{cc} & Cl^- \\ Cl^- & Ag^+ \end{array}\right]^- + \left[\begin{array}{cc} Ag^+ & Cl^- \\ & Ag^+ \\ Cl^- & Ag^+ \end{array}\right]^+ . \qquad (5.79)$$

We note that the crystal section containing the vacancy is now negatively charged, and that the one containing the interstitial defect is positively charged. (The formulation selected for Eq. (1.2b) (on page 16) is even more concise.) If we subtract the perfect AgCl structure from the real structure — i.e. in Eq. (5.78) we subtract[71] $(AgCl)_8$ on both sides and in Eq. (5.79) we subtract[71] $(AgCl)_4$ — we obtain the disorder reaction in the building element notation ($Ag^{\cdot}$: interstitial silver ion, $|Ag|'$: silver ion vacancy):

$$\mathrm{Nil} \rightleftharpoons Ag^{\cdot} + |Ag|' . \qquad (5.80)$$

The old–fashioned charge designations which are exclusively used for the defects, indicate particles and indicate the relative charge with respect to the perfect lattice. According to the arguments detailed in Section 5.2 we know that there is, at equilibrium, a mass action law of the form (F indicates Frenkel reaction)

$$[Ag^{\cdot}]\,[|Ag|'] = K_F(T) = \exp - \frac{\mu^{\circ}_{Ag^{\cdot}} + \mu^{\circ}_{|Ag'|}}{RT} = \exp \frac{\Delta_F S^{\circ}_m}{R} \exp - \frac{\Delta_F H^{\circ}_m}{RT} . \qquad (5.81)$$

Since the rest of this chapter is exclusively devoted to equilibrium concentrations, we can suppress the "equilibrium arc" (as done in Eq. (5.81)). Values for $\Delta_F S^{\circ}_m$ and $\Delta_F H^{\circ}_m$ are listed in Table 5.1 (on page 114).

The building element formulation (5.80) evidently lacks vividness, even though it is thermodynamically the most correct [166]: In accordance with the treatment in

[71]The presentation of Eq. (5.78) is schematic for different reasons. (i) The defect reaction corresponds to the formation of a well–separated "pair". (ii) In view of the precise crystallographic situation (Fig. 5.3a, page 112) it refers to a section parallel to a cube face into which the interstitial particle is projected. The crystallographically adequate description has to involve at least 4 AgCl units for the interstitial defect and at least 6 AgCl units for the vacancy defect (see Section 5.2 and legends to Figs. 5.3a, 5.3b).

Section 5.2 we conceive, in this description, the point defects as species, that we add to the perfect crystal; that is, they must be true relative elements. The more vivid representation [167] in the form of structure elements ($\vee$: vacancy, i: interstitial position, see Section 1.2), which now takes account of the actual structure of the real crystal in Eq. (5.78), but only includes the centres actually affected and defines the charge relative to the perfect lattice, reads[72]

$$\mathrm{Ag_{Ag}} + \mathrm{\vee_i} \rightleftharpoons \mathrm{Ag_i^{\cdot}} + \mathrm{\vee'_{Ag}}. \tag{5.82}$$

The re–arrangement

$$\mathrm{Nil} \rightleftharpoons (\mathrm{Ag_i^{\cdot}} - \mathrm{\vee_i}) + \left(\mathrm{\vee'_{Ag}} - \mathrm{Ag_{Ag}}\right) \tag{5.83}$$

and comparison with Eq. (5.80) reveals the connection. The combinations $(\mathrm{Ag_i^{\cdot}} - \mathrm{\vee_i})$ and $\left(\mathrm{\vee'_{Ag}} - \mathrm{Ag_{Ag}}\right)$ represent our building elements $\mathrm{Ag^{\cdot}}$ and $|\mathrm{Ag}|'$. They express the fact that it is only possible to create a vacancy ($\mathrm{\vee'_{Ag}}$) when a regular silver ion ($\mathrm{Ag_{Ag}}$) is removed at the same time, and that the occupation of an interstitial site (leading to $\mathrm{Ag_i^{\cdot}}$) requires annihilation of the free space there ($\mathrm{\vee_i}$). Strictly speaking, it is only possible to define thermodynamic potentials for such combinations; and only their concentrations then appear in the mass action laws. However, it is evident that the concentrations of structure elements and building elements are identical so that, in the end, we obtain Eq. (5.81) here in the form

$$\mathrm{K_F} = [\mathrm{Ag_i^{\cdot}}]\left[\mathrm{\vee'_{Ag}}\right]. \tag{5.84}$$

In other words — as long as the defect concentrations are small —, the correct result is obtained if we naively formulate mass action laws via the structure elements, i.e. assign (formal) activities for the individual structure elements and set those of regular species at 1 or think of them as being incorporated in K as a constant.

Until now, thermodynamics only allows us to make a statement concerning the product of the two concentrations. In order to determine the individual concentrations of $[\mathrm{Ag_i^{\cdot}}]$ and $\left[\mathrm{\vee'_{Ag}}\right]$, we require further information. This is provided by the electroneutrality condition of the crystal. In the case of pure substances it follows that $[\mathrm{Ag_i^{\cdot}}] = \left[\mathrm{\vee'_{Ag}}\right]$ must apply (locally) and, hence,

$$[\mathrm{Ag_i^{\cdot}}] = \left[\mathrm{\vee'_{Ag}}\right] = \mathrm{K_F(T)^{1/2}} = \exp\frac{\Delta_\mathrm{F}\mathrm{S^\circ_m}}{2\mathrm{R}}\exp-\frac{\Delta_\mathrm{F}\mathrm{H^\circ_m}}{2\mathrm{RT}}. \tag{5.85}$$

Figure 5.36 shows the dependence of defect concentration in pure AgCl in the van't Hoff presentation ($\Delta_\mathrm{F}\mathrm{H^\circ_m} = 140\mathrm{kJ/mol}$, $\Delta_\mathrm{F}\mathrm{S^\circ_m} = 9.4\mathrm{R}$, see Table 5.1, page 114). The slope is given by $-\Delta_\mathrm{F}\mathrm{H^0_m}/2\mathrm{R}$. One can see that the immense influence of the

[72] Thus, building elements are pure relative elements with respect to structure and charge (real crystal section minus perfect crystal section), while the above structural elements are absolute with respect to the structure, but relative with respect to the charge. It will be shown (Chapter 6), that the structure element notation is more particularly suitable for kinetic considerations. The whole solid is represented — concerning structure and charge — by the "sum" of the structure elements SE (perfect solid = $(\Sigma\mathrm{SE})_{\mathrm{perfect}}$, real solid = $(\Sigma\mathrm{SE})_{\mathrm{real}}$). The Σ–symbol here includes the structural arrangement. The "sum" of the defect building elements (BE) follows as: $\Sigma\mathrm{BE} = (\Sigma\mathrm{SE})_{\mathrm{real}} - (\Sigma\mathrm{SE})_{\mathrm{perfect}}$. A precise analysis of the notation is given in Refs. [5,122].

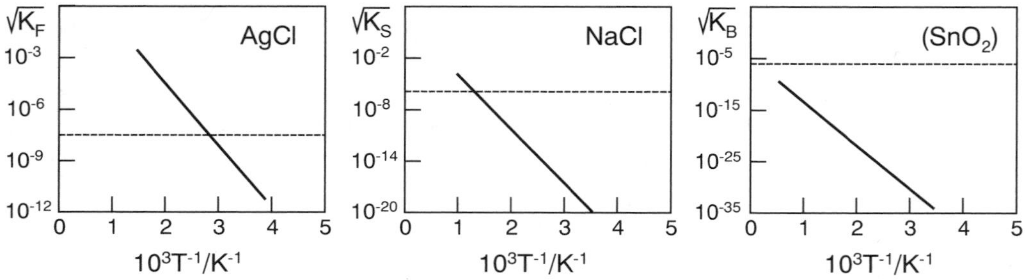

Fig. 5.36: The intrinsic concentration of the Frenkel defects in AgCl, Schottky defects in NaCl and electronic charge carriers in a material with a band gap of 3.5eV. Such a band gap is found in SnO_2. The value of the relevant equilibrium constant decreases from the left to the right. The broken line represents a typical lower impurity limit. When the temperature is not too high, impurity effects (Section 5.6) dominate. In addition, for the case of SnO_2 (r.h.s.) effects of nonstoichiometry (Section 5.5.2) have to be taken into account: Even in ideally pure SnO_2, an intrinsic electronic behaviour would require extreme oxygen partial pressures (that is why SnO_2 is bracketed). In AgCl and NaCl such nonstoichiometry effects are negligible due to the low redox variability.

Frenkel disorder enthalpy causes the defect concentration to increase steeply with temperature. One can also see, however, that the defect concentration at room temperature is so low that other effects, the so–called extrinsic effects, which are discussed in the coming Sections 5.5.2, 5.6 will certainty dominate here[73]. The analogy with aqueous chemistry is also evident: There $K^{1/2}$ corresponds to the intrinsic proton or hydroxide ion concentration; and also there, the charge carrier effects are generally extrinsically controlled at room temperature.

The point defect formation according to the Schottky reaction (formation of "Schottky pair") can be treated analogously:

$$\begin{bmatrix} Na^+ & Cl^- \\ Cl^- & Na^+ \end{bmatrix} + \begin{bmatrix} Na^+ & Cl^- \\ Cl^- & Na^+ \end{bmatrix} \rightleftharpoons \begin{bmatrix} & Cl^- \\ Cl^- & Na^+ \end{bmatrix}^- + \begin{bmatrix} Na^+ & Cl^- \\ & Na^+ \end{bmatrix}^+ + NaCl \qquad (5.86)$$

Here both a cation and anion vacancy are created. This is done by the generation of a monomeric unit which we consider to be placed at a surface site of repeatable growth (half crystal position in Fig. 5.24). In the building element notation this reads

$$\mathrm{Nil} \rightleftharpoons |\mathrm{Na}|' + |\mathrm{Cl}|^{\cdot} + \mathrm{NaCl} \qquad (5.87)$$

and in the structure element notation

$$\mathrm{Na_{Na}} + \mathrm{Cl_{Cl}} \rightleftharpoons \mathrm{V'_{Na}} + \mathrm{V^{\cdot}_{Cl}} + \mathrm{NaCl} \qquad (5.88)$$

[73]Impurities are meant here (see Section 5.6). In AgCl or NaCl the equilibrium with the neighbouring phase (treated in Section 5.5.2) significantly affects — relatively speaking — only the electrons as minority defects, whose concentration is small with respect to that of the silver defects. However, the latter is relevant for the SnO_2 example in Fig. 5.36.

with the mass action law

$$[\mathsf{V}'_{\mathrm{Na}}][\mathsf{V}^{\cdot}_{\mathrm{Cl}}] = \mathrm{K_S(T)} = \exp\frac{\Delta_\mathrm{S}\mathrm{S}^\circ_\mathrm{m}}{\mathrm{R}} \exp -\frac{\Delta_\mathrm{S}\mathrm{H}^\circ_\mathrm{m}}{\mathrm{RT}}. \tag{5.89}$$

Using the experimental values of $\Delta_\mathrm{S}\mathrm{H}^\circ_\mathrm{m} = 240\mathrm{kJ/mol}$ and $\Delta_\mathrm{S}\mathrm{S}^\circ_\mathrm{m} = 9.8\mathrm{R}$ (Table 5.1, page 114), the ionic defect concentrations in the pure material ($[\mathsf{V}'_{\mathrm{Na}}] = [\mathsf{V}^{\cdot}_{\mathrm{Cl}}] = \sqrt{\mathrm{K_S}}$) follow, as given in Fig. 5.36. Owing to the lower value of the mass action constant, the window within which the carrier concentration is intrinsically determined is shifted toward much higher temperatures than in the case of the previous example (given a similar impurity content). Even if we ignore the coarse–grained nature of the crystal structure, the Schottky reaction is not (like the anti–Schottky reaction discussed below and unlike the Frenkel reaction) a truly homogeneous internal reaction, since the lattice molecule and thus surfaces (or other higher–dimensional defects such as interfaces or dislocations) are involved. This is important for the kinetic treatment.

The anti–Frenkel reaction ($\bar{\mathrm{F}}$) is a third fundamental, internal defect reaction: It describes a disorder variant analogous to type F but in the anionic sublattice (and is therefore also termed anion–Frenkel disorder). This typically occurs in alkaline earth halides, particularly the fluorides; here the anions are small enough (fluorides) or polarizable enough (in the case of the higher halides) to take up interstitial positions. In the building element notation and the structure element notation we formulate

Reaction $\bar{\mathrm{F}}$ =
$$\mathrm{Nil} \rightleftharpoons \mathrm{F}' + |\mathrm{F}|^{\cdot} \tag{5.90a}$$

Reaction $\bar{\mathrm{F}}$ =
$$\mathrm{F_F} + \mathsf{V}_\mathrm{i} \rightleftharpoons \mathrm{F}'_\mathrm{i} + \mathsf{V}^{\cdot}_\mathrm{F} \tag{5.90b}$$

respectively, with the mass action law

$$[\mathrm{F}'_\mathrm{i}]\,[\mathsf{V}^{\cdot}_\mathrm{F}] = \mathrm{K}_{\bar{\mathrm{F}}}(\mathrm{T}) = \exp\frac{\Delta_{\bar{\mathrm{F}}}\mathrm{S}^\circ_\mathrm{m}}{\mathrm{R}} \exp -\frac{\Delta_{\bar{\mathrm{F}}}\mathrm{H}^\circ_\mathrm{m}}{\mathrm{RT}}. \tag{5.91}$$

Naturally, the counterpart of the Schottky reaction exists as the last possibility, namely, the additional introduction of a monomeric unit (lattice molecule) into cationic and anionic interstitial positions. This anti–Schottky reaction requires an adequately flexible structure and is usually not a dominant disorder reaction. One exception might be yellow PbO, for which we then formulate:

Reaction $\bar{\mathrm{S}}$ =
$$\mathrm{PbO} \rightleftharpoons \mathrm{Pb}^{\cdot\cdot} + \mathrm{O}'' \tag{5.92a}$$

or

Reaction $\bar{\mathrm{S}}$ =
$$\mathrm{PbO} + 2\mathsf{V}_\mathrm{i} \rightleftharpoons \mathrm{Pb}^{\cdot\cdot}_\mathrm{i} + \mathrm{O}''_\mathrm{i}. \tag{5.92b}$$

(Strictly speaking, it is necessary to distinguish between the different interstitial positions in Eq. (5.92b).) The respective mass action law is

$$[\mathrm{Pb_i^{\cdot\cdot}}]\,[\mathrm{O_i''}] = \mathrm{K_{\bar{S}}} = \exp\frac{\Delta_{\bar{S}}\mathrm{S_m^\circ}}{\mathrm{R}}\exp-\frac{\Delta_{\bar{S}}\mathrm{H_m^\circ}}{\mathrm{RT}}. \tag{5.93}$$

We have now completed our consideration of simple, internal, ionic defect reactions. The mutual internal substitution (anti–site disorder) does not play a role in binary ionic crystals (M^+X^-). It is difficult to imagine defects of the form $M_X^{\cdot\cdot}$ or X_M'' in significant quantities at realistic temperatures. However, the significant occurrence of anti–site defects (see Section 5.2) is possible in less polar materials, such as GaAs, or in multinary materials, such as $YBa_2Cu_3O_{6+x}$, prepared at very high temperatures. Defects of the type $Y_{Ba}^{\cdot}$ are very probable in the latter and are of significant influence [121]. More complex defects, that occur as a result of the association of the defects discussed here, will be ignored for the moment but described in detail later.

In addition to the four ionic disorder reactions (acid–base reactions) discussed, there is the fundamental electronic disorder reaction treated in detail in Section 5.3 (redox reaction) to consider:

$$\text{Reaction B} = \qquad \mathrm{Nil} \rightleftharpoons \mathrm{e'} + \mathrm{h^{\cdot}}. \tag{5.94a}$$

In many binary, ionic crystals (M^+X^-), e.g. in oxides or halides with sufficient electronegativity difference, the valence band (VB) can be assigned to the X orbitals and the conduction band (CB) to the M orbitals (see Section 2.2), so that, in a localized picture, the following simplified presentation reproduces the reaction in absolute notation:

$$\begin{bmatrix} \mathrm{M^+} & \mathrm{X^-} & \mathrm{M^+} & \mathrm{X^-} \\ \mathrm{X^-} & \mathrm{M^+} & \mathrm{X^-} & \mathrm{M^+} \\ \mathrm{M^+} & \mathrm{X^-} & \mathrm{M^+} & \mathrm{X^-} \\ \mathrm{X^-} & \mathrm{M^+} & \mathrm{X^-} & \mathrm{M^+} \end{bmatrix} \rightleftharpoons \begin{bmatrix} \mathrm{M^+} & \mathrm{X^-} & \mathrm{M^0} & \mathrm{X^-} \\ \mathrm{X^-} & \mathrm{M^+} & \mathrm{X^-} & \mathrm{M^+} \\ \mathrm{M^+} & \mathrm{X^-} & \mathrm{M^+} & \mathrm{X^-} \\ \mathrm{X^0} & \mathrm{M^+} & \mathrm{X^-} & \mathrm{M^+} \end{bmatrix} \tag{5.94b}$$

or

$$(\mathrm{X^-_{X^-}}) + (\mathrm{M^+_{M^+}}) \rightleftharpoons (\mathrm{X^0_{X^-}})^{\cdot} + (\mathrm{M^0_{M^+}})'. \tag{5.94c}$$

After subtracting the perfect state, Eq. (5.94b) or Eq. (5.94c) yields Eq. (5.94a) once again. Comparison with Eq. (5.94a) reveals that the building elements $h^{\cdot}$ and e' correspond to the structure element combinations $(X_X^{\cdot} - X_X)$ and $(M_M' - M_M)$. Instead of Eq. (5.94c) we could also have formulated the transition from the valence band to the conduction band as follows:

$$(\mathrm{e_{VB}})^\times + (\mathrm{V_{CB}})^\times \rightleftharpoons (\mathrm{e_{CB}})' + (\mathrm{V_{VB}})^{\cdot} \tag{5.94d}$$

Like Eq. (5.94a) this description[37] would be independent of band structure details, but would lead to a double counting: As has been explained previously Eq. (5.94a)

is to be preferred, since the consideration of the ionic elements as structure elements already includes the elements $(e_{VB})^{\times}$ and $(V_{CB})^{\times}$.
Irrespective of the formulation, there is a mass action law

$$[e'][h^{\cdot}] = K_B = \exp\frac{\Delta_B S_m^{\circ}}{R}\exp-\frac{\Delta_B H_m^{\circ}}{RT}. \tag{5.95}$$

In Section 5.3 it is shown in more detail that, to a good approximation, $\Delta_B H^{\circ}$ corresponds to the band gap. According to our convention $\Delta_B S_m^{\circ}$ formally includes (approximately) constant terms, such as the effective density of states (cf. page 122). Figure 5.36c illustrates $\sqrt{K_B}$ for a solid with a band gap of 3.5eV; one can see that the electronic behaviour is dominated extrinsically even at higher temperatures. Intrinsic behaviour according to $[e'] = [h^{\cdot}] = \sqrt{K_B}$ can only be observed for very pure semiconductors with comparably small band gaps and at sufficiently high temperatures.

5.5.2 External defect equilibria

Table 5.2 lists the internal disorder reactions. All these defect reactions leave the M/X ratio unaltered.
However, every realistic treatment of defect chemistry of pure materials must take account of the fact that there are always slight deviations from the exact stoichiometric composition, i.e. that the defects also interact with the external world. This finds expression in the phase rule, whereby, at fixed temperature and constant partial pressure, all degrees of freedom are only utilized when there are as many phases present as components. Fixing the properties thermodynamically, thus, demands equilibration with a particular M or X activity and thus with a particular M or X_2 partial pressure (cf. also Section 4.3).

This can be generally achieved by bringing the sample into contact with the M mother material (or if existent, with the appropriate metal–rich thermodynamically compatible compound), or by contact with pure X_2 gas[74] (or with the correspondingly phase–compatible X–rich compound). Then the extreme stoichiometries corresponding to the highest possible and lowest possible M–content (lowest possible and highest possible X–content), respectively, are set up. When there is low disorder in the compound MX, M activity and P_{X_2} partial pressure are coupled by[75]

$$K_f^{-1} = P_{X_2}^{1/2} a_M, \tag{5.96}$$

the mass action law of

$$\text{Reaction f} = \quad \frac{1}{2}X_2 + M \rightleftharpoons MX(s). \tag{5.97}$$

[74]Setting a particular partial pressure (e.g. X_2/Ar mixtures) allows the tuning of the component potential.

[75]The standard pressure is included in the mass action constant.

Table 5.2: Simple defect chemistry of $M_{1+\delta}X$.

reaction (r)	mass action law
internal	
Schottky (S)	$[V_X^{\bullet}]\,[V_M'] = K_S$
Frenkel (F)	$[M_i^{\bullet}]\,[V_M'] = K_F$
Anti-Frenkel ($\bar{F}$)	$[V_X^{\bullet}]\,[X_i'] = K_{\bar{F}}$
Anti-Schottky ($\bar{S}$)	$[M_i^{\bullet}]\,[X_i'] = K_{\bar{S}}$
Band-Band (B)	$[h^{\bullet}]\,[e'] = K_B$
$K_r(T) \propto \exp + \frac{\Delta_r S_m^0}{R} \exp - \frac{\Delta_r H_m^0}{RT}$	
external	
reaction with the gas phase (X)	$P_{X_2}^{-1/2}\,[V_X^{\bullet}]^{-1}\,[e']^{-1} = K_X$
electroneutrality condition	
$[V_X^{\bullet}] + [M_i^{\bullet}] + [h^{\bullet}] = [V_M'] + [X_i'] + [e']\ (\pm C)$	

As demonstrated copiously in Section 4.3.5, Eq. (5.96) only applies if μ_{MX} is approximately constant, i.e. in the case of slight deviations from stoichiometry which do not affect the absolute value of the Gibbs energy $\left(\left|\frac{\mu_{MX}-\mu_{MX_{1+\delta}}}{\mu_{MX}^{\circ}}\right| \ll 1\right)$. When $\delta \ll 1$ this is well fulfilled in accordance with the Gibbs–Duhem relationship. This invariance does not hold for the components M and X (cf. Section 4.3.4). It is worth looking at Fig. 4.4 (page 90) once again at this point; it shows how immensely the chemical potential varies within the phase width, even though the change in composition is slight. In other words: The following treatment is concerned with the chemical changes occurring on passage through the phase width.

This variation in δ manifests itself in changes of the electronic and ionic point defect concentrations. We may expect that on passing through the phase width from left to right, i.e. at increasing oxygen potential, the concentration of oxygen vacancies decreases and the concentration of interstitial oxygen ions increases, the metal vacancy concentration rises, the interstitial metal concentration falls and, at the same time, the electron concentration falls and the number of holes increases on account of the oxydation. We will now discuss this situation quantitatively and take yellow lead (II) oxide as an example.

The interaction of the oxide PbO with oxygen can, for example, be formulated as the incorporation of oxygen interstitially. The oxygen incorporated sits in the lattice as O^{2-}, i.e. in our example as O_i'' and, therefore, conduction electrons e' are annihilated (formally low valence lead is oxydized to Pb^{2+}):

$$\frac{1}{2}O_2 + \mathsf{V}_i + 2e' \rightleftharpoons O_i''. \tag{5.98a}$$

Note that this equation (in contrast to the pure redox or acid–base reactions discussed above) represents both a redox reaction and an acid–base reaction. It hence couples electronic and ionic defect concentrations via the chemical potential of the components ($\mu_O = \widetilde{\mu}_{O^{2-}} - 2\widetilde{\mu}_{e^-}$). Note that this corresponds to the coupling of the electronic and ionic Fermi levels that is shown in Fig. 5.37. We could formulate the oxygen incorporation as

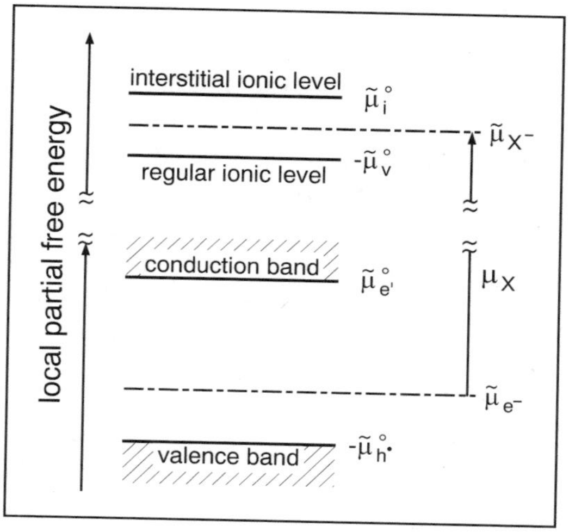

Fig. 5.37: Coupling of the ionic and the electronic energy level pictures for an anti–Frenkel disordered material M^+X^- through $\mu_X = \frac{1}{2}\mu_{X_2} = \widetilde{\mu}_{X^-} - \widetilde{\mu}_{e^-}$. According to the position of the electrochemical potentials the material is in the I–regime on the r.h.s. ($[h^\cdot] > [e']$, $[X_i'] \simeq [\mathsf{V}_X^\cdot]$) of the intrinsic point (cf. Fig. 5.38). Note that $\widetilde{\mu}_{e^-} = \widetilde{\mu}_{e'} = -\widetilde{\mu}_{h^\cdot}$ and $\widetilde{\mu}_{X^-} = \widetilde{\mu}_{X_i'} = -\widetilde{\mu}_{\mathsf{V}_X^\cdot}$, see Sections 5.2, 5.3). Instead of $\widetilde{\mu}_{X_i'}$ (or $\widetilde{\mu}_i$) and $\widetilde{\mu}_{\mathsf{V}_X^\cdot}$ (or $\widetilde{\mu}_\mathsf{V}$) it should more precisely read $\widetilde{\mu}_{X_i'-\mathsf{V}_i}$ and $\widetilde{\mu}_{\mathsf{V}_X^\cdot - X_X}$. (Cf. also Fig. 5.9 on page 127.) From Ref. [168].

$$\frac{1}{2}O_2 + \mathsf{V}_i \rightleftharpoons O_i'' + 2h^\cdot \tag{5.98b}$$

with equal justification. (The introduction of holes approximately corresponds to the oxydation of O^{2-} to O^-, i.e. to the consumption of bonding electrons). Equation (5.98b) does not bring in any new information for the equilibrium situation, but merely represents the coupling of Eq. (5.98a) with the band–band reaction (5.94a). Analogously, reaction (5.98c), according to which oxygen is brought to vacant sites, also has its justification:

$$\frac{1}{2}O_2 + \mathsf{V}_O^{\cdot\cdot} + 2e' \rightleftharpoons O_O. \tag{5.98c}$$

It corresponds to the coupling of reaction (5.98a) with the anti–Frenkel reaction in the form (cf. Eq. (5.90))

$$\mathrm{O_O + V_i \rightleftharpoons O_i'' + V_O^{\cdot\cdot}}. \tag{5.99}$$

In the same manner the coupling of Eq. (5.98a) with the anti–Schottky reaction (5.92) yields a relevant formulation, namely

$$\frac{1}{2}\mathrm{O_2 + Pb_i^{\cdot\cdot} + 2e' \rightleftharpoons PbO + V_i}. \tag{5.98d}$$

Moreover, it is possible to regard the interaction with Pb as a constituent of the neighbouring phase, in, for example, the form

$$\mathrm{Pb_{Pb} + 2e' \rightleftharpoons Pb + V_{Pb}''} \tag{5.100}$$

instead of the incorporation of oxygen. Here the coupling of Eq. (5.98d) with the equilibrium in the gas phase (Eq. (5.97)) is addressed.
To cut a long story short: Provided internal defect–chemical equilibrium is established, it is sufficient to formulate a single reaction, that expresses the interaction of the neighbouring phase with the bulk, e.g. Eq. (5.98c). Naturally, if we have some knowledge of the material we will prefer the formulation that involves the major defects, i.e. those encountered in the majority in the material.
When a specific binary material is considered, we will need to add to the list of all internal disorder reactions (which all occur in that material even though with differing weight) a single external one (which we have already included in Table 5.2, page 162). Without restriction of generality we will select for our model compound MX the formulation with anion vacancies and conduction electrons, i.e.

$$\text{Reaction X} = \qquad \frac{1}{2}\mathrm{X_2 + V_X^{\cdot} + e' \rightleftharpoons X_X}. \tag{5.101}$$

The mass action law reads[75]

$$\mathrm{P_{X_2}^{-1/2}\,[V_X^{\cdot}]^{-1}\,[e']^{-1} = K_X(T) = \exp\frac{\Delta_X S^\circ}{R}\exp-\frac{\Delta_X H^\circ}{RT}}. \tag{5.102}$$

In addition to the temperature, we hence have, a second parameter, the X_2 partial pressure (or the X or M activity). The last condition, we need to consider, is the requirement of electroneutrality

$$\mathrm{[X_i'] + [V_M'] + [e'] = [M_i^{\cdot}] + [V_X^{\cdot}] + [h^{\cdot}]}\,. \tag{5.103a}$$

Our remaining task is to calculate the defect concentration as a function of T and P_{X_2}. The mathematical difficulty is not in the multiplicity of mass action laws, but in the differing mathematical structures of the electroneutrality relationship and the mass action laws: After taking logarithms, all mass action laws are linear, in contrast to the electroneutrality condition Eq. (5.103a). In order to be able to

treat the problem analytically we must simplify Eq. (5.103a) sufficiently. But first let us count the variables. We have 6 independent defect concentrations, but 7 equations. This means that one of the internal ionic defect reactions is redundant. Thus, for instance, the anti–Schottky reaction is represented by combination of the first three ionic disorder reactions. It can be seen immediately that $K_{\bar{S}} = K_{\bar{F}}K_F/K_S$. Therefore, we can leave out one of the four — let us select $\bar{S}$.
Back to the electroneutrality relationship: As can be readily verified by substitution Eq. (5.103a) can be rewritten as

$$[X_i'] \, (K_S/K_{\bar{F}} + 1) + [e'] = [V_X^{\cdot}] \, (K_F/K_S + 1) + [h^{\cdot}] \, . \tag{5.103b}$$

In order to make the logarithmic form of Eq. (5.103b) linear it is necessary that one of the two additive terms on each side of the equation is negligible. This corresponds to the Brouwer approximation [169], which we will now discuss for a somewhat simplified disorder model and specifically for an oxide.
Let us assume that in the oxide $(M^+)_2(O^{2-})$ appreciable disorder only occurs on the anion sublattice[76]. This simplification made does not affect the validity of the approximation at all, it is only intended to simplify the considerations. We can now neglect the mass action constants $K_{\bar{F}}, K_B$ and K_O. The mass action laws[73]

$$K_{\bar{F}} = [O_i''] \, [V_O^{\cdot\cdot}] \tag{5.104}$$

$$K_B = [e'] \, [h^{\cdot}] \tag{5.105}$$

$$K_O = P_{O_2}^{-1/2} \, [e']^{-2} \, [V_O^{\cdot\cdot}]^{-1} \tag{5.106}$$

remain. The condition of electroneutrality reads

$$2 \, [O_i''] + [e'] = 2 \, [V_O^{\cdot\cdot}] + [h^{\cdot}] \, . \tag{5.107}$$

(A factor of 2 appears on account of the double charge of the oxygen ion.) In addition, we do not question, at the moment, the stability limits of the oxide or demand experimental achievability of the external conditions. At extremely low partial pressures of oxygen (let us call it the low (partial) pressure range or N–range) we will find minimal δ values in the oxide $M_2O_{1+\delta}$ $(-1 \ll \delta < 0)$ and there will be an O deficit. The concentration of oxygen vacancies will predominate over the interstitial ions. In addition, the crystal will be essentially reduced, i.e. $[e'] \gg [h^{\cdot}]$. As a result, the electroneutrality condition in this region simplifies itself to a desired form $[e'] = 2[V_O^{\cdot\cdot}]$. It immediately results from Eq. (5.106) that

$$[e'] = 2 \, [V_O^{\cdot\cdot}] = 2^{1/3} K_O(T)^{-1/3} P_{O_2}^{-1/6} \, . \tag{5.108}$$

[76] UO_2 as a binary or La_2CuO_4 as a ternary compound may be named as examples. However, it would not be sensible here to restrict to a particular example, since it is not possible to verify all the facets to be discussed in the same example.

Equations (5.104) and (5.105) then give separately the solutions for the hole and oxygen interstitial concentrations:

$$[h^{\cdot}] = K_B/[e'] = 2^{-1/3}\left[K_B(T)K_O(T)^{1/3}\right] P_{O_2}^{1/6}, \tag{5.109}$$

$$[O_i''] = 2^{2/3}\left[K_{\bar{F}}(T)K_O(T)^{1/3}\right] P_{O_2}^{1/6}. \tag{5.110}$$

The dependence on the partial pressure of oxygen is illustrated in the left hand part of Fig. 5.38 in a double logarithmic presentation. Presentations of this type are known as Brouwer or Kröger–Vink diagrams[77].

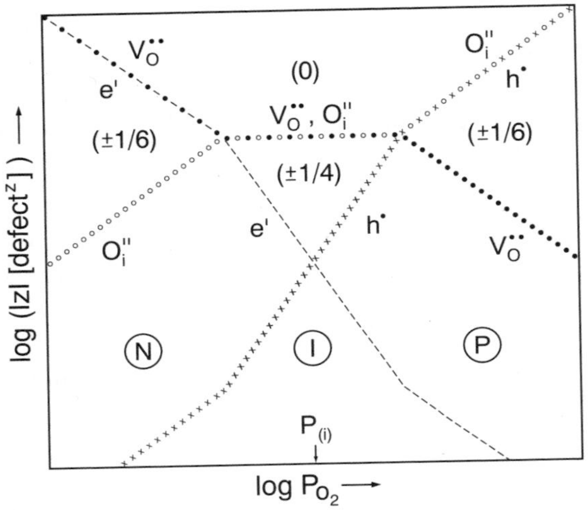

Fig. 5.38: Defect concentrations of our model oxide as a function of the partial pressure of oxygen (Kröger–Vink or Brouwer diagram). The Brouwer approximation is naturally no longer applicable at the crossing points, and the transitions are smoothed. The partial pressure of the intrinsic point (index i) is indicated. Note that the charge number z is included in the definition of the ordinate.

After counting all additional and missing oxygens, the deviation from the Dalton composition (stoichiometric or intrinsic point) in the oxide $M_2O_{1+\delta}$ becomes[78]

$$\delta = [O_i''] - [V_O^{\cdot\cdot}] = \frac{1}{2}\left([h^{\cdot}] - [e']\right). \tag{5.111}$$

It is negative in the N–region[79]. By increasing the partial pressure the intrinsic point at which δ is exactly 0, is passed through. Here $[O_i'']$ and $[V_O^{\cdot\cdot}]$ are of the same magnitude, as are $[e']$ and $[h^{\cdot}]$. Usually, but not always, the ionic defects are the major charge carriers in oxides at their intrinsic points. Thus, in the neighbourhood

[77] A large number of such diagrams are discussed in the standard work of F. A. Kröger [167], see also Refs. [4–7].

[78] In $MX_{1+\delta}$ it holds that $\delta = [X_X]+[X_i'] - [M_M] - [M_i]$. Because of site conservation ($[X_X]+[V_X^{\cdot}] =$ const $= [M_M] + [V_M']$) it holds that $\delta = [X_i'] + [V_M'] - [V_X^{\cdot}] - [M_i]$; on the other hand, because of electroneutrality $\delta = [h^{\cdot}] - [e']$.

[79] In a few oxides (e.g. CuO, see page 176) the electronic concentration predominates at the Dalton–composition. In our example the decision on whether ionic or electronic disorder prevails is already anticipated in the concentration distribution in the N–regime. Since in the N–regime of Fig. 5.38 $[O_i''] \gg [h^{\cdot}]$, the O_i'' become majority carriers in the I–regime, i.e. $[O_i''] \simeq [V_O^{\cdot\cdot}]$, before $[h^{\cdot}] \approx [e']$.

of this point it also follows with very little error[80] that $[O_i''] \simeq [V_O^{\cdot\cdot}]$ or more precisely $\log[O_i''] \simeq \log[V_O^{\cdot\cdot}]$ and then $\delta \simeq 0$. This is the I region which follows the N region (see Fig. 5.38). On account of Eq. (5.104)

$$[O_i''] = [V_O^{\cdot\cdot}] = K_{\bar{F}}^{1/2}(T). \tag{5.112}$$

In this region the ionic point defect concentrations are — relatively speaking[80] — independent of the partial pressure of oxygen. The situation is different for the electronic defect concentration. Equation (5.106) reveals according to

$$[e'] = K_O^{-1/2} K_{\bar{F}}^{-1/4} P_{O_2}^{-1/4} \tag{5.113}$$

a $P_{O_2}^{-1/4}$ dependence for the conduction electron concentrations. On account of the band–band equilibrium (Eq. (5.105)) the hole concentration is then

$$[h^{\cdot}] = K_B K_O^{+1/2} K_{\bar{F}}^{+1/4} P_{O_2}^{+1/4}, \tag{5.114}$$

i.e. characterized by a $P_{O_2}^{+1/4}$ dependence. As P_{O_2} is increased further, the hole concentration becomes relevant in the electroneutrality equation.
Since now for very high partial pressures, i.e. in the P region, the concentration of oxygen in interstitial positions must be much greater than that of the conduction electrons, Eq. (5.107) reduces to

$$[h^{\cdot}] = 2\,[O_i'']. \tag{5.115}$$

In symmetry to the N region we obtain

$$\left.\begin{array}{lcllll}
[O_i''] & = & 2^{-2/3} & K_{\bar{F}}^{1/3} & K_O^{+1/3} & K_B^{+2/3} & P_{O_2}^{+1/6} \\
[h^{\cdot}] & = & 2^{1/3} & K_{\bar{F}}^{1/3} & K_O^{+1/3} & K_B^{2/3} & P_{O_2}^{+1/6} \\
[V_O^{\cdot\cdot}] & = & 2^{2/3} & K_{\bar{F}}^{+2/3} & K_O^{-1/3} & K_B^{-2/3} & P_{O_2}^{-1/6} \\
[e'] & = & 2^{-1/3} & K_{\bar{F}}^{-1/3} & K_O^{-1/3} & K_B^{+1/3} & P_{O_2}^{-1/6}
\end{array}\right\} \text{ and } \delta = [O_i''] - [V_O^{\cdot\cdot}] > 0. \tag{5.116}$$

These results can obviously be generalized:
When mass action laws are valid and Brouwer approximation can be used then power laws of the type

$$c_k(T,P) = \alpha_k P^{N_k} \Pi_r K_r(T)^{\gamma_{rk}}, \tag{5.117}$$

exist for every charge carrier k; here N_k and γ_{rk} are rational numbers and α_k is a constant. Equation (5.117) gives us, in an approximate and sectional form, the desired solution for the internal chemical thermostatics of pure binary ionic solids with simple disorder as a function of the two control parameters, partial pressure

[80] In absolute numbers the ionic and electronic changes are comparable (see Eq. (5.101)). On account of Eq. (5.111) the difference $[h^{\cdot}] - [e']$ is (apart from the factor 2) equal to $[O_i''] - [V_O^{\cdot\cdot}]$. However, in the case of the electronic carriers the relative deviation is enormous. Note that the changes in the logarithm reflect the relative changes.

(of one of the two neutral components)[81] and temperature. (In Eq. (5.117) we have used the molar concentrations (per volume) which are more convenient for the discussion that follows. The K's have been rescaled accordingly via the molar volume. As agreed in Section 5.2 we use [k] for c_k or x_k depending on the context.) On account of the temperature dependence of the mass action constants involved, the T dependence of the defect concentration is characterized by the defect reaction enthalpies:

$$-R\left.\frac{\partial \ln c_k}{\partial 1/T}\right)_P = \Sigma_r \gamma_{rk} \Delta_r H^\circ \equiv W_k. \tag{5.118}$$

Qualitatively we can formulate the following "T theorem":

> Increasing the temperature favours endothermic reactions, such as all internal disorder reactions, since their equilibrium is shifted to the right (cf. Eq. (4.37b)), as the temperature is increased (and vice versa). Since the individual defect concentrations are given by a combination of mass action constants (cf. Eqs. (5.117), (5.118)), the tendency is not generally predictable. In most cases, however, the defect concentrations increase with T (i.e. $W_k > 0$) (but see Fig. 5.59d on page 198).

The partial pressure dependence is obtained from the Brouwer diagram in the form of a polygonal function with the slopes N_k (cf. Eq. (5.117)), which are characteristic[82] for the particular defect model:

$$\left.\frac{\partial \ln c_k}{\partial \ln P}\right)_T = N_k. \tag{5.119}$$

N_k is positive for interstitial oxygen defects and holes ($k = O_i''$ and $h^\cdot$), and negative for oxygen vacancies and conduction electrons ($k = V_O^{\cdot\cdot}$ and e'). This can be generalized to the following intelligible theorem (referred to below as the "P theorem"):

> If the chemical potential of the electronegative component of a (binary) ionic crystal (e.g. oxide) is raised ($P = P_{O_2}$), the concentrations increase (fall) of all those defects which considered individually would increase (reduce) the anion–cation ratio; in addition, the concentration of oxydized states (holes) increases, that of the reduced states (conduction electrons) falls[83].

---

[81] If the compounds are multinary and other concentrations are variable, P^{N_k} or a^{N_k} is to be replaced by a product expression of the form $\Pi_\ell a_\ell^{N_{lk}}$ (see the proton conduction example discussed later, Eq. (5.179)). In contrast to the partial pressure P, the hydrostatic pressure is denoted by p in the whole book and usually considered constant.

[82] It is not just since Karl Popper [170] that it must be realized that the agreement of experimentally found data with those predicted by the (defect) model is a necessary but not a sufficient criterion for its validity. It is naturally essential to disprove (or support) other defect models, that yield the same slope, with additional experimental results.

[83] Even though $N_{V_O^{\cdot\cdot}} \simeq N_{O_i''} \simeq 0$ in the I regime, $[O_i'']$ ($[V_O^{\cdot\cdot}]$) increases (falls) with P_{O_2}, viewed absolutely, parallel to the changes in the electronic concentrations.

It is important and no longer trivial that this applies to each individual carrier on account of the individual mass action laws. Compensation effects[84] do not occur.

The Kröger–Vink diagram (Fig. 5.38) reproduces the internal chemistry within the homogeneity region of the phase (see Chapter 4). The connection between the exact position in the phase diagram defined by δ (see Eq. (5.111)) and the equilibrium partial pressure is determined by an expression of the form $|a|\, P^{|N|} - |b|\, P^{-|N|}$. When intrinsic parameters are used ($x_{(i)}$= concentration of the majority charge carrier at $\delta = 0$; $P_{(i)} \equiv P_{O_2}(\delta{=}0)$; $\mu_{(i)} \equiv \mu_{O_2}(\delta{=}0)$) it follows more precisely[85]:

$$\delta = x_{(i)} \left(\left(\frac{P_{O_2}}{P_{(i)}} \right)^{|N|} - \left(\frac{P_{O_2}}{P_{(i)}} \right)^{-|N|} \right) = 2x_{(i)} \sinh \frac{|N| \left(\mu_{O_2} - \mu_{(i)} \right)}{RT}. \qquad (5.120)$$

$\delta(\mu_{O_2})$ is a monotonic function with a point of inflection at the intrinsic point. We will meet this function again in Chapter 7 as a titration curve.
Naturally, the whole region shown in Fig. 5.38 is not in general observed. In the case of many oxides (usually they should contain reducible cations[86], such as SnO_2 [171,172], ZnO [173]), the position of the equilibria is such that only the N region is observed. The partial pressure limit to the right is generally either set by the formation of a higher oxide or by the realization limit for the necessary high P_{O_2} value. If the total pressure is 1 bar, the upper limit is 1 bar O_2. Higher partial pressures can be realized in high pressure cells. (Note that then also the hydrostatic pressure changes, see Eq. (4.14b).) Conversely, in other oxides it is only possible to observe the P region and the stability limit at lower partial pressures (formation of lower oxides or the metal) is reached before the I region is entered. In these cases there is always oxygen excess (or equivalently metal deficiency as for Cu_2O [174], NiO [175]). Experimentally, it is only possible to realize partial pressures of oxygen down to $\sim 10^{-5}$ bar using O_2/inert gas mixtures; however, very low oxygen activities can be achieved by the use of buffer gas mixtures, such as H_2O/H_2 or CO_2/CO, or via solid

[84] It is a trivial consequence of the mass conservation and charge conservation that an increase of P leads to an increase of the O content (i.e. of $\delta = [O_i''] - [V_O^{\cdot\cdot}] + [V_M''] - [M_i^{\cdot\cdot}]$) in $MO_{1+\delta}$ and a decrease of total electron concentration in conduction and valence bands (cf. $[e'] - [h^{\cdot}]$). The changes with regard to the *individual* defect concentrations as stated above, only follow in conjunction with the individual mass action laws. The "P theorem" applies in the case of simple defect chemistry. In the case of exotic associates (e.g. $M_i^{\cdot\cdot}$ in M^+X^-) the tendencies can compete so that it may become invalid.

[85] On account of Eq. (5.117), Eq. (5.112) (intrinsic case) and Eq. (5.111), δ is of the form $aP^{|N|} - bP^{|N|}$, with the intrinsic quantities being

$$x_{(i)} = \sqrt{ab} = |a| P_{(i)}^{|N|} = |b| P_{(i)}^{-|N|} \quad \text{with} \quad P_{(i)}^{|N|} = \sqrt{\frac{b}{a}}.$$

[86] Note, however, that, in many cases, the oxydized state O^- constitutes the hole, so that a P region can also occur with cations that are not capable of further oxydation (e.g. $SrTiO_3$, see Section 5.6).

phase mixtures, such as Cu/Cu_2O or Cu_2O/CuO (see Section 4.3.4). Finally, many materials (in particular ionic conductors) exhibit only I–regime behaviour.

In order to be able to compare the theoretical predictions with experiments it is worth saying in advance that the conductivity of a material is a sensitive measure, very readily determined experimentally, of the defect concentration. The quantitative justification will given in more detail in Chapter 6. The total conductivity σ can be split up into ionic and electronic contributions ($\sigma_{ion}, \sigma_{eon}$), then, also, into the individual contributions of the defects (k):

$$\sigma(P, T) = \sigma_{ion} + \sigma_{eon} = \Sigma_k \sigma_k(P, T). \tag{5.121}$$

For every individual defect (of charge z_k)

$$\sigma_k = |z_k| \, F u_k(T) c_k(T, P), \tag{5.122}$$

whereby c_k denotes the mole number per volume[87]. (In what follows we will use the mass action constants on this c scale.) Since the mobility u_k does not depend on the defect concentration, the partial pressure dependence of the specific conductivity yields the N value directly:

$$\frac{\partial \ln \sigma_k}{\partial \ln P} = N_k. \tag{5.123}$$

However, account must be taken, with respect to the temperature dependence, as temperature also affects mobility (see Chapter 6) so that

$$-R \frac{\partial \ln \sigma_k}{\partial 1/T} \equiv E_k = W_k + \Delta H_k^{\neq}. \tag{5.124}$$

Here $\Delta H_k^{\neq}$ is the energetic portion of the activation threshold overcome on transfer of the defect k from one position to the next equivalent position (see Fig. 5.39). Additional terms of the order of RT have been neglected in Eq. (5.124). The exponential ansatz for σ is frequently also a good approximation in the case of electronic conductors. Even though $\Delta H_k^{\neq}$ may then be negligible, W_k is usually large enough to dominate the additional temperature effects, which are neglected in Eq. (5.124). There is a more detailed treatment in Chapter 6.
Since the mobility of electrons is normally higher than that of ions, and the values for holes and conduction electrons do not differ very greatly, the N regime is characterized by the occurrence of n–type conduction (i.e. primarily conduction by conduction electrons), and that of the P regime by p–type conduction (i.e. hole conduction). The occurrence of an I regime (here primarily ionic disorder) is generally

[87] We agreed in Section 5.2 that the designation [k] should be applied flexibly to x_k or c_k. In the great majority of cases discussed later, the distinction is without relevance, since we are considering changes or proportionalities. In other cases the meaning is obvious from the context. This is true, in particular, of equilibrium constants. For a limited temperature region and in dilute cases, the enthalpies involved are invariant because $x_k \propto c_k$.

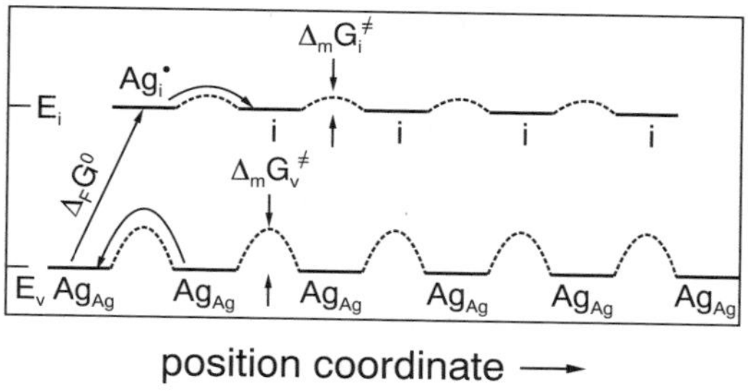

Fig. 5.39: The local (free) energy balance for the conduction process. Formation and migration of defects [176]. More precisely $Ag_i^{\cdot}$ should be replaced by the term $Ag_i^{\cdot} - V_i$ and Ag_{Ag} by $Ag_{Ag} - V'_{Ag}$.

a necessary but not a sufficient condition for a prevailing ionic conduction. The latter usually requires very large ratios of ionic and electronic defect concentrations. In agreement with this, (binary and undoped)[88] fast ionic conductors are (nearly) stoichiometric compounds (i.e. $\delta \simeq 0$). Frequently in the I–regime the conductivity exhibits a transition from n– to p–type via an ionically conducting region (e.g. PbO). If for all partial pressure values the ratio of the concentrations (ionic vs. electronic) is smaller than the reciprocal ratio of the mobilities, or if the electronic concentrations in the I region are even predominant ($[e'] \simeq [h^{\cdot}]$), as in CuO [5,178], the material is an electronic conductor over the whole range.

Let us consider some examples explicitly. SnO_2 is an example of an n–type conductor. Figure 5.40 shows that the N regime occurs over the whole partial pressure range and a more precise analysis proves that the conductivity is, in fact, electronic in nature and of n–type: The drop in σ with P_{O_2} indicates transport by conduction electrons and the slope of $-1/6$ at high temperatures[89] (low partial pressure) is, indeed, that which is expected (see Eq. (5.108)). It follows (without our needing to know the type of ionic defect) immediately from

$$\text{Reaction O} = \qquad \frac{1}{2}O_2 + V_O^{\cdot\cdot} + 2e' \rightleftharpoons O_O \qquad (5.125)$$

with $2[V_O^{\cdot\cdot}] = [e']$. The temperature dependence of the concentration of the conduction electrons is (cf. also Eq. (5.108))

$$E \simeq W_{e'} = -\Delta_O H^\circ/3. \qquad (5.126)$$

Until now we have neglected cationic defects. On account of the isomorphy of the internal defect reaction, it is clear that analogous relationships result. Of course,

[88]Deviations from stoichiometry can occur without redox effects in multinary compounds. This is obviously the case if the SrO content in $SrTiO_3$ is varied. The following example is even more illustrative: The variation of water content in hydroxides such as NaOH(s) (which can be considered thermodynamically as a mixture of the components Na_2O and H_2O) is connected with pure Brønsted acid–base effects. Fig. 5.38 is qualitatively applicable, if we replace O_i'' by OH_i', $V_O^{\cdot\cdot}$ by $V_{OH}^{\cdot}$ and the electronic carriers e' and $h^{\cdot}$ by the basic O'_{OH} (i.e. O^{2-} instead of OH^-) and the acidic $HOH_{OH}^{\cdot}$ (i.e. H_2O instead of OH^-). Of course, the slopes are correspondingly flatter [177] (see also footnote 112 on page 198).

[89]Values at lower temperatures (high partial pressures) are dominated by impurity effects (see Section 5.6).

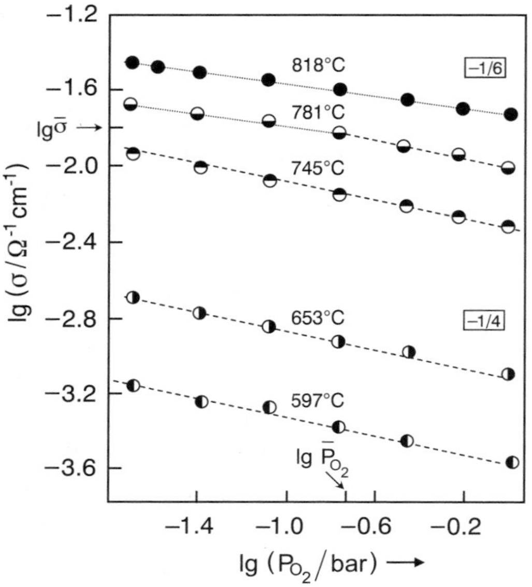

Fig. 5.40: The partial pressure of oxygen as a function of the conductivity of nominally pure SnO_2 (bold lines). At low temperatures (or higher partial pressures) the behaviour is determined by impurities (broken lines). (The parameters $\bar{\sigma}$, $\bar{P}$ are explained in Section 5.6 [172].)

the charges can be different and so can the absolute N_k values. In the case of SnO_2 it is possible, in principle, to assume additional Sn ions, i.e. $Sn_i^{\cdot\cdot\cdot\cdot}$, as ionic defect species instead of $V_O^{\cdot\cdot}$. The corresponding incorporation equation would then be

$$O_2 + Sn_i^{\cdot\cdot\cdot\cdot} + 4e' \rightleftharpoons SnO_2. \tag{5.127}$$

With the mass action law and the condition for pure substances $4\,[Sn_i^{\cdot\cdot\cdot\cdot}] = [e']$ we would get $\sigma = \sigma_{e'} \propto P_{O_2}^{-1/5}$, a somewhat different slope than is observed experimentally[90]. In general, it is assumed, on account of the strong Coulomb effects, that at relatively low temperatures highly charged defects are not formed in large concentrations and are not mobile enough to come to equilibrium with the oxygen. On the other hand, thermogravimetric studies suggest that Sn vacancies as well as oxygen vacancies play a role, but only at very high temperatures.

The counterpart, the pure P regime, is realized in pure La_2CuO_4, in which a comparatively high hole concentration, caused by high oxygen partial pressure, is known to lead to superconduction at low temperatures. Here it has even been possible to determine the positions of the excess oxygen ions in the interstices by neutron diffraction measurements (see Fig. 5.3c on page 113). According to Fig. 5.38 and Eq. (5.116), we expect $N_{h^\cdot} = 1/6$ and

$$E \simeq W_{h^\cdot} = 1/3\Delta_{\bar{F}}H^\circ + 1/3\Delta_O H^\circ + 2/3\Delta_B H^\circ. \tag{5.128}$$

[90]In the case of $Sn_i^{\cdot\cdot}$, etc. lower slopes would result (see Section 5.7). Such lower charges would be the result of association with the conduction electrons (see Section 5.7.1).

The seemingly complicated relationship for the T dependence results because the incorporation equation has been formulated using the oxygen vacancies which are minority charge carriers. For a direct discussion of this compound it is more straightforward to write the interaction in terms of the majority charge carriers, i.e. the oxygen interstitials:

$$\text{Reaction O'} = \qquad \frac{1}{2}\mathrm{O_2} + \mathrm{V_i} \rightleftharpoons \mathrm{O_i''} + 2\mathrm{h^{\cdot}} \tag{5.129}$$

leading of course also to the result $N_{h^{\cdot}} = +1/6$ on account of $2\,[O_i''] = [h^{\cdot}]$. The temperature dependence becomes simpler

$$-R\frac{\partial \ln [h^{\cdot}]}{\partial 1/T} \equiv W_{h^{\cdot}} = \Delta_{O'}H^\circ/3, \tag{5.130}$$

whereby $\Delta_{O'}H^\circ$ refers to the reaction Eq. (5.129). (Combination of the oxygen incorporation reaction used for the Kröger–Vink diagram, the band–band equation and the anti–Frenkel equation, allows us to verify that $\Delta_{O'}H^\circ = \Delta_O H^\circ + 2\Delta_B H^\circ + \Delta_{\bar{F}} H^\circ$ and, hence, that Eq. (5.130) is identical to Eq. (5.128).)
In fact, the experimentally determined conductivity displays the P–dependence predicted in Fig. 5.41. The temperature dependence of the conductivity and, hence,

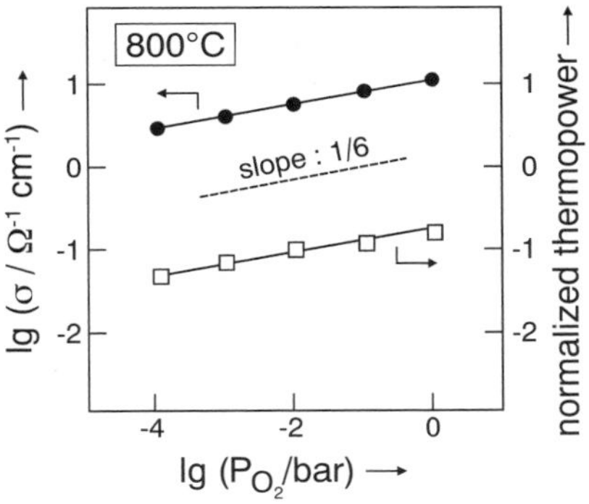

Fig. 5.41: Conductivity (filled–in circles) as a function of the oxygen partial pressure for La_2CuO_4 at high temperatures. (The thermoelectric effect also exhibits the same dependence (open squares); this is not discussed in the text.) [179]

the enthalpy of reaction of reaction O turns out to be very small, which can be attributed to the somewhat roomy structure and the presence of transition metal elements. It might puzzle the reader that we are considering a ternary oxide and treating it as our binary model oxide. However, the experiments indicate that the La/Cu ratio is very close to 1 and does not change during the experiments (low cation mobility). Such a ternary oxide behaves in a quasi–binary manner under these circumstances. This is even the case if the ratio perceptibly deviates from 2:1; then La or Cu defects occur but are frozen in at the measurement temperature. In this case, however, metal defects must be taken into account as a constant contribution in the condition for electroneutrality, if they are present at large concentration (treatment as a doped pseudo–binary, see Section 5.6). In the general case of a

ternary oxide it would, however, be necessary to fix a further component potential by including another phase[81].
NiO and Cu_2O are other examples of p–type conducting oxides. Here it is well–established that the metal vacancies are the dominant ionic defects. The p conduction, that is observed experimentally, reflects the oxydizability of the materials. In the case of nickel oxide we write

$$\frac{1}{2}O_2 + Ni_{Ni} \rightleftharpoons NiO + V''_{Ni} + 2h^{\cdot}, \tag{5.131}$$

and with $2[V''_{Ni}] = [h^{\cdot}]$ it follows that $N_{h^{\cdot}} = 1/6$. On account of the V''_{Ni} and O''_i having the same charge number, the N value is the same as we would have expected for O''_i. In the case of Cu_2O, however, we have to formulate[91]

$$\frac{1}{2}O_2 + 2Cu_{Cu} \rightleftharpoons Cu_2O + 2\,V'_{Cu} + 2h^{\cdot}. \tag{5.132}$$

With $[V'_{Cu}] = [h^{\cdot}]$ it follows that $N_{h^{\cdot}} \simeq 1/8$, which is in reasonable agreement with the experiments ($\sim 1/7$).

PbO is a material displaying the I region. There is no anti–Frenkel disorder, but probably an anti–Schottky disorder is present. Nevertheless, on account of $[Pb_i^{\cdot\cdot}] = [O''_i] = \sqrt{K_{\bar{S}}}$ here, too, we expect

$$N_{Pb_i^{\cdot\cdot}} = N_{O_i^{\cdot\cdot}} = 0 \tag{5.133}$$

and

$$-N_{e'} = +N_{h^{\cdot}} = 1/4, \tag{5.134}$$

which is found experimentally. Figure 5.42 shows the distribution of the total conductivity between electronic and ionic components. In the mixed conduction range the effective N value is represented by a sum weighted by the transference numbers[92]. A more complicated example (Fig. 5.43) exhibiting n–type conduction, ionic conduction and p–type conduction (apparent on the r.h.s. of the figure) is the pyrochlore $Gd_2(Zr_{0.3}\,Ti_{0.7})_2O_7$ [181]. The fact that it is a mixed crystal does not modify our considerations, since (i) the cationic concentrations do not change (as for La_2CuO_4) and (ii) the Zr–content does not cause electrical effects, i.e. does not influence the electroneutrality equation (see Section 5.6).

AgCl is a further useful example (see Fig. 5.44). Here, too, the ionic conductivity is not dependent on the component activity, i.e. on the partial pressure of Cl_2, but

[91] Interstitial defects appear to play a significant role at high P_{O_2} values. However, associates between O''_i and $h^{\cdot}$ occur here (see Section 5.7).

[92] If the contributions of various charge carriers to the total conductivity ($t_j \equiv \frac{\sigma_j}{\sigma}$) are comparable, it is advantageous to apply the relationship $E_{eff} = \Sigma_j t_j E_j$ or $N_{eff} = \Sigma_j t_j N_j$ for the corresponding change in the total conductivity [180]. It follows from $\partial \ln \sigma = \frac{\partial \Sigma_j \sigma_j}{\sigma} = \Sigma_j \frac{\partial \sigma_j}{\sigma} = \Sigma_j \frac{\sigma_j}{\sigma}\frac{\partial \sigma_j}{\sigma_j} = \Sigma_j t_j \partial \ln \sigma_j$.

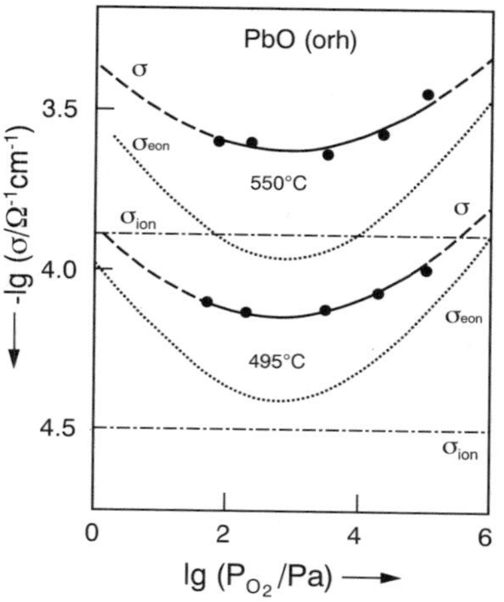

Fig. 5.42: Contributions to conductivity in orthorhombic ("yellow") lead(II) oxide as a function of oxygen partial pressure. The ionic conductivity was determined from EMF experiments (see Section 7.2.2) [117].

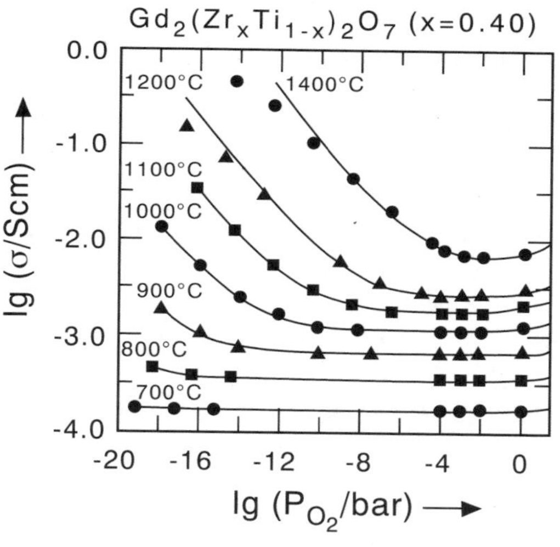

Fig. 5.43: The conductivity of $Gd_2(Zr_{0.4}Ti_{0.6})_2O_7$, as a function of the partial pressure of oxygen, exhibits n–type conductivity, oxygen ion conductivity and there is an indication of p–type conductivity (please note that the figure exclusively refers to the I–regime). From Ref. [181].

the n–type and p–type conductivites react strongly to the slightest deviations from stoichiometry. As we already know, $Ag_i^{\cdot}$ and V'_{Ag} are the majority defects so that

$$[Ag_i^{\cdot}] = \left[V'_{Ag}\right] = K_F^{1/2} . \tag{5.135}$$

Owing to

$$\text{Reaction Cl} = \qquad \frac{1}{2}Cl_2 + Ag_{Ag} + e' \rightleftharpoons AgCl + V'_{Ag} \tag{5.136}$$

and, hence, because of $K_{Cl} = \left[V'_{Ag}\right][e']^{-1}P_{Cl_2}^{-1/2}$ there are dependences of the form

$$\begin{aligned} [e'] &\propto P_{Cl_2}^{-1/2} \\ [h^{\cdot}] &\propto P_{Cl_2}^{+1/2} . \end{aligned} \tag{5.137}$$

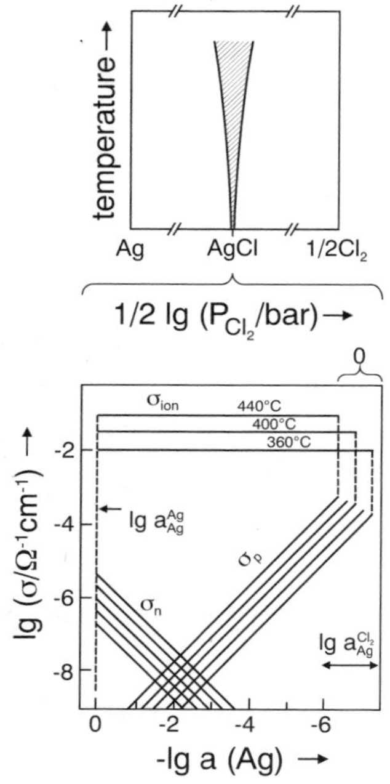

Fig. 5.44: The variation of n–type, p–type and ionic conduction in AgCl as a function of the silver activity (the scale for the partial pressure of chlorine runs in the opposite direction). More exactly: The two logarithmic scales differ (apart from the minus sign) by an additive term which is given by the mass action constant of the formation reaction of AgCl and hence T–dependent. The l.h.s. limitations are set by silver activity 1 (contact with Ag, a_{Ag}^{Ag}) and thus also by the decomposition chlorine partial pressure which is temperature dependent. The r.h.s. limit of the scale corresponds to the maximum ambient Cl_2 pressure (1 bar) corresponding to a T–dependent low silver activity ($a_{Ag}^{Cl_2}$). The upper figure shows that these changes mirror the "internal chemistry" on passing through the phase width. The electronic conductivity was determined from polarization measurements (see Section 7). Data according to Ref. [182].

The incorporation equation can be just as readily formulated with Ag; then the following relationship is obtained using the silver activity a_{Ag}

$$[e'] \propto a_{Ag}^{+1} \qquad \text{and} \qquad [h]^{\cdot} \propto a_{Ag}^{-1}, \tag{5.138}$$

that can, in turn, be converted to Eq. (5.137) making use of the equilibrium between Ag, Cl_2 and AgCl.

In CuO, too, the conductivity is almost independent of oxygen partial pressure for a large parameter window [178]. Here, however, the comparatively high conduction is electronic. It is to be assumed that the roles of ionic and electronic charge carriers in the I regime are exchanged and, thus, that slight deviations from stoichiometry do not affect the comparatively high, internal electronic disorder[93]. In other words:

$$[e'] = [h^{\cdot}] \tag{5.139a}$$

and

$$N_{e'} = N_{h^{\cdot}} = 0. \tag{5.139b}$$

Conversely, in such cases now the ionic defect concentrations are strongly dependent upon the oxygen partial pressure[93].

[93] Cf. also the Fe_3O_4 example in Fig. 6.18 on page 299.

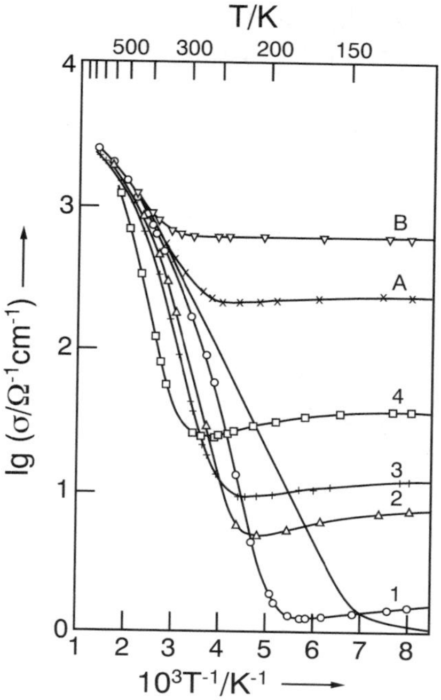

Fig. 5.45: Conductivity of Si over a large temperature range. The labels 1, 2, 3, 4, A, B indicate samples of differing purity and indicate extrinsic effects. From Ref. [128].

The relationships described in Eq. (5.139) correspond exactly with the behaviour of pure silicon. Figure 5.45 illustrates the conductivity values for this semiconductor over a wide temperature range. The values measured at lower temperatures are evidently not intrinsic (compare with Fig. 5.36c) and vary from sample to sample. An analogous situation occurs with impure ionic conductors (compare with Figs. 5.36a,b). These effects will be discussed now. Since we treat the general case of mixed conductors (i.e. ionic plus electronic conductivity), effects as they occur in typical semiconductors are automatically included.

5.6 Doping effects

We will see in this section how impurities can have a profound effect on defect chemistry; this means conversely that well–defined and deliberate contamination, i.e. doping, is an extremely powerful method for influencing material properties.
Let us first consider the ion conductor AgCl. Experimentally it is found that the Frenkel–disordered AgCl is frequently contaminated with higher–valent cations, such as Cd^{2+}. In view of the fact that Cd will exhibit the charge 2+ and the ionic radius of Cd^{2+} is very similar to that of Ag^+, it is to be expected that Cd^{2+} will substitute Ag^+ and form effectively charged defects of the form $Cd^{\cdot}_{Ag}$. The positive excess charge can be predominantly compensated only by a deficiency of Ag^+ ions: Excess

Cl^- ions are not sufficiently stable within the lattice and conduction electrons[94] can only be generated in a minor amount, since redox–effects do not play a dominant role in AgCl. Naturally, on account of the low concentration of excess silver ions their annihilation — not to mention that of the holes — cannot serve to compensate for the excess charge to a significant degree. Yet, owing to the interactions described all these effects which are coupled together occur to some degree.
Let us now quantify the situation. The incorporation reaction can be written as follows:

$$CdCl_2 + 2Ag_{Ag} \longrightarrow Cd^{\cdot}_{Ag} + V'_{Ag} + 2AgCl. \tag{5.140a}$$

The "excess chlorine" offered by the doping with $CdCl_2$, effects the extraction of a regular Ag^+ and hence the formation of a vacancy. The annihilation of this chlorine can (and will in a limited amount) also be carried out by an interstitial silver ion according to

$$CdCl_2 + Ag_{Ag} + Ag^{\cdot}_{i} \longrightarrow Cd^{\cdot}_{Ag} + 2AgCl. \tag{5.140b}$$

Other possibilities for formulating the incorporation reactions release the excess chlorine in the form of a neutral molecule, i.e. via a redox reaction

$$CdCl_2 + Ag_{Ag} \longrightarrow Cd^{\cdot}_{Ag} + 1/2Cl_2 + e' + AgCl \tag{5.140c}$$

or

$$CdCl_2 + Ag_{Ag} + h^{\cdot} \longrightarrow Cd^{\cdot}_{Ag} + AgCl + 1/2Cl_2. \tag{5.140d}$$

Again, it is obvious that when, for example, reaction (5.140a) is written down, other reactions (5.140b), (5.140c), (5.140d) are redundant and can be obtained directly by combination of reaction (5.140a) with the native equilibrium disorder reactions ((F), (Cl), (B), see Section 5.5). Hence, it is sufficient to select one reaction. For the purpose of clarity we will use the majority charge carrier formulation (here reaction (5.140a)). The above doping reactions are formulated in the form of irreversible reactions. The foreign atoms are generally incorporated at temperatures (often at temperatures above the melting point) that are far higher than the measurement temperature at which the doping centres are then immobile. In this sense, reactions (5.140a – 5.140d) bring about a defined but constant doping concentration, hence $[Cd^{\cdot}_{Ag}]$ appears merely in the form of a constant in the electroneutrality condition; mass action laws are not allowed to be formulated for such doping reactions. Thermodynamically speaking, the doping reactions alter the conditions in a similar manner to that discussed in Section 5.4 for the existence of metastable, higher–dimensional defects (e.g. grain boundaries): The doping concentrations have to be regarded as ex situ parameters. (At very high temperatures, however, such reactions become reversible. As far as the relatively low measurement temperatures are

[94]It may seem implausible to the chemist that the incorporation of Cd^{2+} can result in a redox effect at all. This is because, as mentioned on page 17, the excess Cl is not — as it is the case with fluid phases — incorporated on doping with $CdCl_2$. Thus it can be thought of being removed as Cl_2 ($\widehat{=}2Cl^- - 2e'$). Hence, the substituting reactand is "CdCl".

concerned the high temperature conditions can be regarded as the prehistory of the samples which defines the ex situ parameters[95].)

The occurrence of $[\mathrm{Cd}^{\cdot}_{\mathrm{Ag}}]$ in the electroneutrality condition means that the Cd doping increases the silver vacancy concentration over its intrinsic value and lowers the interstitial silver concentration correspondingly as implied in Eqs. (5.140a, 5.140b). The mass action laws coupling ionic and electronic effects (see Table 5.2, page 162) show that — as is expressed directly by Eqs. (5.140c, 5.140d) — the electronic concentration increases and the hole concentration drops at constant $\mathrm{Cl_2}$ partial pressure.

This can be generalized to the following fundamental rule of doping. Assuming that the doping ion is introduced irreversibly (with a concentration[96] C) and all other defects are in local equilibrium, the following statement applies for simple defect chemistry (referred to, in what follows, as the "C theorem"):

> If the doping defect introduced is positively charged, then the concentrations of all negatively charged, ionic and electronic defects are increased, while the concentrations of positively charged defects are reduced. The opposite occurs for negatively charged dopants. This applies, not only to a combination of defect concentrations (cf. charge density), but to each individual one. In this sense, there are no compensation effects amongst the mobile defects. More concisely: If $\mathrm{z_k}$ is the charge number of defect k and z that of the doping defect, then it follows for all k that

$$\frac{\mathrm{z_k}\delta \mathrm{c_k}}{\mathrm{z}\delta \mathrm{C}} < 0. \tag{5.141}$$

This "rule of homogeneous doping" (which is complemented by the "rule of heterogeneous doping" describing the effect of irreversibly brought–in higher–dimensional defects, page 247, Eq. (5.264)) is of great importance for material research and will be further quantified below (see Eq. (5.152)). It also means that for defect chemical analysis it is not absolutely necessary to formulate the incorporation reactions, rather it is merely necessary to know the effective charge (and the concentration) of the doping defect introduced. The same considerations show that the designations donor– and acceptor–doping are only meaningful for purely electronic effects and are confusing in the general case. We do better to use the designations positive and negative doping. This effective charge (determined by the absolute charge and the site occupied) can be usually estimated from the requirement that the substitution does not involve too high an energy, i.e. that ionic radius, chemical nature and charge are not too different.

The proof of Eq. (5.141) follows directly from the fact that the incorporation equations and the connecting equilibria must be electroneutral and can be formulated, in each case, with two defects of opposite charges.

[95] cf. page 195 in this context.

[96] As with [k] we use C, depending on context, as a quantity which is related to the number of monomeric units in the perfect state (lattice molecules) ($\equiv \mathrm{x_{dop.}}$) or to the volume ($\equiv \mathrm{c_{dop.}}$).

Let us consider, in detail, the situation in our model substance AgCl. Since the ionic defects are always in the majority, the electroneutrality condition for Cd–doped material is $([\mathrm{Cd}^{\cdot}_{\mathrm{Ag}}] \equiv \mathrm{C})$

$$[\mathrm{Ag}^{\cdot}_{\mathrm{i}}] + [\mathrm{Cd}^{\cdot}_{\mathrm{Ag}}] = [\mathrm{V}'_{\mathrm{Ag}}] = [\mathrm{Ag}^{\cdot}_{\mathrm{i}}] + \mathrm{C}. \tag{5.142}$$

In addition, the Frenkel relationship

$$[\mathrm{Ag}^{\cdot}_{\mathrm{i}}][\mathrm{V}'_{\mathrm{Ag}}] = \mathrm{K_F(T)} \tag{5.143}$$

is still valid. When C is comparatively small, the substance behaves as pure AgCl with $[\mathrm{Ag}^{\cdot}_{\mathrm{i}}] = [\mathrm{V}'_{\mathrm{Ag}}] = \mathrm{K_F^{1/2}}$; $\mathrm{N}_{\mathrm{Ag}^{\cdot}_{\mathrm{i}}} = \mathrm{N}_{\mathrm{V}'_{\mathrm{Ag}}} = 0$ and $\mathrm{W}_{\mathrm{Ag}^{\cdot}_{\mathrm{i}}} = \mathrm{W}_{\mathrm{V}'_{\mathrm{Ag}}} = \frac{1}{2}\Delta_{\mathrm{F}}\mathrm{H}^{\circ}_{\mathrm{m}} \simeq 0.7\mathrm{eV}$. This applies to adequately high temperatures for which $\mathrm{K_F}$ has a very high value. Such a situation for the model material "MX" is shown in Fig. 5.46 (for the working

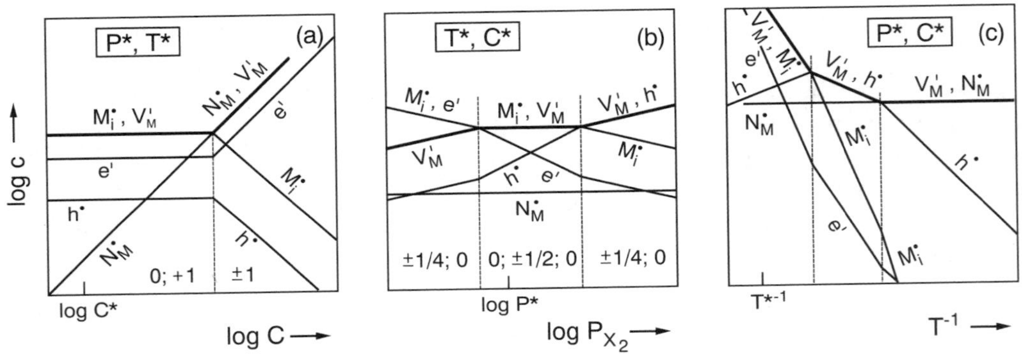

Fig. 5.46: Dependences of the equilibrium concentrations in slightly positively $(\mathrm{N}^{\cdot}_{\mathrm{M}})$ doped Frenkel–disordered "MX" on $\mathrm{P_{X_2}}, \mathrm{T}, [\mathrm{N}^{\cdot}_{\mathrm{M}}]$. At the working point characterized by $\mathrm{P}^*, \mathrm{C}^*, \mathrm{T}^*$, the material still behaves intrinsically. The middle section in the T–dependence characterized by $[\mathrm{V}'_{\mathrm{M}}] \simeq [\mathrm{h}^{\cdot}]$ did not appear (Part c) in the case of AgCl owing to the very small electronic carrier concentration.

point P^*, T^*, C^*). The other extreme case is $\mathrm{C} \gg \mathrm{K_F^{1/2}}$, which is, for a given C–value, always realized at sufficiently low temperatures. In this case the concentration of vacancies must be distinctly greater than in the intrinsic case on account of Eq. (5.141), whilst $[\mathrm{Ag}^{\cdot}_{\mathrm{i}}]$ must be very much lower because of Eq. (5.143). For this extreme case Eq. (5.142) leads to

$$\mathrm{C} \simeq [\mathrm{V}'_{\mathrm{Ag}}] \tag{5.144}$$

i.e. a vacancy concentration, which is independent of temperature (and of the Ag activity), while $[\mathrm{Ag}^{\cdot}_{\mathrm{i}}]$ is now strongly dependent on T:

$$[\mathrm{Ag}^{\cdot}_{\mathrm{i}}] = \mathrm{K_F(T)/C}. \tag{5.145}$$

As in the undoped case,

$$\mathrm{N}_{\mathrm{Ag}^{\cdot}_{\mathrm{i}}} = \mathrm{N}_{\mathrm{V}'_{\mathrm{Ag}}} = 0, \tag{5.146}$$

but now

$$W_{Ag_i^\bullet} = \Delta_F H_m^\circ \gg 0 = W_{V'_{Ag}}. \tag{5.147}$$

The succession of intrinsic regime and impurity regime is shown in Fig. 5.47 as a function of 1/T and in Fig. 5.48 as a function of $[Cd^\bullet_{Ag}]$[97].

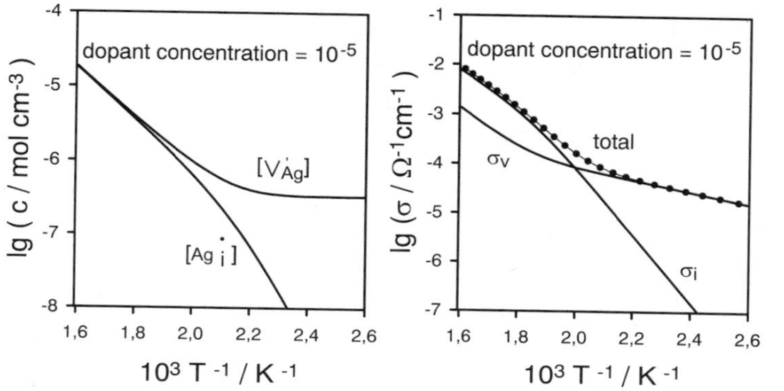

Fig. 5.47: Defect concentrations (a) and conductivities (b) calculated according to Eqs. (5.154, 5.155) for the Cd doping given as a function of temperature[97].

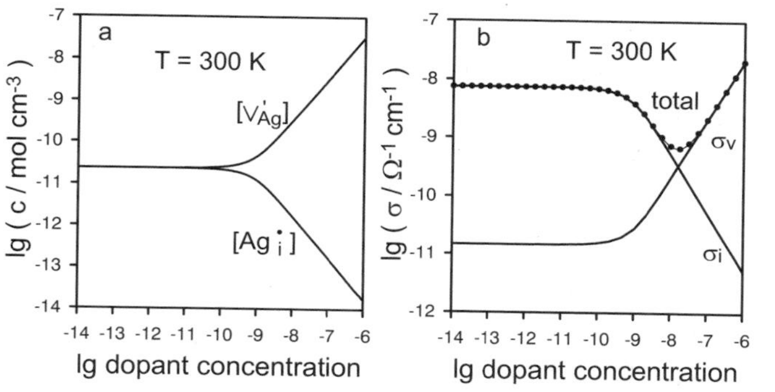

Fig. 5.48: Defect concentrations (a) and conductivities (b) calculated according to Eqs. (5.154, 5.155) as a function of $[Cd^\bullet_{Ag}]$[97].

The transition temperature is naturally very dependent on the C value. Even nominally pure materials behave extrinsically, if the temperature is low enough. High purity AgCl is always found to contain a few ppm of aliovalent ions as impurities (see Fig. 5.36, page 158), whose effects on the ionic conductivity usually become perceptible around room temperature.

The behaviour of the conductivities is more complex than that of the concentrations, due to the temperature dependence of the mobility. On the one hand, the $Ag_i^\bullet$ mobility in AgCl is greater than the V'_{Ag} mobility. On the other hand, the different migration energies have to be taken into account ($\Delta_m H^{\neq}_{Ag_i^\bullet} = 0.05$eV, $\Delta_m H^{\neq}_{V'_{Ag}} = 0.3$eV). On account of Eq. (5.147) the temperature dependence of the partial conductivity

[97]The nuances obtained in the transition regions in comparison to Figs. 5.46a,c come from a precise calculation avoiding the Brouwer approximation (see, for example, Eq. (5.142) instead of Eq. (5.144)).

of strongly Cd–doped AgCl is given by

$$\mathrm{E}_{\mathrm{V}'_{\mathrm{Ag}}} = \Delta_{\mathrm{m}}\mathrm{H}^{\neq}_{\mathrm{V}'_{\mathrm{Ag}}} \quad \text{and} \quad \mathrm{E}_{\mathrm{Ag}_i^{\cdot}} = \Delta_{\mathrm{F}}\mathrm{H}^{\circ}_{\mathrm{m}} + \Delta_{\mathrm{m}}\mathrm{H}^{\neq}_{\mathrm{Ag}_i^{\cdot}}. \tag{5.148}$$

The total conductivity at high temperatures is of interstitial type (intrinsic), at very low temperatures it is dominated by vacancies (extrinsic). Here the concentration of $\mathrm{V}'_{\mathrm{Ag}}$ is then so high if compared to $[\mathrm{Ag}_i^{\cdot}]$ that the higher mobility (we remember that $\sigma_{\mathrm{k}} \propto \mathrm{c}_{\mathrm{k}}\mathrm{u}_{\mathrm{k}}$) of the $\mathrm{Ag}_i^{\cdot}$ is overcompensated. The situation is different in the transition region: If the $\mathrm{V}'_{\mathrm{Ag}}$ concentration just begins to predominate over the $\mathrm{Ag}_i^{\cdot}$ concentration, the conductivity is still of the interstitial type (because $\mathrm{u}_{\mathrm{Ag}_i^{\cdot}} > \mathrm{u}_{\mathrm{V}'_{\mathrm{Ag}}}$). The same cross–over behaviour occurs as a function of doping content. Since the $\mathrm{Ag}_i^{\cdot}$ concentration is greatly reduced in this transition region, as illustrated in Fig. 5.48b, there is a minimum in the $\sigma(\mathrm{C})$ representation before conductivity of the vacancy type dominates. As far as the temperature dependence is concerned (Fig. 5.47b), the effect is mirrored in the appearance of a "knee" ("Koch–Wagner effect" [183]). At high temperatures the conductivity is independent of doping, hence $-\mathrm{R}\partial\ln\sigma/\partial\mathrm{T}^{-1} \simeq \Delta_{\mathrm{F}}\mathrm{H}^{\circ}_{\mathrm{m}}/2 + \Delta_{\mathrm{m}}\mathrm{H}^{\neq}_{\mathrm{Ag}_i^{\cdot}} \simeq \Delta_{\mathrm{F}}\mathrm{H}^{\circ}_{\mathrm{m}}/2$. At low temperatures $-\mathrm{R}\partial\ln\sigma/\partial\mathrm{T}^{-1} \simeq \Delta_{\mathrm{m}}\mathrm{H}^{\neq}_{\mathrm{V}'_{\mathrm{Ag}}}$.

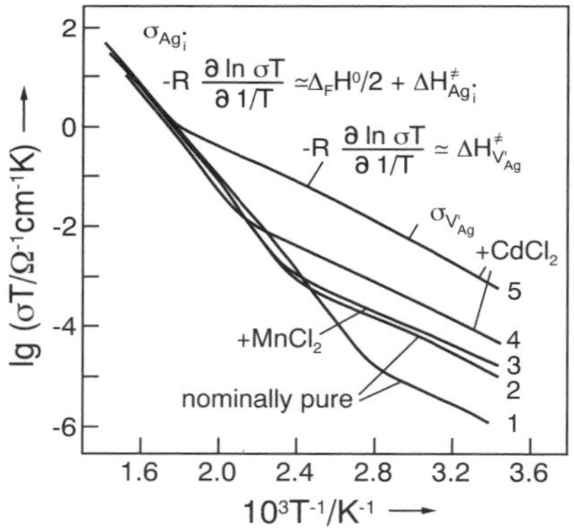

Fig. 5.49: Experimental conductivity data for nominally pure and doped AgCl as a function of 1/T. Here $\lg(\sigma\mathrm{T})$ is plotted instead of $\lg\sigma$ to take account of the slight T–dependence of the prefactor (see Chapter 6). However, this does not alter the slope noticeably. According to Ref. [184].

Figure 5.49 displays experimental results for nominally pure AgCl and for various doping levels, which are in very good agreement with the above considerations. One can see that doping can not only be used to finely control the conductivity (cf. Fig. 5.49: $\sigma(300\mathrm{K})$ varies over several orders of magnitude; see also Fig. 5.45), but, also, that the doping experiments can provide useful thermodynamic and kinetic parameters. Thus, the low temperature slope of AgCl in the $\log\sigma$vs.$1/\mathrm{T}$ representation gives the migration threshold for silver vacancies directly.

Now let us treat in detail defect concentration and conductivity as a function of dopant concentration (Figs. 5.48a,b, cf. also Fig. 5.46a). As already discussed, at very low C values $[\mathrm{V}'_{\mathrm{Ag}}]$ and $[\mathrm{Ag}_i^{\cdot}]$ are independent of C (corresponding to the intrinsic condition of electroneutrality $[\mathrm{V}'_{\mathrm{Ag}}] = [\mathrm{Ag}_i^{\cdot}] \gg [\mathrm{Cd}_{\mathrm{Ag}}^{\cdot}]$). If C becomes comparable with the intrinsic values, $[\mathrm{Ag}_i^{\cdot}]$ falls and $[\mathrm{V}'_{\mathrm{Ag}}]$ rises until, for high C–values, $[\mathrm{V}'_{\mathrm{Ag}}] = \mathrm{C}$

is reached, i.e. in the end we get a straight line for $\lg[V'_{Ag}] = \text{fct}(\lg C)$ with the slope 1, while $[Ag_i^{\cdot}]$ then decreases with a slope -1 on account of Eq. (5.143).
If we apply the Brouwer approximation we also directly obtain the results for the minority charge carriers. From Eqs. (5.135, 5.136) it follows (lower temperatures):

$$[V'_{Ag}] = C \tag{5.149a}$$

$$[Ag_i^{\cdot}] = K_F C^{-1} \tag{5.149b}$$

$$[e'] = K_{Cl}^{-1} P_{Cl_2}^{-1/2} C \tag{5.149c}$$

$$[h^{\cdot}] = K_B K_{Cl} P_{Cl_2}^{+1/2} C^{-1}. \tag{5.149d}$$

(In particular it can be seen that in this case the P dependence of the minority species is also unchanged compared with that for pure AgCl in which the ionic defect concentrations were constant too.)
Obviously Eq. (5.117) can be generalized to

$$c_k(T, P, C) = \alpha_k P^{N_k} C^{M_k} \Pi_r K_r(T)^{\gamma_{rk}} \tag{5.150}$$

to take account of doping effects. One should remember that, in contrast to the in situ parameters P and T, the parameter C cannot be reversibly altered within the window of conditions under consideration (ex situ parameter). The important rational numbers M_k (in Eq. (5.149): $1, -1, 1, -1$)

$$M_k \equiv \left.\frac{\partial \ln c_k}{\partial \ln C}\right)_{T,P} \simeq \left.\frac{\partial \ln \sigma_k}{\partial \ln C}\right)_{T,P} \tag{5.151}$$

reproduce the doping effect quantitatively (Fig. 5.48a). The fact that $M_{V'_{Ag}}$ and $M_{e'} > 0$ and $M_{Ag_i^{\cdot}}, M_{h^{\cdot}} < 0$ reflects the above doping rule (C theorem) (Eq. (5.141)), which can now be formulated more succinctly as

$$\frac{z_k}{z} M_k < 0 \tag{5.152}$$

(z is the effective charge number of the dopant). The right hand side of Eq. (5.151) follows because the mobility is not dependent on doping for dilute defect chemistry.
Now let us abandon the comfortable Brouwer approximation and discuss the situations in generality, i.e. we use the complete electroneutrality condition in accordance with Eq. (5.142). Combination with the Frenkel equilibrium (Eq. (5.143)) leads to the quadratic equation

$$[V'_{Ag}]([V'_{Ag}] - C) = K_F \tag{5.153}$$

with the solutions

$$[V'_{Ag}] = C/2 + \sqrt{C^2/4 + K_F} \tag{5.154a}$$

$$[Ag_i^{\cdot}] = -C/2 + \sqrt{C^2/4 + K_F}. \tag{5.154b}$$

(The extreme cases $\sqrt{K_F} \ll C/2$ or $\sqrt{K_F} \gg C/2$ correspond to the extrinsic and intrinsic limiting cases discussed above.) The conductivity is given by

$$\sigma/F = u_{V'_{Ag}}\left(C/2 + \sqrt{C^2/4 + K_F}\right) + u_{Ag_i^{\cdot}}\left(-C/2 + \sqrt{C^2/4 + K_F}\right). \tag{5.155}$$

The graphs in Figs. 5.47 and 5.48 have been constructed on the basis of Eqs. (5.154, 5.155). Notice that the transition regions are curved compared with the discontinuous change in the slope as predicted by the Brouwer approximations (curved sections). Regarding AgCl, for which $u_{Ag_i^{\cdot}} > u_{V'_{Ag}}$, Eq. (5.155) exhibits a minimum at[98]

$$C_{min} = \frac{2\left(\frac{u_{Ag_i^{\cdot}} - u_{V'_{Ag}}}{u_{Ag_i^{\cdot}} + u_{V'_{Ag}}}\right) K_F^{1/2}}{\sqrt{1 - \left(\frac{u_{Ag_i^{\cdot}} - u_{V'_{Ag}}}{u_{Ag_i^{\cdot}} + u_{V'_{Ag}}}\right)^2}} \simeq \sqrt{\frac{K_F u_{Ag_i^{\cdot}}}{u_{V'_{Ag}}}}. \tag{5.156}$$

Because $-R\frac{\partial \ln C_{min}}{\partial 1/T} \simeq \frac{1}{2}\Delta_F H^\circ_m + \frac{1}{2}\left(\Delta H^{\neq}_{Ag_i^{\cdot}} - \Delta H^{\neq}_{V'_{Ag}}\right) \simeq (0.7 - 0.15)\text{eV} = 0.55\text{eV}$ the minimum is displaced to distinctly higher values as the temperature is increased, essentially because more doping is then required to overcompensate the greater intrinsic disorder ($\Delta_F H^\circ_m$ term). Simultaneously the minimum is shallower because $u_{V'_{Ag}}$ and $u_{Ag_i^{\cdot}}$ approach each other. Figure 5.50 (right side) shows the excellent

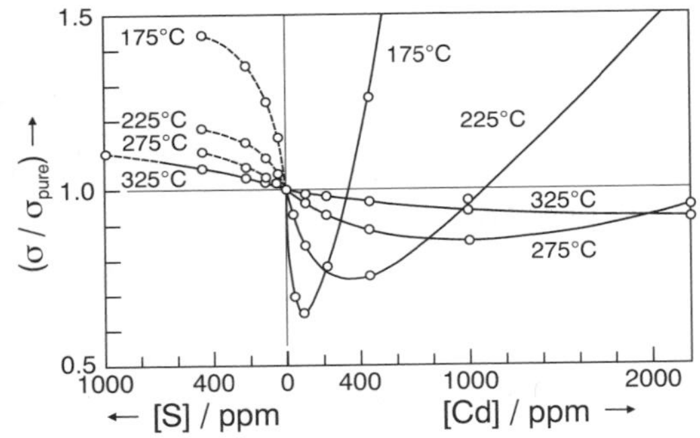

Fig. 5.50: The dependence of the conductivity of AgCl (normalized w.r.t. the intrinsic value) on the S and Cd content. The solid curves (r.h.s.) are calculated according to Eq. (5.155). According to Ref. [185].

agreement in all respects of the experimental data with the simple mass action theory. This has been treated in somewhat more detail here in order to demonstrate that the defect–chemical viewpoint is well based and effective. Nonetheless, for the sake of clarity the Brouwer approximation will always be used in what follows.
We can also aim for the converse effect and dope AgCl, e.g. with Ag_2S, to create effectively negatively charged S'_{Cl} defects, whereby now, according to our rule, Eq. (5.152), $[Ag_i^{\cdot}]$ and $[h^{\cdot}]$ increase, while $[V'_{Ag}]$ and $[e']$ fall. Since the conductivity of AgCl itself is of interstitial type, doping with Ag_2S leads to a monotonic increase

[98] The approximation on the right hand side follows for $u_{Ag_i^{\cdot}} \gg u_{V'_{Ag}}$. The third binomial formula is best used for verification.

in conductivity (Fig. 5.50, left hand side) and not to extreme values. However, the situation is somewhat complicated when doping with Ag_2S, because S'_{Cl} and $Ag_i^{\cdot}$ defects have a — chemically not unexpected — tendency to associate strongly. These complications concerning the internal chemistry will be discussed in detail in the next section.

Let us now analyse as a second example the effect of doping on the electron conductor SnO_2. We remember that this oxide is oxygen–deficient and in pure material the e' and $V_O^{\cdot\cdot}$ concentrations are determined by the partial pressure of oxygen and the temperature. Figure 5.40 (on page 172) showed the (electronic) conductivity as a function of P_{O_2} at various temperatures. The slopes of $-1/6$ (Eq. (5.108)) predicted for pure SnO_2 are evidently only realized at high temperatures and/or low oxygen potentials. Accordingly, in those isotherms in Fig. 5.40, in which a change in slope is detectable (at 781°C), this change from $-1/4$ to $-1/6$ takes place in the direction of lower P_{O_2} values. At sufficiently high P_{O_2} and low temperatures $N_{e'} = -1/4$ always applies. Here impurities evidently determine the behaviour. According to the decisive oxygen incorporation reaction

$$\text{Reaction O} = \qquad \frac{1}{2}O_2 + V_O^{\cdot\cdot} + 2e' \rightleftharpoons O_O \tag{5.157a}$$

with

$$K_O = P_{O_2}^{-1/2}[V_O^{\cdot\cdot}]^{-1}[e']^{-2}, \text{ where } \Delta_O H^\circ < 0, \tag{5.157b}$$

the slope of -1/4 follows immediately when $[V_O^{\cdot\cdot}]$ is constant, i.e. when in the electroneutrality condition a large concentration of an effectively negatively charged impurity defect has to be taken into account. In fact, a precise chemical analysis of the material reveals a considerable concentration of Fe, that, with certainty, exceeds the electron concentration of the pure compound in the P_{O_2}, T regions being discussed. The similar ionic radii of Fe^{3+}(0.6Å) and Sn^{4+}(0.7Å) suggest that there is substitution of Sn^{4+} ions and formation of the singly negatively charged defect Fe'_{Sn}. The condition for electroneutrality is now

$$[Fe'_{Sn}] + [e'] = 2[V_O^{\cdot\cdot}] = C + [e']. \tag{5.158}$$

The iron doping means an increase of $[V_O^{\cdot\cdot}]$ and a reduction of $[e']$, so that at comparatively high impurity content $2[V_O^{\cdot\cdot}] = [Fe'_{Sn}] \equiv C$ and, hence, $N_{e'} = -1/4$. Because $\Delta_O H^\circ < 0$, at high temperatures the impurity concentration (C) is overwhelmed on account of the high degree of defect formation according to Eq. (5.157), so that the Fe content no longer plays a role and the slope is $(-1/6)$. The same applies, in full agreement with the experimental results, on reduction of the partial pressure of oxygen for mass action reasons (see Eq. (5.157)).
The incorporation equation can be written in detail as

$$Fe_2O_3 + 2Sn_{Sn} + O_O \rightarrow 2Fe'_{Sn} + 2SnO_2 + V_O^{\cdot\cdot} \tag{5.159}$$

or

$$Fe_2O_3 + 2Sn_{Sn} + 1/2O_2 + 2e' \rightarrow 2Fe'_{Sn} + 2SnO_2, \tag{5.160}$$

directly mirroring the above tendencies (negative doping).
It is informative to consider the point $(\bar{\sigma}_{e'}, \bar{P})$, at which, at the temperature given, the impurity regime passes over into the regime of native[99] disorder.
At this point $\sigma_{e'}(C \rightarrow \infty) = \sigma_{e'}(C \rightarrow 0)$. Equation (5.157) gives us

$$\sigma_{e'}(C \rightarrow \infty) = 2^{1/2} F u_{e'} C^{-1/2} K_O^{-1/2} P^{-1/4} \tag{5.161}$$

$$\sigma_{e'}(C \rightarrow 0) = 2^{1/3} F u_{e'} K_O^{-1/3} P^{-1/6} \tag{5.162}$$

explicitly for the limiting cases and, hence,

$$\bar{P} = 4K_O^{-2} C^{-6} \tag{5.163a}$$

and, not unexpectedly,

$$\bar{\sigma}_{e'} = F u_{e'} C. \tag{5.163b}$$

It is worthy of note that the value of $\bar{P}$ is independent of the mobility and reacts extremely sensitively to the doping concentration. As expected $\bar{P}$ is displaced to lower values when C is increased. It is evidently possible to determine K_O in this convenient manner, provided that C is known (and vice versa). On the other hand, $\bar{\sigma}_{e'}/(u_{e'}F) = \bar{c}_{e'} = C$ gives the level of doping if $u_{e'}$ is known.
Equations (5.161, 5.162) also reveal to us immediately the T dependences in the impurity and native regimes. If we neglect $\Delta_m H^{\neq}_{e'}$, then we get

$$E_{e'}(C \rightarrow \infty) \simeq -1/2 \Delta_O H^\circ \tag{5.164a}$$

and

$$E_{e'}(C \rightarrow 0) \simeq -1/3 \Delta_O H^\circ. \tag{5.164b}$$

The temperature curve of σ (Fig. 5.51) does, in fact, exhibit a kink at $\bar{\sigma}$. The slopes to the left and right of the kink do, in fact, exhibit a ratio[100] of 3:2 (see Eq. (5.164)) and, thus, permit the determination of $\Delta_O H^\circ$. The deviations occurring at lower temperatures can be explained by association (see next section) and freezing–in effects (see below).
If SnO_2 is doped with larger quantities of Fe_2O_3 — or In_2O_3 as in Figs. 5.51 and 5.52 — the native region is no longer observable in the T and P window and the conductivity is very much less than in the case of the pure substance under the same conditions (cf. Fig. 5.40, page 172).

[99] "Native" refers to the absence of elements except Sn and O. We reserve the expression "intrinsic" for the special case $\delta = 0$.

[100] The reader will notice that in contrast to the assumptions of Eq. (5.164) the slopes of the two Fe–doped curves are different. The reason for this is not clear.

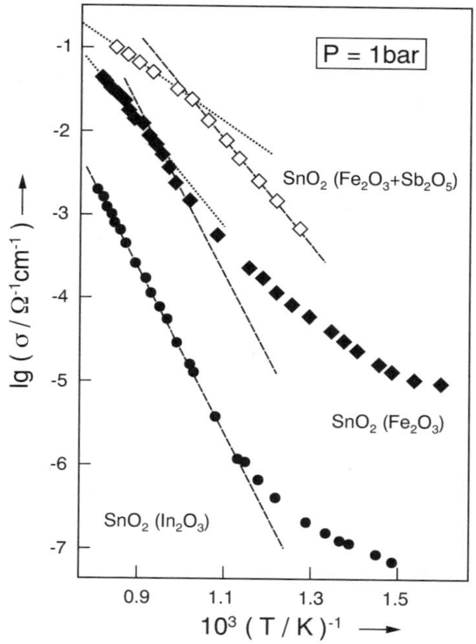

Fig. 5.51: The temperature dependence of the conductivity of weakly negatively (Fe_2O_3) doped SnO_2 reveals a native and an impurity-dominated region. The upper curve also contains Sb^{5+}, but is still negatively doped overall[100]. The lower curve is strongly negatively doped (In^{3+}) and does not exhibit a native region. The flattening out at lower temperatures is attributable to complications concerning the defect chemistry (see next section) [172].

Figure 5.53 shows the conductivity of SnO_2 films, that were prepared by chemical deposition using $SnCl_4$ gas in the presence of H_2O or O_2. As in the first case, it is possible to identify the native region at high temperatures and low partial pressures. However, in the impurity region the conductivity is now almost completely independent of P_{O_2} and T. The reason for this is that contamination with Cl^- from the preparation procedure, leads to a positive doping and thus to an increase in $[e']$ and a reduction of $[V_O^{\cdot\cdot}]$:

$$SnCl_4 + 2O_O \rightarrow SnO_2 + Cl_2 + 2Cl_O^{\cdot} + 2e', \tag{5.165}$$

or in a different (not independent) formulation:

$$SnCl_4 + 2O_0 + 2V_O^{\cdot\cdot} \rightarrow SnO_2 + 4Cl_O^{\cdot}. \tag{5.166}$$

As predicted by Eq. (5.165) the conductivity is now very much greater than in the pure material. When the temperature is sufficiently low or the partial pressure is sufficiently high, the electroneutrality relationship becomes

$$[e'] = [Cl_O^{\cdot}] = C \tag{5.167}$$

and it follows immediately that

$$N_{e'} \simeq 0 \simeq W_{e'} \simeq E_{e'}. \tag{5.168}$$

The slight dependence of the conductivity on the temperature in Fig. 5.53 is a result of the effects of temperature on the mobility of the conduction electrons (see Chapter 6).

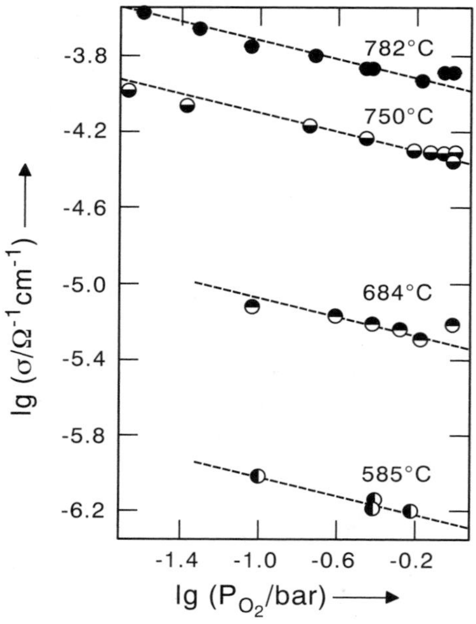

Fig. 5.52: Conductivity as a function of the partial pressure of oxygen for distinctly negatively (In_2O_3) doped SnO_2 [172].

In order to really drive the point home: It is sufficient to know the effective charge of the impurity in order to be able to predict the direction of all the concentration effects. In this manner, it is possible to understand the consequence of the frequently realized donor doping of SnO_2 by Sb, viz. the increase in conductivity caused by the formation of $(Sb^{5+}_{Sn^{4+}})^{\cdot}$. Figure 5.51 shows the temperature dependence of SnO_2 doped by both Fe and Sb. In such mixed cases, it is the $[Fe'_{Sn}] - [Sb^{\cdot}_{Sn}]$ difference that is decisive. It is greater than zero in Fig. 5.51, i.e. the negative doping predominates, though the effect is reduced by Sb according to the comparatively higher σ–values.

We are now able to understand, why a high hole concentration can be induced in La_2CuO_4 by Ba or Sr doping. As already mentioned, this makes the material superconducting at $T \lesssim 40K$. In fact, it was this form of oxydation — and not the oxygen treatment discussed above (increase in oxygen partial pressure) — that led to Bednorz and Müller making their Nobel prize–winning discovery [186]. The incorporation of Sr results in $(Sr^{2+}_{La^{3+}})' = Sr'_{La}$ defects, which increase the hole concentration and decrease the O''_i concentration. Remember, in the pure material the interstitial oxygens compensate for the holes. Figure 5.54 shows the Kröger–Vink diagram for a particular Sr content at constant temperature. The Sr concentration is only exceeded at very high partial pressures (Fig. 5.54 right) and the material then behaves like the pure substance (see also Fig. 5.46b). When P_{O_2} is reduced, $[O''_i]$ and $[h^{\cdot}]$ also fall; however, on account of the electroneutrality condition,

$$[Sr'_{La}] + 2[O''_i] = [h^{\cdot}] \tag{5.169}$$

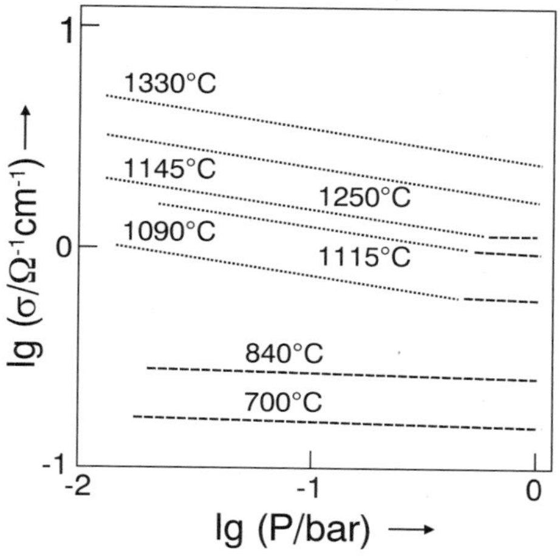

Fig. 5.53: The P_{O_2} dependence of the conductivity [171,172] for positively doped SnO_2 ($SnCl_4$) exhibiting native and impurity–dominated regions.

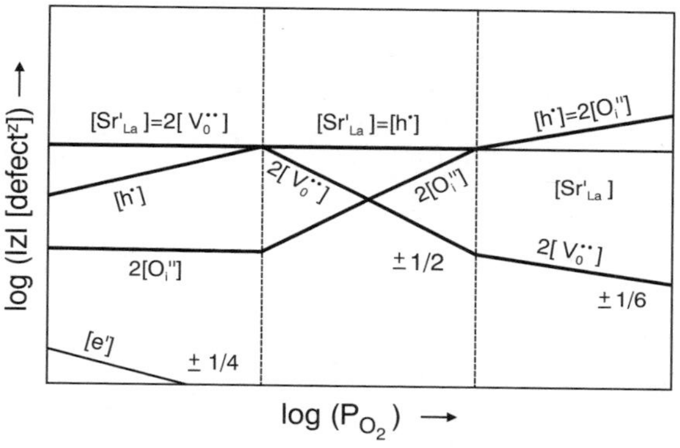

Fig. 5.54: Defect concentrations as a function of oxygen partial pressure at a given negative doping (SrO) for La_2CuO_4 [187].

$[h^{\cdot}]$ cannot fall below the doping level. In the extreme case it follows that

$$[Sr'_{La}] \equiv C = [h^{\cdot}]. \tag{5.170}$$

The (electronic) conductivity is then constant (Fig. 5.54 centre). The interstitial concentration falls steeply as P_{O_2} is reduced and, hence, the partial ionic conductivity in this region too. At the same time the oxygen vacancy concentration, which is increasing on account of the anti–Frenkel equilibrium, becomes important. As P_{O_2} is reduced further, the necessary depression of hole concentration results in the impurity being finally compensated by oxygen vacancies (Fig. 5.54 left). Eventually the material should become n–type conducting at very low partial pressures, but this region probably lies beyond the stability range of the phase. Even moderately low P_{O_2} values lead to the production of reduced phases.

Figure 5.55 shows, in double logarithmic format, the defect chemistry in $(La, Sr)_2CuO_4$ as a function of doping content at constant P_{O_2} and constant T. The C dependences are described quantitatively by Eq. (5.151). Trivially $[h^{\cdot}]$ and $[O''_i]$ are independent of C in the region of very small Sr addition. As C is increased,

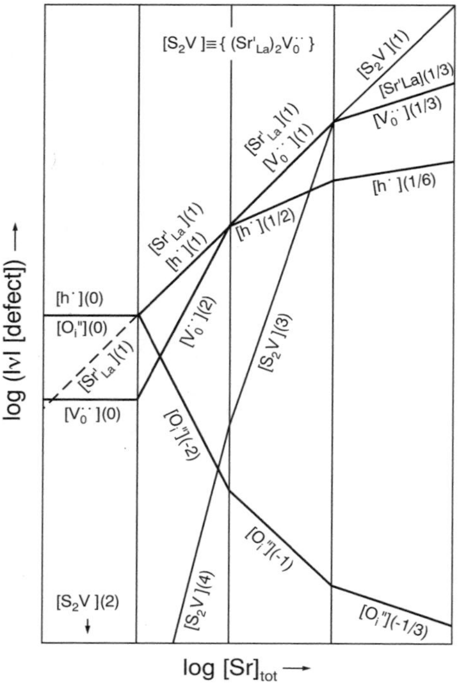

Fig. 5.55: The dependence of defect concentrations on Sr doping of La_2CuO_4 within the framework of the defect model described. Interactions are not included. Only the formation of one associate $S_2V = (Sr'_{La})_2 V_O^{\cdot\cdot}$ is taken into account (see following section). The figures in brackets give the M values (slopes). The parameter $|\nu|$ in the ordinate labelling gives the absolute value of the charge in the case of ionized defects; in the case of S_2V, however, it gives the Sr content per formula unit ($|\nu|=2$) [187].

the impurity defects are initially compensated by holes for the starting conditions shown in Fig. 5.55[101]. Simultaneously the oxygen vacancy concentration increases steeply and takes over the role of compensation. Here the M value for $[h^{\cdot}]$ then falls from 1 to 1/2. The ultimate reason that $[V_O^{\cdot\cdot}]$ dominates in the end is the double positive charge. Associates bringing about a further flattening certainly occur at high concentrations (see the next section). However, the experimentally observed reduction of the electronic conductivity (like the ionic conductivity (see Chapter 6, Fig. 6.19)) with increasing doping finally cannot be explained in this manner. The simple defect concept breaks down at such high defect concentrations:
Structural changes and long–ranging order processes occur, meaning that it is no longer possible to consider the mixed phase as a slightly perturbed version of the initial phase. Formally the mass action constants and mobilities become perceptibly concentration–dependent.
Nonidealities also occur in $YBa_2Cu_3O_{6+x}$, the most popular high temperature superconductor ($x>0.5$). Here anti–site disorder effects are of importance, in addition to associates. Defects of type $Y_{Ba}^{\cdot}$ generated during preparation and then frozen–in, formally take the same role as added aliovalent doping (see also metal vacancies in $SrTiO_3$) [187]. At low temperatures ordering processes are of significance.
Y_2O_3 or CaO–doped ZrO_2 is probably the most important solid electrolyte. The

[101] This naturally varies according to the T, P conditions.

high oxygen ion conductivity is due to the oxygen vacancies formed according to

$$CaO + Zr_{Zr} + O_O \rightarrow Ca''_{Zr} + ZrO_2 + V_O^{\cdot\cdot}. \quad (5.171)$$

Defect concentrations of the order of 10% are required to stabilize the relevant cubic phase. At such concentrations we cannot expect the ionic conductivity to increase proportionally as described by $[V_O^{\cdot\cdot}] = C$ for small doping levels. Rather σ_{ion} flattens and on higher doping it even decreases with increasing Y–content which is only explicable in terms of marked interactions and nonidealities, as in the example above [188].

The alkaline earth (and also lead) titanate or zirconate perovskites are an even more important group technologically. On account of their electrical and dielectric properties they are key materials for the construction of capacitors, resistors, actuators and sensors. In addition, $SrTiO_3$ is a very valuable model compound. The decisive defect–chemical interaction is the incorporation of oxygen at vacancies according to Eq. (5.98c) which is reversible for $T > 700K$. The Schottky equilibrium is only established at very high temperatures ($T > 1300K$)[102]. Since this temperature regime is involved during manufacture, frozen–in metal vacancies (V''_{Sr}) play the role[103] of negative doping. In order to understand the situation it is necessary to know that the mobility of the oxygen vacancies is orders of magnitude greater than that of the V''_{Sr} defects. In practical use $SrTiO_3$ is either appreciably negatively or positively doped [189]. In the latter case La_2O_3 is typically added, yielding a strongly n–conducting $SrTiO_3$ ($[La^{\cdot}_{Sr}] \simeq [e']$); in the first case Fe_2O_3 is often used. While the larger La^{3+} occupies Sr^{2+} sites ($La^{\cdot}_{Sr}$), the smaller Fe^{3+} substitutes Ti^{4+} ions forming effective negatively charged $[Fe'_{Ti}]$ defects, which are largely compensated by oxygen vacancies. Figure 5.56 shows the dependence of the conductivity on oxygen content. As in the case of SnO_2, we can see the transition from the impurity region (with slope $-1/4$) into the native region (with slope $-1/6$), at which the concentration of positive doping is overwhelmed by the concentration of the native defects. In contrast to SnO_2, $SrTiO_3$ exhibits a minimum in the conductivity caused by a further increase, as a result of the formation of holes, which is predicted by Eq. (5.94a). The slope is $+1/4$ as expected. The incorporation of O as O^{2-} first consumes conduction electrons, i.e. "chemically speaking", reduced states ($\sim Ti^{3+}$) are oxydized and the (n–type) conductivity falls. A further incorporation of oxygen results in the increased consumption of bonding electrons, so that holes (p–type conduction) are produced as new charge carriers, chemically speaking the O^{2-} ions are oxydized ($\sim$ redox comproportionation of O and O^{2-} to $2O^-$). The assignments made (Ti^{3+}, O^-) conform roughly to the results of band structure calculations: The

[102] Here it is the Schottky equilibrium with respect to the SrO component that is meant and, hence, the formation of Sr^{2+} and O^{2-} vacancies with the formation of an SrO deficiency with respect to TiO_2. On account of the high vapour pressure of PbO, the same applies significantly at relatively low temperatures to the Pb titanates or zirconates, which are important as actuators.

[103] Associates with $h^{\cdot}$ are also important at high P_{O_2}.

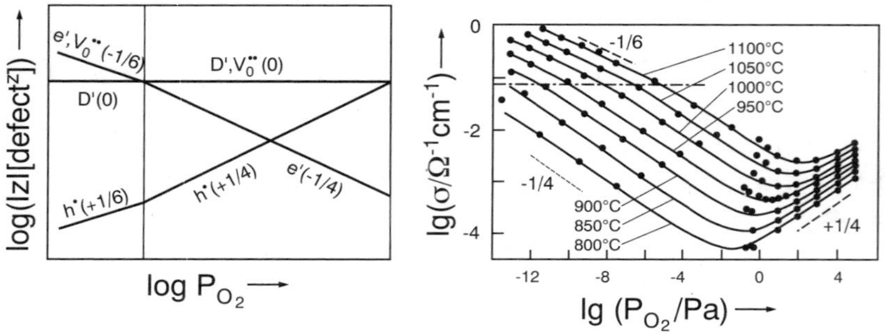

Fig. 5.56: The conductivity of negatively doped $SrTiO_3$ as a function of oxygen partial pressure. Left: model (D′: negative doping), right: experimental data from Ref. [190].

conduction band is provided by the Ti d orbitals, the valence band predominantly by the O 2p orbitals (cf. Section 2.2.2):
We take Eq. (5.157) for the region of n conduction

$$\text{Reaction O} = \qquad \frac{1}{2}O_2 + V_O^{\cdot\cdot} + 2e' \rightleftharpoons O_O. \tag{5.172a}$$

When $2[V_O^{\cdot\cdot}] = C$ this yields

$$[e'] \propto P^{-1/4} \exp + \frac{\Delta_O H_m^\circ}{2RT}. \tag{5.172b}$$

The corresponding relationships for the holes are produced by coupling with the band–band equation. For the sake of symmetry we will write this out explicitly as

$$\text{Reaction O'} = \qquad \frac{1}{2}O_2 + V_O^{\cdot\cdot} \rightleftharpoons O_O + 2h^{\cdot} \tag{5.173a}$$

and

$$[h^{\cdot}] \propto P^{+1/4} \exp - \frac{\Delta_{O'} H_m^\circ}{2RT}. \tag{5.173b}$$

The difference between the two temperature dependences of the conductivities (we neglect the T–dependence of mobility) gives twice the thermal band gap, which is, thus, readily determined in this manner (result: 3.3eV). Figure 5.11 on page 133, illustrates the band gap deduced from optical measurements (compare Section 5.3). The approximate agreement allows us to conclude that the maximum of the valence band in the $\epsilon(k)$ presentation and the minimum of the conduction band are at least approximately at the same k value, i.e. at the same position in the band scheme. Another means of determining the thermal band gap involves following the

T dependence of the conductivity minimum ($\sigma_{\min}$), which is given[104] by

$$-R\frac{d\ln\sigma_{\min}}{d1/T} \simeq 1/2E_g. \tag{5.174}$$

When heavily doped the ionic conduction region is clearly visible (Fig. 5.57). (We note the correspondence with Fig. 5.38, page 166, but do not forget that the centre is not an intrinsic regime.)

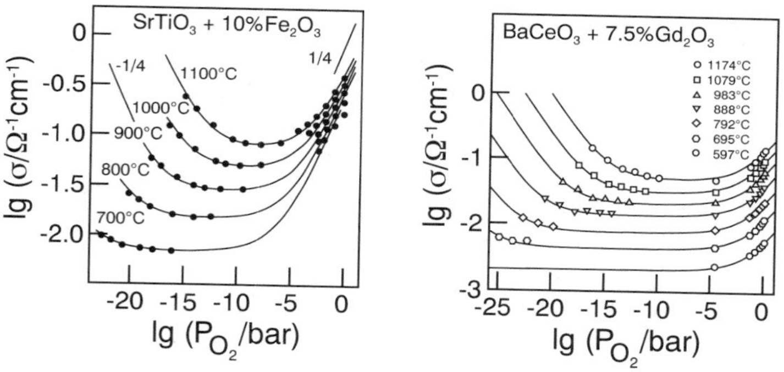

Fig. 5.57: Upon heavy negative doping of the perovskite oxides $SrTiO_3$ (left) and $BaCeO_3$ (right), it is possible to recognize the regions of n–type, p–type and ionic conductions. At this high level of doping it is expected that the thermodynamic parameters differ from those of the undoped material (see footnote 146, page 216). According to Refs. [191, 192].

In the case of Fe–doped $SrTiO_3$ a further complication enters the game, in that Fe^{4+} plays a large role at low temperatures and high oxygen partial pressures, as has been demonstrated in detail spectroscopically[105]. This corresponds to a (naturally exothermic) association between the negatively charged Fe^{3+} defects and the holes:

$$Fe'_{Ti} + h^{\cdot} \rightleftharpoons Fe^{x}_{Ti}. \tag{5.175}$$

The enthalpy of association and corresponding equilibrium constant can be determined optically. (The Fe^{4+} content is responsible for the red colour of the crystals, cf. Sections 5.7 and 6.5.) Fe^{x}_{Ti} does not possess an effective defect–chemical charge; the actual Fe^{3+} concentration ($[Fe'_{Ti}]$) entering the electroneutrality relationship is a function of P_{O_2}, T and the total iron content C, which, as before, is constant. These valence changes are special cases of association reactions, which are dealt with systematically in the next section (see page 204).

[104] Because $\sigma_{\min} = 2\sigma_{e',\min} = 2\sigma_{h^{\cdot},\min} = 2u_{e'}Fc_{e',\min} = 2u_{h^{\cdot}}Fc_{h^{\cdot},\min}$ and $c_{e',\min}c_{h^{\cdot},\min} = K_B$ it follows that $c_{e',\min} = (K_B u_{h^{\cdot}}/u_{e'})^{1/2}$ and $\sigma_{\min} = 2(u_{e'}u_{h^{\cdot}}K_B)^{1/2}$, hence, Eq. (5.174). Note that Eq. (5.174) is not a partial differential of $\sigma(P,T)$. Rather it refers more precisely to $-R\partial\ln\sigma(T,\delta)/\partial 1/T$. Compare the shift of the minimum with temperature in Fig. 5.56.

[105] Note that the lattice is optimized for a four–fold charged ion so that there is a greater driving force for Fe^{4+} (zero effective charge!) than say in aqueous solution.

One further complication with regard to defect chemistry, which is frequently ignored, ought to be discussed here briefly, since it can be clearly demonstrated for the well–understood $SrTiO_3$.
Unexpectedly, ion conduction appears to be dominant at low temperatures, which means at $T < 700K$ in this context, in moderately acceptor–doped $SrTiO_3$; this phenomenon is more pronounced, the quicker the sample is cooled down. This can be interpreted as follows:
The reaction (5.173) becomes frozen in the region $T < 700K$ as a result of the sluggish surface kinetics, and, in addition, at very low temperatures because of the low mobility of the oxygen defects, while the purely electronic equilibrium Eq. (5.175) is still reversible. Accordingly $[V_O^{\cdot\cdot}]$ and, hence, σ_{ion} remain at high levels on cooling, while the holes are captured and the electronic conductivity drops markedly. These effects explain the qualitative change in the behaviour at low temperatures in Fig. 5.58 [193]. Such considerations lie outside the scope of this introduction, but are, nevertheless, important for defect–chemical analyses at "low" temperatures[106] [193–195]. Figure 5.58 reproduces a precise numerical conductivity analysis and shows to what a degree the defect chemistry of a well–investigated material can be understood.
It must be remembered that this freezing–in reduces the number of in situ parameters (here P_{O_2}), while increasing the number of ex situ parameters. The frozen–in metal–oxygen ratio acts as one of these. This is fixed by the P–T–C conditions at the temperature of freezing–in (P_Q, T_Q, C). In this manner the degrees of freedom are increased corresponding to the increased deviation from full equilibrium [102].

Exactly the opposite occurs, namely the conversion of an ex situ parameter to an in situ one, if foreign components become sufficiently mobile. The corresponding incorporation reaction then becomes reversible. Under such conditions it is naturally better to speak of solubility equilibria. Important examples are segregation equilibria of impurities at very high temperatures, another refers to the incorporation of protonic defects in oxides by the dissolution[107] of H_2O. Materials interesting in this respect are CaO–doped ZrO_2 [196] or acceptor–doped perovskites [197], such as the Fe–doped $SrTiO_3$ discussed above. (As before we regard the acceptor dopant

[106] Let us take a simple example. If the concentration of redox–active impurities (e.g. Fe in $SrTiO_3$) is so high that $[h^{\cdot}]$ is small with respect to the concentration of the oxydized (e.g. Fe^{4+}) and the reduced state (e.g. Fe^{3+}), then it follows that the hole concentration in the quenched state is given by $K_A(T)K_A(T_Q)^{-1}P_Q^{1/4}$. K_A is the ionization (dissociation) constant, T_Q the quenching temperature. So, if we quench the samples at T_Q at several partial pressures of O_2 (P_Q), the same partial pressure dependence results, as found in the reversible region. However, the T dependence is different. If the hole concentration were falsely regarded as being frozen–in, i.e. the ionization reaction from impurities or native defects were ignored, so the whole T dependence would be misinterpreted as a marked polaron migration with a substantial migration energy. With respect to achieving a defined state, it is important to choose T_Q in such a manner that the equilibrium is set up in a reasonable time, which, however, is large with respect to the time taken in cooling. A systematic treatment is given in Refs. [194,102].

[107] Similarly there will be a finite solubility of e.g. HF in fluorides.

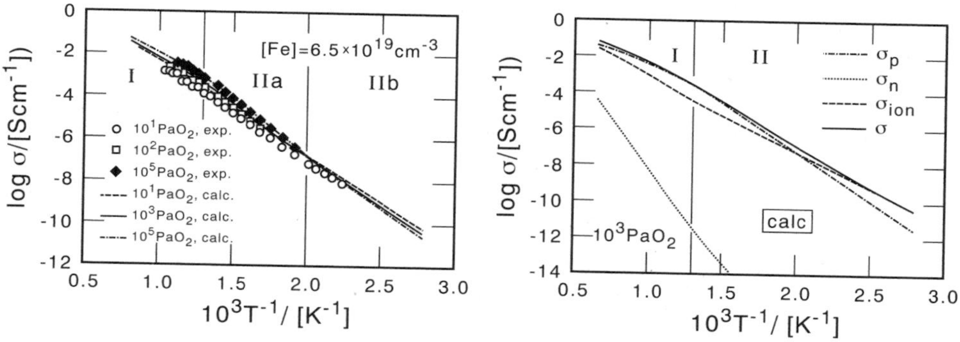

Fig. 5.58: The precise conductivity analysis of Fe–doped $SrTiO_3$ including freezing–in effects can quantitatively describe and predict very subtle changes [193]. Left: experimental and calculated total conductivity. Right: calculated partial conductivities. Regime I: oxygen vacancies equilibrated. Regime II: concentration of oxygen vacancies frozen–in.

as immobile). Such "hydrations" occur to a greater extent in Yb–doped $BaCeO_3$ or Y–doped $Ba_2YSnO_{5.5}$ [198,199][108]. In a manner analogous to the formation of surface OH groups, water molecules can be "incorporated dissociatively" in the interior: The "OH^- part" of the H_2O occupies an oxygen vacancy, while the "H^+ part" combines with a regular O^{2-}; in this way two internal OH^- groups are produced:

$$\text{Reaction } H_2O = \qquad H_2O + V_O^{\cdot\cdot} + O_O \rightleftharpoons 2OH_O^{\cdot}. \tag{5.176a}$$

The change in enthalpy of this reaction is negative (and of the order of -1.5eV for $BaCeO_3$). The OH defect produced (OH^- at an O^{2-} site) bears a single positive charge. Phenomenologically it is also possible to break $OH_O^{\cdot}$ down into O_O^x and $H_i^{\cdot}$[109]:

$$\text{Reaction } H_2O = \qquad H_2O(g) + 2\,V_i + V_O^{\cdot\cdot} \rightleftharpoons O_O + 2H_i^{\cdot}. \tag{5.176b}$$

The partial pressure of water (P_{H_2O}) is an in situ control parameter in such cases, that is to be included in Eq. (5.150) as a further degree of freedom in addition to P_{O_2}. Naturally the problem can also be formulated via the redox reaction

$$\text{Reaction } H = \qquad H_2(g) + 2O_O \rightleftharpoons 2OH_O^{\cdot} + 2e' \tag{5.177a}$$

or[110]

$$H_2(g) + 2V_i \rightleftharpoons 2H_i^{\cdot} + 2e' \tag{5.177b}$$

[108] The $Ba_2YSnO_{5.5}$ saturated with water corresponds to the oxide–hydroxide $Ba_2YSnO_5(OH)$.

[109] Structurally the proton is not in a true interstitial position, since the proton "dives in" to the electronic clouds of the O^{2-}. It is better to speak of an excess proton.

[110] Eq. (5.177b) illustrates the possibility of incorporating protons on contact with hydrogen, even if there are no oxygen vacancies in the structure (see also Ref. [200]). A significant incorporation requires marked redox effects.

with P_{H_2} as parameter. Given the oxygen/hydrogen reaction equilibrium

$$2H_2(g) + O_2(g) \rightleftharpoons H_2O(g) \tag{5.178}$$

Eq. (5.177) is redundant. We will favour the acid–base reaction Eq. (5.176), since it illustrates the main effect under normal conditions[111].

The formal treatment and the calculation of the dependence of the charge carrier concentration, in particular of $[H_i^{\cdot}]$ as a function of T, P_{O_2}, P_{H_2O} and the concentration of the negative dopant, is carried out as described above using the Brouwer approximation. Under reversible conditions the result is formally given by

$$c_k(T, C, P_{O_2}, P_{H_2O}) = \alpha_k C^{M_k} P_{O_2}^{N_k} P_{H_2O}^{N'_k} \Pi_r K_r^{\gamma_{rk}}(T). \tag{5.179}$$

However, at low temperatures a constant water concentration has to be taken into account which then, in the form of $[OH_O^{\cdot}]$, adds to the effective C–parameter (as it is the case for the oxygen vacancy concentration[106]).

Let us consider a rare earth–doped $BaCeO_3$, in which high proton conductivities were measured after treatment with water. Yb'_{Ce} or Gd'_{Ce} defects act as negative dopants. Figure 5.57 shows the conductivity isotherms for the proton–free material. Just as with strongly Fe–doped $SrTiO_3$ we can recognize the regions of n–type, p–type and $V_O^{\cdot\cdot}$ conduction. After H_2O treatment the following electroneutrality condition holds when the partial pressure of oxygen is not too low:

$$[Yb'_{Ce}] = [h^{\cdot}] + 2[V_O^{\cdot\cdot}] + [H_i^{\cdot}]. \tag{5.180}$$

In accordance with the rules set out above the diagrams giving the dependence of defect concentrations of T, C, P_{O_2} and P_{H_2O} then follow (Fig. 5.59). The detailed calculation will be left to the reader as a useful exercise. Only a few characteristics will be indicated here.

When the oxygen partial pressure is increased (Fig. 5.59a), the hole concentration increases, $[V_O^{\cdot\cdot}]$ drops; when $[h^{\cdot}]$ is so large that it finally compensates for the dopant, the proton concentration also falls at a fixed P_{H_2O}. This is seen immediately on

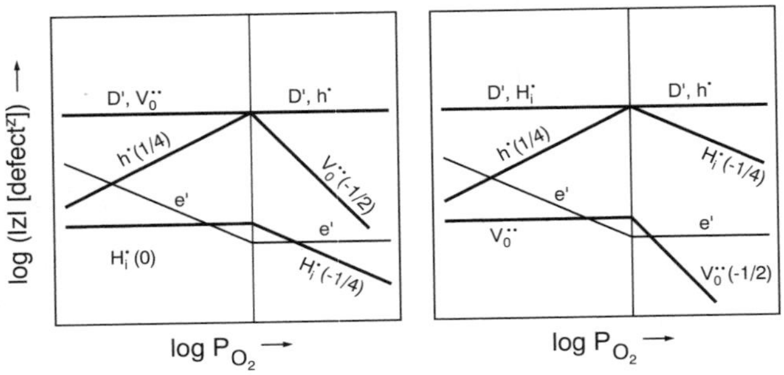

Fig. 5.59a: Dependence of the defect concentrations on oxygen partial pressure for perovskites containing water under reversible conditions. Left: lower, right: higher water partial pressure (D′: negative dopant).

[111] See also Refs. [201,202].

combining Eqs. (5.176) and (5.173a) to give the reaction

$$\frac{1}{2}O_2 + 2H_i^{\cdot} \rightleftharpoons 2h^{\cdot} + H_2O + 2V_i \qquad (5.181)$$

and applying the mass action law to this.
If we raise the partial pressure of water at fixed P_{O_2} (Fig. 5.59b) the proton concentration also increases in accordance with Eq. (5.176a), until it finally compensates

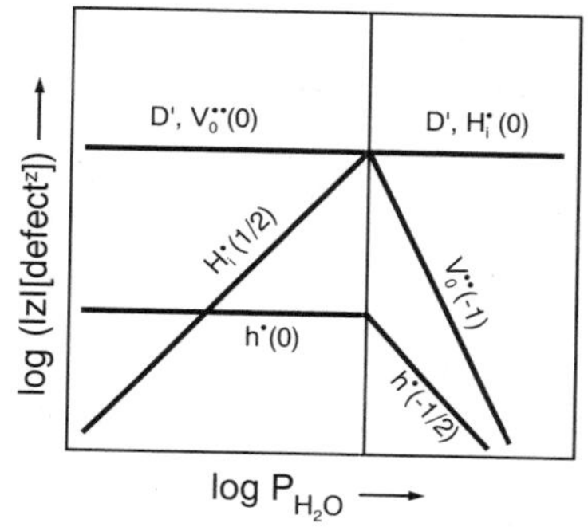

Fig. 5.59b: Dependence of the defect concentrations of water–containing perovskites on the partial pressure of water in the environment (D′: negative dopant) under reversible conditions.

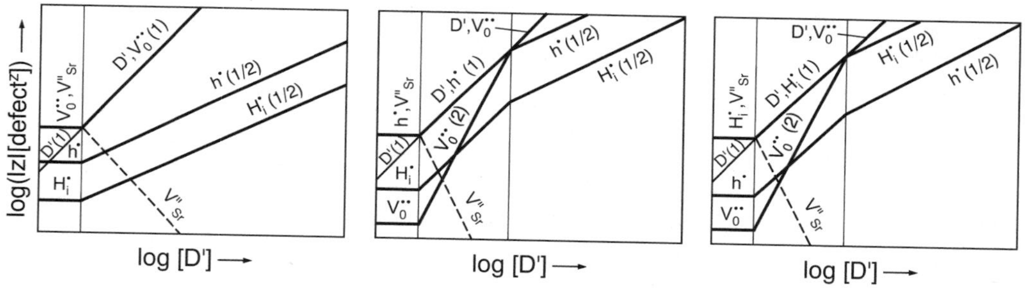

Fig. 5.59c: Dependence of the defect concentrations on the doping concentration in water–containing perovskites for various starting situations (reversible conditions).

the acceptor concentration. Then $[h^{\cdot}]$ has to fall on account of Eq. (5.181).
The dependence on the concentration of the negative doping is shown by Fig. 5.59c for various P_{H_2O}–, P_{O_2}–, T–dependent situations. Both hole and proton concentrations increase (doping rule); the same applies for the oxygen vacancy concentration. On account of the double charge the latter increases steeply so that in the end $[V_O^{\cdot\cdot}]$ always predominates, provided the Boltzmann approximation is still applicable.
The last diagram in the sequence (Fig. 5.59d) shows the temperature dependence for typical parameters ($\Delta_{O^{\cdot}}H^\circ \sim 1.2$eV, $\Delta_{H_2O}H^\circ \simeq -1.5$eV : $\Delta_m H^{\neq}_{V_O^{\cdot\cdot}} \simeq 0.6$eV, $\Delta_m H^{\neq}_{H_i^{\cdot}} = 0.5$eV, $\Delta_m H^{\neq}_{h^{\cdot}} \simeq 0$). While the hole concentration increases continuously, $[H_i]$ falls continuously as the temperature increases. $[V_O^{\cdot\cdot}]$ exhibits intermediate behaviour: The vacancy concentration first increases, goes through a maximum and then falls. This is easily understood from the enthalpy changes in Eq. (5.176a)

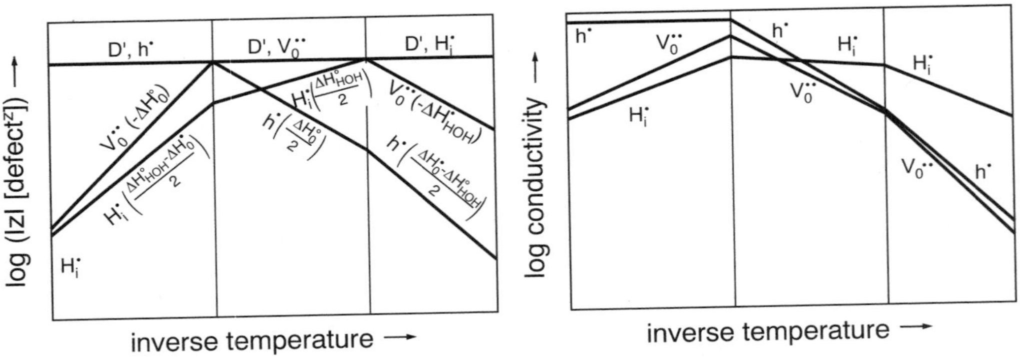

Fig. 5.59d: Dependence of the defect concentrations and the conductivities of water–containing perovskites on the temperature (see text) under reversible conditions (D′: negative dopant).

and Eq. (5.173a). Since Eq. (5.176a) is exothermic, the equilibrium described by it is displaced to the left as the temperature increases, while the equilibrium described by Eq. (5.173a) is shifted to the right on account of the endothermicity. The intermediate position of $[V_O^{\bullet\bullet}]$ follows from the fact that the vacancies are included symmetrically in both equations.

The fair agreement of the experiments with the predicted O^{2-}, H^+ and e^- conductivities even in these complex cases again reveals the power of the defect–chemical approach[112].

A technological conclusion ought to be mentioned here, namely that an optimized proton conductor requires comparatively low temperatures, appreciable but not too high negative doping, and a high water but not too high an oxygen content in the ambient. It must be mentioned once again that both deviations from Boltzmann behaviour at high concentrations and the kinetic problems at low temperatures mentioned in connection with $SrTiO_3$ have been neglected for simplicity's sake.

At the end of this section let us briefly inspect the situations in covalent compounds [42,128] (cf. Section 1.2). Let us consider the silicon network. Extrinsic effects are dominant here too at low temperatures (see Fig. 5.45, page 177). On account of the symmetrical arrangement it is possible to assign regular Si the oxydation number 0. Every regular structure element Si_{Si} possesses 4 bonding electrons (covalency is 4). A phosphorus impurity introduces 5 outer electrons. When four covalent bonds are

[112]It is sensible at this point to recall the defect chemistry of alkali hydroxides (see footnote 88, page 171). Apart from native acid–base effects, the role of CO_2 is worth paying particular attention to. Such a contamination leads, according to

$$CO_2 + 2OH_{OH}^{\times} \rightleftharpoons (CO_3)'_{OH} + HOH_{OH}$$

to an increase of acidity and proton conductivity. Only at high temperatures will this occur reversibly (P_{CO_2} as in situ parameter). At lower temperatures the CO_2 content is frozen and plays the role of a dopant concentration (carbonate content adds to the C–term in Eq. (5.179)) [177].

formed it is easy to delocalize the excess electron (∗). The phosphorus, now being isovalent to Si, then has the effective (formal) charge of +1 ($(P^{+}_{Si^0})^{\cdot}$, for short $P^{\cdot}_{Si}$):

Reaction P: $$^{*}P + Si_{Si} \rightarrow P^{\cdot}_{Si} + e' + Si. \qquad (5.182)$$

Such phenomena have been extensively discussed in Section 5.3. Here it should be pointed out again that the doping reactions of semiconductors are usually divided up into the irreversible incorporation of the defect

Reaction P1: $$^{*}P + Si_{Si} \rightarrow {}^{*}P_{Si} + Si \qquad (5.183a)$$

and the reversible electronic excitation

Reaction P2: $$^{*}P_{Si} \rightleftharpoons P^{\cdot}_{Si} + e'. \qquad (5.183b)$$

The asterisk indicates the excess electron compared to Si (and to the ionized phosphorus). Since Eq. (5.183b) corresponds to a purely electronic effect in a wide band semiconductor, it can be described adequately by means of the band diagram. The relative stability of $P^{\cdot}_{Si}$ corresponds to the low ionization energy of $^{*}P_{Si}$. In physical jargon we speak of shallow donors. The small value of $\Delta_{P2}G^{\circ} = \mu^{\circ}_{e'} - [\mu^{\circ}(^{*}P_{Si}) - \mu^{\circ}(P^{\cdot}_{Si})] = \mu^{\circ}$(electron in conduction band) − μ°(excess electron in the doping atom) corresponds to a small distance between the donor level and the conduction band (see Section 5.3 for a more precise treatment). Analogously doping with Al creates an Al^{-} state ($(Al^{-}_{Si^0})'$), which corresponds to a shallow acceptor. In chemical language:

Reaction Al: $$^{\square}Al + Si_{Si} \rightarrow Al'_{Si} + h^{\cdot} + Si \qquad (5.184)$$

Reaction Al1: $$^{\square}Al + Si_{Si} \rightarrow {}^{\square}Al_{Si} + Si \qquad (5.185a)$$

Reaction Al2: $$^{\square}Al_{Si} \rightleftharpoons Al'_{Si} + h^{\cdot}. \qquad (5.185b)$$

The symbol □ indicates the electron deficit compared to Si (and to the ionized Al). The electron necessary for completion of the 4 bonds is removed from the valence band (i.e. from the regular Si atoms)[113]. In shallow states, it is possible to neglect the concentration of unionized states to a good approximation. If the distances between the levels are greater, we speak of deep impurities and two valence states

[113]It may sound strange: If aluminium forms three covalent bonds, then its formal charge is zero; it exhibits a formal negative charge in the case of four bonds. On the other hand, doping of SnO_2 with trivalent iron (see above) leads to a negative defect (Fe'_{Sn}), while tetravalent Fe is neutral. This is because in the case of covalent bonding aluminium shares its electrons with the partner and does not release them, hence, the triply bonded has one electron less than the quadruply bonded, while, in the case of Fe^{4+}/Fe^{3+}, this is reversed in regard to valency. In any case the negative defect has received an extra electron. These relationships mirror the difference between the concepts "oxydation number" and "covalency".

are of importance (see Fe^{3+}/Fe^{4+} in $SrTiO_3$). These effects are treated in more detail in the next section.

We can use a similar language for ionic conductors. Thus, the "ground state" of Cd doping in AgCl in the ionic diagram (see Fig. 5.61, page 203) corresponds to the undissociated $Cd^{\cdot}_{Ag} - 2V'_{Ag}$ complex and, in the case of the sulfide doping, to the $2Ag^{\cdot}_{i} - S'_{Ag}$ complex. The transition to the free vacancy and interstitial levels corresponds to the dissociation of the complexes according to a Ag^{+} transfer to the lower and energy level, respectively. See the next section for this, too. There defect–defect interactions are treated explicitly. It should be remembered that such considerations form a bridge from pure compounds to mixed crystals (see Chapter 4).

5.7 Interactions between defects

The above examples showed that ideal mass action laws require correction in a variety of cases. Ideal mass action laws for defects only apply when there are random distributions and, hence, at very low defect concentrations. A high disorder — consider, for instance, a solution of 5% SrO in lanthanum copper oxide — can certainly no longer be treated as a simple structural perturbation; rather for accurate analysis, individual treatment of the mixed phases is necessary. Then, of course, tractable models lose their general character, and the reader's expectations of the correction possibilities should not be too high. In this section the intention is — with just one exception — to discuss simple concepts, that are valid, when the ideal mass action laws have "just lost their validity", that is, we are still limiting ourselves, in principle, to more or less low defect concentrations. In many cases, however, such concepts qualitatively indicate the right direction, beyond the region of strict validity.
In a purely formal manner it is possible to correct the Boltzmann approach by replacing concentration by activity and, hence, inserting an activity coefficient term[114]:

$$\mu_k(c_k) = \mu_k^\circ + RT \ln a_k(c_k) = \mu_k^\circ + RT \ln c_k + RT \ln f_k(c_k). \tag{5.186}$$

In what follows we will discuss two methods: One formulating $f_k(c_k)$ explicitly for the interacting defects in simple cases and the other involving the introduction of new species occurring as a result of the interaction. In the second case the introduction of a new set of defects leads to a rescaling such that, to a first approximation, the Boltzmann approach is valid once again, but now for the new set of defects over a greater concentration range. Both methods can be combined to give a better approximation. We start with the last concept.

[114]Please note that higher concentrations demand corrections with respect to the Boltzmann–form, already via statistical effects, that is corrections with respect to the configurational entropy. Cf. page 214ff.

5.7.1 Associates

Let us consider the Frenkel reaction in an ideally pure crystal. If we gradually reduce the temperature, the interstitial ion will eventually fall back into the vacancy (cf. $K_F \rightarrow 0$ for $T \rightarrow 0$). In the same way, the semiconductor's conduction electrons will, via annihilation of holes, become valence electrons ($K_B \rightarrow 0$). In a similar sense, in strongly Cd–doped AgCl, there is a Coulomb attraction between $Cd^{\cdot}_{Ag}$ and the counterdefect V'_{Ag} leading to a mutual trapping and thus to a deviation from random distribution at low temperatures. This can be described approximately by an exothermal production of associates [203,204] of the form

$$\text{Reaction As} = \qquad V'_{Ag} + Cd^{\cdot}_{Ag} \rightleftharpoons (Cd_{Ag}V_{Ag})^{x} \tag{5.187}$$

with the association constant[115]

$$K_{as} = \frac{[(Cd_{Ag}V_{Ag})^{x}]}{[Cd^{\cdot}_{Ag}]\,[V'_{Ag}]}. \tag{5.188}$$

The associate formed corresponds to a mutual neutralization; the complex formed possesses (approximately!) a zero effective charge and does not contribute to conductivity[116]. Structurally the association means that the vacancy is to be found in the immediate neighbourhood of the dopant ion, from which it cannot be released without perceptible effort.

To a large degree, associate formation not only accounts for Coulomb effects, but in particular for specific interactions, as they markedly occur between $Ag^{\cdot}_{i}$ and S'_{Cl} in Ag_2S–doped AgCl. This interaction reflects the large electronic polarizabilities or even distinct covalencies (remember the anomalous behaviour in Fig. 5.50) which also give rise to the extremely low solubility product of the Ag_2S phase in water. The high association is certainly faciliated by the fact that the defects can approach each other very closely (interstitial position!).

The classification as associated pairs and free ions is naturally somewhat arbitrary in the case of purely electrostatic interactions. This distinction was carried out by Bjerrum [203] for solutions of strong electrolytes: The electrostatic interaction requires an increased probability density of the counterions to reside in the neighbourhood of the central ion which drops as a function of the distance in the given direction. On the other hand, the spherical shell increases in volume with distance in the given direction from the central ion. Overall the probability of the counterions residing at a given distance passes through a minimum. This minimum a_B ($\propto 1/\varepsilon$) is 3.6Å for an aqueous solution of M^+X^- at 25°C. In this sense, it is plausible to regard ion pairs with a smaller distance as associates. The effects at larger distances are explained (for instance) with the aid of the Debye–Hückel theory [205]. In this way

[115] Since the associate represents a dipole, that can take up various orientations, the number of orientations enters the mass action constant K_{as} as a correction with respect to the concentration measure, i.e. an entropy correction that can be included in the mass action constant.

[116] Apart from this, the mobility will be also negligible owing to the low Cd–mobility.

a significant part of the interaction is accounted for in a simple manner. However, the concept cannot be applied with equal precision to solids. The notion that the contact pair is effectively neutral at distance $r < a_B$ is less likely to be true for a solid body, one reason lies in the fact that on account of the usually lower dielectric constant, the parameter a_B can be appreciably higher than the lattice constant, another one lies in the rigidity of the lattice[117].

But back to the formal phenomenological treatment. The introduction of associates divides up the constant doping concentration C into the concentration of the associates and that of the free Cd ions:

$$[\mathrm{Cd}^{\cdot}_{\mathrm{Ag}}] + [(\mathrm{Cd}_{\mathrm{Ag}}\mathrm{V}_{\mathrm{Ag}})^{\mathrm{x}}] = \mathrm{C}. \tag{5.189}$$

The individual concentrations can be calculated using Eqs. (5.188, 5.189) and the electroneutrality condition. We have added one more variable by introducing associates, but there is also an additional equation. Although the precise calculation is simple (see results in Fig. 5.60), let us restrict to the Brouwer approximation and

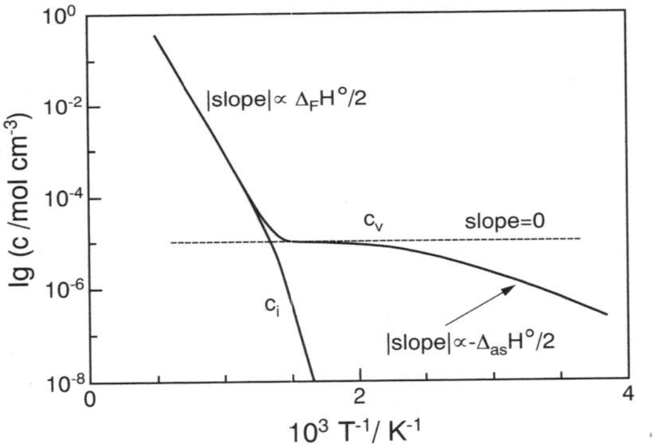

Fig. 5.60: Concentrations of vacancies and interstitial particles in a positively doped Frenkel–disordered material taking account of the association between cationic vacancies and dopant ions. The parameters used were $\Delta_{\mathrm{as}}\mathrm{S}^{\circ}_{\mathrm{m}} = 0$, $\Delta_{\mathrm{F}}\mathrm{S}^{\circ}_{\mathrm{m}} = 10\mathrm{R}$, $\Delta_{\mathrm{as}}\mathrm{H}^{\circ}_{\mathrm{m}} = -40\,\mathrm{kJ/mol}$ and $\Delta_{\mathrm{F}}\mathrm{H}^{\circ}_{\mathrm{m}} = 200\,\mathrm{kJ/mol}$.

the extreme case, where extensive association occurs ($-\Delta_{\mathrm{as}}\mathrm{G}^{\circ}$ high and/or T small). In this case, almost all the Cd is associated ($[\mathrm{Cd}^{\cdot}_{\mathrm{Ag}}] \ll [(\mathrm{Cd}_{\mathrm{Ag}}\mathrm{V}_{\mathrm{Ag}})^{\mathrm{x}}] = \mathrm{C}$). Then, with the condition of electroneutrality, it follows

$$[\mathrm{V}'_{\mathrm{Ag}}] = [\mathrm{Cd}^{\cdot}_{\mathrm{Ag}}] \tag{5.190}$$

and from Eq. (5.188)

$$[\mathrm{V}'_{\mathrm{Ag}}] = \left(\frac{\mathrm{C}}{\mathrm{K}_{\mathrm{as}}}\right)^{1/2} \tag{5.191a}$$

or, in a form that will be required later,

$$\ln[\mathrm{V}'_{\mathrm{Ag}}] = \ln \mathrm{C} + \ln[(\mathrm{K}_{\mathrm{as}}\mathrm{C})^{-1/2}]. \tag{5.191b}$$

---

[117] A more rigorous treatment is given by Allnatt and Lidiard [9]. It is certainly sensible to define crystal–chemical interaction criteria in the individual cases, e.g. via the nearest neighbour sphere.

Equation (5.191) shows how association affects ionic conduction. In extrinsic, non-associated AgCl ($[V'_{Ag}] = C$) the charge carrier concentration would be higher by a factor of $\sqrt{K_{as}C}$. Accordingly, the (free) charge carrier concentration, responsible for ionic conduction, becomes temperature–dependent again and it follows that

$$W_{V'_{Ag}} = -1/2\Delta_{as}H^{\circ} > 0, \tag{5.192}$$

which is the low temperature limit of the more precise calculation represented in Fig. 5.60. The experimental data for $\Delta_{as}H^{\circ}$ in the case of Cd^{2+} doping vary between -0.3 and -0.5 eV [10].
The association tendency is more marked in a less polar environment. The tendency of forming ion pairs limits the ionic conductivity of salt–containing polymers (see page 291).

Analogous processes occur with respect to electronic charge carriers. If the dopant ion is redox active, the analogous process corresponds to a change in the dopant's redox state, which has already been considered in some detail in the previous sections (especially Section 5.3). The association energy is reflected by the distance of the defect levels from the band edges as shown e.g. in Fig. 5.10 (page 128). (The analogous "physical diagram" for the purely ionic defects[118] just discussed is given in Fig. 5.61.) In the case of ionic–electronic interactions involving redox stable

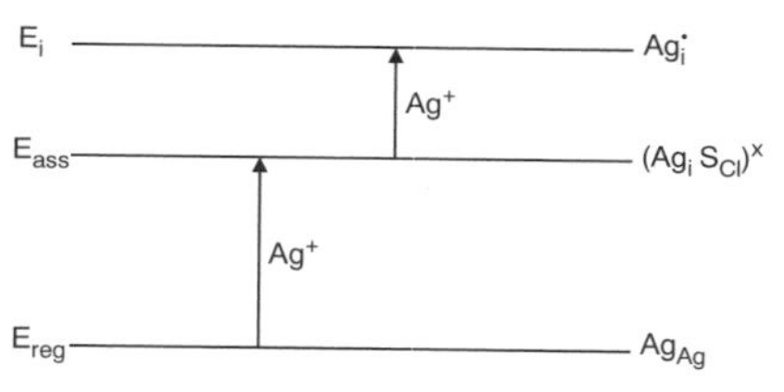

Fig. 5.61: The ionic analogue of the defect level picture in semiconductors. The levels are more accurately electrochemical standard potentials (see Section 5.3). The individual processes correspond to the transition of Ag^+. More precisely the terms $Ag_i^{\cdot}$, $(Ag_iS_{Cl})^x$ and Ag_{Ag} have to be replaced by $Ag_i^{\cdot} - V_i$, $(Ag_iS_{Cl})^x - (V_iS_{Cl})'$ and $Ag_{Ag} - V'_{Ag}$ [206]. A generalized acid–base picture[118] that considers the associates as internal acids or bases is given in Ref. [15].

dopants dipole states may be formed (instead of true valence changes). (Thus, the association of a hole in Al–doped $SrTiO_3$ with a shallow defect Al'_{Ti} should correspond to $(Al'_{Ti}O^{\cdot}_{O})^x \equiv (Al'_{Ti}h^{\cdot})^x$ rather than to $Al^x_{Ti} \equiv \left(Al^{4+}_{Ti^{4+}}\right)^x$.)
If valence changes are possible, the association concept in the electronic case has a clearer meaning than in the purely ionic one and it can, generally, be easily verified, e.g. spectroscopically. This is the case, for instance, for the various states of

[118]Such a picture is also very advantageous in the consideration of proton transfer in water (see also Fig. 5.9). The levels within the fundamental gap characterize the strength of bases and acids (e.g. NH_3 or CH_3COOH). The distance to the fundamental levels is determined by pK_S and pK_B. In the same way the levels of the associates in Fig. 5.61 can be conceived as the levels of internal acids and bases (if Ag^+ is taken as the acidic particle) [15].

oxydation of iron[119] in $SrTiO_3$:

$$Fe'_{Ti} + h^{\cdot} \rightleftharpoons Fe^{x}_{Ti}. \tag{5.193}$$

Fe^{x}_{Ti} can be regarded as an associate between Fe'_{Ti} and a hole. For the formal treatment it makes no difference whether the hole is situated at a oxygen ion neighbouring the Fe–defect (i.e. $Fe^{3+}O^{-}$) or whether the valence of Fe is strictly 4+. While the latter (Fe^{4+} in an O^{2-} environment) is the better approximation in this case (in contrast to the above Al–example), generally the truth may lie in between the extremes (electronic defect distributed over the impurity and the neighbours). The $[Fe^{4+}]/[Fe^{3+}]$ ratio increases as the temperature falls (exothermic association) and as the oxygen partial pressure increases (oxydation). Figure 5.62 displays the calculated dependence of the Fe^{4+} concentration on T and P_{O_2}; this has been confirmed

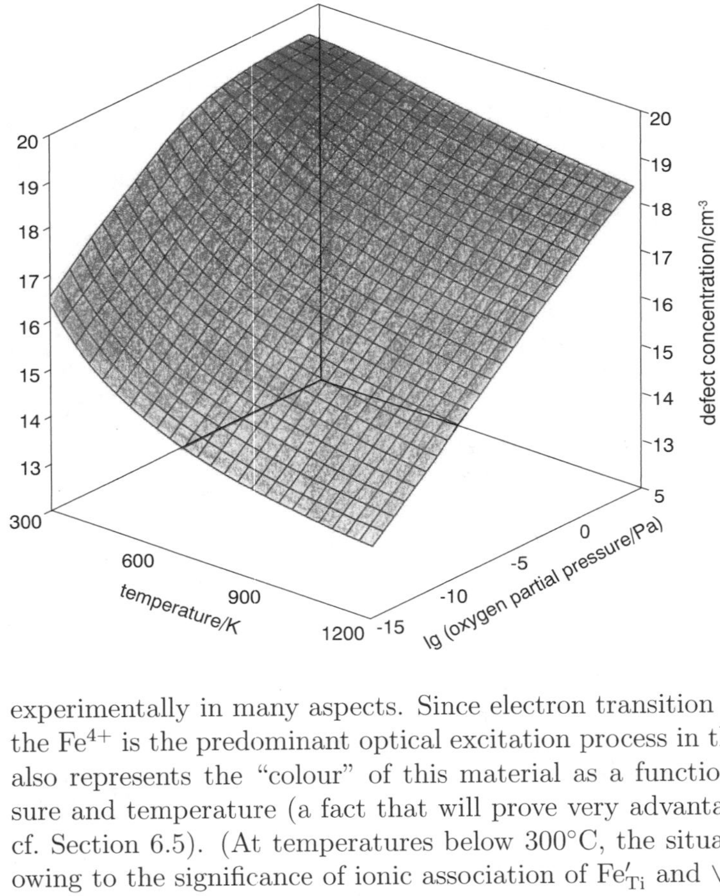

Fig. 5.62: Fe^{4+} concentration in Fe–doped $SrTiO_3$ as a function of T and P_{O_2}. The total iron concentration ($\sim 10^{19}/cm^3$) corresponds to the limiting case for $(T, P_{O_2}) \rightarrow (0, \infty)$ [127].

experimentally in many aspects. Since electron transition from the valence band to the Fe^{4+} is the predominant optical excitation process in the visible range, Fig. 5.62 also represents the "colour" of this material as a function of oxygen partial pressure and temperature (a fact that will prove very advantageous for kinetic studies, cf. Section 6.5). (At temperatures below 300°C, the situation is more complicated owing to the significance of ionic association of Fe'_{Ti} and $V_{O}^{\cdot\cdot}$ [207].)

[119] If the reader is confused by the fact that (under standard conditions) Fe^{2+}/Fe^{3+} is the stable redox couple in H_2O, whereas Fe^{3+}/Fe^{4+} is more stable in $SrTiO_3$, he or she may consider the effective charges in both cases.

Associations between native ionic and electronic defects can also play a role. Particularly at lower temperatures (see Fig. 5.51, page 187) the conductivity effects in SnO_2 (and ZnO) are attributed to the capture of e' at oxygen vacancies [171]. This means that (one or two) electrons are trapped quasi as anion replacements in oxygen vacancies according to:

$$\begin{aligned} V_O^{\cdot\cdot} + e' &\rightleftharpoons V_O^{\cdot} \\ V_O^{\cdot} + e' &\rightleftharpoons V_O^{x} \end{aligned} \tag{5.194}$$

corresponding to two donor states in the band gap. (In the case of MgO even different migration energies (see Section 6.2) for the different defects have been determined [208].) The counterpart occurs at oxygen excess p conductors:

$$O_i'' + 2h^{\cdot} \rightleftharpoons O_i' + h^{\cdot} \rightleftharpoons O_i^{x}. \tag{5.195}$$

Here the interstitial oxygen itself or the immediate environment is oxydized up. Such effects have to be considered for La_2CuO_4 at low temperatures. On account of the higher defect densities, such interactions are already important at higher temperatures in the high temperature superconductors $YBa_2Cu_3O_{6+x}$ or $Bi_2Sr_2CaCu_2O_8$ and can explain the unusually high $N_{h^{\cdot}}$ value of up to 1/2 [209,210].
Provided the Brouwer approximation holds, power laws as given by Eq. (5.150) are also valid in the associated cases. The corresponding (P, T, C) rules, which determine the sign of the change, are valid here (pages 168, 169, 179), too. Since $V_O^{\cdot}$ is an associate between $V_O^{\cdot\cdot}$ and e', that is $[V_O^{\cdot}] \propto [V_O^{\cdot\cdot}][e']$, the P and C dependences predicted for $[V_O^{\cdot\cdot}]$ and $[e']$ add up. In the case of doping dependence, the opposing effects on $V_O^{\cdot\cdot}$ and e' weaken each other: Since $V_O^{\cdot\cdot}$ is double–charged, the first tendency predominates and Eq. (5.152) still applies[120]. The situation is analogous for O_i'. (This would be particularly true of highly unlikely defects such as $V_O^{\cdot\cdot\cdot}$, O_i'''.) In the case of the P law (page 168) it has been established that, on increasing the potential of the electronegative components, the defects, which themselves increase (decrease) the anion–cation ratio or the state of oxydation, are increased (decreased) in their concentration. This applies definitely to $V_O^{\cdot}$ and O_i', since the two effects accumulate (e.g. $N_{V_O^{\cdot}} = N_{V_O^{\cdot\cdot}} + N_{e'}$, because of Eq. (5.194)). $V_O^{\cdot}$, for instance, corresponds to a reduced state and in addition its formation simultaneously means a decrease of the anion–cation ratio[121].
Colour centre formation in alkali halides is another well–investigated interaction reaction [211]. If we treat NaCl with metallic Na, that is, if we bring the NaCl phase to the Na–rich end of the (extremely small) region of homogeneity ($NaCl_{1-|\delta|_{max}}$), an enhanced electron concentration is induced in the Schottky–disordered NaCl as

---

[120] This can be seen on the basis of reactions involving both unassociated defects. It follows, for example, from $[V_O^{\cdot\cdot}]P_{O_2}^{1/2} = K_O^{-1}(T)[e']^{-2}$ the relation $M_{V_O^{\cdot\cdot}} \equiv \left(\frac{\partial \ln [V_O^{\cdot\cdot}]}{\partial \ln C}\right)_{P,T} = -2\left(\frac{\partial \ln [e']}{\partial \ln C}\right)_{P,T} \equiv -2M_{e'}$. Using $[V_O^{\cdot\cdot}][e'] = K_{as}^{-1}(T)[V_O^{\cdot}]$ we see that $M_{V_O^{\cdot}} = M_{V_O^{\cdot\cdot}} + M_{e'} = M_{e'}(-2+1) = -M_{e'} = \frac{1}{2}M_{V_O^{\cdot\cdot}}$ and, hence, $\mathrm{sign}\,(M_{V_O^{\cdot}}) = \mathrm{sign}\,(M_{V_O^{\cdot\cdot}}) = -\mathrm{sign}\,(M_{e'})$.
[121] This no longer applies for exotic defects such as O_i''' or $V_O^{\cdot\cdot\cdot}$.

is an increased anion vacancy concentration according to

$$\mathrm{Na} + \mathrm{V}'_{\mathrm{Na}} \rightleftharpoons \mathrm{Na}^{\times}_{\mathrm{Na}} + \mathrm{e}' \tag{5.196a}$$

and (in combination with the Schottky equilibrium)

$$\mathrm{Na} + \mathrm{Cl}^{\times}_{\mathrm{Cl}} \rightleftharpoons \mathrm{NaCl} + \mathrm{V}^{\cdot}_{\mathrm{Cl}} + \mathrm{e}'. \tag{5.196b}$$

At lower temperatures[122] free electrons are captured according to

$$\mathrm{V}^{\cdot}_{\mathrm{Cl}} + \mathrm{e}' \rightleftharpoons \mathrm{V}^{\mathrm{x}}_{\mathrm{Cl}} \tag{5.197}$$

in the (effectively positively charged) chloride vacancies (formation of $\mathrm{V}^{\mathrm{x}}_{\mathrm{Cl}}$). This treatment endows the crystal with a violet colour. Other colour centres in other alkali halides are characterized by similar characteristic optical absorptions. Since the electron cloud of the trapped electron is, roughly speaking, limited to the vacancy, it is possible to predict the colour by using the electron–in–the–box problem and identifying the box size with the lattice constant (a). The light absorbed raises the electrons from the ground state to the next highest state. According to Chapter 2 (Eq. (2.27)) the quantum number vector $(\mathrm{n_x}, \mathrm{n_y}, \mathrm{n_z})$ changes from (1, 1, 1) to (2, 1, 1) or (1, 2, 1) or (1, 1, 2). Because $\epsilon = \frac{\mathrm{h}^2}{8\mathrm{m}_{\mathrm{e}^-}}\left(\mathrm{n_x^2} + \mathrm{n_y^2} + \mathrm{n_z^2}\right)\mathrm{a}^{-2}$ it follows that

$$\mathrm{h}\nu_{\mathrm{abs}} = \Delta\epsilon = \left(\frac{3\mathrm{h}^2}{8\mathrm{m}_{\mathrm{e}^-}}\right)\mathrm{a}^{-2}. \tag{5.198}$$

This simple calculation yields an absorption maximum for NaBr that only deviates by ca. 20% from the exact result. In particular, Equ. (5.198) explains the result known as the Mollwy–Ivey law [11], stating that in the case of the colour centres of the alkali halides, the wavelengths vary approximately with the square of the lattice constant (Fig. 5.63). Such simple estimates, of course, do not make it possible to calculate the exact size of the vacancies[123], but they did play a historically important role in generating confidence in the defect–chemical concept.
The hole $((\mathrm{Cl_i})'_2(\mathrm{V}^{\cdot}_{\mathrm{Cl}})_2)$ trapped at Cl^- in NaCl under oxydizing conditions mentioned in Section 5.3 is not an associate of two defects that were originally present, rather it refers to the polarization of the perfect lattice by $\mathrm{h}^{\cdot}$ (cf. electron–phonon interaction).

Association of intrinsic defects is also known. The electron–hole pair is an example:

$$\mathrm{Nil} \rightleftharpoons (\mathrm{e}'\mathrm{h}^{\cdot}) \equiv \mathrm{exciton} \rightleftharpoons \mathrm{e}' + \mathrm{h}^{\cdot}. \tag{5.199}$$

---

[122] In fact, the temperature need not be very low owing to the comperatively high energy of the free electrons in NaCl (excess electron in the Na–orbitals!).

[123] About 90% of the wave function of the ground state is localized at the vacancy. The remainder is smeared out over several lattice constants.

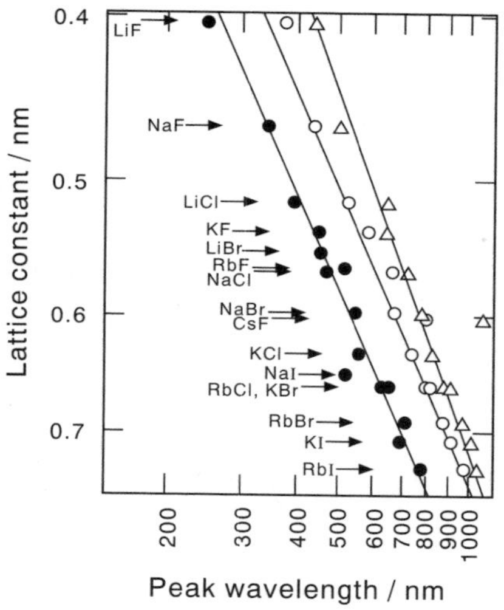

Fig. 5.63: The Mollwy–Ivey law (i.e. $\nu_{abs} \propto a^{-x}$) for vacancy defects "F", "F_2" and "F_3" in alkali halides. The left hand curve refers to the elementary F centres of interest in this context. The two lines displaced to higher wavelengths refer to aggregates of two or three F centres (F_2 and F_3 centres). The plot is double logarithmic. The slope reveals the power law $\nu_{abs} \propto a^{-1.77}$. From Ref. [11].

Here it is not a recombined "zero state", but an excited state characterized by polarization effects, which is separated from the fully dissociated state by an activation threshold and, as such, is capable of migrating. Then excitation energy is, as it were, transported in the form of excited, but neutral, particles. Such excitons are somewhat similar to hydrogen atoms and also possess similar spectra; however, the energy states, in particular, are different, roughly speaking, on account of the fact that the dielectric constant of the medium is not one. (The ionic analogue would be a nondissociated metastable Frenkel pair.)

The Cooper pair is a further electronic associate of considerable importance:

$$\begin{aligned} 2e' &\rightleftharpoons (e')_2 \\ 2h^{\cdot} &\rightleftharpoons (h^{\cdot})_2 . \end{aligned} \tag{5.200}$$

Associates of this type exist in several compounds, in spite of the Coulomb repulsion (e.g. because of phonon effects) but generally at extremely low temperatures; they behave like bosons and provide for superconduction (see also page 293). One consequence of the Coulomb repulsion is that, in many cases, the correlation length can be considerably greater than the lattice constant. The (free) enthalpy of the reaction described by Eq. (5.200) corresponds to the Cooper gap in the energy level presentation.

The intrinsic defects formed by the Schottky reaction or the anti–Schottky reaction can also associate, without forming or annihilating a regular monomer (otherwise it would be a simple return to the zero state of the perfect crystal):

$$\begin{aligned} M_i^{\cdot} + X_i' &\rightleftharpoons (M_iX_i)^x \\ V_M' + V_X^{\cdot} &\rightleftharpoons (V_MV_X)^x . \end{aligned} \tag{5.201}$$

These reactions can play an important role as precursors of phase precipitation or pore formation.
In systems with trapping in which percolative effects play an important role, as can be the case in highly doped systems, the situation soon becomes complex. A numerical treatment of ion dynamics in systems with arbitrarily distributed Coulomb–trap centres is given in Ref. [212]. The possibility of transport of electron carriers exclusively via impurity states corresponds to the formation of a (possibly very narrow) impurity band.
The treatment becomes very specific when we are considering extended defects or ordering effects, which we can regard as agglomerations or as results of co–operative segregation of point defects. Thus, clusters of defects are held responsible for the marked "nonstoichiometry"[124,125] of $Fe_{1-\delta}O$ or UO_{2+x} [213,214]. Oxygen–deficient Magnéli phases [215] are produced when TiO_2 is reduced; these can be thought of as being formed by whole rows of oxygen atoms going into the gas phase and whole rows of Ti going into the interstices[126] (see Fig. 5.64). The Ruddlesdon–

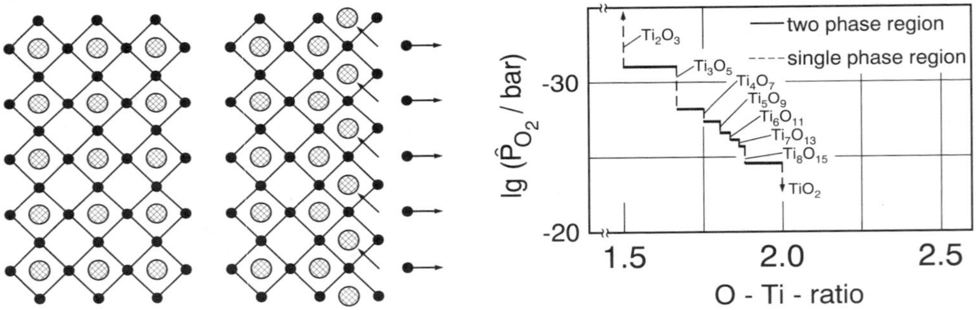

Fig. 5.64: The formation of Magnéli phases (left). Oxygen (dark circles) is removed at low oxygen partial pressure, and ordered interstitial cation configurations are formed (pale circles). The thermodynamic stability range of the resulting phases is shown in the graph of the equilibrium oxygen partial pressure as a function of composition (right) (cf. Chapter 4). According to Refs. [6,216].

Popper phases, which have been demonstrated in SrO–deficient strontium titanates, are a complementary example [217]. Oxygen defects in $YBa_2Cu_3O_{6+x}$ at lower

[124]The true Dalton composition is actually nonexistent in $Fe_{1-\delta}O$ under usual conditions.

[125]Thermodynamically, we can understand the formation of such extended defects under equilibrium conditions as follows: Either there is only a very slight positive ("initial") free enthalpy of formation which is compensated by the low configurational contribution, or there is possibly even a negative value for the formation of the first defect, which, together with the configuration contribution, is then compensated by a repulsive defect–defect interaction, increasing strongly with concentration.

[126]Strictly speaking, in cases in which the free standard enthalpy of formation with respect to the perfect state is negative, we should not use the form "defect".

temperatures can be given as an example of defect order; they arrange themselves in chains which affects the superconducting transition temperatures(see Fig. 5.65). These points take us to the limits of the defect concept (see also Section 4.3.5 and particularly page 99).

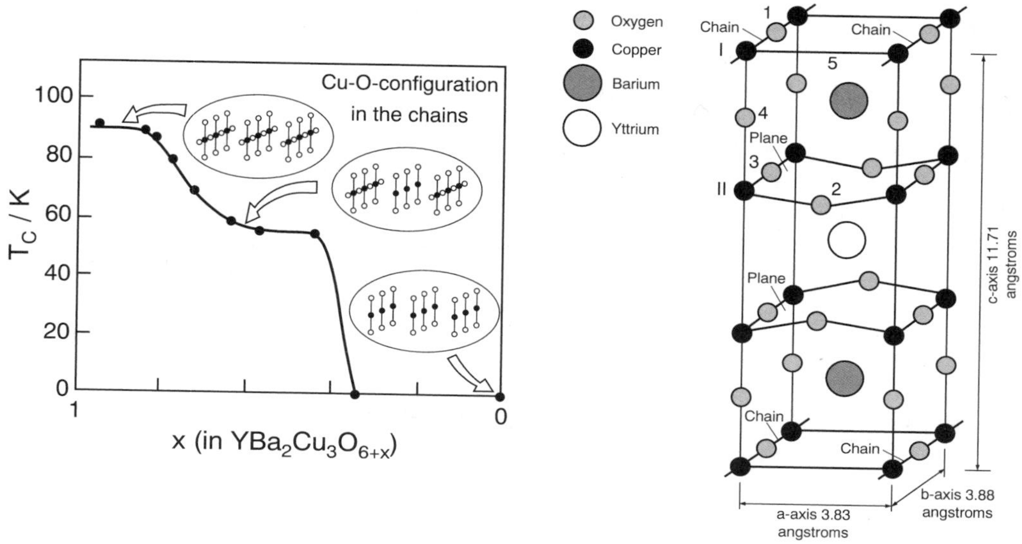

Fig. 5.65: Ordered structures for interstitial oxygen in the Cu–O chains of the high temperature superconductor $YBa_2Cu_3O_{6+x}$ (at x=0.5 every second chain position is occupied at sufficiently low temperatures, while at x=1 all chain positions are occupied) and the effects on the critical temperature T_C. The precise structure for x=1 is shown in the right hand diagram. From Ref. [218].

Let us now return to the general theme of our book and consider our $CdCl_2$–doped AgCl and Eq. (5.191b). If we were not to describe the reduction in conductivity in terms of association as we have done above (see Eq. (5.191)), and we had to measure every silver vacancy by the same yardstick ("V'_{Ag}"), we would use an activity coefficient, which, on account of C = ["V'_{Ag}"], would evidently be given by the second term in Eq. (5.191b)[127]:

$$f_{eff} = (K_{as}C)^{-1/2} < 1. \tag{5.202}$$

[127] Since we can write the incorporation equilibrium of chlorine (see Eq. (5.136)) with the free but also with the "overall" vacancies, their chemical potentials are identical (page 124). Apart from irrelevant questions of normalisation and apart from μ° and RT, the l.h.s. of Eq. (5.191b) corresponds to this potential. The term ln C corresponds to the Boltzmann–term for the "overall" vacancies and hence $\ln(K_{as}C)^{-1/2}$ to the activity coefficient (per RT).

If this seems too abstract you can also see the connection as follows: Let us consider Eq. (5.136) once again and imagine we knew the mass action constant and the correct hole concentration experimentally. Tentatively we may use the value C for $[V'_{Ag}]$ and find that the mass action law is violated and, hence, correct C with f to Cf. On the other hand, we may find the mass action law

This effective activity coefficient is smaller than 1, since $C > [V'_{Ag}]$ (see Eq. (5.191)) and decreases with increasing doping concentration and increasing association strength. In this sense we can also describe the ideal association concept as follows: After rescaling charge carrier types and concentrations ideal mass action laws can be used again to a better approximation (but now for an increased number of carrier types). In this way a large proportion of nonideality is taken account of.

5.7.2 Activity coefficients

The Coulomb interaction over larger distances, that is not included in association, can be described for moderate concentration ranges by means of the Debye–Hückel theory [205]. The negative space charge around a positive defect decays monotonically with the Debye length λ as the characteristic length. In the case of two defects of the same charge[128] ($z_+ = |z_-| = z$), the electrochemical treatment yields[129] (see also following section):

$$\lambda = \sqrt{\frac{\varepsilon RT}{2z^2F^2c_\infty}}. \tag{5.203}$$

At very high concentrations λ approaches zero. Here (if the arrangement were still stable) a dense rigid defect lattice (cf. rigid double–layers in the one–dimensional representation) would be formed. At infinite dilution the Debye length becomes infinitely long and the distribution goes over into a statistical one.
Let us write the correction term[130] in the chemical potential as

$$\begin{aligned} \mu_k^{ex} &= \mu_k - \mu_k^0 - RT\ln c_k = RT\ln\left(a_k/c_k\right) \\ &= RT\ln f_k \end{aligned} \tag{5.204}$$

with a_k and f_k as the activity and activity coefficient, so that the latter, which is less than one on account of the interaction observed, is given by[129]

$$RT\ln f_k = -\frac{z_k^2F^2}{8\pi\varepsilon N_A}\lambda^{-1} \propto c_\infty^{1/2}. \tag{5.205}$$

At higher concentrations a better approximation is arrived at if the expression is corrected by the factor $(1 + a_B/\lambda)^{-1}$, which takes a minimum distance into account (see previous section). It is shown from the theory of electrolytic solutions that the Debye–Hückel theory breaks down at higher concentrations (in the 0.1% region

to be fulfilled to a good approximation by the free concentration $[V'_{Ag}] = (C/K_{as})^{1/2}$. Obviously the result is $f = (K_{as}C)^{-1/2}$.

[128] When there are more than two defects, then the "defect strength" $\frac{1}{2}\sum_k z_k^2 c_{k\infty}$ has to be used instead of z^2c_∞.

[129] See electrochemistry textbooks [151–153,219–221].

[130] Since here the two measures of concentration (x, c) are proportional to each other and we have agreed to include the conversion factors in μ°, we only need to concern ourselves with a single activity coefficient (primarily related to x).

at the latest) for various reasons [222], but, primarily, because the structure of the immediate environment becomes important, that is the "ion atmosphere" of the continuum treatment is becoming "coarse–grained". It is evident that the importance of structure is more pronounced in solids, i.e. those concepts which neglect individual structural details rapidly lose their validity. More accurate descriptions [223–227,222] are extremely complicated, awkward and/or specific.

Nevertheless, it has been demonstrated, at least in the case of AgCl, AgBr, AgI and PbF_2, that at high defect concentrations, at which the Debye–Hückel theory breaks down, the nonideality effects can be described very well on the basis of a simple cube root law [115]:

$$RT \ln f_\pm = -J^c_\pm c_\pm^{1/3} = -J_\pm x_\pm^{1/3} \tag{5.206}$$

with J or $J^c(= JV_m^{1/3})$ as c–independent positive parameters. The index $\pm$ indicates that here, according to $\mu_\pm = \frac{1}{2}(\mu_+ + \mu_-)$, we refer to the arithmetic mean of the individual contributions of the differently charged particles and thus to the mean activity coefficient $f_\pm = \sqrt{f_+ f_-}$. Only these mean parameters are accessible to measurement. (This is also true in the range of validity of the Debye–Hückel approach.) The quantities $c_\pm$ and $x_\pm$ correspond to the respective intrinsic concentrations. Equation (5.206) appears to be intuitively reasonable, since $c_\pm$ scales with the mean defect separation; it does not imply that an ordered defect lattice actually forms, but merely that the energetic effects can, on average, be considered in terms of an ordered defect lattice[131]. In fact, numerical calculations show that, on account of a topological order, the Coulomb energy of a molten salt can be described very well by means of Madelung energies[132] (see Section 2.2.2), even though the distances between the ions are fluctuating [114]. According to these calculations the effective Madelung constants are smaller than in the ordered lattice. In this sense we can describe the energetic effects by superimposing a defect lattice on the perfect lattice. The lattice constant of the former is related to the actual lattice constant by $x^{-1/3}$, that is by $c^{-1/3}V_m^{-1/3}$. If we neglect the factors of the order of 1, and take account of Eq. (2.37a) from Section 2.2.2, the above law[133] (Eq. (5.206)) follows with (φ, φ_d: Madelung constants of the basic lattice and the defect lattice)

$$J_\pm \simeq \frac{2}{3}\frac{\varphi_d}{\varphi}\frac{U_{Mad}}{\epsilon_r}. \tag{5.207}$$

---

[131] In this sense it is a "mean–field–concept": The mean effect of an (inhomogeneous) ensemble of properties is approximated by the effect of an (homogeneous) ensemble of mean properties.

[132] This is not trivial and not exact, since the summation over $1/r_{ij}$ does not average out a random variation in the denominator.

[133] The Madelung energy of the defect lattice is

$$\Delta G_{int} = -n_d \frac{e^2 \varphi_d}{4\pi\varepsilon_r\varepsilon_0 b_d} N_m = -\frac{U_{Mad}}{\varepsilon_r}\frac{\varphi_d}{\varphi}\frac{n_d^{4/3}}{n^{1/3}}$$

with the dielectric permittivity ε_r of the perfect lattice and the mean nearest neighbour distance of the defects b_d. The (effective) Madelung constant was abbreviated by φ in order to avoid confusion.

The estimate via the simple relationship Eq. (5.207) is in approximate agreement with the experimental data. Thus, for silver halides $\varphi \simeq \frac{3}{2}$, $\Delta_F H^\circ \simeq \frac{4}{3} U_{Mad}/\varepsilon$ (cf. Eq. 5.8). Computer experiments for fluids yield $\varphi_d \sim 0.7$ [114]. Hence, we expect $J_\pm$ to range between 0.3eV and 0.6eV ($0.7 < \varphi_d < \varphi$). It is of importance that, in contrast to the description of the perfect crystal lattice energy, the dielectric permittivity of the crystal enters the equation for the quasi lattice energy of the defect lattice. In contrast to the intrinsic concentration at low temperatures (see Eq. (5.85)) the implicit relationship of $x_\pm$ for say AgCl is

$$x_\pm = [Ag_i^{\cdot}] = [V'_{Ag}] = \exp - \frac{\Delta_F G_m^\circ - 2J_\pm x_\pm^{1/3}}{2RT}. \tag{5.208}$$

Equation (5.208) reveals that the effective free enthalpy of formation becomes ever smaller as the temperature increases and, hence, an anomalous increase in defect concentration is the result. More precisely, the site restriction must be taken into account in the mass action law and $x_\pm$ on the left hand side of Eq. (5.208) be replaced by $x_\pm/(1 - x_\pm)$ (see below).

In this manner, the anomalous increase of the ionic conduction of AgCl, AgBr, AgI and PbF_2[134] in the temperature region immediately below the transition temperature to the highly disordered phase can be much better explained than by using the Debye–Hückel theory [115,228]. In the respective highly–disordered high–temperature phase either the cationic (AgI) or the anionic (PbF_2) partial lattice is molten or the material is in the liquid state (AgCl, AgBr) (see Fig. 5.66).

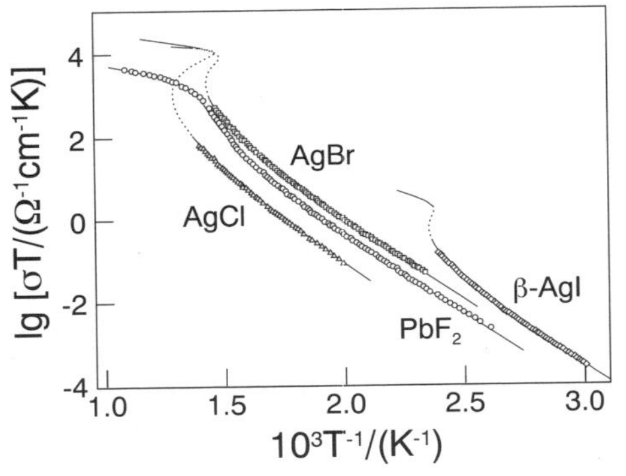

Fig. 5.66: High temperature anomalies in the conductivity can be explained with the cube root law. Even the phase transition temperatures themselves are well predicted. The S curve corresponds to the instability during the first order phase transitions. In the case of PbF_2 (higher order) the high temperature region is also correctly described. In the other cases a jump characterizes the equilibrium conductivity (1st order phase transition). It occurs at the place on the S curve where the G values of the two phases are identical (see Fig. 5.67) [115].

The ratio of the nearest neighbour distance (b) of the regular particles to b_d gives $x_\pm^{1/3} = (n_d/n)^{1/3}$. Differentiating by the defect mole number n_d and dividing by 2 leads to the chemical interaction potential, i.e. to the activity coefficient: $\Delta\mu_\pm = RT \ln f_\pm = -\left(\frac{2}{3}\frac{U_{Mad}}{\varepsilon_r}\frac{\varphi_d}{\varphi}\right) x_\pm^{1/3}$. Owing to the comparatively small distances ε should be smaller than the static dielectric constant. This was clearly seen in materials with high ε–values [207,228].

[134] There are a series of further effects, that can be important in the high temperature region, such as the significance of other defects, volume changes and — in the case of AgI — boundary phase transitions (see following section).

Cube root laws are also helpful when considering the interaction between purely electronic ("gap narrowing"[135]) charge carriers, and also between dopants and electronic charge carriers in semiconductors [229]. It is particularly worth mentioning that cube root corrections came into use very early on to describe the ion–ion interaction in very concentrated liquid electrolytes [230,231]. But let us return to the solid electrolytes and consider the phase transition in the light of the defect–defect interactions.

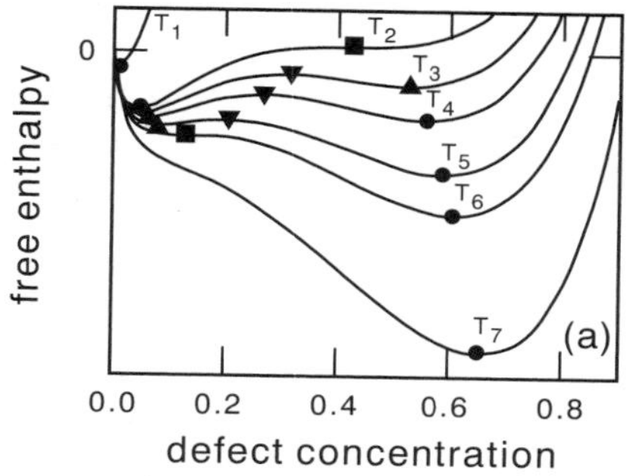

Fig. 5.67: Defect interactions according to the $c^{1/3}$ law. The interaction displaces the left hand minimum to higher concentrations (cf. also Fig. 5.7) and leads to a second minimum, corresponding to a transition into the (possibly virtual) superionic phase. The symbols indicate minima, maxima and points of inflection with horizontal tangents. The temperature increases monotonically from T_1 to T_7. Cf. Refs. [232,233].

Figure 5.67 shows that ion–ion interactions can lead to a second minimum in $G(n_d)$ at high defect concentration. If this second minimum becomes the absolute minimum, a phase change is predicted to occur that can be of first (AgCl, AgBr, AgI) or higher order (PbF_2). Figure 5.66 illustrates the relationship for AgBr, AgCl, AgI and PbF_2. The phase transition temperatures calculated from Eq. (5.206) (AgCl, AgBr: melting point, AgI: α/β transition) and the predicted order are in reasonable agreement with the experimental results for these substances. In the case of PbF_2, which fulfils the conditions for this model (no structural changes) over the whole T–range, there is also reasonable agreement with the data in the high–temperature range.

The agreement with the phase transition temperatures for the other substances is surprising, since structural effects have not been taken into account, so that the treatment gives an upper limit within the framework of the description[136].

Anyway, in this manner, we have built a thermodynamic bridge between defect properties, solid and liquid state[137], that is also mirrored in many semiempirical relationships (e.g. between enthalpy of defect formation and melting point). This, of course, is to be expected, since defect formation contains almost all the relevant energetic and entropic information. Thus, it is understandable that typical binary

[135] The interaction term μ^{ex} influences the "energy levels" (Fig. 5.9, page 127). In this case the latter are no longer given by $\widetilde{\mu}^{\circ}$, but by $\widetilde{\mu}^{\circ} + \mu^{\circ,ex}$, where $\mu^{\circ,ex}$ only counts the nonconfigurational part of the excess values (here the energetic part).

[136] When the phase transition is associated with structural change then G (real high temperature phase) – G (virtual high temperature phase) is less than zero (the virtual high temperature phase is the highly disordered phase with the "ground" structure of the low temperature phase). Concerning phase transitions in ionic conductors, see also Ref. [234].

[137] In this context the reader is also referred to Devonshire's melting theory [235].

ionic conductors (low disorder energy, see Eq. (5.208)) have low melting points[138]. Recently the cube root laws for AgI and PbF_2 have been confirmed by means of molecular dynamics and Monte Carlo calculations [228].
In such a semiquantitative way the high temperature behaviour (including order–disorder transition) can be traced back to the same parameters that are responsible for the formation of single defects, too, namely lattice energy and dielectric constant (see Section 5.2) [236].
Taking these parameters as roughly temperature independent, they determine the "thermal destiny" of ionic crystals: Owing to finite defect formation energies there is a finite carrier concentration at nonzero temperature. Since the concentration increases with temperature, the defects necessarily start interacting at temperatures that are the lower the easier the defects are formed. Finally these interactions "lead" to a phase transformation to a truly disordered state. Fig. 5.68 shows this by using the energy level diagrams.

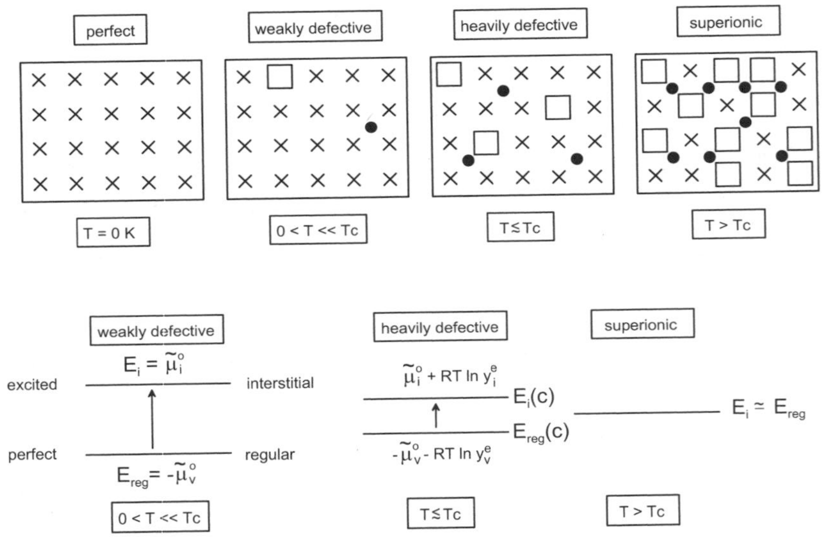

Fig. 5.68: The thermal density of an ionic crystal. The weakly defective state is characterized by a small number of chemical excitations. Attractive defect interactions lead to a level narrowing (increased and even accelerated defect formation) and eventually to a degeneracy (superionic state). From Ref. [236].

It is interesting to note that this behaviour also provides an explanation of Tamanns rule (typically at 2/3 of the melting temperature atomic transport becomes significant).

[138] One should not be confused by the fact that U_{Mad} or ΔH_F° enters the numerator of J in Eq. (5.208). The opposite influence via $x_\pm^{1/3}$ is much more sensitive.

Furthermore, there are obvious analogies between the transition of a weakly disordered ionic conductor into the superionic state and the transition of a semiconductor into the metallic state[139].

In spite of energetic interactions, we have, until now, always used Boltzmann statistics and, thus, neglected activity coefficients resulting from configurational entropy effects. This is not a bad approximation in many instances. Accurate corrections are very complex [224–227,222]. Conversely there are deviations from the Boltzmann relationship simply because the sites available for disorder are exhaustible. This affects the statistics in the form of site restriction. Strictly speaking we ought to apply the Fermi–Dirac form of the chemical potential according to Eq. (5.14)[140] [237] leading to a correction of the form

$$f_k = (1 - c_k/c_k^{max})^{-1} > 1. \tag{5.209}$$

(The nominal maximum concentration c^{max} is usually but not always[141] the reciprocal molar volume, i.e. $x^{max} = 1$ (see Section 5.2)[142]. Degeneracy factors as for $Ag_i^{\cdot}$ in AgCl or for electronic states at the doping atoms, are included in $\mu°$.) This relationship alone is not very useful, as can be seen from the example already described, since interactions occur as $x \to 1$. Combination with a cube root law leads in Eq. (5.208) to $x_\pm/(1 - x_\pm)$ on the left hand side. It is interesting that the activity coefficient described by Eq. (5.209) is greater than 1, because occupation becomes ever more "difficult", on account of the increasing free enthalpy, and the activity rises ("Fermi pressure"). These effects bring about a lower equilibrium concentration than would be calculated "according to Boltzmann".

Such considerations are very important for electrons. Here, the exhaustibility of the states, as already shown in Section 5.3, results for a given energy state in a

[139] Cf. footnote 29, page 125 and Ref. [206].

[140] The Fermi–Dirac type correction results from the exclusion principle that not more than one ion can occupy a given lattice site. Activity expressions of the form $\frac{x}{1-x}$, such as are obtained, for example, for electrons in donor or acceptor states, also belong in this section. In Section 5.3 such expressions were obtained (sometimes artificially), in that, instead of building elements, structure elements were considered; then the term $x/(1 - x)$ appeared as the quotient of two ideal concentration terms (x and $x' = 1 - x$). See also footnotes 37 and 172. This "trick" also works in more complicated cases. In the case of water incorporation in oxides according to $\frac{1}{2}H_2O + V_O^{\cdot\cdot} + O_O \rightleftharpoons 2OH^{\cdot}$ (see Eq. (5.176)) we get the same result, regardless of whether we use separate Boltzmann expressions for $V_O^{\cdot\cdot}$, $OH^{\cdot}$ or O_O or whether we use thermodynamically exact Fermi–Dirac–like expressions for the building elements ($V_O^{\cdot\cdot} - O_O$) and ($OH^{\cdot} - O_O$). Strictly speaking, it is incorrect to apply Boltzmann statistics to O_O. On the other hand, in cases in which corrections for O_O become necessary, also the building element activity, too, must be corrected with respect to interactions.

[141] In the case of proton–conducting perovskites (see previous section) $c_{H^+}^{max}$ is given by the doping concentration.

[142] Since the number of interstitial positions is 2N in AgCl, this leads to $\mu = \text{const} + RT \ln \frac{x}{2-x}$ which is different from Eq. (5.209). As already noted in Section 5.2, "double occupations" are unlikely. In this case $\mu = \text{konst} + RT \ln \frac{x}{2(1-x)}$ (see footnote 24 on page 123) results, i.e. an expression which can be converted to the functional form (5.209). The exact calculation is complex.

chemical potential of type Eq. (5.209) (see Eq. (5.40)). If all electronic states are included (see Eq. (5.44)) and μ° refers to the band edge, the exact result contains the Fermi–Dirac integral $\mathcal{F}_{1/2}$ (Eq. (5.53)). Figure 5.69 shows the increase of the

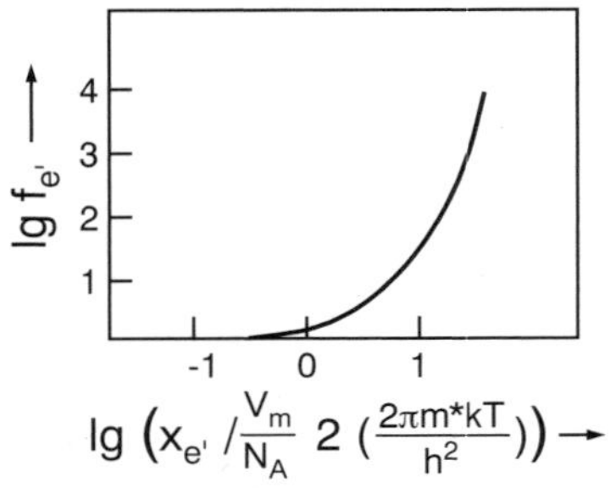

Fig. 5.69: Fermi–Dirac activity coefficient of electronic charge carriers as a function of concentration. According to Ref. [7].

activity coefficient with the electron density becoming comparable with the effective state density or exceeding it[143].

In contrast to Fermi–Dirac statistics ("occupation makes new occupation more unfavourable"), there is a positive sign in Eq. (5.209) in the case of Bose–Einstein statistics (see also Chapter 4) in leading to a value of f that is less than 1. (The designation c_k^{max} is then no longer meaningful. Here "occupation makes new occupation less unfavourable", i.e. f falls with increasing concentration.) The "Bose condensation"[144], associated with this, is of great importance for superconductivity and superfluidity. Exciting actual research programmes deal with Bose–condensates [238,239].

A more precise quantitative treatment of configurational corrections due to charge carrier interactions is complex; it soon leads to unmanageable expressions (cf. e.g. [9,222,227,240]). A simple type of statistical correction has already been considered in the exclusion process for occupation of neighbouring interstitial sites in the silver halides (see page 122)[142,145].

In spite of the simplicity of the approaches described here, they enable efficient descriptions of bulk defect chemistry[146]. The silver halides may serve as models. Their behaviour can be quantitatively described over an enormous temperature range — from low temperatures to the phase transition (low temperature: extrinsic with

[143]Interactions between electronic charge carriers, such as electron–hole interactions, must be taken into account in addition, e.g. via the $c^{1/3}$ law, as discussed above.

[144]The 2001 nobel prize for Physics refers to that phenomenon.

[145]Similarly in silicate chemistry neighbouring Al atoms are not allowed on energetic grounds (no Al–O–Al configurations), i.e. the enthalpy of formation of such a state is considered to be infinity.

[146]It is also worth mentioning that activity coefficients — even though not unity — can be constant with respect to the variation of control parameters. In this respect, and only in this respect, they can be formally included in the mass action constant. Examples are the ideal P_{O_2} dependences (0, −1/4, +1/4) of $V_O^{\cdot\cdot}$, e′, h$^\cdot$ in highly Y_2O_3 doped ZrO_2 or the behaviour described in Fig. 5.57.

association (Section 5.7.1); at about room temperature: extrinsic without association (Section 5.6); high temperatures: intrinsic without interaction (Section 5.5), very high temperatures below the melting point: intrinsic with interaction (Section 5.7.2)).

5.8 Boundary layers and size effects

5.8.1 General

We now have a concept for treating the equilibrium defect chemistry within a solid which we can reliably use and build upon. Yet, we have ignored, until now, the presence of interfaces[147] and their effects on the point defect concentration in their immediate neighbourhood.
In the case of the two prototype substances, water and silicon, mentioned at the beginning, we know that boundary layer effects frequently exceed the volume effects in their importance; one may think, on the one hand, of p–n junctions, transistors, photoelements, varistors, Schottky diodes, and, on the other, of electrode/liquid electrolyte junctions or colloid chemistry. First let us look at five particularly illustrative examples (Fig. 5.70).
The first example (Fig. 5.70a) involves dipping a solid into an aqueous salt solution, thus, creating a solid–liquid interface. Either cations or anions will be preferentially adsorbed on the surface, and the result is an excess surface charge[148]. The countercharge is distributed in the zone of the solution adjacent to the interface; the extent of this zone is determined by the Debye length (see Eq. (5.203)). It is inversely proportional to the square root of charge carrier concentration in the bulk solution. A rigid double–layer is formed in a concentrated solution; in dilute solution a diffuse layer is formed with appreciable extension (typical numbers are several tens of nanometres). Such electrostatic effects are responsible for the kinetic stability of dispersed systems in colloid chemistry[149].
Example 2 (Fig. 5.70b) shows the contact between the metal Al and the semiconductor Si. For simplicity we will consider temperatures at which mutual atomic solubility effects can be neglected. Because of the difference in electronegativity we expect electrons to seek to pass from the less noble aluminium to the more noble Si. If the electrons were not charged, the effect would certainly be very considerable.

[147] We regard the "ground structure" of the interface as invariant, i.e. it is either in equilibrium (e.g. surface of a Wulff crystal) or — as is almost always the case — it refers to a metastable structure element. Cf. Section 5.4 for details.

[148] More precisely, there is an exactly defined parameter set, for which the excess charge disappears. A nonzero excess charge is required in the other cases.

[149] Equivalent colloid particles carry the equivalent surface charges and repel each other. This generates an activation threshold for growth processes (cf. flocculation). The same double–layer repulsion is significant, in addition to other reasons, for the removal of grain boundary phases during the sintering process. See Section 5.4.

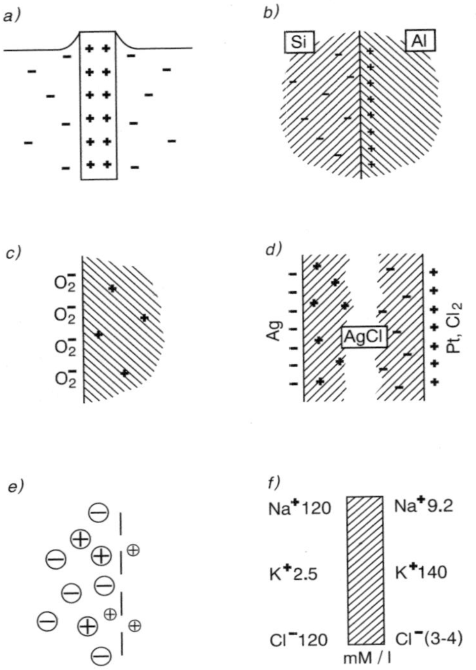

Fig. 5.70: Formation of electrical and chemical potential differences at interfaces, as described in the text. (In diagram d the positive charge, on the right hand side, refers to the charge on the metal side (Pt) not to the charge of the adsorption layer [241].)

However, an electric field is produced on account of the charge, and this limits the amount and the extension in space severely. On account of the high electron concentration in the metallic aluminium the charge is restricted to the surface, while the charged zone in silicon is extended to a degree which depends on the purity (see Eq. (5.203)). Enrichment, depletion or inversion layers are formed depending on the type of conduction in Si (n or p) (see also Fig 5.76, page 226). Such semiconductor boundary layers play a paramount role in devices, such as diodes, transistors and solar cells. Contact potentials in semiconductors also give rise to thermo–voltaic effects.

Example 3 (Fig. 5.70c) illustrates a solid–gas contact, more precisely the contact between the n–type semiconductor SnO_2 with O_2 gas. At low temperatures the adsorbed oxygen cannot penetrate the interior, but remains adsorbed on the surface. It traps electrons from the boundary layer and makes this positively charged, producing a depletion layer of increased resistance. This is the basic principle of the Taguchi sensor, which we will discuss later. Such effects must also be discussed at elevated temperatures when phase equilibrium with oxygen is established.

Example 4 (Fig. 5.70d) illustrates the superposition of asymmetrical contact potentials at the ionic conductor AgCl, to yield a measurable potential difference and, hence, a battery voltage. The cell shown can also be used as a sensor for Cl_2. This is treated in Chapter 7.

Finally example 5 (Fig. 5.70e) illustrates the relevance of membrane potentials for biology. Solutions of differing concentrations are to be found on the two sides of the membrane. A simple concentration equalization does not occur, since, because of its structure, the membrane only allows the effectively small cations to pass. As a result, there is an electrical potential difference on account of charge separation; this is of basic importance in electrophysiology (e.g. propagation of nerve signals)[150]. Figure 5.70f refers to a real electrochemical situation involving a frog muscle cell membrane [243].

In more general terms: The creation of an interface means a break of symmetry of the homogeneous starting situation. Given a sufficient mobility of charge carriers, this necessarily leads to a charging[148].

Our aim in this section is to show how it is possible to calculate the defect concentrations in boundary layers as a function of the control parameters temperature, component activity and doping content, the local coordinate and the materials parameters of both bulk and interface core. Thus, we are going to extend our previous defect chemical analysis to boundary layers, whereby we will, in general, consider a mixed conductor with low defect concentrations [244].

In all the examples described, a difference in chemical potential was bought at the cost of a difference in electrical potential, in such a manner that the electrochemical potential difference between the mobile charge carriers disappears. Let us consider, quite generally, the transfer of a charged (z_A) particle A from place x to place x′, whereby the nature of A can change ($A' \neq A$) or not. For simplicity, let us leave the charge unchanged ($z_A = z_{A'}$) (see Fig. 5.71):

$$A(x) \rightleftharpoons A'(x'). \tag{5.210}$$

According to the general treatment set out in Section 4.3.6 the spatial equilibrium is described by the equality of the electrochemical potentials:

$$\widetilde{\mu}_A(x) = \widetilde{\mu}_{A'}(x'). \tag{5.211}$$

If we split $\widetilde{\mu}$ up into μ and $zF\phi$ and consider dilute states, then the result is

$$\mu_A^\circ + RT \ln c_A(x) + z_A F\phi(x) = \mu_{A'}^\circ + RT \ln c_{A'}(x') + z_A F\phi(x') \tag{5.212}$$

or

$$\frac{c_{A'}(x')}{c_A(x)} = \exp -\frac{\mu_{A'}^\circ - \mu_A^\circ}{RT} \exp -\frac{z_A F(\phi(x') - \phi(x))}{RT} \equiv K_{AA'}\kappa_{xx'}. \tag{5.213}$$

We will assume, in the following, that the "ground structure"[151] of the phase remains intact up to the proper phase boundary layer[152], so that boundary effects on that

[150] It should be noted that the space requirement of Na^+ can be greater than that of K^+ because of a greater hydration. The fixed charges on the membrane surface are important for permeability in addition to the pore size and other factors (see also [242]).

[151] Structure minus the changes brought about by point defects (in the sense of Fig. 1.2, page 14).

[152] This also applies with respect to bond lengths (i.e. the absence of marked elastic effects is assumed [245]). In hard materials such elastic effects can be of longer range [245]. The actual stationary interface core exhibiting a different structure is allowed to be of finite thickness. With respect to the spatial dependence see also Ref. [246].

phase are attributable to point defects. This assumption of an abrupt contact is frequently useful (see semiconductor physics) even though (as may be reflected by changes in the band gap) only approximately correct. In this approximation $K_{AA'} = \exp((\mu_A^\circ - \mu_{A'}^\circ)/RT)$ (in Eq. (5.213)) only differs from unity if either the charge carrier changes its nature or passes from one phase to another. Within the same phase we obtain the following important relationship for the charge carrier $A \equiv A'$:

$$\frac{c_A(x')}{c_A(x)} = \kappa_{xx'}. \qquad (5.214)$$

$$\boxed{A(x) \rightleftharpoons A'(x')}$$

$$\boxed{\begin{array}{c} \tilde{\mu}_A(x) = \tilde{\mu}_{A'}(x') \\ \dfrac{c_{A'}(x')}{c_A(x)} = K\kappa \end{array}}$$

$$K = \exp\left[-(\mu_B^0 - \mu_A^0)/RT\right]$$
$$\kappa = \exp\left[-zF(\phi(x') - \phi(x))/RT\right]$$

$zF\phi(x) = zF\phi(x')$	$A \equiv A'$
conventional chemical reaction	particle transport

$$\boxed{A \rightleftharpoons A'}$$

$$\boxed{\begin{array}{c} \mu_A = \mu_{A'} \\ \dfrac{c_{A'}}{c_A} = K \end{array}}$$

$$\kappa = 1$$

$$\boxed{A(x) \rightleftharpoons A(x')}$$

$$\boxed{\begin{array}{c} \tilde{\mu}(x) = \tilde{\mu}(x') \\ \dfrac{c(x')}{c(x)} = \kappa \end{array}}$$

$$K = 1$$

$$zF\phi(x) = zF\phi(x')$$
no charge or no field

$$\boxed{\begin{array}{c} \mu(x) = \mu(x') \\ \dfrac{c(x')}{c(x)} = 1 \end{array}}$$

$$K = \kappa = 1$$

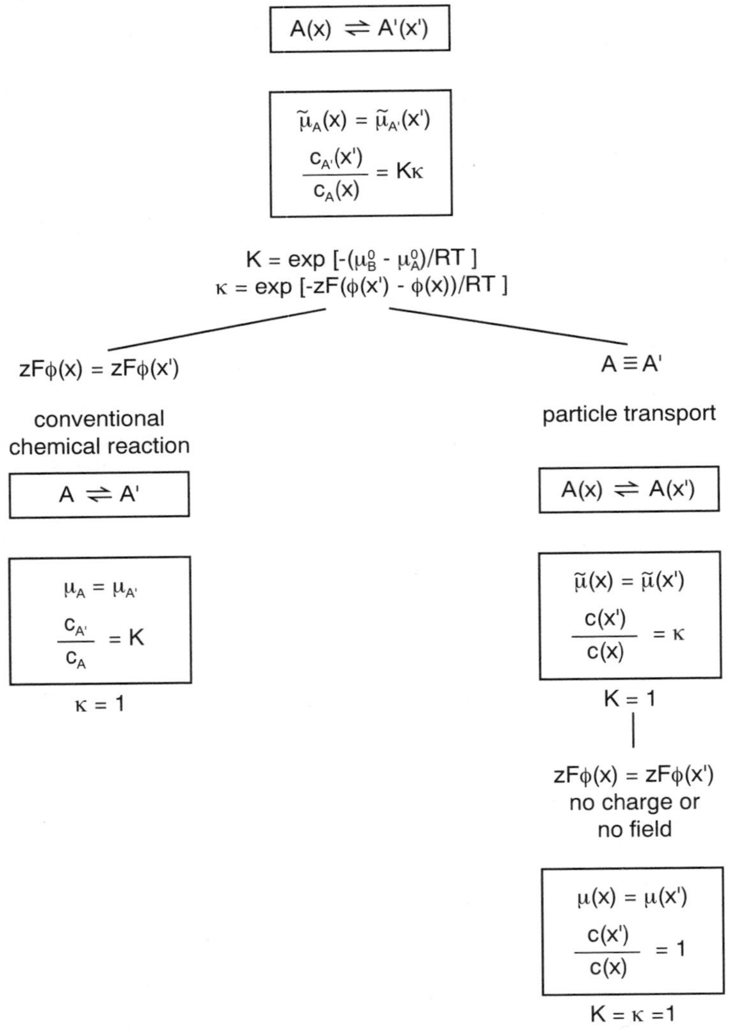

Fig. 5.71: Heterogeneous equilibria including the specific cases of homogeneous chemical reaction (left) and transport equilibrium (right) within a phase [14].

If we introduce the concentration enhancement ζ as the concentration relative to that in the bulk ($x = \infty$), i.e. ($\zeta(x) \equiv c(x)/c(x{=}\infty)$)[153,154], the result is

$$\zeta_A^{1/z_A} = \exp -\frac{[\phi(x) - \phi_\infty]\,F}{RT}. \qquad (5.215)$$

In the bulk ($\phi(x) = \phi_\infty$) there are no concentration inhomogeneities ($\zeta = 1$), in contrast to the neighbourhood of the interface. The homogeneity in the bulk also applies for high concentrations, since the activity coefficient is a one–to–one function of the concentration. In Eq. (5.215) z_A has been brought to the left hand side, in order to show that the resulting expression is independent of A. This leads to the following important conclusion:

A given electrical potential difference influences all (mobile) charge carriers in a strictly defined manner, according to their charge. For instance, if all defects are effectively monovalent, a positive potential difference will deplete all positive charge carriers by the factor $\exp|\frac{\Delta\phi F}{RT}|$, and all negative charge carriers will be enriched by the same factor. This is shown in Fig. 5.72 for the mixed conducting, Frenkel

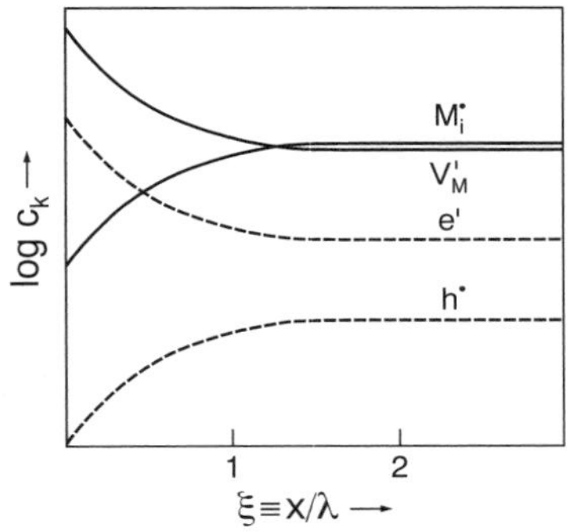

Fig. 5.72: Splitting of the concentrations for positive electrical interface potential. The concentration of anion defects is neglected.

disordered phase MX. Specifically for AgCl it follows that: $\zeta_{Ag_i^{\cdot}}(x) = \zeta^{-1}_{V'_{Ag}}(x) = \zeta_{h^{\cdot}}(x) = \zeta^{-1}_{e'}(x)$. Typical potential differences are of the order of a few 100mV. A value of ± 250 mV corresponds to a concentration influence of a factor $e^{10} \sim 10^4$ at 300 K! In ideally pure AgCl, for $\phi(x{=}0) - \phi_\infty = 250$mV, the number of interstitial ions at the contact is reduced by 4 orders of magnitude with respect to the bulk value of level $K_F^{1/2} \simeq 10^{-10}$; the vacancies are increased by 4 orders of magnitude. Naturally the value of the interfacial potential is determined by chemical interaction and, hence, will be strongly dependent on the neighbouring phase and on temperature.

[153]These abbreviations can be used for the following advantageous formulation of the chemical potential:

$$\mu_k = \mu_{k\infty} + RT \ln \zeta_k.$$

[154]Strictly speaking, relationship (5.215) only presupposes spatial equilibrium with respect to species A, not necessarily local equilibrium with the counterdefect. The spatial equilibrium is compatible with the local one because $\zeta_+ \cdot \zeta_- = 1$ and so $c_+ c_- = c_{+\infty} c_{-\infty}$.

Since, as already discussed, the energy levels in the ideal case correspond to the standard electrochemical potentials (i.e. $\mu^\circ + zF\phi$), these are curved in accordance with the variation of the electrical potential[155]. The Fermi levels as electrochemical potentials are horizontal and constant even over the phase boundary (electrochemical equilibrium). In semiconductor physics it is usual to represent boundary layer effects in the form of band bending. For this reason Fig. 5.73 illustrates the behaviour of the electrochemical, chemical and electrical potentials and of the "energy levels" (in the sense of Fig. 5.9, page 127).

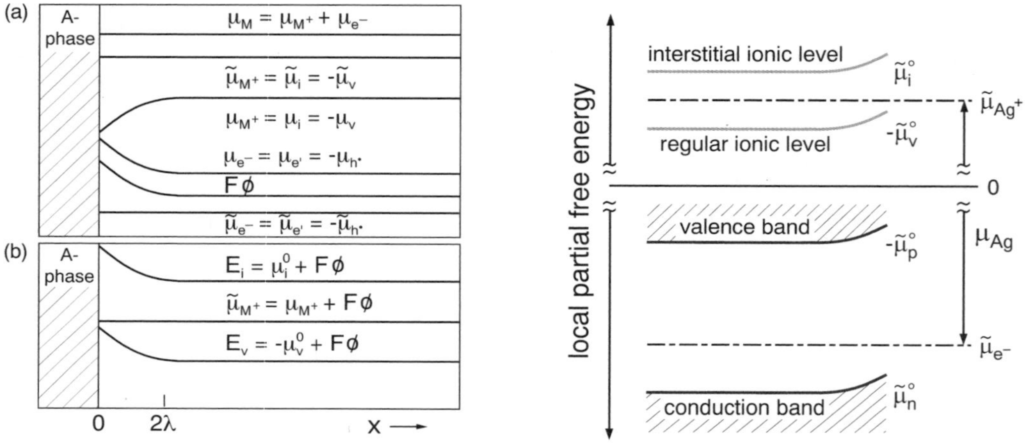

Fig. 5.73: L.h.s.: Bending of the thermodynamic potentials for ions and electrons and of the ionic energy levels in the boundary layers of a cation disordered solid MX [176]. R.h.s.: Bending in the energy level diagram (cf. Figs. 5.9 and 5.37). Note that the electronic level diagram is plotted upside down.

In order to calculate the concentration distribution we must set up a relationship between concentration (or μ) and electrical field. This was, in the bulk, simply the condition of electroneutrality, but this does not apply at the boundary. The general relationship, that provides the complete electrical information, is Poisson's equation. It is obtained as follows from Maxwell's equations:

According to the first Maxwell equation the curl of the electrical field is zero in the absence of time dependent magnetic fields. This means that, in this case, no turbulent electrical fields occur. It is then and only then that we can define a scalar electrical potential, which we require. This is visible from the definition of curl **E**

$$\mathrm{curl}\,\mathbf{E} \equiv \nabla \times \mathbf{E} \equiv \begin{pmatrix} \partial E_z/\partial y & - & \partial E_y/\partial z \\ \partial E_x/\partial z & - & \partial E_z/\partial x \\ \partial E_y/\partial x & - & \partial E_x/\partial y \end{pmatrix} = \mathbf{0}. \tag{5.216}$$

[155] If structural invariance is not maintained, which, to a certain degree, is more or less always the case, then a bending of the μ°–level has to be considered (see Section 5.8.5).

Since every component must disappear, i.e. $\partial E_i/\partial j = \partial E_j/\partial i$, the expression

$$E_x dx + E_y dy + E_z dz = \mathbf{E}\,\mathbf{dr} \equiv d\phi^* \tag{5.217}$$

is a total differential which we denote as $d\phi^*$. Hence, this quantity ϕ^*, whose gradient corresponds to the electrical field, is only dependent on the state of the system. We term $\phi \equiv -\phi^*$ the electrical potential. The second Maxwell equation, which we need, states that charges are sources of the dielectric displacement field and, as we ignore spatial variations of the dielectric constant ε, also of the electrical field; therefore ρ/ε determines its divergence:

$$\operatorname{div}\mathbf{E} \equiv \nabla\,\mathbf{E} = \rho/\varepsilon, \tag{5.218}$$

ρ denotes the charge density, which is obtained from the defect concentration using $\Sigma_k z_k F c_k$. Combining the two relationships (Eqs. (5.216, 5.218)) yields the desired connection, namely the Poisson equation

$$\nabla^2\phi = -\rho/\varepsilon. \tag{5.219}$$

Since, in the bulk, in equilibrium, ϕ is constant for reasons of symmetry[156], i.e. $\phi'' = 0$, we derive the electroneutrality condition which we extensively used in the previous sections. This also applies when ϕ is linear, which is the case for the homogeneous bulk subject to an external electrical field. For the boundary zones, however, Eq. (5.219) must be taken into account explicitly.

The combination with the condition of the constancy of the electrochemical potential leads us to the Poisson–Bolzmann relationship. The one–dimensional form reads

$$\frac{d^2(\phi-\phi_\infty)}{dx^2} = -\frac{F}{\varepsilon}\Sigma_k c_{k\infty} z_k \exp -\left(z_k F\frac{\phi-\phi_\infty}{RT}\right). \tag{5.220}$$

5.8.2 Concentration profiles in the space charge zones

The solution of Eq. (5.220) for semiinfinite boundary conditions leads to the Gouy–Chapman profile [247]. Let us follow the treatment in Ref. [248] and change to the concentration as the variable. First, let us consider the case that only two equivalently (but oppositely) charged defects (subscripts + and -) are relevant ($z_+ = |z_-| = z$). The bulk concentrations are equal for reasons of electroneutrality ($c_{+\infty} = c_{-\infty} \equiv c_\infty$). On combination with Eq. (5.215) we obtain the differential equation for the concentration enhancement (ζ_+ or ζ_-)

$$\frac{d^2\ln\zeta_\pm}{d\xi^2} = \frac{1}{2}(\zeta_\pm - \zeta_\mp) \tag{5.221}$$

[156] Naturally the electrical potential can and does vary periodically at the atomic scale in the crystal. Our ϕ is an adequately coarse–grained mean value. Note also that Poisson's equation is — depending on the charge density — not necessarily a linear differential equation.

(whereby either the upper or the lower subscripts apply). For ease of manipulation the local coordinate has been normalized to the Debye length ($\xi \equiv x/\lambda$)

$$\lambda = \sqrt{\frac{\varepsilon RT}{2z^2F^2c_\infty}}, \tag{5.222}$$

which, under the simplified conditions, automatically evolves from the prefactor in Eq. (5.220), and with which we are familiar from the previous section.
If both defects are mobile and there is electrochemical equilibrium, it follows because $\zeta_+ = \zeta_-^{-1}$ (see Eq. (5.215)) that:

$$\frac{d^2 \ln \zeta_\pm}{d\xi^2} = \frac{1}{2}\left(\zeta_\pm - \zeta_\pm^{-1}\right). \tag{5.223}$$

This differential equation can be integrated using the boundary conditions

$$\zeta_\pm(x{=}0) = \zeta_{\pm 0} \quad \text{and} \quad \zeta_\pm(x \longrightarrow \infty) \equiv \zeta_{\pm\infty} = 1. \tag{5.224}$$

Before we give the general solution, let us consider a helpful approximation (Fig. 5.74). If the splitting effect is very marked, the depleted defect (defect 2) very rapidly

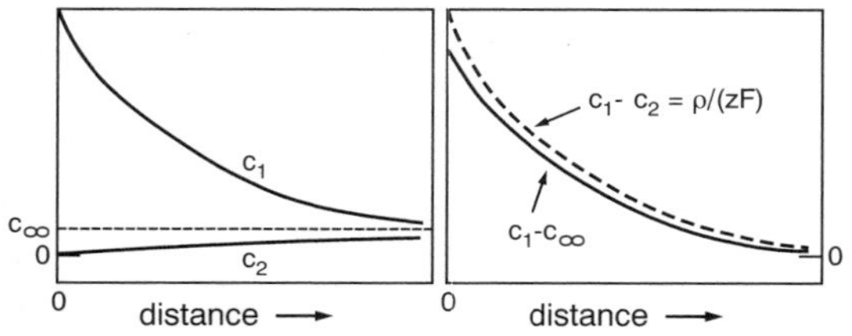

Fig. 5.74: Absolute variation of the charge carrier concentrations (left) and the charge density (right) close to the interface. If the contribution of the depleted charge carrier 2 is ignored in the charge density, this results in an error in the total charge density given by $((1-\vartheta_1)/(1+\vartheta_1)$ (cf. Eq. 5.229) (i.e. $< 6\%$ for $\vartheta > 0.9$) [244].

becomes unimportant for the charge density as long as we are not concerned with too great distances from the boundary layer (i.e. either ζ or ζ^{-1} becomes negligible). Then Eq. (5.223) simplifies to

$$\frac{d^2 \ln \zeta_1}{d\xi^2} = \zeta_1/2. \tag{5.225}$$

The subscript 1 now applies to the enriched positive or negative defect. Using

$$\ln \zeta_1 = \ln a + b \ln(1 + c\xi) \tag{5.226}$$

as an evidently applicable and adequately general test function, the solution follows as

$$\zeta_1 = \frac{\zeta_{10}}{\left(1 + \sqrt{\zeta_{10}}\,\xi/2\right)^2}, \tag{5.227}$$

where $\zeta_{10} \equiv \zeta_1(x{=}0)$.

Two parameters are important in Eq. (5.227), the bulk concentration (appears in λ and ζ_{10}), whose dependence on the control and material parameters can be inferred from the previous section, and the interfacial concentration c_0 which is included in ζ_{10} and parametrizes the interfacial chemistry. The significance of this parameter will be analysed further below (see Section 5.8.4). The concentration of the counterdefect is then given by $\zeta_2 = 1/\zeta_1$. As we can see when $\xi = 2$, i.e. $x = 2\lambda$, the parameter ζ_1 has fallen to the bulk value 1 because we assume $\sqrt{\zeta_{10}} \gg 1$. Hence, the (double) Debye length is a suitable measure of the extent of the space charge zone. The fact that ζ_1 falls to zero as $x \to \infty$ in Eq. (5.227) should not puzzle the reader, since the solution Eq. (5.227) is no longer usable in this region on account of the approximation made. Equation (5.225) can also be integrated without this approximation. The result is more complicated, but applies over the entire local coordinate range and also for smaller splitting effects. We get

$$\zeta_{\pm} = \left(\frac{1 + \vartheta_{\pm} \exp -\xi}{1 - \vartheta_{\pm} \exp -\xi}\right)^2 = \zeta_{\mp}^{-1}. \qquad (5.228)$$

The parameter ϑ, which depends on ζ_0 and hence on c_0, is defined by

$$\vartheta_{\pm} = \frac{\zeta_{\pm 0}^{1/2} - 1}{\zeta_{\pm 0}^{1/2} + 1} = -\vartheta_{\mp}. \qquad (5.229)$$

We term it the degree of influence [249]. As can be seen from Fig. 5.75 $\vartheta = 0$, if the boundary layer defect chemistry does not differ from the bulk defect chemistry, i.e.

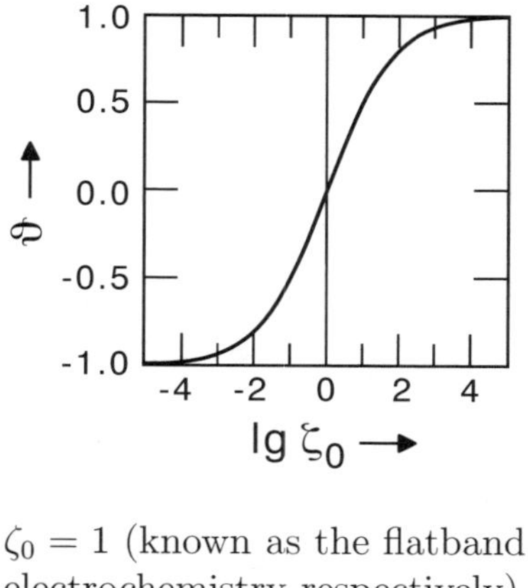

Fig. 5.75: The relationship between the degree of influence and the concentration increase at the interface [244].

$\zeta_0 = 1$ (known as the flatband case or zero charge case in semiconductor physics or electrochemistry respectively), it approaches $+1$ for maximum enrichment ($\zeta_0 \gg 1$) and -1 for maximum depletion ($\zeta_0 \ll 1$). To a good approximation at $x = 2\lambda$ (that is $\xi = 2$) the concentration has fallen to a small value that is approximately independent of c_0, viz. $(1...1.7)c_{\infty}$[157].

The approximation described first (see Eq. (5.227)) has the disadvantage of diminished precision and breaks down completely for small effects. However, it has the

[157] $\zeta(\xi = 2)_{\vartheta=1} = 1.72$, $\zeta(\xi = 2)_{\vartheta=0} = 1$. $\zeta(\xi = 1)$ in contrast, varies between 1 and 4.7.

advantage of being more generally applicable, since it is not limited to intrinsic or native conditions, provided we are concerned with large effects, which will primarily be of interest to us in what follows. Since the counterdefect (defect 2) has been completely neglected in the calculation, Eq. (5.227) is also valid (i) if the nonenriched counterdefect possesses a different absolute charge (ζ_2 is then obtained from $\zeta_2 = \zeta_1^{-z_2/z_1}$) or (ii) if this counterdefect is completely immobilized and does not change with the local coordinate (frozen–in profile). In such cases the Debye length then includes the parameters $c_{1\infty}$ as concentration term and z_1F as molar charge. The first condition (i) is e.g. met in many p–conducting or n–conducting materials, e.g. oxides with bulk electroneutrality relationships $2\,[V_O^{\cdot\cdot}] = [e']$ or $2\,[O_i''] = [h^{\cdot}]$. The second case (ii) is important for enrichment phenomena in doped substances with immobile doping defect. The concentration distribution of mobile minority charge (k) carriers follows from ζ_1 via $\zeta_k(x) = \zeta_1^{-z_k/z_1}(x)$.
The changes in defect concentrations as a result of space charge effects are of direct and crucial importance in questions of reactivity, catalysis and interfacial kinetics, and they are of particular significance with respect to electrical effects (see Section 5.8.5). Figure 5.76 shows the enrichment, depletion and inversion effects with respect

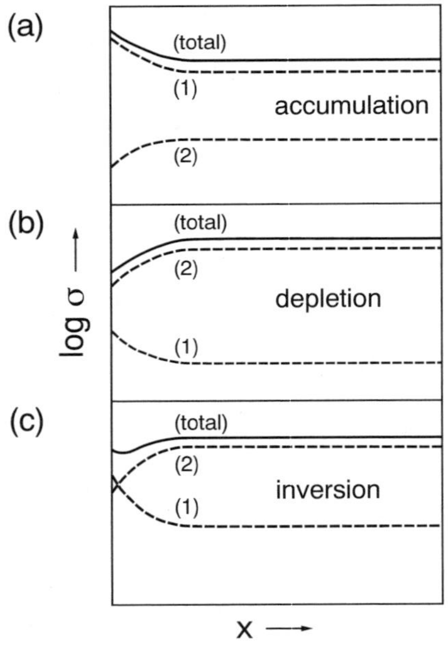

Fig. 5.76: Enrichment, depletion and inversion effects with regard to the partial conductivities and the total conductivity. In the concentration presentation carrier (1) is enriched and the counterdefect (2) is depleted. Note too that, in the case of a mixed conductor (electronic and ionic defects) the space charge profile causes changes in the transference number (σ_{eon}/σ, σ_{ion}/σ).

to the local conductivity established according to the mobilities of the two majority carriers. Figure 5.77 gives an experimental example of a surface space charge profile in AgCl measured by radioanalysis.
Depending on the space charge situation the conductivity type (ionic/electronic) can change as a function of the space charge potential. This is particularly obvious in cases in which an ionic and an electronic defect are the majority charge carriers.

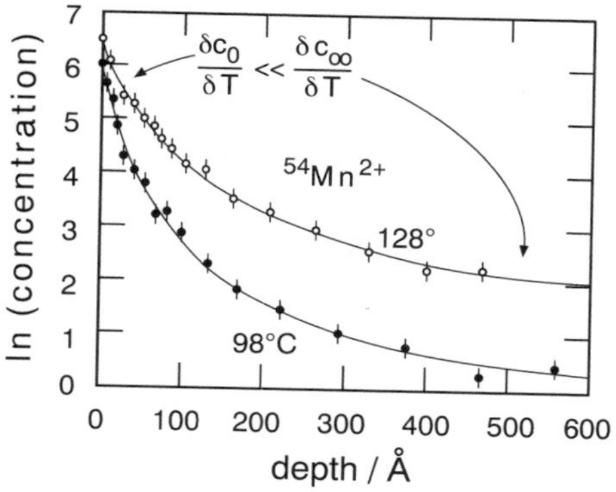

Fig. 5.77: AgCl was doped with $^{54}MnCl_2$ at 98°C or 128°C and equilibrated ($^{54}Mn_{Ag}^{\cdot}$ defects are formed). The quenched tracer distribution was determined at room temperature by etching and radioanalysis. The solid line shows the fit according to Eq. (5.228). Note also the low temperature dependence of c_0 compared to c_∞. From Ref. [250]. A similar profile was measured for $SrTiO_3$ doped with Fe_2O_3 by SIMS [251].

An important case is not included in the above discussion (let us subsume the above situations under the name Gouy–Chapman case), namely the one in which one majority charge carrier (1) — usually the dopant — is immobile, while the other majority charge carrier (2) is depleted on account of the interfacial chemistry. In this situation — here referred to as the Mott–Schottky case (see e.g. [252,253]) — which is very frequent in doped systems, it is not possible for the majority charge carrier that determines the space charge to follow the electric field. A significantly reduced screening will be the consequence. In the following description we neglect the depleted carrier type 2 in the local space charge chemistry. If the doping profile is completely horizontal ($c_1 = c_{1\infty}$), the charge density, as well as ϕ'' and $(\ln \zeta_2)''$ are constant, and a (half) Gauss function directly results for ζ_2, namely ($\xi^* \equiv \lambda^*/\lambda$):

$$\zeta_2 = \exp - \left|\frac{z_2}{z_1}\right| \left(\frac{x - \lambda^*}{2\lambda}\right)^2 = \exp - \left|\frac{z_2}{z_1}\right| \left(\frac{\xi - \xi^*}{2}\right)^2 \tag{5.230}$$

for $x \leq \lambda^*$.

Bulk concentration and molar charge in the Debye length (cf. Eq. 5.222) are referred to charge carrier 1[158].

The maximum of $\zeta_2(x)$ is at λ^*. At exactly this point $\zeta_2 = 1$ becomes unity, i.e. reaches the bulk value. Beyond λ^* — as already in the neighbourhood of λ^* because of the conditions imposed — the function is no longer permitted. In contrast to the Gouy–Chapman case the extent of the space charge thickness (λ^*) is dependent on the interfacial parameters according to

$$\lambda^* = \sqrt{\frac{2\varepsilon}{z_1 F c_{1\infty}}(\phi_\infty - \phi_0)}. \tag{5.231}$$

A comparison with the Debye length λ shows that

$$\lambda^*/\lambda = \sqrt{\frac{4 z_1 F}{RT}(\phi_\infty - \phi_0)} = \sqrt{4\frac{z_1}{z_2}\ln \zeta_{20}}. \tag{5.232}$$

[158] It is $\phi'' = -z_1 F c_\infty/\epsilon$. The boundary conditions to be taken into account by integrating twice are $\phi'(x = \lambda^*) = 0$ and $\phi(x = \lambda^*) = \phi_\infty$. Eqs. (5.230, 5.231) then follow from $\phi(x{=}0) = 0$.

The ratio λ^*/λ, thus, becomes larger, the greater the interfacial effect[159]. The difference $\lambda^* - \sqrt{2|z_1/z_2|}\lambda$ equals the local coordinate of the point of inflection in the Gauss function. This means that as depletion increases ($\vartheta \to 1$) the situation can be described ever more accurately by means of a rectangular function; that is, in the extreme case, the boundary layer is almost cleaned out. In the Gouy–Chapman case above, the space charge effect was concentrated strongly at $x = 0$, even for large surface potentials, and the double–layer could not penetrate so far into the interior because of the stronger screening. Figure 5.78 compares the two cases with each other.

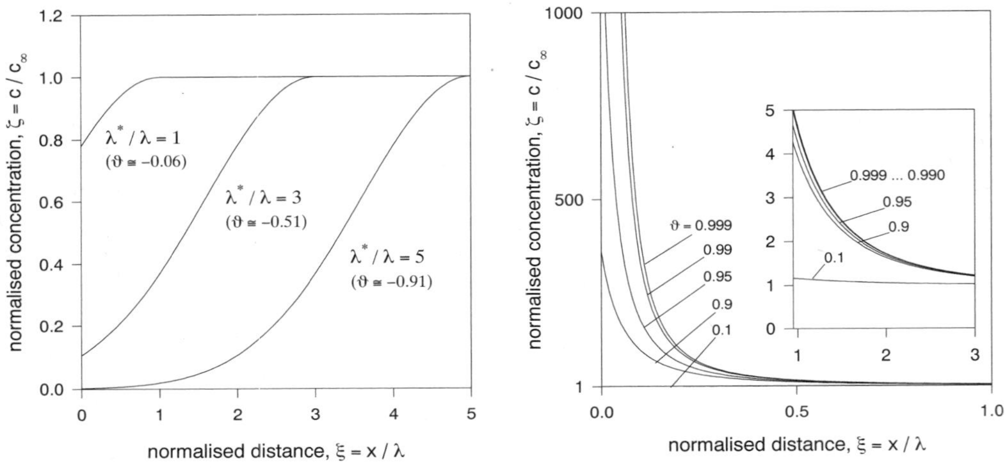

Fig. 5.78: L.h.s.: Depletion profiles in the Mott–Schottky case as a function of the strength of the effect measured by ϑ.
R.h.s.: Enrichment profiles in the Gouy–Chapman case as a function of the strength of the effect measured by ϑ.

5.8.3 Conductivity effects

The integral effect exerted by such profiles, is what is generally measured or perceived in a specific application. A conductance experiment is a well–suited example, because the local specific conductivity is proportional and sensitive to the charge carrier concentration. (See Section 6.6.2 for diffusion through boundary layers, and Section 7.3.3 for the capacitance effects.)

[159] The relationship between $\vartheta_2 (< 0)$ and λ^*/λ is obtained from Eqs. (5.232, 5.229, 5.215) as

$$\frac{\lambda^*}{\lambda} = \sqrt{8\frac{z_1}{z_2}\ln\frac{1+\vartheta_2}{1-\vartheta_2}}.$$

On account of the σ–profile (in x–direction) the direction of measurement is naturally important. When conductivity measurements are performed in the direction of the profile (normal to the active interface, i.e. in the x direction) the most strongly insulating parts of the profile are most perceived, while a measurement perpendicular to the x–direction (parallel to the active interface) senses the highly conducting parts most strongly. Electrical depletion layers are preferably investigated in the first mode, enrichment layers in the second mode.
Let us first consider a measurement parallel to the active interface in Fig. 5.79 (we neglect the additional effects of boundaries normal to the direction of measurement)

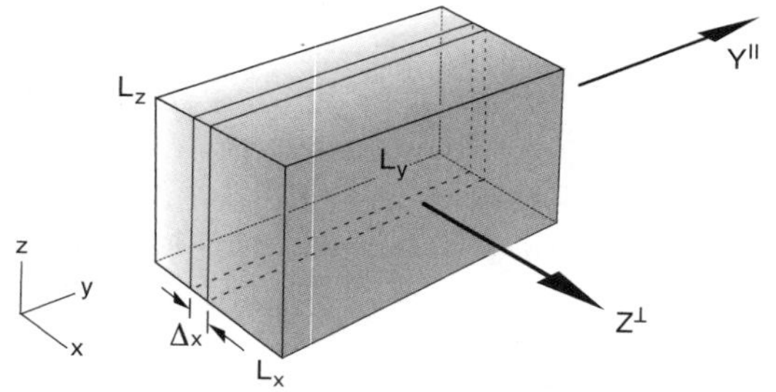

Fig. 5.79: Concerning the conductance normal ($Z^{\perp}$) and parallel ($Y^{\|}$) to the interface (0, y, z). On account of existence of the interface the concentration inhomogeneity occurs (see text).

and cut a very thin layer from the crystal, having the dimensions L_y and L_z in the y and z directions. The thickness in the x direction is chosen to be differentially small, so that variation in conductivity within it can be neglected. The conductance ($dR^{\|-1}(x)$) of this slice is $\frac{L_z dx}{L_y}\sigma(x)$. The total conductance of the crystal is obtained as the sum of these differential parallel slices and given by

$$R^{\|-1} = \frac{L_z}{L_y}\int_0^{L_x} \sigma(x)dx. \qquad (5.233)$$

Let us define an effective conductivity $\sigma_m^{\|}$, which is an apparent specific quantity related to the conductance according to $R^{\|-1} \equiv \frac{L_z L_x}{L_y}\sigma_m^{\|}$, i.e. as σ in the homogeneous case. This quantity then is given as an arithmetic mean according to Eq. (5.233):

$$\sigma_m^{\|} = \frac{1}{L_x}\int_0^{L_x} \sigma(x)dx. \qquad (5.234)$$

In the case of a measurement normal to the active interface the slices are connected in series. The differential resistances $dR^{\perp}(x) = \frac{dx}{L_y L_z}\sigma(x)^{-1}$ sum up according to[160]

$$R^{\perp} = \frac{1}{L_y L_z}\int_0^{L_x} \sigma(x)^{-1}dx. \qquad (5.235)$$

The effective parameter $\sigma_m^{\perp}$ then follows as the harmonic mean $\left(R^{\perp} \equiv \frac{L_x}{L_y L_z}\sigma_m^{\perp -1}\right)$

$$\sigma_m^{\perp -1} = \frac{1}{L_x}\int_0^{L_x} \sigma(x)^{-1}dx. \qquad (5.236)$$

Since in the case of the silver halides for example the local conductivity is the sum of the partial local conductivities[160] and each of these is proportional to the particular concentration, it is easily possible to calculate the electric effects, if we can assume constant mobility. On account of the structural invariance assumed, this is fulfilled if the fields are not too high (see Chapter 6).
Let us first consider a case where both defects are in spatial equilibrium, that is the Gouy–Chapman case (with $|z_1| = |z_2| \equiv z$).
As Eq. (5.233) demonstrates, it is most useful to consider the quantity $R^{\parallel -1}L_y/L_z$ in discussing the parallel electrical conductance; we shall call this $Y^{\parallel}$ $(= \sigma_m^{\parallel} \cdot L_x)$ in the discussion that follows. After partial fraction expansion of Eq. (5.228) and subsequent integration, it is found as expected for the two defects (1 and 2) (e.g. the Frenkel defects in Fig. 5.72) that they can be split additively in the contribution of the bulk (which extends through the whole sample as a background value right up to the phase boundary), given by $Y_{1,2\infty} = L_x z F u_{1,2} c_{\infty}$, and an excess value [249]

$$\Delta Y_{1,2}^{\parallel} = (2\lambda) z F u_{1,2}\left[2c_{\infty}\frac{\vartheta_{1,2}}{1-\vartheta_{1,2}}\right]. \qquad (5.237)$$

It is not surprising that this includes the two parameters λ and ϑ and, hence, c_∞ and c_0. The total excess electrical conductance $\Delta Y^{\parallel}$ is obtained as the sum $\Delta Y_1^{\parallel} + \Delta Y_2^{\parallel}$.

---

[160] A further difficulty must be pointed out, which is not met in the case of Frenkel disorder and which we will not follow up further in detail. If there is no local equilibrium, as with a dislocation–free Schottky–disordered single crystal, the particle fluxes are not convertible, i.e. $\sigma = \Sigma_k \sigma_k$ no longer applies locally and neither does $R^{\perp -1} = \text{const}\left(\int(\Sigma_k\sigma_k)^{-1}dx\right)^{-1}$ but instead $R^{\perp -1} = \text{const}\Sigma_k(\int \sigma_k^{-1}dx)^{-1}$ (the resistor elements belonging to the different carriers are not locally connected). The same difficulty appears in materials in which ions and electrons represent the major carriers. Such difficulties do not occur in the parallel case because $\int$ and Σ are interchangeable. Accordingly the lifetime for Schottky pairs may be substantially larger than for Frenkel–pairs. In semiconductor physics intermediate cases are also considered, corresponding to the ratio of dielectric relaxation time (ε/σ) and recombination lifetime (rate constant of the defect reaction, Chapter 6). The limiting cases are termed life–time and relaxation semiconductor respectively [253].

In the case of large effects $\vartheta_1 \rightarrow 1$ (1 refers again to the enriched defect, in the example referred to in Fig. 5.72 this is V'_M), $1-\vartheta_1$ approaches $2\zeta_{10}^{-1/2}$, and the result

$$\Delta Y_1^{\|} = z_1 F u_1 (2\lambda)\sqrt{c_{10}c_\infty} = u_1\sqrt{2\varepsilon RT c_0} \tag{5.238}$$

is obtained. Equation (5.238) can be interpreted in the following way. The double Debye length (2λ) plays the role of an effective thickness; and the geometric mean of the two extreme concentrations ($\sqrt{c_{10}c_\infty}$), the role of an effective concentration (see Table 5.3). Thus, if the concentration at x=0 is raised by four orders of magnitude

Table 5.3: Effective values[a] for the Gouy–Chapman and the Mott–Schottky cases ($|\vartheta| \rightarrow 1$).

	effective thickness	effective concentration
conductance in Gouy-Chapman-case	2λ	$\sqrt{c_0 c_\infty}$
resistance in Schottky-Mott-case	λ^*	$2c_0 \ln\frac{c_\infty}{c_0}$

[a] For exact range of validity see text. For a more detailed compilation see [254].

with respect to the bulk value this results in an increase of the effective space charge conductivity by two orders of magnitude. If the sample is thin or fine grained, such conductivity increases are certainly measurable. (The generalization for finite boundary conditions, which are important when working in the nano–ranges are given on page 255.)

If we express the result in terms of $\sigma_m^{\|}$ we see that the effective conductivity of the boundary layer must be weighted with the volume fraction (φ), for this is just $2\lambda/L_x$ (Eq. (5.238)), according to

$$\sigma_m^{\|} = \Sigma_\alpha \varphi_\alpha \sigma_\alpha. \tag{5.239}$$

Equation (5.239) generally applies to the parallel connection of various zones α with identical cross sections (dimensions in y and z directions). This relation also makes it possible to include the interfacial core parallel to the space charge effect (see also Eq. 5.261).

It is obvious that in Eq. (5.238) the influence of c_∞ cancels out ($\lambda \propto c_\infty^{-1/2}$), so that the conductivity increase is independent of the bulk concentration. (An increased bulk concentration increases the effective conductivity of the layer but reduces the effective thickness.) This point is of great advantage, since then impurities do not

play any role, as long as the enrichment effect is large enough[161]. It is consistent with the fact that Eq. (5.238) can also be produced by integration of Eq. (5.227) and hence shows that for large effects it is also valid for a more general defect chemistry (see above). In such cases the integral has to be taken simply over ζ_1 (Eq. (5.227)) from x=0 and x=2λ, since the bulk contribution is negligible close to the boundary and the boundary layer contribution is negligible for $x > 2\lambda$. The approximate relationship Eq. (5.238) is also obtained immediately via first integration of the Poisson equation reflecting the integrated charge density which is determined by the enriched defect[162] (see Fig. 5.74).
Depletion effects occur in the conductance only, if the depleted defect 2 possesses such a high mobility (with respect to 1) that this then dominates the conductivity. When the effect is large, $\vartheta_2 \to -1$ leads to

$$\Delta Y^{\|} \simeq \Delta Y_2^{\|} = -z_2 F(2\lambda) u_2 c_{\infty}. \tag{5.240}$$

Equation (5.240) means that the space region of extension 2λ simply drops out of the total balance.
At this point, it is instructive to discuss the behaviour of the minority charge carrier [255]. Let us assume that the adequately mobile minority charge carrier 3 (e.g. the conduction electrons in the example in Fig. 5.72) is also enriched, which is the case if $z_1 z_3 > 0$. We obtain its concentration profile via $\zeta_3 = \zeta_1^{z_3/z_1}$, according to Eq. (5.215). For $z_3 = z_1$ we obviously get, after integration, a result which is analogous to Eq. (5.237), namely

$$\Delta Y_3^{\|} = |z_3| F(2\lambda) u_3 \left[2c_{3\infty} \frac{\vartheta_1}{1-\vartheta_1} \right] \simeq |z_3| F u_3 (2\lambda) \sqrt{c_{30} c_{3\infty}}. \tag{5.241}$$

In contrast to $\Delta Y_1^{\|}$, however, $\Delta Y_3^{\|}$ is sensitively dependent on the bulk value, i.e. on the impurity content ($\lambda \propto c_{1\infty}^{-1/2}$!).
Analogously for the depleted minority defect 4 (holes in Fig. 5.72, $z_1 z_4 < 0$) it follows that

$$\Delta Y_4^{\|} = -|z_4| F(2\lambda) u_4 c_{4\infty}. \tag{5.242}$$

(In the example in Fig. 5.72 the electronic conductivity is given by the sum of the contributions from Eq. (5.241) and Eq. (5.242).)

In the perpendicular mode it is found, after partial fraction expansion, that the integral of $c^{-1}(x)$ also leads to a simple summation[163] of a volume background $R_\infty^{\perp} = \frac{L_x}{L_y L_z} \sigma_\infty^{-1}$ and an excess value $\Delta R^{\perp}$. Here we assume that a single defect dominates

[161] And as long as the impurities do not act on the boundary layer effect (i.e. ϑ).

[162] $\Delta Y^{\|} \propto \int_0^\infty \rho dx \propto \int_\infty^0 \phi'' dx = \phi'_0 - \phi'_\infty = \phi'_0$.

[163] This is not expected from the outset because background and excess contributions are summed in c (that is in the sense of a parallel connection) but not in 1/c.

conductivity. Normalizing the resistance with respect to the area $L_y \cdot L_z$ (see Eq. (5.235)), i.e. defining $Z^\perp \equiv R^\perp L_y L_z$ ($= \sigma_m^{\perp -1} \cdot L_x$) [256], we arrive at the excess quantity

$$\Delta Z^\perp = -\frac{2\lambda}{zFu}\frac{2}{c_\infty}\frac{\vartheta}{1+\vartheta}. \tag{5.243}$$

As $\Delta Y^\parallel$, the quantity $\Delta Z^\perp$ is proportional to the effective thickness (2λ), but it is now inversely proportional to an effective concentration. In the extreme case ($\vartheta_2 \to -1, (1+\vartheta_2) \to 2\sqrt{c_{20}/c_\infty}$) of pronounced depletion the latter again turns out to be of the familiar form $\sqrt{c_{20}c_\infty}$ and it follows that

$$\Delta Z^\perp = \Delta Z_2^\perp = \Delta R^\perp L_y L_z = \sqrt{\frac{2\varepsilon RT}{c_{20}c_\infty^2 F^4 z_2^4 u_2^2}}. \tag{5.244}$$

Again, this result can also be obtained from the approximate profile (Eq. (5.227)). Analogously to the above case it is approximately possible to describe the resistivity $\sigma_m^{\perp -1}$ by a summation over the contributions from local regions α, the individual terms being weighted with the volume proportion φ_α:

$$\sigma_m^{\perp -1} = \Sigma_\alpha \varphi_\alpha \sigma_\alpha^{-1}. \tag{5.245}$$

Again it is possible in this way, additionally to the space charge effect, to include the contribution of the core of the boundary layer (see also Eq. (5.262)).

It now remains for us to calculate the resistance of a Mott–Schottky boundary layer, that is a depleted boundary layer in the extrinsic case with nonmobile dopant. As we saw for very severe depletion, the total effect took the form of an almost stepwise function at λ^* ($\simeq \lambda^* - \sqrt{2}\lambda$). In this extreme step–function approximation it follows ($\Delta Z^\perp \simeq Z^\perp$) that

$$\Delta Z^\perp \simeq \frac{\lambda^*}{|z_2|Fu_2c_2^*} \tag{5.246}$$

with $c_2^* = c_{20}$. This means that the boundary layer concentration now appears as the effective concentration and the parameter λ^* as the effective thickness. The bulk effect does not disappear in the overall result, i.e. in $\Delta Z^\perp$. (On parallel measurement we would have found the excess conductance value $(\sigma_{20} - \sigma_{2\infty}) / \lambda^* \simeq -\sigma_{2\infty}/\lambda^*$.) Even though the assumption of a rectangular profile through c_0 is useful for initial orientation, the neglect of the contributions shown in Fig. 5.78a is far too restrictive. In a much better approximation[164] $c_2^* = c_{20} \cdot 2\ln(c_{2\infty}/c_{20})$ [253] (see Table 5.3).

5.8.4 Defect thermodynamics of the interface

Before we turn to the examples, we have to discuss briefly how the interface chemistry can be related to the parameter c_0, i.e. to ϑ (cf. Section 5.4.3). Let us consider,

[164] While in the zeroth approximation c and, hence, ϕ are taken to be constant in the space charge zone, it is a better approximation for small x to linearize $\phi(x)$, which is actually parabolic.

in order to sketch out the problem, the specific, highly simplified case of the surface of a mixed conducting oxide in the P–regime with fully ionized excess oxygen atoms and defect electrons as majority charge carriers in bulk, space charge zone and boundary core. The structure varies from bulk (x≥0) to core (x<0) in the form of a sharp step–function. We will also assume that the core consists of a single atomic layer centred at x=s<0. Lateral inhomogeneities are neglected. The generalization of this artificial example to other cases is evident (see Fig. 5.80). We will break down the thermodynamic treatment of the problem as a whole into the following individual steps[165]:

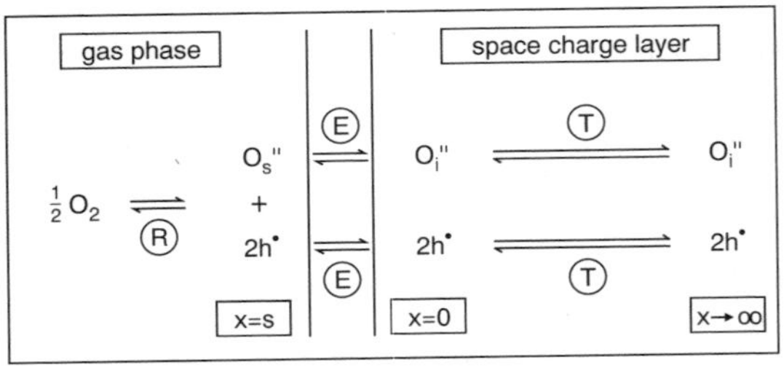

Fig. 5.80: Scheme for the derivation of a defect model for the boundary layer. In contrast to x≥0, the ground structure is changed (that is the standard potentials) at x=s (referring to the surface or adsorption layer). As an example we use the interaction of an oxide with O_2. For simplicity it is assumed that oxygen is doubly charged even at x=s. The kinetics of such processes will be discussed in Chapter 6 (especially Section 6.4).

1) The first step involves solution of the bulk problem, i.e. relating $c_{O_i''\infty}$ and $c_{h^{\cdot}\infty}$ to the material parameters and the control parameters, temperature, component activity and doping content. In view of the treatment already given we can regard this problem as having been solved (see Eq. (5.150)).
2) In order to describe the equilibrium in the space charge zone we consider the transport equilibrium ("T") of the charge carrier within the space charge zone. This problem was subject to discussion above and led to the necessity of determining the two parameters $c_{k\infty}$ and c_{k0}. Hence, the whole problem — at least for large effects[166] — is reduced to the analysis of the parameters c_{k0} or ϑ_k. As an example we assume that O_i'' defects are enriched and $h^{\cdot}$ defects are depleted in the space charge zone (i.e. negative space charge potential).
3) For this purpose we consider the transfer of the two defects from position x=0 to the neighbouring, but structurally different, layer x=s (Fig. 5.80), where the countercharge is situated, that is to the actual surface[167]:

$$O_i''(x=0) + V_s(x=s) \rightleftharpoons O_s''(x=s) + V_i(x=0) \tag{5.247}$$

[165] Here we follow the presentation in Ref. [248]. A more detailed treatment is given in Ref. [257].
[166] Equation (5.228) is not generally valid for our example (different absolute charge numbers). However, for large effects, Eqs. (5.227) and (5.230) apply, respectively.
[167] The definition of the interstitial position largely loses its meaning at the surface. In this context, only the coupling with interfacial chemistry is important. For this reason and for the purpose of precisely describing the coordinate the index "s" is used (and not the bulk symbols or the designation "ad").

$$h^{\cdot}(x=0) \rightleftharpoons h^{\cdot}(x=s). \tag{5.248}$$

These two processes are associated with a change in both the standard chemical potentials and the electrical potential[168] (electrochemical process "E"). It is also necessary to use the activity a, at least for x = s:

$$\frac{a_{O''_s}(s)}{a_{O''_i}(0)} = \exp -\frac{\mu^\circ_{O''_s}(s) - \mu^\circ_{O''_i}}{RT} \exp +\frac{2F(\phi(s) - \phi(0))}{RT} \tag{5.249}$$

and

$$\frac{a_{h^\cdot}(s)}{a_{h^\cdot}(0)} = \exp -\frac{\mu^\circ_{h^\cdot}(s) - \mu^\circ_{h^\cdot}}{RT} \exp -\frac{F(\phi(s) - \phi(0))}{RT}. \tag{5.250}$$

The electrostatic contributions are the same for ionic and electronic defects apart from the charge numbers, while the chemical contributions naturally differ. The parameters $\mu^\circ_{O''_i}$ and $\mu^\circ_{h^\cdot}$ depend on the bulk mass action constants of the reaction O′ (see Eq. (5.129)):

$$-RT \ln K_{O'} = \mu^\circ_{O''_i} + 2\mu^\circ_{h^\cdot} - \frac{1}{2}\mu^\circ_{O_2}. \tag{5.251}$$

The relationships for $\mu^\circ_{O''_s}(s)$ and $\mu^\circ_{h^\cdot}(s)$ are analogous, with a modified constant $K_{O's}$ for the incorporation in the core zone of the surface (see below), provided a single layer model is adequate.

We can conclude three points from these formulations:

Firstly, because of the relevance of $\mu^\circ(s)$ a model for the extended defect "surface" is required in addition to a bulk defect model; this will be referred to as the (surface) core model in what follows.

The simplest reasonable model involves the assumption of a random distribution of surface defects, yet with a limited number of surface sites as assumed in the treatment according to Langmuir (see Section 6.7.1). The associated activity coefficients are obtained from Eq. (5.209). (The standard potentials are naturally related to the surface tension [149], see Section 5.4.4.)

Secondly we must relate the potential difference $\phi(s) - \phi(0)$ to the contact chemistry. This requires clearer ideas concerning the charge distribution and the atomic structure[169]. Let us take the simple case, where there is no charge between x = 0 and x = s < 0 and for more precise characterization a dielectric constant ε_s can be introduced[170]. From the fact that the countercharge of the core charge (Σ) at x=s is distributed between zero and ∞ (cf. is global electroneutrality), the integration

[168] It is more rigorous to formulate the building elements $O''_i(x=0) - V_i(x=0) \equiv O''(x=0)$ and $O''_s(x=s) - V_s(x=s) \equiv O''(x=s)$ explicitly.

[169] The above equilibrium thermodynamics can also be used in the case of a less ordered "ground structure" (say, in the core). The description is then less precise, due to a variability of μ°, but, as is seen in aqueous chemistry, can be largely adequate.

[170] It is advantageous to introduce ε_s as an effective material parameter if the thickness of the interfacial core is extended.

of Poisson's equation for the countercharge gives the expression (Gouy–Chapman case)

$$|\Sigma| = |\varepsilon_s E_s| = \sqrt{2\varepsilon RTc_\infty}\left(\zeta_{10}^{-1/2} - \zeta_{20}^{-1/2}\right) \simeq \sqrt{2\varepsilon RTc_{10}}. \tag{5.252}$$

In the case of large effects $\zeta_{10} \gg 1 \gg \zeta_{20}$ the simplification on the right is obtained, which naturally must correspond to Eq. (5.238). On the basis of the Gauss' law it is also possible to interpret the condition (5.252) as continuity of the dielectric displacement (D = dielectric constant times electric field; $\varepsilon E_0 = \varepsilon_s E_s$)[171]. The electrical potential varies linearly (zero charge density) with x, and the difference $\phi(s) - \phi(0)$ is given by the product of the (constant) field strength and s.
4) The last (and actually redundant) step is the interaction with the neighbouring phase (chemical process "C"). Here we formulate the interaction with oxygen:

$$\text{Reaction } O_s\text{'} = \qquad \frac{1}{2}O_2(g) + V_s(s) \rightleftharpoons O_s''(s) + 2h^{\cdot}(s), \tag{5.253}$$

with the mass action law:

$$\frac{a_{O_s''}(s)a_{h^{\cdot}}^2(s)}{P_{O_2}^{1/2}} = K_{O_s'}, \tag{5.254}$$

$V_s(s)$ does not appear, since $a_{O_s''}(s)$ is, strictly speaking, the activity of the building element $(O_s'' - V_s)$ and also contains configurational effects with respect to $V_s(s)$[172]. Neither do electrical effects occur, since we refer to the same position x=s (i.e. $\phi(x{=}s)$ cancels out). Since both Eq. (5.252) and the charge and mass balances refer to concentrations, we must, as already mentioned, derive activity coefficients from the surface core model.

Let us now once again work out the difference from the situation in which core and bulk are separated in thought and being in equilibrium with the gas phase. Now the transfer equilibria described by Eq. (5.247) and Eq. (5.248) become directly relevant and the core charge will be compensated by space charges of opposite sign. This additional unknown is accounted for by including the Poisson equation as an additional relationship[173]. One of the four chemical equilibria (bulk equilibrium, ionic equilibrium, transfer equilibria for both defects) is redundant so it is not necessary that all four μ° values ($\mu^\circ_{O_s''}(s)$, $\mu^\circ_{O_i''}$, $\mu^\circ_{h^{\cdot}}(s)$, $\mu^\circ_{h^{\cdot}}$) are known. Knowledge of three independent combinations is sufficient.

---

[171] According to Gauss' law it is also possible to express the volume integral $\int \mathrm{div}\mathbf{D}dV$ as the surface integral $\int \mathbf{D}d\mathbf{a}$. If we integrate over a narrow area element with the normal vector parallel to the x direction, then, neglecting the contributions from the sides, we get as difference of the x–components: D(s) − D(0) = surface charge density. Since there is no charge accumulated between x=0 and x=s, D is continuous. Because of Eq. (5.218) this is identical with the condition of global electroneutrality ($\int \varrho dV = 0$).

[172] Formally we obtain the same expression (Eq. (5.254)) for the mass action law if we take the structure elements $O_s''(s)$ and $V_s(s)$ into account separately and set their (virtual) activity equal to the concentration.

[173] Strictly speaking we also have to consider double–layer effects in the separated regions individually.

Analogous considerations apply for the simple clear–cut case of primarily ionic disorder in both bulk (I region) and core. The behaviour is illustrated in Fig. 5.81 in terms of the energy level picture (cf. Fig. 5.9 on page 127, and Fig. 5.73 r.h.s.) for Frenkel or anti–Frenkel disorder in bulk and core (i.e. vacancies and interstitial defects predominate), and is transformable to related cases. (In the P region, the level shift is influenced by electronic effects.) Not only are the energy levels ($\widetilde{\mu}^\circ$) in bulk and core different, but also their separation corresponding to the different mass action constants. In the space charge zone the levels are bent (variation in ϕ) such that $\widetilde{\mu}$ = const. (Fig. 5.81). Knowledge of the sum of the chemical standard

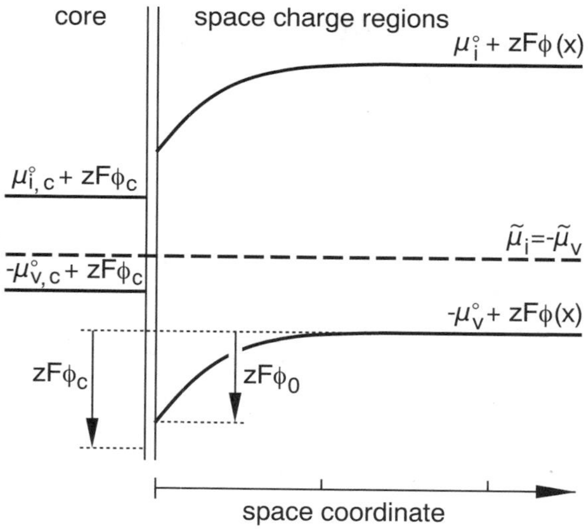

Fig. 5.81: Boundary equilibrium in the level diagram for Frenkel defects (i, V: e.g. $Ag_i^{\cdot}$, V'_{Ag}, z=1) or anti–Frenkel defects (i, V: e.g. O_i'', $V_O^{\cdot\cdot}$, z=-2). The electrical potentials cause the levels ($\widetilde{\mu}^\circ$) to bend in the space charge zone and to shift in the core in order to satisfy that $\widetilde{\mu}$ = const. For the sake of simplicity the electrical bulk potential (ϕ_∞) is set to zero. It can be seen that the level difference on both sides of x = 0 is different. (In the pure material $\widetilde{\mu}$ lies in the middle of the gap between the two bulk standard values ($\mu_i^\circ + zF\phi(x=\infty)$ and $-\mu_v^\circ + zF\phi(x=\infty)$). The index c designates the core region. An analogous situation is met in the case of electronic levels. When considering O_i'' and $h^{\cdot}$ as majority carriers ($\mu_{i,c}^\circ \equiv \mu_{O_s''}^\circ$) — as done in the text — the direction of the level bending is the same for both carriers, with the electronic effect being less pronounced (cf. charge). The coupling is given by $\widetilde{\mu}_{O_i''} + 2\widetilde{\mu}_{h^{\cdot}} = \frac{1}{2}\mu_{O_2}$ [257].

potentials in the core and bulk ("energy level gap") and the relative positions to each other (e.g. the difference between the μ° values of a given defect) are sufficient to define the situation. Again it is not necessary to have knowledge of all four standard potentials.

In the literature the interaction with the neighbouring phase, the differing thermostatics of the core and also the potential jump are usually neglected in the treatment of space charge chemistry. If, in such greatly simplified approximations, we allow s to approach zero, we obtain the results associated in the literature with the model of Pöppel and Blakely [258], while additional neglect of the finite number of sites is associated with the model by Kliewer and Köhler [259]. In the latter case the following relationship applies

$$c_{k0}/c_k^* = \exp\left(-\frac{\Delta_k G_m^K}{RT}\right). \tag{5.255}$$

If we are considering a vacancy (k=v) then $\Delta_k G_m^K$ can be interpreted as the chemical free enthalpy of transfer (without configurational contribution) of the regularly situated ion from the bulk to the interface; for k=i it can interpreted as the corresponding free transfer enthalpy from the interface to an interstitial position in the bulk[174]. In other words, an improved relation is obtained by transferring k explicitly from the bulk to the position $x = s \neq 0$, viz.

$$c_{k0}/c_k^* = \exp\left(-\frac{\Delta_k G_m^{K'}}{RT}\right). \tag{5.256}$$

The assumption that $\Delta G_m^{K'}$ is constant is, in fact, a reasonable approximation for the ionic defects, if pronounced disorder is realized in the core. It is, for example, imaginable that a high density of adsorbed anions and cations is present in the core region. Then the (small) difference of the two (large) concentrations constitutes the surface charge. In this case, the transition of a particular charge carrier into the surface layer (or the incorporation of a component from the gas phase) alters the

---

[174] If we formulate the individual "defect formation reactions" as

$$O_O + V_s \rightleftharpoons V_O^{\cdot\cdot} + O_s''$$

$$O_s'' + V_i \rightleftharpoons O_i'' + V_s$$

and identify the chemical standard enthalpies of reaction as $\Delta_v G^*$ and $\Delta_i G^*$, then we see that for x=0 (κ_{0s} describes the potential jump between x=0 and x=s)

$$c_{v0}/c_v^* = \left(\exp -\frac{\Delta_v G^*}{RT}\right) / (a_{O_s''}\,\kappa_{0s})$$

$$c_{i0}/c_i^* = \left(\exp -\frac{\Delta_i G^*}{RT}\right) \cdot a_{O_s''}\kappa_{0s}.$$

For $c_v(x)$ and $c_i(x)$ the parameter ΔG^* must be replaced by $\widetilde{\Delta G}^*$ $(= \Delta G^* \pm 2F(\phi_0 - \phi(x)))$. In the case of bulk concentration the relevant potential difference is $\phi_0 - \phi_\infty$. Evidently it follows that $\widetilde{\Delta_v G}_m^* + \widetilde{\Delta_i G}_m^* = \Delta_v G_m^* + \Delta_i G_m^* = \Delta_F G_m^\circ$, while the difference $\widetilde{\Delta_v G}_m^* - \widetilde{\Delta_i G}_m^*$ is determined by $\ln(a_{O_s''} \cdot \kappa_{0s})$ and the space charge potential. The latter follows from the electroneutrality of the bulk, i.e. from $c_{v\infty} = c_{i\infty}$. The effect of a neighbouring phase on the defect concentrations can be seen immediately from the above reactions. If O_s'' is stabilized by the neighbouring phase there is a tendency for $[V_O^{\cdot\cdot}(x)]$ to increase and for $[O_i''(x)]$ to get smaller. In Kliewer's approximation κ_{0s} and $a_{O_s''}$ are tacitly taken as constant and these can be included in the ΔG^* values. The result is then a relationship of the form (5.256). The sum of these $\Delta G_m^{K'}$ values is then again $\Delta_F G^\circ$, while the difference is given directly by $\pm 2zF(\phi_0 - \phi_\infty)$. This consideration also applies if there are additional vacancy defects in the core. It is important that the definition of the ΔG^*–values refers to the same defect species.

concentration relatively little[175]; the field contribution, that is actually included in the Equation (5.256), is then also invariant. Hence $c_{O''_s}(s)$ and $a_{O''_s}(s)$ in the mass action law (Eq. (5.249)) and accordingly $c_{O''_i}(x=0)$ act as constant quantities[174], as suggested by Eq. (5.256). But this is then no longer correct for the electronic charge carrier: When $a_{O''_s}(s) = \text{const.}$, Eq. (5.254) reveals a proportionality between $a_{h^\cdot}(s)$ and $P_{O_2}^{1/4}$.

We have seen that the picture is always simple if the charge density is independent of the component potential. This then applies on account of Eq. (5.252) also for the enriched defect at x=0, in our example for O''_i. For mass action reasons, the concentration of the counterdefect ($V_O^{\cdot\cdot}$) is also constant. However, electronic charge carriers — as minority species — are, as in the bulk case, appreciably dependent on the component potential. In our example $N_{h^\cdot,0} = 1/4$, while in the bulk $N_{h^\cdot\infty} = 1/6$. In such cases then, it is in fact possible within the framework of our simple core–space charge model to express $c_k(x=s)$ and in simple cases $c_k(x=0)$ too in the form of the bulk relationship Eq. (5.150)

$$c_{k0} = \alpha_{k0} P^{N_{k0}} \Pi_r K_r^{\gamma_{rk0}}, \tag{5.257}$$

whereby the characteristic coefficients differ, in general, from those in the bulk. From Eq. (5.229) ϑ can be obtained and from relations such as Eqs. (5.227), (5.228), (5.230) the local concentration $c_k(x;P,T)$.
For the usually more important integral effect, as, for example, realized in $\Delta Y^{\|} \propto \sqrt{c_0}$, we conclude that $\partial \ln \Delta Y^{\|}/\partial \ln P = N_0/2$ since $\partial \ln c_0/\partial \ln P = N_0$. Thus, if the N_0 value is identical to the bulk value, the exponent measured in a (parallel) conductivity experiment for the boundary layer contribution is half the value measured for the bulk. This does not apply to the resistance in the Mott–Schottky case (see Table 5.3, cf. also the following section). The dependence on the doping content is expected to be rather involved due to segregation effects.

We have now set out the principal problem, but it is also evident how much knowledge of the interfacial chemistry is needed for establishing $c_k(T, a_M, C; x)$ relationships, i.e. for predicting Kröger–Vink and van't–Hoff diagrams for boundary layers, a knowledge which is usually not available. We will become acquainted with simple examples below.

As already pointed out in the intrinsic case, simplified treatments of the dependence of the component activity (e.g. P_{O_2}) are feasible even in quite general cases (intrinsic or extrinsic) if it is not necessary to consider the core chemistry explicitly. This may be the case if a priori assumptions concerning the core charge density can be made [244,248]. By skipping point 3 (see page 234) we formulate the incorporation

[175] Cf. Eqs. (5.249), (5.250). The relative change of the charge density, however, is not negligible. This is different if the charge density is on a relatively high level, e.g. if essentially only anions are adsorbed.

equation as

$$\frac{1}{2}O_2(g) + V_i(x) \rightleftharpoons O_i''(x) + 2h^{\cdot}(x) \tag{5.258}$$

which is valid for every location, that is, also for x = 0[176]. Since the electrical field drops out, it follows that

$$c_{O_i''}(x{=}0)c_{h^{\cdot}}^2(x{=}0) \propto P_{O_2}^{1/2}. \tag{5.259}$$

Let us consider the Gouy–Chapman case. If O_i'' is the enriched majority charge carrier, then, according to Eq. (5.252), the surface charge density Σ is proportional to $c_{O_i''}^{1/2}(x{=}0)$. Hence, with Eq. (5.259) we obtain

$$\frac{\partial \ln c_{h^{\cdot}}(x{=}0)}{\partial \ln P_{O_2}} = \frac{1}{4} - \frac{\partial \ln |\Sigma|}{\partial \ln P_{O_2}}. \tag{5.260}$$

If the relative change of $|\Sigma|$ with the change of partial pressure is small then — as already mentioned — it follows that $N_{h^{\cdot}0} = \frac{1}{4}$, while in bulk $N_{h^{\cdot}\infty}$ is, in general, $\frac{1}{6}$ (if $2c_{O_i''}(x{=}\infty) = c_{h^{\cdot}}(x{=}\infty)$) or $\frac{1}{4}$ (if $c_{O_i''}(x{=}\infty) \gg c_{h^{\cdot}}(x{=}\infty)$) corresponding to a positive doping ($D^{\cdot}$).
As already demonstrated, a very small relative change in Σ on oxygen partial pressure treatment is most likely if ionic disorder[177] predominates. In other cases, the partial pressure of oxygen may have an appreciable influence on the surface charge density, e.g. as a result of variation in the number (or in the charge) of adsorbed negativated oxygens, in so far as specific segregation effects or crystallographic effects do not keep Σ more or less constant.
The relationships are different in the Mott–Schottky case. There $\partial \ln c_{O_i''}(x{=}0) \propto \partial \ln |\Sigma|$ does not apply. Rather, because of $\Sigma \propto [D^{\cdot}]\lambda^*$ and Eqs. (5.215), (5.231), it follows that $\partial \ln c_{O_i''}(x{=}0) \propto \partial |\Sigma|^2$, i.e. the absolute change is important (more than that: The concentration effect is proportional to $|\Sigma|\partial|\Sigma|$, in complete contrast to $\partial|\Sigma|/|\Sigma|$ in the Gouy–Chapman case). A detailed example will be discussed in the next section.

It is evident that the core region will be usually more complex in structure (e.g. multi–adsorption layers), particularly when solid–solid contacts are considered. In the case of solid homophasic contacts the boundary layer consists of two symmetrical space charge zones separated by the grain boundary core (which in some cases includes an amorphous grain boundary phase). In the case of heterocontacts the symmetry is lost and charge enrichment in one space charge zone is imaginable solely

[176] When equilibrium is complete, Eq. (5.259) is also obtained from the combination of Eq. (5.215) with the bulk mass action law. Note that because of Eq. (5.215) it follows that $c_{O_i''}(x = 0)/c_{O_i''}(x = \infty) = c_{h^{\cdot}}^2(x = \infty)/c_{h^{\cdot}}^2(x = 0)$.

[177] This case appropriately transformed applies to AgCl. It is comparable to the constancy within the I–regime in the bulk. If pure electronic disorder governs the I–regime, the roles of electrons and ions are to be exchanged.

at the expense of depletion in the other.

The abrupt structural model described so far is the simplest possible model, but offers a surprisingly good approximation in many cases. In practice the structural adjustment will be more or less continuous in particular if elastic effects of longer range (see Section 4.3.7) complicate the picture (see page 259 for a more detailed consideration).

We expect particularly interesting phenomena in the nanosize range, when the distance between the two interfaces (e.g. grain boundaries in nanocrystalline material) is of the order of magnitude of the Debye length or smaller. If we ignore structural effects just mentioned (see however Fig. 5.108, page 261) we expect the thermodynamic profiles shown in Fig. 5.82. Nano–size effects will be considered explicitly in Example c.

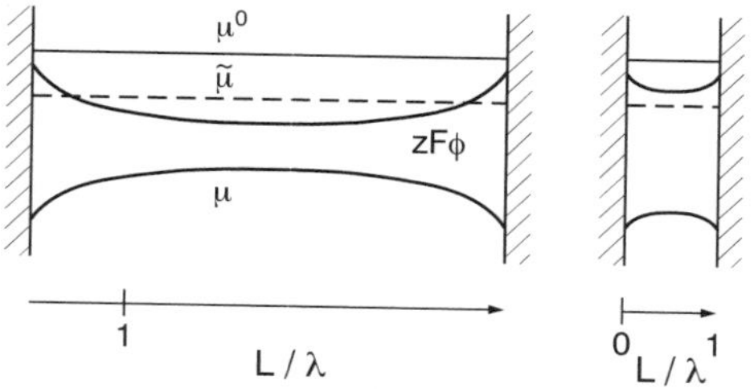

Fig. 5.82: Thermodynamic potentials of thin films (perpendicular to the film plane) of thickness L for $L > 4\lambda$ (l.h.s.) and $L < 4\lambda$ (r.h.s.). Structural invariance is presupposed (c_0 = const). In the left hand example bulk behaviour is established in the interior. This is not the case in the right example; there the sample is charged throughout and a mesoscale effect realized.

We will now consider a range of experimental examples. Without substantial loss of generality — the treatments apply to the general case of a mixed conductor — we shall concentrate our discussion on enrichment effects in ionic conductors, and on the discussion of depletion layers in electronic conductors.

5.8.5 Examples and supplementary comments

a) Heterogeneous solid electrolytes

As discussed, conductivity anomalies due to interfaces can be attributed to core effects (charge carrier concentration and mobility are altered there) and/or to space charge zone effects (primarily concentrations are altered there). In particular, in ionic conductors with low bulk disorder and high defect mobility the last point is likely to play a highly important role.
In recent times intensive research has concentrated on enrichment layers of ionic conducting systems[178]. Systematic investigations have confirmed that (almost[179])

[178] See Ref. [244].

[179] However, a fundamental difference must be remembered. On account of the large mass, delocalization phenomena do not play a role in our treatment of ionic defects (with the proton as a partial exception). Cf. also Section 5.8.5e. In addition (but not independent of this) the elec-

all analogues to electronic boundary layer effects with ionic defects can be verified. It has already been shown in Figure 5.77 that ionic space charges are, in principle, possible in solid ionic conductors. Let us now consider the following relevant interfaces: (i) the contact of an ionic conductor with an electrically insulating solid phase, (ii) the contact of an ion conductor (MX) with another ion conductor (MX′) — which includes both grain boundary as homocontact (MX=MX′) as well as the heterocontact of two conductors of different composition (MX≠MX′) — and (iii) the contact ionic conductor/gas.

Research into such systems was started as a result of the surprising finding [244,260, 261] that, under certain circumstances, the addition of fine particles of insulating oxides such as Al_2O_3 (most effectively γ, or η–Al_2O_3) and SiO_2 increases the ionic conductivity of cationic conductors such as LiI, LiCl, AgCl, AgBr, CuCl, CuBr but also of anionic conductors such as CaF_2, SrF_2 and PbF_2 by several orders of magnitude. Ionic space charge regions, occurring as a result of the surface activity of the insulating phase, offer a qualitative, and in cases in which the necessary parameters are known, also a quantitative explanation (cf. Fig. 5.83). According to the above

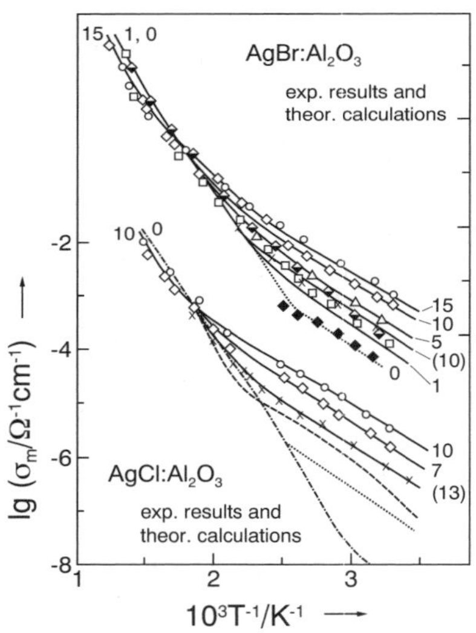

Fig. 5.83: Experimental results (symbols) and theoretical calculations (solid lines) for $AgBr{:}Al_2O_3$ and $AgCl{:}Al_2O_3$ two–phase mixtures. The labels give the volume fraction of Al_2O_3 as a percentage and refer to an Al_2O_3 grain size of 0.06μm (if not in brackets) or 0.15μm (if in brackets). The line marked with dashes for AgCl refers to the nominally pure single crystal, the dotted line to the polycrystal and the broken line to a positively doped single crystal (with respect to the "knee", cf. Fig. 5.88) [249].

treatment the increase in conductivity along interfaces is proportional to the mobility and to the square root of the boundary layer concentration (c_0) of the enriched defect determining the conductivity (Eq. (5.238)). If the surface charge is adequately stabilized by the influence of the neighbouring phase, $\Delta_k H^{K'}$ in Eq. (5.256) is sufficiently small and $1/2\Delta_k H^{K'}$ can be neglected compared with $\Delta_k H^{\neq}$[180], so that the

tron's mobility can be so high, that local equilibrium may be violated during transport processes. As far as the distinction between "relaxation semiconductors" and "lifetime semiconductors" is concerned, see footnote 160.

[180] Even without interaction ("free surface"), $\Delta_k H^{K'}$ is small for AgCl and AgBr (see temperature dependence of the intercept in Fig. 5.77). If the interstitial particle were the enriched carrier in the

temperature dependence is primarily determined by the enthalpy of migration. This is confirmed by Fig. 5.84 for the cationic conductors under discussion (but also holds

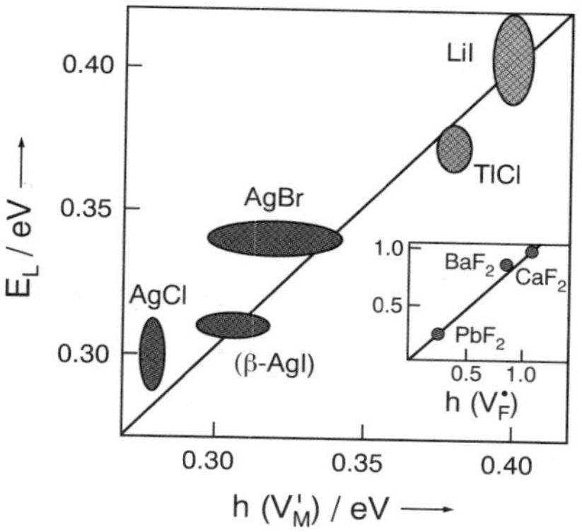

Fig. 5.84: In all cases described the addition of Al_2O_3 to the cation conducting halide results in increased conductivity, whose energy of activation is very close to the energy of migration of the cation vacancy. (An analogous correlation of the activation energy with the vacancy migration energy of the anion vacancy $V_F^{\bullet}$ applies to the fluoride ion conductors CaF_2 and PbF_2 heterogeneously doped with SiO_2. See inset.) On account of the polytypism in AgI (see Section 5.8.5) this datum is given in brackets [244].

for F^- conductors investigated). The activation energy observed is almost identical with the migration energy of the cation vacancies (or fluoride vacancies). This points to the adsorption of cations (or F^- in the case of CaF_2, PbF_2) at the surface as defect–inducing mechanism, a mechanism that is analogous to the example of adsorption from the liquid phase treated at the beginning of this chapter[181]. The surface basicity is also here decisive, and the effect of the oxide is parallel to the pH of zero charge[182]. Variation of the nature of the insulating compound (Al_2O_3, SiO_2, ZrO_2), of its crystallography (α–Al_2O_3, γ–Al_2O_3, η–Al_2O_3), and also of its surface chemistry (e.g. silylation) confirm this picture. Wetting experiments[183] with liquid silver chloride provide evidence concerning the effectiveness of various surfaces [264]. Splitting up the interfacial tension into enthalpy and entropy terms reveals a comparatively low interfacial energy (see Section 5.4) of $3.2 J/m^2$ for the (0001) Al_2O_3 surface, that is more active in adsorption than the $(10\bar{1}0)$ and $11\bar{2}0)$ surfaces (3.8 and $4.6 J/m^2$ respectively). When the surface is covered by hydroxyl groups the interfacial effect increases and the dependency on crystallography is largely lost. Measurement and analysis are greatly complicated by the fact that the effect of a single interface is very small even if local enrichment is very great. For two–phase

case of the silver halides, $\Delta H^{K'}$ would not be negligible compared with the extremely low migration enthalpy of the interstitial defects. Note, that the interaction with the neighbouring phase reverses the space charge potential compared with the free surface value ((100), (111)).

[181] However, the situation is more complex for AgI [262,263].

[182] pH value of a contact solution (in general composed of H_2O, H_3O^+, OH^- and inert counterions) at which the oxide surface is electrically neutral. However, it should be noted that the oxide surface is changed by the neighbouring phase. In the case of an electronic conductor (e.g. oxide electrode) there is, in addition, the possibility of an electronic excess charge. Note that in liquid electrochemistry the point of zero charge is generally referred to the disappearance of this excess charge.

[183] It is to be expected that the energetic interaction of the AgCl melt with the oxide will partially reflect the solid–solid interaction. See Section 5.4.4. The experiments [264] also show the significance of entropic contributions for the interfacial tension and hence for the adsorption.

mixtures with high interface densities, however, as is the case for a dispersion of a fine–grained insulator phase in the ion conductor matrix, substantial effects can be observed. As has been demonstrated by scanning electron microscope studies, the small oxide particles (A) are preferentially located at the grain boundaries of the relatively large ionic conductor grains and form coherent conductivity paths, even at relatively low volume fractions (Fig. 5.85). This means that the percolation

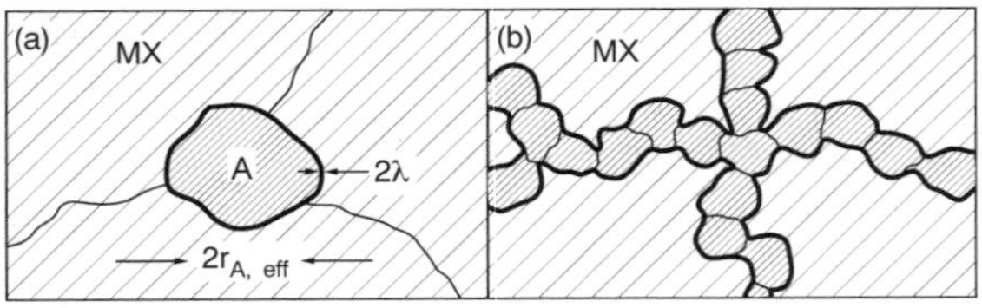

Fig. 5.85: By cation adsorption the insulating A particles induce highly conducting boundary layers in the ionic conductor MX. In (a) the A–grain is isolated (e.g. in the triple–grain junction), and the overall effect is negligible. In (b) coherent paths are formed between the MX grains [249].

threshold, i.e. the volume fraction (φ), at which the first continuous path is produced, occurs at very low values of φ_A. A further increase of the insulator content (φ_A) does not necessarily lead to a higher volume fraction in the grain boundaries, rather in many cases the grain size of the ionic conductor is (provided that φ_A is sufficiently small) reduced accordingly. Given a significantly attractive surface interaction of the two phases, this corresponds to a low energy morphology and is thermodynamically expected. In these cases the interfacial contribution to the specific conductivity σ_m is approximately proportional to the volume concentration. On account of the parallel connection of the pathways, it follows (see Eq. (5.239)) that

$$\sigma_m = (1 - \varphi_A)\beta_\infty\sigma_\infty + \varphi_A\beta_L\sigma_L, \tag{5.261}$$

β_∞ and β_L measuring the number of continuous paths. These factors are independent of φ_A and T under the above conditions. A semiquantitive insight is obtained by considering a primitive cubic morphology. Here $\beta_\infty = 1$ and $\beta_L = 1/3...2/3 \sim 0.5$ in the ideal case. If we take into account a possible blocking effect as a result of grain boundaries perpendicular to the direction of the current we get approximately [256]

$$\widehat{\sigma}_m = \left[\widehat{\sigma}_\infty\widehat{\sigma}_L + \beta_L^{\|}\varphi_L\widehat{\sigma}_L^{\|}\widehat{\sigma}_L^{\perp}\right] / \left[\widehat{\sigma}_L^{\perp} + \beta_L^{\perp}\varphi_L\widehat{\sigma}_\infty\right]. \tag{5.262}$$

Introduction of the complex conductivity into Eq. (5.262) ($\widehat{\sigma}$ instead of σ) also takes account of capacitive effects. This will be treated in more detail in Chapter 7. Here it will suffice to mention that the relevant, high–conductivity paths are parallel to the bulk[184]. More complex distributions must be dealt with using percolation theory, effective medium theory or finite element calculations [265,266]. The result for the

[184]It is the high frequency branch which has to be evaluated in an impedance spectroscopic analysis (see Section 7.3.6).

(stationary) equilibrium conductivity [249] (neglecting or after separating out the blockage effect) turns out to be

$$\sigma_m = (1 - \varphi_A)\sigma_\infty + \beta_L \Omega_A \varphi_A (2\varepsilon_r \varepsilon_0 RT)^{1/2} u_1 \sqrt{c_{10}}. \qquad (5.263)$$

Mobility effects of the adsorbed cations themselves were neglected, but can be taken into account by means of a third term of the form $\beta_c \varphi_c \sigma_c$. (The index "c" stands for "core".) Ω_A is the ratio of the surface to the volume of the surface active A phase. For exactly spherical particles of radius r_A, $\Omega_A = 3/r_A$. Figure 5.83 shows how accurately Eq. (5.263) describes the results for various volume fractions and various particle sizes in the case of AgCl and AgBr, for which the mobilities are well known. The agreement is already fairly good over the whole range[185] if a maximum effect of $c_{v0} = 1/V_m$ is assumed. The agreement is excellent if a slight T dependence of c_{v0} as a result of a nonzero $\Delta_v H_m^{K'}$–value is taken into account (cf. Eq. (5.256)). The values of entropy and enthalpy obtained for AgCl/γ–Al_2O_3 are consistent with a significant interfacial bonding [244].
The behaviour described corresponds to an inversion layer in extremely pure AgCl or AgBr, since in the pure material the bulk conductivity is of the interstitial type. In real silver halides the material is generally extrinsic and is doped with higher valent cations. In these cases there is a pure enrichment effect in the extrinsic temperature range (enrichment with respect to the charge carrier concentration V'_{Ag}). (Consider Fig. 5.76 in this connection.) As discussed above the bulk behaviour is not of perceptible influence on the results. In addition, the fact, as to whether a dopant follows the field or not, is unimportant for $\sigma_m^{\parallel}$ as long as the large–effect–approximation is used and as long as the major counterdefect is enriched and not depleted (see previous section).
In the case of the Li^+ conductor LiI (Schottky–disorder) the conductivity of the anion vacancy as counterdefect is negligible, and the contact with a cation–adsorbing, neighbouring phase results in a pure enrichment effect. The relationships are exactly reversed in TlCl, which is also Schottky defective. Here the most mobile species is the anion vacancy. When doped with Al_2O_3 the composite material is, at sufficiently low temperatures, expected to exhibit strong inversion effects, leading in this case to predominant cation conductivity (Fig. 5.76).
In contrast to enrichment effects which are in parallel with the bulk response and whose detection usually[186] presupposes a high proportion of interfaces (as met in dispersions or thin films), depletion effects of a single interface can be studied by conventional impedance spectroscopy. Such experiments have been described for the contact RuO_2/AgCl[187].

---

[185] On account of the discreteness of the problem this is not necessarily the maximum equilibrium effect in reality (see Ref. [267]).

[186] See, however, microelectrode impedance techniques (Section 7.3.7).

[187] The effect of RuO_2 is similar to that of Al_2O_3; however, RuO_2 is electronically conducting and can simultaneously act as an electrode [268].

As already mentioned analogous phenomena also occur in anti–Frenkel disordered anionic conductors (e.g. CaF_2 or PbF_2). In contrast to the above conductors, the more acidic SiO_2 is more effective here than Al_2O_3. This qualitative aspect and, even more, the precise analysis [267] lead to the conclusion that the adsorption of F^- is the dominant mechanism.
In ionically conducting polymers a similar effect is expected: Owing to the less polar environment salt additions such as $LiClO_4$ in PEO are undissociated to a marked degree. Adsorption of one ion sort by oxide filler particles results in breaking up the ion pairs[31,188]. Even though the increased Li^+ conductivity on increasing the acidity of the filler points towards such a phenomenon [270,271], severe mobility effects may also occur here (e.g. influence on segmental motion, amorphization).

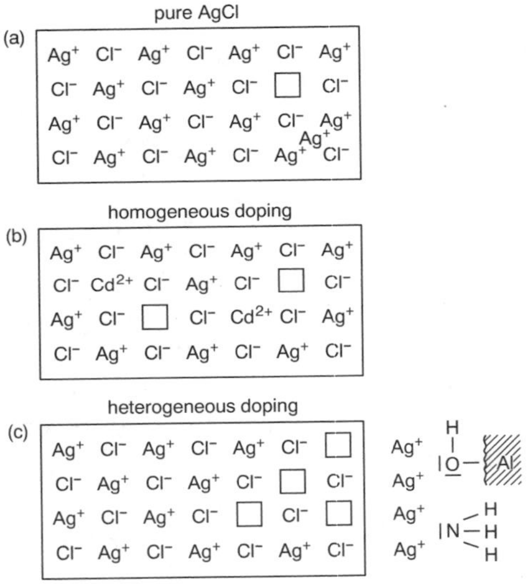

Fig. 5.86: As in homogeneous doping, heterogeneous doping lowers or raises in a predictable manner the concentration of the individual defects (Eq. (5.264)). However, the effect is limited to the region near the interface [244].

Figures 5.86, 5.87, 5.88 compare the principle of this "heterogeneous doping" (introduction of metastable higher–dimensional defects) with that of homogeneous doping (see Section 5.6, introduction of immobile aliovalent point defects) with regard to defect chemistry and conductivity. In both cases the charge of the irreversibly brought in defect (dopant or interface) controls the effect on all the mobile carriers. One important difference is that in homogeneous doping "knees" [183] appear in the van't Hoff representation of silver halides ("Wagner–Koch effect", see dashed line[189] in Fig. 5.83) while this is not detectable in heterogeneous doping: The reductions in conductivity occurring (see local minimum in Fig. 5.87b) are by–passed by the better conducting bulk (see Fig. 5.88).

[188] Such an effect has been clearly demonstrated in the so–called "soggy sand electrolytes", in which SiO_2 admixtures significantly enhance the conductivities of salt–containing nonaqueous liquid solvents (e.g. $LiClO_4$ in MeOH) [269].
[189] Compare here Fig. 5.47b, page 181.

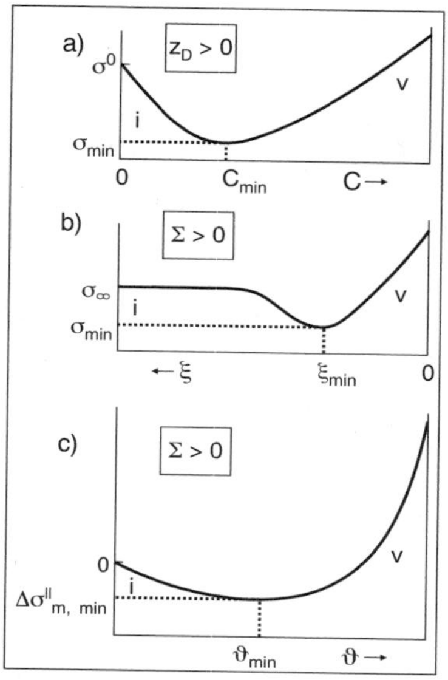

Fig. 5.87: Comparison of homogeneously (a) and heterogeneously (b, c) doped material (example: AgCl): a) conductivity as a function of the concentration of a homogeneous positive dopant ($z_D > 0$) of concentration C (Eq. (5.155)), b) local conductivity as a function of the spatial coordinate at positive charge density of the interface core ($\Sigma > 0$) in the case of heterogeneous doping ($\xi \equiv x/\lambda$); $\xi = 0$ denotes the layer adjacent to the neighbouring phase (Eq. (5.228)), c) integral conductivity increase ($\Delta\sigma^{||}_m$) as a function of the strength of the interfacial interaction (Eq. (5.237)) in heterogeneous doping. (ϑ_{min} $\left(= \left[\left(\frac{u_i}{u_v}\right)^{1/2} - 1\right] / \left[\left(\frac{u_i}{u_v}\right)^{1/2} + 1\right]\right)$ should not be confused with a minimum effect, rather it denotes the ϑ parameter, that brings about minimal total conductivity.)

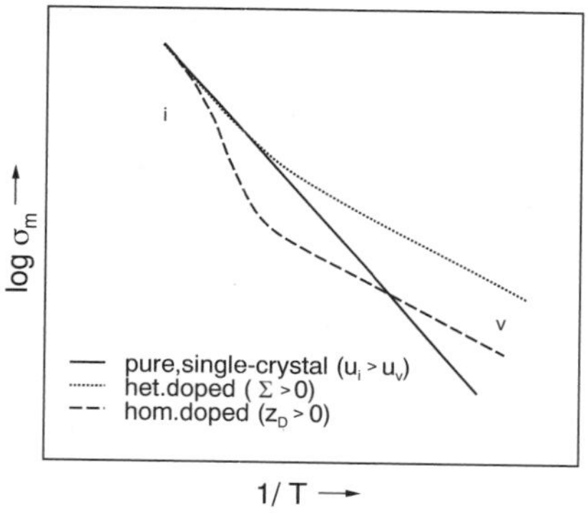

Fig. 5.88: The dependence of the conductivity on temperature for positively homogeneously and heterogeneously doped material (example: AgCl, cf. Fig.5.83) [244]. The "knee" which is characteristic for the homogeneous case, is by–passed in the heterogeneous case.

In view of the potential of heterogeneous doping in materials research and the similarity to the homogeneous effect, let us formulate the rule for "heterogeneous doping" [14] as a counterpart to Eq. (5.141):

$$\frac{z_k \delta c_k}{\delta \Sigma} < 0 \tag{5.264}$$

(Σ is the core charge density.) Equation (5.264) expresses the fact that when the core charge is positive, the concentrations of all positive (negative) charge carriers in the space charge zone are reduced (increased) and vice versa. Owing to Eq. (5.215) compensation effects do not occur.

It is also possible to analyse the behaviour of minority charge carriers [255] and, in particular, their dependence on component potential by using blocking electrodes (as described in Chapter 7). The results for $AgCl{:}Al_2O_3$ are in agreement with an increase in the concentration of conduction electrons in the boundary layers as required by Eq. (5.241). The fact that the behaviour of the electronic minority charge carriers is controlled by the electrical field but that the latter is determined predominantly by the interaction of the majority charge carriers with the neighbouring phase means that there is "foreign control" of the electronic charge carrier in such cases ("fellow traveller effect" [244]). This is valid, in general, for all charge carriers that are not of importance in the Poisson equation. This consideration has important consequences for electronic conductors, in which the ionic point defects very often predominate the defect chemistry — in spite of their low conductivity. At least during the process of preparation ionic profiles can be produced that may have a dominant influence on the electronic distribution.

On account of the high ionic concentration — as already discussed in detail above — the concentrations of the silver defects in the space charge zone are not dependent on the silver activity or the partial pressure of chlorine. In contrast, the local electronic concentration and electronic local specific conductivity are pronounced functions of the component activities, according to $\mathrm{N_{e'}} = \mathrm{N_{e'\infty}} = -1/2$ and $\mathrm{N_{h^\cdot}} = \mathrm{N_{h^\cdot\infty}} = +1/2$:

$$\sigma_{e'} \propto \mathrm{P}_{\mathrm{Cl}_2}^{-1/2} \quad \text{and} \quad \sigma_{h^\cdot} \propto \mathrm{P}_{\mathrm{Cl}_2}^{+1/2}. \tag{5.265}$$

This example allows a comprehensive interpretation of the dependence of $c_k(x)$ as a function of the control parameters. This is shown in detail in Figs. 5.89, 5.90.

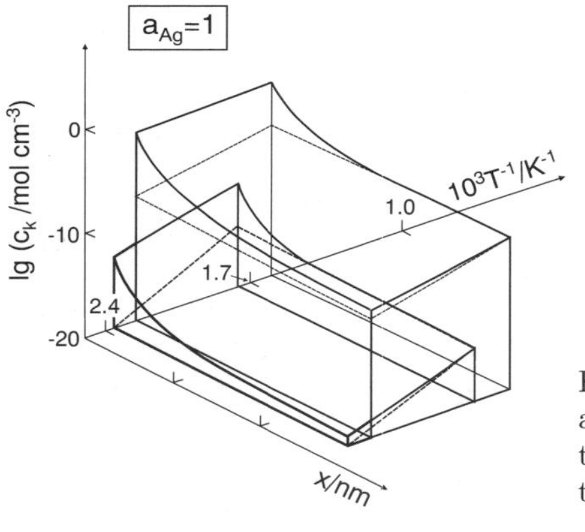

Fig. 5.89: Concentration of silver vacancies (large shape) and conduction electrons (small shape) as a function of the temperature and the distance from the γ–Al_2O_3 interface [255].

Sections of Fig. 5.90 parallel to the $\lg c_k - T^{-1}$ plane represent, as it were, Kröger–Vink diagrams of boundary layers (see Fig. 5.91).

In the case of mixed conductors appreciable stoichiometry effects are to be expected. Examples[190] for this include Ag_2S [273] and CeO_2 [274,275]. It has been demonstrated in these cases that power laws describe the dependence of the deviations

[190] Very recently the possibility of excess stotage at heterophase interfaces has been pointed out. So, the heterogeneous storage of Li leads to increased battery capacity in Li–batteries with nano–crystalline electrodes, e.g. consisting of Li_2O, metal composites in the lithiated state (see also page 491): Li^+ stored at the Li_2O side, e^- at the metal side [272].

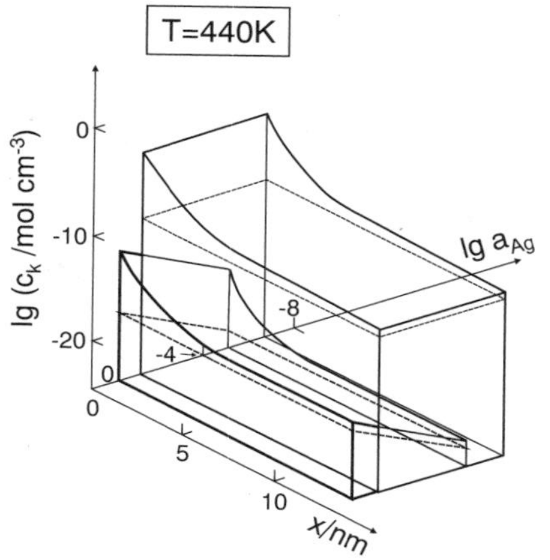

Fig. 5.90: Concentration of silver vacancies (large shape) and conduction electrons (small shape) in AgCl as a function of the silver activity and the distance from the γ–Al_2O_3 interface [255].

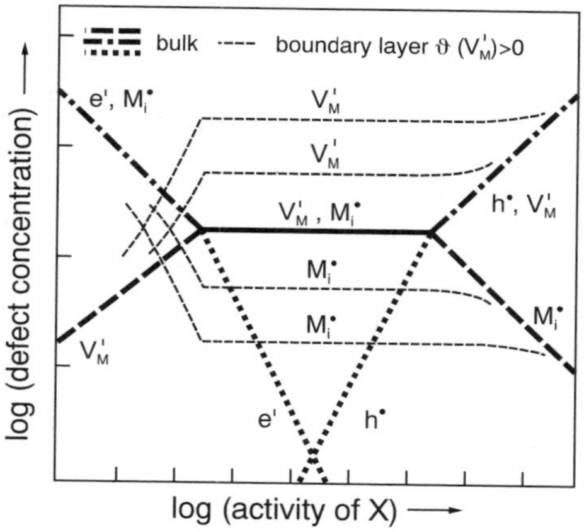

Fig. 5.91: Kröger-Vink diagram of boundary layers for our model substance MX when $\vartheta(V'_M) > 0$. The broken lines refer to the ionic defect concentrations at (two) different distances from the interface. The behaviour of the electrons in boundary regions is not shown. The approach to the bulk values (printed boldface) for extreme abscissa values is attributable to the disappearing Debye length. The mirror symmetry on comparison of V'_M with $M_i^•$ follows from Eq. (5.215) [244].

from the Dalton composition brought about by higher–dimensional defects. It has not yet been clarified how far purely structural effects account for this; in principle, power laws can also be consistent with the space charge relationships according to Eqs. (5.254) and (5.257) (see also Subsection 5.8.5c).

Let us now turn to the contact of two ionic conductors [276]. If the ionic conductors are identical or almost identical, i.e. our contact is a homocontact, then this grain boundary nonetheless, on account of its structural singularity, serves as a sink for cations or anions (see Fig. 5.92). Accordingly the result is a charged core with (almost) symmetrical profiles on both sides. It has been found for the silver halides that the grain boundaries stabilize surface Ag^+ ions, though not so effectively as Al_2O_3. As for $AgCl{:}Al_2O_3$ the resulting increased vacancy conductivity can be easily measured. In that case, however, the fine oxide particles "decorate" the grain–boundaries and enhance their effect. A similar effect can be caused by contamination with NH_3 (Lewis–base). The same can be done in the case of CaF_2 by "activating" the boundaries by contamination with SbF_5–molecules [267], as strong

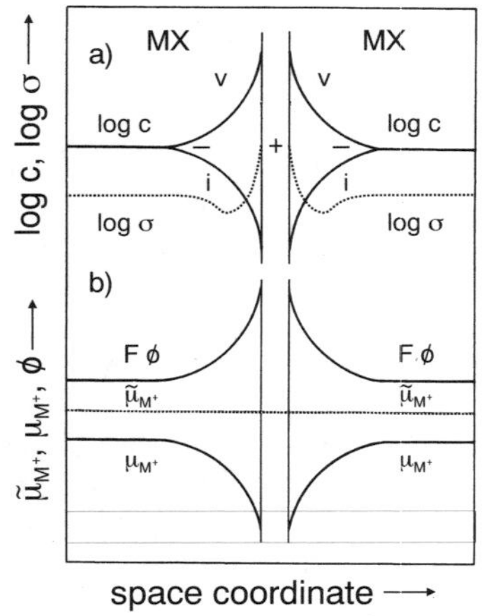

Fig. 5.92: Ionic space charge effects at grain boundaries [256].

Lewis–acids ("fluoride ion attractor" by their strong tendency to form SbF_6^-). The use of "activation" by gas adsorption for chemical sensors will be discussed below. Because the core region of the grain boundary is sandwiched between two space charge regions and because of the concentration profile in the space charge region itself, the situation is marked by strong anisotropy, and even one and the same grain boundary, can — depending on the current direction — block the path of the current (e.g. as a result of an insulating core region if the grain boundary is perpendicular to the current direction) or by–pass the bulk (parallel to bulk, e.g. via conducting space charge zones). Such effects occuring simultaneously in polycrystalline silver halides can be separated by means of impedance spectroscopy using Eq. (5.262) (see Sections 7.3.6 and 7.3.7).

The effects can be much more marked when the neighbouring silver halide grains differ chemically. The enormous conductivity anomalies in the miscibility gaps of the systems AgBr:β–AgI [261] and AgCl:β–AgI [277] are evidence of this (Fig. 5.93). In the case of such heterocontacts the contact equilibrium requires a partial transfer of Ag^+ from one space charge zone[191] to the other in accordance with

$$\text{Reaction F}\alpha\alpha' = \qquad \mathrm{M_M(MX)} + \mathrm{V_i(MX')} \rightleftharpoons \mathrm{M_i^{\bullet}(MX')} + \mathrm{V_M'(MX)}. \qquad (5.266)$$

When the effect is large the vacancies in the boundary layers are in the majority in one phase, the interstitial defects in the other. In this manner V–i junctions can be produced, the ionic analogue to p–n junctions (cf. Figs. 5.94, 5.95). If we neglect the charge accumulation in the core region — which can naturally always occur in

[191] Corresponding to Eq. (5.266) such pronounced conductivity effects do not occur in two phase mixtures consisting of a Frenkel–disordered and a Schottky–disordered alkali halide [278], neither are they expected in systems of two alkali halides. Anomalies occuring there [279,280] are most probably due to charge segregation into grain boundaries or mobility effects.

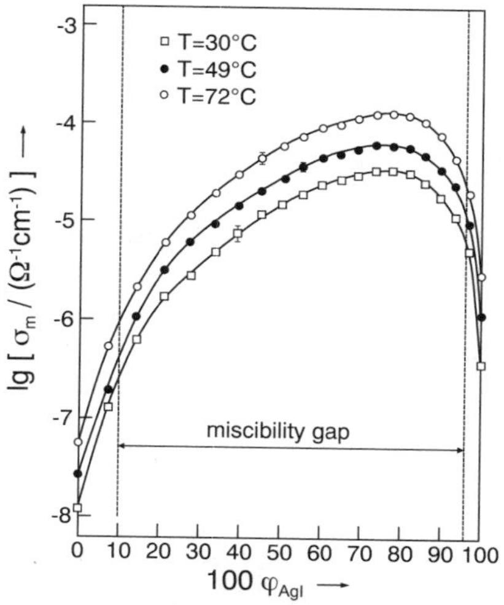

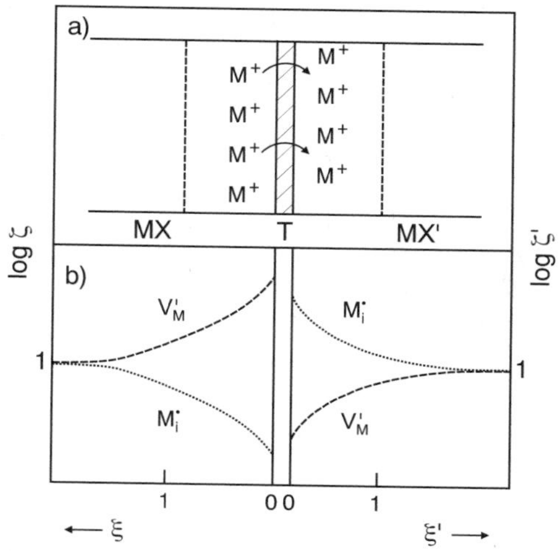

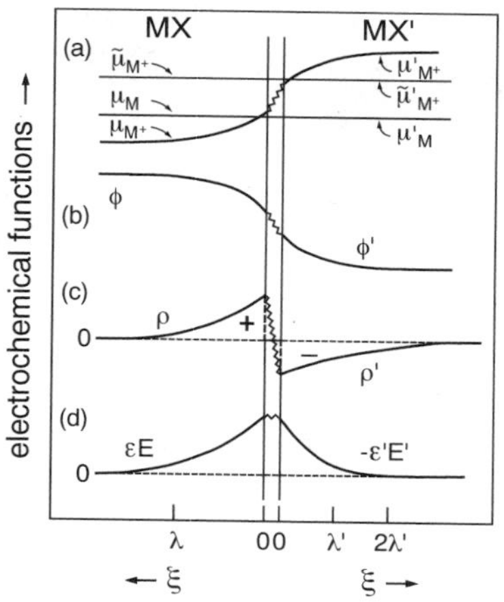

Fig. 5.93: Top left: Conductivity variation in the system β–AgI–AgCl [276].

Fig. 5.94: Top right: Ion redistribution process at the contact MX/MX′: concentration effects [276].

Fig. 5.95: Bottom left: Variation of the potentials, charge densities, and dielectric displacements at the contact of two Frenkel defect ionic conductors [244]. The translation into the energy level picture (cf. Fig. 5.81) is left to the reader as a useful exercise [206].

addition —, then application of our boundary layer model gives[192]:

$$c'_{Ag_i^{\cdot}}(x'=0) = \left[\kappa \frac{\varepsilon}{\varepsilon'} \exp - \frac{\Delta_F^{\alpha\alpha'} G_m^\circ}{RT}\right]^{1/2} = \frac{\varepsilon}{\varepsilon'} c_{V'_{Ag}}(x=0). \tag{5.267}$$

($\Delta_F^{\alpha\alpha'} G_m^\circ$ is the molar standard free enthalpy of the heterogeneous Frenkel reaction (Eq. (5.266)) which can be associated with the standard potentials of the two phases; the dashed parameters refer to MX′, the nondashed parameters to MX, and κ takes account of the potential difference at the point of contact (Fig. 5.71)). The excess

[192] This follows from Eq. (5.252). Instead of $\varepsilon_s E_s$ we write the corresponding r.h.s. expression for the other phase; the result is $\sqrt{\varepsilon' c'_{Ag_i^{\cdot}}(x'=0)} = \sqrt{\varepsilon c_{V'_{Ag}}(x=0)}$ [276].

conductance of the heterojunctions follows from this:

$$\Delta Y^{\parallel} = \left(4R^2T^2\varepsilon\varepsilon'\kappa \exp - \frac{\Delta_F^{\alpha\alpha'}G_m^\circ}{RT}\right)^{1/4} \left(u_{V'_{Ag}} + u'_{Ag_i^\cdot}\right). \quad (5.268)$$

On account of the nonzero conductivities of the homocontacts and the bulk phases the percolation problem is complex, even when the grains are randomly distributed and of the same size. Put in a very simplified manner, we expect and observe that the relevant conduction paths[193] will not only consist of heterocontacts but will be made up of a sequence of homocontacts and heterocontacts. Hence the fraction of heterocontacts (Fig. 5.96 centre) and, thus, both the resistance ratio R_{hetero}/R_{homo} and the capacitance ratio C_{homo}/C_{hetero} will exhibit maxima roughly at the centre of

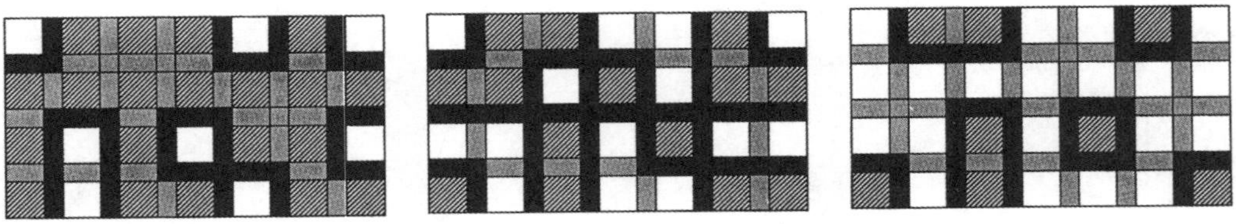

Fig. 5.96: Schematic interfacial network in the two–phase system MX/MX′. Left: MX (hatched), predominates, many homocontacts (dotted); centre: comparable proportions of both phases, many heterocontacts (black); right: MX′ (white) predominates, many homocontacts (dotted) [277].

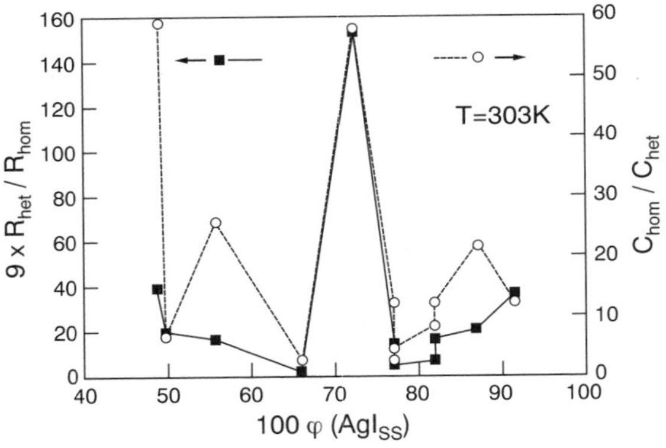

Fig. 5.97: The maximum number of heterocontacts between β–AgI and AgCl is reflected by an extremum in resistance and capacitance plots. The individual contributions were determined by impedance spectroscopy [277].

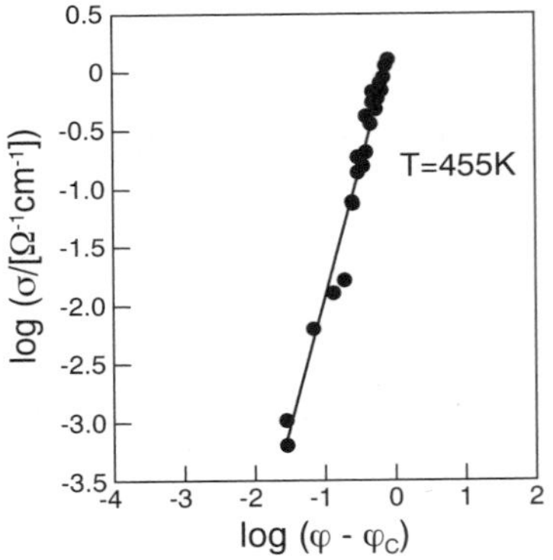

Fig. 5.98: Above the transition temperature into the α–phase the conduction in the AgCl–AgI composite occurs via normal percolation over the α–AgI grains, corresponding to a power law with a percolation threshold $\varphi_c \simeq 0.15$ and an exponent of 2.3 (cf. slope) [277].

[193] $\sigma_{hetero} > \sigma_{homo} \gg \sigma_\infty$.

the miscibility gap (Fig. 5.97). However, at high temperature, at which AgI exists in the highly conducting α–phase (see Chapter 6) there is simply percolation[194] via α–AgI grains that reveals itself in the characteristic power law illustrated in Fig. 5.98[195]. The analysis of the temperature dependences suggest a transition of Ag^+ from β–AgI to AgCl and a defect distribution as illustrated in Figs. 5.94, 5.95 (if MX is identified with AgI and MX′ with AgCl, see also Ref. [282]).
Investigations of CaF_2–BaF_2–CaF_2 . . .–heterolayers grown by molecular beam epitaxy allow for a much more elegant and a more quantitative test. Films with an overall thickness of typically 500nm exhibiting tunable spacings (by almost 3 orders of magnitude) and phase sequences reveal distinct enhancement effects which can be

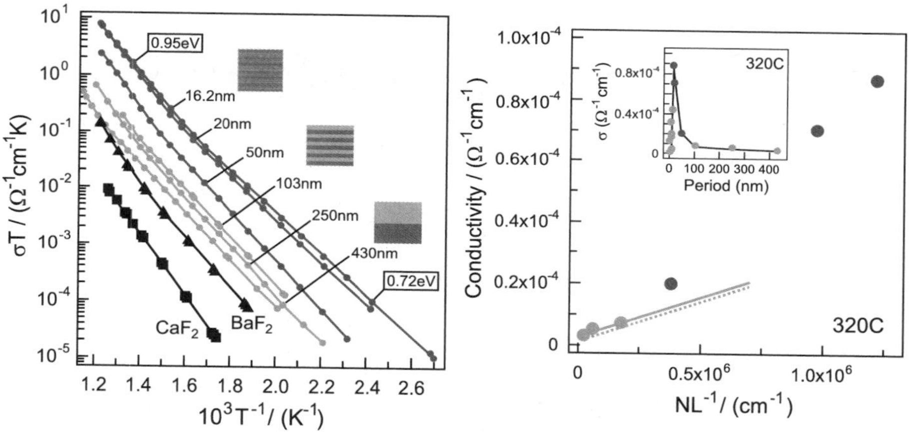

Fig. 5.99: Ionic superlattices of CaF_2 and BaF_2 show significant enhancement of the F^- conductivity which progressively increases with the density of interfaces [283]. The period L/N is varied from ∼nm to ∼ μm while the total thickness L is approximately constant. The temperature and thickness dependence is initially as predicted by semiinfinite space charge theory. The σ–increase does not saturate in the sub–Debye regime, it rather increases more steeply. At extremely small spacings (cf. insert) the conductivity drops with decreasing L due to loss of connectivity.

[194] If we mix an absolutely insulating with a conducting phase, then (d.c.) conductivity only commences when there are percolating pathways of the conducting phase. The volume fraction at which this "insulator–conductor phase transition" occurs is referred to as the percolation threshold. The percolation theory is also concerned with the structure and the effect of the conducting clusters, in particular, in the region of the threshold. Power laws are characteristic here (cf. Fig. 5.98). Correspondingly the percolation clusters are fractal objects (see Section 6.10.3) [281].

[195] This means that the conductivity is determined almost entirely by the α–AgI, and the α–AgI pathways determine the conductivity behaviour, while when $T < T_{\alpha/\beta}$ the overall effect is significantly more complicated. There the conductivity of the homocontacts (and also of the grains) is not small enough for an exclusive conduction over the percolating heterocontacts as discussed in the text. The power law to be seen in Fig. 5.98, that is, characteristic of percolation behaviour, is related to the fractal geometry (see Section 6.10.3) [281].

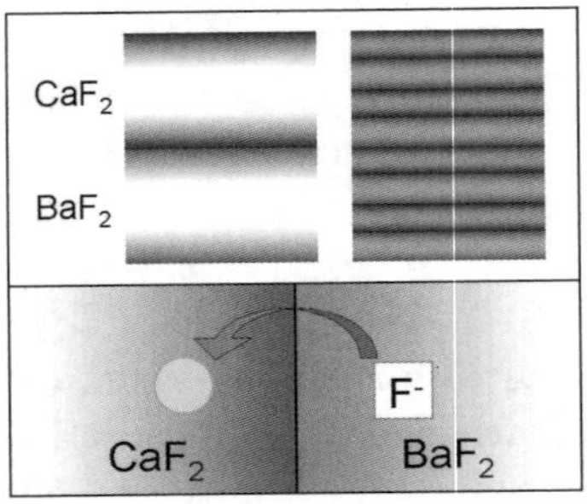

Fig. 5.100: In the sub–Debye regime the bulk disappears and the whole artificial crystal consists of space charges [283]. The transfer of fluoride ions from BaF_2 to CaF_2 explains the results.

largely understood by the space charge theory discussed and suggest a F^- transfer from BaF_2 to CaF_2 (see Figs. 5.99, 5.100) [283].

b) Conductivity effects at the ionic conductor/gas interface

A gas phase can also be used as the defect–inducing second phase [267,284]. Figure 5.101 illustrates the conductivity effect of the adsorption of NH_3 on AgCl interfaces.

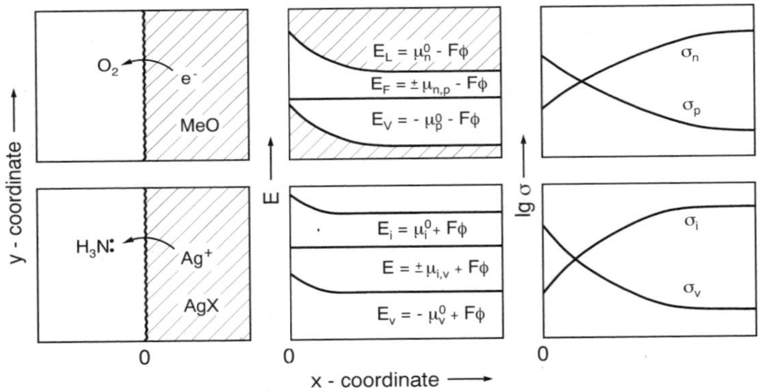

Fig. 5.101: Analogous mechanisms of the acid–base gas sensor, based on the surface ion conductivity change, and the redox gas sensor (see text) [14].

The effect is adequately described in terms of a stabilizing interaction of the NH_3 molecules such as is well known in aqueous chemistry and can be formulated[196] as

$$(\mathrm{NH_3})_\mathrm{s} + \mathrm{Ag_{Ag}} \rightleftharpoons (\mathrm{NH_3}\ldots\mathrm{Ag})^{\cdot}_\mathrm{s} + \mathrm{V}'_\mathrm{Ag}. \tag{5.269}$$

Analogously generalized acids, such as BF_3 (and SbF_5 as already discussed), exert an effect on the conductivity of CaF_2, that can be explained in terms of an enrichment of fluoride vacancies connected with the formation of BF_4^- or SbF_6^-, e.g.

$$(\mathrm{BF_3})_\mathrm{s} + \mathrm{F_F} \rightleftharpoons (\mathrm{BF_4})'_\mathrm{s} + \mathrm{V}^{\cdot}_\mathrm{F}. \tag{5.270}$$

In turn, it is possible, in this manner, to detect acid–base–active gases (see Fig. 5.102). This analogy with semiconductor sensors for the detection of redox–active

[196] A similar effect was recently discussed for copper halides [285].

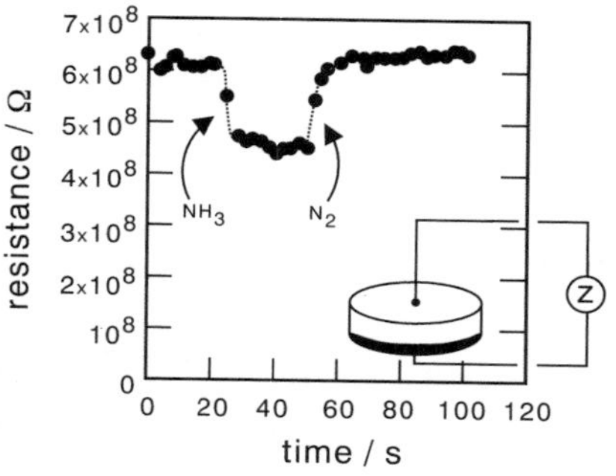

Fig. 5.102: The reversible conductivity change when a AgCl single crystal surface is brought into contact with NH_3 gas can be exploited for sensor purposes. Point electrodes can be used to advantage for rapid detection. [286].

gases is taken up again below (see also Chapter 7). As mentioned above, such a chemical treatment can also be exploited in order to activate grain boundaries in appropriate ceramics.

c) Nanosystems

Last of all let us discuss the important case of nanosystems. In crystals of a few nm in size, boundary zones are not only important, because their relative content in the sample becomes immensely large[197], but also because of local size effects caused by the tiny spacing of interfaces (true size effect). Fig. 5.103 illustrates such a size effect [287]. If the sample thickness is less than four times the Debye length, the centre of the sample itself is no longer electrically neutral, and the bulk value is not reached anywhere in the sample. The calculation is now appreciably more complex on account of the boundary conditions being finite on both sides. Let us limit ourselves to the case of symmetrically contacted films. It is true that $\zeta=1$ no longer applies in the centre of the sample, but $\phi'=0$ no doubt applies there for symmetry reasons. The detailed calculation [288] using the approximation[198] $\Delta\sigma_{\mathrm{m}}^{\|} \propto \mathrm{d}\phi/\mathrm{dx}|_{\mathrm{x}=0}$ which is already known to us (see Section 5.8.3), leads to

$$\Delta\sigma_{\mathrm{m}}^{\|}\cdot \mathrm{L} = \Delta \mathrm{Y}^{\|} \simeq 2\mathrm{u}_1\left[2\mathrm{RT}\varepsilon\left(\mathrm{c}_{10}-\mathrm{c}_1^*\right)\right]^{1/2}. \tag{5.272}$$

Equation (5.272) can also be interpreted such that in relevant cases the effective boundary conductivity is increased additionally compared with the semiinfinite case

[197] The behaviour within the grain boundary is not uniform. Thus, the lines of intersection of grain boundaries and the points of intersection of these lines are energetically distinguished compared with the grain boundaries themselves. The proportions of these regions are size–dependent. Additional structural features are discussed below.

[198] A qualitatively similar thickness dependence also applies for small effects [288], namely

$$\Delta \mathrm{Y}^{\|} = \pm\left(\mathrm{zFu_1c_\infty}\ln\frac{\mathrm{c}_{01,2}}{\mathrm{c}_\infty}\right)2\lambda\tanh\frac{\mathrm{L}}{2\lambda}. \tag{5.271}$$

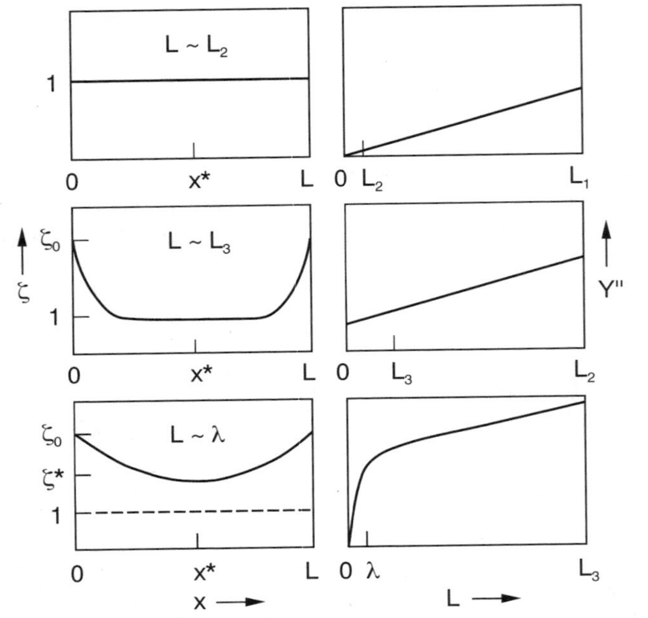

Fig. 5.103: The mesoscale effect on defect concentration discussed in the text, when $L < 4\lambda$, is also mirrored in the dependence of the conductance on thickness (see text). If the boundary layers "overlap", the interfacial effect previously hidden in the intercept is now resolved. It is presupposed that surface concentration and Debye length do not depend on L. (Both can be violated at sufficiently small L because of interaction effects and exhaustability of bulk concentrations.) [244]

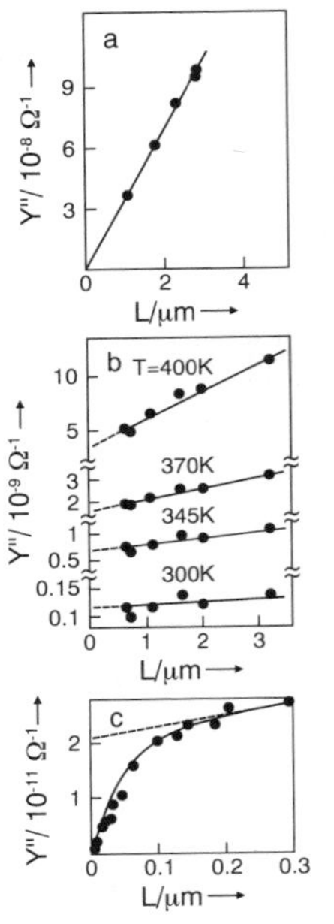

Fig. 5.104: Examples concerning the thickness dependence of the parallel conductance (not the excess value $\Delta Y^{\parallel}$) corresponding to Fig. 5.103: a) LiI on SiO_2, b) AgCl on mica, c) LiI on Al_2O_3 [244].

(see Eq. (5.272)) by a "nanosize" factor[199]

$$g \simeq \frac{4\lambda}{L}\left[\frac{c_{10} - c_1^*}{c_{10}}\right]^{1/2} . \qquad (5.273)$$

Figs. 5.105a and 5.105b are helpful in this context. In Fig. 5.105a the excess film conductance normalized to the interfacial spacing L, i.e. the averaged (excess) film conductivity $\Delta\sigma_{\mathrm{m}}^{\parallel}$, is plotted against 1/L (as in Fig. 5.104 an accumulation effect with a constant boundary concentration is assumed). If the film thickness is decreased, this value increases as the bulk proportion is diminished. (This behaviour at $L > 4\lambda$ may be called a "trivial size effect".) The local increase of the concentration when the space charges overlap leads to the fact that the averaged conductivity does not flatten if $L \simeq 4\lambda$ but substantially later (the difference in the logarithmic plot being described by g(L)). The limit is given if the local concentration even in the centre assumes the boundary value. The transition from a trivial to a true

[199]Compare $\Delta\sigma_{\mathrm{m}}^{\parallel}(L < 4\lambda)$ with $\Delta\sigma_{\mathrm{m}}^{\parallel}(L = 4\lambda)$.

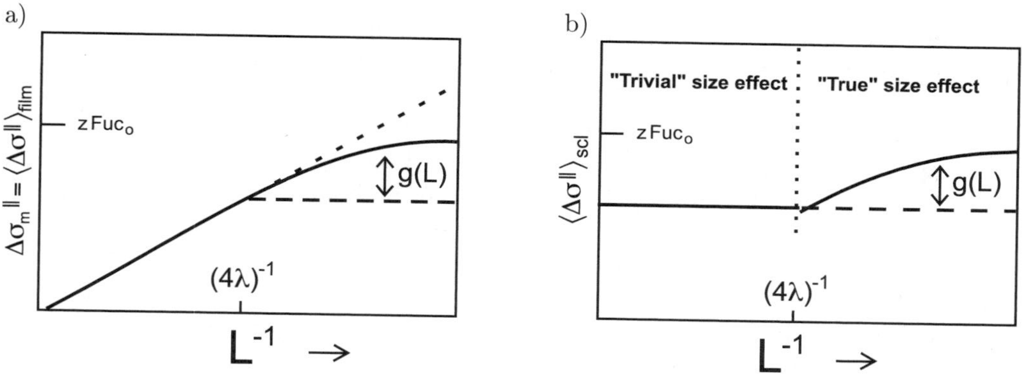

Fig. 5.105: a) Area related conductance per film thickness representing the measured excess space charge conductivity, as a function of inverse spacing. b) Area related conductance per boundary thickness representing the mean specific excess space charge conductivity, as a function of inverse spacing. The boundary concentration is assumed to be invariant. From Ref. [289].

size effect is more clearly elaborated if the excess conductance is normalized with respect to the width of the space charge layers ($= 4\lambda$ for $L \geq 4\lambda$ and $= L$ for $L < 4\lambda$) (see Fig. 5.105b) to yield an averaged space charge conductivity ($\Delta\sigma^{\|}_{sc}$). While this quantity is constant for $L \geq 4\lambda$ ("trivial size effect") it increases by the g–factor for $L \leq 4\lambda$ ("true size effect").

Not too far from $L \simeq 4\lambda$ it should be possible to approximate g by $4\lambda/L$ ($c_1^* \ll c_{10}$). If the sample thickness is half the Debye length, Eq. (5.272) then predicts an increase of about one order of magnitude compared with the value at $L \simeq 4\lambda$ where the bulk has disappeared. The parameter c_1^* denotes the concentration of the enriched charge carrier in the centre of the sample ($\xi = \xi^*$) [288]

$$\xi^* \equiv \frac{L}{2\lambda} = 2\sqrt{\frac{c_{1\infty}}{c_1^*}}\left[elli\left(\frac{c_{1\infty}}{c_1^*}, \frac{\pi}{2}\right) - elli\left(\frac{c_{1\infty}}{c_1^*}, \mathrm{Arcsin}\sqrt{c_1^*/c_{10}}\right)\right] \tag{5.274}$$

and hence depends itself on the bulk and boundary layer effects [288][200]. Equation (5.272) is fundamental to the discussion of the ionic conductivity effects of nanocrystalline samples.

The most direct way of analysing the thin film results is to plot the conductance (normalized to the area) as a function of the layer thickness (see Fig. 5.103). If the boundary effect is zero, $Y^{\|}(L)$ represents simply a straight line starting from the zero–point with the slope σ_∞. If there is an interfacial effect, but the film is relatively thick, the only change with thickness is the increased spacing of the invariant space charge regions. In this case the original straight line is displaced parallel, with $\Delta Y^{\|}$ as intercept, the interfacial conductance being $Y^{\|} = \Delta Y^{\|} + \sigma_\infty L$. In the case of

[200] Elliptical integrals of the first kind are defined as $elli(k, \chi) = \int_0^\chi d\alpha \left(1 - k^2 \sin^2\alpha\right)^{-1/2}$. Note that Eq. (5.274) predicts $c_1^* \to c_{10}$ for $L \to 0$.

very thin films, however, the space charge regions "interfere" and the behaviour is described by Eq. (5.272).
Figure 5.104 shows the occurence of these three types of thickness dependences. In the first example displayed, LiI on SiO_2 [290], we do not expect any great interfacial effect and accordingly the intercept is zero — at least at the resolution shown. Note that under certain conditions LiI, which normally crystallizes in the rock salt structure, can be deposited on the SiO_2 in the (metastable) hexagonal form. This emphasizes that extreme caution is necessary when assuming structural invariance. Films of AgCl or AgBr on mica [291] exhibit finite intercepts as expected for $L > 4\lambda$. Nevertheless, again extreme caution is necessary in the analysis. Particularly if annealing was not adequate, the conductivity may be enhanced as a result of a high concentration of higher–dimensional defects (dislocations). A careful analysis of the cases of AgCl and AgBr led to consistent data for the enthalpy and entropy of transfer of silver ions from the bulk to the surface and for the interaction of the surface with the neighbouring phase. Care is also necessary when analyzing the conductivity of the extremely thin LiI films on Al_2O_3 substrates shown as the third example [292]. It appears to mirror the predicted nanosize effects perfectly. However, even though the fit with the above conditions is very good, the astonishingly large, apparent Debye lengths suggest that the results are probably falsified by island formation and possibly also by dislocations.
Let us consider the conductance data of the $CaF_2-BaF_2-CaF_2-\ldots$ heterostructure already discussed in Section 5.8.5a as a function of the interfacial spacing (Fig. 5.99). Indeed the conductivity (the initial values of which are in good agreement with semiinfinite effects) does not saturate at $L \simeq 4\lambda$ [283] but increases further if $L \lesssim 4\lambda$ ($\simeq$40nm). (The reason for the anomalous steep increase — compare Fig. 5.99 with Fig. 5.105a — is under investigation.) In this regime the heterostructure is expected to be charged at any locus and represents an artificial ion conductor.
Even though Eq. 5.272 is strictly valid only for films, it is nevertheless basic for the behaviour of nano–crystalline samples in a semiquantitative way. Measurements in this field are quite sparse[201]. Results obtained for CaF_2 [293] or CeO_2 [274,294–296], still refer to the regime in which the Debye–length is smaller than or possibly just comparable with crystallite size. Certainly such size effects are of importance for ion exchange membranes (cf. Fig. 5.106), such as Nafion (consisting of a perfluorinated polyaliphatic hydrocarbon backbone with ether side chains carrying sulfonic acid surface groups) or PEEK (based on a polyaromatic backbone). These polymers have nanometre size water channels built into them, in which the protons of the sulfonic acid groups can migrate (see e.g. [297]). For this reason width and distribution of the channels are of high relevance for the overall effect. Another example is the

[201] It might even be that granular systems better obey the relations under considerations, since epitaxial films may exhibit strong stress effects (if the misfit w.r.t. substrate is substantial). Higher–dimensional defects such as misfit dislocations (see CaF_2–BaF_2–CaF_2–... heterostructure) or grain boundaries (polycrystalline films) can however "absorb" the crystallographic misfit stress to a large degree.

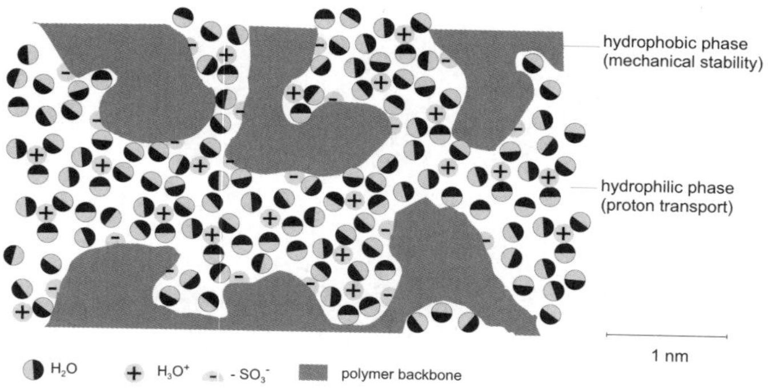

Fig. 5.106: Microstructure of the Nafion ion exchange membrane (cf. text and Ref. [297]).

stacking fault phase at the interface of β–AgI/Al_2O_3 [262], that can be regarded as composed of mesoscopic heterolayers of γ– and β–AgI (see Fig. 5.109 and subsequent Section d).

If discussing nanocrystalline samples we also have to be aware of the fact that the interfacial effect will change with L (core charge interaction) and also the Debye length will change (since the bulk concentration starts being affected by the defect redistribution). In addition the "curvature" of nano–crystals is connected with a modified internal pressure (cf. Eq. (5.73)) and hence "behaves" as a "chemically different phase" as far as the contact behaviour is concerned (standard potential modified by the excess term $(2\bar{\gamma}/\bar{r})\times$ partial molar volume in Eq. (5.73)). Morphology and geometry effects also lead to complications in charge distribution and screening [298].

As already indicated we should not totally ignore the mobility of the charge carriers in the core. As discussed above, such carriers can be formed by "intrinsic core disorder" but also as excess carriers by segregation to the core (space charge effect). Let us assume here that the first effect prevails: If the formation energy in the core is reduced by a factor β (in the simplest approximation $\beta \sim 2/3$ for a surface, see footnote 3 in Section 5.2) and if we assume an invariant packing density, not only the local defect concentrations are appropriately enhanced, but also the characteristic dependences are reduced (by the factor β in the simplest approximation, if we neglect charging). Whether reduced partial pressure or temperature dependences have to be explained by core or space charge effects, depends on the individual cases[202] [130].

So far we have assumed a step function behaviour of μ° and, almost exclusively, attributed the mesoscopic effects to changes in the electrical potential (see Fig. 5.107). But this structural invariance cannot be fulfilled down to the smallest dimensions. More subtle changes in the basic atomic structure (ionic and atomic standard potentials) are to be expected. This becomes evident when we build up solids from

[202] As a rule of thumb space charge effects are very important in high mobility materials, while pure core effects are expected to be significant in terms of conductivity for low mobility materials.

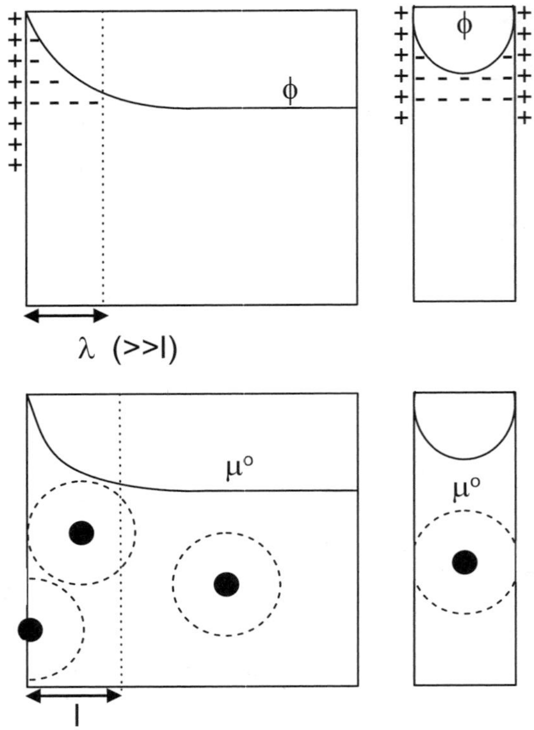

Fig. 5.107: Effects of semiinfinite interfaces and effects of narrowly spaced interfaces on the electrical potential (a) and the chemical standard potential (b) of ionic point defects or polarons. In wide–band materials the confinement occurs already for large spacings corresponding to the electron's delocalization (see Sections 2.2.1.1, 2.2.1.2). From Ref. [289].

oligomeric molecules (clusters). However, the discussion in Section 2.2 (see Fig. 2.4 on page 35), shows that, at least, in the case of markedly ionic crystals the bulk and surface energetics of the massive crystals are reached very soon (in the case of NaCl for cluster sizes above ca. 10 formula units). This is also supported by calculations of defect formation energies [246]. Quantum–chemical calculations for $SrTiO_3$ indicate that in both SrO and TiO_2 terminated surfaces[203] the first layer is seriously perturbed while already the third layer is similar to the bulk [106]. In hard materials it is necessary to take elastic strain effects of comparably long range into account which make μ° space dependent (cf. Section 4.3.7) [245].

Fig. 5.107b displays the local energetic situation. Each ionic or polaronic defect is characterized by a sphere of influence within which the local structure is distinctly modified (cf. Section 5.2). If this sphere comes close to the boundary or is even confined, $\mu^\circ_{\text{defect}}$ must change.

At sufficiently low defect concentrations the variation of the thermodynamic potentials within a grain should be reflected qualitatively by Fig. 5.108, where it is assumed that the Debye length is large in comparison to the boundary region, in which structural effects become important. Hence a variation in the ground struc-

[203] Unless the surface is polar [106].

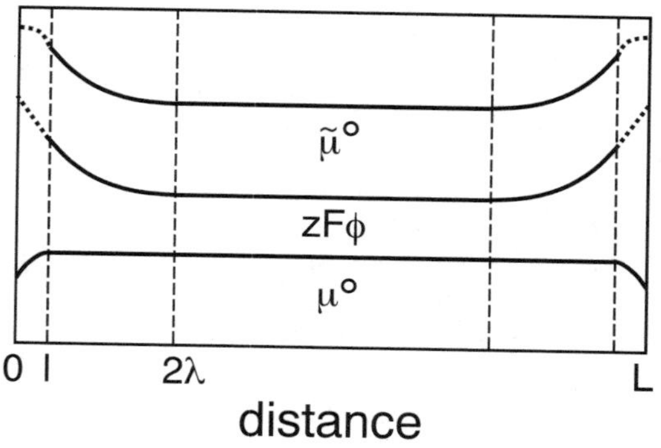

Fig. 5.108: Variation of the thermodynamic potentials within a grain (see text). The core (of thickness |s|) is not shown. In most relevant cases, l and |s| should be comparable magnitude [289].

ture, i.e. in the μ° values, should occur[204] between zero and ℓ. Accordingly, if the dimensions are successively reduced, two mesoscale regimes[205] are to be expected (if $\ell \ll \lambda$). Please note that in the general case the characteristic lengths are themselves functions of the size (L) [289].
Generally nanostructured systems ("nanoionics") are expected to provide systems not only characterized by structural and functional complexity but also (in particular because of the high density of metastable structural elements) by a high information density [299]. The reader is referred to Table 6.7 on page 393 in this respect.

d) Boundary layer phase transitions

Before we consider primarily electron–conducting systems, we turn briefly to a consequence of the boundary layer effects with respect to phase stability:
On account of the different thermodynamics (of core and bulk), phase changes are to be expected at interfaces which take place at temperatures that differ from the bulk phase transition temperatures [300]. Mechanistically this may described, in some cases, as a consequence of the charge carrier interaction (as a result of the altered defect concentrations) in the boundary layers (see Section 5.7.2) [244].
In the case of ionic conductors surface phase transitions are found in the proton conductor $CsHSO_4$ [301] or for the AgI/Al_2O_3 contact (cf. also the LiI example above) [262,302]. As already mentioned, in the last case it is possible to describe the stacking disorder [303] as the limiting case of a multilayer–arrangement formed of β–AgI and γ–AgI, each a few atom layers thick. In this way the extremely high conductivity of the AgI/Al_2O_3 system is understandable (in fact a ionic conductor–ionic conductor–nanocomposite according to Fig. 5.109). In both cases the interface

[204] In fact, even in the abrupt core model μ° is expected to vary at x > 0 (if the structure abruptly changes at x=0) owing to the finite size (see Section 5.2) of the relaxation sphere (of radius r) around each defect (leading to a profile of μ° between x=r and x=0). The situation is similar for polarons but different for delocalized electrons.

[205] A variety of nano–size phenomena, such as capillarity effects, geometrical screening effects, hetero–size charging, edge and corner energies, statistical problems in nanocrystals, etc. are discussed in Ref. [289].

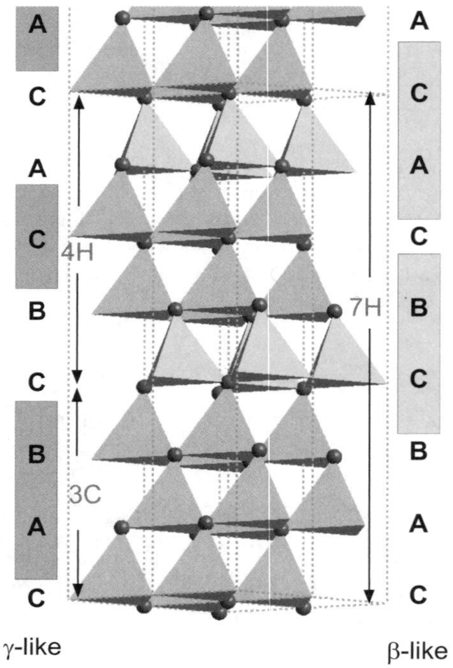

Fig. 5.109: Left: 7H-stacking fault phase formed at the β–AgI/Al_2O_3 contact [262, 289]. The enormous conductivity effects at the interface can be understood if we think of the boundary layer phase as a cation–disordered heterostructure in the sub–nm range. Right: Ion redistribution occuring at each interface of the heterostructure leads, for small enough spacing, to almost predominant disorder. The charge carrier concentrations (v, i) are much higher than in the bulk [263]. According to Ref. [262].

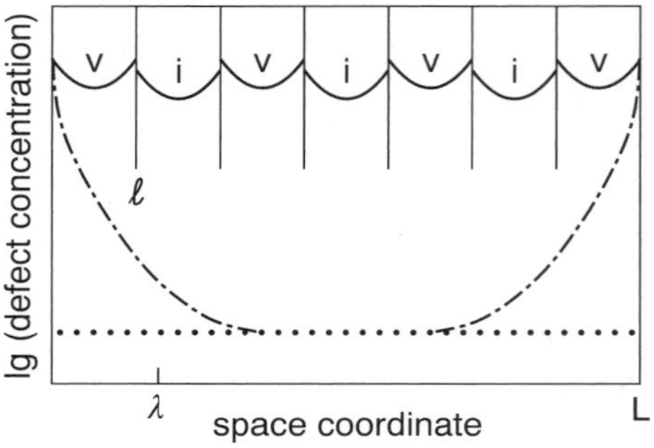

exhibits an appreciably higher conductivity than the bulk in a temperature range at which the latter has not undergone transition to a superionic phase.
Not only the "premature" disordering and rearranging of a partial lattice are known, "surface melting" is too: Thus, the first atomic layers of ice are liquid–like[206] [146] and make sliding possible.

e) Space charge effects in electronic conductors[207]

Let us now turn briefly to space charge effects in electronic conductors and concentrate on the important case of the depletion layers, insofar as this is of importance for our considerations. It was mentioned above that depletion layers in extrinsic conductors can be fairly large on account of lack of significant screening[208]. The resultant grain boundary resistances can be enormously high, so that, as in a zinc oxide varistor, in the substrate material Si_3N_4 or the PTCR[209] material $BaTiO_3$,

[206] Conversely it has been postulated that the upper layer of liquid water (in contact with electrodes) exhibits an ice–like structure [304].

[207] The interested reader is referred here to the voluminous semiconductor literature.

[208] This must also be taken into account for purely ionic conductors.

[209] The PTCR effect refers to anomalies observable in some semiconductors or ionic conductors, in that the resistivity increases significantly with the temperature (PTCR = positive temperature coefficient of the resistance). Trivial but powerful PTCR effects rely on composites of insulating

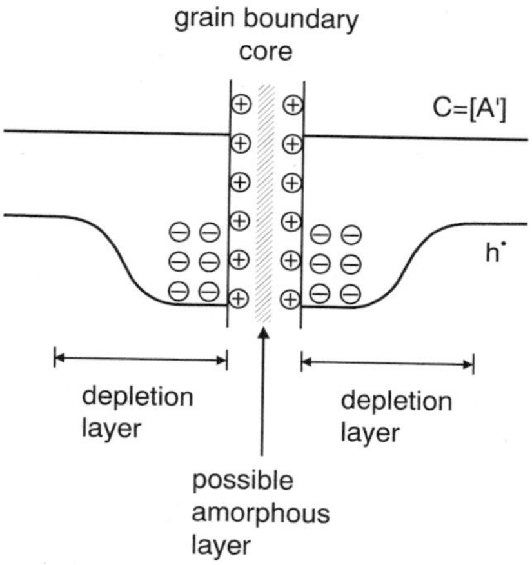

Fig. 5.110: Grain boundaries as double Schottky contacts (schematic). The concentration profile of the mobile oxygen vacancies is similar to that of $[h^{\cdot}]$ but, on account of Eq. (5.215), $[V_O^{\cdot\cdot}]$ falls off more rapidly (double charge).

they completely determine the direct current resistance — if suitably conditioned [59].

Figure 5.110 illustrates a layer depleted of holes at a $SrTiO_3$ grain boundary (see also Fig. 5.23). The positive countercharge is formed in the core region of the grain boundary probably by excess titanium ions (and dopant ions)[210]. Even though, as discussed above, in the Mott–Schottky case the change in the surface charge density (Σ) with the oxygen partial pressure can be very important, under certain circumstances, however, Σ is determined during preparation and is not dependent on P_{O_2}[211]. Such a situation is met in (Fe–doped) $SrTiO_3$. Hence, $\partial \ln[V_O^{\cdot\cdot}]_0/\partial \ln P_{O_2} \simeq 0$, $\partial \ln[h^{\cdot}]_0/\partial \ln P_{O_2} \simeq 1/4$ and the dependence of the space charge resistance on the partial pressure of oxygen (Table 5.3) is approximately

$$-\frac{\partial \ln \Delta Z^{\perp}}{\partial \ln P_{O_2}} = +\frac{\partial \ln c_{h^{\cdot}}}{\partial \ln P_{O_2}} = 1/4. \tag{5.275}$$

This has been verified experimentally after extraction of the grain boundary contribution using impedance microscopy (see Sections 7.3.6, 7.3.7) (see Fig. 5.111). (Similar effects are found at electrode interfaces Pt/$SrTiO_3$.) The current–voltage behaviour corresponds to two Schottky diodes connected in opposite directions [305, 306]. The fact that the current increases steeply at high voltage is associated with nonlinear effects that will be discussed in Chapters 6 and 7. Owing to the same sign the vacancy concentration is also reduced. The effect is — due to the double charge — much more pronounced than for the hole. Recent experiments fully confirm this picture [308]. (The reduction of both $[V_O^{\cdot\cdot}]$ and $[h^{\cdot}]$ leads also to serious

polymer matrix with percolating metal inclusions. Upon healing, the morphology changes and the short–circuit is interrupted.

[210] With respect to literature see Refs. [139,164,305–307].

[211] This is also shown by capacitance measurements, which indicate a constant thickness of the space charge zone λ^* (see Eq. (5.231)) (see Chapter 7.3.3).

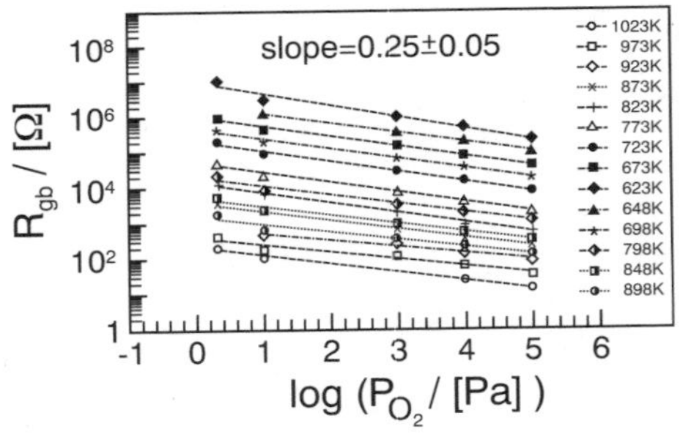

Fig. 5.111: The $P_{O_2}^{1/4}$ dependence of the grain boundary resistance is in agreement with depletion boundary layers. The temperature range extends from 623K to 1023K (in 50 or 25 degree steps) [307].

chemical resistances which will be treated in Chapter 6, Section 6.6.2). Again owing to the sign, the conduction electron concentration is increased. Such inversion layers explain the enhanced n–type conduction in nano–crystalline CeO_2 [295,294] and possibly the current voltage characteristics of certain $SrTiO_3$ grain boundaries [309].
The boundary conductivity can be influenced by electric fields. An example of great practical importance is the so–called varistor which is used for overload protection (see Fig. 5.112): An appropriately doped ZnO polycrystal (with highly conducting

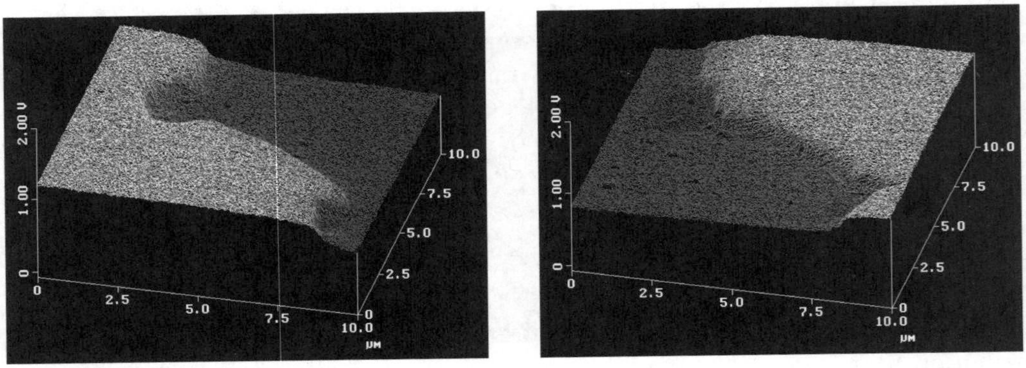

Fig. 5.112: The drop in electrical potential over a ZnO varistor grain boundary made visible by scanning electron microscopy (Kelvin-capacitor method) under potential load (left hand micrograph) and after reversing the potential (right hand micrograph). From Ref. [310].

bulk and insulating grain boundaries) becomes conductive under high voltage for such reasons. In the case of Si_3N_4 substrate materials, the grain boundary also makes the nitride material electrically insulating, but does not prevent thermal conduction by phonons. In this manner it is possible to create a valuable substrate material with a combination of two useful properties. In the case of the PTCR effect in polycrystalline $BaTiO_3$, the conductivity drops very greatly with temperature above 120°C. The reason for this lies in the importance of the dielectric constant for the grain boundary conductivity [311]. Above the phase transition from the

ferroelectric to the para–electric phase, ε falls with temperature and, hence, so does the interfacial conductance[212].
The counterpart to Fig. 5.95, namely the situation of a p–n junction in equilibrium (e.g. contact of donor–doped and acceptor–doped Si) is illustrated in Fig. 5.113. Marked, rectifying properties of such junctions occur primarily in life–time semiconductors[160,207]. In view of the extensive literature on this subject, we will not go in any further detail here.

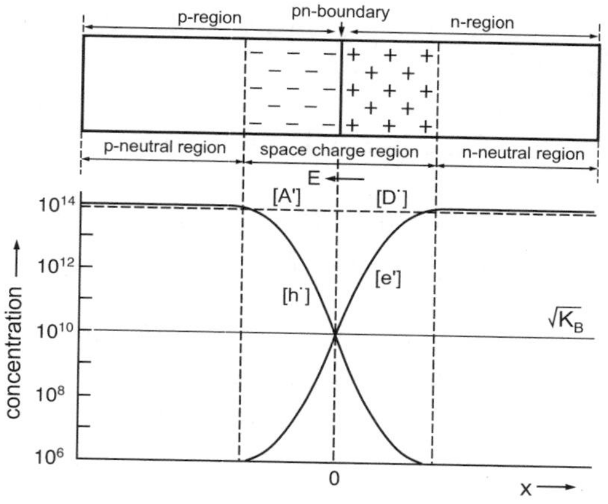

Fig. 5.113: Sketch of the dopant (A′: acceptor, D˙: donor) and carrier distribution in a symmetrical pn–junction in thermal equilibrium.

A further depletion effect important in this context is observed at the interface SnO_2/O_2. Phase equilibrium is not set up at low temperatures, but there is (partial) contact equilibrium. This means that the oxygen is not dissolved in the bulk but withdraws electrons from the boundary zone on adsorption. The increased surface resistance, of the n–type conducting SnO_2, can be used for the detection of oxygen. In the presence of reducing gases the tendency is reversed. The analogy between such sensors for redox–active gases and the ionic conductivity sensors described above for acid–base–active gases (shown in Fig. 5.101) has already been discussed. The dependence of the depletion effect and, hence, the sensitivity of such Taguchi sensors [313] on the partial pressure of oxygen can be estimated from the above treatment. The application aspect will be returned to once again in Section 7.3.1.
If the field effects at the Schottky barrier are high, it is possible for inversion boundary layers to appear here too.
The technique of heterogeneous doping, that is so successful for ionic conductors, has also been suggested for electronic conductors: An admixture of metallic particles to semiconducting oxides should be able to modify the total impedance via the space

[212] The explanation given by Jonker [312] is the following: According to Eq. (5.231) the boundary layer thickness varies with ε. In the neighbourhood of the Curie point the change in ε is very much greater than any change in the stored charge, leading to a distinct change of the boundary concentration c_0. Combination of Eqs. (5.231) and (5.215), taking into account the fact that the interfacial charge Σ is roughly given by the doping concentration and the boundary layer thickness, leads to the result that the logarithm of the boundary layer resistance is proportional to Σ^2/ε.

charge effect [314]. While an enhanced overall conductivity could also simply be due to the formation of metallic paths which prolong the electrode, the reduction of electronic conductivity discovered for various systems on admixture of metallic phases [315] can, however, not be explained on the basis of such an "artifact".

The phenomenon of interface metallization of semiconductors can be partly considered analogous to the surface phase transitions in ionic conductors just discussed. Structural features and electronic interactions are important in both cases (see Section 5.7.2).

If the grain size falls below the effective boundary layer thickness (here $2\lambda^*$) extreme effects, such as discussed for ionic conductors, naturally also occur for electronic conductors. An interesting recent example, refers to the overlap of depletion zones around dislocation cores constituting low angle grain–boundaries in $SrTiO_3$ (cf. Section 5.4 and Fig. 5.20) [316].

While the situation is similar for small polarons, in materials with wide bands we are no longer able to draw parallels[213] between the behaviour of electronic and ionic effects: Tunnelling effects are significant in electronic boundary layers and, energy levels change as a result of wave function confinement, as discussed in Chapter 2. In other words, the effective size of the electron cloud causes variations in μ° even without structural changes, which is in contrast to ionic potential variations (cf. Fig. 5.107 on page 260 and footnote 204 on page 261). The former are approximately described by Eq. (2.29)[214]; the corresponding energy shift is inversely proportional to the square of the size. Modifications of the energy levels are experimentally evidenced, e.g. by modified band gaps (referring to standard potentials in the interaction free case). The colour change of semiconductor particles with size is a striking consequence [318]. Fig. 5.114 displays the evolution of the band gap with size (number of Cd atoms) for CdS–clusters including the value of the macroscopic crystal, as obtained from photoelectron spectra [319]. The band edge shifts agree well with theoretical expectations.

As for typical semiconductor contact phenomena, we will not follow such effects further, since they are treated in detail in the solid state physics literature.

It just remains for us to mention that the investigation of quantum effects in dimensionally reduced electron conductors is an attractive modern field of semiconductor research[215], characterized by the investigation of the sub–band formation in heterostructures[216], of the problems of quantum wires and quantum dots, of "artificial atoms" and "artificial molecules". In this context the term "nanoelectronics" [50,74] is more than just a fashionable slogan, but describes an exciting scientific

[213] Compare footnotes 160 and 179 on pagse 230 and 242.

[214] In addition attention has to be paid to the effect, that in nano–crystals excess electron and holes cannot be sufficiently separated, and Coulomb attraction (cf. level narrowing, Section 5.7.2) which is proportional to the inverse size, plays an important role (cf. exciton radius [317]). This holds generally for defect–defect interactions in nano–sized systems (see Section 5.7).

[215] It is a technologically important challenge to enable the reproducible and systematic preparation of metallic nano–contacts [320].

[216] Also an electronic analogue to the transport in stacking faults (here SiC) is known [321].

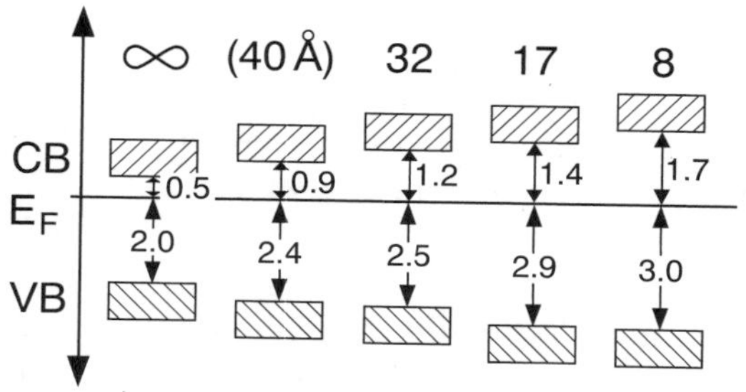

Fig. 5.114: The increase of the bandgap (data in eV) in CdS with cluster size. The clusters the Cd–content of which are indicated in the top row (∞ means bulk, the designation 40Å refers to a cluster for which the number of molecules is not precisely known but is above 32), are chemically complex in terms of the terminal groups; the local environments, however, are comparable with pure CdS. The shifts of valence and conduction band (data in eV) agree with values predicted from an electron–in–the–box treatment (see Section 2.2.1) appropriately adjusted and extended (spherical geometry, finite energy threshold at the surface) [319].

area. If "nanoionics", even though an equally justified expression and of similar basic significance [322,130,323], can be of a comparable technological attraction (cf. page 393) remains to be seen[217]. While nanoionics relies on mass transport, nanoelectronics — and this provides us with a bridge to the next chapter — demands negligible "ionic kinetics" under operating conditions.

[217]Nanocrystalline mixed conductors appear to be promising electrodes for Li batteries (see Section 7.4.3b).

6 Kinetics and irreversible thermodynamics

6.1 Transport and reaction

Until now we have investigated equilibrium states; in particular, we have made it our task to work out the equilibrium concentration of defects in order to describe the complete chemical equilibrium of the solid, as a function of material constants and control parameters. The latter are thermodynamic variables of state, but, in the case of partial equilibrium, also parameters which characterize metastable structural elements, such as irreversibly introduced dopants, frozen–in native point defects (cf. V''_{Sr} in $SrTiO_3$ at $T < 1300K$) or frozen–in interfaces (cf. Sections 5.6 and 5.8). Now we are going to leave (total or partial) equilibria and interest ourselves in the change in defect concentrations with time. It is particularly important in this physico–chemical context to know how rapidly new equilibrium concentrations will be reached if the component activity is varied. As a master example we treat the question how rapidly the conductance of an oxide (such as $SrTiO_3$, see Fig. 5.56, page 192) changes from the initial equilibrium to the new equilibrium on sudden change of the partial pressure of oxygen. Formulated in an application–oriented manner, we ask how rapidly a bulk conductivity sensor responds to oxygen. This question encompasses the kinetics of the surface reaction and the kinetics of oxygen diffusion. If, by the jump in oxygen partial pressure, the existence range of the oxide phase is exceeded, a new phase is formed (see Fig. 4.4, page 90). Such "real" chemical reactions also involve interfacial processes and diffusion steps and can partly be treated in a similar manner. Typical features of processes far from equilibrium will be treated at the end of this chapter.

The key to this chapter is the consideration of the kinetics of a heterogeneous elementary step

$$\text{Reaction R/T} = \qquad A(x) \rightleftharpoons B(x'), \tag{6.1}$$

in which species A can change its nature (R for reaction) and/or its position (T for transport). If $A \equiv B$, a pure transport step is being described. If $A \not\equiv B$ but conversely $x=x'$ can be assumed (more precisely: If possible differences in the position coordinate are unimportant), a homogeneous chemical reaction is addressed. As any heterogeneous reaction the process described by Eq. (6.1) can be formally broken down into a homogeneous reaction ($A(x) \rightleftharpoons B(x)$) and a coupled transport step ($B(x) \rightleftharpoons B(x')$); this is not necessarily mechanistically (and hence kinetically) correct. In addition, transport processes (see Sections 6.2 – 6.6) and chemical reactions (see Sections 6.7 – 6.10) are by no means always monomolecular, as is assumed in Eq. (6.1).

In what follows we will be dealing (in contrast to usual cases in semiconductor physics, cf. Section 6.2.2) with comparatively slow transport processes, so that we

Physical Chemistry of Ionic Materials J. Maier
©2004 John Wiley & Sons, Ltd ISBN: 0-471-99991-1 (HB); 0-470-87076-1 (PB)

will generally presuppose local equilibrium for the coupled internal reactions (see also Section 6.6.1).

6.1.1 Transport and reaction in the light of irreversible thermodynamics

Our general equilibrium condition for Eq. (6.1) is

$$\widetilde{\mu}_A(x) = \widetilde{\mu}_B(x'). \tag{6.2}$$

In the case of transport equilibria of a given particle (A≡B) the gradients in its electrochemical potential disappear in accordance with this. If there are no electrical fields and if the structure is invariant (i.e. constant μ° in $\widetilde{\mu} = \mu^\circ + RT\ln a(c) + zF\phi$), the concentration gradients disappear. In the case of a homogeneous reaction equilibrium ($x = x'$) between two species the condition reduces to the equality of the chemical potentials of A and B, i.e. to a vanishing chemical affinity of the reaction. However, if the electrochemical potentials in Eq. (6.2) deviate from each other, e.g. as a result of a transient perturbation of the equilibrium state, fluxes (J) or reaction rates ($\mathcal{R}$) appear[1]. These fluxes endeavour, when the distance from equilibrium is not too large, to restore this equilibrium. In an analogous manner temperature gradients make for an appropriately directed thermal flow. In this sense the equilibrium state

[1]Some fundamental comments concerning fluxes (J) and rates ($\mathcal{R}$) should be made here. The reaction rate measures the increase of a product concentration (decrease of educt concentration) in a given reaction normalized w.r.t. the stoichiometric number. The flux density measures the number of particles per time unit flowing through an area element perpendicular to the flux direction (flux density = particle density times velocity). Both are related via the continuity equation $\partial c/\partial t = -\mathrm{div}\mathbf{J} + \nu\mathcal{R}$ which expresses that c can increase by transport ($\delta_e c$, cf. Section 4.2) as well as by local generation ($\delta_i c$, cf. Section 4.2; cf. Eq. (6.78)).
By taking seriously Eq. (6.1) one can also conceive the flux term as being constituted by local (hopping) reaction rates. In the case of a reaction series $A \rightleftharpoons B \rightleftharpoons C$ the (conversion) rate from A to B ($\mathcal{R}_{AB}$) means the difference between the forward and back rates ($\vec{\mathcal{R}}_{AB} - \overleftarrow{\mathcal{R}}_{AB}$); $\vec{\mathcal{R}}_{AB}$ ($\overleftarrow{\mathcal{R}}_{AB}$) designates the number of successful conversion events from A to B (B to A). The situation is analogous for the second individual step ($\mathcal{R}_{BC} = \vec{\mathcal{R}}_{BC} - \overleftarrow{\mathcal{R}}_{BC}$). The decrease or increase (with time) of A or C is directly determined by $\mathcal{R}_{AB}$ and $\mathcal{R}_{BC}$, while the increase (with time) of B is given by the difference $\mathcal{R}_{AB} - \mathcal{R}_{BC}$. In pure transport steps this relationship corresponds to the flux divergence. We may write $B(x) \rightleftharpoons B(x') \rightleftharpoons B(x'')$, with J' being proportional to the number of successful hops per unit time (net) from x to x' (corresponding to $\mathcal{R}_{AB}$); analogous considerations apply to J'' (and the transition from x' to x''). The relationship $\partial[B]/\partial t = \mathcal{R}_{AB} - \mathcal{R}_{BC}$ in the reaction chain corresponds in the transport chain to $\partial[B(x')]/\partial t = \frac{J'-J''}{\Delta x}$ (i.e. $-\mathrm{div}\mathbf{J}$), whereby Δx is the distance of the dividing "planes" between x and x', on the one hand, and x' and x'', on the other (cf. also Eq. (6.78)). (Note that a further difference between fluxes and rates comes into play if the stoichiometric ν number deviates from unity; unlike the flux, the reaction rate is usually normalized w.r.t. ν.) Conceiving more precisely transport as a re–arrangement reaction allows for a deeper insight into the energetic situation. Then the reaction coordinate is not the space coordinate but refers to re–arranging the solid from the state $\square(x)A(x')$ to the state $A(x)\square(x')$ if without loss of generality a vacancy ($\square$) mechanism is assumed (cf. also bimolecular description in Section 6.2).

is stable and takes the form of an attractor state. If external constraints in the form of constant forces (or fluxes) prevent the system from taking up the equilibrium state then — again, if there is not too great a deviation from equilibrium — stable stationary states are set up. For instance, if we apply a constant temperature difference across a homogeneous medium this leads, after a certain time (transient behaviour), to a constant heat flow (stationary state). The situation is similar with respect to mass flows for given concentration differences or with respect to electrical currents for given electrical potential differences. The experiment that has been mentioned several times, the use of blocking electrodes to separate electronic and ionic conductivity (see also Section 7.3.4), is a more complex example: If a constant voltage is applied to a cell, the electrodes of which have a blocking effect on the ions (or electrons), then the current falls from an initial value which is proportional to $\sigma_{\mathrm{eon}} + \sigma_{\mathrm{ion}}$ to a value proportional to σ_{eon} (or to σ_{ion}) [324,325]. Initially all charge carriers contribute to the conductivity, but, when the stationary state is set up, the only current to flow is that of the nonblocked species.
We have already seen in Section 4.2 that, in the case of a chemical reaction, the entropy production[2], which is zero at equilibrium, can be described at nonequilibrium as a product of the driving force (affinity[3] $\mathcal{A}$), which we will regard in what follows as being given, and the corresponding rate (reaction rate $\mathcal{R}$). This can be generalized for different processes (k) to

$$\Pi \equiv \mathrm{T}\frac{\delta_{\mathrm{i}}\mathrm{S}}{\delta \mathrm{t}} = \Sigma_{\mathrm{k}} \mathrm{J}_{\mathrm{k}} \mathrm{X}_{\mathrm{k}} \tag{6.3}$$

with J as the generalized rate (or flux[1]) and X as the generalized driving force. In Eq. (6.3) Π and S denote volume densities. It can be shown that when processes are close to equilibrium the integral entropy production in the stationary state — although it is not zero, as is the case for the equilibrium state — must, nevertheless, be minimal[4] [94]. Furthermore, the processes in the neighbourhood of the stationary state always take place in such a manner that Π decreases. Hence, stationary states are stable in such cases and, like the equilibrium state, represent attractor states (see Section 6.10).

[2] We also designate the dissipation function Π which is the product of the entropy density produced per unit time and the temperature, simply as "entropy production". In the literature the forces are frequently defined in such a manner that $\Sigma_{\mathrm{k}}\mathrm{J}_{\mathrm{k}}\mathrm{X}_{\mathrm{k}}$ yields $\delta_{\mathrm{i}}\mathrm{S}/\delta \mathrm{t}$ and not Π (cf. Eq. (6.3)) Since we are considering isothermal processes here these details do not have any consequence for the treatment [94,326,327].

[3] Naturally, the affinity introduced for chemical reactions, can also be used in a generalized sense as $\widetilde{\mathcal{A}} \equiv \widetilde{\mu}_{\mathrm{A}}(\mathrm{x}) - \widetilde{\mu}_{\mathrm{B}}(\mathrm{x}')$ to describe Eq. (6.2). It reduces to $\mathcal{A} = -\Delta_{\mathrm{R}}\mathrm{G}$ for chemical reactions, to the purely entropic driving force $\mathrm{RT}\ln(\mathrm{c}/\mathrm{c}')$ for pure diffusion processes and to the purely electrical driving force $\mathrm{zF}(\phi - \phi')$ for pure electrical conduction. It is evident in view of Eq. (4.72) that generally $\mathrm{T}\delta_{\mathrm{i}}\mathrm{S} = \widetilde{\mathcal{A}}\mathcal{R} \geq 0$ (for a single process), if $\mathcal{R}$ is also used for transport processes.

[4] The limits have been worked out by Landauer [328] for electrical circuits. Care is particularly required in the presence of inductivities.

Let us now look at the relationship between J and X and expand[5] J(X) as

$$J(X) = \alpha + \beta X + \gamma X^2 + \delta X^3 + \ldots \tag{6.4}$$

Since the rate disappears in the equilibrium (X=0), it follows that $\alpha = 0$ and, as the equilibrium is approached, the higher terms become less important, so that we are left with the linear relationship[6,7]

$$J(X) = \beta X. \tag{6.5}$$

(Given the appropriate definition of X, β is positive definite.) This characterizes the regime of linear irreversible thermodynamics[8]. Naturally, Eq. (6.5) is almost a triviality in the sense that, according to Eq. (6.4), any relevant curve can be approximated by a straight line if we are sufficiently close at X=0. The major task will be to make a statement on the range of validity, i.e. on the term "sufficiently close". Empirically, a series of linear flux–force relationships are known. The most important (w.r.t. the context of this book) are collected in Table 6.1. Let us first accept their validity, look at the consequences of the linear laws (see Table 6.2) and investigate the range of validity later. In the case of a normal chemical reaction $A \rightleftharpoons B$ (i.e. $x = x'$ or at least $\phi(x) = \phi(x')$) Eq. (6.5) predicts that the reaction rate is proportional to the affinity (negative gradient of the mean chemical potential with respect to the reaction progress),

$$J_R \equiv \mathcal{R} = \beta_R \mathcal{A}_R \propto -\Delta_R G = -\Delta_R G^\circ - RT \ln Q_R = -RT \ln(Q_R/K_R), \tag{6.6}$$

a relationship which is quite evidently not well supported by experience: For example, in spite of a high affinity ($|\Delta_R G| \gg RT$ because $|\Delta_R G^\circ| \gg RT$) hydrogen is stable (kinetic stability) in the presence of traces of O_2 over a large condition window, i.e. no measurable reaction takes place. Near critical values a slight increase in oxygen concentration can, in spite of an increase of $|\Delta_R G|$ that is quite small (of $\sim |RT\Delta \ln P_{O_2}|$), under otherwise identical conditions, lead to a dramatic change

[5]With respect to symmetry restrictions according to the experimental situation, see Ref. [329].

[6]When various driving forces and fluxes become important, then, according to $J_k = \Sigma_{k'} \beta_{kk'} X_{k'}$, mixed terms also appear, whose coefficients are symmetrical according to the Onsager relation ($\beta_{kk'} = \beta_{k'k}$) [330]. The latter follows from the principle of microscopic reversibility [326]. We shall mostly ignore such cross effects in the treatment (however, see Sections 6.6.1, 6.10.1).
Of course, the choice of the driving force is not unequivocal. This is even the case if only one independent driving force is relevant, as assumed in the text. For simplicity we restrict to one dimension. Instead of X (e.g. if dilute: $(\partial/\partial x)\mu$) it is also possible to introduce X′ as driving force, with $X' \simeq X(\partial X'/\partial X)_{\text{equilibrium}}$ (e.g. $(\partial/\partial x)c = (\partial/\partial x)\mu(\overline{c}/RT)$). However the range of validity is altered.

[7]Strictly speaking Eq. (6.5) describes, in the sense of system theory, a linear signal transfer which is analogue, continuous and time invariant (note that β is constant). In particular in Section 7.3.6, we have to be concerned with time–variant systems.

[8]Specific presentations of irreversible thermodynamics are to be found in Refs. [326,327,331–333].

Table 6.1: Relevant empirical linear flux–force relationships.

$\mathbf{J} = \beta \quad \mathbf{X}$	Linear flux–force relationship
$\mathbf{j} = \mathrm{D}\ (-\nabla \mathrm{c})$	Fick's law
$\mathbf{i} = \sigma\ (-\nabla \phi)$	Ohm's law
$\mathbf{f} = \lambda\ (-\nabla \mathrm{T})$	Fourier's law

$\mathbf{j}$ = particle flux density, $\mathbf{i}$ = electrical current density
$\mathbf{f}$ = heat flux density; D = diffusion coefficient
σ = specific electrical conductivity
λ = specific thermal conductivity

and the harmless mixture may become explosive (according to a typical nonlinear behaviour, see Section 6.10).

In the case of transport $\mathrm{A(x)} \rightleftharpoons \mathrm{A(x')}$ (i.e. $\mathrm{A} \equiv \mathrm{B}, \mathrm{x} \neq \mathrm{x'}$) Eq. (6.5) takes the form

$$\mathbf{J}_{\mathrm{T}} = \beta_{\mathrm{T}}(-\nabla\widetilde{\mu}) = -\beta_{\mathrm{T}}\nabla\mu - \mathrm{zF}\beta_{\mathrm{T}}\nabla\phi, \tag{6.7}$$

expressing the fact that both chemical and electrical fields can act as driving forces. We will now look more closely at specific cases[3,9].
If uncharged particles are concerned, such as, for example, sugar molecules in aqueous solution or germanium atoms in silicon, the process is reduced to a pure diffusion ($\nabla\widetilde{\mu} = \nabla\mu$), and

$$\mathbf{J}_{\mathrm{T}} = \mathbf{J}_{\mathrm{D}} \equiv \mathbf{j} = -\beta_{\mathrm{T}}\nabla\mu \tag{6.8}$$

remains (T stands for transport, D for diffusion).
In the case of the dilute state $\nabla\mu = \mathrm{RT}\nabla\mathrm{c}/\mathrm{c}$, so that[10]

$$\mathbf{j} = -\frac{\mathrm{RT}\beta_{\mathrm{T}}}{\mathrm{c}}\nabla\mathrm{c}. \tag{6.9}$$

A comparison with Fick's law (Table 6.1) permits the identification

$$\beta_{\mathrm{T}} = \mathrm{Dc}/\mathrm{RT}. \tag{6.10}$$

[9]The information flux, which can be regarded as a consequence of information gradients, is another example. The significance of communication barriers (cf. β) in real (mostly nonlinear) processes need not be emphasized.

[10]Equation (6.9) is of course also obtained by the application of Eq. (6.6) to the transport step. Here $\Delta_{\mathrm{R}}\mathrm{G}^{\circ} = 0$ and $-\ln\mathrm{Q}_{\mathrm{R}} = \ln\frac{\mathrm{c}-\Delta\mathrm{c}}{\mathrm{c}} = \ln\left(1 - \frac{\Delta\mathrm{c}}{\mathrm{c}}\right) \simeq -\frac{\Delta\mathrm{c}}{\mathrm{c}}$ (see below) and, in contrast to the chemical reaction, the linear relationship applies over a wide range, since the driving force is small (purely entropic) because $\Delta_{\mathrm{R}}\mathrm{G}^{\circ} = 0$.

Table 6.2: Diagram for the nonequilibrium behaviour in solids close to equilibrium. Conventional chemical reaction ($x \equiv x'$) and particle transport ($A \equiv B$) are described in a general manner. Particle transport also includes the limiting cases of pure diffusion ($zF\Delta\phi = 0$) and pure electrical conduction ($\Delta\mu = 0$) [334].

linear irreversible thermodynamics
particle transport or reaction

$$A(x) \rightleftharpoons B(x')$$

$$\tilde{\mu}_A(x) \neq \tilde{\mu}_B(x')$$
$$J \propto -[\tilde{\mu}_B(x) - \tilde{\mu}_A(x)]$$

$zF\phi(x') = zF\phi(x)$ $A \equiv B$

conventional
chemical reaction   particle transport

$$A \rightleftharpoons B$$

$$\mu_B - \mu_A = \Delta_R G \neq 0$$
$$J_R \propto -\Delta_R G$$
for $\Delta_R G \ll RT$

$J_R = \mathcal{R}$ = reaction rate

$$A(x) \rightleftharpoons A(x')$$

$$\tilde{\mu}(x') - \tilde{\mu}(x) \neq 0$$
$$J_T = -\beta(\delta/\delta x)\tilde{\mu}$$

$(\delta/\delta x)zF\phi = 0$
diffusion $(\delta/\delta x)\mu = 0$
conduction

$$\mu(x') - \mu(x) \neq 0$$
$$J_T = -\beta(\delta/\delta x)\mu$$
$$j = -\frac{RT}{c}(\beta(\delta/\delta x)c$$

$$\phi(x') - \phi(x) \neq 0$$
$$J_T = -\beta z F(\delta/\delta x)\phi$$
$$i = -\beta z^2 F^2(\delta/\delta x)\phi$$

j = flux density i = current density

$$D \equiv \frac{RT}{c}\beta$$ $$\sigma = z^2 F^2 \beta$$

$$\frac{D}{RT} = \frac{\sigma}{cz^2 F^2} = \frac{u}{zF}$$

Nernst-Einstein-relation

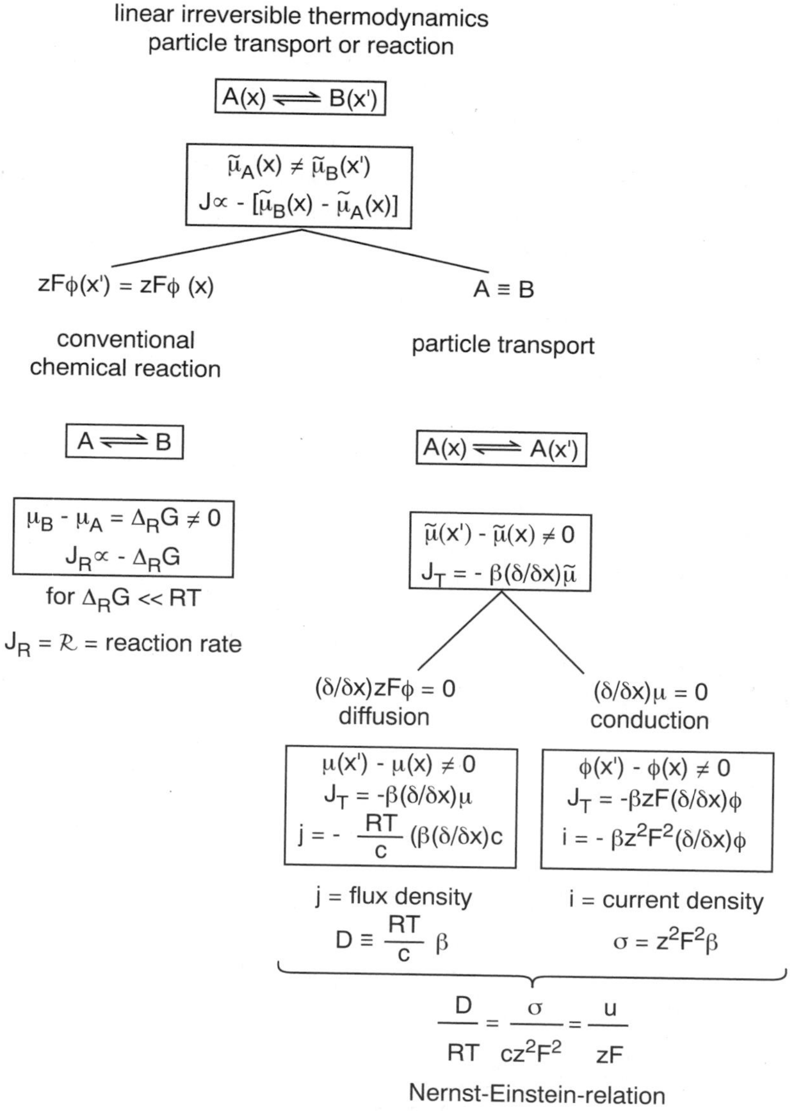

In the case of concentrated solutions Eq. (6.9) and Eq. (6.10) obviously include the so-called thermodynamic factor $\mathrm{w} = \frac{\mathrm{d}\ln \mathrm{a}}{\mathrm{d}\ln \mathrm{c}} = 1 + \frac{\mathrm{d}\ln \mathrm{f}}{\mathrm{d}\ln \mathrm{c}}$ (a: activity, f: activity coefficient, see Section 5.7.2 and also footnote 26 on page 83). The fact that c appears in the numerator of Eq. (6.10) is not inconsistent with the linear laws close to equilibrium, c is simply given by the equilibrium concentration[11] $\widehat{\mathrm{c}}$; $\delta \ln \mathrm{c} \ll 1$ means $\delta \mathrm{c} \ll \mathrm{c} \simeq \widehat{\mathrm{c}}$.

Conversely, if there are no noticeable changes in the chemical potential $\mu(\mathrm{a}(\mathrm{c}))$, the electrical field remains as the driving force. Since $\delta \mathrm{c} \ll \mathrm{c}$, i.e. $\delta \ln \mathrm{c} \simeq 0$, this condition is excellently fulfilled for composition variations in materials with high charge carrier concentrations such as metals (electrons), superionic conductors (ions) and appropriately highly doped systems. It is approximately fulfilled in the I–regime of mixed conductors for the ionic but not for the electronic carriers. The condition $\delta\mu \simeq 0$ is also satisfied in the case of stationary conductivity experiments with reversible symmetrical electrodes. There an invariant composition is maintained: The same number of charge carriers introduced on the one side, is removed on the other[12].

Then, what remains is (E stands here for electrical conduction)

$$\mathbf{J}_\mathrm{T} = \mathbf{J}_\mathrm{E} = -\beta_\mathrm{T} \mathrm{z} \mathrm{F} \nabla \phi \tag{6.11}$$

or after converting particle current densities into charge current density $\mathbf{i} = \mathrm{zF}\mathbf{J}_\mathrm{E}$

$$\mathbf{i} = -\beta_\mathrm{T} \mathrm{z}^2 \mathrm{F}^2 \nabla \phi. \tag{6.12}$$

This is obviously Ohm's law[13].

Evidently, the prefactor in Eq. (6.12) corresponds to the electrical conductivity of the charge carrier under consideration

$$\sigma = \beta_\mathrm{T} \mathrm{z}^2 \mathrm{F}^2. \tag{6.13}$$

Comparison with Eq. (6.10) reveals that (in the case $\mathrm{z} \neq 0$) σ and D are related parameters. The relationship

$$\mathrm{D} = \frac{\sigma \mathrm{RT}}{\mathrm{z}^2 \mathrm{F}^2 \mathrm{c}} \tag{6.14}$$

is known as the Nernst–Einstein equation [335]. Because $\sigma = |\mathrm{z}|\mathrm{Fcu}$ (see Eq. (5.122)) this is formulated more concisely as

$$\frac{\mathrm{D}}{\mathrm{RT}} = \frac{\mathrm{u}}{\mathrm{zF}}. \tag{6.15}$$

---

[11] Even though not always consistent (see Eq. (6.4)), usually in order to obtain a reasonable approximation for nonlinear processes a space dependence of c is implemented in the linear law a posteriori [329].

[12] We neglect the effects of electrical capacitances here. We shall treat such aspects of electrochemical kinetics in Chapter 7. Note that the thermodynamic factor — which comes into play for higher concentrations (see above) — is directly related to chemical capacitance (see Section 6.7.4).

[13] The macroscopic Ohm's law current = resistance × voltage is obtained assuming homogeneous samples. Here the space charge density is zero and, hence, because of Eq. (5.219) $\phi' = \mathrm{const.}$ i.e. $\mathrm{i}_\mathrm{k} = -\sigma_\mathrm{k} \Delta\phi/\Delta \mathrm{x}$ ($\Delta \mathrm{x}$ is the sample thickness here). Since $\mathrm{i}_\mathrm{k} \cdot$ area$\equiv$I, $-\Delta\phi \equiv \mathrm{U}$ and $\mathrm{R} = \frac{\Delta \mathrm{x}}{\mathrm{area}} \sigma^{-1}$, the result is U=RI.

The diffusion coefficient (in "thermal units") and the mobility (in "electrical units") of a charge carrier are accordingly analogous parameters.
To summarize, the partial current density of the species k can be expressed in the form

$$\mathbf{J}_{\mathrm{k}} = -\frac{\sigma_{\mathrm{k}}}{z_{\mathrm{k}}^2 F^2}\nabla\widetilde{\mu}_{\mathrm{k}} \tag{6.16a}$$

or

$$\mathbf{J}_{\mathrm{k}} = -\frac{D_{\mathrm{k}} c_{\mathrm{k}}}{RT}\nabla\widetilde{\mu}_{\mathrm{k}}. \tag{6.16b}$$

The first formulation only has meaning if the particles are charged.

6.1.2 Transport and reaction in the light of chemical kinetics

Now, in order to map out the region of validity of the linear laws we turn to the simple kinetic laws of chemical kinetics [336] for transition rates (Table 6.3). These are not limited to the near–equilibrium conditions (Fig. 6.1), but apply from the start to dilute systems[14]. For simplicity's sake let us consider only one position coordinate, refer to our general[15] reaction (Eq. (6.1)) (only two species and two sites) and formulate:

$$J = \overrightarrow{k} N_A(x) - \overleftarrow{k} N_B(x'). \tag{6.17}$$

The generalized flux (composed of forward and backward contributions, i.e. $\overrightarrow{J} - \overleftarrow{J}$) refers to the net particle number passing from x to x′ in unit time; during that process the particles may or may not change their nature. In the case of a homogeneous

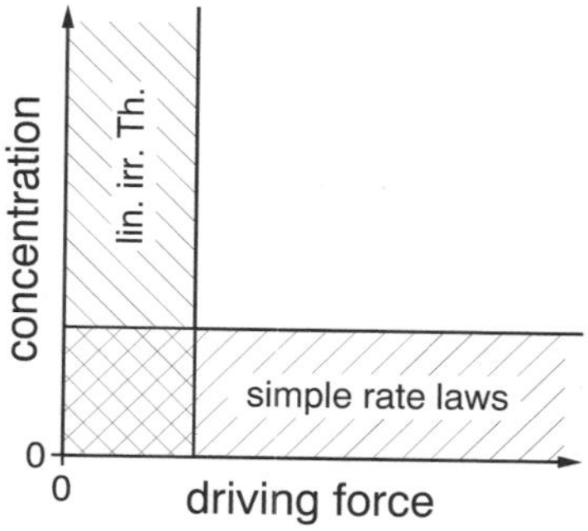

Fig. 6.1: The kinetic relations only apply to dilute states, but then in the nonlinear region too, while the relations of linear, irreversible thermodynamics break down there, but also apply to regions of higher concentrations.

[14]a) Another restriction is that the volume should be sufficiently large. b) The concentration terms in the kinetic formulations are frequently converted to activities in order to enlarge the range of applicability. Even though this is compatible with the mass action law, it must be remembered that the interactions and, hence, the activity coefficients depend on the distance from equilibrium.

[15]The pseudo–monomolecular rate law (6.1) is justified mechanistically in Sections 6.2.1 and 6.7.3.

Table 6.3: Diagram of the kinetic treatment of the nonequilibrium behaviour in the dilute states [334]. Reaction, diffusion and electrical conduction appear as special cases ($\widehat{c}$: equilibrium concentration) of the electrochemical reaction $A(x) \rightleftharpoons B(x')$. See the text for a more precise treatment.

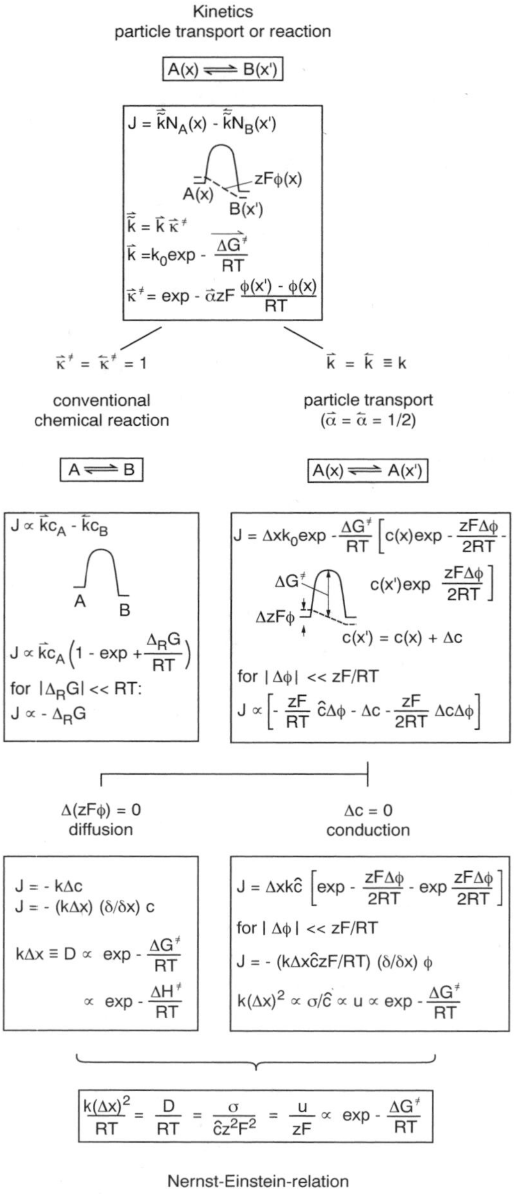

chemical reaction the local coordinate[16] is replaced by the reaction coordinate and the area density by the volume density[1]. Then the relation between J and $\mathcal{R}$ involves the jump distance Δx (in addition to the stoichiometric number). The rate constants $\widetilde{k}$ are independent of the particle density but exponentially dependent on the barrier to be overcome[17,18], e.g. for the forward jump we have

$$\overrightarrow{\widetilde{k}} = k_0 \exp -\frac{\overrightarrow{\widetilde{\Delta G}}^{\neq}}{RT} \propto \exp -\frac{\overrightarrow{\widetilde{\Delta H}}^{\neq}}{RT}. \tag{6.18}$$

The tilde indicates that the actual chemical barrier ($\overrightarrow{\Delta G}^{\neq}$) is modulated by the electrical field, as shown in Fig. 6.2, namely according to $\overrightarrow{\widetilde{\Delta G}}^{\neq} - \overleftarrow{\widetilde{\Delta G}}^{\neq} = \widetilde{\Delta_{R/T} G}^{\circ} = \Delta_{R/T}G^{\circ} + zF(\phi(x') - \phi(x))$. It should be noted that Fig. 6.2 does not represent

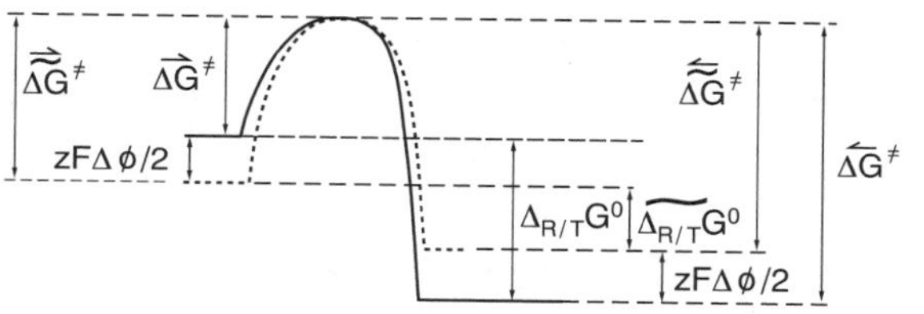

Fig. 6.2: The free–enthalpy profile of the electrochemical reaction (continuous line) is distorted by the potential drop $zF\Delta\phi$ (dotted profile) such that the activation energy for the reaction from left to right (right to left) is increased (reduced) by $zF\Delta\phi/2$. (More precisely the symmetry factor $\overrightarrow{\alpha}$ or $\overleftarrow{\alpha} = 1-\overrightarrow{\alpha}$ must be substituted for $\frac{1}{2}$, see also Chapter 7.) In the pure transport step the chemical barrier is symmetrical and $\Delta_{R/T}G^{\circ} = 0$.

a snapshot of the local chemical potential distribution[1], but rather the partial free enthalpy profile probed by a test particle (given occupation), i.e. the "local standard term". The parameter k_0 corresponds to the attempt frequency. Multiplication by the probability of passing the barrier (exponential term in Eq. 6.18) yields then the rate constant. If the transition state theory is applied, then k_0 is a weak function of the temperature[19].

More precisely the transition state corresponds to a saddle–point. This is shown in Fig. 6.3; the "fastest route" is that path from the set of all parallel processes which exhibits the most favourable transition state. On the other hand, the same transition state forms the most difficult serial part of that specific path (bottleneck)[19]. Again, we first consider the chemical reaction. The barrier is now a purely chemical threshold ($\widetilde{\Delta G} \rightarrow \Delta G$, $\widetilde{k} \rightarrow k$, $J \rightarrow \mathcal{R}$). Note that Eq. (6.17) can be appropriately

[16] Generally, Eq. (6.1) — even for the pure transport case — can be advantageously viewed as a rearrangement reaction with a corresponding reaction coordinate.

[17] This follows from transition state theory (cf. Refs. [9,10,337,338]).

[18] In order to avoid complex notations the index "m" is omitted from the molar thermodynamic functions. Thus ΔG, ΔH, ΔS refer to 1 mol if not stated otherwise.

[19] Compare here Refs. [9,10,337,338].

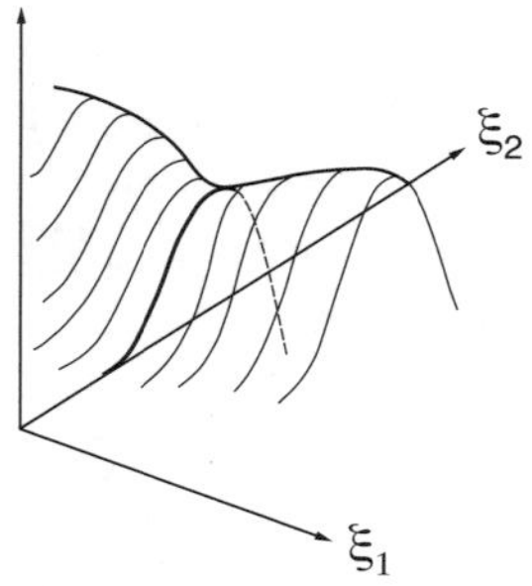

Fig. 6.3: The potential energy as a function of two reaction coordinates. The ordinate is obtained by occupying each state (ξ_1, ξ_2), and reflects a "local standard value" (cf. energy level). The route over the saddle–point ("pass") is marked. If the landscape is weighted with the frequency of such routes or with the width of the "passes" (i.e. if one takes account of entropy contribution), one obtains an effective energy landscape, i.e. G–landscape, the saddle point of which is decisive for k.

rearranged to

$$\mathcal{R} \propto \vec{k}N_A \left(1 - \frac{\overleftarrow{k}\, N_B}{\vec{k}\, N_A}\right). \qquad (6.19)$$

In Eq. (6.19) we used the proportionality sign, simply because (in contrast to the fluxes) the reaction rate is volume related. Since $\mathcal{R} = 0$ in equilibrium, it follows that $K_R = \vec{k}/\overleftarrow{k}$. Because of $\Delta_R G = +RT \ln Q_R/K_R$ (cf. Section 4.2) we obtain

$$\mathcal{R} \propto \vec{k}N_A \left(1 - \exp + \frac{\Delta_R G}{RT}\right), \qquad (6.20)$$

and recognize that the relationship between rate and driving force is not linear. Since N_A depends on $\mathcal{R}$ and $\Delta_R G$ in a very individual manner for each reaction, no simple flux–force relation can — as experience also teaches us — be given in the general case. However, when $|\Delta_R G| \ll RT$ the exponential expression can be linearized[18], N_A then approaches the equilibrium concentration, and we obtain a linear relationship of the form

$$\mathcal{R} \propto -\vec{k}\widehat{N}_A \Delta_R G/RT = -\overleftarrow{k}\widehat{N}_B \Delta_R G/RT = \mathcal{R}_0 \mathcal{A}_R/RT. \qquad (6.21)$$

Here the exchange rate $\mathcal{R}_0$ (the rate of the forward and back reactions at equilibrium) appears as the relevant "permeability" parameter for the reaction in the linear regime[20]. However, the limitation introduced, viz. $|\Delta_R G| \ll RT$, is now extremely restrictive. RT is only of the order of 10kJ even at 1000K, while the chemical free standard enthalpy of reaction can certainly reach MJ in its order of magnitude. Equation (6.21) is only obeyed in the immediate neighbourhood of the equilibrium and is, thus, of little utility for typical chemical reactions[21]. If conversely $|\Delta_R G| \gg RT$ then, depending on the sign of $\Delta_R G$, $\mathcal{R}$ is determined by the forward reaction ($\vec{k}N_A$) or the back reaction ($\overleftarrow{k}N_B$) alone and the explicit dependence on

[20] If again we generalize the affinity concept (see footnote 3) by applying it to the reaction $A(x) \rightleftharpoons B(x')$, i.e. also to the pure transport ($A \equiv B$), we notice that the conductivity corresponds to the exchange rate of the transport (apart from less important factors). Cf. also Section 6.3.4. Note that in generalized cases the rate coefficients may change with distance from equilibrium.

[21] If we start from the equilibrium state and disturb it slightly, the relaxation in the equilibrium state can be expressed by linear relationships (see Section 6.7.3).

the affinity disappears. As we will see below this is different if the barrier is purely electrical or even purely entropic in nature.
The following basic relationship for the ratio of the individual reaction rates ($\vec{\mathcal{R}} \propto \vec{k} N_A, \overleftarrow{\mathcal{R}} \propto \overleftarrow{k} N_B$) has already been used in Eqs. (6.19) and (6.20):

$$\overleftarrow{\mathcal{R}}/\vec{\mathcal{R}} = Q_R/K_R = \exp + \frac{\Delta_R G}{RT}\,. \tag{6.22}$$

It most clearly shows that neglect of one partial rate ($\overleftarrow{\mathcal{R}}$ or $\vec{\mathcal{R}}$) is associated with nonlinear behaviour ($|\Delta_R G| \gg RT$). The inherent nonlinearity of chemical kinetics is the key to our understanding of biology. As has been discussed, it does not just make possible the durability of biological structures, it also leads to the variety of nature and the appearance of dissipative structures [94,339,340], which is after all what we are in a thermodynamic sense. As is well known, nonequilibrium defects play a fundamental role in evolution. In our context, the fact that the nonlinearity of chemical reactions is indispensable for the individuality and complexity of interfacial reactions is of particular interest. Typical nonlinear phenomena are discussed in more detail at the end of this chapter.

Let us now turn to transport processes. Here it is not the type of species (A) that changes but only its position:

$$J_A = \overset{\rightarrow}{\tilde{k}}_A N_A(x) - \overset{\leftarrow}{\tilde{k}}_A N_A(x'). \tag{6.23}$$

First let us consider pure diffusion, that is, we neglect electrical effects. It then follows that

$$\overset{\rightarrow}{\tilde{k}}_A = \vec{k}_A = \vec{k}_0 \exp - \frac{\Delta \vec{G}_A^{\neq}}{RT} \propto \exp - \frac{\Delta \vec{H}_A^{\neq}}{RT}. \tag{6.24}$$

On account of the absence of electrical field effects and the symmetry of the (standard) chemical barrier ($\Delta_R G^\circ = 0$), the rate constants for the forward and back reactions are identical:

$$\vec{k}_A = \overleftarrow{k}_A \equiv k_A \quad \text{and} \quad \Delta \vec{G}_A^{\neq} = \Delta \overleftarrow{G}_A^{\neq} \equiv \Delta G_A^{\neq}. \tag{6.25}$$

In contrast to the chemical reaction, a linear relationship follows without further approximation, which is of the form

$$J_A = -k_A \Delta N_A = -(k_A \Delta x)\frac{\Delta N_A}{\Delta x}. \tag{6.26}$$

This is Fick's law; it is obviously valid for dilute states under very general conditions[22]. Since usually the jump distance is small in comparison to the total sample thickness length, $\Delta N/\Delta x$ corresponds to the differential quotient. If we

[22] Note that this also implies that the proportionality of flux and *chemical potential* gradient (see Eq. (6.8)) — even though applicable for higher concentrations — is restricted to the proximity to equilibrium (see Eq. (6.10)).

change over from N to volume concentration, by thinking of the particles as smeared out over the distance Δx[23] ($c = N/\Delta x$), the equivalence of diffusion coefficient (see Table 6.3) and rate constant becomes obvious

$$k_A(\Delta x)^2 = D_A = D_{A0}\exp -\frac{\Delta G_A^{\neq}}{RT} \propto \exp -\frac{\Delta H_A^{\neq}}{RT}. \tag{6.27}$$

Equation (6.27) also reveals the temperature dependence of D_A: D_{A0} is proportional to the square of the jump distance and only exhibits a weak temperature dependence. Hence D_A as k_A is basically thermally activated via $\Delta H_A^{\neq}$.
D_A and k_A are proportional to Γ_A, the jump frequency of the particle. Let us extend the treatment to three–dimensional space and consider a cubic crystal. Since, there, each particle is equally mobile in each of the three directions it follows that[24]

$$\frac{D}{(\Delta x)^2} = k_A = \frac{1}{6}\Gamma_A \tag{6.28}$$

($\frac{1}{6}$ instead of $\frac{1}{3}$ because particles can jump with positive and negative displacements). Γ is clearly activated by the same $\Delta H^{\neq}$–value that we discussed above. The prefactor $\Gamma_0 \equiv 6k_0$ represents the attempt frequency[25].

---

[23] This smearing out takes place on both sides of the maximum (saddle–point) around initial and final state. The volumes so created are displaced by $\Delta x/2$ from the volume contained between x' and x. The correspondence between the flux and the total generation rate applies when there is restriction to one elementary process. In the real case there are analogous processes, that take place in other spatial directions, i.e. in ^{1}D not only from x to $x + \Delta x$ but also from x to $x - \Delta x$. The reaction rates from x to $x + \Delta x$ and $x - \Delta x$ sum up. If we split off $N(x \pm \Delta x)$ as $N \pm \Delta N$, the N–terms cancel and the difference of the ΔN–terms remains. Accordingly we obtain as the balance the difference between the fluxes being given by the second Fick's law (see Eq. (6.66)): $\Delta J \propto \Delta\Delta c \propto \nabla^2 c$. So, the total growth $\dot{c}$ ($\propto \Delta J$) at a given position corresponds to the generation rate in the case of a chemical reaction, but to the divergence of J in the case of transport (see Eq. (6.78)). If we consider the rates of individual steps, the role of $\mathcal{R}$ and J is analogous. See also footnote 1.

[24] In the isotropic case we can simply attribute one third of the jump attempts to the x coordinate and one half of these to the partial flux in the given direction. A consideration that more precisely takes account of the statistical nature of the diffusion process and also refers to its long range behaviour, involves the random walk model. Starting from the origin the particle is displaced by the vector $\mathbf{r}_1$ and then by $\mathbf{r}_2$, etc., that is a total of $\Sigma\mathbf{r}_i$ from the origin. The square of the displacement (i.e. distance) is $(\Sigma\mathbf{r}_i)^2 = \Sigma\Sigma\mathbf{r}_i\mathbf{r}_j = \Sigma\Sigma r^2\cos\alpha_{ij}$, whereby α_{ij} represents the angle between $\mathbf{r}_i$ and $\mathbf{r}_j$. If there are n steps, this gives $nr^2 + 2\sum_{i=j+1}^{n}\sum_{j=1}^{n-1} r^2\cos\alpha_{ij}$. The first term is made up of the summation of the products of vectors with the same index (diagonal elements), the second term stems from the nondiagonal elements. In order to obtain the mean square displacement, averaging over $\cos\alpha_{ij}$ is necessary. For the uncorrelated defect jump $\overline{\cos\alpha_{ij}} = 0$ and the mean square displacement will be nr^2. On the other hand, the mean square displacement for the random walk problem equals 6Dt (see footnote 25). Equation (6.28) follows with $\Gamma \equiv n/t$.

[25] The fact that the mean square displacement corresponds to the parameters 2Dt (or 6Dt in three dimensions) can be understood as follows: Let us consider a random walker who decides by tossing a coin whether to make one step to the right or to the left depending on whether the outcome is heads or tails. According to the treatment on page 122 Section 5.2, when a coin is tossed

Let us now consider pure electrical conduction as our final case. Here $N_A(x) \simeq N_A(x') \simeq \widehat{N}_A$, i.e. we can neglect concentration changes. However, on account of the applied field $(-\Delta\phi/\Delta x)$, the barrier is no longer symmetrical. (Generally the potential difference can be a consequence of inherent fields (e.g. the equilibrium space charge zone) or of externally applied fields (bias).) In accordance with Fig. 6.2, let us assume that the electrical potential variation is approximately linear (cf. Poisson equation; see Section 5.8), and that the transition state is situated in the middle, so the barrier increases by $zF\Delta\phi/2$ on one side and is reduced by the same amount on the other. If we incorporate the purely chemical term according to Eq. (6.24) in k_A then we obtain

$$\vec{\overline{k}}_A = k_A \exp - \frac{z_A F\Delta\phi}{2RT}, \tag{6.29}$$

$$\overleftarrow{\overline{k}}_A = k_A \exp + \frac{z_A F\Delta\phi}{2RT} \tag{6.30}$$

and, hence, a sinh function relates particle flux density and voltage:

$$J_{TA} = \frac{i_A}{z_A F} = k_A \widehat{N}_A \left(\exp - \frac{z_A F\Delta\phi}{2RT} - \exp \frac{z_A F\Delta\phi}{2RT} \right). \tag{6.31}$$

As in the case of the chemical reaction we only find a linear relationship if we neglect higher order terms in the expansion of the exponential functions. However, in contrast to the chemical reaction, this is a very reasonable approximation in many cases. On account of the decomposition voltage of a particular crystal, the voltage applied is generally restricted to an upper limit on the order of magnitude of 1V[26]. Let us take a typical sample dimension to be in the mm–range and a typical jump distance $\Delta x \simeq 1$nm, then the average drop over the jump distance is less than $\Delta\phi \simeq 10^{-6}$V, while even at 100K, RT/F itself is of the order of magnitude of 10^{-2}V. In boundary layers, however, the linear law may fail: The voltage drop may amount to several 100 mV over a fairly short distance. The same holds true in the case of very thin films or in cases in which very high voltages are applied. It is then necessary to take into account the high field law (Eq. (6.31)). Since normally one of the exponential terms in Eq. (6.31) dominates, the Tafel law [342] follows

n times the probability of any arbitrary but defined sequence (e.g. heads, heads, tails...) occurring is $\left(\frac{1}{2}\right)^n$. (This also makes a statement w.r.t. the time behaviour, since the interval between coin tosses defines the time frequency.) In order to find out the probability of m heads and n–m tails it is necessary to multiply this by the number of possible combinations. The result $\omega_{n,m} = \left(\frac{1}{2}\right)^n \binom{n}{m}$ is the probability that the random walker has made m steps to the right and (n–m) steps to the left, that is has come to 2m–n. By the use of the (detailed) Stirling formula it can be shown that $\omega_{n,m}$ goes over into a Gaussian distribution if n and m are large numbers, as it is described by Eq. (6.68), see, for example, Ref. [341]. The mean square displacement then follows from the definition $\int_{-\infty}^{+\infty} x^2\omega(x,t)dx$ to 2Dt in ^{1}D and analogously to 6Dt in ^{3}D.

[26] This corresponds to the order of magnitude of a free enthalpy of decomposition into the electrolysis products of $\sim$ 100 kJ/mol (z = 1).

rather than Ohm's law. The voltage is then a linear function of the logarithm of the current. However, quite different mechanisms may lead to such situations (see Section 6.10). Here we will assume that $|z_A F\Delta\phi| \ll RT$ and we get

$$i_A = -\frac{k_A \widehat{N}_A z_A^2 F^2}{RT}\Delta\phi \simeq -\frac{k_A (\Delta x)^2 \widehat{c}_A z_A^2 F^2}{RT}\frac{d\phi}{dx}, \qquad (6.32)$$

that is Ohm's law ($i_A = -\sigma_A d\phi/dx$). Comparison with Eq. (6.27) via k_A again yields the Nernst–Einstein equation (Eq. (6.14)). Once again it should be emphasized that σ_A is proportional to the equilibrium concentration $\widehat{c}_A$, whose functionality we have discussed sufficiently in Chapter 5, and to the mobility (proportional to k_A), which is thermally activated via the hopping barrier.

If both concentration and electrical potential influences are of significance, then we must return to the general Eq. (6.23). Since, at this point, we are considering particles at infinite dilution and, hence, c_A can be replaced by $\exp\left(\frac{\mu_A - \mu_A^\circ}{RT}\right)$, we find again after linearizing, our thermodynamic relationship

$$J_{TA} \propto -\Delta\widetilde{\mu}_A \qquad (6.33)$$

which is identical to Eq. (6.21) if we replace the affinity by $-\Delta\widetilde{\mu}_A$.

If we assume that $|z_A F\Delta\phi| \ll RT$, an instructive approximation[27], namely

$$J_{TA} \propto -\frac{z_A F}{RT}\widehat{c}_A\Delta\phi - \Delta c_A - \left(\frac{z_A F}{2RT}\Delta c_A \Delta\phi\right) \qquad (6.34)$$

is obtained from Eq. (6.23). The term in brackets illustrates, in the form of a correction, the occurrence of mixed effects and extends beyond the region of linear, irreversible thermodynamics. Even though illustrative, for it emphasizes the joint effect of both driving forces, the correlation is without quantitative significance since neglible by presupposition ($z_A F\Delta c_A \Delta\phi/2RT \ll \Delta c_A$).

Let us summarize the significant results:

First of all we have obtained an overview of the relationship between fluxes and driving forces, and have particularly investigated the validity range of linear relationships.

Fick's law is valid — with respect to deviation from equilibrium — over a wide range of conditions (but it is restricted to the dilute state), Ohm's law is limited in validity to the bulk and not too small samples, but it is then generally applicable. In the case of transport processes involving both electrical and concentration effects, the fundamental relationship

$$\mathbf{j}_k = -\frac{\sigma_k}{z_k^2 F^2}\nabla\widetilde{\mu}_k \qquad (6.35)$$

[27] It is striking that this term is quadratic in the driving forces and hence breaks the symmetry (about the origin) of the linear law. This is a consequence of the one–sided change of the driving force in Eq. (6.34). Employing a symmetrical master equation ($c_A \rightarrow \widehat{c}_A - \Delta c/2$; $c'_A \rightarrow \widehat{c}_A + \Delta c/2$) the even power terms disappear and a cubic mixed term reflects, as the decisive higher order approximation, the effect of large driving forces. For details see Ref. [329].

is relevant, and extensive use will be made of it in this chapter.
Furthermore, the relationships between the transport coefficients have become evident, enabling an interpretation in the atomistic picture. Rate constants of the hopping process, mobility and the diffusion coefficient of the hopping particle (point defect) are closely related parameters. Equation (6.35) emphasizes the importance of the specific conductivity as a transport parameter, which extends beyond its role as a valuable measurement parameter and an electrical material property. Equation (6.32) demonstrates that (close to equilibrium) it is proportional to the equilibrium concentration of the defect under consideration and its mobility. The proportionality to $\widehat{c}$ was exploited extensively in Chapter 5 for experimental verification of defect chemistry.
Typical chemical reactions (unless they correspond to very small perturbations) are better treated by master equations than by irreversible thermodynamics. This is also the case for a charge transfer process through an interface. Such a transfer is, due to the structural variations, generally related to a change in the standard potential; moreover, at electrodes, a true electrochemical reaction may occur. In these cases, the purely chemical barrier is asymmetrical because $\overrightarrow{\Delta G}^{\neq} - \overleftarrow{\Delta G}^{\neq} = \Delta_R G^\circ \neq 0$. In the case in which the total rate of the process is controlled by chemical effects (e.g. interfacial reactions), we must resort to detailed reaction kinetics. The Butler–Volmer equation and its analogues concerning tracer and chemical processes, which are relevant in such cases, are discussed in detail in Sections 6.7.3 and 7.3.3.

6.2 Electrical mobility

6.2.1 Ion mobility

The actual kinetic parameter, which is discussed in what follows, is the (electrical) mobility u_k of the defect k; it is proportional to the diffusion coefficient of the defect and also proportional to the jump frequency or to the jump rate constant, and as these activated via the migration barrier. More precisely, according to Eq. (6.15), $u_k T \propto D_k \propto k_k \propto \Gamma_k$. The parameters $u_0 T$ and D_0 which are the prefactors to the Boltzmann–term $\exp(-\Delta G^{\neq}/RT)$, are proportional to the attempt frequency $\Gamma_0 \propto k_0$; the Boltzmann–term corresponds to the number of successful jumps. Typical attempt frequencies which are of the order of $10^{13} s^{-1}$ are, owing to a lack of detailed insight, often identified with the Debye frequency[28] (here we refer to the vibrational properties discussed in Chapter 3). The slight difference in the T–dependence of u and D (or k, Γ) reflected by Eq. (6.15) can normally be neglected, with respect to the exponential activation; T dependences on the same order of magnitude have already been neglected in Eq. (6.18) in k_0.

[28] According to the transition state theory [337] the prefactor is given by kT/h. When we set $T = \Theta_D$ we find the direct connection with the discussion above. For $\Theta_D \simeq 500K$ we get $10^{13}/s$.

Overall, we resume

$$uT = \text{const}\Gamma_0 \left(\Delta x\right)^2 \exp \frac{\Delta S^{\neq}}{R} \exp -\frac{\Delta H^{\neq}}{RT}. \tag{6.36}$$

Let us now consider, in what follows, the three elementary jump mechanisms [343]: In the case of the vacancy mechanism in Fig. 6.4a a regular particle (A_A) hops into the vacancy of effective charge number z_v which is of opposite sign to the ionic charge of particle A, and leaves behind a vacant position ($V_A^{z_v}$):

$$\text{Reaction V} = \qquad A_A(x) + V_A^{z_v}(x') \rightleftharpoons V_A^{z_v}(x) + A_A(x'). \tag{6.37}$$

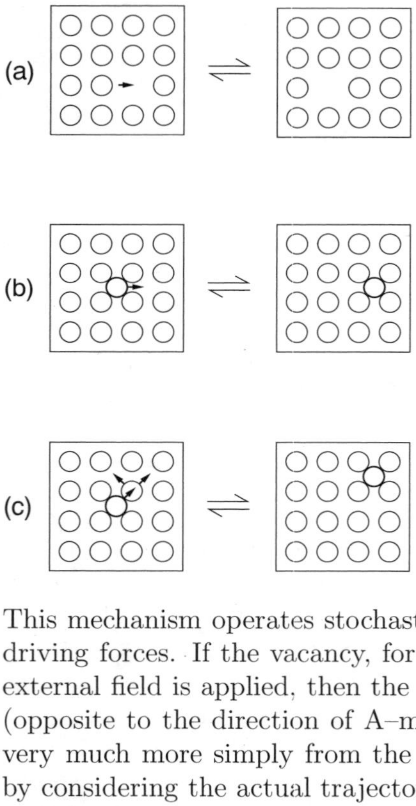

Fig. 6.4: Elementary jump mechanisms in crystals: a) vacancy mechanism, b) direct interstitial mechanism, c) (collinear or noncollinear) indirect interstitial mechanism (interstitialcy mechanism).

This mechanism operates stochastically in the absence of correlations and external driving forces. If the vacancy, for instance, is effectively positively charged and an external field is applied, then the vacancy migrates to the side of the negative pole (opposite to the direction of A–migration). This mass transport can be described very much more simply from the point of view of the vacancy defect, rather than by considering the actual trajectories of the substantial A–particles involved.
Two mechanisms are of importance in the case of the migration of an interstitial particle: (i) The defect (charge number z_i) jumps directly from one interstitial site to the next (Fig. 6.4b)

$$\text{Reaction i} = \qquad A_i^{z_i}(x) + V_i(x') \rightleftharpoons V_i(x) + A_i^{z_i}(x'), \tag{6.38}$$

or (ii) it ($A_i^{z_i}$) pushes a neighbouring regular particle $\underset{\sim}{A}$ (in a collinear or noncollinear way) into a vacant interstitial position and takes its former place (Fig. 6.4c):

$$\text{Reaction ic} = \qquad A_i^{z_i}(x) + \underset{\sim}{A_A}(x^*) + \vee_i(x') \rightleftharpoons \vee_i(x) + A_A(x^*) + \underset{\sim}{A_i^{z_i}}(x'). \qquad (6.39)$$

It is expected that, in the case of large ions in close–packed structures, the second (interstitialcy or indirect interstitial) mechanism is associated with a lower transition barrier compared to the direct interstitial mechanism which involves pushing aside neighbouring A particles (see Fig. 6.4). We can see that the electrochemical information is then transmitted from particle to particle, and a single substantial particle under consideration does not move any great distance, just as in the vacancy mechanism[29].

It is also important, and this is clear immediately when rate equations are applied to the bimolecular reactions (6.37) and (6.38) or to the even more complex reaction (6.39), that — more rigorously — the concentrations of the regular structural elements are also involved in the mobility; we expect a factor $(1 - c/c_{max})$ to appear, which takes account of the probability of finding a jump partner, and which is only constant in the dilute state[30]. In such a dilute state one can take the effect of the regular constituents (e.g. $[A_A(x)]$ and $[A_A(x')]$ in Eq. (6.37)) as constant and incorporate their concentrations in the rate constants. Then one indeed obtains first order reactions as assumed in Section 6.1.2. Further simplifications can be made if the above jump reaction is composed of several elementary reactions. (Such kinetic aspects are taken up again in Section 6.7.)

The energetic barriers ($\Delta H^{\neq}$), which determine the thermal activation of the mobilities are subject to large variations; typical values for ionic conductors are between 0.2eV and 2eV. Figure 6.5 shows the mobilities as a function of temperature for

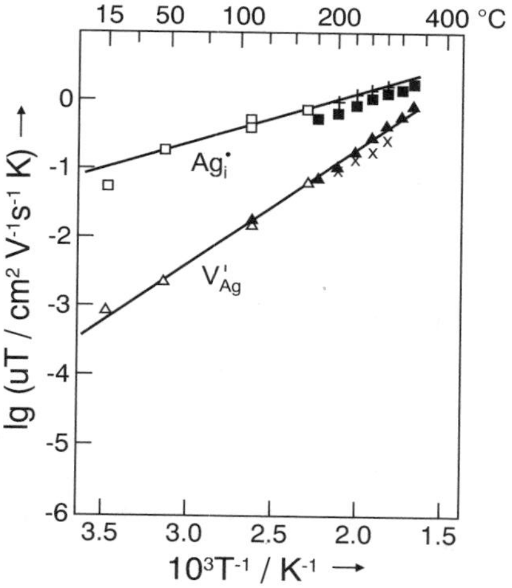

Fig. 6.5: The mobility of the Frenkel defects in AgBr as a function of the temperature (according to Eq. (6.36)). From Ref. [344].

[29] It is advantageous to conceive the transport processes as re–arrangement reactions. Then the reaction path connects the initial state, i.e. crystal in which A at site x and $\vee$ at site x′, with the final state, i.e. a crystal in which $\vee$ at site x and A at site x′.

[30] Compare the site restriction in the thermodynamic treatment (Section 5.7.2). Notice also that the above rate equations are — strictly speaking — restricted to $c \ll c_{max}$.

both Frenkel defects in AgBr. Table 5.1 (page 114) gave a collection of $\Delta S^{\neq}$ and $\Delta H^{\neq}$ values for a series of halides. It is an obvious and very useful rule of thumb that it is advantageous as far as ionic mobility is concerned, either for the migrating ion itself (cf. Ag^+ for example, and compare AgCl with KCl) or for its counterion (cf. LiI and compare it with LiCl) to have a high polarizability, i.e. to be soft and deformable[31,32].

In special cases enthalpies of migration are $\lesssim 0.1$eV and the defects are very mobile (e.g. interstitial mobility of silver halides). If such high mobilities are accompanied by high charge carrier concentrations (low enthalpy of formation, extreme disorder)

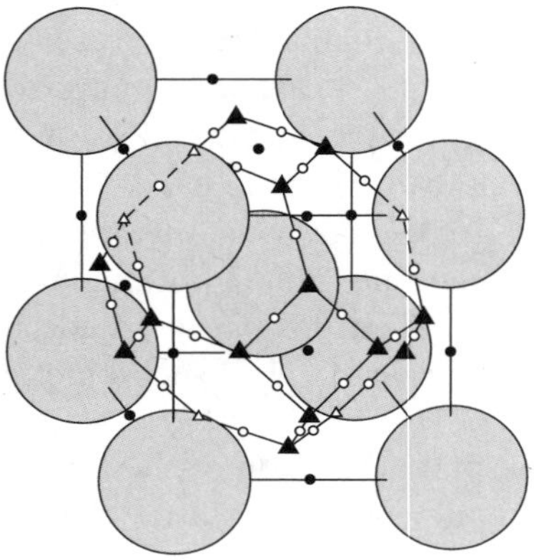

Fig. 6.6: Crystal structure of α–AgI. The iodide ions are represented by large spheres. The multiplicity of (not precisely) energetically identical sites for Ag^+ ions and the low activation energy for site exchange compared with the thermal energy when T > 146°C lead to a "melting" of the Ag^+ partial lattice. According to Ref. [347].

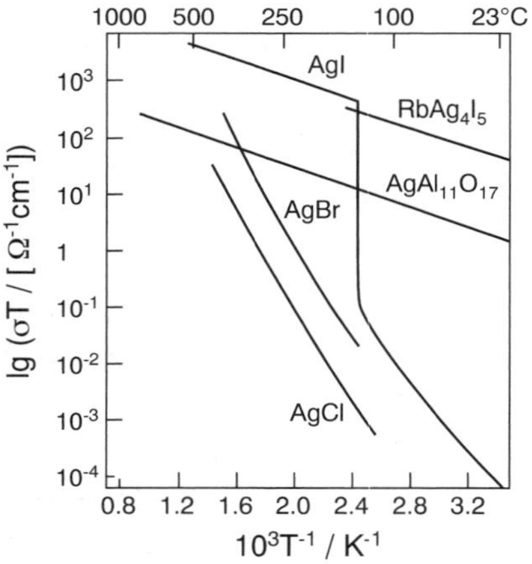

Fig. 6.7: The temperature dependence of the specific conductivity for a series of selected silver ion conductors. ($AgAl_{11}O_{17}$ reads more correctly $Ag_{1+x}Al_{11}O_{17+x/2}$). The "superionic conductors" are characterized by flat slopes and high absolute values.

[31]Such dynamic softness can also be associated with intermediate valence change. So it is conceivable that a transition metal ion (e.g. Cu) could intermediately change its valence during transport and, thus, reduce the activation energy (cf. coordination).

[32]More details on ion conductors can be found in Refs. [13,345,346].

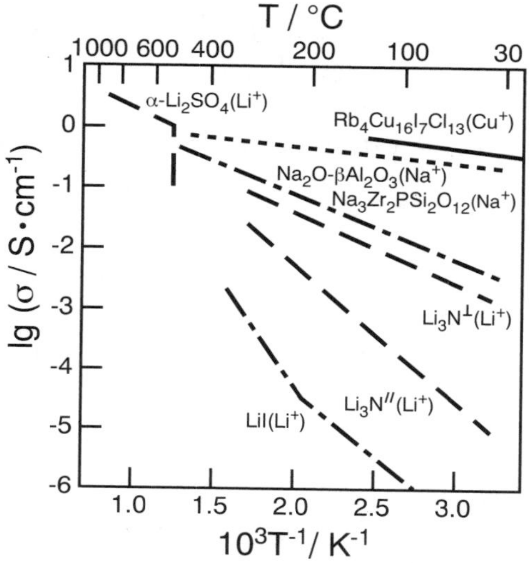

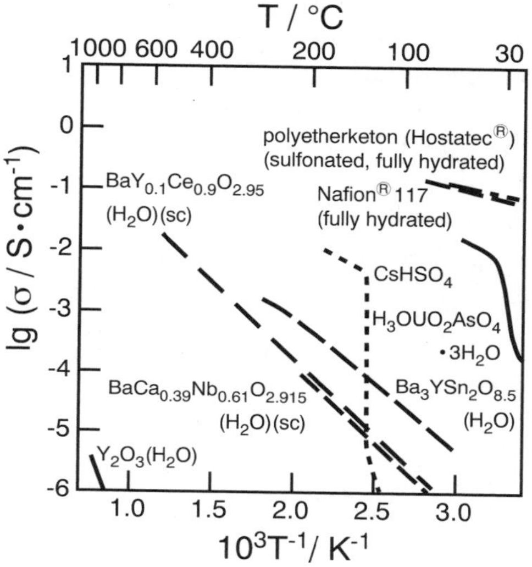

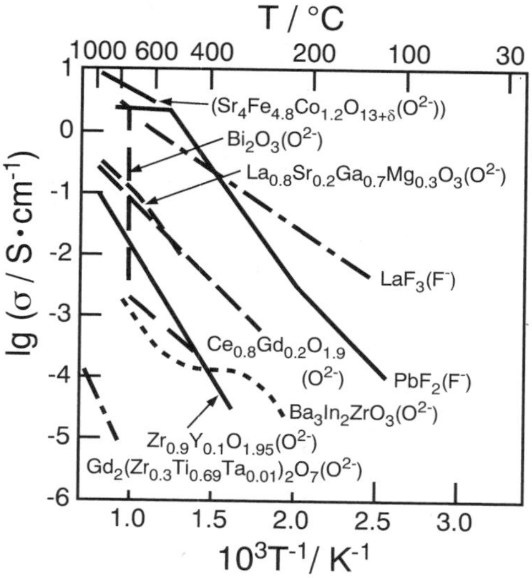

Fig. 6.8: Specific conductivity as a function of the temperature for a selection of compounds. Top left: cation conductors (Li^+, Na^+, Cu^+). With respect to Ag^+ conductors see Fig. 6.7. Top right: proton conductors. Bottom: anion conductors (O^{2-}, F^-) [348]. Very high O^{2-} conductivities are met in some perovskites, typically in doped ferrates [349]. These materials (cf. $Sr_4Fe_{4.8}Co_{1.2}O_{13+\delta}(O^{2-})$) are, however, mixed conductors and hence excellent materials for permeation membranes (cf. also Chapter 7). For more detailed data collections see Refs. [345,346].

we speak of superionic conductors[33]. Naturally the crystallography is all important in superionic conductors and conductivity effects are sometimes markedly anisotropic[34]. In the isotropically conducting α–silver iodide (Fig. 6.6) the silver ions have effectively more "interstitial positions" available than there are Ag^+ ions [347,352] connected by very flat potential profiles. Consequently, the silver ions are more or less statistically distributed; we say that the silver sublattice is "molten" (see also Section 5.7.2). It is illuminating that the entropy of α–AgI on transition to the liquid state (also iodine sublattice molten) increases by the same order of magnitude as on transition from the weakly disordered β–phase into the α–phase (14.5 and 11.3 $JK^{-1}mol^{-1}$ respectively). The high conductivity is isotropic but, on account of the phase change, it is only present above 146°C. (At lower temperatures the silver ions have to "decide on particular positions".) In the case of $RbAg_4I_5$ high conductivities are still maintained down to room temperature[35]; however, the compound becomes unstable with respect to elementary iodine, which is a disadvantage for its use as a battery electrolyte (see Chapter 7) [353,354]. Figure 6.7 gives an overview of various silver ionic conductors. A conductivity record of $0.3\Omega^{-1}cm^{-1}$ at

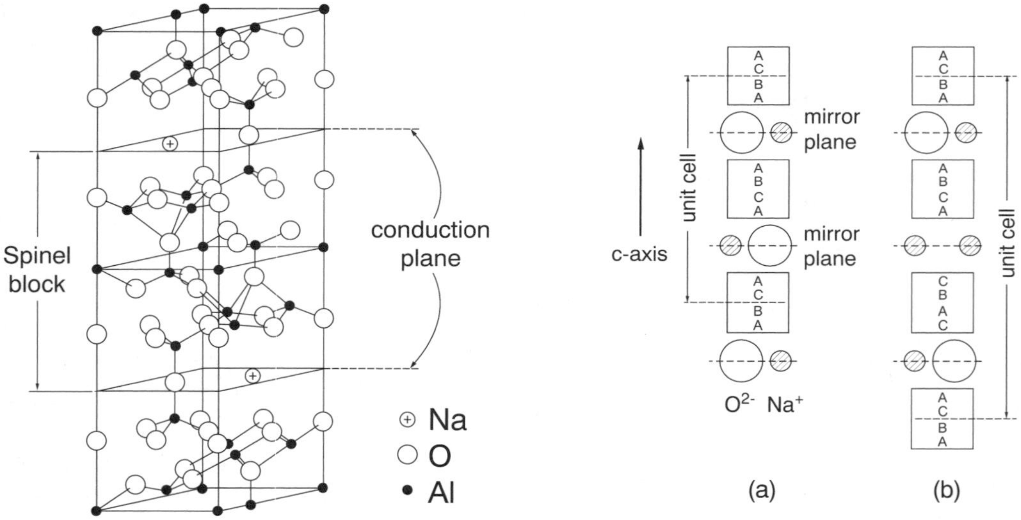

Fig. 6.9: Crystal structure of β–alumina ($Na_2O \cdot 11Al_2O_3$) (left). This is shown schematically on the right (a) and compared with the more complex β″–alumina structure (b) [355]. From Refs. [48,356].

[33]A reduction of the effective enthalpy (slope) is frequently associated with a reduction of the prefactor (intercept, cf. Fig. 6.7). This is referred to as the compensation rule or Meyer–Neldel rule (see also Refs. [212,350]). A temperature dependence of the activation energy is also relevant in this context [351].

[34]Concerning phase transformations in ion conductors cf. Ref. [234].

[35]Similarly Bi_2O_3 shows a very high oxygen conductivity (see Fig. 6.8) only at rather high temperatures, which one attempts to stabilize to lower temperatures by compositional variation.

25°C is held by $Rb_4Cu_{16}I_7Cl_{13}$ (cf. Fig 6.8, [357]). Note, in particular, the absolute value and slope of these superionic conductors in comparison to the ionic conductors (AgCl, AgBr, β–AgI) describable in terms of the dilute defect model (see Sections 5.5.1–5.7).

The family of β– and β''–alumina containing foreign oxides are excellent ionic conductors for a number of cations, such as, for instance, $(Na_2O)_{1+x}(Al_2O_3)_{11}$ ("sodium β–alumina") ($x > 0$) for Na^+ ions [355,356]. $Na_{1.2}Al_{11}O_{17.1}$ is a typical composition. The Na_2O is situated in the layers between the spinel blocks (Fig. 6.9). At higher temperatures the cations (see Fig. 6.9) are distributed over different crystallographic sites in these "conduction planes" and can be regarded as being in a partial "molten" state like the Ag^+ ions in α–AgI. The β''–aluminium oxides also contain divalent ions situated on Al^{3+} sites, the charge deficiency is then balanced by further Na^+ ion incorporation (e.g. $Na_{1+x}Mg_xAl_{11-x}O_{17}$), leaving the oxygen content essentially invariant. Here, too, the excess sodium ions are very mobile. The crystal structure is more complex (see Fig. 6.9). The increased conductivity compared with β–Al_2O_3 is only partially attributable to the higher Na^+ concentration [356] as the lower activation energies show. In a similar manner dissolutions of Ag_2O, K_2O, PbO, H_2O, etc. can lead to Ag^+, K^+, Pb^{2+}, H^+ conductors (see Fig. 6.8) [355,358].

The conductivity being confined to two dimensions represents a drawback to the use of polycrystalline material [359] (see also Section 6.6). Its highly brittle nature is another disadvantage of this ionic conductor. Crack formation in Na^+–conducting ceramics led to quite spectacular failure in the Na–S batteries (described in more detail in Chapter 7).

Li_3N is another excellent two–dimensional conductor, but unfortunately it is very reactive [360]. Nasicon is an acronym for a family of excellent sodium ion conductors

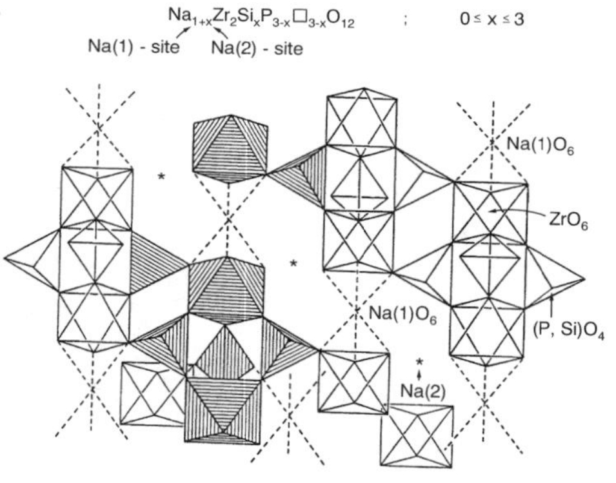

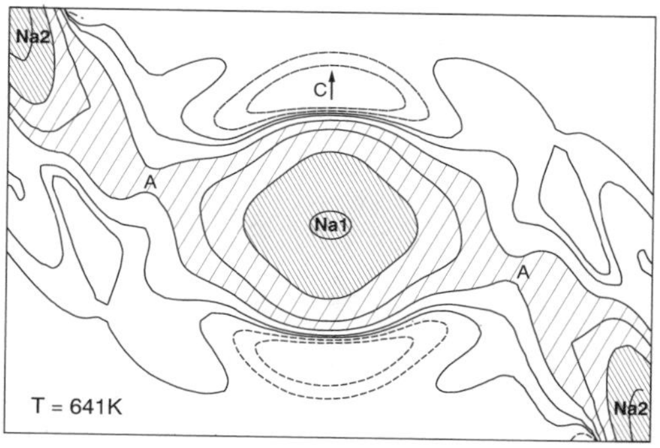

Fig. 6.10: (Left) Schematic structure of "Nasicon" ($Na_{1+x}Zr_2P_{3-x}Si_xO_{12}$). The number of sodium vacancies per formular unit is (3 - x) that of the Na(2) sites is x. To the right is a conductivity channel obtained from temperature–dependent X–ray measurements. Both Na(1) and Na(2) positions are necessary for charge transport. The migration bottleneck is largest when x $\simeq$ 2 (right). From Refs. [361,362].

(Na–SuperIonicCONductor) with the empirical formula $Na_{1+x}Zr_2P_{3-x}SiO_{12}$ [363]. The high conductivity takes place along channels but is isotropic overall on account of their three–dimensional distribution [363,361,364,365] (Fig. 6.10). When $x \simeq 3$ all Na positions are occupied and the conductivity is low. When $x < 3$ there is underoccupation. At $x \simeq 2$ the migration bottleneck is widest for this composition and the conductivity is at a maximum, almost reaching the value obtained with β''–Al_2O_3 (see Fig. 6.10). In addition Nasicon is less brittle and cheaper to synthesize than the aluminium oxides above; unfortunately highly conducting P–containing compounds are not stable with respect to elementary sodium, which would be desirable for application in high power batteries [365].
One–dimensional conductors with conduction channels parallel to each other, as realized in the hollandite structure in Fig. 6.11, are essentially of academic interest. Generally the conductivity depends very strongly on the specific structural environment[36]. This also applies to ionically conducting glasses and polymers. Polar

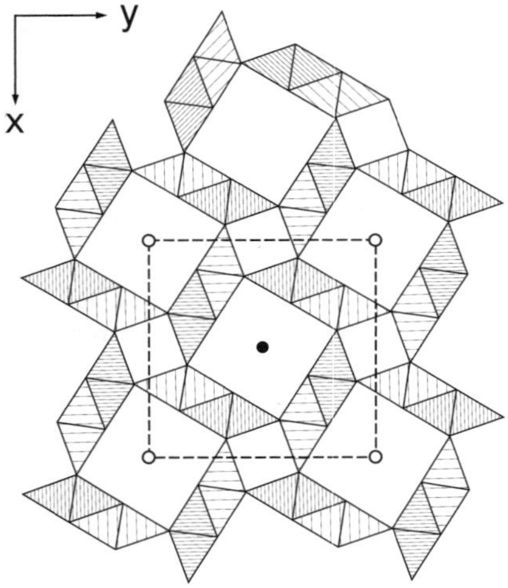

Fig. 6.11: The hollandite structure (section along the c axis) is primarily adopted by compounds of the composition $Ba_xMn_{8-x}O_{16}$. It exhibits channels within the framework of the MnO_6 octahedra, in which the Ba ions are to be found. $K_{1.6}Al_{1.6}Ti_{6.4}O_{16}$ with the same structure has a K^+ conductivity in the c direction (z axis) of the order of $10^{-2}\Omega^{-1}cm^{-1}$ at 400°C. From Ref. [346].

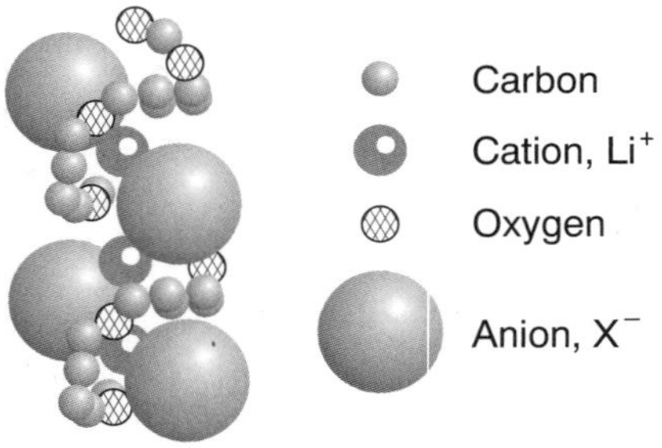

Fig. 6.12: Schematic illustration of the structure of a PEO–LiX complex (Li^+X^- dissolved in poly ethylene oxide [367]. (The relative magnitudes are not representative.). From Ref. [368].

[36]Thus, for Li_2SO_4 a paddle wheel mechanism has been proposed (with respect to the ability of the SO_4 tetrahedron to rotate), in which the rotation of the anion directly effects the cation migration [366].

polymers, such as polyethylene oxide, can solvate ions in a similar manner to aqueous solvents. Here correlation effects are very significant (see Section 6.6.1) [369]. In particular owing to the often covalent environment the tendency of Li^+ and X^- to associate is strong. The reduction of this might be one explanation of the beneficial effect of adding second–phase particles to polymers (see Section 5.8.5). The motion of the polymer segments plays an important role in the charge transport kinetics (see Fig. 6.12) [370]. The multiplicity of possibilities and structural aspects of that field is not treated further here.

It ought to be mentioned at this point that proton mobility[37] occupies a special position in the field of ion transport. As naked elementary particles the protons polarize their surroundings very strongly, so that they do not migrate in this "naked" state. Rather they "are conveyed" by their surroundings. In compounds such as hydronium uranyl arsenate (see Fig. 6.8) it is a vehicle such as H_2O that transports the protons in the form of H_3O^+. In the case of the H_2O–containing oxides (see Fig. 5.5) discussed in Chapter 5, in which internal hydroxyl OH groups are formed, it is oxygen vibrations combined with hydrogen bond formation. The succession of forming and breaking hydrogen bonds being responsible for the detachment and further transport of the protons (see Fig. 6.13) and constituting proton migration, also characterizes the transport in many proton conducting polymers [374]. Proton tunnelling[38] can play a role over short distances and at low temperatures [375].

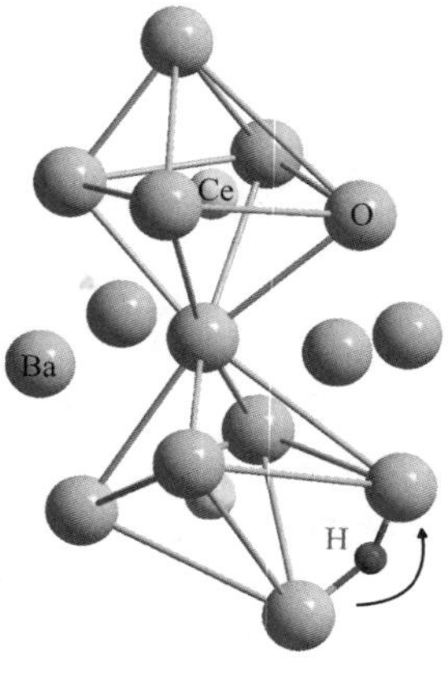

Fig. 6.13: Proton conduction in water–containing $BaCeO_3$. As oxygen ions move towards each other on account of lattice vibrations, the activation energy for the proton jump is lowered, and the proton changes partner. The initial situation was shown in Fig. 5.5 (page 116) (however with the proton localized in the centre of the figure). From Ref. [120].

[37] Compare, for example, [202,371–373].

[38] The delocalization of the wave function corresponds to the ability of the electrons to tunnel through local barriers. Application of Schrödinger's equation for a quantum mechanical object of mass m and energy ϵ to a single potential well of height V and width a leads to a finite probability of an electron to traverse the barrier even if $E < V$. This probability decreases exponentially with increasing $V - E$, a^2 and m [23]. The effect is closely related to the energy level interaction on forming bonds and bands (cf. Section 2.1.1). According to the above, the tunnelling probability for the proton is much less.

6.2.2 Electron mobility

Until now we have mainly treated electrons and holes analogously to the ionic defects. As far as the mobility is concerned, quantum mechanical effects cause severe differences. There is no energy of activation ($\Delta H^{\neq} = 0$) in the case of perfect band conduction[38] and, formally speaking, the temperature dependence of the mobility is effectively determined by the prefactor. The determining process for the finite mobility is scattering by lattice vibrations and/or imperfections. The $T^{-3/2}$ relation for acoustic phonon scattering is a typical law[39] in this context (see Chapter 3) (see Fig. 6.14). Unless the electronic charge concentration has been fixed by dop-

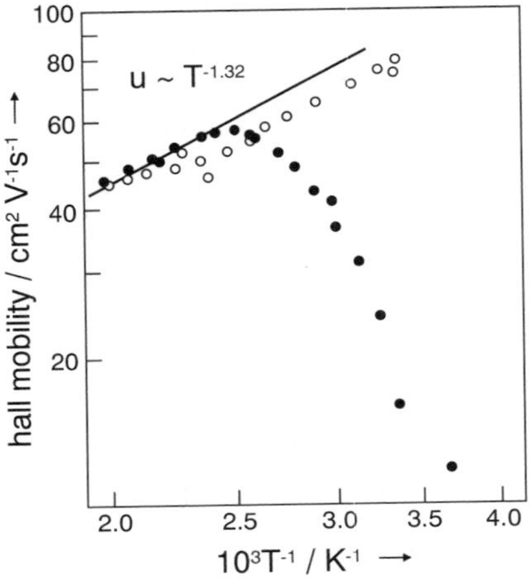

Fig. 6.14: The mobility of the excess electrons in various SnO_2 samples determined by means of the Hall effect and conductivity. The high temperature behaviour points to acoustic phonon scattering. Both samples differ in purity. According to Ref. [376].

ing, this weak T dependence is generally negligible compared to the (exponential) temperature dependence of the concentration[40]. The reader should note that the pre–exponential terms that stem from the concentration are weakly temperature–dependent ($\propto T^{+3/2}$), too, as a result of the Eq. (5.49).
When orbital overlap is not too marked and the bands are not very broad (cf. β, band width, Chapter 2) there is significant interaction between the lattice and the electrons. The electrons (or holes) then polarize their environment (see Section 5.3). The "electron + distortion field" state is known as a polaron. The semiconductor InSb is a typical example of a solid containing "large polarons". Here the effective mass is increased very slightly, the mobility is not greatly reduced, and the band model for transport is a good approximation, in short the polarization effect is not too strong. Typical mobilities are of the order of 10^0 (alkaline earth titanates) and

[39] No elastic approximation is possible on interaction with optical phonons and the relationships are specific. In the case of scattering at shallow defects the mobility is typically proportional to $T^{+3/2}$. See, for example, [128,129].

[40] In the case of metals too there is scatter due to the lattice vibrations (when $T \gg \theta_D$ then $u \propto 1/T$) and due to the defects (constant resistance contribution). The latter contribution dominates the resistance at low temperatures.

$10^2 cm^2/Vs$ compared to typical values of $(10^2 \ldots 10^4) cm^2/Vs$ for band conduction. It is often found that $u_0 \propto T^{-1/2}$ for large polarons. Activation energies are typically small ($\Delta H^{\neq} \sim 0.1$ eV).

The polarization effect is considerable in markedly ionic crystals[41], such as NaCl or AgCl. The electron (or hole) can be regarded as largely localized, and transport at high temperatures actually takes place by thermally activated hops similar to those described for ions (however, tunnelling[38] is significant at low temperatures). Activation energies can be as high as 0.5 eV (or more) for such "small polarons". Effective mass and mobility (typically $10^{-4} \ldots 10^{-2}$ cm^2/Vs) are greatly reduced in comparison to band conduction. The electron jump requires the presence of an excess electron, i.e. of a reduced state or of an electron hole, that is of an oxydized state. In the hole mechanism a regular electron from the neighbouring atom jumps into the "a vacancy of the electron shell" of the central atom. In a manner completely analogous to that discussed for ions, the concentration of the jump partners is also important, so that terms of the form $c\,(c_{max} - c)$ must be taken into account.

As described in detail, electrons and holes can be localized at dopant ions (cf. Section 5.7.1). If the latter are close neighbours, orbital overlap causes band formation and polaron band conduction[42] (see above). In general, since σ is proportional to u and c, a distinction must be made between doping effects on the mobility and those that result from the defect concentration (disorder and ionization equilibria). It is frequently not simple to differentiate between ionization effects from the defect levels with intermediate band conduction and polaron processes from impurity to impurity (see e.g. [377]).

Thermally activated hopping and tunnelling[38] are of great importance for the electronic conductivity of amorphous, inorganic semiconductors, but also for conducting polymers. A very extended, one–dimensional delocalization, in the sense discussed in Chapter 2, may be expected in conjugated hydrocarbons, but is not possible on account of the Peierls distortion [378] (see page 44). Therefore, high conductivities[43], e.g. in polyacetylene (see Fig. 6.15) require doping, i.e. partial oxydation or reduction. In many cases the conductivity can be described by simple hopping models[44].

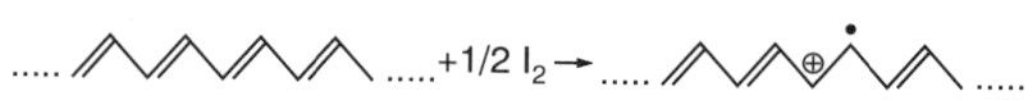

Fig. 6.15: The doping of polyacetylene with iodine leads to the formation of a hole. Iodide (or polyiodide) is the counterion.

[41] Transition metal oxides take up an intermediate position.

[42] According to Mott's conception [47] the transition from delocalization to localization (e.g. metal–insulator transition, transition from isolated states within the band gap to impurity bands) occurs when the mean distance exceeds a certain critical value ($\propto c^{1/3}$) — the distance of the effective Bohr's radius (more precisely 4 × Bohr's radius). The critical behaviour exhibits similarities to that treated in Section 5.7.2.

[43] The internal production of radicals, carbanions or carbocations in polymers also leads to intrinsic polarons or solitons. The charged states lie between the HOMO-LUMO states on account of interactions with the environment, i.e. in the band gap (see, for example, [379]).

In the case of heavy doping, the conductivity can become metallic and comparable to that of the solid metal elements ("one–dimensional metals").

Superconductivity is an important extreme case in which quantum mechanics is of direct and paramount importance[45]. In classical superconductors, "associates" (see also page 207) of the type $(e')_2$ or $(h^{\cdot})_2$, are stabilized by phonon coupling. These Cooper pairs behave as bosons (spin zero), they are at low temperatures correlated by a common wave function, the dynamics of which are no longer hindered by the phonons[46]. In this state, the superconducting state, the nominal mobility is infinite (see Fig. 6.16). The great importance of high temperature superconductors [382]

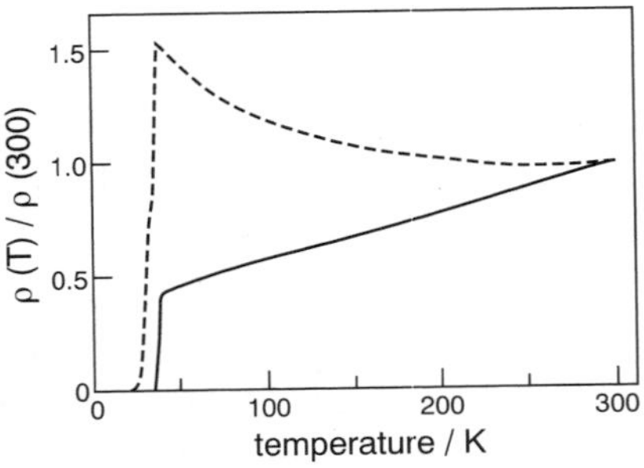

Fig. 6.16: The resistivity of SrO–doped La_2CuO_4 disappears at temperatures less than $T_C \simeq 40K$. The $(h^{\cdot})_2$ "associates" (Cooper pairs) are responsible for this. The two curves are based on samples, that are annealed at differing partial pressures of oxygen [381]. Compare here Sections 5.5, 5.6. From Ref. [382].

(such as $YBa_2Cu_3O_{7-\delta}$) lies in the occurrence of this phenomenon at temperatures that can be reached with liquid nitrogen. Additional electronic correlations are being discussed as possible coupling mechanisms[47].

6.3 Phenomenological diffusion coefficients

The diffusion coefficients of the migrating particles discussed above (proportional to the mobilities of ionic or electronic defects, see Eq. (6.15)) are not directly accessible experimentally to measurement and are sometimes termed "microscopic" diffusion coefficients. It is a complication, in principle, that, on account of the requirement of electroneutrality, we cannot simply move the charge carrier, under consideration, through the solid. However, there are various possibilities for measuring phenomeno-

---

[44] Variable range hopping leads to $\sigma \propto \exp - \left(\frac{T_0}{T}\right)^{\gamma}$, with γ typically 1/4 [47]. On account of the scatter of the energy levels in amorphous systems and of the temperature–dependent availability of phonons, the hopping rate, thus, depends on T as described.

[45] Compare, for example, Ref. [380].

[46] Sometimes one is tempted to assume that, while the distribution of scientists (given a finite number of jobs to do) should obey Fermi–Dirac statistics, the distribution of lawyers seems to be governed by Bose–Einstein statistics: An indefinite number can be concerned with one and the same problem with the tendency to attract even more (cf. page 216). A partial analogy to superconductivity lies in the fact that lawyers sticking tightly together do not experience serious resistance.

[47] Compare, for example, Ref. [383].

logical ("macroscopic") diffusion coefficients, which can be interpreted in terms of the microscopic parameters[48].
Let us consider the three principal solutions (Fig. 6.17) to this "electroneutrality dilemma". The three respective diffusion coefficients to be derived have to be carefully differentiated.

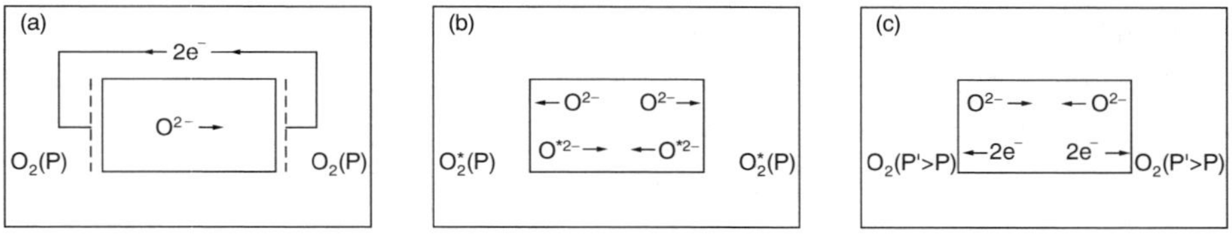

Fig. 6.17: Three different types of electroneutral charge carrier movements: a) stationary conductivity experiment, b) tracer experiment c) chemical diffusion experiment. From Ref. [384].

6.3.1 Ion conduction and self–diffusion

It is naturally possible to maintain a flux of ions (or electrons) within the solid if an equivalent electron flux passes through an external circuit (Fig. 6.17a). This is the principle of a stationary conductivity experiment. We are not interested here in how we avoid or correct for interfacial and other polarization effects, or in how we separate ionic and electronic conductivities (see Chapter 7). (It is clear that we measure the sum of electronic and ionic conductivities when applying reversible electrodes.) Neither do we consider the transients (but see Chapter 7). Since the experiment we are referring to, involves reversible electrodes and sufficiently small currents so that the composition is not affected, the stationary current is given by the electrical field term (Ohmic term) alone ($\mathbf{i} \propto -\sigma\nabla\phi$, see Eq. (6.12)). For simplicity's sake let us consider an ion–conducting oxide (M_2O) and assume that we have established that the ionic conductivity is solely due to the oxygen ions. As we already know, in order to calculate the defect mobilities or the defect diffusion coefficients, we require knowledge of the defect concentrations. If we do not explicitly have this knowledge, we can nevertheless, in analogy with Eq. (6.14), formally convert the oxygen ion conductivity into a diffusion coefficient, according to

$$D^{Q}_{O^{2-}} \equiv \frac{RT}{4F^2}\frac{\sigma_{O^{2-}}}{c_{O^{2-}}}. \tag{6.40}$$

[48]We shall not consider additional complications as a result of different reference systems [7, 333] and will always refer to the lattice of immobile counterions. Neither do we intend to give a systematic overview on the variety of relevant measuring techniques. Besides the ones mentioned, there are various further electrochemical (Chapter 7), thermoelectric, magnetoelectric (Hall effect), spectroscopic (NMR), etc. methods which are relevant as such or in appropriate combinations.

It thus refers[24,25] immediately to the transport of charge (hence, the index Q)[49], and corresponds to the self–diffusion coefficient of the ions. According to our presumed lack of knowledge of the defect concentration, we obtain a diffusion coefficient that averages over all oxygen ions, that is both the immobile, regular particles and the rapid defect particles (for which we derived Eq. (6.14)). Since, under normal circumstances, point defects are only statistically identifiable, this ionic self–diffusion coefficient has a meaning in its own right. $D^Q_{O^{2-}}$ is, obviously, considerably smaller than the diffusion coefficient of a mobile defect. (Formally it is possible in the same manner to define a mean ionic mobility via $2FD^Q_{O^{2-}}/RT$.) If we take into account the actual mechanism of migration in $\sigma_{O^{2-}}$ via

$$\sigma_{O^{2-}} = \sigma_{O_i''} + \sigma_{V_O^{\cdot\cdot}} = 2F\left(u_{O_i''}c_{O_i''} + u_{V_O^{\cdot\cdot}}c_{V_O^{\cdot\cdot}}\right), \tag{6.41}$$

we obtain, using Eq. (6.40) and Eq. (6.41), the desired relationship between D^Q and the microscopic parameters. If, for the sake of simplicity, we assume that the ion flux only occurs via interstitial defects, we can simply state:

$$D^Q_{O^{2-}} = \frac{c_{O_i''}}{c_{O^{2-}}}D_{O_i''} = x_{O_i''}D_{O_i''}. \tag{6.42}$$

Eq. (6.41) and subsequently Eq. (6.42) were obtained from the conclusion that the absolute values of ion flux and of defect flux must be equal to each other ($\sigma_{O^{2-}} \propto j_{O^{2-}}/\phi'$, $\sigma_{O_i''} \propto j_{O_i''}/\phi'$).

Determination of the defect diffusion coefficient (D_k) according to Eq. (6.42) requires knowledge of the defect concentration (c_k). As can be seen from the above relationships, the temperature dependence of $D^Q_{O^{2-}}$ is given by the temperature dependence of $\sigma_{O^{2-}}$ and, in contrast to D_k and u_k, the enthalpy of defect formation is also included.

Doping experiments, thermoelectric and Hall effect experiments[50] provide powerful methods for separating the charge carrier concentration and mobility.

6.3.2 Tracer diffusion

A second fundamental experiment (Fig. 6.17b) consists of exchanging the ambient oxygen surrounding an oxide (which is usually an isotope mixture primarily made up of $^{16}O_2$) with a gas enriched with either $^{17}O_2$ or $^{18}O_2$. The consequence of a difference between the tracer content of the gas and that of the solid phase is a tracer exchange in the solid. As in conductivity experiments the chemical composition is not altered. Obviously this tracer diffusion is a counterdiffusion of the two oxygen isotopes[51].

[49] The formally obtained diffusion coefficient, described by Eq. (6.40), is also termed the "conductivity diffusion coefficient" D^σ. We prefer the index Q, since we proceed similarly (Section 6.7) for the effective rate constant of the interfacial processes. Please note that a conductance experiment with reversible electrodes is not a diffusion experiment (concentrations do not change).

[50] Compare, for example, Ref. [385].

[51] It is important to avoid stoichiometric gradients (see Section 6.3.3). There can be errors, if this point is not heeded, particularly when metal isotopes are exchanged in a compound by application of a metal film.

Since storage phenomena (here of the tracer content; cf. capacitive phenomena, Section 6.7.4) necessarily occur, this is — like the process in experiment c but unlike the one in experiment a (Fig. 6.17) — a true diffusion process. The driving force for the exchange comes solely from the configurational entropy. The isotope exchange involves the whole oxygen ensemble of the crystal and not, as is the case in experiment a, just the point defects, even though the mechanism of ion diffusion necessarily involves defects. Let us consider the flux of the ^{18}O particles (that is $\mathbf{j}_{^{18}O} = \mathbf{j}_{^{18}O^{2-}} = \mathbf{j}_{^{18}O_i''} = -\mathbf{j}_{^{16}O} = -\mathbf{j}_{^{16}O^{2-}} = -\mathbf{j}_{^{16}O_i''}$, if we assume here exclusively interstitial particles to be relevant[52]) and designate the parameters referring to ^{18}O with an asterisk. Then we get

$$\mathbf{j}^*_{O_i''} = -\frac{\sigma^*_{O_i''}}{4F^2}\nabla\mu^*_{O_i''}. \tag{6.43}$$

Electrical field effects do not occur and $\nabla\widetilde{\mu}^*_{O_i''}$ can be replaced by $\nabla\mu^*_{O_i''}$. Since the isotope distribution in the oxygen sublattice is purely random (mass effects are neglected), it follows that

$$\nabla\mu^*_{O_i''} = RT\nabla c^*_{O_i''}/c^*_{O_i''}. \tag{6.44}$$

The same relationships can also be attributed to the ions on account of local equilibrium and the random distribution of all labelled oxygens[53], hence

$$\mathbf{j}^*_{O^{2-}} = \frac{\sigma^*_{O^{2-}}}{4F^2}\nabla\mu^*_{O^{2-}} = \frac{\sigma^*_{O^{2-}}RT}{4F^2 c^*_{O^{2-}}}\nabla c^*_{O^{2-}}. \tag{6.45}$$

Naturally[54] $\sigma^*_{O^{2-}}/c^*_{O^{2-}} = \sigma_{O^{2-}}/c_{O^{2-}}$, whereby the parameters without asterisks represent the total values, and so Eq. (6.45) defines the tracer diffusion coefficient

$$D^*_{O^{2-}} = \frac{\sigma_{O^{2-}}}{4F^2}\frac{RT}{c_{O^{2-}}} \tag{6.46}$$

which again reflects the self–diffusion coefficients of the O^{2-} ions.
Since the tracer diffusion represents the migration of the isotope, and since it is not primarily the migration of the defect that is observed in the conductivity experiments, the total ion oxygen concentration ($c_{O^{2-}}$) appears in Eq. (6.46) in a natural

[52]For simplicity, we refer here to the interstitialcy mechanism. If only jumps from one interstitial site to the other were possible, the regular sites would not be affected at all.

[53]$\sigma_{O_i''} = \sigma_{O^{2-}}$, $\mathbf{j}_{O_i''} = \mathbf{j}_{O^{2-}}$. $\nabla\mu_{O_i''} = \nabla\mu_{O^{2-}}$ results from the equivalence of:
a) $O^{2-}(\text{gas}) \rightleftharpoons O^{2-}(\text{solid})$
b) $O^{2-}(\text{gas}) + V_i \rightleftharpoons O_i''$.
Please note that gradients in activity coefficients disappear in the ionic formulation due to the ideality of the tracer distribution (thermodynamic factor is unity).

[54]It is presupposed that $c_{16_{O_i''}}/c_{16_{O^{2-}}} = c_{18_{O_i''}}/c_{18_{O^{2-}}}$. The fact that this relationship is equivalent to $c_{O_i''}/c_{O^{2-}}$ trivially follows from
$\frac{c_{O_i''}}{c_{O^{2-}}} \equiv \frac{c_{^{16}O_i''}+c_{^{18}O_i''}}{c_{^{16}O^{2-}}+c_{^{18}O^{2-}}} = \frac{c_{^{16}O_i''}}{c_{^{16}O^{2-}}}\left(\frac{(c_{^{16}O^{2-}}/c_{^{18}O^{2-}})+1}{(c_{^{16}O^{2-}}/c_{^{18}O^{2-}})+1}\right) = \frac{c_{^{16}O_i''}}{c_{^{16}O^{2-}}} = \frac{c_{^{18}O_i''}}{c_{^{18}O^{2-}}}$. Consequently the above relationship in the text follows with $\sigma_{16,18O^{2-}} = \sigma_{16,18O_i''} \propto c_{16,18O_i''}$.

manner. Moreover, this point leads to a slight difference in the parameters $D^*_{O^{2-}}$ and $D^Q_{O^{2-}}$ if considered in a better approximation. To illustrate this, we consider a vacancy mechanism:
It can be seen from Fig. 6.4 that after a jump process has taken place a vacancy experiences the same environment and, hence, to a good approximation a second jump is not correlated with the first[55]. (This must be refined on closer examination for concentrated defects.) However, the situation is different from the point of view of the (labelled) ion in the case of tracer diffusion. After a site exchange of the tracer particle and the vacancy, a further vacancy is normally not available, and a return jump of the tracer to the original position is favoured. This "back–orientation" results in a deviation from the uncorrelated migration and, thus, in a deviation from the self–diffusion coefficients of the ions. This is represented by a correction factor[56], which is to be taken into account when calculating the diffusion coefficient from the jump rate (see Eq. (6.28)). It should be noted that this correlation factor (F) does not reflect the longer waiting period on account of the low concentration of jump partners (this is taken into account in Eq. (6.46) via the concentration term), but only the deviation from a homogeneous angle distribution[56].
Typical correlation factors for the case of the vacancy mechanism are $F_v = 0.5$ for the diamond lattice, $F_v \simeq 0.65$ for the primitive cubic lattice, 0.73 for the CsCl lattice (body–centred cubic) and 0.78 for the NaCl lattice (face–centred cubic). In the case of an interstitial mechanism, tracer correlation coefficients are unity on account of the overall availability of jump partners, in contrast to the interstitialcy mechanism (for which F is between 0.6 and 1.0).
In general, the ratio between the tracer diffusion coefficient and the conductivity diffusion coefficient is referred to as the Haven ratio (H):

$$D^*_{O^2} = H_{O^{2-}} D^Q_{O^{2-}}. \tag{6.47}$$

H contains further corrections[57], such as the possibility that different mechanisms contribute to the conductivity experiment and the tracer experiment[58]. Of special interest are situations in which more than one valence state is of relevance (cf. Section 6.6). Since usually $H_{O^{2-}}$ is of the order of 1, the use of this parameter for mechanistic elucidation is restricted to very well investigated systems. The effects are frequently within the error margins of the measurement techniques.
Figure 6.18 shows the dependence of the tracer coefficients for iron (^{59}Fe) in $Fe_{3-\delta}O_4$ on the oxygen partial pressure for different temperatures. In accordance with Eq. (6.46) it reflects the behaviour of the ionic conductivity in this primarily electronically disordered oxide. In agreement with Eq. (5.117) power laws are obtained.

[55]The term $\overline{\cos\alpha_{ij}}$ in footnote 24 measures this correlation.

[56]If we repeat the derivation given in footnote 24 for the tracer atoms, then $\overline{\cos\alpha_{ij}}$ is only zero in the case of the interstitial mechanism, otherwise a correlation factor F is obtained, with the consequence that $D^* = \frac{1}{6}F\Gamma r^2$, whereby $F = 1 + \frac{2}{n}\sum_{i=j+1}^{n}\sum_{j=1}^{n-1}\cos\alpha_{ij}$. In the random walk game (footnote 24), the probability of the decision "right or left" is no longer equal. The increased probability of a return jump leads to a reduction in the mean distance from the starting position

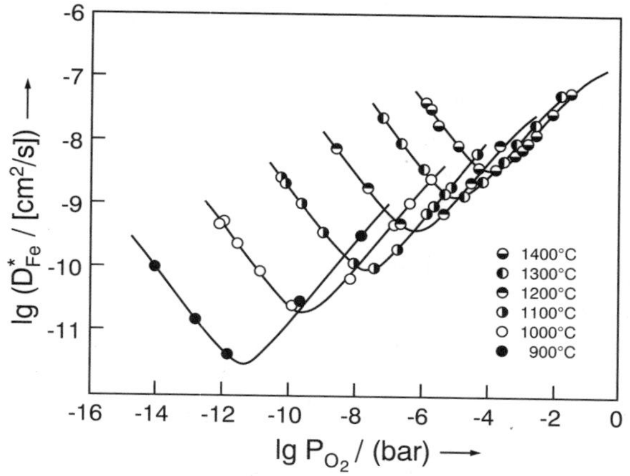

Fig. 6.18: Tracer diffusion data (^{59}Fe) for various temperatures as a function of the partial pressure of oxygen for $Fe_{3-\delta}O_4$. In agreement with the special defect chemistry (Frenkel disorder in the Fe–sublattice with high electronic disorder) a dependence of the form $\alpha P^{-2/3} + \beta P^{+2/3}$ is fulfilled. From Ref. [387].

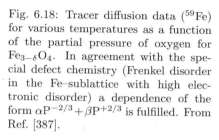

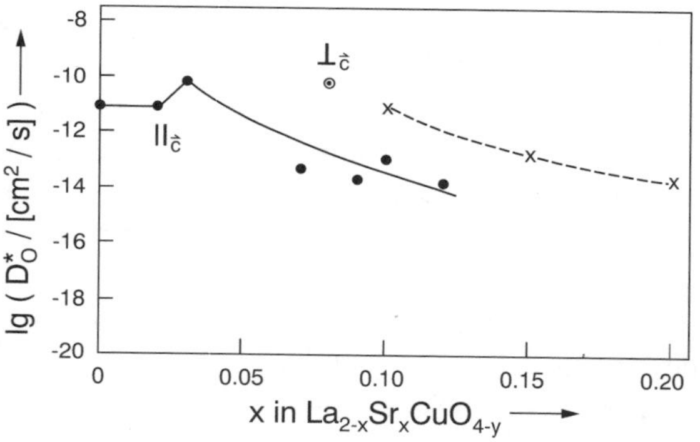

Fig. 6.19: The tracer diffusion coefficient of oxygen in SrO–doped La_2CuO_4 as a function of the doping concentration (cf. Fig. 6.52). The invariance of the single–crystal data (•) for low x–values corresponds to the native disorder (cf. Fig. 5.55). While for x<0.05 the shape of the curve corresponds to Fig. 5.55, at higher Sr concentrations obviously structural changes have a perceptible influence. The considerable anisotropy (see Fig. 5.3c) is also shown. The transport is primarily perpendicular to the c direction in the polycrystalline material (x) [388].

The minimum in Fig. 6.19 indicates the change from interstitial to vacancy dominance (cf. the "P law" in Chapter 5). Note again that the tracer experiment yields element–specific information.

A more complex example is La_2CuO_4. The doping dependence of oxygen tracer diffusion coefficients in La_2CuO_4 and their anisotropy are illustrated in Fig. 6.19. Whilst the behaviour at low x–values is in accordance with simple defect chemistry, the interactions and structural changes at high doping concentrations lead to deviations from the ideal mass action laws (see Fig. 5.55).

In the above considerations we neglected isotope effects on the mobility, which is a

---

and, hence, to a lower diffusion coefficient. So this effectively refers to a "hesitating random walker".

[57] See Refs. [7–12,386] for details. Note that conductivity means dc conductivity in our context. In the case of ac conductivity at high frequencies correlation effects have to be taken account of (see Section 6.6.1).

[58] The alkali hydroxides provide a striking example (see Section 6.6.1d).

reasonable assumption for oxygen but certainly incorrect for protons. Conversely, these isotope effects are useful means to draw conclusions on the conduction mechanism in proton conductors (see Section 6.2.1).

6.3.3 Chemical diffusion

The most important diffusion coefficient for chemistry and materials science is the chemical diffusion coefficient which characterizes the diffusion kinetics of composition changes. This is formally a diffusion of neutral components, and, for ionic compounds, a charge–neutral ambipolar diffusion of at least two chemically different charged particles[59,60]. A relevant example is the change in stoichiometry of the oxide "M_2O" (Fig. 6.17c) in the sense of

$$M_2O_{1+\delta} + \frac{\Delta\delta}{2}O_2 \rightleftharpoons M_2O_{1+\delta+\Delta\delta}. \tag{6.48}$$

Here the driving force is a gradient of the partial pressure of oxygen. When there is only disorder in the oxygen sub–lattice, the diffusion[59] of "O" takes place as a result of coupled transport of O^{2-} and $2e^-$ in opposite directions. The external wire of the conductivity experiment (cf. Fig. 6.17a) is, as it were, internalized and instead of an imposed difference in electrical potential we have one in the chemical potential. This also causes internal gradients in the chemical potential. In contrast to experiments a and b in Fig. 6.17 the situation involves a complex mixture of ionic and electronic transport properties. Since stoichiometric effects only affect the defect budget, it is clear from the start that the chemical diffusion coefficient D_O^δ is going to be of the order of magnitude of the defect diffusion coefficients and, hence, very much larger than the ionic self–diffusion coefficients[60,61]. In addition, it must represent a combination of ionic and electronic defect diffusivities. Even if only one ionic or one electronic defect is dominant, and even if metal defects are neglected, a range of mechanisms is possible: In the simplest case we have to consider a diffusion in the same direction of $V_O^{\cdot\cdot}$ and $2e'$ or of O_i'' and $2h^{\cdot}$; in various cases[62], e.g. in doped material, diffusion may occur in opposite directions, e.g. of $V_O^{\cdot\cdot}$ and $2h^{\cdot}$. Let us first assume that $\sigma_{ion} = \sigma_{O_i''}$ and $\sigma_{eon} = \sigma_{h^{\cdot}}$. (The treatment of La_2CuO_4 is a concrete example of this (see Chapter 5).)

[59] In binary compounds it always involves redox effects. Compare [7,335,389–394].

[60] See Section 6.6 for more complicated situations.

[61] The chemical diffusion coefficient is very frequently denoted by $\tilde{D}$. Since, in our case, the tilde has a very clearly defined meaning (cf. μ and $\tilde{\mu}$, k and $\tilde{k}$, etc.), namely extension by an electrical potential term, we shall use the superscript δ which refers to the composition inhomogeneities that always occur. This has also been found advantageous in Section 6.7.

[62] This may also happen in pure material if ionic disorder predominates (I–regime: e.g. $[O_i''] = [V_O^{\cdot\cdot}] \gg \frac{1}{2}[h^{\cdot}]$ if $\sigma_{V_O^{\cdot\cdot}} \gg \sigma_{O_i''}$), or also in N– or P–regimes if the majority carriers are not both sufficiently mobile (e.g. $[O_i''] \simeq \frac{1}{2}[h^{\cdot}] \gg [V_O^{\cdot\cdot}]$ but $\sigma_{V_O^{\cdot\cdot}} \gg \sigma_{O_i''}$).

Application of our general transport equation permits us to quantify the relationships. For the separate ionic and electronic flux densities we have:

$$\mathbf{j}_{O_i''} = -\frac{\sigma_{O_i''}}{4F^2}\left(\nabla\mu_{O_i''} - 2F\nabla\phi\right), \tag{6.49a}$$

$$\mathbf{j}_{h^\cdot} = -\frac{\sigma_{h^\cdot}}{F^2}\left(\nabla\mu_{h^\cdot} + F\nabla\phi\right). \tag{6.49b}$$

According to the requirement of electroneutrality the fluxes are coupled by

$$2\mathbf{j}_{O_i''} = \mathbf{j}_{h^\cdot} \tag{6.50}$$

Elimination of the electrical potential from Eq. (6.49) and taking account of Eq. (6.50) leads to

$$\frac{1}{2}\mathbf{j}_{h^\cdot} = \mathbf{j}_{O_i''} = \mathbf{j}_{O^{2-}} = \mathbf{j}_O = -\frac{1}{4F^2}\frac{\sigma_{O_i''}\sigma_{h^\cdot}}{\sigma_{O_i''} + \sigma_{h^\cdot}}\left(\nabla\mu_{O_i''} + 2\nabla\mu_{h^\cdot}\right). \tag{6.51}$$

In our case $j_{O_i''}$ is naturally identical to the flux density of the neutral component "O" (j_O). Evidently the harmonically averaged conductivity expression in Eq. (6.51) corresponds to an effective, ambipolar conductivity σ_O^δ, expressing the fact that both ionic and electronic charge carriers are necessary and the respective resistors so–to–speak "connected in series". The expression in brackets obviously represents the chemical potential gradient of the component "O" ($\mu_O = \mu_{O_i''} + 2\mu_{h^\cdot}$). The result is a force–flux relationship of the expected form, viz.

$$\mathbf{j}_O = -\frac{1}{4F^2}\sigma_O^\delta\nabla\mu_O = -\left(\frac{1}{4F^2}\sigma_O^\delta\frac{\partial\mu_O}{\partial c_O}\right)\nabla c_O. \tag{6.52}$$

The expression in brackets[63] is — in one dimension — equivalent to $-j_O/(\partial c_O/\partial x)$ and, on account of conservation of mass and charge, identical to $-j_{O^{2-}}/(\partial c_{O^{2-}}/\partial x)$ as well as to $-j_{e^-}/(\partial c_{e^-}/\partial x)$. It is hence precisely the chemical diffusion coefficient we have been seeking:

$$D_O^\delta = \frac{1}{4F^2}\sigma_O^\delta\frac{\partial\mu_O}{\partial c_O} = \frac{RT}{4F^2}\frac{\sigma_O^\delta}{c_O}\frac{\partial\ln a_O}{\partial\ln c_O}. \tag{6.53}$$

---

[63] We recognize that — while σ_O^δ measures the rate of change of stoichiometry (j_O) at a given gradient of *chemical component potential* ($\partial\mu_O/\partial x$) and represents, as it were, the component permeability — the diffusivity D_O^δ measures this at a given gradient in the *component concentration*. Thus, the latter parameter involves, in addition to the chemical or ambipolar resistance $R^\delta \propto 1/\sigma^\delta$, the chemical capacitance C^δ [391,392] (cf. Section 4.2), i.e. $(\partial\mu_O/\partial n_O)^{-1} \propto (\partial\mu_O/\partial c_O)^{-1} \propto c_O^\delta$ (ambipolar concentration, term in brackets in Eq. (6.54a)), which describes how much c_O is affected by μ_O and how much "stoichiometric change" is to be brought about. We note the isomorphy (n: mole number) between $\partial n/\partial t = (\partial n/\partial\mu)(\partial\mu/\partial t)$ and $\partial Q/\partial t = (\partial Q/\partial\Delta\phi)(\partial\Delta\phi/\partial t)$ with $(\partial n/\partial\mu)$ as differential chemical and $(\partial Q/\partial\Delta\phi)$ as differential electrical capacitance. This language also reveals the role of $D^\delta \propto (R^\delta C^\delta)^{-1}$ with regard to an inverse chemical time constant (cf. Section 7.3.4).

The thermodynamic factor $w_O \equiv \partial \ln a_O / \partial \ln c_O$ appearing on the right hand side is naturally orders of magnitude different from 1 in ionic crystals, for after all the oxygen is neither present as O nor randomly distributed as such. Because[64] $\partial\mu_O = \partial\mu_{O^{2-}} - 2\partial\mu_{e^-}$ and $\partial c_O = \partial c_{O^{2-}} = -\frac{1}{2}\partial c_{e^-}$ it is possible to break D_O^δ up into ionic and electrical terms:

$$D_O^\delta = \frac{RT}{4F^2}\frac{\sigma_{O^{2-}}\sigma_{e^-}}{\sigma_{O^{2-}} + \sigma_{e^-}}\left(\frac{1}{c_{O^{2-}}}\frac{\partial \ln a_{O^{2-}}}{\partial \ln c_{O^{2-}}} + 4\frac{1}{c_{e^-}}\frac{\partial \ln a_{e^-}}{\partial \ln c_{e^-}}\right). \tag{6.54a}$$

Again, the thermodynamic factors differ greatly from 1 in this notation because the carriers considered are regular, ordered particles. Further simplification is only possible by formulating the problem at the level of the defects. Because[65] $\partial\mu_{O_i''} = \partial\mu_{O^{2-}}, \partial\mu_{e^-} = -\partial\mu_{h^\cdot}$ and $\partial c_{e^-} = -\partial c_{h^\cdot}, \partial c_{O^{2-}} = \partial c_{O_i''}$ we obtain from Eq. (6.54a) or more directly from Eq. (6.51)

$$D_O^\delta = \frac{RT}{4F^2}\frac{\sigma_{O_i''}\sigma_{h^\cdot}}{\sigma_{O_i''} + \sigma_{h^\cdot}}\left(\frac{1}{c_{O_i''}}\frac{\partial \ln a_{O_i''}}{\partial \ln c_{O_i''}} + 4\frac{1}{c_{h^\cdot}}\frac{\partial \ln a_{h^\cdot}}{\partial \ln c_{h^\cdot}}\right). \tag{6.54b}$$

Now, however, the thermodynamic factors are 1 for weak disorder and it follows that

$$D_O^\delta = \frac{RT}{4F^2}\frac{\frac{1}{c_{O_i''}} + \frac{4}{c_{h^\cdot}}}{\frac{1}{\sigma_{O_i''}} + \frac{1}{\sigma_{h^\cdot}}} \equiv \frac{RT}{4F^2}\frac{\sigma_O^\delta}{c_O^\delta}. \tag{6.54c}$$

The ambipolar concentration c_O^δ is introduced as an abbreviation that occurs in the form of the harmonic mean of defect concentrations weighted by the square of the charge numbers. The isomorphy with Eq. (6.14) is obvious. A simple rearrangement using the Nernst–Einstein equation (Eq. (6.15)) yields the following identical formulations[66]

$$D^\delta = \frac{(D_{h^\cdot}c_{h^\cdot})\left(D_{O_i''}c_{O_i''}\right)}{(D_{h^\cdot}c_{h^\cdot}) + 4\left(D_{O_i''}c_{O_i''}\right)}\left(\frac{1}{c_{O_i''}} + \frac{4}{c_{h^\cdot}}\right) \tag{6.54d}$$

or (with the transference number $t_k \equiv \sigma_k/\sigma$)

$$D_O^\delta = t_{h^\cdot}D_{O_i''} + t_{O_i''}D_{h^\cdot} = \frac{F^2}{RT}\left(\frac{c_{h^\cdot}D_{h^\cdot}D_{O_i''}}{\sigma} + 4\frac{c_{O_i''}D_{O_i''}D_{h^\cdot}}{\sigma}\right) = \frac{F^2}{RT}\frac{D_{O_i''}D_{h^\cdot}}{\sigma}\left(c_{h^\cdot} + 4c_{O_i''}\right). \tag{6.54e}$$

[64] The equivalence of $\partial c_{O^{2-}}$ and $-\partial c_{e^-}/2$ does not require the electronic and ionic defects to be the majority carriers, and is also fulfilled in doped material. There are however important modifications necessary, if the dopant can change its valence. This is treated in Section 6.6.1c.

[65] One should keep footnote 53 on page 297 in mind. In contrast to tracer diffusion we have

$$\partial \ln a_{O^{2-}} = \partial \ln a_{O_i''},\ \partial a_{O^{2-}} \neq \partial a_{O_i''},\ \text{but}\ \partial c_{O^{2-}} = \partial c_{O_i''},\ \partial \ln c_{O^{2-}} \neq \partial \ln c_{O_i''}.$$

[66] If the two migrating defects are also the majority charge carriers, the concentration terms in Eq. (6.54d) disappear. If their charge numbers are also identical, the result is very simple and clear: D^δ is then the harmonic mean of the two defect diffusion coefficients.

In the case of stronger disorder it is necessary to take account of appropriate activity coefficients (see Section 5.7.2, and also Section 6.6).
In important specific cases, it is useful, for further simplification, to start out from Eq. (6.54c). We can see that two independent conditions must be considered which refer to the ratio $c_{h^\cdot}/c_{O_i''}$, on the one hand, and to the ratio $c_{h^\cdot}u_{h^\cdot}/c_{O_i''}u_{O_i''}$, on the other.
First let us treat the case in which either the ionic or the electronic defect belongs to the majority charge carriers, that is, in our example, $c_{h^\cdot}/c_{O_i''} \gg 1$ or $c_{h^\cdot}/c_{O_i''} \ll 1$ as usually the case for sufficiently doped materials (cf. C–theorem, page 179). This condition is also fulfilled for pure materials in the I regime of the Brouwer diagram (see Section 5.5), where either ionic disorder (PbO: $[O_i''] = [Pb_i^{\cdot\cdot}]$) or the electronic disorder predominates (CuO: $[e'] \simeq [h^\cdot]$). In the case of predominantly ion–conducting materials, on account of the low mobility of ions, $c_{O_i''}u_{O_i''} \gg c_{h^\cdot}u_{h^\cdot}$ normally implies automatically $c_{O_i''} \gg c_{h^\cdot}$ and it follows that D^δ is represented by the diffusion coefficient of the (most rapid) electronic defect. In order to go beyond our example and to include the case of n–conducting compounds we write

$$D_O^\delta = D_{h^\cdot/e'}, \tag{6.55}$$

(i.e. either $D_O^\delta = D_{h^\cdot}$ or $D_{e'}$ depending on the defect chemistry)[67]. Thus, for example, in Y_2O_3–doped ZrO_2[68] $D_O^\delta = D_{h^\cdot}$ at high and $D_O^\delta = D_{e'}$ for low partial pressures of oxygen. In the extreme case of predominantly electronically conducting compounds the situation is not so clear cut, since a negligible ionic conductivity is very far from meaning that the charge carrier concentration concerned is also negligible. In the case of "electron rich electron conductors", however, as is the case, for instance, in positively doped SnO_2 ($\sigma \simeq \sigma_{e'}, C = [e'] \gg [V_O^{\cdot\cdot}]$) or (weakly)[69] negatively doped La_2CuO_4 ($\sigma \simeq \sigma_{h^\cdot}, C = [h^\cdot] \gg [O_i'']$) then

$$D_O^\delta = D_{O_i''/V_O^{\cdot\cdot}} \tag{6.56}$$

appears as a counterpart to Eq. (6.55). In the remaining, third special case of the "ion–rich electronic conductors", the result is

$$D_O^\delta = \frac{4c_{O_i''/V_O^{\cdot\cdot}}}{c_{h^\cdot/e'}} D_{O_i''/V_O^{\cdot\cdot}} = \frac{\sigma_{O_i''/V_O^{\cdot\cdot}}}{\sigma_{h^\cdot/e'}} D_{h^\cdot/e'}. \tag{6.57}$$

Here, D_O^δ is not only dependent on T but on P_{O_2} and C, too. Equation (6.57) is for instance realized in pure ($[V_O^{\cdot\cdot}] = [V_{Sr}''] \gg [e']$ or $[h^\cdot]$) or acceptor–doped $SrTiO_3$ ($2\,[V_O^{\cdot\cdot}] = [A'] \gg [h^\cdot] \gg [e']$).
Unification can be obtained for all three extreme cases by writing

$$D^\delta = t_1 D_3 = \begin{cases} D_3 & \sigma_1 \gg \sigma_3 \\ \frac{\sigma_1}{\sigma_3} D_3 & \sigma_1 \ll \sigma_3 \end{cases} \tag{6.58}$$

---

[67] For the case that e' and $h^\cdot$ are important simultaneously cf. Section 6.6.1c.
[68] However, it must be free of traces of redox–active impurities (see Section 6.6).
[69] $V_O^{\cdot\cdot}$ becomes significant when doping is heavy.

if 1 refers to the most conducting of the majority charge carrier types (1,2) and 3 to the most conducting of the minority charge carrier types (3,4) (i.e. $c_1, c_2 \gg c_3$ by definition). Here the expression charge carrier type means either ions or electrons. For $c_1 \gg c_3$ Eq. (6.58) follows directly from Eq. (6.54e) since $D_1\sigma_3 \propto D_1D_3c_3 \ll D_1D_3c_1 \propto D_3\sigma_1$. In the case when the majority charge carrier (1) is also more conducting compared with the minority charge carrier, D^δ is merely temperature–dependent (via of $\Delta_3 H^{\neq}$) but not dependent on partial pressure or doping (case 1 in Eq. (6.58)). In the case, however, that the minority defect is the more conducting (case 2 in Eq. (6.58))

$$D^\delta \propto \exp -\frac{\Delta_1 H^{\neq}}{RT}\left(P^{N_1-N_3}C^{M_1-M_3}\Pi_r K_r^{\gamma_{r1}-\gamma_{r3}}\right). \tag{6.59}$$

results[70] (cf. Eq. 5.150). In particular the temperature dependence is given by

$$-R\frac{\partial \ln D^\delta}{\partial 1/T} = \Delta_1 H^{\neq} + \Sigma_r\left(\gamma_{r1}-\gamma_{r3}\right)\Delta_r H^\circ. \tag{6.60}$$

Let us now turn to the cases in which both the decisive mobile ionic and the decisive electronic defect are the majority carriers. Thus, in the case of a pure material we refer to the P or N regime, see Section 5.5. In both regimes we can assume that the conductivity is determined by the electronic defects. It is sufficient to consider the case of the P regime where $2[O_i''] = [h^\cdot]$. It follows immediately from Eq. (6.54) that

$$D^\delta = 3D_{O_i''}, \tag{6.61}$$

a result that, for an order of magnitude consideration, can also be subsumed under Eq. (6.58) (case 1) if we now replace defect number 3 by the poorer conducting of the two majority charge carriers.

We can see from Eq. (6.54e) that D^δ lies between $D_{h^\cdot/e'}$ and $D_{O_i''/V_O^{\cdot\cdot}}$. Since the first is generally the largest, this means that the electronic influence on the ambipolar diffusion is an accelerating one, and that the ionic influence acts as a brake. You can see that both diffusion coefficients and, in particular, the ionic defect diffusion coefficient must be sufficiently large[71] in order to produce a high D^δ. The lower value $D^\delta = D_{O_i''/V_O^{\cdot\cdot}}$ is — as demonstrated above — found for the case of an electron–rich electronic conductor.

In the case of multinary compounds the composition changes described by the chemical diffusion need not always be a redox effect (as here). Thus, the incorporation of water in the perovskite proton conductors (see Section 5.6) comes about as a result of the coupled diffusion of $2H^+$ and O^{2-}, that is an acid–base effect. Mechanistically this corresponds to the counterdiffusion of oxygen vacancies and proton defects ($H_i^\cdot$

[70] The relation is more complex in the case of trapping effects (Section 6.6.1).

[71] A high D^δ in Eq. (6.54c) does not necessarily imply a high permeability (σ_O^δ) on account of c_O^δ which may have a compensating influence. For a high σ^δ it is sufficient that σ_{eon} and σ_{ion} be large. Because $\sigma_O^\delta = \sigma_{eon}\sigma_{ion}/\sigma$, this parameter reaches, for a given σ, a maximum of $\sigma/4$ at $\sigma_{eon} = \sigma_{ion}$.

or more precisely $OH_O^{\cdot}$). The corresponding relationships are analogous, instead of μ_O and c_O, $\mu_{H_2O} = 2\mu_{H_i^{\cdot}} - \mu_{V_O^{\cdot\cdot}}$ and c_{H_2O} go into the equations. Such acid–base effects play a large role in solid state reactions (see Section 6.9).
In the case of purely morphological changes (plastic deformations as a result of significant changes in hydrostatic pressure or as required to minimize the surface free energy), it is necessary for the whole substance to migrate (i.e. migration of all components involved). In the case of diffusion–controlled creep with comparable concentrations, the slowest component is decisive in accordance with the discussion above, such as the Zr^{4+} ions in ZrO_2 [134] (see also Sections 5.4 and 6.9.2).
Some experimental examples of chemical diffusion are presented in detail in Section 6.5.

6.3.4 A comparison of the phenomenological diffusion coefficients

The derivations above give the following interrelation of the three phenomenological oxygen diffusion coefficients (see Eqs. (6.47), (6.54), see Fig. 6.20)

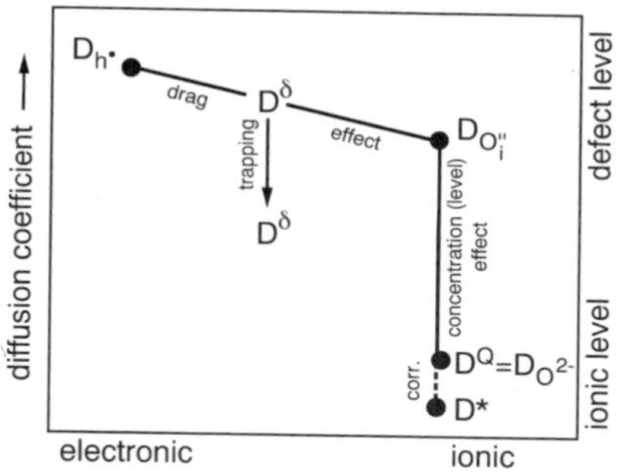

Fig. 6.20: The relationships between the different diffusion coefficients discussed in the text. See Section 6.6.1 with respect to trapping [393]. Holes and interstitial oxygen ions are assumed as mobile defects.

$$D^Q_{O^{2-}} : D^*_{O^{2-}} : D^\delta_O = 1 : H_{O^{2-}} : \frac{\sigma_{h^\cdot/e'}}{\sigma}\left(\frac{c_{O^{2-}}}{c_{O_i''/V_O^{\cdot\cdot}}} + \frac{4c_{O^{2-}}}{c_{h^\cdot/e'}}\right). \tag{6.62}$$

These relationships will be generalized later for the presence of several charge carriers and, in particular, for the occurrence of association reactions.
It is worth emphasizing once again that in contrast to D^Q and D^* the chemical diffusion coefficient is first of all addressing the (small) ensemble of (rapid) defects[66] and secondly is made up of ionic and electronic contributions. In other words: It is enhanced with respect to $D_{O_i''/V_O^{\cdot\cdot}}$ on account of the drag effect of the electrons, while $D_{O_i''/V_O^{\cdot\cdot}}$ itself is usually much larger than D^Q and D^* because of the concentration effect. Thus, it is not correct to ascribe the ratio D^δ/D^* (instead of $D^\delta/D_{O_i''/V_O^{\cdot\cdot}}$) solely to an acceleration by the electrons as is frequently done[72].In the case of a

[72]Only the ratio $D^\delta/D_{O_i''/V_O^{\cdot\cdot}}$ can be regarded as an acceleration factor in this sense. We should remind ourselves again that the parameter D^δ is comparatively large, because it measures the

predominantly electronically conducting solid the ratio of D^δ_O to D^Q_O is reduced to the thermodynamic factor w_O (cf. the term in brackets in Eq. (6.62)). If the electronic defect concentration is also greater than that of ionic defects, then $w_O = x^{-1}_{O_i''/V_O^{\cdot\cdot}}$ remains.

It is certainly illuminating in this latter case[73] in which it is only the ionic jump process that is relevant for the chemical diffusion, to derive D^δ and D^* using chemical kinetics. Let us consider the jump of an O^{2-} ion (at x) into a vacancy (at $x' = x+\Delta x$) and the associated back reaction. According to Eqs. (6.17, 6.37) it follows the master equation[74]

$$j_{O^{2-}}/\Delta x = \vec{k} c_{O^{2-}}(x) c_{V_O^{\cdot\cdot}}(x') - \overleftarrow{k} c_{O^{2-}}(x') c_{V_O^{\cdot\cdot}}(x). \tag{6.63}$$

In the chemical diffusion experiment the value of $c_{V_O^{\cdot\cdot}}$ is variable, while $c_{O^{2-}}$ can be regarded as being approximately constant[75]. Since $\vec{k} = \overleftarrow{k} \equiv k$ it follows that

$$j_{O^{2-}} = j_O = +k c_{O^{2-}} (\Delta x)^2 \frac{\Delta c_{V_O^{\cdot\cdot}}}{\Delta x} = -k c_{O^{2-}} (\Delta x)^2 \frac{\Delta c_{O^{2-}}}{\Delta x}. \tag{6.64}$$

Conversely, in the tracer experiment $c_{V_O^{\cdot\cdot}}(x) = c_{V_O^{\cdot\cdot}}(x')$, and the tracer concentration is the variable quantity. Thus

$$j^*_{O^{2-}} = -k c_{V_O^{\cdot\cdot}} (\Delta x)^2 \frac{\Delta c^*_{O^{2-}}}{\Delta x}. \tag{6.65}$$

The products in front of the gradient term in Eqs. (6.64) and (6.65) obviously constitute D^δ and D^*, which are coupled by $x_{V_O^{\cdot\cdot}} = c_{V_O^{\cdot\cdot}}/c_{O^{2-}}$ in our specific case as we have already shown.

If we also include the facts already used in Section 6.1, namely that in a stationary measurement of the ionic conductivity (and hence of D^Q) it is possible to neglect gradients in both $c_{O^{2-}}$ and $c_{V_O^{\cdot\cdot}}$ whilst (on account of the field) differences in k value become relevant (see also Section 6.1), the informative Table 6.4 can be constructed. After linearization all that remains of Δk (more precisely speaking, of $\Delta\tilde{k}$), apart

(ambipolar) mobility of the defects while D^* (as D^Q by definition) refers to ions (regular and defective).

[73]That is, we consider the case of the "electron–rich electron conductor". The presentation follows Ref. [384]. Also more general cases can be tackled by chemical kinetics [9,329]. In the case of mixed conduction a more sophisticated (cf. Chapter 7) experiment must be selected to determine $D^Q_{O^{2-}}$ (i.e. the ionic conductivity) from a conductivity measurement.

[74]The reader will have noticed that in spite of the high concentration, the regular term (here $c_{O^{2-}}$) has been included in the master equation (6.63) in an ideal way (cf. Fig. 6.1). This is allowed, since (i) this quantity is constant anyway in the cases of the electrical and chemical experiments, while (ii) it reflects the ideality of the tracer O^{2-} distribution in the tracer experiment.

[75]More accurately, since $c(x') = c(x) + \Delta c$ and $\vec{k} = \overleftarrow{k} \equiv k$, we may state

$$\vec{k} c_{O^{2-}}(x)[c_{V_O^{\cdot\cdot}}(x) + \Delta c_{V_O^{\cdot\cdot}}] - \overleftarrow{k}[c_{O^{2-}}(x) + \Delta c_{O^{2-}}] c_{V_O^{\cdot\cdot}}(x) \simeq k[c_{O^{2-}}(x)\Delta c_{V_O^{\cdot\cdot}} - c_{V_O^{\cdot\cdot}}(x)\Delta c_{O^{2-}}].$$

In the case of chemical diffusion it is true that $|\Delta c_{V_O^{\cdot\cdot}}|$ and $|\Delta c_{O^{2-}}|$ are identical, but there is no doubt that $|\Delta c_{V_O^{\cdot\cdot}}/c_{V_O^{\cdot\cdot}}| \gg |\Delta c_{O^{2-}}/c_{O^{2-}}|$ and, hence, Eq. (6.64).

Table 6.4: Comparison of ion conductivity, tracer diffusion and chemical diffusion in electron–rich electron conductors, using
$\mathrm{j} \propto \mathcal{R} = \overrightarrow{\mathrm{k}} \mathrm{c}_{\mathrm{O}^{2-}}(\mathrm{x}) \mathrm{c}_{\mathrm{V}_\mathrm{O}^{\cdot\cdot}}(\mathrm{x}') - \overleftarrow{\mathrm{k}} \mathrm{c}_{\mathrm{O}^{2-}}(\mathrm{x}') \mathrm{c}_{\mathrm{V}_\mathrm{O}^{\cdot\cdot}}(\mathrm{x})$
(Δk refers to the difference of the electrochemical rate constants $\overrightarrow{\overrightarrow{\mathrm{k}}}$ and $\overleftarrow{\overleftarrow{\mathrm{k}}}$. The tilde has been supressed for simplicity.)

Experiment	Simplification ($\mathcal{R}$)
Ionic conductivity	$-(\Delta \mathrm{k}) \mathrm{c}_{\mathrm{O}^{2-}} \mathrm{c}_{\mathrm{V}_\mathrm{O}^{\cdot\cdot}}$
Tracer exchange	$-\mathrm{k}(\Delta \mathrm{c}_{\mathrm{O}^{2-}}) \mathrm{c}_{\mathrm{V}_\mathrm{O}^{\cdot\cdot}}$
Chemical diffusion	$-\mathrm{k} \mathrm{c}_{\mathrm{O}^{2-}} (\Delta \mathrm{c}_{\mathrm{V}_\mathrm{O}^{\cdot\cdot}})$

from less important constants, is the mobility ($\mathrm{u} \propto \mathrm{k}$) and the electrical potential difference. The resulting transport parameter then is the ionic conductivity ($\propto \mathrm{k}\mathrm{c}_{\mathrm{V}_\mathrm{O}^{\cdot\cdot}}$). The same parameter also appears in the case of tracer exchange, while the transport parameter for chemical diffusion reduces to $\mathrm{k} \propto \mathrm{u}_{\mathrm{V}_\mathrm{O}^{\cdot\cdot}} \propto \mathrm{D}_{\mathrm{V}_\mathrm{O}^{\cdot\cdot}}$, in complete agreement with the results of irreversible thermodynamics for this special case. The relationships for the interstitialcy mechanism are analogous but less clear cut [384]. Before we discuss a few experimental examples we must first discuss the problems of evaluating concentration profiles.

6.4 Concentration profiles

While conductivity measurements are evaluated by determining a flux, in the case of tracer diffusion and chemical diffusion it is simpler to use concentration changes for analysis[76]. Let us consider the volume section in Fig. 6.21 and assume one–dimensional geometry. It is evident that the increase in concentration can only occur in two ways: (i) by more flowing into the volume than out of it ($\propto -(\mathrm{j}_{\mathrm{x}+\Delta\mathrm{x}} - \mathrm{j}_\mathrm{x})/\Delta\mathrm{x}$); (ii) by the presence of internal sources or sinks, i.e. (defect) chemical reactions in the

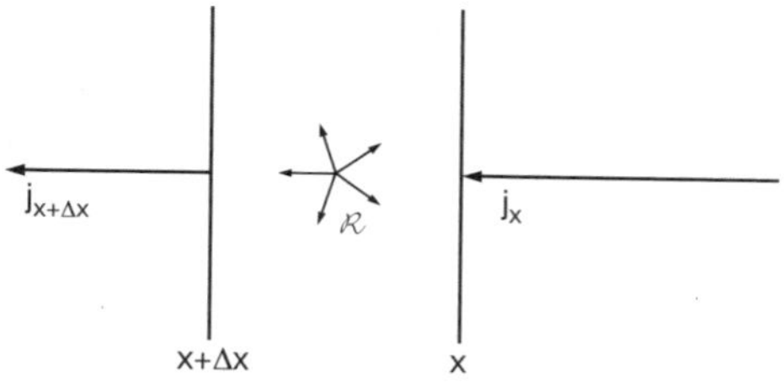

Fig. 6.21: The increase in concentration between x and Δx comes from internal sources/sinks and from the difference between import and export rate.

[76] See, however, permeation measurements in Chapter 7.

form of production or annihilation reactions which release or consume the defects under consideration ($\propto \nu_k \mathcal{R} = \nu_k(\vec{\mathcal{R}} - \overleftarrow{\mathcal{R}})$). We will discuss the latter effects in Section 6.6 and ignore them completely for the moment. In the three–dimensional extension[77] it follows that

$$\frac{\partial c_k}{\partial t} = -\nabla \mathbf{j}_k = -\nabla(D\nabla c_k). \tag{6.66}$$

This relationship applies to both tracer and chemical diffusion if the requirements of Fick's ("first") law are fulfilled. In what follows we assume that D remains approximately constant[78]. Then the well–known "Fick's second law" results which reads for one dimension,

$$\frac{\partial c}{\partial t} = D\frac{\partial^2 c}{\partial x^2}. \tag{6.67}$$

The term "Fick's second law", is somewhat misleading, since it is not a new phenomenological law; rather Eq. (6.67) has been obtained from Eq. (6.9) by coupling with material conservation.

Even though the diffusion problems generally involve solving one and the same differential equation, the solutions can be tremendously different, on account of the individual boundary and initial conditions that characterize the specific physical experiment. There are a series of mathematical procedures for the solution of Eq. (6.67). The Laplace transform (see page 464) is a very general procedure, which transforms Eq. (6.67) into a simple algebraic equation containing the boundary and initial conditions in a straightforward way. The solution must then be back–transformed with usually substantial mathematical effort or by simply referring to tables (see Section 7.3.6). Most solutions of Eq. (6.67) have been worked out in the literature. Reference [395] is particularly useful, and contains solutions of the thermal conduction equation ("Fourier's second law") for very many cases, i.e. differing boundary and initial conditions. Owing to an analogous linear flux–force relationship, and the respective continuity equation, the differential equation for the temperature change is fully isomorphic to the diffusion equation Eq. (6.67) (see Table 6.1, page 272). Reference [395] also gives information for approaching the case of variable transport coefficients. Here an analytical solution is not, in general, possible, and one must resort to graphical (e.g. Boltzmann–Matano method) or numerical methods (e.g. finite difference methods). A further complication, in particular for fluid systems, is related to the reference system [7,333]. In our simple cases we will take the immobile partial structure as the internal reference[79].

[77]See footnote 23 on page 280 for the derivation of Eq. (6.66) via jump considerations.

[78]More precisely $|\nabla D \nabla c| \ll |D\nabla(\nabla c)|$, i.e. the relative change in the diffusion coefficient has to be small with respect to the relative change in the concentration gradient.

[79]A local displacement of the material can also occur in the solid state (see also Kirkendall effect [396]). Also compare with Fig. 6.62, page 368.

Experimentally speaking it is useful to differentiate boundary conditions according to whether the external diffusion sources[80] remain constant or whether they can be exhausted, whether the concentration at the boundary remains constant or the total amount is kept constant; another simple variant is to keep the diffusion flux at the boundary constant; in addition the spatial position of the diffusion source as well as the geometry and dimensionality of the set–up are decisive.
Perhaps the most clearcut case is that of an exhaustible source in the centre of a very extended one–dimensional system (Fig. 6.22). A relevant experimental example

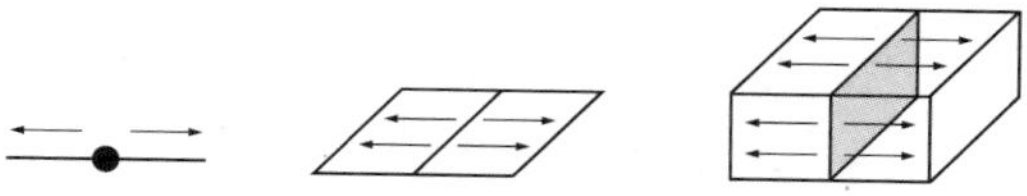

Fig. 6.22: Two–sided infinite quasi one–dimensional diffusion configurations as discussed in the text.

is the application of a tiny spot of gold to the centre of a very thin silver wire at elevated temperatures. In the initial stages of the diffusion, the solution turns out to be a Gaussian curve[81]:

$$c(x,t) = \frac{s_0}{(4\pi Dt)^{1/2}} \exp\left(-\frac{x^2}{4Dt}\right). \tag{6.68}$$

As can be readily verified, Eq. (6.68) is not only a solution of the differential equation Eq. (6.67) but also fulfills the initial condition that for $x \neq 0$ $c(t=0)$ is zero while for $x=0$ it is constant (s_0)[82], as well as the "boundary conditions" (for the infinitely extended sample) $c(x = \pm\infty, t) = 0$. In addition, Eq. (6.68) has the intuitively expected form (see Fig. 6.23a). It exhibits a maximum at $x = 0$ and is symmetrical about the concentration axis. On account of the requirement $c \to 0$ for $x \to \pm\infty$ the solution must also exhibit two inflection points. These are at

$$|x_{ifn}| = \sqrt{2Dt}. \tag{6.69}$$

The area under the curve is constant with time and yields the total quantity[82] of the gold (s_0). Since a major proportion (ca. 68%) of the material introduced is to be found between the inflection points, it is possible to obtain a rough estimate[83] of the duration (t_{eq}) of the diffusion experiment by setting $2x_{ifn}$ equal to the total sample length (L):

$$L^2 \simeq 2Dt_{eq}. \tag{6.70}$$

[80] The term "source", normally used here, refers to the location of higher concentration, which is setting the boundary conditions and must not be confused with internal source and sink terms.

[81] This also follows from the random walk problem (see footnote 24 on page 280).

[82] The initial condition is written $s_0\delta(x)$ with mathematical precision, whereby $\delta(x)$ represents the delta function. $\int_{-\infty}^{+\infty} s_0\delta(x)dx$ yields s_0. In addition it holds that $\int_{-\infty}^{+\infty} \exp\left(-y^2\right) dy = \sqrt{\pi}$. Cf. also footnote 108 on page 464 in Chapter 7.

[83] Equation (6.68) naturally loses its functional validity for finite boundary conditions. Conversely, the simpler relations derived for infinite boundary conditions are always valid for sufficiently small times.

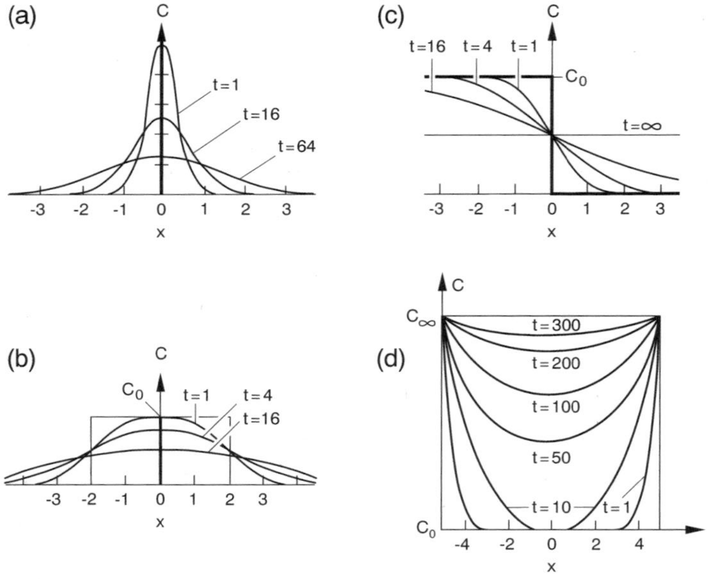

Fig. 6.23: Concentration profiles for various initial and boundary conditions (see text):
(a) exhaustible point source[80] (infinite boundary conditions)
(b) extended, exhaustible source (infinite boundary conditions)
(c) infinitely extended (one-sided) source (infinite boundary conditions)
(d) inexhaustible (two-sided) source (finite boundary conditions)[84].

(Note that the parameter 2Dt (or 6Dt in ^{3}D) corresponds to the square of the mean displacement in stochastic processes[77].)
The case of two–sided finite initial boundary conditions is discussed more precisely below. If we apply a diffusion source[80] on one side of a very thin, long diffusion channel (corresponding to one–sided finite or semiinfinite boundary conditions), Eq. (6.68) applies once again (comparatively small times), provided s_0 is replaced by double the value. Figure 6.24 shows a tracer diffusion example.

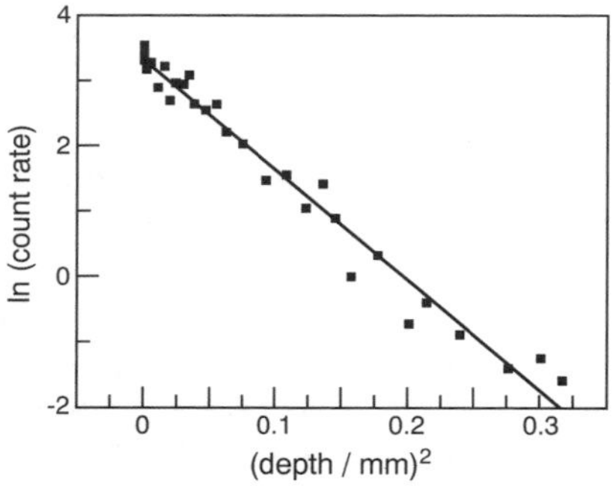

Fig. 6.24: Co tracer diffusion (^{57}Co) in CoO at 1100°C in air, detected by measuring the γ–radiation of successively removed layers. Evaluation according to Eq. (6.68). From Ref. [397].

In many cases the diffusion source is not exhaustible (see, for example, Fig. 6.23d), but is maintained at a constant concentration. The total amount diffusing in initially follows a $\sqrt{t}$ law. For longer times the particles experience the finite size, and an exponential law is approximately valid (compare following section).

Figure 6.23 illustrates a variety of situations[84]. We will meet further cases, in Chapter 7, in which the fluxes and hence the gradients are kept constant at the boundary. Naturally the one–dimensional solutions are not restricted to one–dimensional systems (cf. Fig. 6.22). In ^{2}D a suitable pseudo one–dimensional experiment is the application of a diffusion source in the form of a thin strip (e.g. gold) onto a thin film (e.g. a thin sheet of silver) (chemical diffusion). It can also be a thin strip of the same but now radioactive material (different isotope), or the exposure of a slit–shaped opening of the otherwise sealed thin oxide film to a (radioactive or chemically modified) gas atmosphere (tracer diffusion). The analogue in ^{3}D is the sandwich technique or the planar application of the diffusion source onto the surface (as already considered in Fig. 6.24). If the diffusion source is a gas phase, again sealing is necessary unless the aspect ratio is very favourable, i.e. if the extension is sufficiently small in the direction of diffusion compared with the other directions in space[85]. Otherwise the three–dimensional solution has to be considered.
The mathematical complexity of the problem can be reduced by formulating the diffusion problem in coordinates adapted to the symmetry. Thus, in the case of the cylinder geometry illustrated in Fig. 6.25, the diffusion problem can be described with a single positional variable if cylindrical coordinates are used instead of cartesian coordinates (see below Eq. (6.76)).

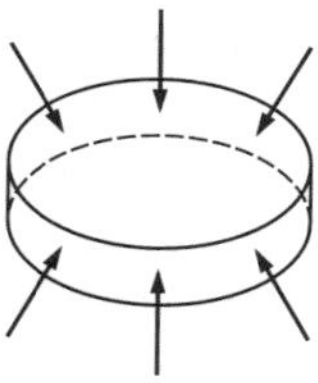

Fig. 6.25: Radial diffusion in a cylindrical sample. For symmetry reason the radial distance from the centre is the only parameter. In cases in which top and bottom of the sample are not negligible in area compared with the lateral surfaces (as here), these planes need to be covered (e.g. with a glass seal).

In the following section we will discuss in–situ profiles (as a function of space and time) that develop as a result of chemical diffusion (see Fig. 6.33). Most frequently the analysis is carried out ex situ (quenching the nonequilibrium state, successive removal of the material and analysis of the respective concentration by chemical analysis, mass spectroscopy, radiometry etc). Removal techniques used in the context of tracer diffusion involve particle bombardment or etching. Figure 6.26 illustrates the ^{18}O profile obtained by SIMS measurements[86] on $La_{0.8}Sr_{0.2}Mn_{0.8}Co_{0.2}O_{3-x}$ for the purpose of determining D^*_O; such Sr–doped La manganates[87] or cobaltates are prominent candidates for cathodes in fuel cells (see Section 7.4.2). As a preview of Section 6.7, it is to be emphasized that the boundary conditions used so far presuppose an infinitely rapid surface reaction. The fact, that the normalized concentration

[84] See also Fig. 7.11, page 418 and Fig. 7.26, page 450 for further examples.

[85] Attention has to be paid to the fact that transport coefficients can be very anisotropic.

[86] Secondary ion mass spectroscopy: The analysis is carried out by mass spectrometry of sputtered ions produced by particle bombardment.

[87] These compounds also exhibit interesting magneto–resistive properties [398].

does not exactly reach unity in Fig. 6.26, reveals that this condition is not exactly fulfilled here.

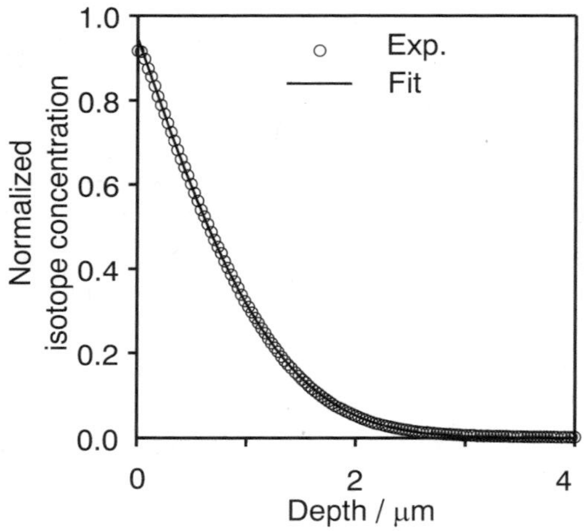

Fig. 6.26: ^{18}O–diffusion profile in $La_{0.8}Sr_{0.2}Mn_{0.8}Co_{0.2}O_{3-x}$ determined by SIMS analysis. In contrast to Fig. 6.53 (page 347) this is almost purely diffusion–controlled. However, the fit detects — and takes account of (cf. Section 6.7) — small surface reaction contributions (the interface would reach a value of 1 for pure diffusion kinetics). Only 20% of the data points recorded have been plotted. From Ref. [399].

6.5 Diffusion kinetics of stoichiometry change

At this stage we are going to treat explicitly the important example of diffusion–controlled alteration of oxygen stoichiometry as an example (Fig. 6.17c). In other words: The rate of dissolution of oxygen in the oxide $M_2O_{1+\delta}$ via chemical diffusion is considered, presupposing that the surface reaction is comparatively rapid.
We assume that a conductance measurement[88] is used to monitor the stoichiometric change. We consider a thin rectangular sheet of isotropic material, which has been equilibrated under a partial pressure of oxygen P_1 ($\widehat{=}M_2O_{1+\delta_1}$). We suddenly alter the partial pressure to the value P_2 and follow the relaxation process to the final state ($\widehat{=}M_2O_{1+\delta_2}$) (Fig. 6.27). The initial homogeneous profile is described by c_1, (here c denotes the concentration of the charge carrier measured in the conductance

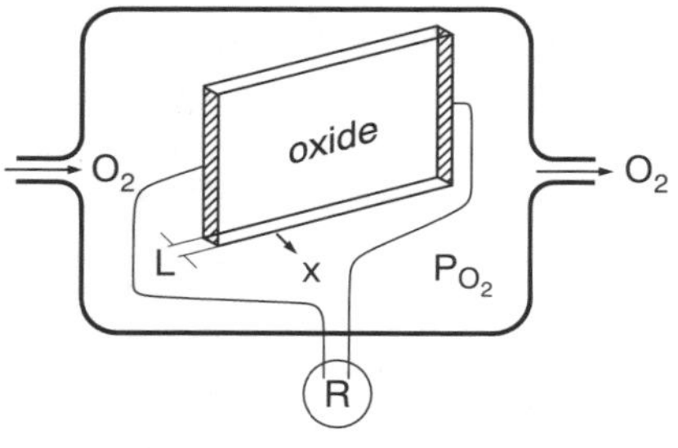

Fig. 6.27: The partial pressure of oxygen over an equilibrated sample is suddenly changed and the conductance is measured as described.

[88]The conductance measurement only serves for detection of the diffusion profile here. The primary experiment is of type (c) in Fig. 6.17.

experiment[89]), the final profile by c_2. The initial condition is, thus, $c(x,t{=}0){=}c_1$ if x is the coordinate normal to the large sheet surfaces. We will neglect the oxygen entering through the small side planes. On account of the rapid surface reaction assumed the value c_2 will be found at the surface immediately after changing the gas partial pressure from P_1 to P_2 at t=0. This leads to the boundary conditions $c(x = \pm L/2; t > 0) = c_2$, if we set the zero–point in the centre of the sheet. The solution is a modulated Fourier series [395]:

$$\frac{c\,(x,t) - c_1}{c_2 - c_1} = 1 - \sum_{0}^{\infty} \frac{4(-1)^i}{\pi\,(2i+1)} \cos\left[\pi\,(2i+1)\,\frac{x}{L}\right] \exp\left[-\pi^2\,(2i+1)^2\,\frac{D^\delta t}{L^2}\right]. \quad (6.71)$$

The profiles set up are illustrated in Fig. 6.28 (see also Fig. 6.23d); they exhibit the weighted cosine–shaped spatial dependence predicted by Eq. (6.71).

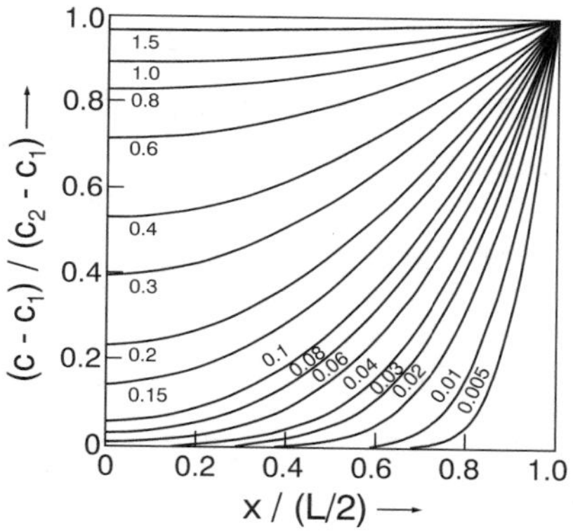

Fig. 6.28: Profiles for half the sample on rapid exchange of the surface concentration on both sides by $x = \pm L/2$. The value $x = 0$ refers to the centre of the sample. The parameter on the curves is $D^\delta t/\,(L/2)^2$. You can see that an approximate equilibrium is reached for the value 2. The mean square displacement is then equal to the sample thickness (see Eq. (6.70)).

We will see below that these profiles can be experimentally verified as a function of space and time using an optical in–situ technique. (Compare Fig. 6.28 with Fig. 6.33 on page 318.) If one would like to carry out such a time and space resolved analysis by conductivity measurements, a great number of microelectrodes would have to be positioned along the diffusion direction.

It is much easier to employ integral techniques. In the experiment sketched in Fig. 6.27, the conductance of the sample is measured parallel to the profile by applying electrodes to the two opposite side planes.

If the conductivity is dominated by one charge carrier then, according to Eq. (5.234), the mean or effective conductivity is

$$\sigma_m^{\parallel}(t) = \frac{1}{L} \int_{-L/2}^{+L/2} zuFc(x,t)dx. \quad (6.72)$$

[89] On account of charge conservation a Fick's "first" and "second" law apply to both the relevant ionic and the relevant electronic defects.

Since only the trigonometric functions in Eq. (6.71) are subject to integration, we obtain for the mean conductivity $\sigma_{\rm m}^{\|}$

$$\frac{\sigma_{\rm m}^{\|}(t) - \sigma_1}{\sigma_2 - \sigma_1} = 1 - \frac{8}{\pi^2}\sum_0^{\infty}\frac{1}{(2i+1)^2}\exp\left(-\frac{t}{\tau_{\rm d}}(2i+1)^2\right) \tag{6.73}$$

which is only time–dependent. The time constant τ^δ serves as an abbreviation of

$$\tau^\delta = \frac{L^2}{\pi^2 D_{\rm O}^\delta}. \tag{6.74}$$

For small times ($t \ll \tau^\delta$) a $\sqrt{t}$ law of the form[90]

$$\frac{\sigma_{\rm m}^{\|}(t) - \sigma_1}{\sigma_2 - \sigma_1} \simeq \frac{4}{\pi^{3/2}}\sqrt{\frac{t}{\tau^\delta}} \tag{6.75a}$$

is obtained as an approximation. In this regime we extract the chemical diffusion coefficient from the slope of the $\sqrt{t}$ plot. It is simpler and usually more reliable[91] to determine $D_{\rm O}^\delta$ from the long–term behaviour. When $t = \tau^\delta$ the first exponential term in Eq. (6.73) is $e^{-1} \simeq 0.36$, the second is already only $e^{-9} \simeq 10^{-4}$, and the third $e^{-25} \simeq 10^{-11}$, so that the infinite sum can be well approximated by the first term. It is helpful to use a somewhat different normalisation by referring to the final state:

$$M_\sigma \equiv \frac{\sigma_{\rm m}^{\|}(t) - \sigma_2}{\sigma_1 - \sigma_2} = \frac{8}{\pi^2}\exp -t/\tau^\delta \tag{6.75b}$$

when $t > \tau^\delta$. In this case plotting $\ln M_\sigma$ against t gives a straight line and $D_{\rm O}^\delta$ can be calculated from the slope. Figure 6.29 illustrates such an evaluation for $YBa_2Cu_3O_{6+x}$: Fig. 6.29a displays the "raw data" (specific resistance plotted against time), in Fig. 6.29b it is plotted in a linearized form according to Eq. (6.75b).

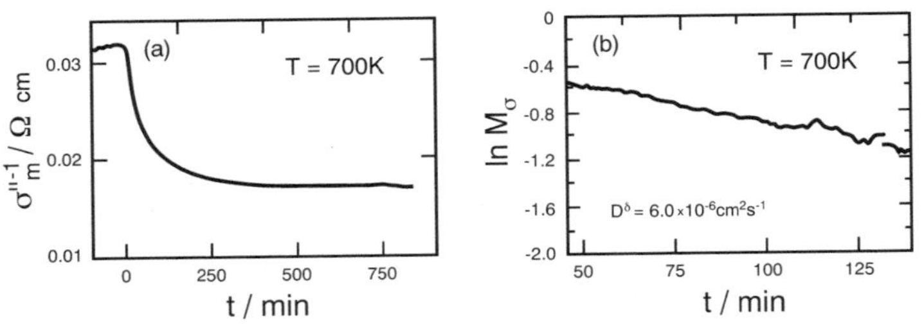

Fig. 6.29: In the logarithmic plot (b) the measurement curve of the specific resistance of $YBa_2Cu_3O_{6+x}$ as a function of time (a) linearizes if the time is not too short; the slope is determined by D^δ and L [400].

---

[90] In general, for short times: $1 - 8\pi^{-2}\sum_0^{\infty}(2i+1)^{-2}\exp\left[-(2i+1)^2 t/\tau^\delta\right] \simeq 4\pi^{-3/2}\sqrt{t/\tau^\delta}$. This solution is obtained directly from the diffusion equation on the assumption of semiinfinite boundary conditions, which are approximately correct for short times.

[91] Other effects may interfere, particularly for short times (e.g. interfacial processes, gas diffusion).

The chemical diffusion coefficient[92] obtained is considerable. Such a high value explains the rapid stoichiometric change which occurs on annealing this material in an oxydizing atmosphere, and is required in order to get a suitable composition which is superconducting at low temperatures ($x > 0.5$).
We can see from Fig. 6.28 that the diffusion experiment is almost complete when the parameter $4D^\delta t/L^2$ has acquired the value 2. This confirms, in an explicit manner, for finite boundary conditions our rule of thumb that a reasonable estimate of the diffusion period t_{eq} results when the square of the mean displacement has reached the value of the thickness of the sample. In this case $t_{eq}/\tau^\delta \simeq \pi^2/2 \simeq 5$ and the relative conductivity change in Eq. (6.75b) has sunken to less than 1%.
It is informative, at this point, to remember the orders of magnitude of possible equilibration times (see Table 6.5), since chemical diffusion is an important serial

Table 6.5: Equilibration times for one–dimensional diffusion in a 1 mm thick sample after a (small) jump in external chemical component potential.

D^δ / cm^2s^{-1}	τ^δ_{eq} (L = 1mm)	
10^{-20}	5×10^{17}s	~4 x Earth's age
10^{-10}	5×10^{7}s	~duration of PhD work
10^{-8}	5×10^{5}s	~1w
10^{-6}	5×10^{3}s	~1h
10^{-5}	5×10^{2}s	~10 min
10^{-4}	50s	~1min
10^{-3}	5s	(fluid phases)
10^{-2}	0.5s	

step in (almost) all solid reactions (see below). Let us select a typical diffusion distance of L=1 mm. Values of $D^\delta \simeq 10^0 cm^2/s \ldots 10^{-2} cm^2/s$, which are typical of fluid phases, form the upper limit for the solid state. In this case the equilibration time is of the order of a second or less. Rapid chemical diffusion processes, such as the oxygen treatment of $YBa_2Cu_3O_{6+x}$ discussed above, exhibit values of the order of $10^{-6}cm^2/s$. Here the equilibration times are in the readily measured order of magnitude of 1 h. A value of about $D^\delta = 10^{-10}cm^2/s$ corresponds to an equilibrium

[92]See the next section for the more complicated interpretation of D^δ in the case of various valence states in $YBa_2Cu_3O_{6+x}$.

time which is typical of the time required for a Ph.D. thesis[93]. Much lower D^δ values are by no means uncommon. A value of 10^{-20}cm^2/s corresponds to a time exceeding the age of the Earth! Clearly a spatial equilibrium with respect to all components is not achievable in such cases. It is worth recalling that low diffusion coefficients are a characteristic of our environment; otherwise structuring would not be possible and almost all contours, including our own, would dissolve. It becomes evident that it is extremely important in many cases to include consideration of (partially) frozen–in states. Figure 6.30 displays a selection of compounds with relatively high chemical diffusion coefficients.

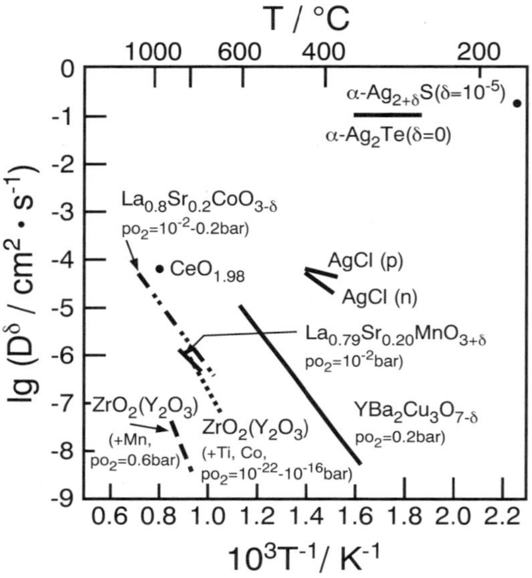

Fig. 6.30: Chemical diffusion coefficients for silver or oxygen in a selection of materials with high D^δ values under specified conditions [401]. It should be noted that a high D^δ value does not necessarily imply a high permeability $(\sigma^\delta)^{63,71}$: Sr–doped $LaCoO_3$ exhibits a much better oxygen permeability (cf. also Fig. 7.1, page 400) than Sr–doped $LaMnO_3$, while the D^δ data cited here are obviously similar. Note also that the content of redox–active impurities in ZrO_2 (Y_2O_3), even though low ($\sim 10^{-4}$) compared with the Y content, has a significant depressive effect on the chemical diffusion coefficient (see Section 6.6.1c) as a result of buffer effects [393, 402].

Let us return to the measurements. The concentration dependence of D^δ (see Eq. (6.54)) shows us that the treatment according to Fick's laws is not well fulfilled for significant gradients. If we wish to avoid numerical or graphical analyses[94] it is necessary, in such cases, to narrow the interval $[P_1, P_2]$ appropriately.

It is also necessary to take account of the fact that generally D^δ possesses tensor character. Figure 6.31 shows the pronounced anisotropy of D^δ in the high–temperature superconductors $YBa_2Cu_3O_{6+x}$ and $Bi_2Sr_2CaCu_2O_{8+x}$, which is expected in view of the anisotropy of the structure (cf. Fig. 5.65, page 209).

One of the conditions necessary for the conductivity method described above to function is that the partial pressure exponent of the conductivity of the dominating charge carrier differs from zero, which is frequently not the case for ionic

[93] The Ph.D. student involved would be well advised to increase the temperature, to reduce the sample thickness, to study a different material, or to change the supervisor.

[94] In such cases the equilibrium concentration entering D^δ is usually replaced by the local nonequilibrium value. Even though it is a helpful a posteriori correction, it is not precise in view of Eq. (6.4). For a consideration of nonlinear corrections see Ref. [329].

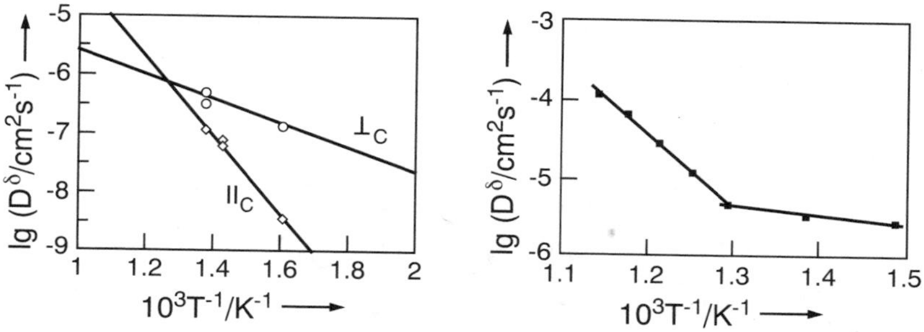

Fig. 6.31: Anisotropy of the chemical diffusion coefficients of $YBa_2Cu_3O_{6+x}$ (left) ($\sim 10^{-2}$ bar O_2) and $Bi_2Sr_2CaCu_2O_{8+x}$ (right) (0.2 bar O_2). The steeper branch of the curve represents the diffusion in the c direction, the flatter one the diffusion coefficients in the ab plane. Left: two different orientations of the single crystal. Right: parallel connection of various paths in a polycrystalline sample [403].

conductors[95]. Another parameter that can be used to analyse the defect–chemical relaxation, is weight change. However, this presupposes large differences in stoichiometry, which are not expected for ionic conductors. In the case of binary compounds, this method is therefore restricted to materials which tolerate significant redox effects and hence also not applicable to typical ionic conductors.
A very elegant method that can definitely be employed for ionic conductors, such as ZrO_2 (Y_2O_3), consists of spectroscopically observing the concentration changes of optically detectable defects. Traces, for example, of Fe^{4+} in $SrTiO_3$ or of Ni^{3+} in the case of ZrO_2 are readily detected optically. For this purpose the materials should be weakly doped (see Section 5.7.1). As a result of the defect chemical equilibrium the distribution of the redox states (that is the ratio Fe^{3+}/Fe^{4+} or the Ni^{3+}/Ni^{2+} ratio) is an unambigous function of the local oxygen activity for a given temperature and given total doping level (see Section 5.7.1). If we measure the integral optical absorption[96], an expression, corresponding to Eq. (6.75), is obtained, whose time–dependence yields D^{δ}. Here it is to be noted that (unlike the conductance experiment) the measurement should be performed perpendicular to the interface[97] in order to obtain a measure proportional to the concentration. (Since internal defect chemical reactions necessarily play a role here, the analysis of D^{δ} will be delayed

[95] See Chapter 7 for selective measurement of minority charge carriers and for the determination of chemical diffusion coefficients by electrochemical methods.

[96] ESR, NMR and Mössbauer techniques are also amongst the methods suitable [8,404,405].

[97] The decrease in the intensity (I) on passing through a virtually homogeneous region of thickness dx is determined by $\ln \frac{I(x+dx)}{I(x)} = -\alpha dx \propto -c dx$. On passage through the inhomogeneous sample perpendicularly to the interface, the integral of c over x has to be considered in the integral measurement. When the profiles are locally resolved (parallel to the interface) the local absorption coefficient α (and hence c(x)) is measured directly (more precisely $\alpha(c)$ times thickness in the direction normal to the x axis).

until the next section, where it will be shown that the evaluation described here is correct.) The elegance of the method is that the in–situ profiles can also be resolved spatially, even at high temperatures[97]. Figure 6.32 shows a sequence of images

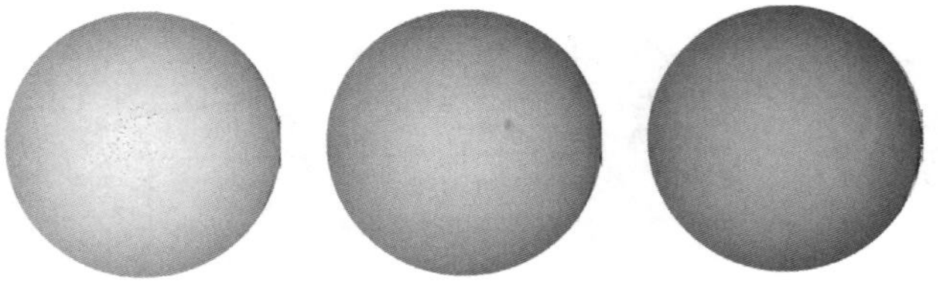

Fig. 6.32: In situ snapshots of the local profiles of Fe^{4+} concentration in $SrTiO_3$ as a function of time, using the digital camera technique (CCD–camera, i.e. charge coupled device) [406].

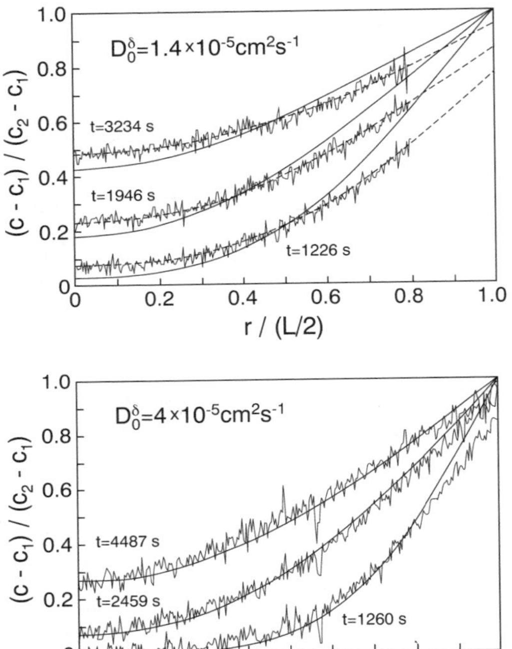

Fig. 6.33: The concentration profiles (top: 893 K, bottom: 848 K) obtained as a function of space and time. The profiles are obtained by fitting with a constant D^δ. The introduction of a concentration dependence only alters the result slightly. It can be seen that the assumption of a pure diffusion control (solid lines) is better fulfilled for diffusion from fresh surfaces (bottom) than is the case for the relaxed surfaces (top). In this case the fit only becomes perfect if the surface control is taken into account (broken line). The analysis of the above set of curves then yields — instead of 1.4×10^{-5}cm^2/s — more precisely $D^\delta = 2.0 \times 10^{-5}$cm^2/s with a rate constant $\bar{k}^\delta = 2 \times 10^{-4}$cm/s. The fact that a radial geometry was analyzed (see abscissa) top and a cartesian geometry below does not affect the result.

obtained by a CCD–camera, and Fig. 6.33 a time–sequence of the corresponding stoichiometry profiles for $SrTiO_3$, which can be used for an exact determination of D^δ. Since here disc–shaped crystals were employed through the lateral surface of which oxygen diffused in, while the profiles developing under the glass–protected circular surface were "observed" (see Fig. 6.25), the appropriate, geometry adapted diffusion law[98] reads (r: distance from the centre):

$$\frac{\partial c}{\partial t} = \frac{1}{r}\frac{\partial}{\partial r}\left(rD^\delta \frac{\partial c}{\partial r}\right). \tag{6.76}$$

[98]See, for example, Ref. [407] for conversion from Cartesian to curvilinear coordinates.

As can be seen, even in Fig. 6.26 deviations from the conformity to a second Fick's law reveal the influence of the surface reaction. This point will be developed in Section 6.7, while the interpretation of the D^{δ} values is described in the next section. Not only dissolution of gaseous components is of importance. Of special significance is the introduction of Li in Li_xMO_2 (M= Ni,Co,V, see Section 7.4.3) as a so–called intercalation process ("guest–host–reaction") which can occur there over a wide range of compositions with invariant morphology. The latter point is relevant if this reaction is made use of in electrochemical applications (electrode function in lithium–batteries, see Section 7.4).

6.6 Complications of matter transport

In this section we will consider complications resulting from internal interactions and the occurrence of boundary layers.
We have already referred briefly to complications caused by anisotropies (tensor character of the diffusion coefficient) as well as to complications caused by possible concentration dependences[99]. For more details of these problems and those dealing with effects which are caused by mechanical stress phenomena the reader is referred to the literature (see e.g. Refs. [8,408,409]). Nonidealities which can be traced back to concentration effects in the conductivity have been treated in Chapter 5 and are not followed up here either.

6.6.1 Internal interactions

a) Conductivity and jump relaxation

When considering the mobility in the case of higher concentrations, the finite number of jump partners becomes important, as well as interactions. While the first can be taken account of in a straightforward way by introducing a concentration term for the jump partners (Section 6.2), the occurrence of interactions requires the consideration of correlations.
The jump relaxation model of Funke is a concept of wide validity [410]. In certain aspects it may be compared to the Debye–Falkenhagen theory [411] of liquid electrolytes. The interaction of the point defects expresses itself in a relatively flat defect potential[100], that is superimposed on the lattice potential, as shown in Fig. 6.34.
In toto, the neighbouring defects act in a repulsive way on a jump of the carrier under consideration from A to B. As long as the defect environment has not adjusted to the new situation, the potential minimum at B is higher than that at A, causing

[99] The problems of the reference system were also referred to briefly, but will not be followed up here.

[100] The defect potential is obtained by calculating the energy of interaction for various displacements of the central particle (cf. Section 5.7.2).

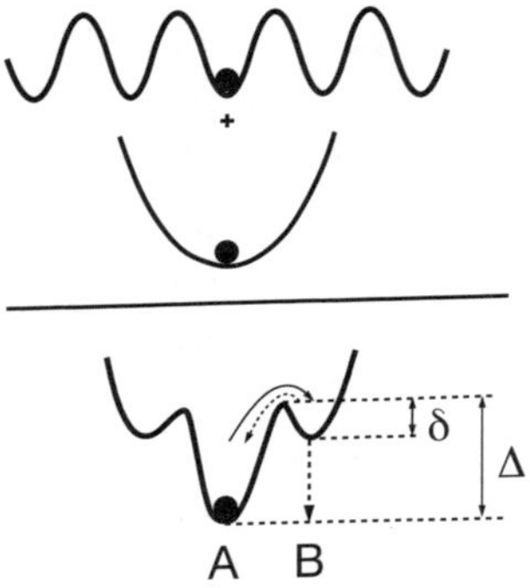

Fig. 6.34: A jump within the true potential (lattice potential (top) + defect potential (centre)) requires the surmounting of a relatively high activation energy and a subsequent relaxation of the environment, before the site B takes on the original potential surroundings of A (see text). From Ref. [412].

the back jump to be greatly favoured as long as the relaxation of the surroundings has not taken place. A successful jump of a particle requires the relaxation of the environment[101].
First let us consider the two following materials as artificial limiting cases. (i) The relaxation of the environment is extremely slow in material 1, so that almost all jumps are unsuccessful. Thus, the particle essentially carries out slight displacements from its original position, which can only be observed at extreme time resolution, i.e. at a high frequency of measurement (compare Section 7.3.6). The d.c. conductivity is virtually zero. (ii) The relaxation is infinitely fast in material 2; the situation is then as that treated until now and the conductivity is independent of the measurement frequency. The two processes (jump between A and B and relaxation of the environment) are not independent of each other in realistic materials[102] and the frequency–dependent conductivity mirrors the interaction. Figure 6.35 shows an example. The plateau towards lower frequencies corresponds to the d.c. conductivity. Only the successful jumps are recorded in this range, while at higher frequencies the unsuccessful ones also contribute. At very high frequencies vibra-

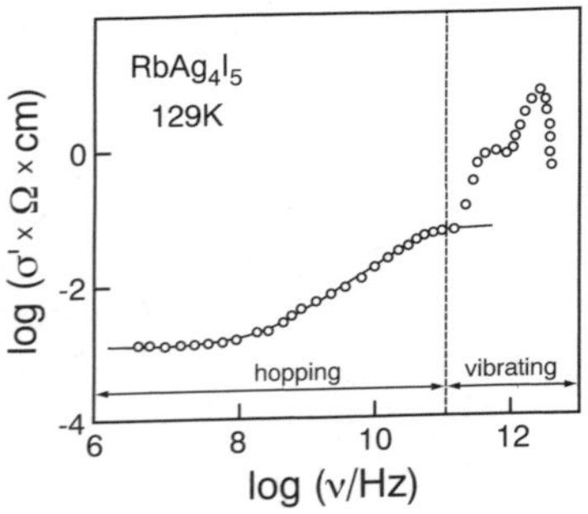

Fig. 6.35: Frequency–dependent conductivity of $RbAg_4I_5$ at 129K. (More precisely σ' represents the real part of the complex conductivity, cf. Chapter 7.) The continuous line is reproduced from the jump-relaxation model Refs. [410, 413,414]. The structure in the high frequency range results from the excitation of oscillatory silver ion motion and optical phonons. By plotting the real part (σ', cf. Chapter 7) capacitive effects are separated out to a good approximation. From Ref. [410].

[101] Here, it is primarily the relaxation of the defect environment that is meant. It is assumed that the structural relaxation of the perfect lattice takes place in the time or frequency range of the lattice vibration (and is, thus, of the same order of magnitude as the attempt frequency, see Chapter 3).

[102] It has been found in many experimental examples that the tendency of an ion to jump back and the tendency of the neighbours to reorganize themselves are proportional to each other [412].

tional processes are solely observed. In the intermediate region ($10^8 - 10^{11}$Hz in Fig. 6.35) the time–dependence of relaxation is very important and the result is a time–dependent correlation factor (here with respect to the conductivity experiment, cf. Section 6.3.2)[103]. The treatment quantitatively explains the high frequency behaviour of the jump process in many materials, as confirmed in measurements of various crystalline or amorphous ionic conductors (see Fig. 6.35). We will return to this again in Chapter 7[104].
Detailed simulations [416] indicate the importance of inhomogeneity effects[105] for the dispersion (cf. static distribution of thresholds and time constants, see also Section 7.3.6). This is of special importance in the case of amorphous systems. While the jump relaxation model explicitly addresses dynamic inhomogeneity effects in the sense of Fig. 6.34, possible static inhomogeneities have to be included in the effective defect potential and the relation between jump and relaxation rate. A competent recent review is given in Ref. [418].
Formally the mobility becomes, in general, a complicated function of the composition, when we face high–defect concentrations. This must be reflected in both σ and D^* (see Fig. 6.19).

An interaction effect, known as the "mixed alkali effect" [419–421], depends on the presence of two different mobile ion types[106]. Thus, we observe, in mixed crystals of $(Na,K)_2O$–β-Al_2O_3 or $(Na,Ag)_2O$–β-Al_2O_3 at comparable concentrations, conductivities, which are lower than for the pure components by orders of magnitude (see Fig. 6.36) [420]. The qualitative explanation is simple. Each ion in the pure components has a preferred environment. In the mixed crystal the individual neighbourhoods conflict with each other in the conduction process [425]. Such losses in mobility can come into being, in general, as a result of repulsive (tendency to segregation) or attractive (tendency to agglomeration) interactions of the two carrier species. The excess enthalpies, which are expected to dominate the activity coefficients for the mixing of the separate components, indicate such interactions (see Chapter 4) [421]. The sensitivity of the effect is remarkable. Even small quantities of K_2O reduce the conductivity of (Na_2O)–β-Al_2O_3 by orders of magnitude. Analo-

[103] The effect corresponds to the relaxation effect in the Debye–Hückel–Onsager theory for liquid electrolytes [219]. It remains to be investigated in how far quasi–Madelung models are helpful in this context for detailing the phenomenon in solids [115,231].

[104] Cf. Ref. [415].

[105] Conductivity effects in highly doped alkaline earth fluorides and lead fluorides have been explained by a concentration dependent distribution of migration enthalpies [417].

[106] On the other hand, significant enhancement effects are known in the homogeneous solutions Ag(I,Br) or $(Ca,Sr)F_2$ when compared to the end–numbers. As far as an increased carrier concentration is concerned, this phenomenon may be conceived as the extreme case of the effects discussed in Section 5.8.5 for an atomistic separation distance, resulting in a diminished formation enthalpy. In addition, a mobility variation can also be expected. Note that, in contrast to the mixed alkali effect, here the counterions (with respect to the mobile ones) are mixed. A conductivity enhancement due to a "mixed counterion effect" has also been noticed in Refs. [422, 423].

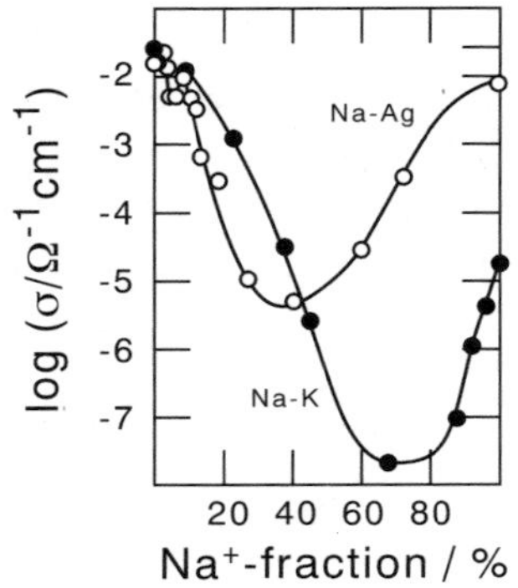

Fig. 6.36: Mixed alkali effect in $(Na,K)_2O$–β-Al_2O_3 and $(Na,Ag)_2O$–β-Al_2O_3. From Ref. [424].

gous effects in glasses are also marked and require more subtle interpretations [426, 427].

Continuous (percolating) pathways are generally required for appreciable d.c. conductivities (cf. page 396). In the case of electron conduction the conductivity is appreciable when the corresponding orbitals overlap (cf. Section 2.2). As discussed in Section 5.3, electronic charge carriers can be generated by impurities. If the distance between the impurities is less than a critical distance, direct migration from dopant to dopant is possible (formation of an impurity band, see Section 6.2).
Transport properties become particularly difficult to treat in inhomogeneous and heterogeneous media (see also Section 5.8). The reader is referred to Ref. [281] and Sections 5.8, 6.10 and 7.3.7.

b) Chemical diffusion and local relaxation

An essential simplifying assumption that we make in this book in most cases is the assumption of local equilibrium. In the case of chemical diffusion it means that, irrespective of existing gradients, the defects are locally equilibrated with respect to the local chemical component potential. This requires sufficiently sluggish transport and sufficiently fast relaxation rates, as shown to be established in various cases (see Kirkendall effect [428], see following section). As set out in Ref. [8] the validity of this assumption may be questioned, especially in multinary compounds (cf. also Section 6.6.1a). Generally, the gradients should not be too large, otherwise the system is driven too far from thermodynamic equilibrium and the effective local chemical potential depends on the gradients. In all these cases the precise description becomes complicated and involved (see also previous section). The interested reader is referred to Refs. [8,9,429,430,409] as well as to Section 5.8.3 as far as analogous questions involving semiconductors are concerned (cf. footnote 160 on page 230). For simplicity's sake we presuppose the validity of local equilibrium in the following considerations.

c) Chemical diffusion and conservative ensembles

Now let us turn to the resulting consequences of internal charge carrier interactions on chemical diffusion. The respective transport coefficient (chemical diffusion coefficient) includes, as we know from the formal treatment in Section 6.3.3, both an effective ambipolar conductivity and an effective ambipolar concentration, the latter being correlated with the thermodynamic factor.

If defects are not randomly distributed, it is not possible to neglect the activity coefficients. In better approximation then $c^{1/2}$ or $c^{1/3}$ laws (see Section 5.7.2) may be introduced in the thermodynamic factors of the defects (Eq. 6.54a)).

However, it may be necessary to make corrections already in the starting flux equations. The detailed formulation of the linear, irreversible thermodynamics also includes (symmetrical) coupling terms (cross–terms) (cf. Section 6.10). They take into account that the flux of a defect k may also depend on the gradient of the electrochemical potential of other defects. This concept has been worked out, in particular, for the case of the ambipolar transport of ions and electrons [390].

A more explicit procedure is to introduce chemical interactions directly: There is a certain analogy with the equilibrium situation where it is possible to avoid the use of activity corrections over wide ranges by considering associates to correct for the interactions (see Section 5.7). The relevance of this method of treatment is particularly evident if such associates can also be detected experimentally, e.g. by spectroscopic techniques in the case of ionic defects with differing charge states. This leads to a rescaling of defect concentrations and defect fluxes, which can now be taken to be ideal again to a good approximation (i.e. with negligible activity corrections values and negligible coupling terms in the Onsager relationships). However, it is now necessary to include source and sink terms from the start, on account of the internal dissociation and association reactions (see Fig. 6.21). This leads us, in the case of local equilibrium, to the concept of the "conservative ensemble" [393,431, 432], which will now be described.

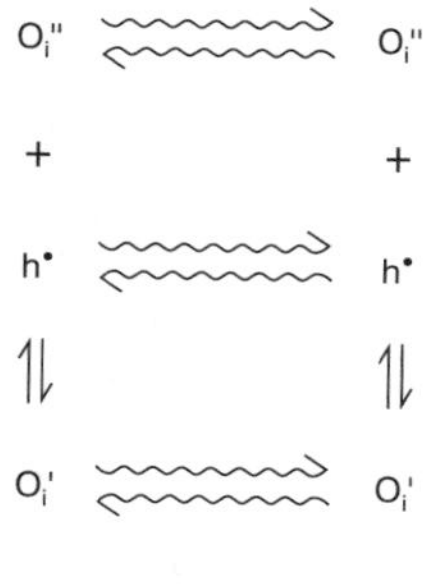

Fig. 6.37: If we can also allot individual existence to each defect state for transport, then the internal mass and charge transport is a reaction–diffusion problem.

Figure 6.37 uses the example of associate formation between the ionic defect O_i'' and the electronic defect $h^{\cdot}$ to emphasize that the strict treatment requires the solution of coupled diffusion–reaction relationships, describing the general (electro–)chemical reaction scheme with individual diffusion or rate constants as parameters (cf. Section 6.1). Source terms (q) must be taken into account in the relevant continuity equations, e.g. for defect B that can be created by

$$\nu_A A \rightleftharpoons \nu_B B. \tag{6.77}$$

In the case that Eq. (6.77) describes the decisive elementary reaction, the continuity equation (6.77) for B (see Section 6.4 as well as footnote 1 in Section 6.1) is

$$\frac{\partial c_B}{\partial t} = -\mathrm{div}\mathbf{j}_B + q_B \equiv -\mathrm{div}\mathbf{j}_B + \nu_B \mathcal{R} = -\mathrm{div}\mathbf{j}_B + \nu_B \left(\vec{k} c_A^{\nu_A} - \overleftarrow{k} c_B^{\nu_B}\right). \tag{6.78}$$

In the general case the problem depends on the rate constants of the association and dissociation processes. Since we assume local equilibrium (interaction much faster than transport) the situation simplifies considerably (only $K_{as/dis}$ is included, not the rate constants themselves). It would be wrong to believe that the second term in Eq. (6.78) can be neglected as a consequence of local equilibrium. It is true that the bracketed term in the formulation

$$\frac{\partial c_B}{\partial t} = -\mathrm{div}\mathbf{j}_B + \nu_B \vec{k} c_A^{\nu_A} \left(1 - \frac{c_B^{\nu_B}}{c_A^{\nu_A}} \frac{\overleftarrow{k}}{\vec{k}}\right) \tag{6.79}$$

is close to zero because $c_B^{\nu_B}/c_A^{\nu_A} \simeq (\widehat{c_B^{\nu_B}}/\widehat{c_A^{\nu_A}}) = \vec{k}/\overleftarrow{k} = K$; however, the second part is not negligible in comparison with the flux divergence on account of the high values of the individual rate constants (see prefactor). Expressed in another manner: The high partial rates of the forward and back reaction are similar; the difference between the forward and back reaction rate is small in itself, but not with respect to the difference between the rate of influx and outflux. An analogy may serve to illuminate this point. Let us consider a salt crystal in contact with saturated solution and assume rapid dissolution and precipitation reactions, i.e. local solution equilibrium. We now dilute the solution continuously. Precisely because local equilibrium is established, the salt crystal dissolves; under stationary conditions ($\partial c/\partial t = 0$) the rate of dissolution even compensates the dilution flux.

It can now be shown that as far as the diffusion of certain ensembles, namely of the "conservative ensembles", is concerned, the source terms disappear. The resulting chemical diffusion coefficients then have to refer to this ensemble.

Let us consider an anti–Frenkel disordered material, taking into account both O_i'' and $V_O^{\cdot\cdot}$, but initially neglecting the occurrence of variable valence states. (In addition, we will assume a quasi one–dimensional situation.) Doing this we reduce the problem to a relatively trivial case. For it is immediately evident that the source terms disappear on consideration of the total ion flux (or current)

$$j_{O^{2-}} = j_{O_i''} - j_{V_O^{\cdot\cdot}} \quad \text{or} \quad i_{O^{2-}} = i_{O_i''} + i_{V_O^{\cdot\cdot}} \tag{6.80}$$

and the total electron flux

$$j_{e^-} = j_{e'} - j_{h^{\cdot}} \quad \text{or} \quad i_{e^-} = i_{e'} + i_{h^{\cdot}}, \tag{6.81}$$

In this sense $(c_{O_i''} - c_{V_O^{\cdot\cdot}})$ and $(c_{e'} - c_{h^{\cdot}})$ refer to "conservative ensembles". This is self–evident since all changes take place within these ensembles. Because

$$O_O + V_i \rightleftharpoons V_O^{\cdot\cdot} + O_i'' \tag{6.82}$$

and

$$\mathrm{Nil} \rightleftharpoons e' + h^{\cdot} \tag{6.83}$$

it follows that $q_{V_O^{\cdot\cdot}} = q_{O_i''}$ and $q_{e'} = q_{h^{\cdot}}$ and, hence, $q_{O^{2-}} = q_{O_i''} - q_{V_O^{\cdot\cdot}} = 0$ and $q_{e^-} = q_{e'} - q_{h^{\cdot}} = 0$, if q represents the increase in concentration with time resulting from the defect chemical reactions[107]. Since $\sigma_{O^{2-}} = \sigma_{O_i''} + \sigma_{V_O^{\cdot\cdot}}$ and $\sigma_{e^-} = \sigma_{e'} + \sigma_{h^{\cdot}}$, it follows that the flux equations, formulated in terms of O^{2-} and e^-, are the same as those given in Section 6.5. The self–diffusion coefficient of the ions is then

$$D_{O^{2-}} = \frac{RT}{4F^2}\frac{\sigma_{O^{2-}}}{c_{O^{2-}}} = \frac{c_{O_i''}}{c_{O^{2-}}} D_{O_i''} + \frac{c_{V_O^{\cdot\cdot}}}{c_{O^{2-}}} D_{V_O^{\cdot\cdot}} \tag{6.84}$$

and the chemical diffusion coefficient reads

$$D_O^{\delta} = \frac{1}{4F^2}\frac{\sigma_{O^{2-}}\sigma_{e^-}}{\sigma}\frac{d\mu_O}{dc_O} = \frac{RT}{4F^2}\frac{\sigma_{O^{2-}}\sigma_{e^-}}{\sigma}\left(\frac{\partial \ln a_{O^{2-}}}{\partial c_{O^{2-}}} + 4\frac{\partial \ln a_{e^-}}{\partial c_{e^-}}\right). \tag{6.85}$$

On account of the local equilibrium $(\mu_{O^{2-}} = \mu_{O_i''} = -\mu_{V_O^{\cdot\cdot}})$ it follows for the ions that $\partial \ln a_{O^{2-}} = \partial \ln a_{O_i''} = -\partial \ln a_{V_O^{\cdot\cdot}}$ and analogously for the electrons that $\partial \ln a_{e^-} = \partial \ln a_{e'} = -\partial \ln a_{h^{\cdot}}$. Simultaneously $\partial c_{O^{2-}} = \partial c_{O_i''} - \partial c_{V_O^{\cdot\cdot}}$ and $\partial c_{e^-} = \partial c_{e'} - \partial c_{h^{\cdot}}$. In the case of dilute solutions (defect activity = defect concentration) the re–arrangement leads to:

$$D_O^{\delta} = \frac{RT}{4F^2}\frac{(\sigma_{V_O^{\cdot\cdot}} + \sigma_{O_i''})(\sigma_{e'} + \sigma_{h^{\cdot}})}{\sigma}\left(\frac{1}{c_{O_i''} + c_{V_O^{\cdot\cdot}}} + \frac{4}{c_{e'} + c_{h^{\cdot}}}\right) \tag{6.86}$$

which is more general than Eq. (6.54). Now the sum of the defect concentrations appears in the denominator. (When there is local equilibrium, this result can be generalized to cation defects.)[108]

If we permit interactions between ions and electrons, the situation becomes qualitatively more interesting and quantitatively more complex (see Fig. 6.37). In such cases we introduce variable valence states by permitting[109] O^- and O^0 as valence

[107] $q_k \equiv \dot{c}_k|_{chem} = \Sigma_r \nu_{rk}\mathcal{R}_r$ ($\mathcal{R}_r$: rate of reaction r which leads to the production of k; ν_{rk}: stoichiometric coefficient of k in reaction r) (see Section 4.2, page 72).

[108] The appearance of the sum of the defect concentrations is the expression of the local convertibility. This is not fulfilled for Schottky defects in an ideal single crystal!

[109] The objection that such valences are unrealistic can readily be rebutted by the consideration that unusual valences are already realized by the existence of e′ and h˙. Here we are dealing with the interaction of these "unusual" valences. Examples of such associates have also been described already in Section 5.7.1. The question of the circumstances, under which such associates carry individual mobilities, is more subtle.

states, i.e. O_i'', O_i', $O_i^\times$, $V_O^{\cdot\cdot}$, $V_O^{\cdot}$, $V_O^\times$ as defects. Then the electronic and ionic ensembles (O^{2-} and e^-) are no longer conservative but merely the combinations {O} and {e} as defined by Fig. 6.38. As can be derived from the reaction scheme, it follows that:

$$q_{\{O\}} \equiv q_{O^{2-}} + q_{O^-} + q_{O^\times} = q_{O_i''} - q_{V_O^{\cdot\cdot}} + q_{O_i'} - q_{V_O^{\cdot}} + q_{O_i^\times} - q_{V_O^\times} = 0 \qquad (6.87a)$$

and

$$q_{\{e\}} = q_{e'} - q_{h^\cdot} - q_{O_i'} + q_{V_O^{\cdot}} - 2q_{O_i^\times} + 2q_{V_O^\times} = 0. \qquad (6.87b)$$

This result is immediately understandable because the concentration of the oxygen ensemble so–defined (see Fig. 6.38, third arrangement in the upper row)

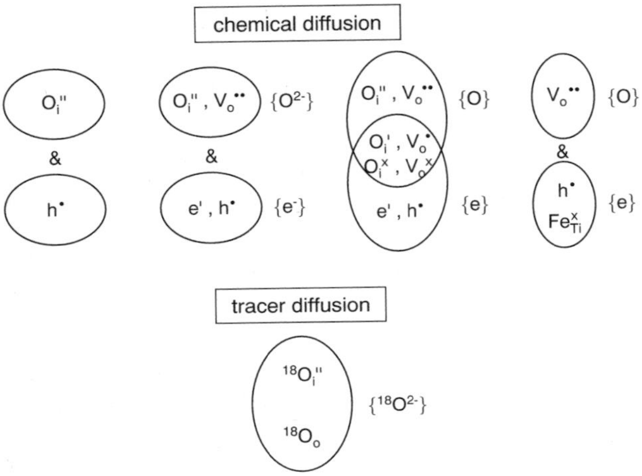

Fig. 6.38: Conservative ensembles in the case of complex defect chemistry for various examples discussed in the text. The diagram on the far right describes the situation for chemical diffusion in Fe–doped $SrTiO_3$ [433].

$$c_{\{O\}} = c_{O_i''} + c_{O_i'} + c_{O_i^\times} - c_{V_O^{\cdot\cdot}} - c_{V_O^{\cdot}} - c_{V_O^\times}, \qquad (6.88a)$$

$$c_{\{e\}} = c_{e'} - c_{h^\cdot} - c_{O_i'} + c_{V_O^{\cdot}} - 2c_{O_i^\times} + 2c_{V_O^\times} = -2c_{\{O\}} \qquad (6.88b)$$

obviously represents the deviation from the stoichiometric composition (δ), which is unchanged by the internal reaction[110]. Equation (6.87b) then follows as a consequence of the condition of electroneutrality $((-2c_{\{O\}} = c_{\{e\}})$.
For D^δ, now a structurally different expression is obtained after a lengthy calculation [393,432]:

$$D_O^\delta = \frac{1}{4F^2}\left[2\sigma_{O^-} + 4s_{O^0} + \frac{(\sigma_{O^{2-}} + 2\sigma_{O^-})(\sigma_{e^-} - \sigma_{O^-})}{\sigma}\right]\frac{d\mu_O}{dc_O} \qquad (6.89)$$

with the abbreviations $\sigma_{O^-} = \sigma_{O_i'} + \sigma_{V_O^{\cdot}}$ and $s_{O^0} = \frac{F^2}{RT}\left(D_{V_O^\times}c_{V_O^\times} + D_{O_i^\times}c_{O_i^\times}\right)$. In a purely formal manner, the conductivity factor in Eq. (6.89) now permits oxygen permeation even for a zero partial electronic conductivity (as a result of the migration

[110] This applies, as long as reactions that couple anion and cation partial structures, do not occur. Otherwise $c_{\{M\}} + c_{\{O\}}$ (i.e. δ) is to be considered.

of neutral defects[111] but also as a result of an ambipolar migration[112] of $2O^-$ and O^{2-}). The term $d\mu_O/dc_O$ takes another form as well when reformulated in terms of charge carriers, e.g. in terms of the fully ionized particles as

$$\frac{d\mu_O}{dc_O} = RT\left(\frac{\chi_{O_i''}}{c_{O_i''}} + 4\frac{\chi_{h^\cdot}}{c_{h^\cdot}}\right) = RT\left(\frac{\chi_{V_O^{\cdot\cdot}}}{c_{V_O^{\cdot\cdot}}} + 4\frac{\chi_{e'}}{c_{e'}}\right). \tag{6.90}$$

The χ_k–terms refer to differential defect fractions and are defined via the corresponding conservative ensembles according to

$$\chi_k = \frac{\partial c_k}{\partial c_{\{k\}}}, \tag{6.91}$$

whereby $c_{\{k\}} = c_{\{O\}}$ or $c_{\{e\}}$ in the case of the excess particles (O_i'', e'), or $-c_{\{O\}}$ and $-c_{\{e\}}$ in the case of missing ($V_O^{\cdot\cdot}$, $h^\cdot$) particles, respectively. These can be calculated from the defect chemistry.

In the case that the carriers used for the explicit formulation in Eq. (6.90) (O_i'', e') are the decisive carriers, and that $c_{O_i''}$, $c_{e'}$ are approximately the concentrations when ignoring the association, the χ–terms can also be considered to be activity corrections.

If we exclude associates totally, we return to our simple result, namely to Eq. (6.86), for then

$$\chi_{O_i''}^{-1} = \frac{\partial c_{O_i''} - \partial c_{V_O^{\cdot\cdot}}}{\partial c_{O_i''}} = 1 - \frac{\partial c_{V_O^{\cdot\cdot}}}{\partial c_{O_i''}} = 1 - \frac{c_{V_O^{\cdot\cdot}}}{c_{O_i''}}\frac{\partial \ln c_{V_O^{\cdot\cdot}}}{\partial \ln c_{O_i''}}, \tag{6.92}$$

Locally $\partial \ln c_{V_O^{\cdot\cdot}} = -\partial \ln c_{O_i''}$ on account of the Frenkel equilibrium, so that

$$\chi_{O_i''}/c_{O_i''} = \frac{1}{\left(c_{O_i''} + c_{V_O^{\cdot\cdot}}\right)}. \tag{6.93}$$

In recent years it has been found that these considerations are not merely academic, but constitute necessary and considerable corrections in many important cases.

Thus for $YBa_2Cu_3O_{6+x}$, even at fairly high temperatures, it is necessary to take account of variable valences for the ionic defects; this certainly affects the concentration term in D_O^δ and, probably, the conductivity term in D_O^δ, too. The situation is similar for mixed conducting copper ion conductors. A more recent example refers to the transport of different valence states of hydrogen in oxides [434]. The literature should be consulted here for more details [187].

Fe–doped $SrTiO_3$, which has already been discussed above, is a well investigated example, in the case of which the above effects are not of relevance for σ_O^δ but rather appreciably influence $\partial\mu_O/\partial c_O$ and hence c_O^δ. The internal source and sink reaction is the conversion of Fe^{3+} to Fe^{4+}, which we formulate, with advantage, as

[111] When $\sigma_{e^-} = \sigma_{O^{2-}} = \sigma_{O^-} = 0$, $\sigma_O^\delta = 4s_{O^0}$ remains in the square brackets.

[112] When $\sigma_{e^-} = s_{O^0} = 0$ then $\sigma_O^\delta = \frac{\sigma_{O^-}\sigma_{O^{2-}}}{\sigma_{O^-}+\sigma_{O^{2-}}}$ remains exhibiting an analogous form to σ^δ in Eq. (6.54).

Reaction As = $$\mathrm{Fe'_{Ti} + h^{\cdot} \rightleftharpoons Fe^{\times}_{Ti}} \tag{6.94}$$

in the high partial pressure regime. In this case the mass action constants and mobilities, required to test the relationships quantitatively, are well–known.
If we plot the D^{δ}_{O} values obtained by a variety of methods simultaneously with those calculated using Eq. (6.54c), then there is very good agreement only at high temperatures (see Fig. 6.39), but deviations of several orders of magnitude at low

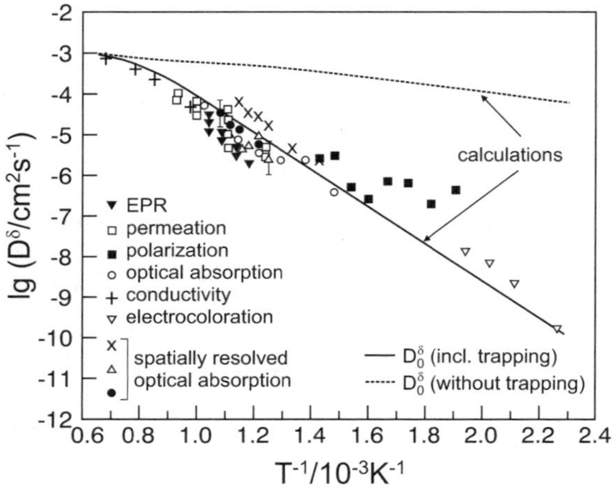

Fig. 6.39: The chemical diffusion coefficient of oxygen in $SrTiO_3$ as a function of the temperature. The broken line only includes the doping effect with respect to σ^{δ} as determined by the ionization reaction (i.e. the effects of doping with an acceptor, whose concentration is equivalent to that of Fe'_{Ti}, but which is not redox active). The continuous line now also contains the correction due to the χ–terms (i.e. additional effect with respect to c^{δ}). The calculation applies to a doping content of $10^{19}cm^3$ and an oxygen partial pressure of 10^5Pa [435].

temperatures, even if the effects of the ionization reaction on the conductivities, are taken into account in Eq. (6.54c). The reasons for the deviations are more sophisticated: Equation (6.94) represents an internal source/sink for the holes. The free holes no longer represent a conservative ensemble, in contrast to the total number of the free and trapped holes ($-c_{\{e\}}$), for which it is possible to formulate a Fick's second law directly. Since the mobility of the trapped holes ($Fe^{\times}_{Ti}$) can be assumed to be zero, the effect is only reflected in the concentration term by the appearance of a $\chi_{h^{\cdot}}$ factor, that reduces the chemical diffusion coefficients (see also Fig. 6.20 on page 305). This can readily be understood qualitatively. As has been demonstrated (see Eqs. (6.52, 6.89)) D^{δ} is essentially the quotient of an effective conductivity (σ^{δ}) and an effective concentration (c^{δ}); the first term remains unchanged, the second is greatly increased as a result of the presence of $Fe^{\times}_{Ti}$ defects. These are immobile, but can make mobile holes available any time by dissociation

$$\begin{array}{lcl} \mathrm{V^{\cdot\cdot}_{O}} & \rightsquigarrow & \mathrm{V^{\cdot\cdot}_{O}} \\ \mathrm{2h^{\cdot}} & \leftsquigarrow & \mathrm{2h^{\cdot}} \\ & & \uparrow\downharpoonleft -\mathrm{2Fe'_{Ti}} \\ & & \mathrm{2Fe^{\times}_{Ti}} \end{array} \tag{6.95}$$

and, thus, enlarge c^{δ}. In other words, trapping does not simply alter the resistive component ($1/\sigma^{\delta}$) in D^{δ} ($\propto \frac{\sigma^{\delta}}{c^{\delta}}$), it also strongly increases the capacitive component ($C^{\delta} \propto c^{\delta}$, see Section 6.7.4) of D^{δ}. However, the precise interpretation is not quite so simple: χ is a differential parameter, as the capacitance to be taken into account is a differential capacitance $(\partial c_O/\partial\mu_O)$. The electroneutrality relation yields $2[V_O^{\cdot\cdot}] \simeq [Fe'_{Ti}] = C - [Fe_{Ti}^{\times}]$ and $2\partial[V_O^{\cdot\cdot}] = -\partial([h^{\cdot}] + [Fe_{Ti}^{\times}])$. Owing to conservation of mass and ionic equilibrium the correction follows as[113]

$$\chi_{h^{\cdot}} = \frac{\partial[h^{\cdot}]}{\partial[h^{\cdot}] + \partial[Fe_{Ti}^{\times}]} = \frac{(1+K_{as}[h^{\cdot}])^2}{(1+K_{as}[h^{\cdot}])^2 + CK_{as}}. \tag{6.96}$$

The concentration $[h^{\cdot}]$ is obtainable from the set of defect chemical equations as a function of P, T, C; the quantity $\chi_{h^{\cdot}}$ itself is, thus, a sensitive function of these three parameters.

Instead of Eq. (6.54e) we find

$$D^{\delta} = \frac{\sigma_{h^{\cdot}}}{\sigma} D_{V_O^{\cdot\cdot}} + \frac{\sigma_{V_O^{\cdot\cdot}}}{\sigma} \chi_{h^{\cdot}} D_{h^{\cdot}}. \tag{6.97}$$

We can see from Eq. (6.97) that in principle D^{δ} can now even lie below the two defect diffusion constants (see Fig. 6.20). The values derived according to Eq. (6.97) (without any adjusted parameter!) agree very well with the experiments. If the concentration of holes is much less than the iron content, we obtain[114] from Eq. (6.94) the simple result that the electronic term in Eq. (6.90) is given by the sum of the reciprocal iron concentrations $(4/[Fe'_{Ti}] + 4/[Fe_{Ti}^{\times}])$. If $[Fe_{Ti}^{\times}]$ dominates, the result is simply $D^{\delta} = 3D_{V_O^{\cdot\cdot}}$. This applies well to the linear section of Fig. 6.39.

$ZrO_2(Y_2O_3)$ is an example, in which the importance of internal redox reactions is perhaps even clearer. Conventionally, i.e. without taking internal reactions into account, the result is $D^{\delta} = D_{e'}$ or $D_{h^{\cdot}}$ (Eq. (6.55)), depending on the oxygen partial pressure region. In the absence of the above chemical capacitance phenomenon it is not evident that, in view of the high Y–doping further minority doping would exert any effect. However, experimentally we find great differences from material to material for the same Y content and astonishingly strong dependences of D^{δ} on oxygen partial pressure and temperature. This apparent contradiction is again resolved by taking account of the role of redox–active impurities with respect to the chemical capacitance. Rather than using Eq. (6.55) we must write, for high[115] P_{O_2}

$$D^{\delta}(T,P,C) = \chi_{h^{\cdot}}(T,P,C)\, D_{h^{\cdot}} \tag{6.98}$$

[113] Do not confuse C, the total doping content, with C^{δ}, the capacitance.

[114] It is instructive to derive this directly from $d\mu_O/dc_O$. When the hole concentration is negligible, then $dc_O = -d[V_O^{\cdot\cdot}] = -\frac{1}{2}d[Fe'_{Ti}] = \frac{1}{2}d[Fe_{Ti}^{\times}]$ as a consequence of electroneutrality and mass conservation. In addition because of the internal equilibria $d\mu_O = d\mu_{O^{2-}} - 2d\mu_{e'} = -d\mu_{V_O^{\cdot\cdot}} - 2\left(d\mu_{Fe'_{Ti}} - d\mu_{Fe_{Ti}^{\times}}\right)$. Hence, it follows that $d\mu_O/dc_O = d\mu_{V_O^{\cdot\cdot}}/d[V_O^{\cdot\cdot}] + 4d\mu_{Fe'_{Ti}}/d[Fe'_{Ti}] + 4d\mu_{Fe_{Ti}^{\times}}/d[Fe_{Ti}^{\times}] \simeq RT/[V_O^{\cdot\cdot}] + 4RT/[Fe'_{Ti}] + 4RT/[Fe_{Ti}^{\times}]$. When $[Fe_{Ti}^{\times}] \gg [Fe'_{Ti}]$ then $\sigma_O^{\delta}/4F^2 \simeq \sigma_{V_O^{\cdot\cdot}}/4F^2$ results as a consequence of electroneutrality and, hence, $D^{\delta} = 3D_{V_O^{\cdot\cdot}}$. (It is interesting to compare this with Eq. (6.61).)

[115] In the range of very low partial pressures, in which conduction electrons are the electronic charge carrier, there is, for not too high temperatures, evidence of native redox reactions of the type $V_O^{\cdot\cdot} + e' \rightleftharpoons V_O^{\cdot}$.

(whereby $D_{h^\cdot}$ is a weak function of temperature). In Ni– or Ti–doped material in which D^δ values vary appreciably with the concentration of redox active dopants, even when the Ni or Ti content is completely negligible with respect to the Y content and hence is not relevant at all in the electroneutrality condition, the results can be quantitatively described by Eq. (6.98) [402]. (At very high levels of redox–active impurities an altered mobility (i.e. $D_{h^\cdot}$), associated with polaron bands (see also Section 6.2)) is to be expected. Native trapping ($V_O^{\cdot\cdot} + e'$) must be taken into account in highly reduced material.)

It is instructive to consider χ as a function of the parameter $r \equiv K_{as}^{-1}[h^\cdot]^{-1}$. In the case of the association of an acceptor A' and a hole to the oxydized form A^x, r is identical to the redox ratio $[A']/[A^x]$. As we can see from Eq. (6.96), and as is illustrated in Fig. 6.40, the maximum effect (minimum χ) does not lie at r=0,

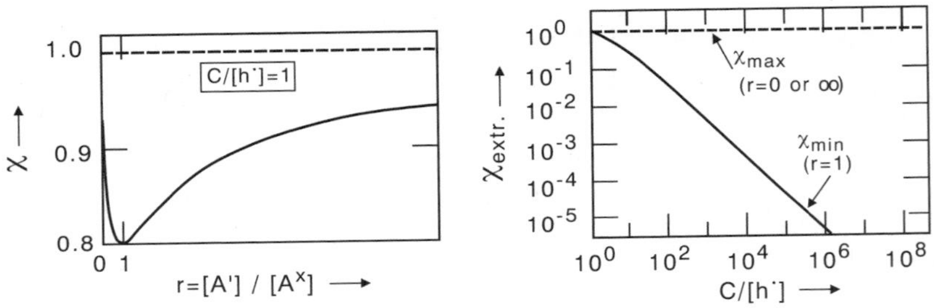

Fig. 6.40: The χ factor of the holes as a function of the redox ratio r (left), if $[h^\cdot]$ can be regarded as independent of r[116]. Maximum ($\chi = 1$) and minimum ($\chi = \chi_{min}$) values as a function of doping (right) [436].

as perhaps expected at first sight, but at r=1 where the ratio[116] of oxydized and reduced species is unity. This shows that impurities are particularly relevant whose energy levels lie about in the centre of the band gap (see Section 5.3) (as with Mn as a deep dopant in the case of YSZ [405]), and also reveals the similarity to static buffer effects in acid–base chemistry in aqueous solutions. A maximum pH buffer capacity requires r = [acid]/[base] = 1, i.e. $\log c_{H^+} = \log K_s$ where K_s = acid–base constant) [436].

The maximum effect at r = 1 on the left hand side of Fig. 6.40 is small just because a small concentration of redox–active impurities has been selected for the purpose of better illustration. If we plot the maximum correction for r = 1, as a function of doping (r.h.s. of Fig. 6.40), we can see the immense consequences.

These considerations also demonstrate clearly that high D^δ values, e.g. as required for bulk conductivity sensors, demand materials that are free from redox centres,

[116]Strictly, this only applies if $[h^\cdot]$ itself is not dependent on K_{as}, i.e. if the redox-active doping is not in the majority (see also YSZ). For this reason the minimum in $SrTiO_3$ is somewhat displaced (see Ref. [436] for more precise details).

while the minimization of drift phenomena in boundary layer sensors demands just the opposite [437].
Another complication, caused by the presence of internal equilibria, namely their impact on the evaluation of electrochemical measurement methods will be discussed in Chapter 7.

Let us now turn to the tracer (D^*) and the charge diffusion coefficients (D^Q). If we allow for variable valence states of the oxygen defects, modifications in these parameters result simply because of the different weight of the different charges. It is not difficult to show[117] that for D^* (we ignore correlation factors here)

$$D^*_O = \frac{RT}{4F^2}\frac{\sigma_{O^{2-}} + 4\sigma_{O^-} + 4s_{O^0}}{c_O} = \frac{1}{c_O}\Sigma_k[k]D_k, \tag{6.99}$$

whereby k indicates the individual ionic defects (O_i'', O_i', O_i^x, $V_O^{\cdot\cdot}$, $V_O^{\cdot}$, V_O^x). This can also be re–formulated as

$$D^*_O c_O = [O^{2-}]D^*_{O^{2-}} + [O^-]D^*_{O^-} + [O^0]D^*_{O^0}. \tag{6.100}$$

The situation is different for the charge diffusion coefficient: Firstly, the un–ionized defect does not appear at all in the conductivity experiment, secondly, if we convert $\sigma_{ion} = \sigma_{O^{2-}} + \sigma_{O^-}$ formally to D^Q_O by using z=2 as the effective valence[118]

$$D^Q_O = \frac{RT\,(\sigma_{O^{2-}} + \sigma_{O^-})}{4F^2 c_O}. \tag{6.101}$$

is obtained.
The ratios of the various phenomenological transport coefficients are described by

$$D^Q_O : D^*_O : D^\delta_O = 1 : \frac{\sigma_{O^{2-}} + 4\sigma_{O^-} + 4s_{O^0}}{\sigma_{O^{2-}} + \sigma_{O^-}} : \frac{\left[2\sigma_{O^-} + 4s_{O^0} + \frac{(\sigma_{O^{2-}}+2\sigma_{O^-})(\sigma_{e^-}-\sigma_{O^-})}{\sigma}\right]\frac{\partial \ln a_O}{\partial \ln c_O}}{(\sigma_{O^{2-}} + \sigma_{O^-})} \tag{6.102}$$

When the conductivity is predominantly electronic ($\sigma_{e^-} \simeq \sigma$) the complicated ratio D^δ_O/D^*_O is reduced to the thermodynamic factor $\partial \ln a_O/\partial \ln c_O$.

[117] The total tracer flux is obtained by summation. Note that both $\partial\phi/\partial x$ and $\partial\mu_{e^-}/\partial x$ are zero.

[118] The separation of the total conductivity into ion and electron conductivity is considered more precisely in Chapter 7. It can also be seen here that the sum $(\sigma_{O^{2-}} + \sigma_{O^-})$ is relevant for ion conductivity, while $(\sigma_{O^{2-}} + 2\sigma_{O^-})$ is important for stoichiometry change. The (electrical) ion current is $i_{ion} = -2Fj_{O^{2-}} - Fj_{O^-} = \frac{\sigma_{O^{2-}}}{2F}\nabla\widetilde{\mu}_{O^{2-}} + \frac{\sigma_{O^-}}{F}\nabla\widetilde{\mu}_{O^-}$. On account of $\mu_O = \widetilde{\mu}_{O^{2-}} - 2\widetilde{\mu}_{e^-} = \widetilde{\mu}_{O^-} - \widetilde{\mu}_{e^-}$ it follows that $i_{ion} = \frac{\sigma_{O^{2-}}+2\sigma_{O^-}}{2F}\nabla\mu_O + \frac{\sigma_{O^{2-}}+\sigma_{O^-}}{F}\nabla\widetilde{\mu}_{e^-}$. In the pure conductivity experiment, reversible electrodes (Pt, O_2, for example) are used and partial pressure gradients are absent. The first term then disappears. It will be shown in Chapter 7 that the integration of $\nabla\widetilde{\mu}_{e^-}$ (i.e. $\Delta\widetilde{\mu}_{e^-}$) leads to the voltage; hence, $\sigma_{O^{2-}} + \sigma_{O^-}$ corresponds to the ionic total conductivity; analogously $\sigma_{eon} = \sigma_{e'} + \sigma_{h^\cdot}$.

d) Co–operative processes during tracer–diffusion

The previous paragraph has shown that discrepancies between conductivity and tracer experiment will occur particularly if particles contribute also in an effectively uncharged form. A more complex actual example may be briefly mentioned: In the case of the alkali hydroxides a highly correlated ring mechanism is expected for the protons, which manifests itself in mass transport but not in conductance (see Section 6.3).
The first one can be also be investigated by PFG–NMR[119] (instead of an isotope exchange experiment). The comparison with conductivity reveals Haven–ratios for NaOH (see Eq. (6.47)) which are higher than $10^3 - 10^4$ [438].

6.6.2 Boundary layers and grain boundaries

As described at the end of Chapter 5, boundary layers exert a double influence, even in the simple core–charge space model described there: A structural one, which is more or less restricted to the core region (if elastic effects at long range are neglected[120]) and the space charge effect which may be much larger in spatial extent. In the structurally altered core regions of the interface the mobilities and, hence, the defect diffusion coefficients are clearly modified. In the space charge region the mobility, parallel to the interface, is invariant if structural changes can be neglected. The influence of electrical fields must be considered perpendicular to the interface, but may, however, be negligible in the case of comparatively high migration thresholds. Generally, such anisotropy of the mobility may be significant in spite of the isotropy of the perfect structure. This is even more relevant for the conductivity and, hence, D^Q, on account of the anisotropy of the concentration, as described in Chapter 5.
The relationships are more complex for chemical diffusion and tracer diffusion.
Let us consider first the chemical transport of oxygen in an oxide along an interface (x–y plane, see Fig. 6.41, right) as a consequence of a sudden chemical potential change in the ambient. As we do for the bulk we assume local equilibrium[121]. Since no gradients occur in the electrochemical potentials or in the chemical potential perpendicular to the interface at the start of the experiment, the flux, at a given position, is completely determined by[122] $\sigma_O^\delta \partial\mu_O/\partial y$. It would be wrong, however, to assume that the flux lines would be parallel to the interface for the whole experiment, and that the effective diffusion coefficient $D_{O,m}^{\delta\|}$ could be obtained from an arithmetic mean of the local $D_O^\delta(y)$ values, as it was for D^Q in the conductance experiment with

[119] PFG–NMR: nuclear magnetic resonance with pulsed field gradients. In this way a spatial coordinate is defined in the sample.

[120] This is often a reasonable approximation for soft materials.

[121] This condition of sufficiently fast local reaction may be especially violated in the case of interfacial transport. Concerning this important point the reader is referred to Refs. [8,430,439] (see Section 6.6.1b).

[122] In the case of the tracer–experiment $\sigma^*_{O^{2-}} \partial\mu^*_{O^{2-}}/\partial y$ is the relevant quantity.

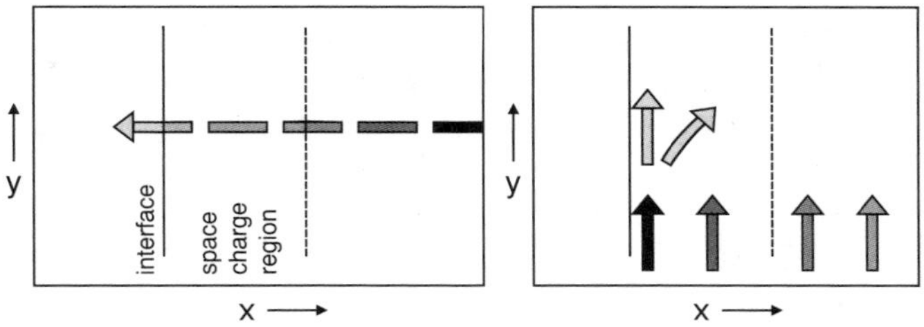

Fig. 6.41: Diffusion across (left) and along (right) an interface (see text).

reversible electrodes (Section 5.8). The diffusion experiment is not a steady state experiment, and gradients also occur in the x direction as a result of the differing local fluxes, so that the diffusion problem becomes multidimensional and, hence, complicated (see Fig. 6.41 right), even if space charge effects are negligible.

The situation is similar in the case of tracer diffusion[122] which we will consider now more explicitly. Let us consider the grain boundary diffusion of a tracer; for simplicity we neglect space charges and restrict ourselves to fast grain boundary transport. (Diffusion along dislocation paths would be another example.) The tracer flux is not confined to the core, but spreads sideways into the bulk region. It is only when the boundary conditions are extreme, with high ratios of grain boundary diffusion coefficient D_{gb} to bulk diffusion coefficient (D_{∞}) and very small grain boundary thickness d_{gb} (more accurately $d_{gb}^2 \ll D_{bulk}t \ll D_{gb}t$), that there is a relatively simple result, known as Fischer's solution [440]. In such a case the bulk transport takes place approximatively perpendicular to the grain boundary transport (the latter being quasi–stationary, cf. Section 6.7). For further details and a full treatment the reader is referred to the relevant literature [441]. So long as ions and electrons are strictly coupled, the treatment of tracer diffusion and the treatment of chemical diffusion are analogous. Inhomogeneity effects can be particularly involved in the latter case on account of the different impact on σ_{ion} and σ_{eon}.

A greatly simplified treatment can be achieved, if it is possible to assume that the lateral diffusion into the bulk is absolutely negligible with respect to the core diffusion (Fig. 6.42, left). The bulk transport then takes place on a completely separate

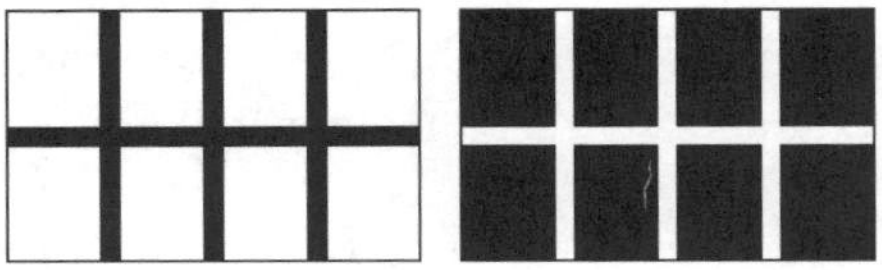

Fig. 6.42: Schematic diagram for the diffusion through a ceramic with highly permeable grain boundaries (left) or scarcely permeable grain boundaries (right). The black and white zones refer to the local permeabilities, not to the concentrations.

time scale. A suitable experiment here would be to follow an equilibration experiment, as discussed above, in which we suddenly change the ambient gas composition (e.g. increase of oxygen content or of tracer content) and follow the homogenization towards the new composition (or tracer content) throughout the whole sample. The rate–determining step to be considered is then the diffusion of O (or of the isotope) from the very rapidly equilibrated grain boundaries into the interior of the grain, so that the effective diffusion length of the ceramic is the grain size (ℓ) and not the sample size (L)[123] (see Fig. 6.42, left). Considering, e.g., the chemical experiment, we have

$$\mathrm{D}_{\mathrm{m}}^{\delta} = \frac{\mathrm{L}^2}{\ell^2}\mathrm{D}_{\mathrm{bulk}}^{\delta} \propto \frac{\mathrm{L}^2}{\ell^2}\frac{\sigma_{\mathrm{bulk}}^{\delta}}{\mathrm{c}_{\mathrm{bulk}}^{\delta}}. \tag{6.103}$$

Such relationships appear to determine the oxygen diffusion in $YBa_2Cu_3O_{7-\delta}$ ceramics under certain conditions and are generally most important for low temperature transport (see Fig. 6.31). Donor–doped $BaTiO_3$ or $SrTiO_3$ is another example. There bulk diffusion is very slow because of the very low $\mathrm{V_O^{\cdot\cdot}}$ concentration (see Sections 5.6 and 6.5) and the low mobility of the metal vacancies. If we start from a perovskite ceramic of very high n–conductivity, prepared under reducing conditions, and expose this to a high partial pressure of oxygen, then in–diffusion of oxygen takes place via the grain boundaries. Even after a long period of exposure there is only equilibrium in the upper layers of the grains. The ceramic produced, consisting of highly conducting grains with insulating "skins", exhibits very high effective (electrical) capacitances and finds application as a powerful dielectric ceramic[124] [189].

In the case of chemical diffusion perpendicular to the interface, the situation is complicated for other grounds (see Fig. 6.41, right). The problem is one–dimensional for reasons of symmetry, but structural inhomogeneities (core versus bulk) and electrical field effects in the direction of transport are of importance here. Apart from transfer reactions through the interfacial core (see Chapter 7) the space charge zones have a profound influence on diffusion. The electrical and chemical issues are naturally closely interwoven and the whole problem is complex. In the instationary case the coupling of the partial fluxes according to Eq. (6.50) is violated and internal net fluxes occur also in the absence of external fluxes[125]. This complication disappears in the stationary state. Nonetheless, the chemical diffusion coefficient is not a suitable transport coefficient any longer, since, even at equilibrium, concentration gradients occur in the space charge zones; hence, considering oxides, ∇c_{O} is not a suitable driving force. In addition, j_{O} itself (in contrast to $j_{\mathrm{O}^{2-}}$ and j_{e^-}) is not a properly defined parameter in the space charge regions, unless we refer to the stationary state. (There $j_{\mathrm{O}} \propto \sigma_{\mathrm{O}}^{\delta}\nabla\mu_{\mathrm{O}}$ still applies.) Even if we select the initial parameters so that no equilibrium space charge exists ($\vartheta = 0$, Section 5.8) space charges build up during chemical diffusion in the general case. This space charge is

---

[123] Eq. (6.103) is obtained by equating the equilibration times (cf., for example, Eq. (6.70)). It is, as such, only valid if all surfaces are exposed to the new partial pressure. Otherwise an additional geometrical factor has to be implemented.

[124] Capacitors of this type in technological use are frequently based on tantalum. Here the surface of the (usually phosphorus–doped) nanocrystalline tantalum grains are oxydized electrochemically (formation of insulating oxide). High storage capacities are reached on account of the high internal interfacial density and the low thickness of the oxide skin (cf. also Section 7.4.3).

[125] A detailed treatment is presented in Refs. [442–444].

required by Gauss' theorem[126] in order to adjust the nonzero internal field, existing to maintain electrical neutrality, to the zero field outside. Such space charge effects have been referred to in the discussion of tracer profiles of $SrTiO_3$ [445] or drift effects in Taguchi–sensors [446].

We will now make a short investigation of the effect of equilibrium space charges (cf. also Section 5.8) on transport in our model oxide (with just O_i'' and $h^{\cdot}$ as defects). Even though $2\partial c_{O_i''} - \partial c_{h^{\cdot}}$ is now nonzero in contrast to the bulk and is given by the variation of the field gradients, it still remains valid, as before, that both fluxes — $j_{O_i''}$ and $j_{h^{\cdot}}$ — are necessary for chemical diffusion. In the stationary state this is expressed by the fact that we must consider the ambipolar conductivity[125] σ^{δ}. More precisely, spatial integration of its reciprocal values is necessary, which can be split into contributions stemming from integration of local ionic and electronic resistivities [447]. If the two main charge carriers are of similar mobility then it is possible to neglect, in the ambipolar diffusion, that charge carrier, whose concentration is elevated in the space charge zone. In a simple Gouy–Chapman situation (one charge carrier enriched, the other correspondingly depleted) the ambipolar transport, in such cases, is always slowed down, with respect to the bulk, independent of the sign of the space charge effect[125,127,128]. A slowing–down effect was already hinted in Fig. 6.41 and explains the barrier effect of the Schottky boundary layer in $SrTiO_3$ on chemical diffusion; there, even both decisive charge carriers are depleted (see Section 5.8). If the optical measurements described in Section 6.5 (see Fig. 6.32) are carried out with a grain boundary perpendicular to the direction of measurement, colour profiles are obtained, such as is shown in Fig. 6.43 for a low–symmetry tilt grain boundary ($\sim \Sigma\,13$). In the case of the highly symmetrical $\Sigma\,3$ grain boundary[129] (Fig. 6.44, right) it is not possible to resolve any colour discontinuities. Note that the colour difference basically reflects the difference in stoichiometry. Figure 6.44 illustrates such profiles. In the case of the low–symmetry grain boundary the space charge potential determined is 450 mV, which is in good agreement with the analysis of the electronic conductivity behaviour (see Section 5.8).

---

[126] The Maxwell equation $(\partial/\partial x)D = \rho$ (with $D = \epsilon E$) becomes $\Delta D = \Sigma$ at the interface (cf. footnote 171 on page 236).

[127] We can compare the transport to the bicycle journey in which — in order to simulate the atomic situation at least half–correctly — the (strongly demanded) vehicles must be changed at kilometre intervals. The ambipolar diffusion can then be compared with the journey of a Swabian married couple (Swabians because the coupling is very pronounced: Neither lets the other out of sight because of anxiety that the other might spend money unnecessarily.) Let us assume that in the neighbourhood of the boundary the couples are about to cross (e.g. between Swabia and Badonia) the number of available men's bicycles is doubled and that of ladies' bicycles is halved, the journey is delayed even though, in all, more bicycles are available. The same applies if ladies' bicycles are preferred. If you don't know Swabia and Badonia you may consider Scotland and England.

[128] In the case of differing mobilities, it is possible for local accelerations to occur (if the slower one is increased in its concentration).

[129] Compare here Fig. 5.23 (page 143).

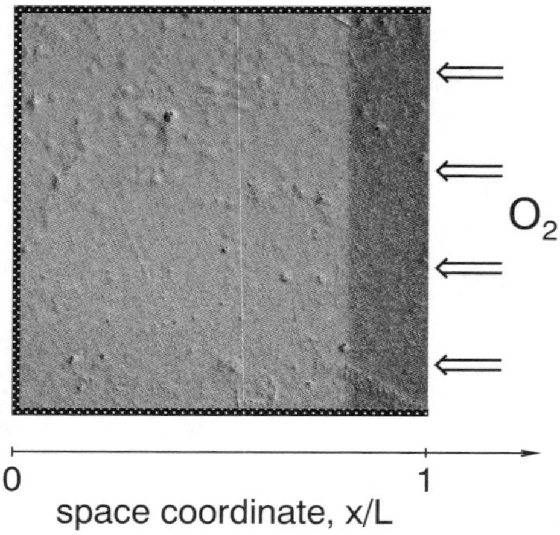

Fig. 6.43: The figure shows the colour change over a grain boundary of low symmetry (of a bicrystal of Fe–doped $SrTiO_3$) during the in–diffusion of oxygen through the right surface (partial pressure jump from 10^4Pa to 10^5Pa, T=873K, Fe concentration: 2.15×10^{18}cm^{-3}). The others are not activated (24° tilt grain boundary, axis of rotation: [001], near Σ 13). According to Ref. [447].

In the experiment described oxygen diffuses in from one side, while the other sides, also the opposite side (left), are sealed. If this left hand side is also open and if one creates a flux from right to left by applying different oxygen partial pressures, we have realized a permeation experiment. If we measure the (permeation) flux leaving the bicrystalline sample on the side of low potential we obtain the calculated graphs shown in Fig. 6.45. Again we can clearly recognize the impeding influence of the space charge zone.
Even if, in a polycrystalline sample, the grain boundary resistance dominates completely, the thermodynamic factor (or chemical capacitance) is still determined by the bulk, provided the density of boundaries is not too great. In other words, the largest part of the stoichiometric change ($d\mu_O/dc_O$) takes place there (Fig. 6.42, right). In this case[130] we obtain, for hardly permeable boundaries, the surprisingly simple relationship

$$D_m^\delta \propto \frac{\ell}{d_{gb}} \frac{\sigma_{gb}^\delta}{c_{bulk}^\delta} \tag{6.104}$$

as a counterpart to Eq. (6.103) [443]. Note that $d_{gb}/\ell \propto \varphi_{gb}^{\perp}$. The proportionality factor in Eq. (6.104) is $\frac{RT}{4F^2}$.
The tracer diffusion is naturally influenced too by boundaries to be crossed. However, the situation is simpler here[131] and the effects are different. Since the tracer diffusion does not produce chemical or electrical effects, the tracer transport across a grain boundary, is — apart from pure tracer concentration effects — characterized by the equilibrium situation and analogous to stationary ionic conduction as described in Section 5.8. It ought, in particular, to be added that, if the dominating charge carriers O_i'' and $h^\cdot$ have the same mobility and the space charge potential is positive (negative), tracer diffusion is accelerated (retarded), while in chemical dif-

[130] This can be derived from the equivalent circuit shown in Section 7.3 (Fig. 7.33) for negligible C_{gb}^δ and dominating R_{gb}^δ.
[131] We neglect isotope effects on the mobility which is certainly incorrect in the case of the proton.

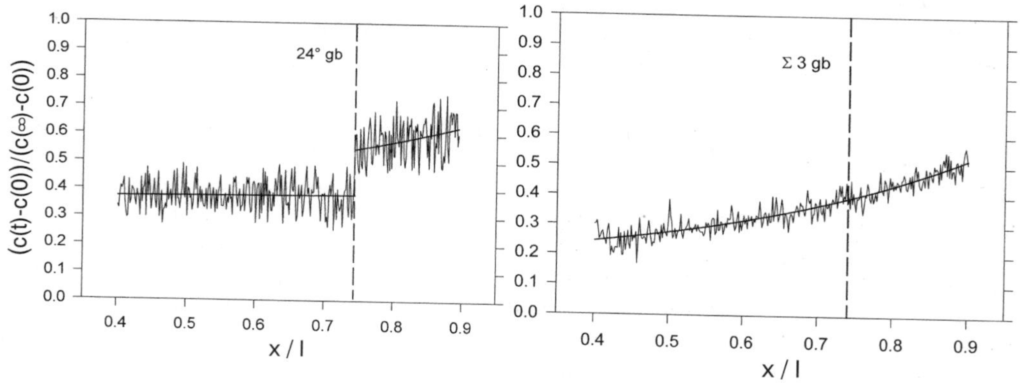

Fig. 6.44: Normalized vacancy profiles for low symmetry ($\sim \Sigma\,13$) and high symmetry tilt grain boundaries (Σ 3, 70.5° tilt grain boundary) as obtained from the in–situ measurements shown in Fig. 6.43. In the case of the low–symmetry grain boundary (left) the fit reveals a space charge potential of 450 mV. In the case of the $\Sigma\,3$ grain boundary (right) the influence of grain boundary does not show up (corresponding to an upper limit of 300 mV for the potential), the continuous line corresponds to the bulk calculation with an effective surface rate constant of 3.6×10^{-5}cm/s (see following section). From Ref. [447].

fusion a retarding effect was to be expected for both signs. Just as for D^Q and D^δ, D^* is also subject to microstructural and bulk–crystallographic anisotropies, that endow the D values with tensor character (cf. again Fig. 6.31). In polycrystalline materials percolation processes via favourably orientated grains are very important. As far as the one–dimensional dislocations are concerned, possible blocking effects retreat, in contrast to interfaces, into the background, while they very frequently provide rapid diffusion paths as a result of core and space charge effects ("pipe diffusion").
It should be added that, as in the case of the conductivity experiments (cf. Section 7.3.7), current–constriction effects can occur in the diffusion experiments [449], if

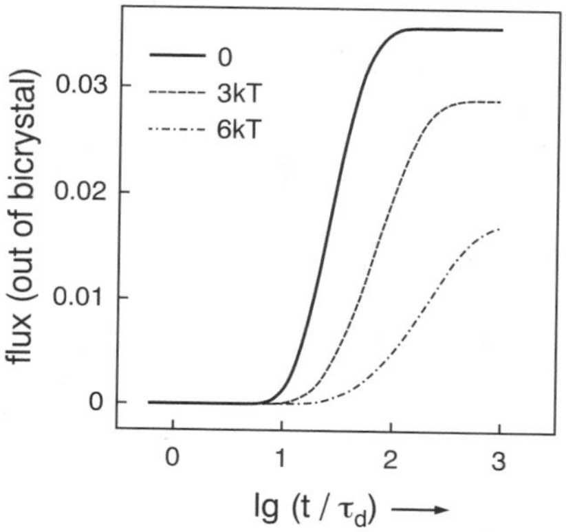

Fig. 6.45: Normalized flux emerging from a bi–crystal with perpendicular grain boundary calculated for a given chemical potential gradient (Gouy–Chapman case) (arbitrary units). Any core resistance was ignored. The decrease with increasing space charge potential (3kT, 6kT: space charge effect in energy units) reveals the influence of the equilibrium space charge [448].

lateral inhomogeneities are present. In this way resistivities occur that can be easily misinterpreted in terms of sluggish surface steps. A concise treatment of proper surface kinetics will be given now.

6.7 Surface reactions

6.7.1 Elementary processes

We encountered the fact, that the surface reaction is of importance for the overall rate, several times in Sections 6.4 and 6.5. This itself is usually made up of very many individual processes. Let us consider our master–example, namely oxygen incorporation, once again. Figure 6.46 shows a selection of possible individual steps.

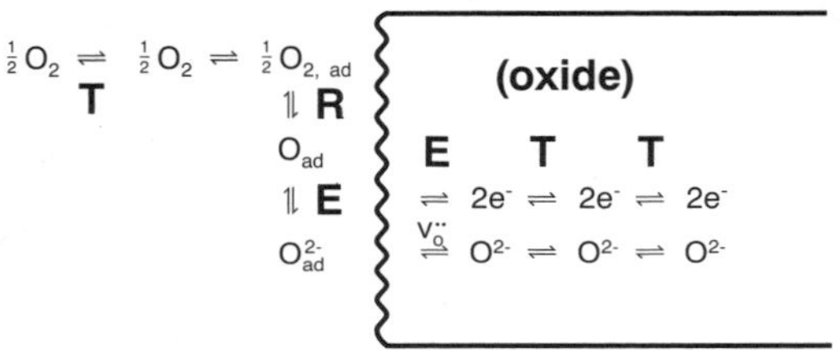

Fig. 6.46: A selection of possible elementary steps for the incorporation of oxygen in an oxide. The surface reaction, in particular, is made up of complex individual steps. In reality the ionization degree of adsorbed atoms lies in between zero and the bulk value (here -2) (cf. mechanism on page 360). T: transport, R: chemical reaction, E: electrochemical reaction.

First the oxygen has to be transported within the gas phase to the sample — this is generally a comparatively rapid process[132]. Then it must be adsorbed, dissociated, ionized and enter the condensed phase; there it crosses the space charge zone, the actual internal diffusion then follows (which itself can consist of serial and parallel steps, particularly when grain boundary effects must be taken into account too). In addition, surface diffusion is of great importance, although it has been "suppressed" for simplicity in Fig. 6.46. As already frequently mentioned, each of these steps can be understood as an electrochemical reaction (E) of the form

$$A(x) + ... \rightleftharpoons B(x') + ... \qquad (6.105)$$

Pure adsorption represents a proper chemical reaction (R), in which electrical fields can be neglected, bulk diffusion (T) represents a pure transport case (A≡B, i.e. same μ°–values), while the transfer reaction refers to the general case, in which chemical standard potentials and electrical potentials change. The role of the electrons depends essentially on which of the three experiments described in Fig. 6.17 is being discussed.

Of these elementary steps, we will now discuss adsorption[133] in its very simplest form.

[132] This step can also be rate–determining at high levels of turnover, such as can occur in fuel cells (cf. Chapter 7). Diffusion–limiting currents are then set up for high driving forces (cf. Chapter 7).

[133] A detailed consideration is given in Ref. [450].

Let us formulate the adsorption of a gas G on a free surface position V_{ad} as

$$\mathrm{G} + \mathsf{V}_{\mathrm{ad}} \rightleftharpoons \mathrm{G}_{\mathrm{ad}}, \tag{6.106}$$

so that in the simplest case we get[134]

$$\mathcal{R}_{\mathrm{ad}} = \vec{\mathrm{k}}\mathrm{P}_{\mathrm{G}}[\mathsf{V}_{\mathrm{ad}}] - \overleftarrow{\mathrm{k}}[\mathrm{G}_{\mathrm{ad}}]. \tag{6.107}$$

The concentration is best formulated in terms of the degree of coverage Θ_{G} — this is the fraction of occupied sites with respect to the total number of available sites[135] — so that an equivalent formulation of Eq. (6.107) with rescaled k or $\mathcal{R}$ values is:

$$\mathcal{R}_{\mathrm{ad}} = \vec{\mathrm{k}}_{\mathrm{ad}}\mathrm{P}_{\mathrm{G}}(1 - \Theta_{\mathrm{G}}) - \overleftarrow{\mathrm{k}}_{\mathrm{ad}}\Theta_{\mathrm{G}}. \tag{6.108}$$

In equilibrium (arc–symbol) it follows that

$$\frac{\widehat{\Theta}_{\mathrm{G}}}{1 - \widehat{\Theta}_{\mathrm{G}}} = \frac{\vec{\mathrm{k}}_{\mathrm{ad}}}{\overleftarrow{\mathrm{k}}_{\mathrm{ad}}}\widehat{\mathrm{P}}_{\mathrm{G}} = \mathrm{K}_{\mathrm{ad}}\widehat{\mathrm{P}}_{\mathrm{G}}. \tag{6.109}$$

This is the relationship for the Langmuir equilibrium [451]. (Strictly speaking V_{ad} and G_{ad} should not be considered as thermodynamically independent elements, rather $\mathrm{G}_{\mathrm{ad}} - \mathsf{V}_{\mathrm{ad}}$ constitutes the relevant building element and the Fermi–Dirac like expression $\Theta_{\mathrm{G}}/(1 - \Theta_{\mathrm{G}})$ can be regarded as its activity reflecting the exhaustible number of the surface positions (see Eqs. (5.18, 5.134)). The result of this more rigorous thermodynamic procedure is the same, viz. Eq. (6.109).) The relationships are more complex when there is lateral interaction or multilayer adsorption. Then, the kinetic and thermodynamic parameters are formally dependent on the occupation. In the literature $\Delta_{\mathrm{r}}\mathrm{H}^\circ$ is frequently taken as proportional to Θ [450]. On account of the relationships discussed in Section 5.7.2 we may presume that a correction of the type $\Delta_{\mathrm{r}}\mathrm{H}^\circ = \mathrm{const}\Theta^{1/2}$ ($\Theta^{1/2}$ is correlated with the mean distance within the surface) might also be of utility.

The simple reason that the neighbouring reactants are much closer in the adsorbed state than in the gas phase, leading to a facilitated reaction, is already a significant point for the efficiency of heterogeneous catalysts; another is the loosening of bonding or even the complete dissociation as a result of interaction with the substrate (see Section 6.8). If a polyatomic gas such as O_2 is considered, the adsorption is usually more complicated than described by Eq. (6.106) (cf. page 355). The dissociation of complex molecules can be even associated with the desorption of a molecular fragment.

As adsorption means bonding to the solid phase[136] and, hence, is associated with a change in electron density, a mechanistic demarcation with respect to ionization is particularly difficult.

[134] Equation (6.107) neglects the complexity of surface structure (see Section 5.4), the occurrence of multilayer adsorption and other complications.

[135] $\Theta_{\mathrm{G}} \equiv [\mathrm{G}_{\mathrm{ad}}]/([\mathrm{G}_{\mathrm{ad}}] + [\mathsf{V}_{\mathrm{ad}}])$.

[136] More accurately, one distinguishes, according to the strength of the bond, between chemisorption (strength of a typical intramolecular chemical bond) and physisorption (strength of a typical intermolecular bond). These details are not of relevance for our context.

Since the ionization reaction comprises charging, it is necessary to take electrical fields into account[137]. In the case of oxygen adsorption, species such as O_2^- or O^- ($O_2^{\delta-}$, $O^{\delta-}$) play a significant role as ionized states the concentration of which depend on the experimental conditions, and the nature of the substrate. The complexity is additionally increased by the lateral inhomogeneity of the surface (see Fig. 5.23). This also makes the treatment of surface diffusion more difficult (e.g. because of the occurrence of lateral fields).
The transfer of the more or less ionized particles into the solid phase is another important elementary step. Here, it can be difficult to make a demarcation between solid phase and adsorbate layer. If we assume for simplicity that completely ionized oxygen[138] is being transferred into the solid phase according to

$$O''_{ad} + V_O^{\cdot\cdot} \rightleftharpoons O_O + V_{ad}, \tag{6.110}$$

the corresponding rate equation reads

$$\mathcal{R} = \overrightarrow{\overline{k}}\Theta[V_O^{\cdot\cdot}] - \overleftarrow{\overline{k}}(1-\Theta). \tag{6.111}$$

Here Θ refers to the ionized particles. Even when the preceding processes are rapid and appear in the form of equilibrium terms (see below), the coupling between Θ in Eq. (6.111) and Θ_G in Eq. (6.104) can still be complex because of the field effects.
In this section we are primarily interested in the kinetics of stoichiometric change and the kinetics of the tracer experiment (see Figs. 6.17b,c)[139]. In both cases a relaxation experiment, close to equilibrium, is usually performed, from which we extract effective rate constants (in the sense of linear thermodynamics), which we designate with $\bar{k}$. For precise distinction we use $\bar{k}^\delta$ for the rate constant of the composition change and $\bar{k}^*$ for the rate constant of the tracer exchange, while the comparable parameter which is formally derived from the electrical experiment[139], is denoted by $\bar{k}^Q$. These effective rate constants naturally depend on the microscopic k values of the rate–determining step, but also contain information concerning the other elementary processes of the reaction scheme. Before we turn to such issues (see Section 6.7.3) it is necessary to make several general observations concerning more complex mechanistic matters.

6.7.2 Coupled reactions

It is possible to represent the whole scheme in Fig. 6.46 approximately in the form of a network of elementary steps in parallel and series[140]. If the rate constants

[137]Related effects that refer to electrochemical methodology will be treated in more detail in the "Electrochemistry" chapter (Section 7.3.3). Cf. also Fig. 6.2, Eq. (6.18) as well as Section 6.7.3.

[138]In reality adsorbed oxygen will be partially oxydized at the surface and take up additional electrons while being incorporated in the bulk (cf. mechanism on page 360).

[139]The kinetics of the conduction experiment include electrical capacitance effects and are studied in Chapter 7. The phenomenological effective rate constant $\bar{k}^Q$ is only formally derived from σ_{ion} and does (as D^Q) not reflect a genuine kinetic parameter.

[140]Note that reorganization processes have naturally always to be taken into account.

differ sufficiently, the network concentrates on the rate–determining step; this is the slowest series step in the most rapid parallel path.
If two elementary processes are running in parallel (i.e. reaction from E to F via two different mechanisms) according to

$$\begin{aligned} \mathrm{E} \ &\underset{\overleftarrow{\mathrm{k}}_1}{\overset{\overrightarrow{\mathrm{k}}_1}{\rightleftharpoons}} \ \mathrm{F} \\ \mathrm{E} \ &\underset{\overleftarrow{\mathrm{k}}_2}{\overset{\overrightarrow{\mathrm{k}}_2}{\rightleftharpoons}} \ \mathrm{F}, \end{aligned} \tag{6.112}$$

the overall rate is determined by the fastest one. Let us quantify this for the limiting case of situations close to and far from equilibrium. In the steady state it is sufficient to refer to a purely resistive contribution (cf. β^{-1} in Eq. (6.5) for situations close to equilibrium). Hydrodynamic and electrodynamic analogues are then the parallel

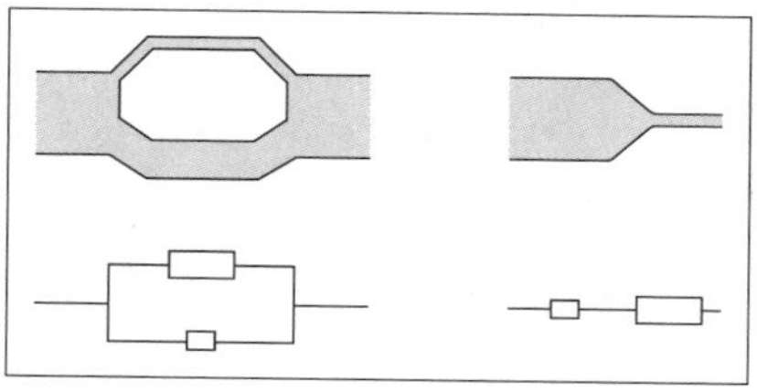

Fig. 6.47: Hydrodynamic and electrical analogues of chemical parallel and series steps.

connection of two pipes, or the parallel connection of two electric resistors (Fig. 6.47). As the driving force is the same for both processes, it follows in the linear range, i.e. close to equilibrium, that

$$\mathrm{J} = \mathrm{J}_1 + \mathrm{J}_2 = \beta_1 \mathrm{X} + \beta_2 \mathrm{X} = (\beta_1 + \beta_2)\,\mathrm{X} = \beta_{\mathrm{eff}} \mathrm{X}, \tag{6.113}$$

and, hence, $\beta_{\mathrm{eff}} = \beta_1 + \beta_2$.
Far from equilibrium we can neglect the back reactions in Eq. (6.112) and it follows that

$$\mathrm{J} \propto \mathcal{R} = \frac{\mathrm{d[F]}}{\mathrm{dt}} = [\mathrm{E}](\overrightarrow{\mathrm{k}}_1 + \overrightarrow{\mathrm{k}}_2); \tag{6.114}$$

as in the previous consideration (but without restriction to steady states), the step with the fastest transport coefficient determines the rate ($\overrightarrow{\mathrm{k}}_{\mathrm{eff}} = \overrightarrow{\mathrm{k}}_1 + \overrightarrow{\mathrm{k}}_2$).
In the case of a series of reactions[141]

$$\mathrm{E} \ \underset{\overleftarrow{\mathrm{k}}_1}{\overset{\overrightarrow{\mathrm{k}}_1}{\rightleftharpoons}} \ \mathrm{F} \ \underset{\overleftarrow{\mathrm{k}}_2}{\overset{\overrightarrow{\mathrm{k}}_2}{\rightleftharpoons}} \ \mathrm{G} \tag{6.115}$$

[141] Compare again, here, footnote 1 on page 269.

the slowest elementary step is naturally decisive. In the stationary state the flux is the same for both processes while the forces are additive. The evident hydrodynamic and electrical analogues are given in Fig. 6.47. Close to equilibrium

$$X = X_1 + X_2 = \beta_1^{-1}J + \beta_2^{-2}J = \left(\beta_1^{-1} + \beta_2^{-1}\right) J = \beta_{\text{eff}}^{-1} J \tag{6.116}$$

applies, with $\beta_{\text{eff}}^{-1} = \beta_1^{-1} + \beta_2^{-1}$. The effective value is hence determined by the smallest β–value, and the total affinity is approximately that of the rate–determining step. This is shown by Fig. 6.48 for the surface reaction controlled oxygen incorporation

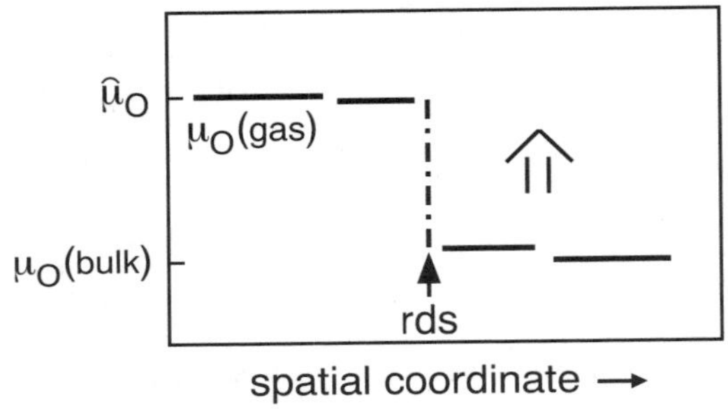

Fig. 6.48: Surface reaction controlled oxygen incorporation in an oxide. The oxide is exposed to a sudden change in the outer oxygen potential to the value $\widehat{\mu}_O$. The affinity of the incorporation is $-(\mu_O(\text{bulk}) - \widehat{\mu}_O)$. The driving force drops predominantly over the rate determining step (rds) (cf. Eq. (6.116)) [452].

in an oxide, a process which will be used as a master example in the following text. We will examine the more general behaviour on the basis of the complex reaction chain

$$A \underset{\overleftarrow{k}_a}{\overset{\overrightarrow{k}_a}{\rightleftharpoons}} B \underset{\overleftarrow{k}_b}{\overset{\overrightarrow{k}_b}{\rightleftharpoons}} C \underset{\overleftarrow{k}_c}{\overset{\overrightarrow{k}_c}{\rightleftharpoons}} \ldots \rightleftharpoons J \underset{\overleftarrow{k}_j}{\overset{\overrightarrow{k}_j}{\rightleftharpoons}} K \underset{\overleftarrow{k}_k}{\overset{\overrightarrow{k}_k}{\rightleftarrows}} L \underset{\overleftarrow{k}_l}{\overset{\overrightarrow{k}_l}{\rightleftharpoons}} \ldots \rightleftharpoons Y \underset{\overleftarrow{k}_y}{\overset{\overrightarrow{k}_y}{\rightleftharpoons}} Z. \tag{6.117}$$

The process K$\leftrightarrows$L is taken as rate–determining, i.e. characterized by very small k values. Since all other processes have comparatively high rate constants, the steps

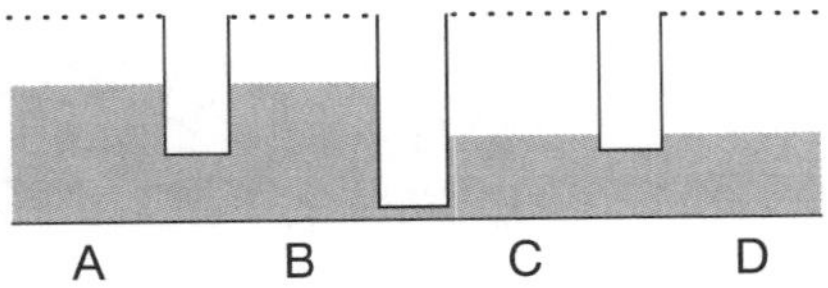

Fig. 6.49: Hydrodynamic analogues of the kinetic situation characterized by a rate–determining step (bottleneck).

before and after are in a quasi–equilibrium state[142,143] (see Fig. 6.49). Thus, for example $\frac{[Z](t)}{[Y](t)} = \frac{\overleftarrow{k}_y}{\overrightarrow{k}_y} \equiv K_y$, but with absolute concentrations that change with time. Hence, it also follows that $\frac{d[Z]}{dt} = K_y \frac{d[Y]}{dt}$, i.e. changes with time in the preceding or succeeding steps are proportional to each other. In Fig. 6.49 we then see these levels rise or fall with identical heights. If we define d[Z]/dt as the reaction rate (chain closed at Z), such proportionalities and the conservation of mass give, in each case, the relationship between $\mathcal{R}$ and the rate of the rate–determining step. Let us first consider its forward reaction ($\overrightarrow{\mathcal{R}}_{lim} = \overrightarrow{k}_k[K]$). [K] can be related to [A] by means of the preceding mass action laws. If, for simplicity, we neglect activity coefficients, then it follows that

$$[K] = K_j[J] = K_a K_b \ldots K_j[A] = \Pi_{r<k} K_r [A]. \tag{6.119}$$

The situation for the back reaction $\overleftarrow{\mathcal{R}}_{lim} = \overleftarrow{k}_k[L]$ is analogous, here it is possible to connect [L] with [Z] via the K value for the succeeding reactions, so that overall for $\mathcal{R}_{lim} = \overrightarrow{\mathcal{R}}_{lim} - \overleftarrow{\mathcal{R}}_{lim}$

$$\mathcal{R}_{lim} = \overrightarrow{k}_k(\Pi_{r<k} K_r)[A] - \overleftarrow{k}_k(\Pi_{r>k} K_r^{-1})[Z]. \tag{6.120}$$

The relationships are very simple in the stationary state, since then all individual reaction rates (fluxes) are the same, i.e. there is no storage of matter at any locus of the chain. Then $\mathcal{R}_{lim}$ is the uniform overall reaction rate $\mathcal{R}$.

---

[142] In order to prove this, let us consider once again any elementary reaction E⇌F of the reaction chain. According to Eq. (6.19) it follows that

$$\mathcal{R}_{EF} = \overrightarrow{k}[E]\left(1 - \frac{[F]/[E]}{K_E}\right) \tag{6.118}$$

is the difference of the turn–over rate of E into F ($\overrightarrow{\mathcal{R}}_{EF}$) and the turn–over rate of F into E ($\overleftarrow{\mathcal{R}}_{EF}$). If the reaction chain is initiated from the far left, i.e. starting from A, then initially $\overrightarrow{\mathcal{R}}_{EF} \gg \overleftarrow{\mathcal{R}}_{EF}$. Since the reaction to Z is hindered by the rate determing step (K⇄L), the back reaction becomes increasingly important within the time window of the KL–step, until $\overrightarrow{\mathcal{R}}_{EF} - \overleftarrow{\mathcal{R}}_{EF} \equiv \mathcal{R}_{EF} \ll \overrightarrow{\mathcal{R}}_{EF}, \overleftarrow{\mathcal{R}}_{EF}$. Then the bracketed term in Eq. (6.118) approaches zero (because $\mathcal{R}_{EF}/(\overrightarrow{k}[E]) \to 0$), and it follows that $[F]/[E] \simeq K_f \equiv \widehat{[F]}/\widehat{[E]}$, i.e. $\mathcal{A}_f \simeq 0$. A quasi–equilibrium is established. The situation is analogous for the preceding, but also for the succeeding steps. If the whole chain is close to equilibrium ($\overrightarrow{k}[E]$ then becomes $\overrightarrow{k}\,\widehat{[E]} = \mathcal{R}_0$=exchange rate), we can formulate the rigorous criterion: All those steps are in quasi–equilibrium, whose reaction rates ($\mathcal{R}$) have, by virtue of the coupling, been depressed considerably below the exchange rates ($\mathcal{R}/\mathcal{R}_0 \to 0$) (cf. Eq. (6.118)). Quasi–equilibrium is definitely established for all (specifically) fast processes in the steady state. There the rates are all equal and reduced to the rate of the rate determing step (rds). The rate of the rds is necessarily small because the k–values are small. The smallness of the rate of the other steps (with their significant k–values) forced by the coupling means, according to Eq. (6.118), that $[F]/[E] \simeq \widehat{[F]}/\widehat{[E]}$.

[143] If the microscopic rate constants include electric fields, as is the case in some of the later examples, $\overrightarrow{k}$ and $\overleftarrow{k}$ also change on approaching equilibrium, and their contributions to the exchange rate are designated by $\widehat{\overrightarrow{k}}$ and $\widehat{\overleftarrow{k}}$.

For the whole system it is only meaningful to assume a stationary (nonequilibrium) state in an open system (influx = outflux). Then the constancy of the rates is not in contradiction with the formulation $\frac{d[Z]}{dt} = K_y \frac{d[Y]}{dt}$, since both time dependences disappear.
If the reaction chain is closed at A and Z, not all the steps can be stationary, since Z changes at the expense of A. This is also true if the chain is closed at one side (as is the case for the dissolution of oxygen in an oxide to be described below). Then A is in immediate contact with the gas phase, and the concentration of Z changes with time accordingly. Nonetheless a part of the chain can behave in a stationary manner, in the sense that the storage of matter can be neglected there[144]. Let the chain be (quasi–)stationary up to the reaction of M to L and let the back reaction of the rate determining step (rds) be negligible (i.e. the reaction chain be far from equilibrium); then — with $\mathcal{R}_s$ as uniform rate of this part of the chain — $[\dot{M}]_{\text{from left}} = \mathcal{R}_s = [\dot{L}]_{\text{from left}} = \vec{k}_k[K]$ and using Eq. (6.119) it follows that

$$\mathcal{R}_s = \vec{k}_k(\Pi_{r<k}K_r)[A] = \vec{k}_k K_{eff}[A] = \vec{k}_{eff}[A], \tag{6.121}$$

whereby $\vec{k}_{eff}$ contains the mass action constants of the preceding equilibria in addition to the rate constant of the rate–determining step.
Generally such one–step approximation follows if we cut the chain right after the irreversible step (i.e. at L): Then the quasi–equilibrium of the preceding steps demands the kinetics of an effective reaction[145] A → L. The result for the temperature dependence of the effective rate $\vec{k}_{eff}$ is

$$-R\frac{\partial \ln \vec{k}_{eff}}{\partial 1/T} = \Delta_k H^{\neq} + \Sigma_{r<k}\Delta_r H^{\circ}. \tag{6.122}$$

Let us emphasize: Instationarity (transient behaviour) is characterized by the (positive or negative) accumulation of a particular species. This can be regarded as a chemical capacitance effect in an analogy with electrical phenomena[63]. If these are negligible for a given species (e.g. $[O_{ad}]$), then the species concerned exhibits

[144] This quasi–stationarity principle ought to be fulfilled in the oxygen incorporation experiment for elementary steps of the surface reaction, if these take place in narrow spatial regions, in which the increase in particle numbers, as a result of stoichiometric changes, can be neglected compared with the bulk. It is fulfilled, precisely speaking, if the rate of storage is small compared with influx and outflux rates. Accordingly, the concentration increase may be locally perceptible, as it is the case in Fig. 6.50 (r.h.s.) at the boundaries. The amount of material stored, however, is small. A similar principle applying to homogeneous kinetics is referred to as Bodenstein's principle [453]: When intermediates are reactive, their accumulation is generally small, since — once they have been formed — they are transformed immediately. Hence, their concentration is small and, thus, without the whole reaction chain being in the stationary state, their (absolute) time dependence is also. The aforementioned discrepancy concerning $\dot{n}$ and $\dot{c}$ does not apply in a homogeneous reaction.

[145] Generally then $\mathcal{R} = d[L]/dt = \vec{k}_k[K] \propto [A]$. If there were a further reversible step, then the total change of [L] would not be proportional to [K] in the transient case.

(quasi–)stationary behaviour[144,146], even if the overall process does not. This, together with the principle of quasi–equilibrium, simplifies the kinetic analysis considerably and is of great importance for Section 6.7.3.

Before we turn to a detailed kinetic analysis, we consider the incorporation reaction of our master experiment grosso motto as a serial coupling of surface reaction (incorporation into the first layer of the solid, i.e. at x=0) and a transport step:

$$A \rightleftharpoons B(0) \rightleftharpoons B(x>0). \tag{6.123}$$

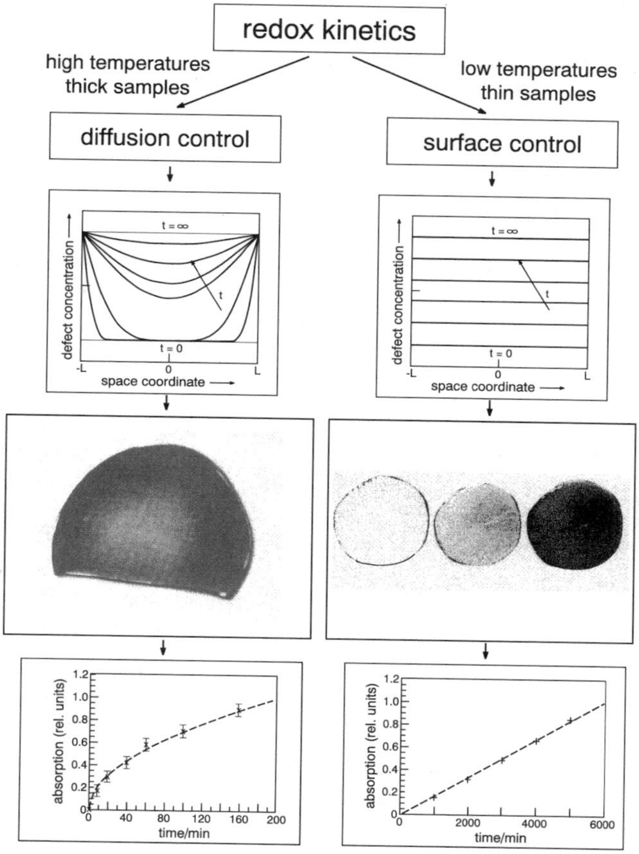

Fig. 6.50: Results of the optical analysis of the incorporation of oxygen in $SrTiO_3$.
Left column: diffusion control, right column: surface control.
Upper row: concentration profiles.
Middle row: ex situ photography of the crystal, quenched on the way to equilibrium.
Lower row: time course of the (integral) optical absorption in the diffusion direction. [127].

[146]The following coffee–analogue may serve as a vivid example even though it fails in several respects. It is particularly illustrative if we refer to the old–fashioned method of making coffee: A sufficient amount of hot water is poured into the coffee filter, containing coffee powder, which allows it to pass as the rate determining step (while extracting the relevant ingredient). Then it drops into the coffee pot. Since the mass redistribution is fast, it occurs via horizontal profiles. Even though the level steadily increases in the pot, the situation in the filter soon becomes stationary. In the very beginning modifications in the filter (in particular concerning the capacitive effects of filter and powder with respect to water) will be significant for the overall process.

If the surface reaction is rapid, then the surface concentration [B](x=0) is in equilibrium with the preceding steps, e.g. in the case of oxygen incorporation with the oxygen in the gas phase (here equilibrium between A and B(x=0)). This is the case, which we treated in Section 6.5. The other extreme case occurs if diffusion is much more rapid than the surface reaction. Then [B](x=0) is in (spatial) equilibrium with [B](x>0) and only homogenous concentration profiles appear within the sample[146]. In realistic cases, the rate determining step is located within the complex reaction A⇌B(x=0), and the description is only this simple if we can neglect the mass increase for the total interfacial regime[144].
These limiting cases, as laid out schematically in the first row of Fig. 6.50, are elegantly confirmed by optically detecting oxygen diffusion into $SrTiO_3$, as described in Section 6.5. Figure 6.51 displays results of in–situ space–resolved measurements

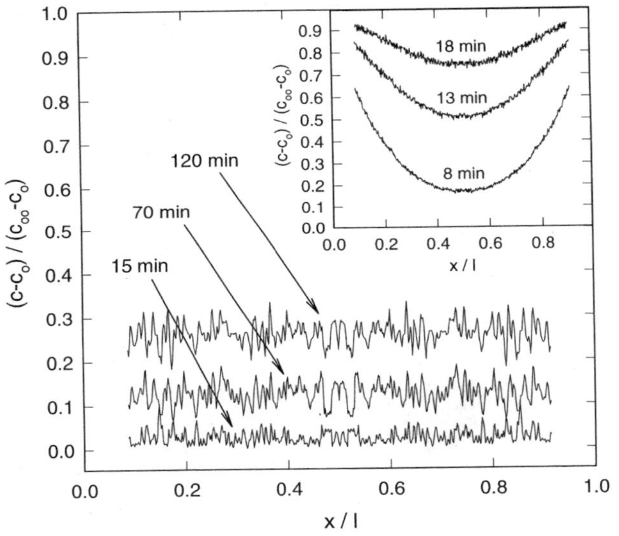

Fig. 6.51: Normalized concentration profiles (see Section 6.5) for Fe–doped $SrTiO_3$ samples (Fe content: $4.6 \times 10^{19}cm^{-3}$) on rapid change of the gas atmosphere from 10^4 to 10^5Pa at 923K. The main diagram shows reaction control, the inset diffusion control. Both cases involve incorporation through a polished (100) surface. In the case of diffusion control, a thin, porous Cr layer was used to activate the surface. According to Refs. [444,454].

on two samples, which only differ in surface treatment. Such measurements, thus, permit direct determination of the effective rate constants of the surface reaction ($\bar{k}$). The middle and bottom rows in Fig. 6.50 show analogous results for the integral measurements (but here at two different temperatures on the same sample) and will be discussed in more detail below.
Naturally it is also possible to evaluate intermediate cases with respect to D and $\bar{k}$ values. In such cases we observe profiles in the interior whose ordinate intercepts do not correspond to the final concentration[147]. Such a behaviour has already been displayed by Fig. 6.33 (upper part) for the kinetics of stoichiometric change and by Fig. 6.26 for the tracer incorporation kinetics. Figure 6.52 shows a tracer profile

[147]The general solution is obtained by solving the second Fick's law with the condition that $j = -D(\partial/\partial x)c = \bar{k}(c - c(t = \infty))$ at the boundary, i.e. the flux there is proportional to the deviation of the boundary concentration from the final value [395].

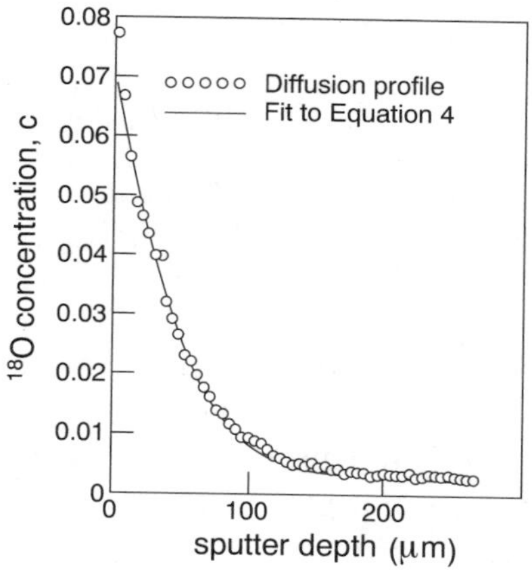

Fig. 6.52: A quenched tracer profile, produced in $La_{1.91}Sr_{0.09}CuO_{4-\delta}$ by means of three minutes of contact with an ^{18}O atmosphere at 500°C, measured by means of SIMS. The continuous line takes into account contributions from the surface reaction (see Section 6.7). Note that the concentration is not normalized. The natural abundance of ^{18}O is 0.2% (cf. bulk value). From Ref. [455].

of a high–temperature superconductor material, the Sr–doped La_2CuO_4, revealing pronounced surface resistances. Figure 6.53 refers to the fuel cell cathode material $La_{0.8}Sr_{0.2}CoO_{3-\delta}$. In comparison with the poorer ion–conducting manganate in Fig. 6.26 the surface exchange is much more apparent in the kinetics of the cobaltate. From such measurements D^* and $\bar{k}^*$ have been obtained.

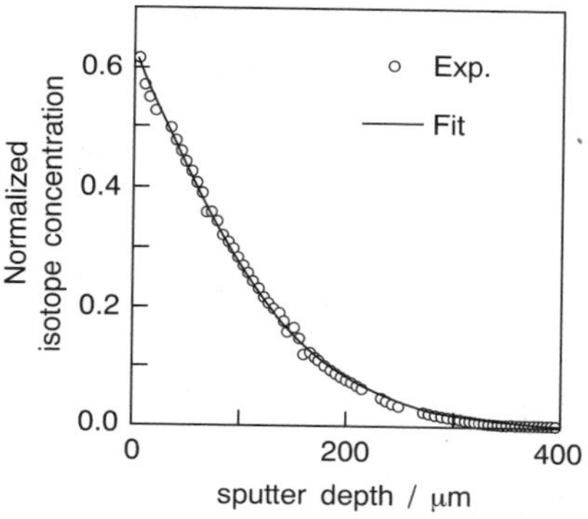

Fig. 6.53: In contrast to Fig. 6.26 (page 312), and in agreement with the high D^* values, the kinetics in $La_{0.8}Sr_{0.2}CoO_{3-x}$, under the conditions given, are strongly influenced by the surface reaction. For pure diffusion control the normalized surface concentration would be unity. From Ref. [399].

6.7.3 Phenomenological rate constants

a) Introductory remarks

Depending on the particular experiment (see Fig. 6.17) we not only need to distinguish between the various diffusion coefficients (D^δ, D^*, D^Q) but also the corresponding $\bar{k}$ values ($\bar{k}^\delta$, $\bar{k}^*$, $\bar{k}^Q$). In contrast to the diffusion case, in which the concentration gradients provide the driving force and the D's are defined as the proportionality constants between the fluxes and these gradients, the respective kinetic parameter of the reaction case, $\bar{k}$, is understood to be the ratio of the flux to the concentration deviation from the equilibrium state, $\delta c \equiv c - \widehat{c}$, taken at the first layer of the bulk[148] adjacent to the surface proper (i.e. at x=0):

$$j = -\bar{k}\delta c(x{=}0). \tag{6.124}$$

This applies, in general and independent of mechanism, for small deviations from the equilibrium state, and corresponds to the linear approximation of the Taylor series of J in δc.

As already referred to, Fig. 6.51 shows space–resolved in situ profiles of the chemical incorporation experiment for the measurement of D^δ and $\bar{k}^\delta$, whilst Fig. 6.50 (lower row) refers to the more easily carried out integral measurements. Let us have a closer look at Fig. 6.50. On sudden alteration of the gas atmosphere from P_1 to P_2, corresponding to the equilibrium concentration c_1 and $c_2 = c(t{=}\infty)$, there is a $\sqrt{t}$ law typical for diffusion processes (Section 6.5). However, the process changes qualitatively for lower temperatures or large sample thicknesses. The linear behaviour observed in such cases — as described below — indicates surface control. This is confirmed by the ex situ observations of the sample with respect to colour distribution (centre row in Fig. 6.50). In both cases the colour changes from pale yellowish to deep red–brown as the partial pressure of oxygen is increased, at the l.h.s. via pronounced spatial profiles, at the r.h.s. via the homogeneous intermediate states. The linear law comes about as follows: If the concentration change from equilibrium is small[149], it is always possible to formulate a relaxation law of the form

$$\frac{dc}{dt} = \frac{d\delta c}{dt} = -\bar{k}_R\delta c = \bar{k}_R\left(c(t{=}\infty) - c(t)\right) \tag{6.125}$$

for the local reaction (cf. Eq. (6.5)); the solution is an exponential law and, as an approximation for small times, we obtain

$$\delta c \propto \exp -\bar{k}_R t \simeq 1 - \bar{k}_R t. \tag{6.126}$$

In this form, Eq. (6.126) applies for simple homogeneous kinetics. If, as is the case for oxygen incorporation, a reservoir of thickness L is coupled, a uniform concentration profile establishes in the sample under the condition of reaction control as directly confirmed by the space–resolved profiles in Fig. 6.51 (quasi–equilibrium). The deviation δc then determines the flux into this volume and, hence, the amount

[148] However, it should be remembered that, in this example, the surface reaction may also include space charge effects. δc then must be referred to $x \simeq 2\lambda$. In this case the interpretation of $\bar{k}$ will be more complex and the quasi–stationary assumption less justified due to the thicker effective boundary. The reader should also pay attention to the fact that we use a different zero–point to the previous discussion, in which the solid ranged from $-L/2$ to $+L/2$ with the origin being in the centre.

[149] This just expresses the linear relation between flux and driving force. Chemical reactions between freshly mixed reactands do not usually fall within the range of applicability of linear, irreversible thermodynamics. However, in this case we only disturb an already established equilibrium (i.e. $|\Delta G| \ll RT$).

of material to be distributed over L, with the consequence that $\bar{k}_R$ in Eq. (6.126) is to be replaced by $\bar{k}/L$[150]. A linear relation similar to Eq. (6.126) is valid for the integral measurement.

There was an exponential solution too for the diffusion control (see Section 6.5) (but with L^2/D instead of $L/\bar{k}$ as time constant). However, the exponential solution only applied to the long–time range there, while here it is valid for small times as well. Hence, short time approximation yields linearity, and not a $\sqrt{t}$ approximation, as was obtained for diffusion control (compare Eq. 6.126 with Eq. 6.75a). It should be remembered that two significant presuppositions had to be made in the derivation of Eq. (6.126) and thus have to be considered in the evaluation of the measurements (see below): Namely small deviations from the equilibrium state, and negligible time delay of the transfer from the gas phase to the locus, to which Eq. (6.126) refers[151]. The detailed analysis can be found in subsection b, where we will also discuss the complexity of the surface reaction.

It is of interest to pause here briefly and to consider a first–order reaction, just to demonstrate, on the one hand, that in this special case Eq. (6.126) applies for any distance from equilibrium and, on the other hand, that the effective constant[152] $\bar{k}_R$ in Eq. (6.126) is correlated with the microscopic rate constants. In the case of the (closed) elementary reaction $A \rightleftharpoons B$ the rate is $\mathcal{R} = \overrightarrow{k}[A] - \overleftarrow{k}[B]$, whereby $\mathcal{R} \equiv d[B]/dt$. Conservation of mass requires that $[A](t) + [B](t) = \text{const}$, from which it follows that $\mathcal{R} = \text{const} - (\overrightarrow{k} + \overleftarrow{k})[B]$. Equilibrium is reached when $t \rightarrow \infty$, ($[B](t=\infty)$ is termed $\widehat{[B]}$), $\mathcal{R}$ becomes zero and the constant is obviously $(\overrightarrow{k} + \overleftarrow{k})\widehat{[B]}$. The result[153]

$$\begin{aligned} \frac{d[B]}{dt} = \frac{d}{dt}\left([B] - \widehat{[B]}\right) &= -(\overrightarrow{k} + \overleftarrow{k})\left([B] - \widehat{[B]}\right) \\ &= (\overrightarrow{k} + \overleftarrow{k})\left([A] - \widehat{[A]}\right) \end{aligned} \tag{6.127}$$

shows that for this closed first–order reaction the effective value $\bar{k}_R$ in Eq. (6.125) is to be identified with the sum of the forward and backward rate constants.

[150] On account of conservation of mass $\partial n/a = L\partial c$. This is more directly evidenced by Fig. 6.21: Since $j_0 = -\bar{k}\delta c$ describes the flux into the sample's volume and $j_L = 0$ the flux out of it, the space invariant quantity $\dot{c}$ is given by $\dot{c} = (\delta c)^{\cdot} = -\mathrm{div}\mathbf{j} = -\frac{j_L - j_0}{\Delta x} = -\bar{k}\frac{\delta c}{L}$ and the exponential law discussed follows, viz. $\frac{c - c(t=\infty)}{c(t=0) - c(t=\infty)} = \exp -\frac{\bar{k}t}{L}$.

[151] This is valid for the case of quasi–stationarity, but also for the case of quasi–equilibrium between the loci addressed. In both cases it is important that the surface layers do not absorb oxygen to a perceptible extent. In the first case the negligible chemical capacitance is an immediate presupposition, in the second case the transient phase leading to quasi–equilibrium would otherwise be comparatively long.

[152] The designation $\bar{k}$ can be understood better when it is demonstrated (see Eq. (6.132)) that this parameter "averages" over the forward and back reaction and over various elementary steps.

[153] Introduction of the approximation $\overrightarrow{k} \gg \overleftarrow{k}$ and $[A](t=0) \gg \widehat{[A]}$ demonstates that Eq. (6.127) is indeed fulfilled for the commencement of the reaction too, independent of the distance from equilibrium: Then $\mathcal{R}$ reduces to the initial forward rate $\overrightarrow{\mathcal{R}} = \overrightarrow{k}[A](t=0)$. This changes obviously if the reaction order is different.

It is instructive for later treatment to consider this from a slightly modified view point [384]. For this purpose we write $\mathcal{R}$ in the form

$$\mathcal{R} = \vec{k}\,\widehat{[A]}\left(\frac{[A]}{\widehat{[A]}}\right) - \overleftarrow{k}\,\widehat{[B]}\left(\frac{[B]}{\widehat{[B]}}\right). \tag{6.128}$$

At equilibrium $\mathcal{R}$ is zero and the terms in brackets are unity. The quantity

$$\mathcal{R}_0 = \vec{k}\,\widehat{[A]} = \overleftarrow{k}\,\widehat{[B]} = \sqrt{\vec{k}\,\overleftarrow{k}\,\widehat{[A]}\,\widehat{[B]}} \tag{6.129}$$

represents the "exchange rate", and is a measure of the dynamics at equilibrium[143] (cf. exchange current densities in Section 6.1). Hence, it follows that

$$\mathcal{R} = \mathcal{R}_0\left(\frac{[A]}{\widehat{[A]}} - \frac{[B]}{\widehat{[B]}}\right). \tag{6.130}$$

Conservation of mass and the mass action law[154] permit us to express this in the form equivalent to that of Eq. (6.127):

$$\mathcal{R} = -\frac{\mathcal{R}_0}{\langle\widehat{c}\rangle}\delta[B] \equiv -\frac{\mathcal{R}_0}{\langle\widehat{c}\rangle}\left([B] - \widehat{[B]}\right). \tag{6.131}$$

The parameter $\langle c\rangle$ is an abbreviation of the harmonic mean of the equilibrium concentrations $\widehat{[A]}$ and $\widehat{[B]}$. Taking Eq. (6.129) then gives[155]

$$\frac{\mathcal{R}_0}{\langle\widehat{c}\rangle} = \bar{k}_R = \vec{k} + \overleftarrow{k}. \tag{6.132}$$

Usual elementary reactions are not so simple and Eq. (6.126) is only valid sufficiently close to equilibrium[156]. Nonetheless the range of applicability may be wider than assumed, since many reactions behave as if they were monomolecular under certain conditions (so–called pseudo–monomolecular reactions). In addition, relevant processes are usually not just single step reactions, in which B only increases at the expense of A and vice versa. The relations between $\bar{k}$ and $\vec{k}$, $\overleftarrow{k}$ are then more involved. Even under those conditions the exchange rate remains the decisive permeability parameter [452] containing $\vec{k}$ and $\overleftarrow{k}$ of the rate determing step in a symmetrical way.

---

[154] Conservation of mass requires $\widehat{[A]} + \widehat{[B]} = [A] + [B]$. It follows that $[A]/\widehat{[A]} = 1 + \widehat{[B]}/\widehat{[A]} - [B]/\widehat{[A]} = 1 + K - [B]/\widehat{[A]}$. Using the abbreviation $\langle\widehat{c}\rangle = \frac{\widehat{[A]}\,\widehat{[B]}}{\widehat{[A]} + \widehat{[B]}}$ for the harmonic mean we obtain Eq. (6.131).

[155] Express $\mathcal{R}_0/\langle\widehat{c}\rangle$ as a function of $\vec{k}$, $\overleftarrow{k}$ and $\widehat{[B]}/\widehat{[A]} = \vec{k}/\overleftarrow{k}$.

[156] For a reversible second–order reaction analogous calculation

$$A + A' \rightleftharpoons B + B'$$

yields an expression of the form

$$\mathcal{R} = d[B]/dt = (\vec{k} - \overleftarrow{k})P^{(2)}([B]),$$

b) Detailed analysis

Let us discuss some experimental results before we analyse these cases and let us turn to our optical experiment, which we used to measure the effective rate constant of oxygen incorporation / excorporation, i.e. $\bar{k}^{\delta}$. (Conductivity relaxation measurements yield analogous results, see Fig. 6.27.)

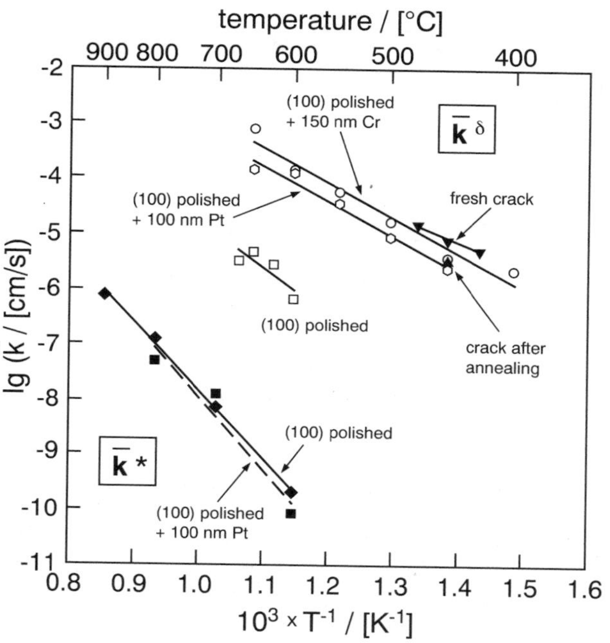

Fig. 6.54: Effective rate constants of the surface reaction for the oxygen incorporation experiment (jump from 10^4 to 10^5Pa) obtained for Fe–doped $SrTiO_3$ (iron content: $4.6 \times 10^{19}cm^{-3}$) under various conditions. The low values for the tracer experiment are shown for comparison (10^5Pa).

Figure 6.54 displays $\bar{k}^{\delta}$ values for $SrTiO_3$ as a function of temperature obtained from both locally resolved and integral measurements. These values are naturally strongly dependent on the surface crystallography and surface structure. In particular, freshly created surfaces resulting from crack formation exhibit considerable $\bar{k}^{\delta}$ values, which rapidly diminish by ageing (relaxation, see Section 5.4). More-

---

whereby $P^{(2)}$ is a second–order power function in [B]. $P^{(2)}$ can be represented by the zeros in the form $([B] - \widehat{[B]})([B] - b)$, so that close to equilibrium

$$\mathcal{R} \cong (\vec{k} - \overleftarrow{k})(\widehat{[B]} - b)([B] - \widehat{[B]}) \propto ([B] - \widehat{[B]}).$$

In general, the relaxation law (Eq. (6.126)) can be proven in the following way: We express the concentrations in the ratio Q/K by means of the deviation (δc) from equilibrium value ($c/\widehat{c} = 1 + \delta c/\widehat{c}$). According to Eq. (6.22) the affinity is given by the logarithm of the concentration ratios. Near equilibrium the logarithm of $(1 + \delta c/\widehat{c})$ can be approximated by $(\delta c/\widehat{c})$. The sum $\Sigma_k \nu_k \ln(1 + \delta c_k/\widehat{c}_k)$ occurring in the affinity becomes $\Sigma_k \nu_k \delta c_k/\widehat{c}_k = (\delta c_k/\nu_k)\Sigma_k \nu_k^2/\widehat{c}_k = \delta\xi \Sigma_k \nu_k^2/\widehat{c}_k$. Note that $\delta\xi = \delta c_k/\nu_k$ is invariant with respect to the nature of the component (see Section 4.2). The affinity and, hence, $\mathcal{R}$ according to Eq. (6.21) are thus proportional to the concentration deviation from equilibrium (ξ: reaction progress variable).

over, the kinetics are sensitively affected by catalytic layers. (Also UV illumination has a distinct effect [456].) We will return to this in Section 6.8. The same figure demonstrates that the corresponding tracer constants ($\bar{k}_0^{\delta}$) are distinctly lower.

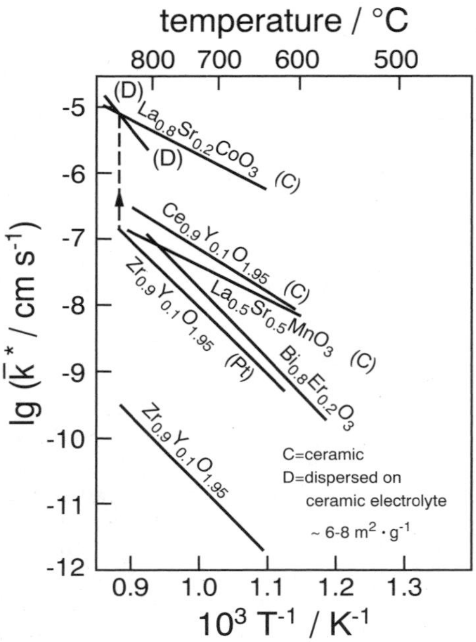

Fig. 6.55: Effective rate constants for oxygen tracer incorporation in various oxides. For $ZrO_2(Y_2O_3)$ the low $\bar{k}^*$ value, in contrast to $CeO_2(Y_2O_3)$ and $Bi_2O_3(Er_2O_3)$, can be considerably increased by a platinum layer. The exchange rate of $ZrO_2(Y_2O_3)$ increases on dispersion of $LaMnO_3(SrO)$ on the ceramic electrolyte (increase in the density of three–phase boundaries). Please note the high values for $LaCoO_3(SrO)$. From Ref. [457].

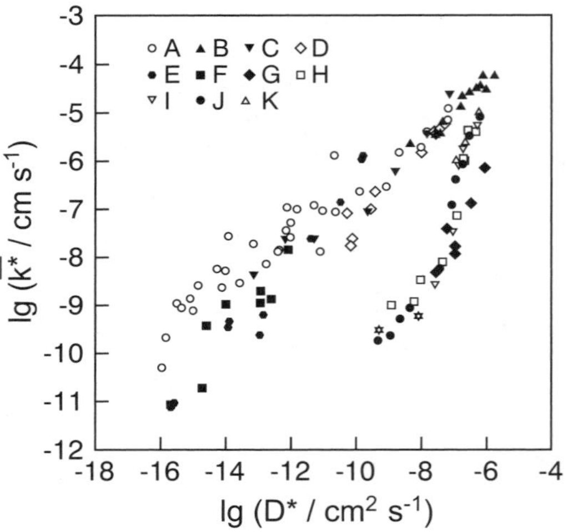

Fig. 6.56: Collection of tracer data for acceptor doped oxides (P_{O_2}=1bar, T=600°C ... 1000°C) in the $\bar{k}^*$ vs. D^* representation. Typical ionic conductors constitute the lower branch while typical mixed conductors belong to the upper one. From Ref. [458].
A – $La_{1-x}Sr_xMn_{1-y}Co_yO_{3\pm\delta}$ [458],
B – $Sm_{1-x}Sr_xCoO_{3-\delta}$ (x=0.4, 0.5, 0.6) [459],
C – $La_{0.6}Sr_{0.4}Fe_{0.8}Co_{0.2}O_{3-\delta}$ [460],
D – $CaZr_{0.9}In_{0.1}O_{2.95}$ [461],
E – $SrCe_{0.95}Yb_{0.05}O_{2.975}$ [461],
F – $La_{1-x}Sr_xYO_{3-x/2}$ (x=0.1, 0.2) [462],
G – $La_{0.9}Sr_{0.1}Ga_{0.8}Mg_{0.2}O_{2.85}$ [463],
H – $Zr_{0.85}Y_{0.15}O_{1.925}$ [464],
I – $Ce_{0.9}Gd_{0.1}O_{1.95}$ [465],
J – $Zr_{0.81}Y_{0.19}O_{1.905}$ (single crystal) [466],
K – $Ce_{0.69}Gd_{0.31}O_{1.845}$ (single crystal) [467].

Figures 6.55 and 6.56 show the rate constants of oxygen tracer exchange ($\bar{k}_O^*$) for a series of different oxides. The profiles were determined by ex situ SIMS analysis[86] (cf. Figs. 6.52, 6.53). From the examples referred to in Figs. 6.54, 6.55, 6.56 we can see that $\bar{k}^*$ and $\bar{k}^{\delta}$ are evidently very different parameters and that there appears to be a correlation between k^* and D^* for many oxides. Furthermore, it is necessary to distinguish between $\bar{k}^*$ and the kinetic parameter $\bar{k}^Q$ obtained from the electrical measurements (cf. Sections 7.3.3 and 7.3.6). In cases in which a reliable comparison is possible $\bar{k}^*$ and $\bar{k}^Q$ have been found to be very similar.

Since in all three cases the $\bar{k}$ values are based on different experiments, it is naturally possible that they depend on different rate–determining steps. In the case of the stoichiometry experiment (Fig. 6.17) electrons are explicitly involved (they are consumed); in the tracer experiment, however, they are only implicitly involved (nominally the electrons are only exchanged between the isotopes), while in the conductivity experiment the electrons stem from the electrode. In addition, the particular material conditions vary (consider, e.g., the presence of electrodes in the electrical experiment). Most importantly, we expect conceptual differences stemming from the different "chemical capacitances" necessarily involved in the experiments. Let us look at the latter point more closely.

The driving force for the chemical experiment is the variation in the chemical component potential; for the tracer experiment, the variation in the chemical potential of the isotope; and for the electrical experiment the variation in electrical potential. Now we concentrate on the first two experiments. The detailed analysis of the electrical experiment will be reserved for Chapter 7.

In the case of pure diffusion control the concentration at the surface ($c(0)$) is in immediate equilibrium ($\overline{c}(0)$) with the gas phase. The same applies to the corresponding potential (μ_O or μ_O^*). When interfacial reactions are inhibited, deviations from the final equilibrium value occur that we will write as $\delta\mu(x{=}0)$, $\delta\mu^*(x{=}0)$, or simply $\delta\mu(0)$, $\delta\mu^*(0)$.

The situation becomes particularly clear when the surface inhibition is the result of transport through a thin interfacial layer of thickness $|s|$ being impeded. Now the deviations $\delta\mu(0)$, $\delta\mu^*(0)$ can (in contrast to the concentration effects[157]) also be considered as spatial differences (Δ) across this layer of thickness $\Delta x = |s|$ (see Fig. 6.48). Here we neglect space charges. Then the situation is obviously analogous to the diffusion case, and, hence introducing Λ as the proportionality factor, we obtain

$$j_O^\delta = -\Lambda_O^\delta \Delta\mu_O = -\Lambda_O^\delta \delta\mu_O(0) \tag{6.133}$$

$$j_O^* = -\Lambda_O^* \Delta\mu_O^* = -\Lambda_O^* \delta\mu_O^*(0). \tag{6.134}$$

In a similar manner it is possible to define a Λ^Q parameter for the electrical experiment if the voltage drop over the layer is taken as the driving force (see Chapter 7). $\Delta\mu$ stands for $\mu(x{=}0) - \mu(x{=}s)$. The coordinate $x{=}0$ refers to the locus in the sample immediately adjacent to the layer[148], while $x{=}s$ refers to the locus in the layer immediately adjacent to the gas phase. The latter is, for the case assumed, in immediate equilibrium with the gas phase. Apart from a geometric factor, the constants of proportionality correspond to the effective conductivities (σ_O^δ, $\sigma_{O^{2-}}^*$) in the layer. The component or tracer fluxes can then, also, be regarded as driven by the local concentration differences[157], whereby the $\bar{k}$ value is defined:

$$j^\delta = -\Lambda^\delta \left(\partial\mu_O/\partial c_O\right) \delta c_O(0) = -\bar{k}^\delta \delta c_O(0) \tag{6.135}$$

[157] Note that when there is inhibition as a result of slow transport through the layer it is true that $\Delta\mu = \delta\mu$ but $\Delta c \neq \delta c$, since at equilibrium $\Delta\mu$ disappears but not Δc.

$$j^* = -\Lambda^* \left(\partial\mu^*_{O^{2-}}/\partial c^*_{O^{2-}}\right) \delta c^*_{O^{2-}}(0) = -\bar{k}^* \delta c^*_{O^{2-}}(0). \tag{6.136}$$

For $\partial\mu_O/\partial c_O$, we obtain according to the treatment in Section 6.3, RTw_O/c_O and for $\partial\mu^*_{O^{2-}}/\partial c^*_{O^{2-}}$ simply $RT/c^*_{O^{2-}}$; $\bar{k}^*$ being isotope independent. It is to be remembered that Λ refers to the layer (x=s), while $\partial\mu/\partial c$ is associated with the locus x=0 (and hence refers to the bulk[148]). As it was the case for D^Q it is necessary to introduce $\bar{k}^Q$ by definition. If we neglect correlation factors, we obtain a result corresponding to the tracer value, when we define $\bar{k}^Q_{O^{2-}}$ as $RT\Lambda_{O^{2-}}/c_{O^{2-}}$. In the case of our homogeneous interfacial layer slowing down the mass transport, Λ is proportional to the effective conductivity of this layer.
The case of the "electron–rich electronic conductor" is particularly transparent (see also Section 6.3.4). The rate–determining step in all three cases is then the ion transport ($\sigma^\delta_O = \sigma_{O^{2-}}$) and it is evident[158], according to Section 6.3, that $\bar{k}^\delta$ and $\bar{k}^*$ are related to each other in the same manner as D^δ and D^*. In addition, if the charge carrier density is high, we can neglect space charge effects[159]. The relative concentration variation at x=0 with respect to x=s on account of rigid double–layer effects (see Sections 5.8 and 7.3.3) is also negligible. So if only one ionic defect, say $V_O^{\cdot\cdot}$, is dominant, it evidently follows that

$$\bar{k}^\delta/\bar{k}^* = \widehat{c}_{O^{2-}}/\widehat{c}_{V_O^{\cdot\cdot}} = \widehat{x}_v^{-1} \quad \text{and} \quad \bar{k}^\delta/\bar{k}^Q = 1 \tag{6.137}$$

as already found for the diffusion coefficients. (Note that we have neglected correlation effects.)
The effective rate constant determined in a chemical relaxation experiment is, thus, considerably larger than that found in a tracer experiment, since this averages over all O^{2-} ions. This already provides us with a possibility of explaining the behaviour in Fig. 6.54. (As we will see below, this is by no means specific for this mode of surface hindering, and rather a common feature.) In addition, there is evidently a correlation between $\bar{k}^*$ and $\bar{D}^*$ (Fig. 6.56), even though this may not be simple. If we assume, e.g. that the conductivity difference between bulk and layer is attributable to the changed concentration and that the mobilities are similar, then elimination of the latter yields

$$\bar{k}^* = aD^*, \tag{6.138}$$

with the parameter a reflecting the concentration ratio.
However, when the material is varied, Eq. (6.138) is purely formal. In other words: One should exercise due care when trying to interpret Fig. 6.56 by Eq. (6.138), since the factor a is not invariant if the material is changed. The correlation between $\bar{k}^*$ and D^* on changing the control parameters, such as the doping content (C) for a given substance, is clearer. If the doping dependence for both x = s and x = 0

[158]Note that Λ^*/c^* is isotope–independent.

[159]The description of chemical transport is additionally facilitated by the following: Approximately constant carrier concentrations (here of the electrons, i.e. if $\delta\Delta\mu_{e^-} = 0$), lead to the result that the electrical potential differences correspond to their equilibrium values ($\delta\Delta\phi = 0$), whenever the electrochemical potential differences for these carriers can be neglected ($\Delta\widetilde{\mu}_{e^-} = \delta\Delta\widetilde{\mu}_{e^-} = 0$).

is determined by the same characteristic exponent $M_{V_O^{\cdot\cdot}}$ (see Section 5.6) then the prefactor in Eq. (6.138) is given by $\left[\frac{C(s)}{C(0)}\right]^{M_{V_O^{\cdot\cdot}}} |s|^{-1}$. Hence, a is not constant here either. However, if the variation of k^* and D^* is varied via the oxygen partial pressure, then indeed $k^* \propto D^*$.

Let us now turn to the proper interfacial kinetics, by focusing on the chemical processes on the surface.
Equations (6.133, 6.134) are still valid. For convenience we replace the flux by the rate ($\mathcal{R} = j/\Delta x$); the driving force corresponds to the affinity $\mathcal{A}$. The prefactor assumes the meaning of an exchange rate (per RT), more precisely of the exchange rate of the rate determining step. This follows immediately from linear irreversible thermodynamics, i.e. for proximity to equilibrium, by using the quasi–equilibrium and the quasi–stationarity principles, as treated in Section 6.7.2. According to this, the flux at x=0 is determined by the approximately constant rate of the surface reaction steps and thus equal to the rate of the rate determining step (rds), i.e. proportional to $\mathcal{R}_{0,rds}\mathcal{A}_{rds}$ according to Eq. (6.21). Since $\mathcal{A}_{rds}$ represents the overall affinity to a good approximation ($\mathcal{A} = \mu(\text{gas}) - \mu(x{=}0,t) = \mu(x{=}0,t{=}\infty) - \mu(x{=}0,t) = -\delta\mu$), we obtain Eqs. (6.135, 6.136) with $\Lambda = \text{const}\mathcal{R}_{0,rds}$. Also here the relations between $\bar{k}^\delta$ and $\bar{k}^*$ given by Eq. (6.137) for the special case of the "electron–rich electron conductor" are fulfilled, since the electronic steps are not perceptible here. The application of chemical kinetics gives more detailed information, in particular with respect to the range of validity.

Let us assume initially that the adsorption is rate determining. We further assume that oxygen is dissociatively adsorbed. If the reaction occurs in a single step

$$O_2 + 2V_{ad} \rightleftharpoons 2O_{ad} \tag{6.139}$$

the corresponding rate equation is

$$\mathcal{R} = \vec{k}P_{O_2}[V_{ad}] - \overleftarrow{k}[O_{ad}]^2. \tag{6.140}$$

(We presuppose fast redistribution of V_{ad} if necessary.)
Formally the same (but with different constants) expression results for a rate determining dissociation according to

$$O_{2,ad} \rightleftharpoons 2O_{ad} \tag{6.141}$$

if the proper adsorption is in equilibrium.
In the case that the adsorption mechanism involves a pre–dissociation of O_2 before the rate determining step we have

$$\mathcal{R} = \vec{k}P_{O_2}^{1/2}[V_{ad}] - \overleftarrow{k}[O_{ad}]. \tag{6.142}$$

This is possible if a significant dissociation into reactive atomic oxygen occurs in the gas phase (very high temperatures, very low oxygen partial pressures) or if there is dissociative adsorption of O_2 at special adsorption sites followed by slow redistribution.

Even though the rate equation (6.142) does not apply in most cases, we will work with it for the purpose of simplicity. At this point we are not interested in details (generalization is straightforward, see below, page 358), and Eq. (6.142) has the advantage that both $\mathcal{R}$ and j are related to a single oxygen. So we formally refer to

$$\frac{1}{2}\mathrm{O}_2 + \mathrm{V}_{\mathrm{ad}} \rightleftharpoons \mathrm{O}_{\mathrm{ad}}. \tag{6.143}$$

For simplicity we first consider an electron–rich electronic conductor, again, and the case of low coverage. Then as far as the chemical experiment is concerned, only the variation in $[\mathrm{O}_{\mathrm{ad}}]$ in Eq. (6.142) is relevant. We formulate[74]

$$\mathcal{R}^\delta = \vec{\mathrm{k}}\mathrm{P}_{\mathrm{O}_2}^{1/2}\widehat{[\mathrm{V}_{\mathrm{ad}}]}\left(\frac{[\mathrm{V}_{\mathrm{ad}}]}{\widehat{[\mathrm{V}_{\mathrm{ad}}]}}\right) - \overleftarrow{\mathrm{k}}\widehat{[\mathrm{O}_{\mathrm{ad}}]}\left(\frac{[\mathrm{O}_{\mathrm{ad}}]}{\widehat{[\mathrm{O}_{\mathrm{ad}}]}}\right) \simeq \vec{\mathrm{k}}\mathrm{P}_{\mathrm{O}_2}^{1/2}\widehat{[\mathrm{V}_{\mathrm{ad}}]} - \overleftarrow{\mathrm{k}}\widehat{[\mathrm{O}_{\mathrm{ad}}]}\left(\frac{[\mathrm{O}_{\mathrm{ad}}]}{\widehat{[\mathrm{O}_{\mathrm{ad}}]}}\right) \tag{6.144}$$

according to the procedure developed above (cf. Eq. (6.128)). At equilibrium $\mathcal{R}^\delta = 0$ with the result $\vec{\mathrm{k}}\mathrm{P}_{\mathrm{O}_2}^{1/2}\widehat{[\mathrm{V}_{\mathrm{ad}}]} = \overleftarrow{\mathrm{k}}\widehat{[\mathrm{O}_{\mathrm{ad}}]}$. This quantity is the exchange rate ($\mathcal{R}_0$) of this reaction. It can obviously be written as

$$\mathcal{R}_0 = \vec{\mathrm{k}}\mathrm{P}_{\mathrm{O}_2}^{1/2}\widehat{[\mathrm{V}_{\mathrm{ad}}]} = \overleftarrow{\mathrm{k}}\widehat{[\mathrm{O}_{\mathrm{ad}}]} = \sqrt{\vec{\mathrm{k}}\overleftarrow{\mathrm{k}}\mathrm{P}_{\mathrm{O}_2}^{1/2}\widehat{[\mathrm{V}_{\mathrm{ad}}]}\widehat{[\mathrm{O}_{\mathrm{ad}}]}}, \tag{6.145}$$

whereby

$$\frac{\widehat{[\mathrm{O}_{\mathrm{ad}}]}}{\mathrm{P}^{1/2}\widehat{[\mathrm{V}_{\mathrm{ad}}]}} = \mathrm{K}_{\mathrm{ad}}. \tag{6.146}$$

The introduction of $\mathcal{R}_0$ simplifies Eq. (6.144) to

$$\mathcal{R}^\delta = \mathcal{R}_0\left(1 - \frac{[\mathrm{O}_{\mathrm{ad}}]}{\widehat{[\mathrm{O}_{\mathrm{ad}}]}}\right) = -\frac{\mathcal{R}_0}{\widehat{[\mathrm{O}_{\mathrm{ad}}]}}\delta[\mathrm{O}_{\mathrm{ad}}]. \tag{6.147}$$

In the case of tracer (^{18}O) diffusion $[\mathrm{V}_{\mathrm{ad}}]$ is unaffected, even in the case of significant coverage and it follows that

$$\mathcal{R}^* = \mathcal{R}_0^{18}\left(1 - \frac{[^{18}\mathrm{O}_{\mathrm{ad}}]}{\widehat{[^{18}\mathrm{O}_{\mathrm{ad}}]}}\right) = -\frac{\mathcal{R}_0}{\widehat{[\mathrm{O}_{\mathrm{ad}}]}}\delta[^{18}\mathrm{O}_{\mathrm{ad}}]. \tag{6.148}$$

(The index 18 refers to the tracer, while the total oxygen content is referred to otherwise.)

We note that $\mathcal{R}_0^{18}/\widehat{[^{18}\mathrm{O}_{\mathrm{ad}}]} = \overleftarrow{\mathrm{k}} = \mathcal{R}_0/\widehat{[\mathrm{O}_{\mathrm{ad}}]}$. In the case of low coverage both experiments, thus, give analogous expressions with respect to the variation of the degree of coverage.

However, since we obtain the $\bar{k}$ value from the time–dependence of the oxygen concentration at position x=0 of our bulk profile, we must convert $\delta[O_{ad}]$ to $\delta c_O(x=0)$ and $\delta[^{18}O_{ad}]$ to $\delta c_O^{18}(x=0)$. The reactions that follow are in quasi–equilibrium according to our treatment in Section 6.7.2. This means for the tracer incorporation that the tracer fraction in the bulk corresponds to that in the adsorption layer. On account of the local invariance of the parameters without asterisks it follows that

$$\mathcal{R}^* = -\bar{k}\left([^{18}O_{ad}] - \widehat{[^{18}O_{ad}]}\right) = -\bar{k}\,\widehat{[O_{ad}]}\left(\frac{[^{18}O_{ad}]}{[O_{ad}]} - \frac{\widehat{[^{18}O_{ad}]}}{\widehat{[O_{ad}]}}\right)$$
$$= -\bar{k}\,\widehat{[O_{ad}]}\left(\frac{[^{18}O_O]}{[O_O]} - \frac{\widehat{[^{18}O_O]}}{\widehat{[O_O]}}\right) = -\bar{k}\left(\widehat{[O_{ad}]}/\widehat{[O_O]}\right)\delta[^{18}O_O] \equiv -\bar{k}^*\delta c_O^{18}/\Delta x \tag{6.149}$$

and, hence,

$$\bar{k}^*/\Delta x = \bar{k}\,\widehat{[O_{ad}]}/\widehat{[O_O]} = \mathcal{R}_0/\widehat{[O_O]}. \tag{6.150}$$

The elementary distance Δx appears here, since $\bar{k}$ was defined as $-j/\delta c$ and not as $-\mathcal{R}/\delta c$.

In the case of the chemical diffusion we explicitly exploit the equilibrium between the succeeding ionization and transfer reactions. So it follows that[160]

$$O_{ad} + V_O^{\cdot\cdot} + 2e' \rightleftharpoons V_{ad} + O_O. \tag{6.151}$$

From this

$$\frac{\delta[O_{ad}(s)]}{\widehat{[O_{ad}(s)]}} \simeq -\frac{\delta[V_O^{\cdot\cdot}(O)]}{\widehat{[V_O^{\cdot\cdot}(O)]}} \tag{6.152}$$

is obtained, with the consequence

$$\mathcal{R}^\delta = \bar{k}\delta[O_{ad}] = \bar{k}\left(\widehat{[O_{ad}]}/\widehat{[V_O^{\cdot\cdot}]}\right)\delta[V_O^{\cdot\cdot}] = -\bar{k}^\delta\delta c_O/\Delta x. \tag{6.153}$$

Thus, in contrast to $\bar{k}^*$ it follows for $\bar{k}^\delta$ that

$$\bar{k}^\delta/\Delta x = \bar{k}\,\widehat{[O_{ad}]}/\widehat{[V_O^{\cdot\cdot}]} = \mathcal{R}_0/[V_O^{\cdot\cdot}] \tag{6.154}$$

and again

$$\bar{k}^\delta : \bar{k}^* = \widehat{x}_{v,i}^{-1}. \tag{6.155}$$

In the case that electrons are also of importance, the r.h.s. of Eq. (6.152) must be complemented by the term $-\delta[e']/\widehat{[e']}$. Then more generally the ratio of the $\bar{k}$–values

[160] In the case of the chemical experiment $V_O^{\cdot\cdot}$ and $2e'$ refer to the same locus, and the electrical potential cancels in the balance of the electrochemical potentials (in contrast to $\bar{k}^Q$, see footnote 161).

takes the form[161]

$$\bar{k}^\delta \simeq \widehat{w}_O \bar{k}^* \quad \text{and} \quad \bar{k}^* \simeq \bar{k}^Q, \tag{6.156}$$

which also followed from the Eqs. (6.135), (6.136) for identical Λ–values. With the help of Eq. (6.109) the P_{O_2} and the C dependences of the $\bar{k}$–values can be extracted from Eqs. (6.150) and (6.154) according to the treatment in Section 5.5. In simple cases — e.g. in the case of the electron–rich electron conductor — one finds approximately Brouwer– and Arrhenius–dependences of the form ($\bar{N}$, $\bar{M}$, $\bar{\gamma}$ are rational numbers):

$$\bar{k} \propto k_{eff}(T) P^{\bar{N}} C^{\bar{M}} \Pi_r K_r^{\bar{\gamma}_r}(T). \tag{6.157}$$

(If we consider the more common adsorption mechanisms described by Eq. (6.140), the corresponding expression for k^δ formally differs from Eq. (6.154) only by a constant factor of 4 ($dn_{O_2} = dn_O/2$; $\delta c^2/c^2 \simeq 2\delta c/c$). Owing to the different form of the exchange current, the characteristic exponents in Eq. (6.157) are distinctly different.)

Similar expressions result in the case that the transfer reaction is rate determining. Let us, for simplicity, assume that the transfer step is identical with Eq. (6.151). The rate equation then reads in the case of a chemical experiment performed with an electron–rich electron conductor,

$$\mathcal{R}^\delta = \widehat{\bar{k}}\,\widehat{[V_O^{\cdot\cdot}]}\,\widehat{[e']}^2 \left(\frac{[V_O^{\cdot\cdot}]}{\widehat{[V_O^{\cdot\cdot}]}}\right) - \widehat{\bar{k}}\,\widehat{[V_{ad}]}\,\widehat{[O_O]}, \tag{6.158}$$

as merely $[V_O^{\cdot\cdot}]$ changes. Hence

$$\mathcal{R}^\delta = \mathcal{R}_0 \frac{\delta[V_O^{\cdot\cdot}]}{\widehat{[V_O^{\cdot\cdot}]}} \tag{6.159}$$

and again $\bar{k}^\delta = \mathcal{R}_0 \Delta x / \widehat{[V_O^{\cdot\cdot}]} = j_0 / \widehat{[V_O^{\cdot\cdot}]}$, with j_0 as exchange current density ($j_0 = \mathcal{R}_0 \Delta x$).

More generally we have to consider the variation $\delta\left([V_O^{\cdot\cdot}][e']^2\right)$ instead of $\delta[V_O^{\cdot\cdot}]$, and $\bar{k}^\delta = -\mathcal{R}_0 w_O / c_O$ will be the result [468].

In the case of tracer incorporation it holds that

$$\mathcal{R}^* = \widehat{\bar{k}}\,\widehat{[V_O^{\cdot\cdot}]}\,\widehat{[^{18}O_{ad}]} - \widehat{\bar{k}}\,\widehat{[V_{ad}]}\,\widehat{[^{18}O_O]} \left(\frac{[^{18}O_O]}{\widehat{[^{18}O_O]}}\right) = \mathcal{R}_0 \delta[O_O] / \widehat{[O_O]}. \tag{6.160}$$

---

[161] Because of the succeeding equilibrium we can, for the chemical experiment, more generally state $\delta \ln [O_{ad}] + \delta \ln [V_O^{\cdot\cdot}] + 2\delta \ln [e'] = 0$ [468]. In the case of the electrical experiment $[O_{ad}]$ is a variable, too, and a function of the overvoltage η, which appears in the subsequent quasi–equilibrium and enters the pseudo–equilibrium constant. (Note that e′ now stems from the electrode material.) With $\delta \ln [O_{ad}] + \delta \ln \widetilde{K}_\eta = 0$ we obtain $\bar{k}^Q = \bar{k}^*$ after linearization.

In the case of the electrical experiment it is the k–values which vary. The corresponding rate equation,

$$\mathcal{R}^{\mathrm{Q}} = \mathcal{R}_0 \left(\frac{\overline{\overrightarrow{\mathrm{k}}}}{\overline{\widehat{\overrightarrow{\mathrm{k}}}}} - \frac{\overline{\overleftarrow{\mathrm{k}}}}{\overline{\widehat{\overleftarrow{\mathrm{k}}}}} \right), \tag{6.161}$$

corresponds to the Butler–Volmer equation, which we will encounter in Section 7.3.3. Close to equilibrium this leads, as expected, to $\bar{\mathrm{k}}^{\mathrm{Q}} = \bar{\mathrm{k}}^*$ [468].

It is interesting to inspect the validity ranges of the above relations. As expected for each process ϵ ($\epsilon = \delta, *, \mathrm{Q}$), relations of the form

$$\bar{\mathrm{k}}^{\epsilon} = \mathrm{j}_0^{\epsilon} \left(\mathrm{w}^{\epsilon} / \widehat{\mathrm{c}^{\epsilon}} \right)_{\mathrm{bulk}} \tag{6.162}$$

are fulfilled (where $\mathrm{w}^* = 1 \simeq \mathrm{w}^{\mathrm{Q}}$) only approximately. (Stoichiometric numbers stemming from different molecularities (see previous page) are neglected in Eq. (6.162).) For the electrical experiments linearization is a pre–requisite in any case. This is in agreement with the treatment in Section 6.1. In the case of the chemical experiment the situation is variable. Linearization is necessary, if, besides the ionic carrier the electronic one is of importance, too. In the case of the electron–rich material a relation of the form of Eq. (6.162) followed without approximation for the transfer control, while for the adsorption control linearization (see Eq. (6.152)) was necessary. Had we discussed interstitials instead of vacancies, in contrast to Eq. (6.152) the assumption of proximity to equilibrium would not have been necessary because $[\mathrm{O}_\mathrm{i}''] \propto [\mathrm{O}_{\mathrm{ad}}]$. This is consistent with the statements on pages 349f. More precisely we can formulate that the restriction to proximity to equilibrium is not necessary, if the rate determining step follows first order kinetics (monomolecular or pseudo–monomolecular) and if there are proportionalities between the relevant point defect concentrations in the succeeding equilibria. Also in the case of tracer exchange the linear range can be exceeded if the rds is of higher order.
Eq. (6.156) shows that $\bar{\mathrm{k}}^{\delta}$ should be distinctly greater than $\bar{\mathrm{k}}^*$ or $\bar{\mathrm{k}}^{\mathrm{Q}}$. (Compare the right column in Fig. 6.57.) In addition, $\bar{\mathrm{k}}^*$ and D^* are not independent of each other, the intercorrelation, however, is complex. If we analyse the ratio between $\bar{\mathrm{k}}^{\delta}$ and $\bar{\mathrm{k}}^*$

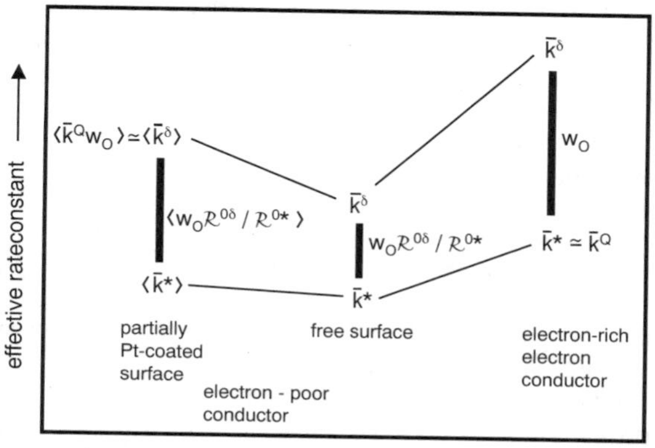

Fig. 6.57: Relations between the $\bar{\mathrm{k}}$'s for high electron concentration (r.h.s. column) and for lower electron concentrations of the oxide. Free surface: centre column, partial Pt coverage: left column. The brackets indicates averaging over the heterogeneous surfaces. There is a close correspondence to Fig. 6.20. From Ref. [452].

in various oxides, we recognize that the ratio $\bar{k}^\delta/\bar{k}^*$ is usually great compared to 1, but in some cases distinctly smaller than predicted by Eq. (6.156) [469,470].

So far we assumed that all experiments are characterized by the same rate determining step ($\mathcal{R}_0^\delta = \mathcal{R}_0^* = \mathcal{R}_0^Q$). Different mechanisms, however, lead to different Λ and $\mathcal{R}_0$ values and in particular to "interfacial Haven ratios" (Λ^*/Λ^Q) that differ from unity[162].
Such mechanistic differences can be important in electron–poor materials[163] (e.g. solid electrolytes). It must be remembered that tracer exchange can in principle also take place as a result of direct substitution of the ^{16}O species by the ^{18}O species and, in such a case, does not involve free electrons (direct electron exchange between the isotopes). In other words: Such a direct tracer exchange is always a possible parallel route which can be followed in the case of a lack of electronic carriers. In these cases we expect that $\Lambda^\delta/\Lambda^* \neq 1$. How far the results of the detailed evaluation of the Λ–ratio for Fe–doped $SrTiO_3$, as well as the different influence of Pt on $\bar{k}^\delta$ and $\bar{k}^*$, may suggest such mechanistic deviations (see Fig. 6.54) is still to be investigated. It is clear that electrical fields as they occur in the case of $SrTiO_3$, complicate the picture. An extensive discussion of the mechanism of oxygen incorporation into $SrTiO_3$ is to be found in Refs. [456,471,469,472]. According to the present status of knowledge the mechanism comprises the following steps[164] [469]

$$\begin{aligned} O_2 &\rightleftharpoons O_2^- + h^{\cdot} \\ O_2^- + e' &\rightleftharpoons O_2^{2-} \\ O_2^{2-} &\rightleftharpoons 2O^- \\ 2O^- + 2V_O^{\cdot\cdot} &\rightleftharpoons 2O_O + 2h^{\cdot} \end{aligned}$$

with the second (e' transfer) or the third step (dissociation) being rate determining.

Examples of electrode kinetics (i.e. referring to the steady state electrical experiment) are given in Chapter 7 (Section 7.3).
In addition to electrical, tracer or chemical transport through solids, a fourth experiment has to be considered for which surface processes are decisive, namely heterogeneous catalysis, which is addressed in Section 6.8.

[162] In the case of the bulk transport such mechanistic differences clearly occur and are reflected by $\sigma^\delta \neq \sigma_{ion}$.

[163] Such a difference in the behaviour of electron–rich and electron–poor materials can already be inferred from Fig. 6.56. One might argue that electrons have to be supplied by the material in the case of an ionization, e.g. according to $O + 2e' \rightleftharpoons O''$ while this might not be the case for a mechanism like $O \rightleftharpoons O'' + 2h^{\cdot}$. This argument, however, only holds far from equilibrium, whereas close to equilibrium forward and backward reactions are of equal importance.

[164] Note that in the first step a regular e^- is consumed whilst the second ionization of O_2 requires an electron of high energy (e') even though present in much lower concentration (p–type $SrTiO_3$).

6.7.4 Reactivity, chemical resistance and chemical capacitance

Let us now discuss some terms of obvious importance in solid state chemistry which are usually used in a purely intuitive manner. In order to refer to rates and not to fluxes, we use here the quantities Λ_R and $\bar{k}_R$ which we obtain from Λ and $\bar{k}$ by division through Δx. The role that Λ_R plays for the interface is analogous to that played by σ for the bulk. Just as we call the latter (specific) conductivity we could designate the first as (specific) "exchange reactivity". (We must note that this parameter is determined by the exchange rate of the rate–determining step normalized to RT; see also β in Section 6.7.2.) It combines thermodynamic and kinetic parameters in a suitable manner (equilibrium concentrations and microscopic k values) in a similar way to σ[165]. It is important to keep in mind that this exchange reactivity (as the conductivity) contains information concerning the forward and back rates. The "reactivity" usually associated with the ability of a substance to react is a property applying far from equilibrium and is identifiable with the forward rate leading to equilibrium (compare $\vec{\mathcal{R}} = \vec{k}[A]$ in Eq. (6.128)) per RT. Such a reactivity declines with time and approaches at equilibrium the exchange reactivity[165] (exchange rate per RT).

Two other very useful terms can be introduced, namely "chemical resistance" (R^δ) and "chemical capacitance" (C^δ) (cf. also Section 7.3.4). As shown by Eqs. (6.74) and (6.126) const.D^δ/L^2 and $\bar{k}^\delta/L$ play the role of inverse time constants. As for electrical processes (R^Q, C^Q) (cf. Section 7.3) the latter can be broken down into chemical resistances and chemical capacitances. The former are given by the reciprocal reactivities (reciprocal exchange rates) or in the bulk case by $1/\sigma^\delta$. The chemical capacitance becomes[63] $\partial n/\partial \mu \propto (\partial c/\partial \mu) L \propto c^\delta L$ and denotes the storage capacity, or, more exactly, the increase in mole number of the component considered as the chemical potential is increased. In the bulk case the effective resistances are proportional to L and hence $\tau^\delta \propto L^2$, while for the surface reaction a proportionality of τ^δ to L^1 has to be taken into account. For tracer diffusion the corresponding parameters take a trivial meaning ($c_{O^{2-}}$, $\sigma_{O^{2-}}$ instead of c^δ, σ^δ). The reader should pay attention to the fact that in the bulk R^δ is characterized by a series switching of the individual elements, while R^Q is characterized by a parallel switching.

The increased (differential) capacitance is finally also largely the reason for the diminution of the diffusion coefficient, which takes place when we go from the case of chemical diffusion without trapping to chemical diffusion with trapping; it is also the decisive factor when we compare defect diffusion (e.g. chemical diffusion in the case of the electron–rich electron conductor ($D^\delta = D_{V_O^{\cdot\cdot}/O_i''}$)) with tracer diffusion (see Fig. 6.19). The situation is similar for the $\bar{k}$ values. As for electrical effects the capacitive components in the chemical and tracer experiment are responsible for the time behaviour.

[165]Suitable quantities being normalized with respect to concentration (hence analogous to the equivalent conductivity or mobility) are $\vec{k}$, $\sqrt{\vec{k}\overleftarrow{k}}$ or $\bar{k}_R$ depending on the situation.

The special role of D^Q and $\bar{k}^Q$ is worth mentioning in this context in that they are derived directly from the resistive parameter[131]. The concentration term involved (included by definition) is not effective as chemical capacitance, since $\delta\mu=0$ and the transient behaviour is merely determined by dielectric effects (i.e. by electrical capacitances). This alters, if chemical changes must also be taken into account in the electrical experiments. Then the conceptual boundaries between the electrical and the chemical experiment disappear (see Section 7.3.4).
In this context the construction of generalized equivalent circuits which include chemical and electrical effects is helpful [473].

6.8 Catalysis

By definition a catalyst (cat) influences the reaction rate but not the position of the equilibrium, i.e.

$$\frac{d\mathcal{A}}{d[\mathrm{Kat}]} = 0 \neq \frac{d\mathcal{R}}{d[\mathrm{Kat}]}. \tag{6.163}$$

In the case that only a single elementary reaction is relevant, on whose transition state the catalyst acts, K and k can stand instead of $\mathcal{A}$ and $\mathcal{R}$. If the presence of the catalyst reduces the free enthalpy threshold of the forward reaction by Δ, then since $K = \vec{k}/\overleftarrow{k}$ it is evident that this must also apply to the back reaction[166].
In the case of heterogeneous catalysis the educts are adsorbed on the surface of the catalyst, and are not incorporated as in the case of the experiments considered in the previous sections but react on the surface — the products being desorbed eventually. Heterogeneous catalysts have, as a rule, a particularly pronounced effect on the reaction path; in addition to changing the chemical bonds, they are helpful because they can enhance the density of reactants via adsorption on the solid surface. The famous ammonia synthesis can serve as an example [473,474]:

$$N_2 + 3H_2 \rightleftharpoons 2NH_3. \tag{6.164}$$

In the gas phase the reaction rate between the elements N_2 and H_2 is negligibly small, primarily because of the kinetic stability of the nitrogen triple bond in $|N \equiv N|$. The situation is different in the presence of a Fe catalyst. Here the N_2 is relatively easily dissociated. Nonetheless this reaction is rate–determining in the

---

166 More precisely we have to consider the uncatalysed ($\vec{k}_0, \overleftarrow{k}_0$) and catalysed ($\vec{k}, \overleftarrow{k}$) reactions in parallel: $\mathcal{R} = \vec{k}[A] - \overleftarrow{k}[B] + \vec{k}_0[A] - \overleftarrow{k}_0[B]$. It clearly follows then at equilibrium that $\frac{\widehat{[B]}}{\widehat{[A]}} = \frac{\vec{k}+\vec{k}_0}{\overleftarrow{k}+\overleftarrow{k}_0} = \frac{\vec{k}_0(1+\exp+\Delta/RT)}{\overleftarrow{k}_0(1+\exp+\Delta/RT)} = K$. Usually $\vec{k}$ and $\overleftarrow{k}$ will contain the catalyst concentration as factors which also cancel in the quotient (appearing as prefactors of the exponential functions).

overall heterogeneous reaction scheme which is represented in a simplified form by

$$\begin{array}{rcl} N_2(g) & \rightleftharpoons & N_{2,ad} \\ \frac{1}{2}N_{2,ad} & \longrightarrow & N_{ad} \\ H_2(g) & \rightleftharpoons & H_{2,ad} \\ \frac{1}{2}H_{2,ad} & \rightleftharpoons & H_{ad} \\ N_{ad} + H_{ad} & \rightleftharpoons & NH_{ad} \\ NH_{ad} + H_{ad} & \rightleftharpoons & NH_{2ad} \\ NH_{2ad} + H_{ad} & \rightleftharpoons & NH_{3ad} \\ NH_{3ad} & \rightleftharpoons & NH_3(g). \end{array} \tag{6.165}$$

In order to avoid complications with respect to various adsorption sites, free adsorption sites have been left out of the formulation (even though decisive for the kinetic interpretation), i.e. the adsorbed species have been regarded as building elements. The relatively rapid diffusion processes, in the gas phase and on the surface, are not included. The effectivity of Fe is qualitatively understandable on the basis of the formation of a bond to nitrogen of moderate strength. Too low an affinity would mean that nitrogen would not be adsorbed or dissociated at all. If the affinity is too strong, subsequent steps would be inhibited [474,475]. This finding, that medium affinities are favourable to catalytic efficiency, is frequently met. Hill–like shapes in the plots of the reaction rate against e.g. heat of adsorption are known as volcano curves (see Fig. 6.58).

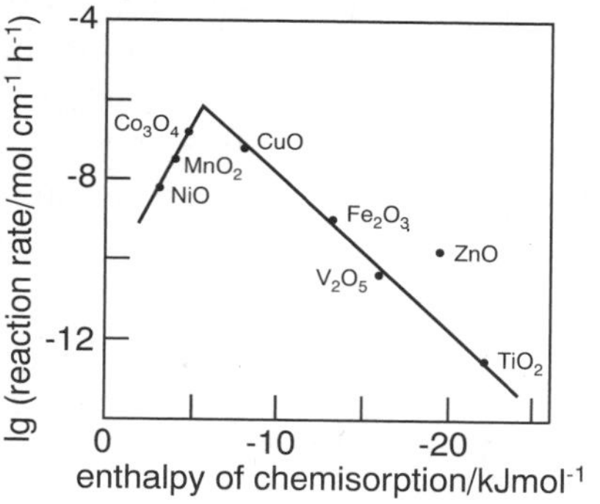

Fig. 6.58: "Volcano" curve for the combustion of H_2 at metal oxides with O_2 excess (T=573K, $[H_2]$ = 10^{-4}mol/l). From Ref. [474].

Such dependences are also consistent with the catalytic efficacy of silver or platinum in the case of many oxygenation processes, such as the incorporation of oxygen in oxides that we have taken as our master–example. It is also striking that many good catalysts are mixed conductors; this also applies to electrocatalysts which are treated in Chapter 7. This suggest that not only the concentration but also the mobility of electronic (redox centres) and ionic defects (acid–base centres, see below) is important[167]. When this is the case, the local conditions at the reaction

[167]The chemical capacitance of mixed conductors can also be of assistance, in that they may supply oxygen via stoichiometric changes. These changes can be nullified later by interaction with the ambient.

centre are much less stringent, since missing partners can diffuse to the site where they are needed. Such mixed conductors are generally very effective catalysts for the oxydation of hydrocarbons. What is desirable, in practice, in addition to acceleration of the reaction, is control of the ratios of the rate constants with respect to competing reactions and to achieve necessary selectivities, e.g. oxydizing CH_4 selectively to HCHO, CH_3OH or HCOOH; many highly active nonspecific catalysts oxydize right through to CO_2.

A certain degree of selectivity control appears possible by using electrically polarized solid electrolytes. More specifically, it has been found that effective oxydation catalysis of hydrocarbons occurs if platinum serves as the anode of a potentiometric oxygen pump

$$CH_x, O_2,^{\oplus} Pt|ZrO_2|Pt^{\ominus}, O_2 \qquad (6.166)$$

(see Section 7.3.1).

Applying a voltage across ZrO_2 effects a flux of O^{2-}. It is now not only that the oxygen, that is transferred by the current, reacts very rapidly, but the reactivity of the hydrocarbons with the oxygen particles present in the gas–chamber is generally increased. The altered surface chemistry of the metal is favoured as an explanation [476]. However, the precise mode of action has not yet been fully elucidated particularly with respect to stating the question of whether the catalytic activity is thermodynamically[168] or kinetically controlled.

Let us briefly return to our model material $SrTiO_3$ once again. It was shown in Chapter 5 that when (Au, Cr)–electrodes are used, only at high temperatures can we expect incorporation equilibrium of oxygen on fully ionized vacancies, with the release of an equivalent number of holes. At lower temperatures the surface step of this reaction becomes extremely sluggish (see Section 5.6). Figure 6.59 now demonstrates that when using $YBa_2Cu_3O_{6+x}$ electrodes the reversibility of the reaction is guaranteed to appreciably lower temperatures. That is, this material has a strong catalytic effect with respect to the surface reaction (as already seen for Pt, Section 6.7.3); the precise mechanism is not known — presumably it involves acceleration of the dissociative adsorption, including ionization[169] of the O_2 molecule. Moreover, the carrier depletion ($V_O^{\cdot\cdot}, h^{\cdot}$) in the space charge regions is presumably lower than in the case of Au, Cr electrodes (cf. Section 5.8.5).

In most cases catalytic effects are explained either in terms of electron transfer processes or acid–base effects of the regular surface groups. However, it must not be overlooked that not only the surface as an extended two–dimensional defect (see Section 5.4), but the point defects therein [478–481] can play a particularly important

---

[168] Note that an electrical potential applied to an oxygen solid electrolyte corresponds to an equivalent modification of the local oxygen activity (Section 7.2.1). The above catalytic effect has been termed NEMCA effect (Non–Electrochemical Modification of Catalytic Activity) [477].

[169] There is, however, still the oxygen–oxygen double bond to break. In this connection we should note that, for instance in electrochemical reactions, it is possible to achieve reversibility for H_2 at relatively low temperatures, the temperature necessary is higher for O_2 and is usually very high for N_2 (cf. also ammonia synthesis).

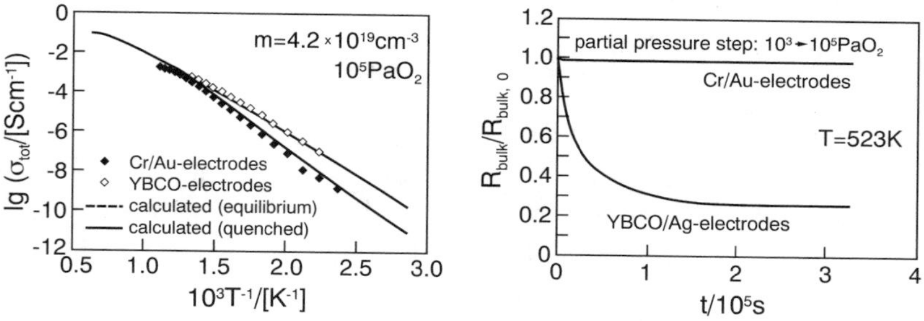

Fig. 6.59: The surface step of oxygen incorporation is greatly accelerated by $YBa_2Cu_2O_{6+x}$ electrodes ("YBCO"). This makes it possible to still attain equilibrium with the gas phase at lower temperatures. L.h.s.: temperature behaviour of the conductivity on cooling. R.h.s.: time change of the resistance on a sudden change in the ambient P_{O_2} (see Section 5.6) [193].

role here in the sense of being potential acid–base partners [15] for all the elementary steps; point defects represent centres of increased, local energy and are generally mobile. Very many reactions are known in organic chemistry (generally homogeneous, i.e. in solution, for instance) that are catalysed by acidic or basic centres. Examples include ester saponifications or dehydrohalogenation reactions. Figure 6.60 displays

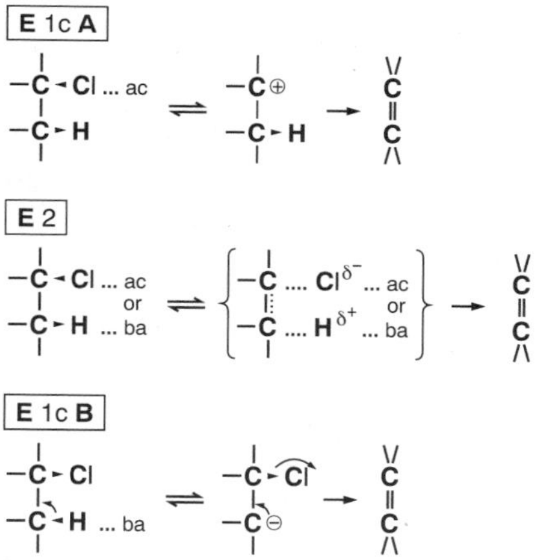

Fig. 6.60: Possible mechanisms of acid–base catalyzed dehydrohalogenation of tertiary butyl chloride (ba: base, ac: acid) [478].

possible acid–base mechanisms for the catalysis of HCl elimination from t–butyl chloride $(CH_3)_3CCl$ ($\equiv$ t–BuCl) in solution. Now let us assume that when this reaction is catalysed heterogeneously the adsorption is the rate–determining step. Since, in these cases, every adsorbed molecule reacts rapidly, it can be assumed, if the reactand pressure is not too great, that the number of free adsorption sites remains approximately constant[170]. Hence, in the case of ideal adsorption the forward

[170] We can compare the situation with the wine racks of a wine drinker. If the wine consumption is relatively low, the wine racks are always filled if wine is purchased regularly. However, if the person

reaction rate of the reaction

$$t\text{–BuCl} \rightleftharpoons \text{isobutene} + \text{HCl} \tag{6.167}$$

should be of first order with respect to t–BuCl partial pressure. Since all three substances involved are gaseous at the temperature of the reaction, the reaction can be followed simply from the increase in pressure. The presence of pure AgCl as catalyst does not have a marked effect on the reaction. However, if it is homogeneously doped with $CdCl_2$, the rate[171] changes almost proportionally with $C \equiv [Cd^{\cdot}_{Ag}]$ [479]

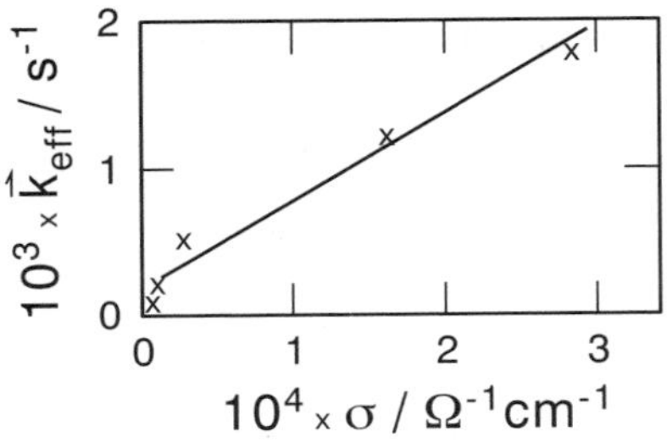

Fig. 6.61: The effective rate constant of elimination of HCl from t–BuCl in the presence of $CdCl_2$–doped AgCl as catalyst as a function of the ion conductivity and, hence, of the Cd content. From Ref. [479].

(Fig. 6.61). An obvious interpretation is that the number of adsorption sites (ad) which is to be included in the effective microscopic rate constant k_{eff} according to

$$\mathcal{R} = k_{eff}[\text{t–BuCl}] \propto [\text{ad}][\text{t–BuCl}] \tag{6.168}$$

also increases proportionally. It is readily understandable that the negatively charged silver vacancies, which are also created to the same extent (see Section 5.6), can act as basic centres and attack a proton of the hydrocarbon, and so facilitate the adsorption and possibly also the elimination itself. The acidic effect of the dopant ion ($Cd^{\cdot}_{Ag}$: interaction with the electronegative chlorine atom) is also conceivable and would be compatible with the results. On account of the low absolute value, effects of electronic defects are very probably excluded. A similar effect is expected and has been observed for heterogeneous doping. There k_{eff} has been indeed found to be proportional to the surface excess charge (see Section 2.2.2) [244].

Finally, homogeneous[172] catalysis can also take place in the solid. Recombination centres may serve as an example: Redox levels (R) in the band gap can accelerate

is a drunkard then so much wine will be consumed that provision will be the rate–determining step and the racks will always be as good as empty. However, the wine rack can also look very empty even when consumption is very low, namely, if there is a significant "pilferage rate". This corresponds to the case of rapid desorption.

[171]It is self–evident that the total turnover rate of the number of particles reacted per time unit is proportional to the area ("Wenzel's law"). However, this no longer occurs for fractal geometries (see Section 6.10).

[172]The term "homogeneous" is naturally to be taken with a pinch of salt.

electron–hole pair formation according to

$$\begin{array}{l} e^-(VB) + R \rightleftharpoons R^- \\ \underline{R^- \rightleftharpoons e^-(CB) + R} \\ e^-(VB) \rightleftharpoons e^-(CB) \quad \text{i.e. Nil} \rightleftharpoons e' + h^{\cdot}. \end{array} \tag{6.169}$$

The name "recombination centres" given to such states reflects the fact that they also accelerate the back reaction, that is electron–hole recombination. Their efficacy depends on the fact that they make possible indirect transitions (i.e. transitions with change of the wave number vector, see Chapter 2) by taking up energy and momentum. In this way the establishing of the local equilibrium is accelerated [128, 129].

A special role is played by auto–catalysis, in which the reaction rate is accelerated by the product. The resulting "upward–spiral" is fundamental to structure formation. This will be treated in Section 6.10.

6.9 Solid state reactions

6.9.1 Fundamental principles

In this chapter we consider processes that involve the formation of a new phase. From the phenomenological point of view alone this field is very rich in variations [4,7] not only with respect to the chemical nature of the reactants but also with respect to their state of aggregation and their spatial distribution. Accordingly a variety of reaction patterns is observed. In this complex field one finds extreme cases such as explosive decomposition reactions generating gaseous phases, fast precipitations of solid phases on reaction of two liquids, as well as slow and soft solid state reactions, e.g. the formation of spinel at a well–defined contact of the two parent oxides. On account of the importance of heterogeneities considering formation, disappearance or displacement of interfaces, the quantitative treatment of individual cases is generally either not possible or so specific that it makes it difficult to see the wood for the trees.

Here, however, where we wish to emphasize the fundamentals we will initially concentrate on the morphologically simple formation of a coherent oxide layer by the oxydation of a metal, and assume that the process is diffusion–controlled [7,482]. Let us imagine that the metal in Fig. 6.62 is saturated with oxygen (cf. also Fig. 4.4 on page 90) and now increase the oxygen potential somewhat over the saturation value. In this manner we avoid large affinities. (Section 6.10 is devoted to special processes far from equilibrium.) Let us now consider Fig. 6.62. Initially a thin layer of reaction product forms. The further progress of reaction depends on whether the reaction layer is gas–tight or not. According to an intuitive rule attributed to Pilling and Bedworth [483] porous oxide layers occur, in particular, when the molar volume of the oxide is appreciably smaller or greater than the molar volume of the

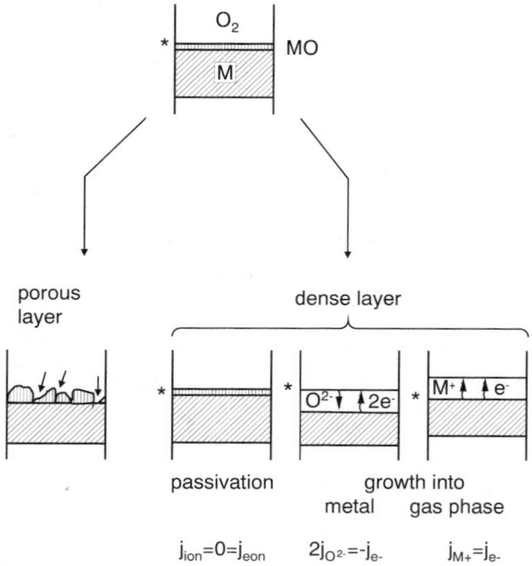

Fig. 6.62: Reactive growth of an oxide layer by metal oxydation (see text) [334].

metal. In the first case, the reaction layer is interrupted by voids, in the second it expands and cracks and pores are formed. In all these cases it is not difficult to see qualitatively that the continued rapid gas diffusion (through pores, for instance) permits oxydation to proceed, even though quantitative treatment is complex, on account of the three–dimensionality and heterogeneity of the process.

However, if closed oxide layers do form, then chemical transport through the oxide layer becomes a necessary condition for the progress of the reaction[173]. The quantitative treatment was carried out by C. Wagner [484]. Here, there are, in principle, two possibilities. Either the metal formally migrates through the layer, as a result of the ambipolar diffusion of metal ions and electrons, or there is a migration of "O" by ambipolar diffusion of oxygen and electrons. In both cases then — if we ignore grain boundary transport — the basis is the chemical bulk diffusion dealt with in the previous chapter. In the first case (see Fig. 6.62) the oxide grows[174] out into the gas phase, in the second case back into the metal. Conversely, if it is possible to observe the reaction front it is possible to decide which is the fastest ionic charge carrier.

Even if the driving force is high, no oxydation takes place if either the ionic or electronic conductivity is vanishingly small. The passivation of Al in air by a thin scale (tarnishing layer) of Al_2O_3, in spite of high affinity, is a well–known example of this.

When there is finite growth there are naturally various transport possibilities depending on the type of disorder. To be specific, let us consider the oxydation of

[173]Chemical transport is usually necessary also in the case of porous oxydation layers, in order to ensure complete oxydation.

[174]In the nomenclature of Ref. [7], we speak of reactive growth. In the case of additive growth, reaction and crystal growth are decoupled.

zinc. In this case it is presumably zinc ions in interstitial positions that provide the predominant contribution to the ionic conductivity[175]. If the temperature is sufficiently high, these are fully ionized.
The electron flow is determined by conduction electrons, and analogously to Eq. (6.52), we have to write

$$\dot{j}_{Zn} = \dot{j}_{Zn^{2+}} = \dot{j}_{Zn_i^{\cdot\cdot}} = \frac{1}{2}\dot{j}_{e'} = \frac{1}{2}\dot{j}_{e^-} = -\frac{\sigma_{Zn_i^{\cdot\cdot}}\sigma_{e'}}{\sigma}\frac{1}{4F^2}\frac{\partial}{\partial x}\mu_{Zn} \tag{6.170}$$

for the total mass transport[176]. Since the zinc flux in ZnO is locally constant[177] — for Zn is not enriched anywhere — it is possible to write Eq. (6.170) in integrated form as

$$\dot{j}_{Zn} = -\frac{1}{L}\frac{1}{4F^2}\int_{\mu_{Zn}(0)}^{\mu_{Zn}(L)} \frac{\sigma_{Zn_i^{\cdot\cdot}}\sigma_{e'}}{\sigma} d\mu_{Zn}. \tag{6.171}$$

In this one–dimensional problem the x axis represents the direction of oxide growth, x=0 represents the interface to the metal (ZnO/Zn), x=L the interface to the gas phase (ZnO/O_2). Space charge effects will be neglected. L(t) is the total thickness of the oxide at time t. Each zinc particle entering the oxide and, thus, migrating through the oxide to the gas phase is converted there to a ZnO lattice molecule, as a result of a rapid surface reaction. As a consequence j_{Zn} is directly related to the growth in thickness of the layer dL/dt and, hence, to the rate of reaction. Because $\dot{n}_{Zn} = \dot{n}_{ZnO} = \dot{V}_{ZnO}/V_m$ (V_m= molar volume of ZnO) and V(t) = L(t)·area we obtain

$$\frac{dL}{dt} = \dot{j}_{Zn}V_m = -\frac{1}{4F^2}\frac{1}{L}\left(V_m\int_{\mu_{Zn}(O)}^{\mu_{Zn}(L)} \frac{\sigma_{Zn_i^{\cdot\cdot}}\sigma_{e'}}{\sigma} d\mu_{Zn}\right) \tag{6.172}$$

as a result. However complicated the integral expression may be, it is independent of L. For it is determined by the value of the integral evaluated at the two interfaces. Because of the fact that we presupposed fast and quasi–stationary interfacial reactions, chemical potentials and specific conductivities are determined by the local interactions, which do not change during growth[178]. We obtain formally a rate law of order -1

$$\dot{L} = \kappa_d L^{-1} \tag{6.173}$$

[175] We adopt here the popular view of the literature even though there are arguments for $V_O^{\cdot\cdot}$ as decisive carrier. This point does not affect the treatment that follows, because the charge is the same (see also footnote 181).

[176] As everywhere we neglect structural (e.g. also significant elastic) effects.

[177] This does not apply to the very short initial time period. In other words: We are referring to a quasi–stationary case, as discussed in Section 6.7.2, and hence neglect "chemical" capacitance effects (cf. page 362).

[178] This applies strictly only when a certain minimum thickness has been exceeded (see Fig. 6.67).

with respect to the dependence on the layer thickness. However, such an analogy with homogeneous kinetics is misleading rather than helpful[179]. For this reason we will select the letter κ to denote the proportionality factor in Eq. (6.173) and not the k usual in the literature. Unfortunately the term "effective rate constant" has been generally accepted for this parameter, even though it contains concentration terms and also the driving force in addition to the mobilities being the proper kinetic parameters. Equation (6.173) states that the diminution of the rate of reaction with increasing layer thickness is a result of the increase in the transport distance.
The integration (Eq. (6.173)) gives the famous square root growth law [482], that has been confirmed in many instances[180]:

$$L = \sqrt{2\kappa_d t}. \tag{6.174}$$

The validity of this relation for zinc oxydation is confirmed in Fig. 6.63 for the pure and doped metal.

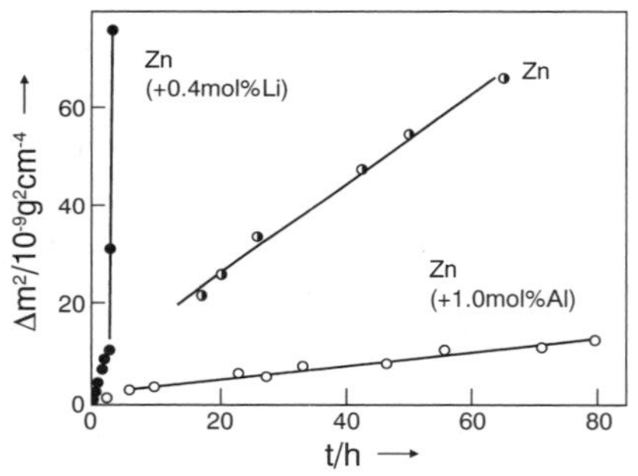

Fig. 6.63: Square of the mass increase ($\propto L^2$) as a function of time during corrosion of pure and doped zinc in air at 390°C. At these temperatures $Zn_i^{\cdot}$ is probably in the majority and not $Zn_i^{\cdot\cdot}$. The discussion is largely unaffected by this (cf., however, footnote 184). From Ref. [485].

A further example is the reduction of the oxygen pressure during the oxydation of titanium to TiO_2 at various temperatures, as displayed in Fig. 6.64 (the variation of the driving force can be neglected).

The dependence of the rate on temperature, oxygen content and doping demands a discussion of $\kappa_d(T, P, C)$.
Let us first conduct this discussion in semiquantitative form. Here we place the ambipolar conductivity even though it is dependent on μ_{Zn} in front of the integral and correct the error caused by regarding the expression in brackets as the corresponding mean value,

$$j_{Zn} = -\frac{1}{4F^2}\left\langle \frac{\sigma_{Zn_i^{\cdot\cdot}}\sigma_{e'}}{\sigma} \right\rangle \frac{\Delta\mu_{Zn}}{L} = -\frac{1}{4F^2}\left\langle \sigma^{\delta}_{Zn} \right\rangle \frac{\Delta\mu_{Zn}}{L}, \tag{6.175}$$

so that we have achieved a formal separation into driving force $\Delta\mu_{Zn}/L = (\mu_{Zn}(L) - \mu_{Zn}(0))/L$ and the kinetic parameters. Over the whole P_{O_2} range ZnO is an (excess)

[179] The change in layer thickness does not constitute a concentration change; furthermore, since the diffusion process is close to equilibrium it is always influenced by the back reaction and κ_d (or D^{δ}) contains rate constants of both directions.
[180] Note here the formal similarity to Eq. (6.69).

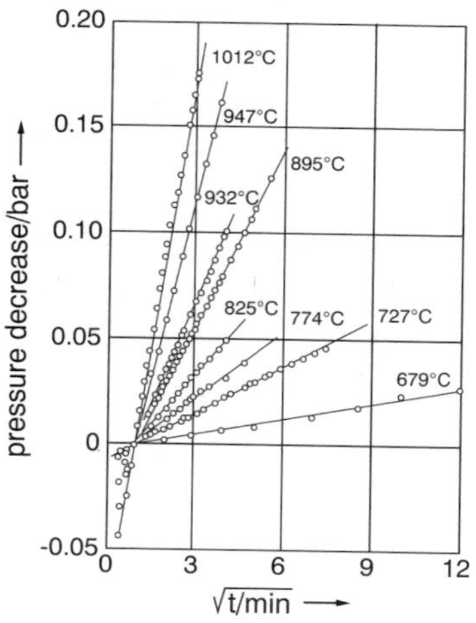

Fig. 6.64: Pressure drop ($\propto$ dL/dt) as a function of the square root of time in the corrosion of titanium to TiO_2. According to Ref. [486].

electron conductor so that $\langle\sigma^{\delta}_{Zn}\rangle = \left\langle\frac{\sigma_{Zn_i^{\cdot\cdot}}\sigma_{e'}}{\sigma}\right\rangle$ simplifies to $\langle\sigma_{Zn_i^{\cdot\cdot}}\rangle$. This means that the ionic conduction determines the specific reaction rate, which directly leads to a characteristic, strong dependence on temperature, as shown in Fig. 6.64 for the formation of (predominantly electronically conducting) TiO_2.

Figure 6.63 has demonstrated that a square root law is also obeyed for the time dependence of the corrosion of pure and of Al– or Li–doped Zn, but also demonstrated the large difference in the corresponding κ_d values. While Li–doped Zn oxydizes much more rapidly than the pure metal, the opposite is true for Al–doped Zn. All this is very simply explained in terms of the defect chemistry. We only have to assume that the doping atoms are partially incorporated into the oxide[181] and substituting Zn^{2+} (formation of Li'_{Zn} or $Al^{\cdot}_{Zn}$). According to our doping rule (Eq. (5.141)) the Li–doping increases the ionic defect concentration ($d[Zn_i^{\cdot\cdot}]/d[Li'_{Zn}] > 0$), while the reverse happens for Al–doping ($d[Zn_i^{\cdot\cdot}]/d[Al^{\cdot}_{Zn}] < 0$). This changes the ionic conductivity and, hence, the rate of scaling in the observed manner[182].

In order to discuss the P_{O_2} dependence we will have to analyse the integral in Eq. (6.172) more precisely. First let us consider the integration limits. The chemical potential μ_{Zn} at the metal contact (x=0) is fixed at μ°_{Zn} on account of the phase equilibrium with the Zn bulk. This is the maximum possible value. At the same time, the oxygen potential takes up its minimum possible value there[183], namely

$$\frac{1}{2}\mu_{O_2}(x{=}0) = \mu^{\circ}_{ZnO} - \mu^{\circ}_{Zn}. \tag{6.176}$$

---

[181] In the case of the metal diffusion assumed here, this requires a mobility of the dopant. As is mentioned, instead of zinc interstitials $V_O^{\cdot\cdot}$ may be the relevant ionic charge carrier. In this case, all the above discussion is quite analogous; however, mobility of the dopant is unnecessary.

[182] An explanation by altered driving forces may apply to Li–doped Zn, but can be ruled out in view of the size of the effect.

[183] This is treated in detail in Chapter 4 and is obtained from the local reaction

Because $\Delta_f G^\circ(\mathrm{ZnO}) = \mu^\circ_{\mathrm{ZnO}} - \mu^\circ_{\mathrm{Zn}} - \frac{1}{2}\mu^\circ_{\mathrm{O_2}}$ the corresponding partial pressure of oxygen is calculated from the free enthalpy of formation of the oxide (cf. Section 4.3.5) as

$$\mathrm{P_{O_2}(x{=}0) = P^\circ \exp \frac{2\Delta_f G^\circ_{ZnO}}{RT} \equiv P^*_{O_2}} . \tag{6.177}$$

($\mathrm{P^\circ}$ is the standard value of the partial pressure.) At the gas contact (x=L) $\mathrm{P_{O_2}}$ varies and $\mu_{\mathrm{O_2}}$ takes a value, that is determined by $\mathrm{P_{O_2}}$, and is usually very high. Simultaneously μ_{Zn} takes a very low value which is given by

$$\begin{aligned}\frac{1}{2}\mu_{\mathrm{O_2}}(\mathrm{x{=}L}) &= \mu^\circ_{\mathrm{ZnO}} - \mu_{\mathrm{Zn}}(\mathrm{x{=}L}) \\ &= \frac{1}{2}\mu^\circ_{\mathrm{O_2}} + \frac{\mathrm{RT}}{2}\ln\left(\mathrm{P_{O_2}(x{=}L)/P^\circ}\right).\end{aligned} \tag{6.178}$$

Using this relation we can write κ_d (see Eqs. (6.172, 6.173)) as a function of the partial pressure of oxygen:

$$\kappa_d = \frac{\mathrm{RTV_m}}{8\mathrm{F}^2} \int_{\ln \mathrm{P^*_{O_2}}}^{\ln \mathrm{P_{O_2}(x=L)}} \sigma_{\mathrm{Zn_i^{\cdot\cdot}}} \, \mathrm{d} \ln \mathrm{P_{O_2}} . \tag{6.179}$$

Since we refer to local equilibrium in the case of diffusion processes (linear regime) we can exploit the dependence in the N range derived in Section 5.5.2. For pure ZnO we found that[184,185]

$$\sigma_{\mathrm{Zn_i^{\cdot\cdot}}} = (2\mathrm{Fu_{Zn_i^{\cdot\cdot}}})2^{-2/3}\mathrm{K'^{-1/3}_O P^{-1/6}_{O_2}} . \tag{6.180}$$

Then it follows after integration that

$$\kappa_d \propto \left(\mathrm{u_{Zn_i^{\cdot\cdot}} V_m}\right) \mathrm{K'^{-1/3}_O} \left(\mathrm{P^{*-1/6}_{O_2} - P^{-1/6}_{O_2}}\right) . \tag{6.181}$$

Since $\mathrm{P_{O_2}}$(x=L) corresponds to the external partial pressure of $\mathrm{O_2}$, the specification x=L is suppressed in Eq. (6.181) and also in what follows. In the case of Al–doped ZnO we obtain ([e′]=const)

$$\kappa_d \propto \left(\mathrm{u_{Zn_i^{\cdot\cdot}} V_m}\right) \mathrm{K'_O} \left(\mathrm{P^{*-1/2}_{O_2} - P^{-1/2}_{O_2}}\right) . \tag{6.182}$$

$$\frac{1}{2}\mathrm{O_2 + Zn \rightleftharpoons ZnO}.$$

Although μ_{O} and μ_{Zn} vary distinctly in ZnO this does not apply for the chemical potential of ZnO, i.e. $\mu_{\mathrm{ZnO\ in\ “ZnO”}} \simeq \mathrm{const} = \mu^\circ_{\mathrm{ZnO}}$ (see Chapter 4)). We can make this plausible by remembering the stoichiometric changes in $\mathrm{Zn_{1+\epsilon}O}$ on addition of δZn, δO, δZnO. When δ is of the order of magnitude of ε, the impact is only negligible for the last change. In the first case we obtain $\mathrm{Zn_{(1+\varepsilon)+\delta}O}$, in the second $\mathrm{Zn_{\frac{1+\varepsilon}{1+\delta}}O \simeq Zn_{(1+\varepsilon)-\delta}O}$, but in the third, $\mathrm{Zn_{\frac{1+\varepsilon+\delta}{1+\delta}}O \simeq Zn_{(1+\varepsilon)-\varepsilon\delta}O}$, that is, a change of second order.

[184] From the reaction $\mathrm{Zn_i^{\cdot\cdot} + \frac{1}{2}O_2 + 2e' \rightleftharpoons ZnO + V_i}$ it follows that $\mathrm{[Zn_i^{\cdot\cdot}][e']^2P^{1/2} = K'_O}$ and hence, using $[\mathrm{e'}] = 2[\mathrm{Zn_i^{\cdot\cdot}}]$, Eq. (6.180) (see Chapter 5). At lower temperatures the associate $\mathrm{Zn_i^{\cdot}}$ predominates [173], and the exponent follows in pure ZnO as $-1/4$ instead of $-1/6$.

[185] The standard pressure is included in $\mathrm{K'_O}$.

Since now $P^*_{O_2} \ll P_{O_2}$ and, hence, $P_{O_2}^{-1/2}$ in Eq. (6.182) (or $P_{O_2}^{-1/6}$ in Eq. (6.181)) can be neglected, the scaling constant is, as observed experimentally, almost independent of the partial pressure of oxygen. This is ultimately attributable to the fact that during integration over $\sigma_{Zn_i^{\cdot\cdot}}$ the region, in which the ionic conductivity is large, is dominant, and this is the region near the metal contact (see for example (Eq. (6.180)) and not the region near the contact to the gas phase (with variable external partial pressure). In the case of heavily Li–doped ZnO $\sigma_{Zn_i^{\cdot\cdot}}$ is independent of oxygen partial pressure; it follows that $\kappa_d \propto \ln(P_{O_2}/P^*_{O_2})$ and, hence, there is a weak dependence on P_{O_2}.
On the other hand, in the case of NiO the defect chemistry is dominated by V''_{Ni} and $h^{\cdot}$ and the ambipolar conductivity is determined by $\sigma_{V''_{Ni}} \propto P_{O_2}^{1/6}$, since $\sigma_{h^{\cdot}} \gg \sigma_{V''_{Ni}}$. It follows a $P_{O_2}^{1/6}$ dependence with respect to oxygen partial pressure at the site x=L. Here then $P_{O_2}^{1/6} \gg P_{O_2}^{*1/6}$ and $\kappa_d \propto P_{O_2}^{1/6}$. In this case the scaling rate can be appreciably increased by raising the O_2 content (e.g. pure oxygen instead of air).
In the case of primarily ionically conducting product layers, as formed when silver is halogenated, it is the electronic conductivity which determines the rate; this also exhibits a power law dependence on P_{X_2}. If the electronic conductivity is of the p–type, there is appreciable P_{X_2} dependence of κ_d, while, on the other hand, κ_d is independent of P_{X_2} in the case of the n–type.
Please note that there is no complete freedom to choose the specific regime. One can select P on one side, but on the side of contact to the metal P is fixed ($P = P^*$). If silver is chlorinated with a relatively high chlorine partial pressures, the electronic conductivity changes from n– to p–type within the scale, while the electronic conductivity at the contact is always of n–type (see Fig. 5.44 on page 176). Then the situation is more complicated.

If one charge carrier dominates, in general, we obviously obtain approximately

$$\kappa_d(T, P, C) \propto u_j C^{M_j} \Pi_r K_r^{\gamma_{rj}} \begin{cases} P^{N_j} & \ldots N_j > 0 \\ \text{const} & \ldots N_j < 0 \\ \ln \frac{P}{P^*} & \ldots N_j = 0 \end{cases} , \tag{6.183}$$

j here indexes the most rapid defect in the slower ensemble.
If the outer oxygen partial pressure exceeds the equilibrium value established at a contact of the oxide under consideration and a higher oxide, then successive layers, made up of different oxide phases, should form depending on the kinetics. Figure 6.65 gives an example. At temperatures of 600 °C and above, Fe is in local phase equilibrium with wuestite ("FeO"[186]), this with magnetite (Fe_3O_4) and the latter with haematite (Fe_2O_3) which coexists with air. The scale layer in Fig. 6.65 exhibits all four solid phases (air, 600 °C). The ratio of the individual layer thicknesses is, in the case of diffusion control, determined by the defect parameters, the external

[186] "FeO" is a phase with marked iron deficit. The Dalton composition Fe_1O_1 is unstable under the experimental conditions.

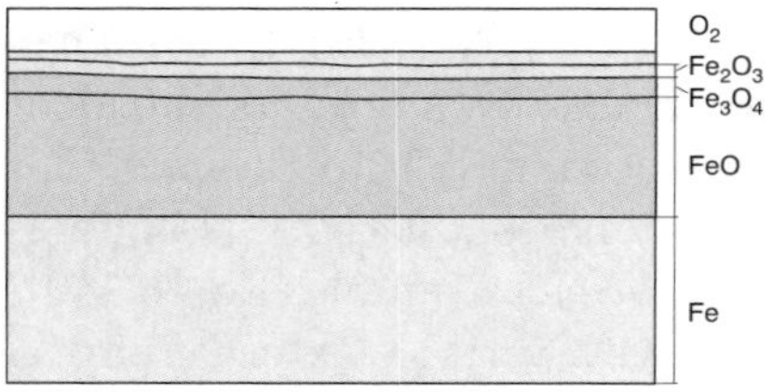

Fig. 6.65: Oxide sequence in the scale layer of a pure iron sample oxydized in air for 16 hours at 600 °C. From Ref. [487].

conditions and the time. In many cases a quasi–stationary state is established in so far as the layer thickness ratios no longer change [488]. Trivially, the thermodynamically stable state under such conditions of oxygen excess corresponds to the existence of Fe_2O_3 alone. Figure 6.66 shows the effective rate constant of the oxy-

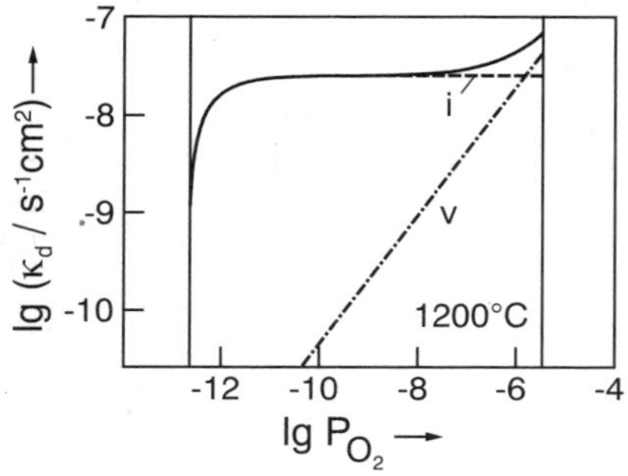

Fig. 6.66: The dependence of κ_d on oxygen partial pressure in the oxydation of "FeO" to Fe_3O_4 (cf. also Fig. 6.18 on page 299). With increasing P_{O_2} the mechanism changes from interstitial to vacancy domination (Frenkel disorder in Fe sublattice). The partial pressure limits left and right are determined by the equilibrium pressures corresponding to the coexistence of Fe_3O_4 with "FeO" and of Fe_3O_4 with Fe_2O_3 respectively. From Ref. [8].

dation of "FeO" to Fe_3O_4. The curve shape describes the change from interstitial mechanism to vacancy mechanism (corresponding to the most conducting carrier of the slower ensemble, i.e. ionic ensemble) with increasing oxygen potential, as was already indicated in the tracer experiment (see Fig. 6.18).

As in the cases mentioned in the previous sections, for solid state reactions, too, the interfacial reaction proper can determine the overall rate.
In addition to adsorption, dissociation or ionization of the oxygen (i.e. charge transfer from zinc to oxygen) it is also possible for phase formation to be rate–determining (nucleation, early stages of layer formation) [489]. Whatever the decisive elementary step is, its rate will after a very short transient, to a good approximation, not be explicitly dependent on the layer thickness. If the surface reaction is stationary[187] then $\dot{L}$ is time independent and we obtain a linear growth law

$$L(t) = S(P_{O_2}, T, C)\, t. \qquad (6.184)$$

Because of the sometimes complex situation at the boundary layer, it is not possible to make simple, general predictions of the dependence of S on P_{O_2}, T, C. Only a

[187]If $\mathcal{R}$ is the rate for the surface reaction then $\ddot{L} = \frac{d\dot{L}}{dt} = \frac{\partial \dot{L}}{\partial L}\frac{dL}{dt} + \frac{\partial \dot{L}}{\partial \mathcal{R}}\frac{d\mathcal{R}}{dt}$. For the conditions discussed this expression vanishes: The first term is zero because $\frac{\partial \dot{L}}{\partial L} = 0$, the second vanishes if $\frac{d\mathcal{R}}{dt} = 0$ (stationary state). In the case of the diffusion–controlled reaction the first term was decisive and the second was always zero.

few more or less qualitative remarks will be made (the reader is also referred to Section 6.4). Upon bringing two reacting phases into contact, such as say O_2 with a nonnoble metal, considerable heats of reaction can be set free, so that, it is generally necessary to consider nonlinear phenomena and thermal transport processes. It was not without reason that we assumed, in the above experiments, that the partial pressure of oxygen was increased just above the existence region of the metal, which was already saturated with oxygen. The reactive adsorption step and the reorganization are followed by the nucleation of the oxide phase, that either takes place homogeneously or occurs heterogeneously with the aid of crystal defects. In the second case the surface energy of the defect is of prime significance. In the first case the formation of a critical nucleus is necessary, whose size and formation probability are considered in Section 5.4. The processes occuring in this time window can also be treated, at least in simple cases, according to the methods of chemical kinetics, namely by treating each cluster as a separate species [490].

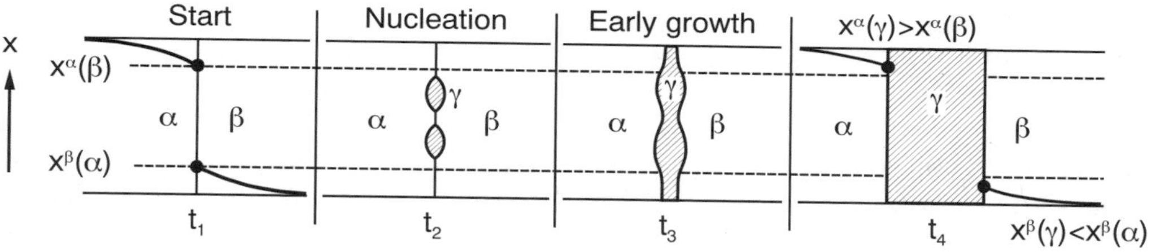

Fig. 6.67: Stages in the reaction of two solid phases α and β to yield phase γ. The initial interdiffusion of the components at contact α/β leads to a supersaturation with respect to phase γ (see right hand diagram). Here concentrations are assumed that correspond at contact to a virtual phase equilibrium between α and β (i.e. $x^\alpha(\beta)$ and $x^\beta(\alpha)$ are the concentrations that would be arrived at if no product formation were to occur ($G_\gamma \to \infty$)). The situation then corresponds to a strong supersaturation with respect to the formation of γ. This is shown in the right hand diagram. The supersaturation (e.g. $x^\alpha(\gamma) - x^\alpha(\beta)$) can be regarded as the driving force. From Ref. [8].

The nucleation stage (see Fig. 6.67) is followed by the stage of early growth in which complications such as lateral mass transport, electronic tunnelling, space charge effects, elastic effects, etc. may play an important role. In contrast to the actual surface reaction, the morphology and layer thickness have a great influence on the rates of these processes. So, $t^{1/3}$ and $\log(t)$ laws are typical for growth dominated by space charge effects and tunnelling (cf. Ref. [491]) and may be even more complicated if adequate attention is being paid to ionic space charges (cf. Section 5.8)[188]. In the course of the further growth diffusion control, which ought to come always into effect when the product layer is sufficiently thick, normally results in the formation of a uniform layer thickness (see below). This is shown in Fig. 6.67 for the reaction of two solid phases α and β to yield phase γ. The stages of nucleation and, in particular, of early growth, involve generally very complex elementary processes, particularly on account of the morphological complexity with which they are associated, the

[188] Cf. also [8,492].

sometimes explicit but often hardly quantifiable thickness dependences and the high–dimensionality of the situation [489,490,493,494].

An example, in which the change in mechanism from surface control to diffusion control can be clearly recognized, is that of spinel formation[189] from the parent oxides (Fig. 6.68) [7]. At low layer thickness, diffusion is sufficiently rapid, but it

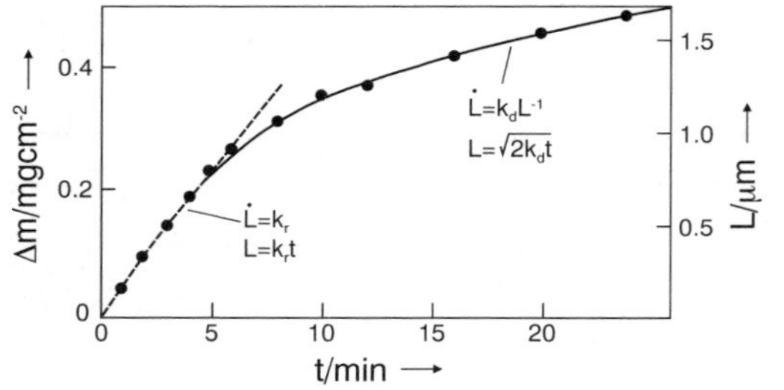

Fig. 6.68: Progress in spinel formation (Zn_2SiO_4 from ZnO and crystobalite at 1350°C in air). Growth of layer thickness or spinel mass as a function of time. Change from reaction to diffusion control. From Ref. [7].

becomes rate–determining at higher values of L and responsible for the change to $\sqrt{t}$ behaviour. In such a ternary, also the pure diffusion process can be complex. In the case that the diffusion occurs via ambipolar conduction of ions and electrons as above, the gas phase is of significance: If A^{2+} and e^- are the migrating partners in the formation of AB_2O_4 from AO and B_2O_3, then there is decomposition to $(A^{2+}/2e^-)$ and $\frac{1}{2}O_2$ at the AO phase boundary. The component $(A^{2+}/2e^-)$ migrates through the product layer and reacts on the other side with B_2O_3 under uptake of oxygen to yield spinel AB_2O_4. Effectively the oxydic compound "AO" has migrated from the AO phase boundary to the B_2O_3 phase boundary: The A component through the product layer, the O component via the gas phase. However, in the case of the formation of multinary compounds it is not absolutely necessary that electrons be involved mechanistically in the chemical diffusion. Thus, as a purely acid–base reaction, the formation of spinel can also occur through the ambipolar diffusion of different ions[190], say in addition to cations (e.g. A^{2+}) O^{2-} ions may migrate through the solid (corresponding to an internal transport of "AO") or also as a result of purely cationic counterdiffusion as illustrated in Fig. 6.69.

In this context we should mention the "electrochemical vapour deposition" (EVD) process which has recently acquired technological interest; it can be used to apply oxydic components of high temperature fuel cells in layer form in an elegant manner[191]. The abbreviation EVD is derived from the fact that elements introduced in the form of halides are converted to the corresponding oxides in a reactive ambipolar diffusion step (such as discussed above).

[189] In the spinel structure AB_2O_4 the oxygen ions form a cubic close–packed structure with the A ions occupying 1/8 of the tetrahedral interstices and the B ions half of the octahedral ones (cf. Section 2.2.7).

[190] Compare also the dissolution of H_2O in solid oxide phases by ambipolar migration of H^+ and O^{2-} (Section 6.3.3).

[191] Compare Ref. [495].

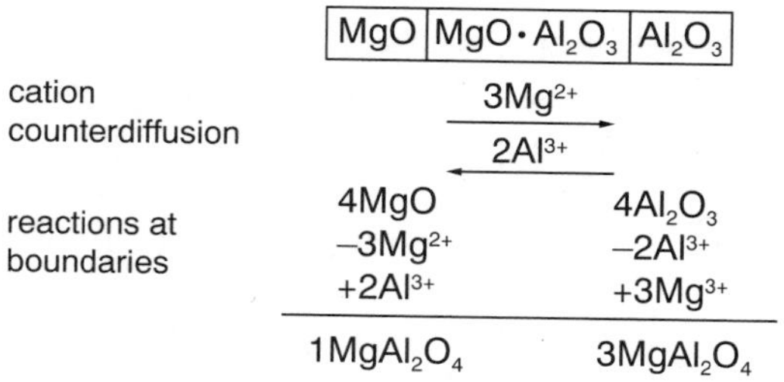

Fig. 6.69: Spinel formation ($MgAl_2O_4$) via ambipolar diffusion of Mg^{2+} and Al^{3+}. From Ref. [7].

The initial morphology of the reactants exerts considerable influence on the reaction kinetics, even in purely diffusion–controlled reactions. The oxydation of metal spheres of radius r_M [496], which are not in contact with each other, is a comparatively simple case. The solution is the Carter equation [497]:

$$\left[(1+(z-1)\,\alpha)^{2/3}+(z-1)\,(1-\alpha)^{2/3}-z\right]r_M^2=2\cdot(1-z)\,\kappa_d t. \tag{6.185}$$

Here α is the relative proportion of the metal that has been reacted to the oxide, and z the ratio of the molar volumes of the oxide and the metal. The validity of Eq. (6.185) has been confirmed for the oxydation of nickel spheres from (diffusion–controlled) initial stages to complete oxydation as shown in Fig. 6.70.

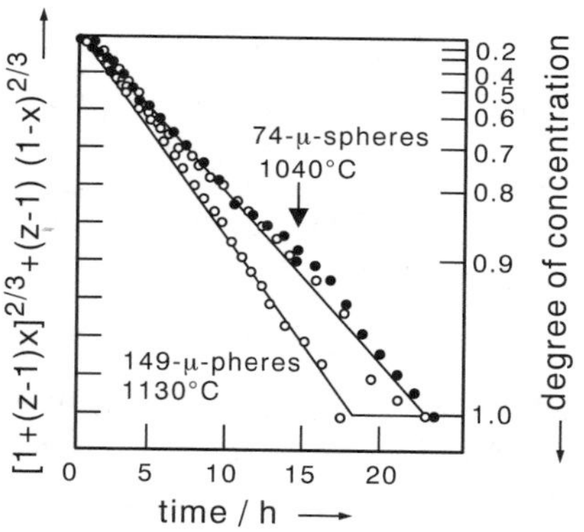

Fig. 6.70: Description of the course of reaction for the oxydation of nickel spheres for differing sphere diameters and temperatures according to the Carter equation. The continuous lines represent the theoretical course. The left scale is linear and ranges from 1.32 to 1.52. From Ref. [497].

A strict analytical treatment of complex real solid reactions is hopeless even in the diffusion–controlled case, above all because of the complex distribution of homo– and hetero–boundaries. In addition, the temperature distribution is not always homogeneous which is particularly true of strongly exothermic reactions. Nevertheless, the simplified treatment above provides a great deal of worthwhile qualitative information concerning the parameters to be taken into account in powder reactions: a) It is important that there be good contact between the reacting phases; the area of contact should be large. Small grain size is important[192]. Intermediate milling and pressing is recommended.

[192] In many cases it is possible to reduce the diffusion lengths drastically via phase formation from a homogeneous state (fluid state, molecular beam synthesis [498] or sol–gel synthesis).

b) The reaction is generally accelerated at elevated temperature. There are naturally thermodynamic limits here (see Chapter 4). Frequently, at high temperature fluid interphases or subsidiary phases occur, which provide preferred transport paths.
c) The influence of doping is important[193]. The direction of the influence depends on the defect chemistry as discussed above. In addition the simultaneous effect of a melting point depression can be advantageous.
d) Generally speaking, an increase in driving force should be striven for, if possible, but may be of much less influence on the kinetics than expected at first glance.

6.9.2 Morphological and mechanistic complications

The situation becomes complex when various intermediates occur during a reaction between powders. Let us consider the formation of $BaTiO_3$ from $BaCO_3$ and TiO_2, which is still a relatively simple synthesis reaction. Owing to the heterogeneity of the initial state and according to the phase equilibria, BaO–rich and TiO_2–rich titanate phases occur. Figure 6.71 indicates the course of compound formation, which is qualitatively explained as follows [499]. $BaTiO_3$, stable on 1:1 contact, is not formed immediately as a durable phase. Rather, any $BaTiO_3$ formed reacts on contact with $BaCO_3$ to yield Ba_2TiO_4. This reaction is evidently much faster than the reaction with TiO_2 to $BaTiO_3$ or even to the TiO_2–richer phases $BaTi_3O_7$ and $BaTi_4O_9$. Ba_2TiO_4 hence appears as an intermediate in considerable amounts, and reacts with TiO_2 or TiO_2 rich phases to give the stable end–product $BaTiO_3$ only if the contact with $BaCO_3$ has been interrupted or if the latter has been consumed. The sequence of the individual steps is naturally a function of the spatial distribution. If, e.g., the BaO rich phases are separated spatially from the TiO_2 rich phases, a multiphase nonequilibrium product is obtained.

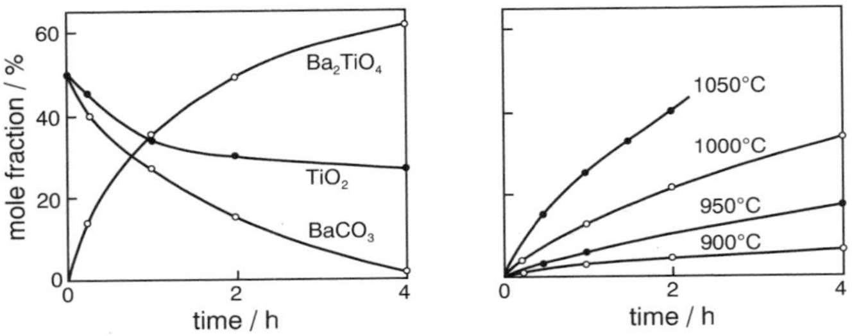

Fig. 6.71: The formation of $BaTiO_3$ according to Ref. [499]. Left: initial reaction course of an equimolar mixture of $BaCO_3$ and TiO_2 in air at 900°C. Right: course of $BaTiO_3$ formation from an equimolar mixture of Ba_2TiO_4 and TiO_2 between 900°C and 1050°C. According to Ref. [4].

A continuous homogenization and milling during reaction is recommended in order to accelerate the equilibration process. Thus, complex morphologies and contact problems are ubiquitous in such cases, for purely practical reasons, because of the

[193] In addition to homogeneous doping (under discussion), heterogeneous doping naturally also should have a considerable effect (see Section 5.8.5).

use of powdered educts[194].

On the other hand, plane interfaces are not always stable even when well wetted. Let us investigate the consequence of a morphological perturbation of the plane M/oxide interface [439,500–502] as shown in Fig. 6.72. Since, during the course of

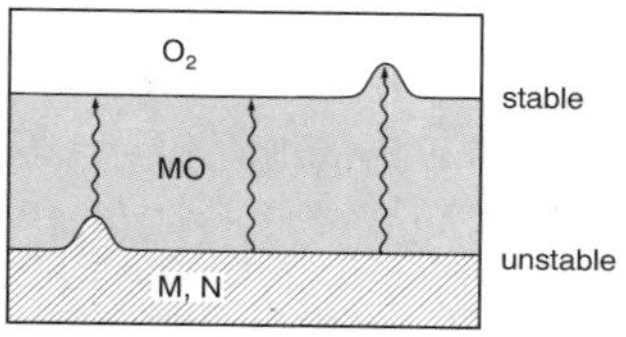

Fig. 6.72: Stability analysis of a nonplanar interface of an alloy with a scale layer ($\sigma_{\text{ion}} = \sigma_{M^{2+}}$). If the transport of the cations through MO is rate–determining, the metal will be more strongly corroded at the site of thinner oxide layers and the unevenness will be levelled off. The perturbation is amplified if the transport of cations through the alloy is rate–determining. The noble alloy component N is not oxydized. N crystals remain dispersed in the oxide matrix [500].

the diffusion–controlled corrosion, further mass transport of either M or O through the oxide at the site of the thicker (thinner) oxide layer takes place more slowly (quickly), the film thickness becomes homogenized as the process continues and a planar morphology is stable[195]. However, the corrosion of an alloy may be different. Here the mass transport in the alloy is also important. If we neglect oxygen transport for simplicity, it is only necessary to consider two cases, namely that the transport of M through the alloy is faster or slower than through the oxide. While the first case leads again to homogenization, the nonplanar morphology is obviously amplified in the latter (Fig. 6.72). Morphological questions are treated in more detail in Ref. [8]. The basic tool in answering such questions lies in the application of local stability criteria of the type discussed. It is not, in general, possible to work out the precise reaction path a priori. Figure 6.73 schematically illustrates experimental interface morphologies for the reaction of $(Fe_yCr_{1-y})_2O_3$ and $(Fe_xMn_{1-x})_3O_4$ as a function of the composition parameters x and y. Important examples of unusual morphologies are whisker growth and, in particular, growth processes in the biosphere (see following section).

The formal counterpart of the formation reaction of a solid is a decomposition reaction. The basic principles are analogous; the relationships, in reality, are no less complex, indeed they are frequently more so as a result of uncontrolled phase generation. In particular, decomposition reactions usually take place as pure surface reactions with complex morphological evolution (e.g. the formation of very porous products). Morphological instabilities play an important role here [8]. When reactivity is high, the reaction products are no longer morphologically related to the starting materials (extreme case: detonation, e.g., of lead azide) while in "gentle"

[194]Plane interfaces are mostly not desired for kinetic reasons, not to mention the fact that powdered educts are usually much cheaper.

[195]In the sense of the treatment in Section 6.10 the lateral "thickening" ΔL plays the role of a Ljapunov function [503]. The interfacial surface is morphologically stable because $\Delta L > 0$ and $d\Delta L/dt < 0$ (more generally $(d\Delta L/dt)\Delta L < 0$).

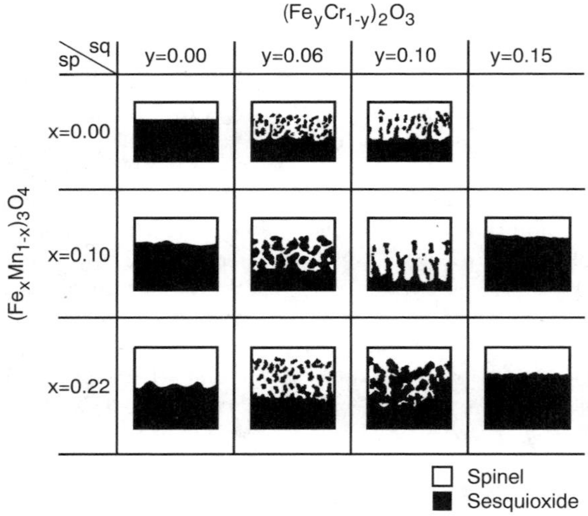

Fig. 6.73: Morphology of the solid state reactions between $(Fe_xMn_{1-x})_3O_4$ spinel, cubic) and $(Fe_yCr_{1-y})_2O_3$ sesquioxide, hexagonal). Compare [501] with respect to the theoretical interpretation. From Ref. [8].

decompositions, such as the mild dehydration of hydrates, pronounced morphological "memory effects" can occur, which are attributable to the low mobility of the structural elements which preserve the original morphology (see topotaxis). A relevant example is the removal of CuF_2 from a $3Cu(OH)_2 \cdot CuF_2$ crystal in an aqueous solution. Hexagon–shaped starting crystals are transformed to $Cu(OH)_2$ needles, which are arranged in six–fold symmetry [504]. (In this context the significance of elastic effects in precipitation reactions should be paid attention to [409].)

Another point concerns the phenomenon of kinetic decomposition of multinary compounds [505,506]. We consider a ternary semiconducting oxide (say a spinel) as an example and expose it to an oxygen gradient that is selected such that both partial pressures P_2, P_1 lie within the stability window of the oxide. Even though the spinel will not be converted into a higher or lower oxide at these oxygen potentials (redox reaction), it can nevertheless decompose, as a result of the differing transport coefficients of the mobile constituents, into the binary oxides (acid–base reaction) with respect to which it would be stable in the absence of the potential gradient. For thermodynamic reasons it is necessary that the free energy of gradient formation at least compensates for the free energy of decomposition[196]. Regardless of

[196] Thus, for spinel formation, we demand that $|\frac{1}{2}RT\ln(P_2/P_1)| > |\alpha\Delta G^\circ_{sp}|$ (whereby ΔG°_{sp} refers to the formation from the oxides); α is then 1 if only A^{2+} is mobile, since, here, for the formation of B_2O_3, the combination $A^{2+}/2e^-$ migrates through the sample to the oxydizing side and reacts to form AO at this surface. Every oxygen, required to maintain the gradient, is equivalent to one AO and, hence, to one AB_2O_4, while if only B^{3+} is mobile, B_2O_3 is "formed" and one O is only equivalent to a third AB_2O_4. In the first case 1O is effectively transferred, while if only B^{3+} is mobile, an effective "transfer" of 3O (B_2O_3) is involved with the spatial distribution of the oxides being reversed; in the last case $|\alpha|$ is thus 1/3. The more precise calculation starting from the usual equations (cf. Eq. (6.171)) shows that, for a primarily electronically conducting compound, $\alpha = \frac{1+\beta/2}{1-3\beta/2}$ with $\beta = \frac{u_{B^{3+}}}{u_{A^{2+}}}$. A detailed treatment is given in Ref. [8]. As we recognize, α changes

mechanistic details, such a situation is not possible in the case of binary oxides, for thermodynamic reasons. Let us, for simplicity, consider an oxide MO that co–exists with the metal M. Then the only decomposition reaction is that into the elements. On account of the laws of thermodynamics the free energy of gradient formation $\frac{1}{2}\mathrm{RT}\ln(\mathrm{P}_2/\mathrm{P}_1)$ must exceed the enthalpy of decomposition ($\mathrm{MO} \longrightarrow \mathrm{M} + 1/2\mathrm{O}_2$). This means trivially that at least one of the partial pressures must lie outside the stability window (cf. Section 4.3.4). This decomposition in potential gradients is of importance for the long–term behaviour of many multinary solids during practical application (cf. Chapter 7).

The kinetics of those solid state "reactions", which involve only a morphological change but (approximately) no "compositional" change, possess a particular importance in ceramic technology. While the driving force for normal solid state reactions is determined by the chemical potentials of the components, in the present case it is a reduction of the interfacial free energy (see Section 5.4). Important examples already discussed are grain growth (in a certain sense this may be regarded as "dimerization of the two ^{3}D giant molecules"), sloppily formulated as

$$\begin{array}{c}\text{small grain} + \text{small grain} \longrightarrow \text{big grain}\\ \text{or approximately}\\ \text{surface} \longrightarrow \text{bulk,}\end{array} \qquad (6.186)$$

and the sintering process (in a certain sense the condensation of two ^{3}D giant molecules, i.e. thermal or chemical "glueing" of powders to yield a mechanically stable solid containing comparatively stable grain boundaries)

$$\begin{array}{c}\text{grain} + \text{grain} \longrightarrow \text{double grain}\\ \text{or approximately}\\ \text{surface} \longrightarrow \text{grain boundary.}\end{array} \qquad (6.187)$$

Ideal sintering necessitates spatial displacement of all components and requires high temperatures. (In order to achieve dense samples (see e.g. Fig. 5.29) significant pressure is usually applied.) So diffusion–controlled creep of an oxygen ion–conducting oxide MO is determined by the cation diffusion and is an ambipolar process as described above (see also Section 6.3.3).
Some thermodynamic aspects of ceramic processing have been discussed in Section 5.4. The reader is directed to the extensive ceramic or metallurgy literature for further information [133,161,507].

---

sign at $\beta = 2/3$. As expected, there is a pole here and not a passage through zero, it is such that for $\beta < 2/3$ the value of α is always greater than 1 (as realized for $\beta = 0$); for $\beta > 2/3$ the value of $|\alpha| = -\alpha$ is above 1/3 (as realized for $\beta \to \infty$). A passage through zero would be contrary to the laws of thermodynamics.

6.10 Nonlinear phenomena

6.10.1 Irreversible thermodynamics and chemical kinetics far from equilibrium, and the special role of autocatalysis

Until now we have largely concentrated on linear relationships between fluxes and forces, which are valid quite generally for diffusion but less so in the case of electrical conduction. We have noticed, for chemical reactions, such as those occurring at interfaces or in the case of electrochemical transfer processes, that the rate equations usually cannot be mapped by linear irreversible thermodynamics, and we have made considerable use of nonlinear formulations. In this chapter we shall specifically consider those phenomena that do not occur even qualitatively close to equilibrium. It is a necessary criterion of the state of equilibrium that entropy production[197] is zero (see Chapter 4). It is always positive at nonequilibrium, and at small displacements from equilibrium there is a restoring force to re–establish that state (see Section 6.1.1). Now it is possible to force maintenance of nonequilibrium states by means of external constraints — think of the chemical diffusion problems discussed above. If we apply a chemical potential gradient across the sample and keep it constant, e.g. by an outer gas stream, then for $t \gg L^2/D^\delta$ the concentration changes with time disappear (that is the process becomes stationary) corresponding to a linear concentration profile (see footnote 198). Another example is the blocking of the ion flux in a mixed conductor (such as Ag_2S) by electrodes (such as Pt) that only allow electrons to pass. In the long run ($t \gg L^2/D^\delta$) of a potentiostatic experiment, the total current has fallen to the value of the purely electronic component (and again approximately linear profiles are established). In Chapter 7 we will exploit this technique as an elegant method of separating partial conductivities. Note that in the latter case ion and electron currents are interdependent due to the condition of electroneutrality, and the driving forces due to the $\nabla\phi$ components in the $\nabla\widetilde{\mu}$ terms (cf. Section 6.3.3).

The so–called coupled kinetic phenomena deserve special consideration in that a certain flux can be caused by different driving forces (that cannot be combined as above). Matter transport or charge transport can be caused not only by chemical or electrical gradients but also by temperature gradients. The inverse is true for the heat flux. This reciprocity is reflected in cross–terms that appear in the flux–force equations and obey the Onsager relations [330,508] (see footnote 6 on page 271). Let us consider a more specific example, namely the thermoelectric effect. It occurs in the chain A|B|A consisting of two different electrical conductors, A

[197] As already described in Section 6.3 (see footnote 2, page 270), we use Π as an abbreviation for $T\delta S/\delta t$, and it is hence referred to somewhat imprecisely as entropy production. The factor T is uninteresting to us with respect to the definition of the fluxes. In addition, in continuous systems Π is obtained as an integral of the product of forces and fluxes. The fluxes usually refer to volume–based changes so that $\Sigma_k J_k X_k$ represents the local entropy density more precisely.

and B, for which the two interfaces are held at different temperatures. In this case the temperature gradient causes not only a heat flux but also a charge flux leading to an electrical potential difference between the two ends of the chain. If the temperature gradient is kept constant, a steady state is eventually obtained in which the electrical current has approached zero, yet, at the expense of a nonzero constant electrical potential difference (Seebeck–effect). This thermovoltage can be used for the precise measurement of temperature. According to Onsager's reciprocity theorem the inverse phenomenon can also be quantified, namely the generation of a temperature difference by an electrical current (Peltier effect). A related process, involving the coupling of matter and heat effects, is thermodiffusion.
As can be demonstrated quite generally [94,326,508,509], in the linear range such a stationary state, set up by constant forces, is a state of minimum entropy production[197,198,199] (Fig. 6.74; cf. also Section 4.2). This state ($\Pi = \Pi_{min}$) is automatically stable: Deviations of $\delta\Pi$ are necessarily positive (see below Eq. (6.192) and footnote 205) and, as can be proven [94,509], the development with time is now

[198] Let us consider the following example to explain the minimal principle (see Fig. 7.11, page 418). The oxide sample of thickness L is equilibrated at an oxygen partial pressure P_1 (oxygen concentration c_1). The oxygen partial pressure is increased on one side to P_2, so that assuming a rapid surface reaction there the concentration is c_2. When t=0 the profile is described by a step function, when t=∞ by a straight line with the above boundary values c_1 and c_2. All intermediate profiles also possess these boundary values. However, they are curved. When the boundary concentrations (and, hence, the external gradient) are fixed, the integral $\int D(\nabla c)^2 dx$ is larger in the case of the curved concentration profiles than for the linear profile. (Note that this integral represents $\int |J\nabla c| dx$ and is, according to footnote 197, the quantity to be minimized.) This can be understood in the following way: If, to a first approximation, we replace the curved profile by two straight segments that meet in the centre, then the integral in the linear case is less by approximately $4\delta^2/L$. (δ is the concentration difference in the centre between the linear and the sectionally linear profiles. The gradient in the stationary state is $\Delta c/L$, Δc = concentration difference over the sample, L = sample thickness; the gradients of the two segments are $\frac{\Delta c+2\delta}{L}$ and $\frac{\Delta c-2\delta}{L}$. In the completely linear case the integral yields $(\Delta c)^2/(4L)$ and in the composite case $(\Delta c+2\delta)^2/(2L)+(\Delta c-2\delta)^2/(2L)$. The difference is $4\delta^2/L$.) This can obviously be continued in higher approximations without changing the result qualitatively.

The following procedure is more precise [444]: It is plausible that the problem of finding a function c(x) that minimizes the integral $\int c'^2 dx$ is equivalent to the problem of minimizing the integral $\int(1+c'^2)dx$ or the integral $\int\sqrt{1+c'^2}dx$. The first point is trivial, the latter can be verified using Euler's equation of variational calculus. (Let $I(c')$ be the integrand, then Euler's equation demands that $\frac{d}{dx}\frac{\partial}{\partial c'}I(c')=0$. It follows both for $I=c'^2$ or c'^2+1 or for $I=\sqrt{c'^2+1}$ that $c' = \text{const} = (c_2-c_1)/L$. See, for example, Ref. [341] for details of variational calculus.) The last integral, however, represents just the length of the c(x) curve ($\sqrt{1+dc/dx}\,dx = \sqrt{dx^2+dc^2} = |ds|$, where ds represents the line element of the curve). The shortest curve, which couples the two endpoints given, is a straight line.

[199] Even though there is symmetry between individual forces and fluxes in the linear range [331], we prefer to set up the steady state by virtue of constant forces in order to avoid misunderstandings, as the following example shows: In the case of the potentiostatic ion blockage in Ag_2S by Pt electrodes the dissipation ($\hat{=} U\cdot I \propto U^2/R$) is at minimum in the steady state, since I has fallen to I_{eon}, i.e. the resistance has increased to R_{eon} (see Section 7.3). Note, however, that in the galvanostatic mode the dissipation has reached a maximum value ($R_{eon}I^2 > RI^2$). The reason for this apparent contradiction lies in the constraints imposed to the boundary of the system (see Ref. [510].

ENTROPY PRODUCTION CLOSE TO AND FAR FROM EQUILIBRIUM

EQUILIBRIUM

$\Pi \hat{=}$ entropy - production

$\Pi = T d_i S/dt \geq 0$

$\dot{\Pi} \leq 0$

$\delta \Pi^{eq} \geq 0$

(prehistory unimportant, stable)

STEADY STATE CLOSE TO EQUILIBRIUM

$\Pi \geq \Pi^{ss} > 0$

$\dot{\Pi} \leq 0$

$\delta \Pi^{ss} \geq 0$

(prehistory unimportant, stable steady states)

STEADY STATE FAR FROM EQUILIBRIUM

$\Pi \geq 0$

$_x \dot{\Pi} \leq 0$

$\delta_x \Pi^{ss} \gtrless 0$

(prehistory important, unstable steady states possible)

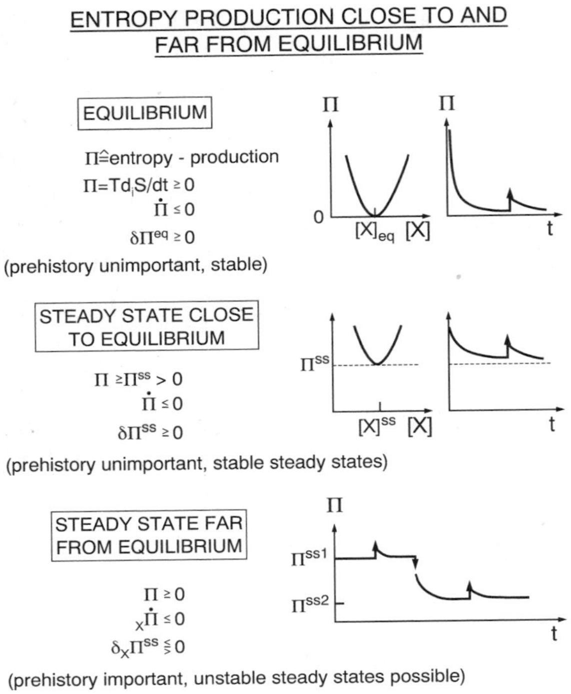

Fig. 6.74: The stability of stationary states. Entropy production close to equilibrium and distant from equilibrium[197]. In the equilibrium case the statement $\delta \Pi \geq 0$ is trivial, since Π cannot fall below zero. The fact that $\delta \Pi$ is positive in proximity to equilibrium is proven by footnote 205. Far from equilibrium, $\delta \Pi$ can become negative, as described in the text.

such (viz. $\dot{\Pi} < 0$) that Π_{min} is re–established. As it is sometimes expressed: In the range of linear effects Π plays the role of a Ljapunov function[200]. Because of this stability overshooting or oscillating behaviour is not possible for linear systems.

All of this no longer applies in the interesting case of large deviations from equilibrium; here only that part of the entropy production which is due to the variation of the forces[201] will approach a minimum value[202] ($d_X\Pi$, not $d_J\Pi = d\Pi - d_X\Pi$). On the other hand, perturbations of this stationary state can be associated with an increase or also with a decrease in this quantity, as we will see (see also Fig. 6.75). In the first case, the displacement will be corrected; in the second case, it will be amplified. Then "dangerous" fluctuations can occur, which destabilize a stationary state, as is shown in Fig. 6.75. New stationary states can be reached, rendering possible a qualitative progress; an oscillatory behaviour becomes possible as well. In other words almost all the interesting variety in our world, particularly in the

---

[200]The existence of a Ljapunov function automatically brings about stability (cf. also footnote 195, page 379). Such a function is characterized by the fact that the excess in its values in the neighbourhood of the stationary state under consideration (singular point) is of the opposite sign to its time derivative [503].

[201]There is no difference in the linear range because $d\Pi = d_X\Pi + d_J\Pi = 2d_X\Pi = 2d_J\Pi$.

[202]The range of applicability of this principle is a matter of controversy (cf. e.g. Ref. [511]).

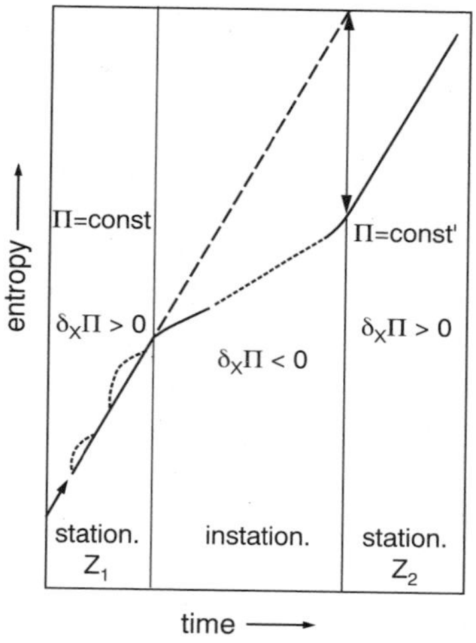

Fig. 6.75: A negative fluctuation of entropy production destroys the stationary state (Z_1). An unstable transition phase leads to a new stationary state (Z_2). The relevant entropy (internal entropy plus entropy produced) in the diagram is approximately $S_{1,2}(t) = S^0_{1,2} + \text{const}\Pi_{1,2}t$, whereby Π is taken as a constant. Although the entropy increases, $S_2 > S_1$, the stationary entropy associated with the respective state has decreased, i.e. S°_2 is less than S°_1 corresponding to a greater degree of order. The quantity $\text{const}\Pi_{1,2}t$ is exported (cf. metabolism). The increase in order in the evolution history of many systems (e.g. living things) is thus not in contradiction to thermodynamics, but demands open systems far from equilibrium. What is also important, and is indeed observed in biological systems, is the decrease of the dissipation (more accurately the force–dependent dissipation) until the stationary state is reached. According Ref. [339].

biosphere[203], requires large deviations from equilibrium [94,339,340].
In the case of an elementary process $A \rightleftharpoons B$, it follows, on account of the treatment in Section 6.3, that

$$\frac{\vec{\mathcal{R}}}{\overleftarrow{\mathcal{R}}} = \frac{K}{Q} = \exp -\frac{\Delta_R G}{RT}, \qquad (6.188)$$

i.e. whenever we can neglect the forward or back reaction, we are necessarily in the nonlinear flux–force range ($|\Delta_R G| \gg RT$).
Let us now investigate the sign of the variation in Π brought about by changing the force. As the entropy production[197] is given by the product of the forces and the fluxes according to

$$\Pi = \Sigma_k J_k X_k, \qquad (6.189)$$

it follows for the partial variation of Π [94] that

$$\begin{aligned} \delta_X \Pi &= \Sigma_k J_k \delta X_k \\ &= \Sigma_k J_k^{(ss)} \delta X_k + \Sigma_k \delta J_k \delta X_k. \end{aligned} \qquad (6.190)$$

In Eq. (6.190) J_k was split up into the stationary value $J_k^{(ss)}$ and the deviation from this. As we refer to the steady state, $\delta_X \Pi$ must disappear in a linear approximation, i.e. $\Sigma_k J_k^{(ss)} \delta X_k$ must vanish and the second order term

$$\delta_X \Pi = \Sigma_k \delta J_k \delta X_k \qquad (6.191)$$

[203] One could introduce a "biological time" which is scaled according to entropy production: Two living organisms are the same "age" if they have produced the same amount of entropy (cf. Eq. (6.189), the integral over $(\Sigma_k J_k X_k)\delta t$) [512]. The biological clocks of beings with greater dissipation would tick faster and the life span would be shorter. This, however, is a very global and somewhat naive way of looking at the complex reality.

remains as a measure of the variation. As already mentioned, in the linear range $\delta_X\Pi$, and because $\delta_X\Pi = \delta_J\Pi = \frac{1}{2}\delta\Pi$, $\delta\Pi$ itself is automatically positive; if cross terms are neglected, this is immediately clear because of

$$\delta_X\Pi = \Sigma_k\delta\left(\beta_k X_k\right)\delta X_k = \Sigma_k\beta_k\left(\delta X_k\right)^2 \tag{6.192}$$

and is also readily verified when cross terms are included[204,205]. In the nonlinear range, $\delta_x\Pi$ can be positive or negative. Since, as already mentioned, chemical reactions very rapidly leave the linear range and are ultimately responsible for the variety of our world, we will inspect $\delta_X\Pi$ more closely. The quantity δJ_k corresponds to the variation in reaction rate. A mechanism, in which a system is very quickly led away from the initial state, is autocatalysis[198]:

$$A + X \longrightarrow 2X. \tag{6.193}$$

This corresponds to a growth rate of X with "consumption" of the fodder A. We can apply it to bacterial multiplication, to the growth of towns[206], but also to impact ionization in solids, i.e., the creation of an electron–hole pair by energy loss of an electron that is already present. (The importance of this phenomenon for point defects is discussed further below. Since examples from this field are still rare, the patience of the reader will possibly be tested temporarily by the biological vocabulary.) It is immediately evident that this reaction with the rate[207]

$$\mathcal{R} \propto d\left[X\right]/dt \propto \left[A\right]\left[X\right] \tag{6.194}$$

is self–accelerating, i.e.

$$\delta\left[X\right]^{\cdot}/\delta\left[X\right] > 0. \tag{6.195}$$

[204] The Onsager relationships $\beta_{ik} = \beta_{ki}$ apply [330].

[205] The statement, that for linear systems $\sum_{i,k}\beta_{ik}X_iX_k$ is greater than or equal to zero, follows from the Second Law; it merely makes a statement concerning β_{ik} and applies for all X_i and X_k. Accordingly, the form $\sum_{i,k}\beta_{ik}\delta X_i\delta X_k$ must be positive definite, too, for the same β_{ik}. (This no longer applies far from equilibrium since β_{ik} formally depends on the forces.) The statement under discussion with respect to β_{ik} is that all β_{ii} and the β_{ik}–determinant must be greater than or equal to zero (see e.g. [341]). In the case of two forces this means that $\beta_{11} \geq 0$, $\beta_{22} \geq 0$ and $\beta_{12}^2 = \beta_{21}^2 \leq \beta_{11}\beta_{22}$. The fact that the pure transport coefficients (β_{ii}) are positive is, thus, a consequence of the 2nd law. However, the mixed coefficients do not have to be positive.

The inequalities can be readily proved as follows for two processes: The quadratic form $\beta_{11}X_1^2 + (\beta_{12}+\beta_{21})X_1X_2 + \beta_{22}X_2^2$ obtained from $J_1 = J_{11}X_1 + \beta_{12}X_2$ and $J_2 = \beta_{21}X_1 + \beta_{22}X_2$ for $\Pi = J_1X_1 + J_2X_2$ can be rearranged to $\beta_{11}[X_1^2 + ((\beta_{12}+\beta_{21})/\beta_{11})X_1X_2 + (\beta_{22}/\beta_{11})X_2^2]$ or, after quadratic extension, to $\beta_{11}[(X_1 + ((\beta_{12}+\beta_{21})/(2\beta_{11}))X_2)^2 + X_2^2(\{4\beta_{11}\beta_{22} - (\beta_{12}+\beta_{21})^2\}/(4\beta_{11}^2))]$. If β_{11} and $\{\ldots\}$ are greater than or equal to zero (this also implies that $\beta_{22} \geq 0$), then this applies to the whole form.

[206] Let us take two neighbouring and identical, initially empty bars (A and B) as a further relevant example. The random decision of the first guest is in favour of bar A. The next guest also decides for bar A, because it already has a customer and he prefers this to empty bar B (autocatalysis: the fact that a customer is already present attracts the next customer). Only when there are very many customers does this become a disadvantage, further customers avoid the crowded bar A (back reaction!) so that an even distribution is to be expected in the end (see also footnote 211).

[207] Compare Ref. [339]

In other words: There is positive feedback. The back reaction becomes important close to equilibrium:

$$\mathrm{A + X \rightleftharpoons 2X}. \tag{6.196}$$

Because of the quadratic dependence

$$\overleftarrow{\mathcal{R}} \propto [\mathrm{X}]^2 \tag{6.197}$$

negative feedback predominates there. Since the affinity (see Section 4.2) is given by

$$\mathcal{A} = \mathrm{RT}\left(\ln \mathrm{K} - \ln\left([\mathrm{X}] / [\mathrm{A}]\right)\right) \tag{6.198}$$

it follows, at constant [A] that

$$\delta\mathcal{A} \propto -\delta\,[\mathrm{X}] / [\mathrm{X}]\,. \tag{6.199}$$

In the case that only the forward reaction counts, the variation in $\mathcal{R}$

$$\delta\mathcal{R} \propto \delta\,[\mathrm{X}] \tag{6.200}$$

is in the opposite direction to that in $\mathcal{A}$ and

$$\delta_{\mathrm{X}}\Pi \propto \delta\mathcal{A}\delta\mathcal{R} < 0. \tag{6.201}$$

We can conclude from Eq. (6.194) that the state of vanishing X concentration is a stationary state ($\mathcal{R} = 0$; the formation of any X implies the pre–existence of X). Yet, the slightest variation in [X] immediately leads to a spontaneous multiplication. In this sense the state is unstable as displayed by Eq. (6.201). When the back reaction is not taken into account (the back reaction becomes ever more important at large values of [X]) it is not possible to reach a stable state. Only taking the back reaction into account leads to a further stationary state (here the equilibrium state), but this is now stable:

$$\begin{aligned}\delta\mathcal{R} &= \delta\left(\left\{\vec{\mathrm{k}}\,[\mathrm{A}] - \overleftarrow{\mathrm{k}}\,[\mathrm{X}]\right\}[\mathrm{X}]\right) \\ &= [\mathrm{X}]\,\delta\,\{\ldots\} + \{\ldots\}\,\delta\,[\mathrm{X}]\,.\end{aligned} \tag{6.202}$$

Close to equilibrium $\{\ldots\} \equiv \left\{\vec{\mathrm{k}}[\mathrm{A}] - \overleftarrow{\mathrm{k}}[\mathrm{X}]\right\} \simeq 0$, and so

$$[\mathrm{X}]\,\delta\,\{\ldots\} = -\overleftarrow{\mathrm{k}}\,[\mathrm{X}]\,\delta\,[\mathrm{X}] \tag{6.203}$$

predominates, i.e. indeed

$$\delta_{\mathrm{X}}\Pi > 0. \tag{6.204}$$

The "counteraction" of the changes of rate and driving force characteristic of autocatalysis (cf. Eqs. (6.194) and (6.198)) can be formally expressed in terms of a negative transport coefficient (cf. Eq. (6.192) and note that this is not possible in the linear regime[208]). A destabilizing relationship between flux and driving force occurs in solids when the current density in the system decreases with increasing voltage (negative differential conductivity), such as effected by the impact ionization process described in more detail below [513].

---

[208]The second law ($\Pi > 0$) requires that β_{ii} and, hence, the conductivities, be positive (footnote 205).

6.10.2 Nonequilibrium structures in time and space

In the previous section we discussed a typical growth reaction ($A+X \rightarrow 2X$). However, in realistic systems X will not only grow, it can also die out, e.g. according to the following decay mechanisms:

$$X \xrightarrow{k'} Z \tag{6.205}$$

(see Table 6.6).
Far from equilibrium overall

$$\mathcal{R} = \frac{d\,[X]}{dt} = (k\,[A] - k')\,[X] \equiv W\,[X]\,. \tag{6.206}$$

Whether X increases or decays, that is, whether W is positive or negative, clearly depends on the ratio $k\,[A]\,/k'$.

Table 6.6 shows the qualitative turnover with increasing [A] corresponding to a nonequilibrium phase transition at the critical selection value[209] k'/k. All species with positive W increase at the expense of A, all others disappear. If the "fodder concentration" [A] is not maintained constant by addition but introduced at the start as a given quantity, competition occurs [339,509]. Here, too, the species with positive W survive, but the value of W is now time–dependent as a result of the exhaustibility of A. The "high jump bar" is, as Eigen put it [339], continually raised; in the end only one species survives. In real systems this configuration is never completely invariant, rather defects continuously occur such as mutations in biological systems as a result of the effects of radiation. These new mutations either die off or drive out the mother species[210]. We can now see how readily "biological" phenomena, such as multiplication, selection and mutation, can occur in simple reaction schemes. Schemes that are not much more complicated (see Table 6.6) permit oscillation over time. Spatial oscillations or generally spatial structure formations occur when the position coordinates come into play, owing to the coupling with diffusion processes [340]. Here instabilities can lead to a breaking of spatial symmetry. Finally, if there are further deviations on account of sequential passage through points of bifurcation[211], a fractal pattern may form characterizing deterministic chaos [516].

[209] In Table 6.6 termed k_2/k_1.

[210] This pronounced sensitivity (see also [514]) is a result of the very simple kinetics in our examples (cf. the model of the "hypercycle" [515]).

[211] The self–stabilizing and initially arbitrary symmetry break, which effects that a random decision between two equivalent alternative determines almost inevitably the next decisions (one–sided reduction of the energy of an initially symmetrical double minimum potential leads to a bifurcation), is a fundamental structure generating mechanism that is characteristic for the irreversibility of history. For the same reason the course of history is singular and nonrepeatable. See also the example in footnote 206. Another example is probably the preferred chirality of protein building blocks in the living world (the necessary amplification of an initially "random" occurrence).

Table 6.6: Kinetics far from equilibrium [334].

autocatalytic reaction	$A + X \longrightarrow 2X$	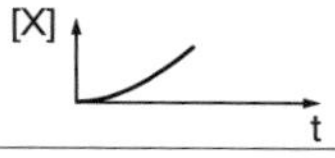	growth, positive feedback
autocatalytic + decay reaction	$A + X \overset{k_1}{\rightleftharpoons} 2X$ $X \overset{k_2}{\longrightarrow} Z$	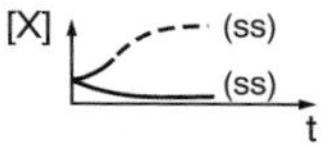	growth or death
no selection pressure	$[A] = \text{const}$ $W_c = k_1[A] - k_2$	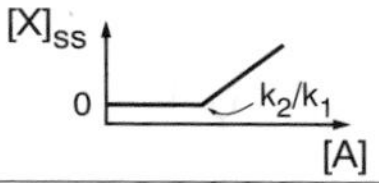	nonequilibrium phase transformation
autocatalytic + decay + selection pressure (+ perturbation)	$A + X \longrightarrow 2X$ $X \longrightarrow Z$ $[A] \neq \text{const}$ $W_c = f(t)$	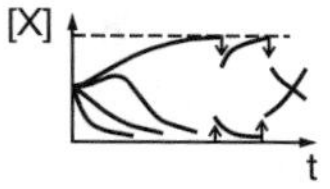	competition selection mutation
Lotka-Volterra reaction scheme	$A + X \longrightarrow 2X$ $X + Y \longrightarrow 2Y$ $Y \longrightarrow Z$	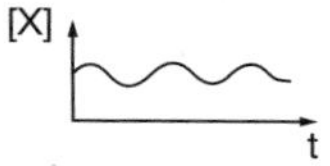	structurally unstable oscillation
Brusselator reaction scheme	$A \longrightarrow X$ $2X + Y \longrightarrow 3X$ $B + X \longrightarrow Y + D$ $X \longrightarrow Y$		limit cycle oscillation
or catalytic CO oxidation	$(CO + O \rightleftharpoons CO_2;$ $Pt_{hex} \rightleftharpoons Pt_{sq})$	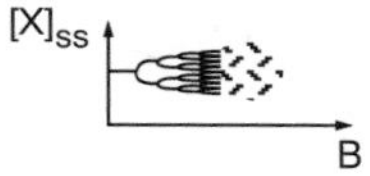	symmetry breaking bifurcation deterministic chaos
above reaction scheme + diffusion	(e. g. $X \rightleftharpoons X$)	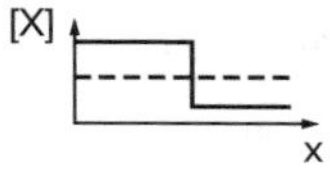	compartmentation, dissipative structure in space

Even though the discussion of biological effects and specifically the immense importance of respective nonequilibrium defects (mutations) for our global history[212] do form part of the subject matter of defect chemistry, it would take us much too far from our chosen theme. Let us now discuss some important implications for the inorganic field.

Table 6.6 lists some reaction schemes, that can lead to temporal or spatial oscillations; all include an autocatalytic elementary step. The catalytic oxydation of CO on Pt surfaces is a very well investigated process, where periodic patterns, that are characterized by surface reconstruction[213], can occur over time [147] (see Fig. 6.76).

[212] Think of the occasional, catastrophic results of the smallest of mutations when inherited, even with respect sometimes to events of historic importance.

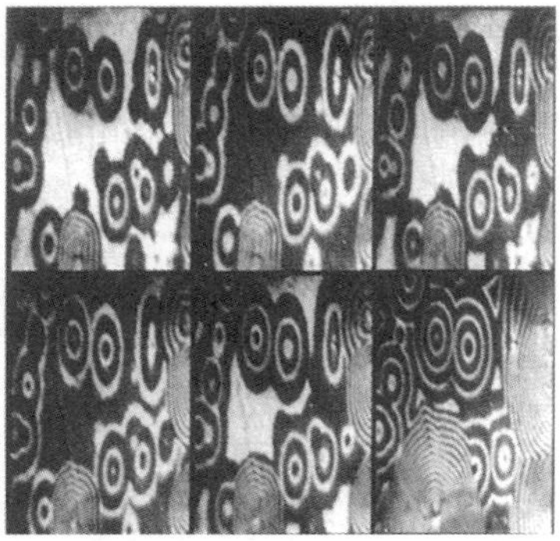

Fig. 6.76: Series of photoemission electron micrographs of a Pt(110) surface during catalytic oxydation of CO (each figure: 0.2mm×0.3mm). From Ref. [517].

Impact ionization is an "autocatalytic" process occurring in solids, that has already been mentioned, in which an excited electron in the conduction band drops down to a lower energy level generating another electron–hole pair. This can, with a pinch of salt, be described as

$$e' \rightleftharpoons 2e' + h^{\cdot}. \tag{6.207}$$

The deviation from equilibrium can frequently be brought about by means of an external voltage. Figure 6.77 shows a nonequilibrium insulator–conductor phase

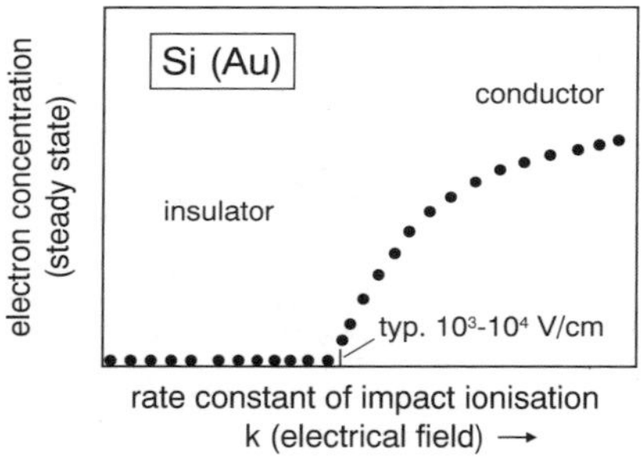

Fig. 6.77: Nonequilibrium phase transition (insulator–conductor transition) in Au–doped silicon as a result of voltage variation in the nonlinear range [513,518, 334].

transition in Au–doped Si, effected by a (voltage–dependent) rate constant elevated in this way over the critical value. This is in complete conformity with the third graph in Table 6.6. Whereas impact ionization acts as a multiplication reaction, the electron–hole recombination acts as a decay mechanism (see reaction Eq. (6.205)). Figure 6.78 shows us the oscillations over time of the current in doped germanium on irradiation with far IR light at high fields. As the voltage is increased further, successive doubling of the period ensues and finally deterministic chaos sets in. This can be seen in the phase portrait (picture a: I against $\dot{I}$), from the time evolution of the current as well as from the voltage behaviour of the current minima (right hand image). Finally Fig. 6.79 shows EBIC micrographs (scanning electron microscopy with electron beam–induced currents) of a nonequilibrium spatial structure in p–Ge. The current filaments (elevated zones) represent zones of altered electron concentration which appear as true dissipative structures under nonequilibrium conditions

[213]Under certain conditions spatially periodic patterns or chaotic states are generated [147].

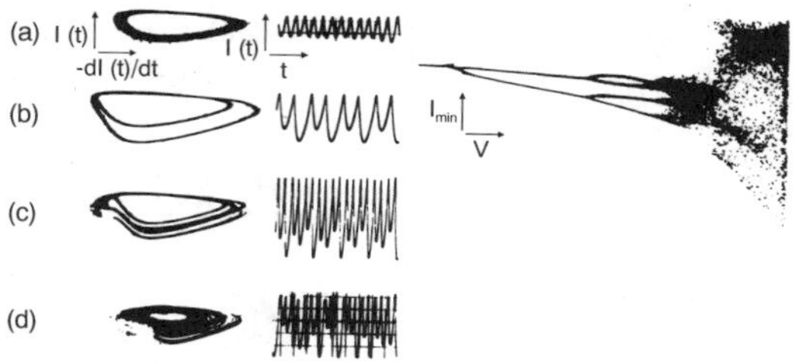

Fig. 6.78: Periodic structures and chaotic behaviour in Ge as a result of voltage variation in the nonlinear range (I: current, V: potential). From Ref. [519].

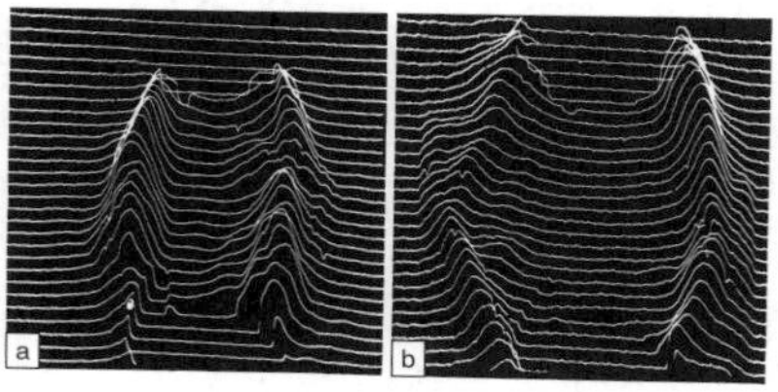

Fig. 6.79: Current filaments in p–Ge at various voltages, made visible by the EBIC technique (electron beam–induced currents). From Ref. [520].

and disappear again when the current is switched off. Compare with Table 6.6 in all cases.

The Liesegang phenomenon [521], which is a striking nonequilibrium effect (not only) in laboratory chemistry, describes the formation of periodic precipitation bands of internal reaction products; the effect can involve periodicity in time and space and depends on the interference of diffusion, supersaturation, nucleation and growth. The collapse and reformation of supersaturation is of importance, in the sense of an overshoot of a partial process.

Other examples of oscillations with time are voltage oscillations[214] depending on mechanical instabilities at the electrode contact on applying a constant current to an Ag/AgI interface (Fig. 6.80) [439,523]. A more popular example is the pulsating

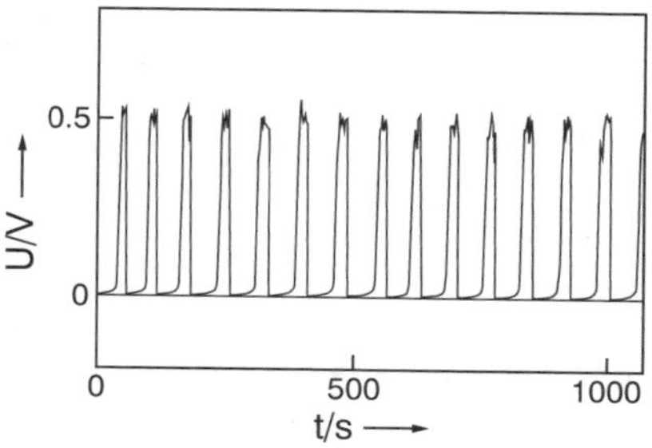

Fig. 6.80: Potential oscillations at the Ag/AgI contact under anodic load and galvanostatic conditions [523]. From Ref. [8].

mercury drop ("beating mercury heart") [524]: An iron nail is at a certain distance from a drop of mercury, and the Hg drop is covered by a $K_2Cr_2O_7$ solution. At the surface mercury is oxydized; the resulting change in surface tension lets the drop

[214]Similar oscillations have been known for a long period in the electrochemical corrosion processes. More recently several reports become available involving proton conducting oxides [522].

reach the iron nail. The oxydation is reversed there, the drop leaves the nail and the whole game starts over again. The periodicity is not created, for example, by a periodic motion of the nail backwards and forwards; it intrinsically emerges with a frequency determined, amongst other things, by the distance of the mercury from the nail. Without supply and removal of educt and product[215] the oscillation will decay and eventually die out.

It is important to remember the reasons for the possibility of such oscillating behaviour. It depends ultimately on the fact that various steps, that are coupled to each other can nevertheless, phase–wise, act more or less independently of each other. Otherwise the whole system would tend to approach equilibrium in a concerted and monotonic manner. In the case of homogeneous reactions, which we considered earlier, autocatalytic processes favoured a particular step to such an extent that competing reactions could be "ignored" for a certain time: In the heterogeneous case such inhibition and triggering processes can also be controlled by spatial distances and the difference of diffusion coefficients (see Hg droplet).

Differing ranges (differing diffusion coefficients) of activator and inhibitor states are of fundamental importance for the biological morphogenesis [525] and are, for example, able to explain the formation of patterns, such as thorn production in plants and the markings of animal coats as well as the development of nerve networks (see Fig. 6.81). In the inorganic field, too, typical growth patterns far from equilibrium

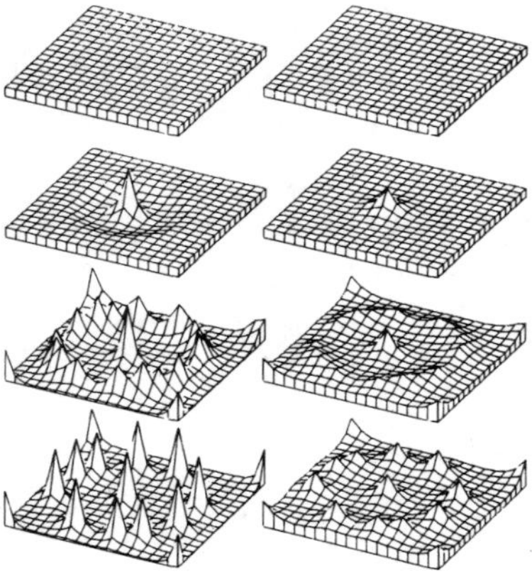

Fig. 6.81: Morphogenesis as a result of the interaction between short–range activation and long–range inhibition. The left hand column shows the evolution of the activator concentration (a) in two dimensions, the right hand column the inhibitor concentration (h). The basic equations are of the form $\dot{a} = \alpha + \beta a^2/h - \gamma a + D_a a''$ and $\dot{h} = \varepsilon a^2 - \mu h + D_h h''$ [525]. From Ref. [340].

— such as dendritic or seaweed–like shapes — are observed. In many cases their formation can be explained in terms of the anisotropy of the interfacial tension [526]. Morphology and the development of morphology within geology and astrophysics are important, not adequately appreciated aspects of solid state research, for which such considerations can be expected to be very fruitful. Nonlinear conditions generally prevailed during the earth's evolution, leaving us the products in a frozen state.

[215]Essentially the process is driven by the difference in the redox potentials of Fe and $K_2Cr_2O_7$.

Table 6.7 closes this section. It stresses the significance of complex nonequilibrium structures for a future "soft" materials science, which provides a high functional complexity on account of a high information content on the basis of metastability [206].

Table 6.7: Increase of complexity and information content at the expense of stability and reproducibility when moving from simple materials such as Si or ZrO_2 crystals ("hard" materials science) to polymers or nano–structured systems ("soft" materials science) or even to biological systems [206].

"Hard" Materials Science	"Soft" Materials Science	Biology
structural simplicity compositional simplicity functional simplicity	structural complexity compositional complexity functional complexity	
monophase - homogeneous (single-) crystalline	nano-integrated systems high density of metastable structure elements: high information content aperiodic - inhomogeneous - heterogeneous	
strong (thermally, mechanically, chemically)	weak	
in or near equilibrium	not near equilibrium	far from equilibrium dissipative structures
reproducible with high precision	reproducible with low precision	self-reproducing structures

6.10.3 The concept of fractal geometry

As mentioned above, reactions for which only the forward (or only the back) reaction needs to be taken into account, are necessarily processes far from equilibrium. Diffusion–limited aggregation (DLA) is a clear example of great importance. Here, every particle that becomes attached to an initial nucleus remains by definition attached ($\vec{\mathcal{R}} \gg \overleftarrow{\mathcal{R}}$). If the transport is rate–determining and if it is possible to ignore surface diffusion processes or other rearrangement processes during diffusion, then protruding positions will be statistically preferred and the result will be fractal structures [527]. Figure 6.82 shows a DLA cluster, that was produced by computer simulation. Typical real DLA examples are the formation of dust particles from gas phase reactions, structures formed at dielectric breakthrough or during the electrolytic deposition of metals from the solution under suitable conditions (see Fig. 6.83). The positive feedback, namely the formation of new extremities, favoured by the presence of "extremities", is most significant for the kinetics.

A related (but more complicated) example is shown in Fig. 6.84. It shows the cauliflower–like shape of a fluoropatite–gelatine composite grown under conditions that mimic a typical biomineralization process as occuring during tooth or bone formation [531].

Let us now explain at the end of this section the important concept of fractal geometry that we have already mentioned several times.

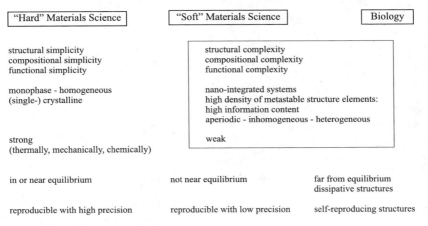

Fig. 6.82: Simulated DLA cluster on a triangular lattice [528]. From Ref. [529].

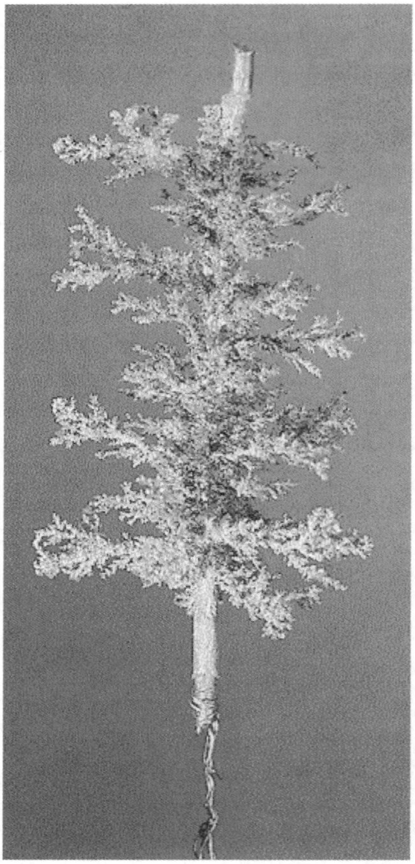

Fig. 6.83: Electrode tree (Wood's metal). Its impedance (cf. Chapter 7) corresponds to its fractal geometry. From Ref. [530].

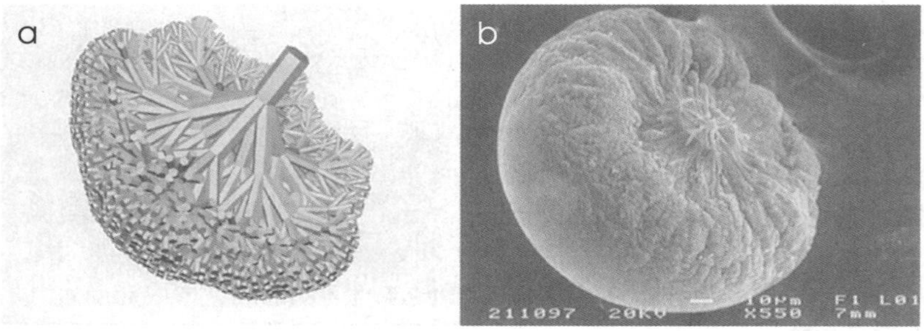

Fig. 6.84: Product of fractal growth of a fluoroapatite–gelatine composite. Computer simulation (a) and experimental observation (b). From Ref. [531].

Typical solid state contacts are, in general, not ideal planes. In particular, the fundamental work of Mandelbrot [532] permits us to go to another again ideal limit, that of fractal geometry. Let us imagine a rough interface, whose individual segments are, when inspected at greater resolution, structured just as the overall structure. If this is fulfilled over a certain size scale (the atomic structure naturally sets a limit), this structure is referred to as fractal and self–similar over this size scale.
It is useful, even though not usual, to designate this as zoom symmetry or enlargement symmetry. It is evident that the length, surface area or volume of these structures is a function of the scale used, since the degree to which the details of the structure contribute differs. Let us consider Koch's curve, shown in Fig. 6.85

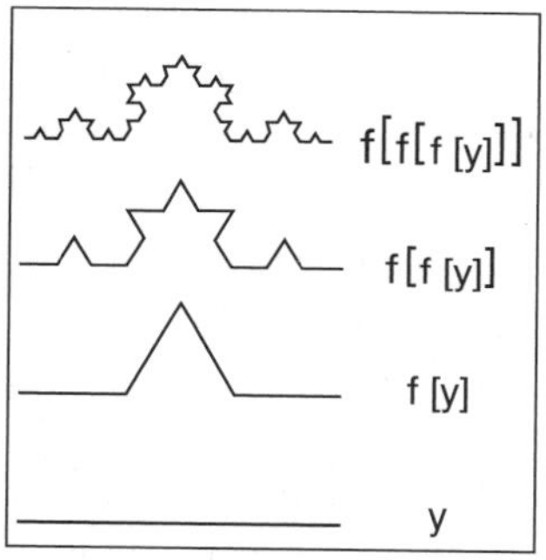

Fig. 6.85: Creation of Koch's curve by successive application of the mapping $y \to f[y]$ (see text).

at various resolutions. Initially let us select the measure ε_0 so that it is identical to the size of the object y in the figure, then the length of the fractal object becomes $\ell(\varepsilon_0) = 1 \cdot \varepsilon_0$; if the measuring rod (scale fineness) $\varepsilon_0/3$ is selected then measurement gives the length $\ell(\varepsilon_0/3) = 4 \cdot \varepsilon_0/3$. In general, we obtain $\ell(\varepsilon/3) = 4/3\ell(\varepsilon)$. Hence, the length is not a whole multiple of ε, rather[216] $\ell(\varepsilon) \propto \varepsilon^{1-\mathrm{d}}$ with d being what is known as the Hausdorff dimension $\mathrm{d} = \log 4/\log 3$. (In the real experiment the scale may for instance be the effective size of (i.e. the area required by) an adsorbed molecule.) If $\mathrm{d} = 1$ then $\ell(\varepsilon)$ is constant.
Let us consider the same point in a somewhat different manner and refer to Fig. 6.86. In addition to Koch's curve (c), the figure displays also two other zoom–

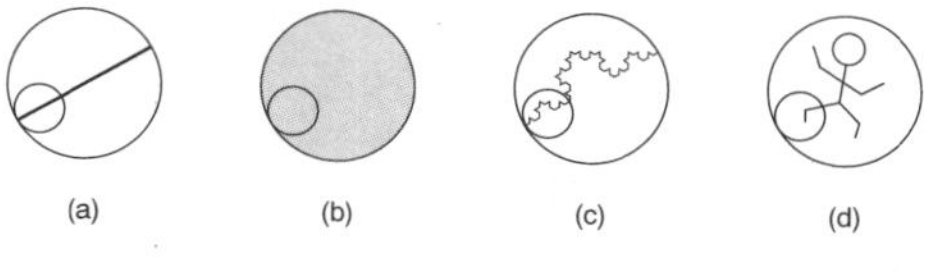

Fig. 6.86: Examples of fractal (c), nonfractal self–similar (a, b) and nonself–similar (d) objects. The inner circle gives the portion to be enlarged.

invariant objects: the line (a) and the plane (b). Let us think of the objects as being homogeneously constructed from a thin metal (constant thickness) of given specific weight, and calculate the mass as a function of the size of the object. The

[216]Proof: $\ell(\varepsilon) = \alpha\varepsilon^{1-\mathrm{d}}; \ell(\varepsilon/3) = \alpha(\varepsilon/3)^{1-\mathrm{d}} = \left(\frac{1}{3}\right)^{1-\mathrm{d}}\ell(\varepsilon)$. Since, on the other hand $\ell(\varepsilon/3) = 4/3\ell(\varepsilon)$ it follows $4/3 = 3^{\mathrm{d}-1}, 4 = 3^{\mathrm{d}}$ and $\mathrm{d} = \log 4/\log 3$. The proportionality constant α is obtained as $\varepsilon_0^{\mathrm{d}}$ because of setting $\ell(\varepsilon_0) = \varepsilon_0$.

latter is measured by the width (L) of the section in Fig. 6.86. It is evident that for the straight line M(bL) = bM(L), i.e. M(L) $\propto$ L, and for the plane M(bL) = b^2 M(L), i.e. M(L) $\propto$ L^2 (Fig. 6.86). In the case of our fractal curve, M is also a homogeneous function (of degree d)

$$M(bL) = b^d M(L), \tag{6.208}$$

but d is not an integer. That is, if we increase the projected length three–fold for the same morphology, the mass increases by 3^d. As we know for Koch's curve M(3L) = 3^dM(L) = 4M(L). Once again, for the Hausdorff dimension we get $\log 4/\log 3 \simeq 1.26$. The fact that the value lies between 1 and 2 expresses the fact that the true length tends to infinity for increasingly smaller scales, but that the two–dimensional space is not entirely filled. We obtain a power law as far as the dependence of the mass on the length is concerned,

$$M(L) = AL^d, \tag{6.209}$$

since $M(bL) = A(bL)^d = b^d AL^d = b^d M(L)$. In this sense the ideal straight line or plane is self–similar, but, because d is an integer, it is not fractal. The stick figure illustrated in Fig. 6.86 (d) as a whole is not self–similar (and hence not fractal), rather the mapping M(L) $\longrightarrow$ M(bL) is complex and depends on the specific case. There is no "zoom symmetry", in contrast to Eq. (6.209) and no universal mapping law. The power law which is valid for the self–similar structures expresses precisely the scale invariance or "zoom symmetry" of the structure ($(b^d)^e = b^{de} = b^{d'}$).

Power laws, found in surface chemical measurements, are frequently the expression of fractal effects. So, not infrequently, we find that reaction rates (amounts or volume changes per unit time) are not proportional to the geometric contact area itself (Wenzel law) but to a noninteger power of the same. Expressed differently, the rate is not proportional to the square of the width ($\propto$L) but to L^{d_c}, whereby d_c is the reaction dimension. The integer variants $d_c = 0$ and 1 are simply explained by assuming that only corner atoms or edge atoms are effective. Noninteger values can be explained in terms of fractal structures; however other explanations are also possible. Table 6.8 lists some experimental exponents for chemisorption of simple gases on dispersed metal catalysts.

Power laws attributable to fractal geometry are also known for electrode impedances [534]. This is (not surprisingly) the case for the example in Fig. 6.83 (see also Chapter 7).

A further example is provided by percolation theory. Already Section 5.8.5 gave an example for this. If one mixes ideally conducting ($\sigma = \infty$) and ideally insulating ($\sigma = 0$) particles[217] randomly, the overall conductivity is zero as long as the percolation threshold (φ_{crit}: critical volume fraction) is not reached, i.e. as long as the mixtures do not exhibit noninterrupted (i.e. percolating) paths of the conducting material which connects the two electrodes. The cluster which just allows percolation, that is, which forms at the threshold to the conductive mixture exhibits a

[217] Cf. also Section 7.3.7.

Table 6.8: Reaction dimensions for some chemisorption reactions on dispersed metals. According to Ref. [533].

catalyt	adsorbate	d_c
Pt - SiO_2	H_2	1.67 ± 0.05
Pt - SiO_2	CO	1.60 ± 0.20
Pt - Al_2O_3	H_2	1.91 ± 0.03
		1.84 ± 0.07
Ag - Al_2O_3	O_2	2.03 ± 0.14
Rh - Al_2O_3	O_2	1.90 ± 0.10
	H_2	1.99 ± 0.09
Ni - SiO_2	H_2	2.13 ± 0.12
Co_3O_4	N_2	1.90
Fe - charcoal	CO	1.60 ± 0.10

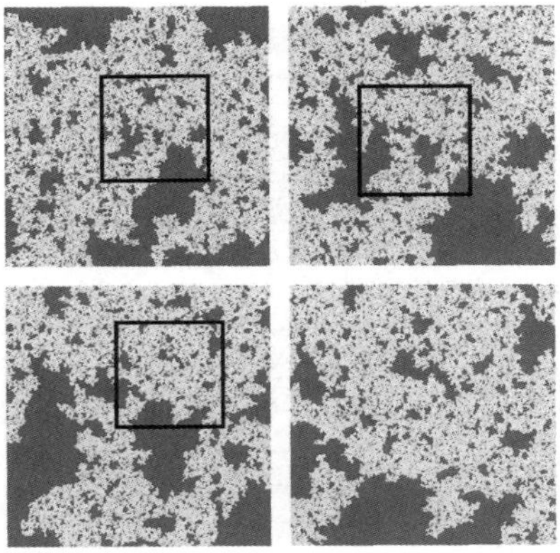

Fig. 6.87: Self–similarity of a large perculation cluster at the critical concentration. The windows indicate the section which is enlarged in the respective succeeding figure. From Ref. [281].

fractal structure (see Fig. 6.87), its mass increases with φ according to a power law, viz. $(\varphi - \varphi_{\text{crit}})^{\beta}$, whereby β is related to the fractal dimension [281,532].
Figure 6.88 is not a poorly digitized landscape photograph, but the result of computer calculations that were totally based on the use of fractal geometry [532]. How much more our complex world resembles fractal structures than structures with flat plane surfaces! However, here we must not overlook the fact that fractal geometry, like "flat" geometry, simply represents an extreme case, that is naturally not met in pure form in nature (as is the case for straight lines or planes).
In addition, technical relevance and structural complexity often behave antagonistically. For reasons of ease of manufacture, reliability and controllability we generally strive to realize shapes that are as simple as possible, such as are considered in the final chapter. In future, however, if we intend to master the generation of more

Fig. 6.88: Computer–generated "mountain landscape". However, the algorithm for creation is not based on real physical–chemical generation laws. From Ref. [532].

complex systems and functions, building more complex structures probably at the expense of sharp reproducibility will become necessary, as it is the rule rather than exception in biology (cf. again Table 6.7 on page 393).

7 Solid state electrochemistry: Measurement techniques and applications

By electrochemistry we mean the study of the electrical behaviour of matter, in relation to changes[1] in composition. In contrast to the previous chapters in which we dealt with electrochemical issues only with regard to the electrical properties of the defects, in particular conductivity effects, in this final chapter, we specifically consider electrochemical systems that are connected to an external circuit (even though, under certain circumstances, the current flowing may be very small). We are interested, in particular, in the mutual conversion of chemical and electrical signals, which is enabled by the occurrence of charged defects. This electrochemistry chapter is, in this sense, not merely a chapter specifically devoted to techniques and applications; it also forms the logical conclusion to this book.
We arrange our treatment as follows: First, we shall consider electrochemical cells in which the electrodes are virtually disconnected (open circuit cells, that is cells with negligible external current); in practice, this implies measurement of the cell voltage with a voltmeter of very high input resistance. The external current can be neglected ($I \simeq 0$), but not necessarily internal partial currents. Second, we investigate the behaviour of cells to which we have applied an external current ($I > 0$), i.e. the current direction — if we are considering asymmetrical cells — opposes that of the short circuit current. Last of all we shall consider cells that (partially or totally) discharge through an external circuit ($I < 0$). There are interesting technological applications for all three cases, and we shall introduce the most important ones. We wish to ignore the purely electronic effects dealt with in detail in the textbooks of semiconductor physics and electrical engineering, and will concentrate on ionic effects or on phenomena (see Fig. 7.1), in which the simultaneous presence of ionic and electronic carriers is of importance.

7.1 Preliminary remarks: Current and voltage in the light of defect chemistry

First we discuss how to connect the parameters current (I) and voltage (U) with the transport parameters and potentials. Here let us consider a general electrochemical cell, i.e. an arrangement of phases between two Cu leads (Fig. 7.2). We concentrate the steady state and neglect displacement currents. The electrical current is obtained

[1] Electrochemistry text books, even recent ones, refer almost exclusively to liquid electrochemistry, but are, nevertheless, to be recommended for background reading (see e.g. [153,219]). Reference [5], dedicated specifically to solid state electrochemistry, concentrates on bulk processes.

Physical Chemistry of Ionic Materials J. Maier
©2004 John Wiley & Sons, Ltd ISBN: 0-471-99991-1 (HB); 0-470-87076-1 (PB)

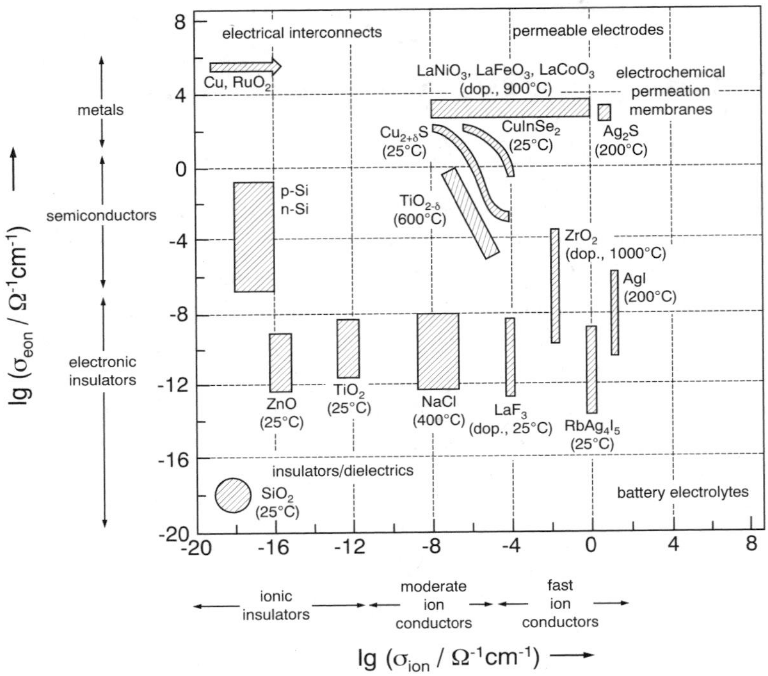

Fig. 7.1: Examples of materials and applications as a function of ion and electron conductivity. The electrical insulators are bottom left, materials with high component permeability ($\sigma^{\delta} \propto$ harmonic mean of σ_{eon} and σ_{ion}) at the top right. Battery electrolytes are bottom right, while purely electronic conductors (electrical interconnects) are to be found at the top left. Between these four extremes there are materials for sensors, electrochromics, diodes, transistors and many other applications. The conductivity values only serve for a first orientation. In various cases, the windows can be substantially extended, e.g. heavy doping. The presentation is based on Refs. [535,536]. Polymers are not included in the figure because there are no reliable data on the relevant transference numbers. They are, however, impressive examples of how widely the conductivity properties can be varied. On the one hand there are (doped) polymers with electronic conductivities as high as Cu or RuO_2 (cf. PA (I_2) in Fig. 6.15). On the other hand there are (doped) polymers exhibiting appreciable ionic conductivities (cf. PEO (LiX) in Fig. 6.12) that typically range between 10^{-8} and $10^{-4}\Omega^{-1}\mathrm{cm}^{-1}$.

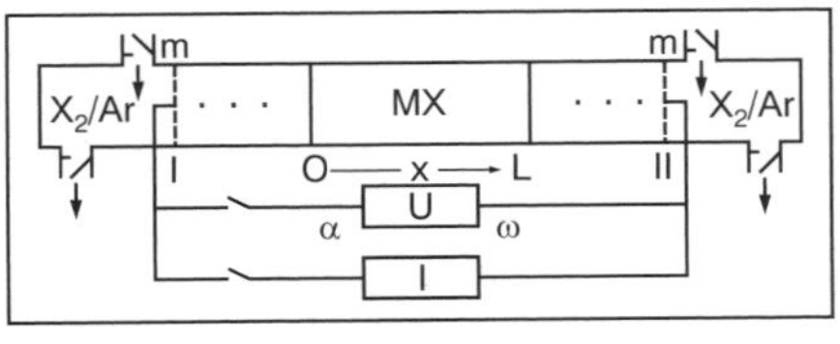

Fig. 7.2: General electrochemical cell with the mixed conductor MX in the centre. The set–up, including the X_2 neighbouring phase, is not only important for fixing the composition, it is also required for the (electro–)chemical polarization measurements. The "current-collecting" metal is denoted by "m".

by multiplying the total current density by the area[2]. The individual current density of the charge carrier k is found, according to the discussion in the previous chapter, to be proportional, to a good approximation, to the corresponding specific conductivity and the corresponding electrochemical potential gradient[3], thus

$$\mathrm{I} = \mathrm{a}\Sigma_\mathrm{k}\mathrm{i}_\mathrm{k} = -\mathrm{a}\Sigma_\mathrm{k}\frac{\sigma_\mathrm{k}}{\mathrm{z_k F}}\left(\partial/\partial \mathrm{x}\right)\widetilde{\mu}_\mathrm{k}. \tag{7.1}$$

At the electrode the internal ionic flux (in the steady state) is converted into a chemical reaction rate (electrode reaction)[4].
The electrical voltage is the difference of the electrical potentials in the two Cu leads (phases α, ω) at the measuring instrument[5]

$$\mathrm{U} = \phi_\omega - \phi_\alpha \equiv \Delta_{\alpha\omega}\phi. \tag{7.2}$$

Since drops of the electrical potential at the phase boundaries do not necessarily disappear at equilibrium and since electrical potential gradients are not directly connected with the current if the composition is altered, it is more sensible to express U in terms of the electrochemical potential of the charge carrier[5]. The chemical potential of the electrons is constant, to a very good approximation, in a metal like copper ($\delta \ln \mathrm{a_{e^-}}(\mathrm{c_{e^-}}) \simeq 0$ since $\delta \ln \mathrm{c_{e^-}} \simeq 0$), and we obtain the desired relationship,

$$\mathrm{U} = -\Delta_{\alpha\omega}\widetilde{\mu}_{\mathrm{e}^-}/\mathrm{F} = -\frac{1}{\mathrm{F}}\int_{(\alpha)}^{(\omega)}\left(\frac{\mathrm{d}\widetilde{\mu}_{\mathrm{e}^-}}{\mathrm{dx}}\right)\mathrm{dx}. \tag{7.3}$$

This formulation is useful since, in equilibrium or in the case of linear nonequilibrium effects, we have information concerning the behaviour of the electrochemical potential. All differences in $\widetilde{\mu}_{\mathrm{e}^-}$ within a phase and from phase to phase, disappear in the absence of a current, if the electrons concerned are mobile enough (see Eq. (7.1)). However, gradients in $\widetilde{\mu}_{\mathrm{e}^-}$ can certainly occur in electrochemical cells

[2] We consider the quasi one–dimensional case. In the case of inhomogeneities perpendicular to the current direction it is necessary to consider the integral $\int \mathbf{i}\mathrm{d}\mathbf{a}$. Moreover Eq. (7.1) is not valid for the displacement current; i.e. Eq. (7.1) is not valid for short times during which capacitive charging occurs (see Section 7.3).

[3] Cross coefficients are neglected.

[4] According to the continuity equation $\partial \mathrm{c}/\partial \mathrm{t} = -\mathrm{div}\mathbf{j} + \nu\mathcal{R}$. If we refer to the steady state ($\partial \mathrm{c}/\partial \mathrm{t} = 0$) and to the ions, one obtains directly that $\mathrm{j_{ion}} \propto \mathcal{R}$ (no outer ionic flux). If we refer under steady state conditions to the electrons, the continuity condition demands that the difference in outer and inner electronic current is given by the rate of the electrode reaction.
In the bulk the steady state condition reduces to $\mathrm{div}\mathbf{j} = 0$ (constant flux) if there are no local generation and annihilation processes. In the transient case stoichiometric changes may occur in the bulk which have to be electroneutral. This follows from the continuity equation for charge and total current ($\partial \mathrm{Q}/\partial \mathrm{t} = 0$, $\mathrm{div}\mathbf{i} = 0$) (see footnote 7).

[5] The question of whether $\Delta_{\alpha\omega}\phi$ or $\Delta_{\alpha\omega}\widetilde{\mu}_{\mathrm{e}^-}$ is measured primarily, is not trivial and depends, ultimately, on the method of measurement.

in which the central phase, MX, is a solid electrolyte[6] resulting in the open circuit cell voltages (see Eq. (7.3)) that we shall discuss in the next section. In metals (and frequently between metal contacts too) the $\widetilde{\mu}_{e^-}$ values are constant, to a good approximation, even when current is flowing, on account of the very high conductivity ($|\nabla\widetilde{\mu}_{e^-}| \propto i/\sigma_{e^-} \simeq 0$). In addition to what happens in the solid electrolyte, differences in $\widetilde{\mu}_{e^-}$ occur at the electrode/solid electrolyte contact and in the phases connected particularly if their electrical conductivities are poor, hence

$$U = I\Sigma(\text{contact resistances}) + U_{MX},$$
$$U_{MX} = -\frac{1}{F}\int_0^L \left(\frac{d\widetilde{\mu}_{e^-}}{dx}\right) dx. \tag{7.4}$$

In Eq. (7.4) we have referred the voltage U_{MX} to the first and last layer of the solid electrolyte (x=0,L). In addition to breaking U down into various local components it is also sensible to make a break down into the open circuit voltage E and the overvoltage η created by the current:

$$U(I) = E + \eta(I). \tag{7.5}$$

In the case of different serial processes i, the stationary[7] overvoltage η can always be formally represented as a product of the current and the associated resistances,

$$\eta(I) = \Sigma_i\eta_i = I\Sigma_i\bar{R}_i(I), \tag{7.6}$$

whereby the resistances are only current–independent close to equilibrium. In contrast to the differential resistances ($\partial\eta/\partial I$), which we term R, the integral[8] resistances appearing in Eq. (7.6) are denoted by $\bar{R}$. In the linear regime R=$\bar{R}$.) As

[6]This shows that only local electrode equilibrium but not global equilibrium applies in electrochemical cells. Differences in $\widetilde{\mu}_{e^-}$ can also occur in open circuit cells if the electrolyte exhibits nonzero electron conductivity (Section 7.2.2). However, an internal current then flows.

[7]If we take I(t) as the excitation, then the whole voltage change is obtained according to $\left.\frac{\partial U}{\partial I}\right)_t \frac{dI}{dt} + \left.\frac{\partial U}{\partial t}\right)_I$ as an implicit change caused by the current–time function plus an explicit change. The latter is effected by capacitance. In the steady state the capacitors block, and the current is a pure Faradaic current (see footnote 2).
The difference between Faradaic current density ($\mathbf{i}_F$) and capacitive current density ($\mathbf{i}_C$) can also be seen from the continuity equation for the charge ($\partial Q/\partial t \propto -\mathrm{div}\mathbf{i}_F$) where $\partial Q/\partial t$ can be expressed as $-\mathrm{const.div}\mathbf{i}_C$.

[8]Since a voltage, namely E, exists when I=0, the precise definition of the integral resistance is η/I where η means the stationary overvoltage. Corresponding to the way in which the elements are connected (cf. decomposition of η and I) $\bar{R}$ can be decomposed into individual contributions. It is generally better to start from a differentially defined resistance that is arrived at from $\partial\eta/\partial I = \partial U/\partial I$. The situation is partly analogous for the capacitance. Viewed differentially (C: differential capacitance) it measures the charge increase on account of the increase in voltage ($\partial Q/\partial U$). Here charge and potential contributions need to be broken down into local contributions. The more precise definition of an integral capacitance ($\bar{C}$) is not $Q/\Delta\phi$ but $Q/(\Delta\phi - \Delta\phi_{pzc})$, since it is only at $\Delta\phi_{pzc}$ that zero charge is achieved (pzc: point of zero charge).

demonstrated in Section 6.1, the restriction "close to equilibrium" depends on the situation; for very thin samples or contact processes deviations from ohmic behaviour may occur at very small currents, in the case of the bulk resistances of thicker samples deviations from ohmic behaviour require extremely high voltages.

In addition to internal charge transport in transient cases[7] the external current can also be compensated by dielectric effects (displacement current by local polarization). This transient charging current is characterized by electrical capacitances. They are considered in more detail in Section 7.3.3. Capacitive effects[7,8] (charge storage) are generally responsible for time dependences. Apart from dielectric effects, storage phenomena can also occur if the stoichiometry changes by virtue of the current flow. Such chemical capacitances (see Section 6.7.4) will be treated in more detail in Section 7.3.4 (cf. also Section 7.4).

Before we start with the more specific treatment two points shall be considered briefly, (i) the connection with global thermodynamics, and (ii) the correlation with terms usually used in surface science.
(Ad i) We have noticed that when we apply an external current to the cell ($I > 0$), then $\eta > 0$ and, hence, according to Eq. (7.5), U is always larger than when $I = 0$, while in the case of current extraction ($I < 0$) in galvanic cells the voltage is decreased ($\eta = U - E < 0$).
This behaviour, since it corresponds to the second law, can be put more clearly in thermodynamic language (Chapter 4) [512]. At electrochemical equilibrium (open circuit, no internal fluxes) the equilibrium voltage E just compensates the reaction affinity (see Section 4.2). The actual driving force for current flow is the electrochemical affinity $\widetilde{\mathcal{A}} = -\Delta\widetilde{G}$, which represents the deviation from electrochemical equilibrium. It is obtained, accordingly, from the difference between the chemical affinity and the voltage multiplied by the charge transferred, and is thus proportional to $E - U = -\eta$. The entropy production (cf. Section 6.1) then is $\Pi = \widetilde{\mathcal{A}}\widetilde{\mathcal{R}} \propto \eta I > 0$. This no longer applies in the same manner when permeation cells are concerned (cf. Section 7.2.2) since short–circuit currents occur ($\Pi \neq 0$), which do not manifest themselves in an external current ($I=0$).
(Ad ii) Speaking about electrical potential, a more careful distinction is necessary. Strictly speaking, the parameter ϕ represents the local inner potential (Galvani potential). However, only the outer (Volta) potential ψ is susceptible to absolute measurement and this differs from ϕ by the surface potential (dipole potential) χ which is often considerable. Polarization effects resulting from the alignment of polar groups or the deformation of nonpolar groups, i.e. charge displacements, make up this dipole potential, while excess charges contribute to the change in the Volta potential. The relationship between $\phi, \psi, \chi, \mu_{e^-}, \widetilde{\mu}_{e^-}$ and the work function w is summarized in Fig. 7.3. The last parameter, also accessible to measurement, describes the (free) energy that is necessary to transport an electron reversibly from

the bulk to an unbonded state[9] immediately in front of the surface.
We are now prepared to treat individual electrochemical systems in detail.

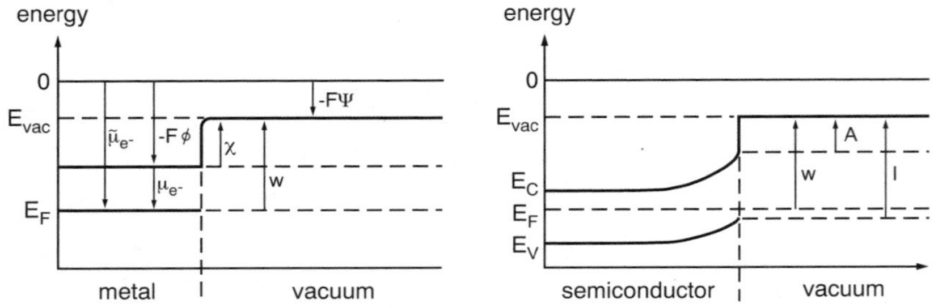

Fig. 7.3: The relationship between the parameters work function (w), Galvani potential (ϕ), Volta potential (ψ) and surface potential (χ) and work function, electron affinity (A) and ionization energy (I) for metal (left) and semiconductor (right). E_{vac} is the energy of the electron in vacuum immediately in front of the surface. Note that the difference between I and A corresponds to the band gap (internal redox disproportionation, cf. Section 2.1.3). A and I refer to the band edges (standard states) while w refers to E_F, that is, also takes account of the configuration entropy. Please do not confuse the symbol for energy with the symbol E used in the text for EMF. Also remember that at finite temperatures local entropy effects are to be considered and hence the free energy must be addressed [537].

7.2 Open circuit cells

7.2.1 Equilibrium cells: Thermodynamic measurements and potentiometric sensors

Absolutely symmetrical cells naturally exhibit a zero voltage in the currentless state at electrochemical equilibrium. Let us consider a simple asymmetrical cell, namely an oxygen concentration cell with the oxygen ion conductor made from Y_2O_3–doped ZrO_2

$$\text{Cell O1} = \qquad \mathrm{Pt, O_2(P_1)|ZrO_2(Y_2O_3)|O_2(P_2), Pt,} \qquad (7.7)$$

in which the two partial pressures of oxygen, P_1 and P_2, on the two sides (x=0 and x=L), are maintained at constant but different levels, with total pressure being the same. Let P_1 be greater than P_2. We measure the voltage with a high resistance voltmeter so that the measurement current is negligible ($U \simeq E$ according to Eq. (7.5)). Hence, there is also no net internal current. But also the partial

[9]The locus (typically 10^{-6}cm distance from the surface) is characterized by the following: Image forces as well as effects of the dipole layer are negligible, but the volta–potential is fully perceived.

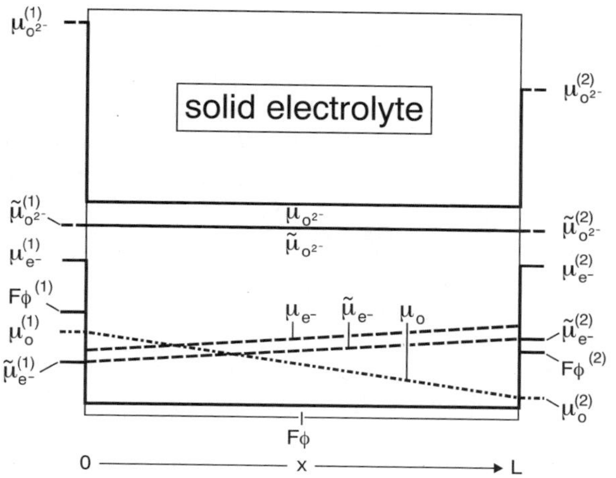

Fig. 7.4: Potential profiles in a solid electrolyte conducting oxygen ions and at the contact with the metal (left and right of the frame) in the stationary state of an EMF measurement (oxygen concentration cell). By virtue of the high carrier density $\mu_{O^{2-}}$ is constant in the electrolyte and μ_{e^-} in the metal and have been set arbitrarily here. All other potentials are obtained from the equilibrium conditions (see Eq. (7.8)). In order to define $\widetilde{\mu}_{e^-}$, μ_{e^-} and μ_O in the solid electrolyte, it is necessary that an (very small) electron conductivity be taken into account (see Section 7.2.2 and Fig. 7.7). Likewise in order to define μ_O, $\widetilde{\mu}_{O^{2-}}$ and $\mu_{O^{2-}}$ in the contact metal it is necessary there to take account of a low oxygen conductivity. The difference in oxygen potential is assumed, for simplicity, to be so small that the specific conductivities can be considered to be constant (cf. Eq. (7.1)). (Space charge zones are regarded as negligible.)

currents disappear: Since the electronic conductivity within the solid electrolyte is negligible[10], the internal pure electronic current is as well (because $i_{e^-} \propto \sigma_{e^-}|\nabla\widetilde{\mu}_{e^-}|$), and since $i_{O^{2-}} = i - i_{e^-} = -i_{e^-}$, the ion current equals zero, too. Now, since $|\nabla\widetilde{\mu}_{O^{2-}}| \propto i_{O^{2-}}/\sigma_{O^{2-}}$ and $\sigma_{O^{2-}} \neq 0$, the electrochemical potential gradient disappears for the oxygen ions; however, this is not the case (because $\sigma_{e^-} \simeq 0$) for the electrochemical potential of the electrons, whose difference between x=0 and x=L ($\Delta_{0L}\widetilde{\mu}_{e^-}$) is, after all, responsible for the cell voltage (see Eq. (7.3)). This is shown in Fig. 7.4.

The incorporation equilibrium applies at each phase boundary[11]

$$\begin{array}{lll} \frac{1}{2}O_2(L) + 2e^-(L) \rightleftharpoons O^{2-}(L) & \text{i.e.} & \frac{1}{2}\mu_{O_2}(L) + 2\widetilde{\mu}_{e^-}(L) = \widetilde{\mu}_{O^{2-}}(L) \\ \frac{1}{2}O_2(0) + 2e^-(0) \rightleftharpoons O^{2-}(0) & \text{i.e.} & \frac{1}{2}\mu_{O_2}(0) + 2\widetilde{\mu}_{e^-}(0) = \widetilde{\mu}_{O^{2-}}(0). \end{array} \tag{7.8}$$

[10]In fact the finite nature of the electronic conductivity is responsible for small internal electronic and ionic currents; this is treated below.

[11]The case of kinetic inhibitions will be discussed later. The designations x=0 and x=L are used in a loose manner: naturally $e^-(L)$ and $e^-(0)$ strictly refer to the respective electrode phases, $O_2(L)$ and $O_2(0)$ to the respective gas phases and $O^{2-}(L)$ and $O^{2-}(0)$ to the respective boundary layers of the solid electrolytes, as can clearly be seen in Fig. 7.4. Precise distinctions will be made whenever this is necessary.

Calculating the difference at the two phase boundaries yields[12]

$$\begin{aligned} EF = -\Delta_{0L}\widetilde{\mu}_{e^-} &= -\frac{1}{2}\Delta_{0L}\widetilde{\mu}_{O^{2-}} + \frac{1}{4}\Delta_{0L}\mu_{O_2} \\ &= +\frac{1}{4}\Delta_{0L}\mu_{O_2} \end{aligned} \tag{7.9}$$

and, hence, the Nernst equation

$$E = \frac{RT}{4F}\ln\frac{P_2}{P_1}. \tag{7.10}$$

When formally constructing the overall cell reaction from the difference of the two electrode reactions, it should be noted that in contrast to purely chemical reactions, the spatial separation of the reactants by an electrolyte is crucial: the different electrical potentials correspond to sites 0 and L; O^{2-} truly cancels in the balance since $\widetilde{\mu}_{O^{2-}}$ is constant, while the difference between the electron terms leads to the voltage, so that the affinity of the overall (chemical) reaction

$$O_2(P_2) \rightleftharpoons O_2(P_1) \tag{7.11}$$

yields the cell voltage

$$\Delta_r G = -z_r FE, \tag{7.12}$$

a result that is expected according to the global laws of thermodynamics[13,14]. If we know the partial pressure on one side, Eq. (7.10) can be used to calculate the partial pressure on the other side. This is the basic principle of a potentiometric gas sensor. As a so–called lambda probe, it finds wide application in automobiles and in metallurgical technology. The oxygen activity in metals, being an important process parameter in metallurgical processes, is measured by means of lambda probes; in automobile applications (Fig. 7.5) the measured signal is directly used to regulate the optimal content which is of particular importance for the functioning of exhaust gas catalysers [539]. Figure 7.5b shows the variation in voltage that is naturally similar to a titration curve. The voltage signal can be used directly to regulate chemical process variables. The point $\lambda = 1$ corresponds to the stoichiometric air–fuel ratio.

[12] As far as the comparison between the cell voltage and the work function difference of the electrodes is concerned, one has to recall that a vanishing of the ionic contribution in Eq (7.9) presupposes the presence of an electrolyte (electrochemical equilibrium).

[13] A much better formulation of the total (electrochemical) reaction, to which then the detailed equilibrium condition is directly applicable, is:

$$O_2(L) + 4e^-(L) \rightleftharpoons O_2(0) + 4e^-(0).$$

The e^- terms are included since their electrochemical potentials are different. By applying the balance of the $\widetilde{\mu}$'s, Eq. (7.12) follows with $z_r = 4$.

[14] As in Chapter 6 the index "m" is omitted in Chapter 7 to avoid complex notation, i.e. ΔG, ΔH, ΔS refer to 1 mol if not stated otherwise.

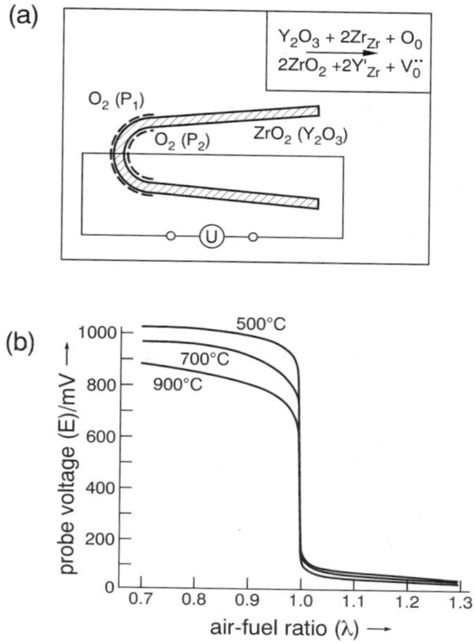

Fig. 7.5: a) Sketch of an oxygen concentration cell with ZrO_2 (Y_2O_3) (termed YSZ) as the ceramic solid electrolyte, which finds application in O_2 sensors (λ probes), pumps and fuel cells (in the latter the partial pressure on the anode side is kept very low by means of gases such as H_2, cf. Section 7.4.2).
b) λ probe voltage as a function of air–fuel ratio. If the mixture is too fat ($\lambda < 1$) or too lean ($\lambda > 1$), the cell voltage deviates strongly from the value at the stoichiometric point ($\lambda = 1$). From Ref. [538].

Instead of using a gas on the reference side, the oxygen activity can be maintained as described in Chapter 4 by means of a metal–metal oxide mixture, e.g. as in

$$\text{Cell O2} = \qquad \mathrm{Cu, Cu_2O|ZrO_2\,(Y_2O_3)\,|O_2, Pt}. \tag{7.13}$$

The same principle can also be applied to the construction of potentiometric sensors for other gases. Thus, cells with proton–conducting solid electrolytes (water–containing In–doped $CaZrO_3$ [540], cf. Section 5.6) are used to measure H activity in aluminium and hence to control the brittleness associated with hydrogen content. It is not necessary that the gas to be detected and the mobile ion refer to the same element. Let us consider a chlorine sensor based on AgCl. AgCl in contact with Cl_2 gas fixes a defined silver activity and cells of the type

$$\mathrm{Pt, Cl_2(P_1)|AgCl|Cl_2(P_2), Pt} \tag{7.14}$$

or more simply (if the minimum value of P_1 is taken)

$$\mathrm{Ag|AgCl|Cl_2, Pt} \tag{7.15}$$

constitute chlorine or silver activity cells. The contact Ag/AgCl corresponds to the metal/metal oxide contact in Eq. (7.13). In the same manner as in the above sample we derive

$$\mathrm{E} \propto -\Delta_{0\mathrm{L}}\mu_{\mathrm{Ag}} \propto \Delta_{0\mathrm{L}}\mu_{\mathrm{Cl_2}}. \tag{7.16}$$

In more detail it follows on the cathode side[15] (x=L) that

$$\frac{1}{2}Cl_2 + Ag^+ + e^- \rightleftharpoons AgCl \quad \text{i. e.} \quad \frac{1}{2}\mu_{Cl_2} + \widetilde{\mu}_{Ag^+} + \widetilde{\mu}_{e^-} = \mu^\circ_{AgCl} \qquad (7.17)$$

and on the anode side (x=0) that

$$Ag^+ + e^- \rightleftharpoons Ag \qquad \text{i. e.} \quad \widetilde{\mu}_{Ag^+} + \widetilde{\mu}_{e^-} = \mu^\circ_{Ag}. \qquad (7.18)$$

In the cell reaction which is the difference of the above partial reactions the Ag^+ contribution disappears (AgCl is a pure Ag^+ conductor), while the e^- contribution gives the cell voltage:

$$\frac{1}{2}Cl_2 + Ag \rightleftharpoons AgCl\left[+e^-(0) - e^-(L)\right]. \qquad (7.19)$$

The expression in square brackets takes account of the difference between the chemical and electrochemical equilibrium[13]. Because $\mu_{Cl_2} = \mu^\circ_{Cl_2} + RT\ln P_{Cl_2}$ and $\frac{1}{2}\mu_{Cl_2} + \mu^\circ_{Ag} = \mu^\circ_{AgCl} - \Delta_{0L}\widetilde{\mu}_{e^-}$ it follows that

$$EF = -\Delta_{0L}\mu_{Ag} = +\frac{1}{2}\Delta_{0L}\mu_{Cl_2} = -\Delta_f G_{AgCl} = -\Delta_f G^\circ_{AgCl} + \frac{1}{2}RT\ln P_{Cl_2}. \qquad (7.20)$$

The standard free enthalpy[14] of formation $\Delta_f G^\circ_{AgCl}$ stands for $\mu^\circ_{AgCl} - \mu^\circ_{Ag} - \frac{1}{2}\mu^\circ_{Cl_2}$.

There is an equivalent treatment, in which the individual electrical potential jumps[16] are calculated and summed up; this will be discussed in more detail now. In the case of heterogeneous equilibrium it follows for the exchangeable species that $\Delta\widetilde{\mu}_k = 0$ if Δ refers to the change at the interface, and hence that the electrical potential jump is given by $\Delta\phi = -\Delta\mu_k/z_k F$. Let us consider cell (7.15) again in this light:
Ag^+ is reversibly exchanged at the Ag/AgCl boundary (i.e. in the equilibrium $\Delta\widetilde{\mu}_{Ag^+} = 0$ hence $\Delta\mu_{Ag^+} = -F\Delta\phi$). On the left hand side of the cell we break down μ_{Ag^+} in silver into $\mu^\circ_{Ag} - \mu_{e^-}(Ag)$. Then the potential jump $F\Delta\phi^{(I)}$ ($\equiv$ potential jump I $= -\Delta\mu^{(I)}_{Ag^+}$) at this phase boundary is given by the expression $\mu^\circ_{Ag} - \mu_{Ag^+}(\text{AgCl, left}) - \mu_{e^-}(Ag)$. On the right hand side ($Pt, Cl_2|AgCl$), there is an analogous potential jump $F\Delta\phi^{(II)}$ ($\equiv$ potential jump II) determined by $\frac{1}{2}\mu_{Cl_2} + \mu_{e^-}(Pt) - \mu^\circ_{AgCl} + \mu_{Ag^+}(\text{AgCl, right})$[17]. Since the cell voltage is probed at the same metals — for simplicity let us assume it is Pt — it is necessary to add

[15] We connect with the formulation in terms of defects by using $\delta\mu_{Ag_i^\cdot} = -\delta\mu_{V'_{Ag}} = \delta\mu_{Ag^+}$ and $\delta\mu_{e^-} = \delta\mu_{e'} = -\delta\mu_{h^\cdot}$ (see Chapter 5).

[16] Since ϕ is a function of state it follows that $\Delta_{0L}\phi = \int_{\phi(0)}^{\phi(L)} d\phi = \Sigma_i\Delta_i\phi$. Hence, the total potential difference is obtained as the sum of the potential jumps at the interfaces, plus any potential differences within the phases themselves.

[17] The first three terms express the negative chemical potential of Ag^+ on the AgCl–surface determined by the contact with the gas phase and the Pt–electrode.

the potential jump at the Pt/Ag boundary (potential jump III), which is given by $\mu_{e^-}(\mathrm{Ag}) - \mu_{e^-}(\mathrm{Pt, left})$. Summation of the potentials yields Eq. (7.20) as before if $\mu_{Ag^+}(AgCl)$ is identical in both expressions, that is, if there is no gradient of this parameter within the silver chloride. This condition is fulfilled with markedly ionically disordered ionic conductors ($\nabla\mu_{Ag^+} = \nabla\mu_{Ag_i^\cdot} \simeq \mathbf{0}$ because $\nabla c_{Ag_i^\cdot}/c_{Ag_i^\cdot} \simeq \mathbf{0}$). If this is only poorly fulfilled, a further potential drop over the ion conductor has to be taken into account; but Eq. (7.20) is the result in any case.
The description leads to the clear conclusion that we can divide the total cell voltage into two half–cell voltages (electrode potentials), provided we use a good ion conductor as an electrolyte. The two partial voltages $\Delta\phi_{\mathrm{left}}$ (composed of potential jump I + potential jump III) and $\Delta\phi_{\mathrm{right}}$ (potential jump II) refer to the half cells Pt|Ag|AgCl and AgCl|Cl_2, Pt.
Even if highly disordered ion conductors are used, care is necessary in the case that the two electrolyte phases of the half cells are not identical (e.g. with the combination of the half cells Pt/Ag/AgX and AgCl/Cl_2,Pt, where X $\neq$ Cl), then the spatial profile of the chemical potential of the ions within the solid is not stationary because of the mutual solubilities of the halides.
There are two possibilities of avoiding this complication. On the one hand, a reversible potential is obtained if both phases are given the opportunity to equilibrate. If there is phase equilibrium (for X=I the coexisting composition[18] $AgI_{ss}|AgCl_{ss}$ is produced; for X = Br the solid solution Ag(Cl,Br)); however, the total cell voltage no longer corresponds to the half cell combination with pure phases. A second possibility is to interpose an electrolyte (e.g. "Ag–alumina", cf. Section 6.2.1) which prevents interdiffusion but connects the μ_{Ag^+} values via the two potential jumps created[19], so that now the difference of the half cell potentials is obtained. In other cases irreversible concentration changes take place at the contact, leading to a diffusion potential. One should pay attention to this point, especially in the case of solid substances[20]. The precise treatment is similar to that for mixed conductors, which are treated at the end of this section [543].

Let us return to the equilibrium chains. The more complex cells

$$\text{cell C1} = \quad (\mathrm{Au}, \mathrm{O_2},)\mathrm{CO_2}, \mathrm{Na_2CO_3}|\beta''\text{–}\mathrm{Al_2O_3}(\mathrm{Na_2O})|\mathrm{SnO_2}, \mathrm{Na_2SnO_3}(, \mathrm{O_2}, \mathrm{Au}) \qquad (7.21a)$$

---

[18]The subscript "ss" stands for solid solution.

[19]$F\Delta\phi$(AgCl/Ag-alumina) = $-\Delta\mu_{Ag^+}$(AgCl/Ag-alumina) and $F\Delta\phi$(Ag-alumina/AgX) = $-\Delta\mu_{Ag^+}$(Ag-alumina/AgX). Since the potentials in Ag–alumina are constant the sum yields $-\Delta\mu_{Ag^+}$(AgCl/AgX).

[20]Obviously the arbitrary inclusion of ion conductors in the circuit may cause considerable effects. In such cases the difference between the chemical potentials of the ions may not be thermodynamically defined but may exhibit appreciable values. The time–dependence of irreversible contributions is often not very great, so that pseudostationary cell voltages are measured. The glass electrode and the Daniell element Cu |$CuSO_4$| |$ZnSO_4$| Zn are examples from aqueous electrochemistry. Such considerations are very important for the performance and selectivity of potentiometric sensors [541,542]. In the case that there are several electrode processes, the phenomenon of mixed potentials must be taken into account (see footnote 59).

$$\text{cell C2} = \quad (\mathrm{Au, O_2,)CO_2, Na_2CO_3}|\beta''\text{–}\mathrm{Al_2O_3(Na_2O)|TiO_2, Na_2Ti_6O_{13}(, O_2, Au)} \quad (7.21\mathrm{b})$$

make a very simple, precise and elegant detection of CO_2 in the gas phase[21] possible [547,548]. They correspond, on the oxide level, to the cells considered at the start. In order to see this, let us abbreviate Na_2O to $\mathtt{N}$, CO_2 to $\mathtt{C}$, and SnO_2 to $\mathtt{S}$. Then with

$$\mathtt{C, NC}|\ldots|\mathtt{NS, S} \tag{7.22}$$

cell C1 takes on a form analogous to cell O2 (cell (7.13)), if we regard the oxygen potential, on the right hand side of cell O2, as being provided by a two–phase mixture (e.g. Ni/NiO)). The chemical cell reaction of cell C1 is $\mathtt{C} + \mathtt{NS} \rightleftharpoons \mathtt{S} + \mathtt{NC}$ or

$$\text{Reaction C} = \qquad \mathrm{Na_2CO_3 + SnO_2 \rightleftharpoons CO_2 + Na_2SnO_3}. \tag{7.23}$$

The comparison, however, is flawed in one important point: the voltage across the Na^+ conductor corresponds to the difference in the Na potentials, not the Na_2O potentials. Put another way: According to the Gibbs phase rule, it is necessary for a third phase to be present on each side, in order that the conditions are thermodynamically defined. Yet, this is the case for the cells described by Eq. (7.21) due to oxygen present on both sides in the gas phase. The particular advantage of this type of cell is that O_2 does not appear in the overall thermodynamic balance, and, hence, the total sensor signal is independent of the partial pressure of oxygen. A second feature of the cell is that in the parameter range of interest, the phase mixture on the right hand side is stable with respect to CO_2, since $\Delta_C G < 0$. The result is an extremely simple experimental set–up: the three pellets can be exposed to the measurement atmosphere without any protection. On account of the overall thermodynamics the EMF of the cell simply is

$$\mathrm{E} = -\frac{1}{2\mathrm{F}}\Delta_\mathrm{C}\mathrm{G} = -\frac{1}{2\mathrm{F}}\Delta_\mathrm{C}\mathrm{G}^\circ - \frac{\mathrm{RT}}{2\mathrm{F}}\ln \mathrm{P_{CO_2}}. \tag{7.24}$$

Figure 7.6 shows how well this relationship is fulfilled, and how rapidly and without drift the cell reacts to changes in CO_2 partial pressure. We shall demonstrate once more, using the electrode reaction and transport processes, that Eq. (7.24) also follows from local considerations and that all activities are fixed. Let us consider Cell C1:

$$\text{measuring electrode}: \ \mathrm{Na_2CO_3(0) \rightleftharpoons 2Na^+(0) + 2e^-(0) + CO_2(0) + \frac{1}{2}O_2(0)} \tag{7.25}$$

$$\begin{aligned} \text{electrolyte}: \ \mathrm{2Na^+(0)} &\rightleftharpoons \mathrm{2Na^+(L)} \\ \mathrm{2e^-(0)} &\not\rightleftharpoons \mathrm{2e^-(L)} \end{aligned} \tag{7.26}$$

$$\text{reference electrode}: \ \mathrm{2e^-(L) + 2Na^+(L) + SnO_2(L) + \frac{1}{2}O_2(L) \rightleftharpoons Na_2SnO_3(L)}. \tag{7.27}$$

[21] Cf. also Refs. [544–546].

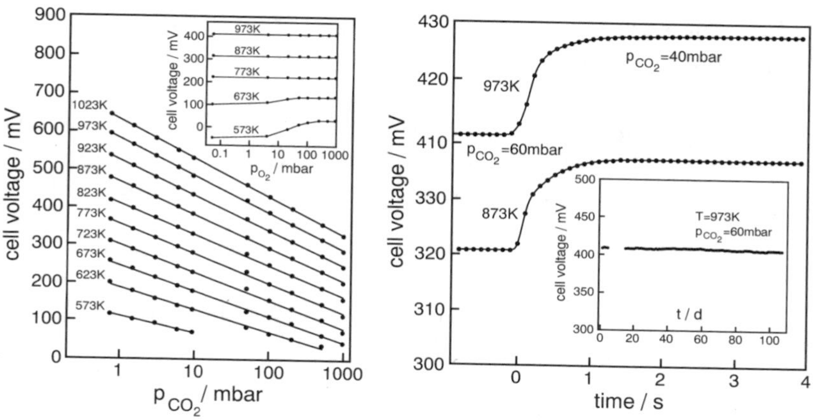

Fig. 7.6: a) Shows how excellently the cell C2 (Eq. (7.21b)) fulfils the criteria for an EMF sensor: it exhibits Nernst behaviour, fast response time and zero-drift. In addition, when T$\geq$ 500°C the signal is independent of P_{O_2}. The detection limit is very low and fixed by the decomposition CO_2–pressure of Na_2CO_3. The high sensitivity is not displayed by the figure [549].

Since the solid electrolyte is a Na^+ ion conductor under the conditions of measurement ($\sigma_{Na^+} \gg \sigma_{e^-} \simeq 0$), the cell voltage is obtained from the difference of the chemical potentials of Na on both sides. Because of the above equation μ_{Na} can be expressed using $\mu^\circ_{Na_2CO_3}(0)$, $\widetilde{\mu}_{Na^+}(0)$, $\widetilde{\mu}_{e^-}(0)$,$\mu_{CO_2}(0)$, $\mu_{O_2}(0)$ on the measurement side and $\mu^\circ_{Na_2SnO_3}(L)$, $\mu^\circ_{SnO_2}(L)$, $\widetilde{\mu}_{Na^+}(L)$, $\widetilde{\mu}_{e^-}(L)$, $\mu_{O_2}(L)$ on the reference side; when the difference between the μ_{Na} values is taken, the μ_{O_2}'s cancel out (the same partial pressure in the open cell) as the other $\widetilde{\mu}_{Na^+}$'s do due to the high Na^+ conductivity in the solid electrolyte. The difference between the $\widetilde{\mu}_{e^-}$'s gives the cell voltage, and Eq. (7.24) is obtained once again.

Conversely, the EMF of a suitable electrochemical cell can naturally be used to determine unknown thermodynamic data of a reactant by means of the free enthalpy of reaction

$$\Delta_r G = -z_r F E. \tag{7.28}$$

The T–dependence delivers the reaction entropy

$$\Delta_r S = -\frac{\partial \Delta_r G}{\partial T} = z_r F \frac{\partial E}{\partial T} \tag{7.29}$$

and from using Eq. (7.28) also the reaction enthalpy

$$\Delta_r H = -z_r F \left(1 - T\frac{\partial}{\partial T}\right) E. \tag{7.30}$$

It is important that the experimental conditions are well–defined, that no other reactions contribute, that the electrode reactions are reversible and that the measurement current is negligible.
Many thermodynamic formation data have been determined in this manner. Thus, the investigation of the formation cell for AgCl (see Eq. (7.15)) provided reliable standard enthalpies of formation for this compound, in accordance with

$$\Delta_f G^\circ_{AgCl} = -EF + \frac{1}{2} RT \ln P_{Cl_2}; \tag{7.31}$$

$\Delta_f S^\circ_{AgCl}$ and $\Delta_f H^\circ_{AgCl}$ then follow from the T dependence according to Eqs. (7.29,7.30).

Table 7.1: Examples of equilibrium cells which provided thermodynamic data of the compounds underlined [6,550].

Mg, MgF_2 | CaF_2 | ThF_4, Th,	(Ni, Mn)O, Ni | ThO_2 (+Y_2O_3) | Ni, NiO
Th, ThF_4 | CaF_2 | AlF_3, Al,	Co, SiO_2, Co_2SiO_4 | ZrO_2 (+CaO) | Co, CoO
U, UF_3 | CaF_2 | AlF_3, Al,	Co, Al_2O_3, (Co, Mg) Al_2O_4 | ZrO_2 (+CaO) | Co, Al_2O_3, $CoAl_2O_4$
Th, ThF_4 | CaF_2 | NiF_2, Ni,	Ni, NiO | ZrO_2 (+CaO) | (Cu, Ni), NiO
Al, AlF_3 | CaF_2 | PbF_2, Pb,	Ag | AgI | (Ag, Te)

Table 7.1 reproduces a selection of equilibrium cells based on fluorine, silver and oxygen ion conductors, which have been used to determine relevant thermodynamic data (such as were treated in Chapter 4) [6,550]. A cell analogous [365] to the cell in Eq. (7.21)

$$O_2, CO_2, Na_2CO_3|\beta''\text{–}Al_2O_3(Na_2O)|Na_2ZrO_3, ZrO_2, O_2 \tag{7.32}$$

has been employed to determine the unknown formation data for Na_2ZrO_3 from the EMF, the CO_2 partial pressure and from tabulated formation data of CO_2, Na_2CO_3 and ZrO_2. The use of $CaCO_3/CaO$ buffer mixtures made possible a further extension of the range of CO_2 partial pressures in the same way as it is possible to establish very low P_{O_2} values by metal/metal oxide two phase mixtures. The combination with C_p measurements is particularly useful. Since the integration over $C_p d \ln T$ gives the entropy increase of the phase (see Chapter 4) and the C_p data can be obtained over an extended temperature range that includes low temperatures, combination with EMF measurements, which are only available at elevated temperatures, yields consistent data over wide ranges. In a similar manner it is possible to determine the corresponding data as well as the mixing behaviour for more complex compounds such as the Nasicon system (see Section 6.2) [100]. Moreover, the comparison of entropy values obtained from C_p values (see Fig. 4.2, page 85) and those obtained electrochemically then allows an estimate to be made of zero point entropies that are caused by kinetic inhibitions of ordering processes in such multinary systems at low temperatures [100].

7.2.2 Permeation cells and chemical polarization: Measurement of transport parameters and chemical filters

Let us return to the oxygen concentration cell considered at the beginning (see Eq. (7.7)) and let us permit an electronic conduction contribution, that is, we replace the solid electrolyte by a mixed conductor[22]. Under these conditions the state $P_1 \neq P_2$ is no longer stable, on account of the occurrence of chemical diffusion. However, if the oxygen partial pressures are maintained (by continuous addition or removal of gas), a cell voltage will be produced that is invariant with time (at the expense of the external flow of gas) and whose value we now wish to calculate.
How large is the value of the stationary cell voltage of the nonequilibrium cell [551, 552]

$$O_2(P_1)|MO|O_2(P_2)\,? \tag{7.33}$$

In spite of a vanishing external current, there is now a flow of internal ionic and electronic partial currents ($j_{O^{2-}} \propto \sigma_{O^{2-}}\nabla\widetilde{\mu}_{O^{2-}}$ and $j_{e^-} \propto \sigma_{e^-}\nabla\widetilde{\mu}_{e^-}$) in contrast to the above, coupled such that formally a neutral oxygen flux (ambipolar coupled ions and electrons) results, as discussed in detail in Section 6.3.3.
Coupling the two currents ($j_O = j_{O^{2-}} = -\frac{1}{2}j_{e^-}$) yields

$$\sigma_{e^-}\nabla\widetilde{\mu}_{e^-} = -\sigma_{O^{2-}}\nabla\widetilde{\mu}_{O^{2-}}/2. \tag{7.34}$$

Furthermore, the electrochemical potentials of the ions and electrons are interrelated via the local equilibria:

$$\widetilde{\mu}_{O^{2-}} - 2\widetilde{\mu}_{e^-} = \mu_O = \frac{1}{2}\mu_{O_2} = \frac{1}{2}\mu^\circ_{O_2} + \frac{1}{2}RT\ln P_{O_2}. \tag{7.35}$$

Figure 7.7 shows (in the first approximation) the spatial variation of the potentials involved.
Combination with Eq. (7.34) and rearrangement leads to $\nabla\widetilde{\mu}_{e^-} \propto (\sigma_{O^{2-}}/\sigma)\nabla\mu_{O_2}$; the EMF is obtained[23] by subsequent integration

$$\begin{aligned} E &= -\frac{1}{F}\int_0^L \nabla\widetilde{\mu}_{e^-}dx = \frac{1}{4F}\int_{\mu_{O_2}(O)}^{\mu_{O_2}(L)} \frac{\sigma_{O^{2-}}}{\sigma}d\mu_{O_2} \\ &= \frac{RT}{4F}\int_{\ln P_1}^{\ln P_2} t_{O^{2-}}\,d\ln P_{O_2} \\ &\equiv \frac{RT}{4F}\langle t_{O^{2-}}\rangle \ln\frac{P_2}{P_1} = \langle t_{O^{2-}}\rangle E_{Nernst}. \end{aligned} \tag{7.36}$$

[22]This represents the general case which includes the special case of solid electrolytes. All real solid electrolytes exhibit a minor but nonzero electronic conductivity.

[23]Measurement with ionic probes yields an analogous result (σ_{eon} instead of σ_{ion}) (cf. Section 7.3.2).

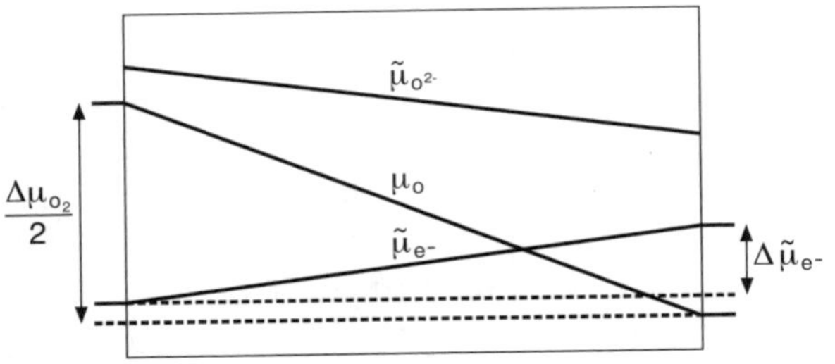

Fig. 7.7: The electrochemical potentials of oxygen ions and electrons in a mixed conductor with an ionic transference number of 2/3 in an oxygen gradient (arbitrarily chosen). The gradient is selected to be so low that the conductivity internally is almost constant. The cell voltage ($\propto \Delta\widetilde{\mu}_{e^-}$) corresponds to $\Delta\mu_O/(3F)$, i.e. $\Delta\mu_{O_2}/(6F)$ according to Eq. (7.36). The details of the variation of $\mu_{O^{2-}}$ require knowledge of the degree of disorder. Since the mobility of the electrons is generally much higher than that of the oxygen ions, it follows that, when σ_{ion} and σ_{eon} are of similar orders of magnitude (here $\sigma_{ion} = 2\sigma_{eon}$), the ionic defect concentration is much greater than that of the electrons and $\mu_{O^{2-}} \simeq$ const. as in Fig. 7.4. Then, in contrast to the solid electrolyte (and in distinction to Fig. 7.4) the bulk is not field–free ($\nabla\phi \neq \mathbf{0}$). (Possible electrode effects are neglected.)

For the sake of simplicity the mean ionic transference number[24] ($\langle t_{O^{2-}}\rangle \equiv \int t_{O^{2-}} d\mu_{O_2} / \int d\mu_{O_2}$) has been introduced into Eq. (7.36); it expresses the reduction of the EMF with respect to the Nernst value (see Eq. (7.10)), which is set up in a pure solid electrolyte. For more precise treatments, such as done for MgO in Fig. 7.8, it is necessary to solve the integral[25] or to measure the change of EMF with P_1

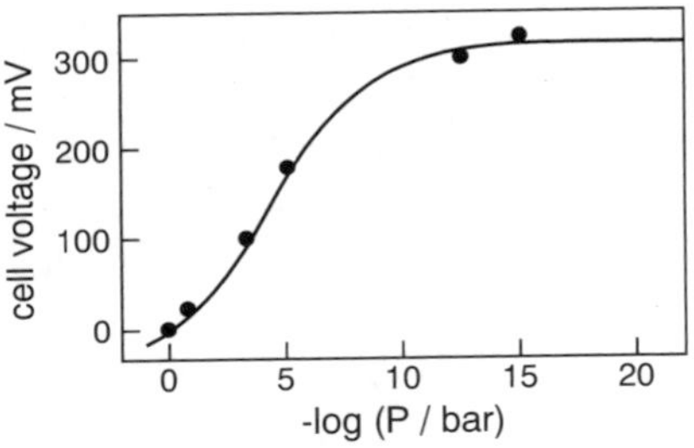

Fig. 7.8: EMF of a concentration cell with MgO as the "electrolyte" as a function of the variable oxygen partial pressure P at 1300°C. The partial pressure on the reference side is 1 bar. The solid line was calculated in accordance with footnote 25. Conductivity data (symbols) are from Ref. [553]. The conductivity is primarily electronic. It is only in the region of the point of inflection that ionic and electronic conductivities are comparable. From Ref. [554].

---

[24]The transference number t_k is more generally defined as the ratio of partial current to total current which usually reduces to σ_k/σ.

[25]It is easy to show that $t_{ion} = (1+\sigma_{eon}/\sigma_{ion})^{-1}$ can be written in the form $t_{ion} = [1+(P_\ominus/P)^N + (P/P_\oplus)^N]^{-1}$ if $\sigma_{e'} \propto P^{-N}$, $\sigma_{h^\cdot} \propto P^N$, and σ_{ion} is independent of the partial pressure. The partial pressures $P_\ominus$ and $P_\oplus$ represent the domain boundaries between the N- and I- and the I– and P–regimes respectively (see Fig. 5.38 in Section 5.5), i.e. $\sigma_{e'}(P_\ominus) = \sigma_{ion}(P_\ominus)$ bzw. $\sigma_{h^\cdot}(P_\oplus) = \sigma_{ion}(P_\oplus)$. When $P_\oplus \gg P_\ominus$ [554] it follows that

$$E = \frac{RT}{F}\left(\ln\frac{P_\oplus^N + P_1^N}{P_\oplus^N + P_2^N} + \ln\frac{P_\ominus^N + P_2^N}{P_\ominus^N + P_1^N}\right).$$

on integration.

at constant P_2 (or vice versa)[26]. The formulation of a mean transference number, however, is not purely academic; it does good service whenever the P_{O_2} range, covered by an individual experiment, is small with respect to the total partial pressure range of interest. In this case the $t_{O^{2-}}(P_{O_2})$–function is obtained in the form of a histogram.
The concentration cell experiment is a powerful method of separating ion and electron conductivities. The requirements are the reversibility of the electrode reaction and the occurrence of appreciable ionic transference numbers. When t_{ion} values are lower than 1% the absolute values of the EMFs normally lie within the fluctuation range of temperature effects.
The need to separate the gas chambers gives rise to serious experimental difficulties. An elegant solution is shown in Fig. 7.9a in which the separation is achieved by

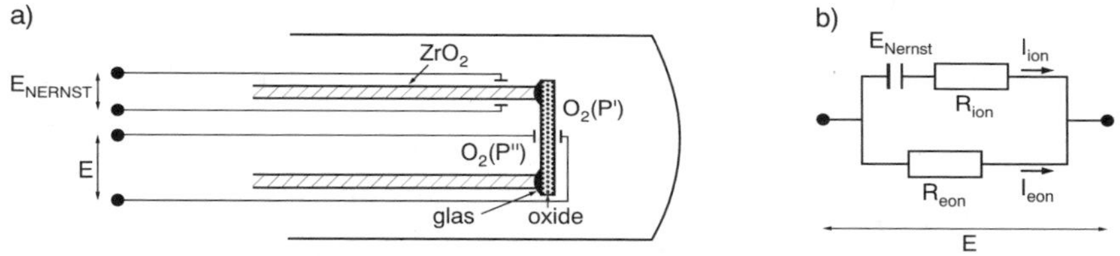

Fig. 7.9: a) The attachment of the measurement sample to a ZrO_2 tube (using a suitable glass solder) permits the direct determination of the ionic transference number from the cell voltages at the tube and the sample.
b) Heuristic equivalent circuit[23] referring to Eq. (7.36).

attaching the sample to a YSZ tube (Y_2O_3–doped ZrO_2) using a suitable glass solder. The use of a solid electrolyte (YSZ) as the separation material makes it possible to control the partial pressure at the same time and, hence, to determine the parameter $\langle t_{ion}\rangle$ directly from the E/E_{Nernst} ratio[27]. Figures 7.10a and 7.10b refer to experimental determinations of the transference number of the ion conductor $Ba_3In_2ZrO_8$ and the mixed conductor $SrTiO_3$.
Equation (7.36) no longer applies if ions are mobile in different valence states (e.g. O^{2-}, O^-, O). It is then necessary to take into account j_{O^-} and j_{O^0} as well as $j_{O^{2-}}$. An analogous calculation which now makes use of the couplings $\widetilde{\mu}_{O^{2-}} = \widetilde{\mu}_{O^-} + \widetilde{\mu}_{e^-} = \mu_O + 2\widetilde{\mu}_{e^-}$ leads to [431]

$$E/E_{Nernst} = \left\langle \frac{\sigma_{O^{2-}} + 2\sigma_{O^-}}{\sigma} \right\rangle = \left\langle t_{ion} + \frac{\sigma_{O^-}}{\sigma} \right\rangle, \tag{7.37}$$

[26] A first order correction possibility is given in Ref. [555] (see also Section 7.3.4).
[27] Fig. 7.9b displays an equivalent circuit which yields Eq. (7.36) if the voltage at the capacitor is identified with the Nernst voltage. This equivalent circuit will become important in the next section. (The following relations are valid: (a) $E_{Nernst} + I_{ion}R_{ion} = I_{eon}R_{eon}$, (b) $E = E_{Nernst} + I_{ion}R_{ion}$, (c) $I_{eon} = -I_{ion}$. Using (a), (b), (c) $E = \frac{R_{eon}}{R_{eon}+R_{ion}} E_{Nernst} = t_{ion}E_{Nernst}$ is obtained.) Cf. also Figs. 7.23, 7.33.

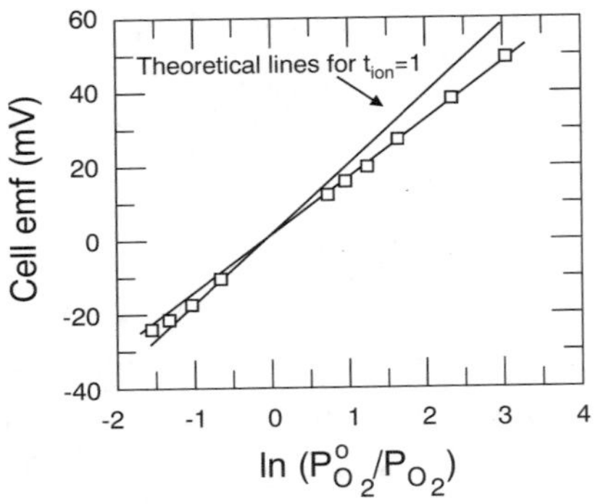

Fig. 7.10a: Comparison of the cell voltage of $Ba_3In_2ZrO_8$ with the Nernst value — i.e. with the value obtainable from the λ probe — reveals primarily ionic conduction. From Ref. [556].

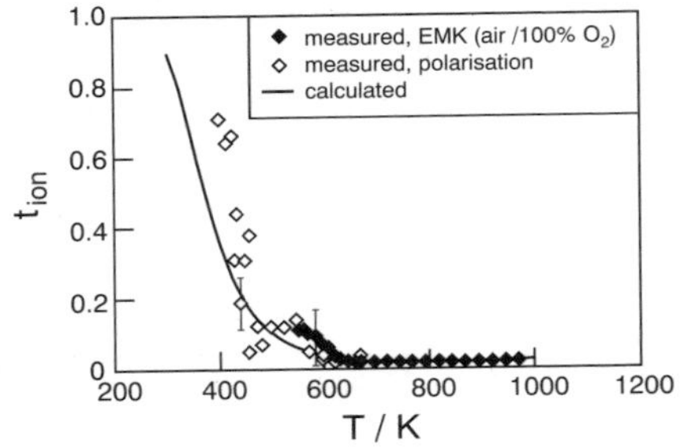

Fig. 7.10b: Ionic transference number of Fe–doped $SrTiO_3$ from EMF measurements compared with polarization measurements (see Section 7.3.4) and independent defect–chemical calculations [193].

whereby once again the arithmetic mean has been taken over μ_{O_2}. Note that $\sigma_{O^{2-}} + 2\sigma_{O^-}$ is not the total ionic conductivity ($\sigma_{ion} = \sigma_{O^{2-}} + \sigma_{O^-}$) but represents $\sigma_{ion} + \sigma_{O^-}$. Consideration of the limiting cases shows that Eq. (7.37) is meaningful. The limit $\sigma_{O^-} \ll \sigma_{O^{2-}}$ has already been discussed, the converse case yields $E/``E_{Nernst}" = 2t_{ion}$, that is a value that could exceed unity, provided the transferred charge included in "E_{Nernst}" is assumed to be (-2); as the real charge is then -1 we have to cancel a factor of 2 in $\frac{RT}{4F} \ln \frac{P_2}{P_1}$ and we get a true Nernst equation, now for a monovalent ion conductor.

Interestingly Eq. (7.37) says that EMF losses may occur without any electrons flowing explicitly. This is in agreement with the fact that an overall currentless permeation is possible if the O^{2-} flux is compensated by a reverse flux of $2O^-$ particles, which we can also regard as electronic vehicle transport coupled to the ion transport. The counterdiffusion of Cu^{2+} and $2Cu^+$ in mixed valance copper conductors is a more realistic case. Nonetheless, in both cases the electronic conductivity is certainly so high, in general, that measurements of this type are not helpful. Short circuit currents that are purely ionically effected are, however, possible in multinary compounds. For this purpose let us consider an oxide proton conductor (such as $SrCeO_3$ discussed in Section 5) in a chemical potential gradient. Here it is necessary to fix two potentials, e.g. P_{H_2O} and P_{O_2}, so that P_{H_2} is also implicitly specified. Internal permeation and, hence, deviations from E_{Nernst} are produced by fluxes of O^{2-} and $2H^+$ in the same directions.

In the general case that O^{2-}, H^+, OH^-, H_3O^+ and e^- are all mobile it can be derived that[28] [557]:

$$\begin{aligned} E &= \frac{RT}{4F}\left[t_{O^{2-}} + 2t_{OH^-} - 2t_{H_3O^+}\right]\Delta \ln P_{O_2} - \frac{RT}{2F}\left[t_{H^+} - t_{OH^-} + t_{H_3O^+}\right]\Delta \ln P_{H_2} \\ &= \frac{RT}{4F}\left[t_{O^{2-}} + t_{H^+} + t_{OH^-} + t_{H_3O^+}\right]\Delta \ln P_{O_2} - \frac{RT}{2F}\left[t_{H^+} - t_{OH^-} + t_{H_3O^+}\right]\Delta \ln P_{H_2O}. \end{aligned} \tag{7.38}$$

We see that an OH^- conductivity makes itself known in a different manner to a H^+ conductivity and, hence, can, in principle, be separated; the deeper reason is that when OH^- migrates it is not merely hydrogen but also oxygen that is transferred. It was possible to conclude from such measurements that direct H^+ migration dominates in the proton conductors under investigation [558].

It is also possible to measure the stationary oxygen permeation flux (see Chapter 6) instead of the cell voltage. This provides us with information concerning the ambipolar conductivity and, hence, concerning the conductivity of the minority charge carrier (e.g. $\sigma^\delta \simeq \sigma_{e^-}$ for $\sigma_{O^{2-}} \gg \sigma_{e^-}$)

$$j_O \propto \sigma^\delta \nabla \mu_O = \frac{\sigma_{O^{2-}}\sigma_{e^-}}{\sigma}\nabla \mu_O, \tag{7.39}$$

which is not otherwise very readily accessible. In the case of surface control the parameter Λ^δ discussed in Chapter 6.7.3 is measured.
If the partial pressure on both sides is not maintained constant, the differences in P_{O_2} level out (we switch off the gas flows in Fig. 7.2). We designate this as chemical depolarization. Its transient behaviour permits calculation of chemical diffusion coefficients or effective rate constants of the surface reaction. Similarly D^δ and k^δ can be obtained from the transient of the chemical polarization (i.e. one–sided step–like change in the partial pressure of oxygen starting from the homogeneous initial situation). Figure 7.11 shows the stoichiometry profiles for a diffusion–controlled chemical polarization. These profiles are obtainable via $D^\delta \partial c/\partial x$ and $c(x,t)$ by solution of the second Fick's law with the initial condition $c(x,0) = c_1$ and the boundary conditions $c(0,t) = c_2$ and $c(L,t) = c_1 \equiv c(x,0)$ (see e.g. [431]).

At the end of this section we shall discuss the technological application of permeation cells, namely the use as chemical filters or pumps. If, as mentioned, a relatively

[28] Equation (7.38) is obtained from Eqs. (7.1) and (7.3) via the couplings $\nabla\tilde{\mu}_{OH^-} = \nabla\mu_{H_2O} - \nabla\tilde{\mu}_{H^+}$; $\nabla\tilde{\mu}_{H^+} = 1/2\nabla\mu_{H_2} - \nabla\tilde{\mu}_{e^-}$, $\nabla\tilde{\mu}_{H_3O^+} = \nabla\mu_{H_2O} + \nabla\tilde{\mu}_{H^+}$ and $\nabla\tilde{\mu}_{O^{2-}} = \frac{1}{2}\nabla\mu_{O_2} + 2\nabla\tilde{\mu}_{e^-}$. Let us consider a somewhat simpler case in order to avoid cumbersome expressions and only take e^-, OH^-- and H^+ into account. Equation (7.1) then demands that $\sigma_{H^+}\nabla\tilde{\mu}_{H^+} - \sigma_{e^-}\nabla\tilde{\mu}_{e^-} - \sigma_{OH^-}\nabla\tilde{\mu}_{OH^-} = \mathbf{0}$. Application of the coupling conditions yields: $(\sigma_{e^-} + \sigma_{H^+} + \sigma_{OH^-})\nabla\tilde{\mu}_{e^-} = \frac{\sigma_{OH^-}+\sigma_{H^+}}{2}\nabla\mu_{H_2} - \sigma_{OH^-}\nabla\mu_{H_2O}$. Equation (7.3) leads, for instance, to $E \simeq -\frac{\sigma_{OH^-}+\sigma_{H^+}}{\sigma}\frac{1}{2F}\Delta\mu_{H_2} + \frac{\sigma_{OH^-}}{\sigma}\frac{1}{F}\Delta\mu_{H_2O} = -\frac{\sigma_{H^+}-\sigma_{OH^-}}{\sigma}\frac{1}{2F}\Delta\mu_{H_2O} + \frac{\sigma_{H^+}+\sigma_{OH^-}}{\sigma}\frac{1}{4F}\Delta\mu_{O_2}$. In addition there are also cases in which hydrogen seems to migrate in different effective valence states [434].

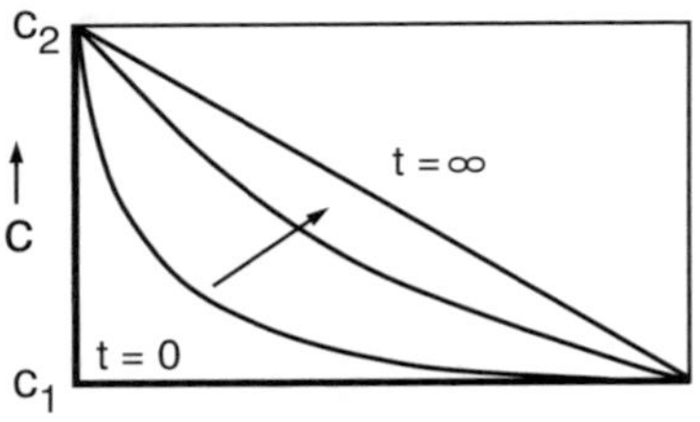

Fig. 7.11: Stoichiometry profiles during diffusion–controlled chemical polarization of a homogeneous initial state (see text). The stationary profile is linear ($\partial c/\partial t = 0 = \partial^2 c/\partial x^2$) corresponding to Fick's law.

reducing situation is created on one side of a mixed conductor, "O" diffuses through the solid[29] to the reducing side in an ambipolar manner. This has been suggested for application in the metallurgical industry for the deoxydation of metals [559]. Direct contact with the reducing agent is not necessary. The selectivity of some oxides in conducting O^{2-} can also be exploited in order to separate O_2 from other gases. This principle is of interest for the preparation of pure hydrogen from the thermolysis of water (separation of the oxygen–hydrogen mixture) or generally for the production of very pure gases, for chemical reaction engineering in ceramic reactors as well as for isotope separation. Relevant materials are, for instance, perovskite–related mixed conductors, such as $Sr(Fe, Co)O_{3-\delta}$ (see e.g. Refs. [560,561] and also Fig. 7.1 (top right)).

7.3 Cells under current load

7.3.1 Electrochemical pumps, conductivity sensors and other applications

In this section we will explore what happens when we force a current to flow through symmetrical or asymmetrical cells (in the last case in the opposite direction to that of short circuit flow), i.e. we consider electrochemical polarization in the generalized sense. (For reasons of presentation charging processes in galvanic elements will be discussed, together with the discharge processes, in the next section.) Conductivity sensors, electrolysis cells, electrochemical reactors and pumps are all important technological applications of cells under a current load.

The last application provides a direct connection to the previous section in which we discussed permeation cells. However, instead of a mixed conductor we use here a solid electrolyte: Ion transport is brought about by application of an external voltage, that is, the electrical potential gradient replaces (or is superimposed on) the chemical potential gradient. Here, too, there are applications in metallurgy, namely the removal from metals of hydrogen, which causes brittleness. Selectively pumping off ("electrochemical filtration") a component from a gas mixture makes efficient enrichment or purification possible. The regulation of gas partial pressures

[29]The same can be achieved by external or internal short–circuiting (admixtures of metal particles) of a solid electrolyte. See the following section for the use of external voltages.

by electrochemical pumping in and out of gases is particularly important in scientific measurement technology.
In the case of electrochemical reactors, specific reactants can be supplied or removed by current flow through the electrolyte. In this manner reactions can be made possible and controlled, or a spatial separation of the reactants can simply be brought about. Figure 7.12 shows how it is possible, in principle, to dehydrogenate saturated

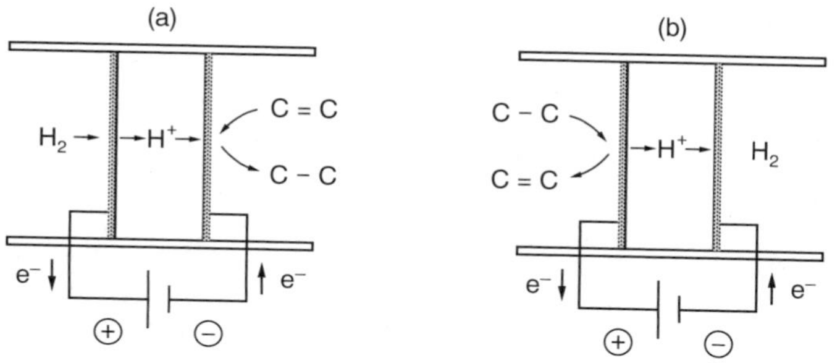

Fig. 7.12: Supply of protons makes it possible to hydrogenate or dehydrogenate hydrocarbons without direct contact with H_2 gas [562]. From Ref. [563].

hydrocarbons or to hydrogenate unsaturated ones using proton conductors. The possibility of ammonia synthesis by this route has been demonstrated [564]. In principle, it is possible to tune yield and selectivity by variation of electrochemical parameters. Additional catalytic effects have been detected in the oxygenation of hydrocarbons [476]. If oxygen is electrochemically pumped through say ZrO_2 (Y_2O_3) into a reaction mixture containing hydrocarbons and oxygen, then, under suitable conditions, very many more oxygen particles react than have been brought in by the current[30].
Electrolysis with solid electrolytes is of potential importance. The generation of hydrogen (and oxygen) by high temperature electrolysis of water might possibly play a significant role in future hydrogen technology. On account of water formation being exothermic the decomposition voltage falls with increasing temperature; kinetic inhibitions are reduced, too. Water–containing β''–Al_2O_3 is a promising proton conducting electrolyte (cf. Section 6.2.1) in this respect.
Electrolysis with the aid of photoelectrodes[31] is a further technologically significant variant (Fig. 7.13). If we irradiate the contact between TiO_2 and a suitable aqueous electrolyte with light of a suitable wavelength, electron–hole pairs are produced and separated by the boundary layer field[32]. Appropriate redox systems can store the energy captured by this process (see Fig. 7.13). In the case of photoelectrolysis of H_2O under ideal conditions, O_2 is evolved at the photoanode as a result of oxydation by the holes, while the electrons pass via the external circuit to the cathode where they generate H_2. In this manner the energy of the light is transformed into chemical energy. The band gap of most oxides (typically 3eV) is too large for visible light to

[30] This "NEMCA effect" has already been discussed in Section 6.8.
[31] Cf., for example, Ref. [565].
[32] Otherwise they would recombine and the irradiation with light would only have a thermal effect.

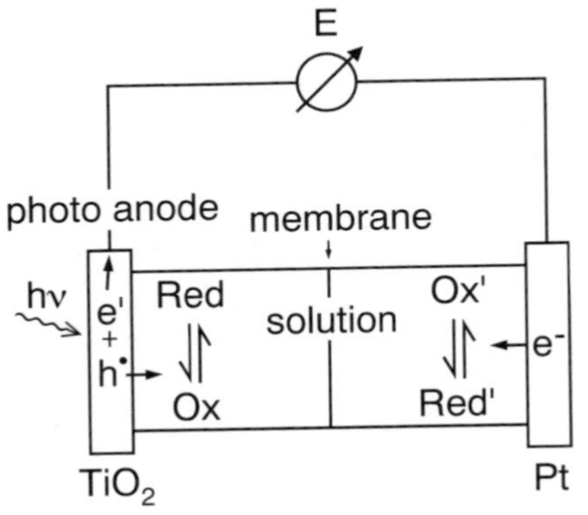

Fig. 7.13: Irradiation with light alters the concentrations of the redox partners, thus, converting solar energy into chemical energy.

come effectively into question. A remedy is to provide an additional electrical auxiliary voltage or to seek more suitable compounds or chemical modifications. Some chalcogenides have favourable band gaps[33]. However, the holes produced frequently lead in such cases to self–decomposition ($S^{2-} + 2h^{\cdot} \rightarrow S$). A variant of great current interest involves small TiO_2 spheres coated with suitable dye adsorbates [567,568], in which electronic excitation takes place with energy level differences better suited to excitation by sunlight. As is so often the case, in this attractive field the lack of suitable materials is the bottleneck.

Bulk and boundary conductivity sensors have already been discussed in Chapter 5 in relation to equilibrium defect chemistry, potentiometric sensors in the previous section. Nevertheless, we wish — in view of the importance of this application — to sketch out some of the fundamentals of electrochemical (composition) sensors[34]. The fact that a variation in the chemical composition (c_k) elicits a physical signal is the rule rather than the exception. This is merely a necessary sensor criterion. In addition, it is important that a sensor signal exhibits adequate sensitivity[35], is sufficiently selective, stable and as free from drift as possible[36], and displays an ad-

[33] A further possibility is to combine different photosystems [566].

[34] Strictly speaking we must distinguish between measurement principles (e.g. thermal, optical, mechanical, electrochemical sensor) and measurement purpose (temperature, pressure, composition sensor). However, in practice, the assignment of names has not been particularly rigorous: The term "thermal sensors" had been applied both to temperature sensors and to sensors that depend on temperature measurement (e.g. pellistors, in which carbohydrates are catalytically oxydized and the measurement effect is detected via heat production), electrochemical sensors are normally electrochemically analysing sensors, while chemical sensors are generally synonymous with analysers of chemical composition.

[35] The sensitivity of the signal $S^{(k)}$ to the concentration of component c_k is given by $\partial S^{(k)}/\partial c_k$, while the selectivity or cross sensitivity is given by $\partial S^{(k)}/\partial c_{k' \neq k}$. See Ref. [569] for a definition with respect to analysis procedures.

[36] The time–dependence of the signal $S^{(k)}(t) = f(c_k(t), t)$ is given by

$$\dot{S}^{(k)} = \frac{\partial S^{(k)}}{\partial c_k}\frac{dc_k}{dt} + \frac{\partial S^{(k)}}{\partial t}.$$

equately short response time (not to mention other criteria, which primarily affect the price). Moreover (except for temperature sensors), the signal should not be too sensitive to temperature. Here we concentrate primarily on the above–mentioned three sensor types (see Table 7.2); these rely on the use of mixed conductors, electron conductors and ion conductors, respectively.

Table 7.2: Solid state sensors for redox–active gases relying on full or partial equilibria[38].

	solid phase	$\nabla\tilde{\mu}_{ion}$	$\nabla\tilde{\mu}_{eon}$
bulk conductivity sensor	mixed conductor	0	0
surface conductivity sensor	electronic conductor	$\neq 0$	0
potentiometric sensor	ionic conductor	0	$\neq 0$

The first type is the bulk conductivity sensor, which is in global and local equilibrium with the gas atmosphere (see Table 7.2). Our prototype example in Section 6.5 was[37] $SrTiO_3$: All driving forces disappear at equilibrium: $\nabla\mu_O = \nabla\widetilde{\mu}_{e^-} = \nabla\widetilde{\mu}_{O^{2-}} = \mathbf{0}$. This type of sensor is very selective on account of the diffusion step involved; in the case of oxides as sensors for oxygen no other gases except hydrogen or water are soluble. (Thermodynamic gas phase interactions of, say, CO with O_2, produce an alteration in O_2 partial pressure and must not be seen as an influence on selectivity.) Response time and selectivity are determined by the $\bar{k}^\delta$ and D^δ values with respect to O_2 and interfering gases (cf. Sections 6.5 and 6.7). In particular, the oxide should be free from redox–active trap centres, since these drastically reduce D^δ_O (and $\bar{k}^\delta_O$) as described in Section 6.6.1. From a thermodynamic point of view, drift effects will not occur with this equilibrium sensor, as long as it is thermodynamically stable with respect to the atmosphere. However, as shown in Chapter 6, such sensors exhibit a high temperature sensitivity and thus may require reference samples which are otherwise not needed according to the principle of operation.

The semiconductor boundary sensor (surface conductivity sensor in Table 7.2) avoids the disadvantage of the high temperatures that were important for the bulk sensors

It is merely an explicit time–dependence that is not desired (see second term); in a case where c_k is time–dependent, $S^{(k)}$ should naturally also be time–dependent. If, for example, $S^{(k)} = \alpha c_k$ then $\partial S^{(k)}/\partial t = 0$ means that the sensitivity is not time–dependent (see footnote 35).

The response time (τ_R) and drift can be meaningfully separated, if a (pseudo)stationary state ("∞"), whose τ_R value varies on a much greater time scale, is reached relatively quickly. Thus, the response time is defined, for example, by $|(S(\tau_R) - S(\text{"}\infty\text{"}))/S(\text{"}\infty\text{"})| = 1\%$ and the drift by $\partial S/\partial t$ when $t \gg \tau_R$ [570].

[37]Siemens is in fact currently (i.e. at the time of writing) testing the use of $SrTiO_3$ as a rapid, cylinder–selective oxygen sensor for automobiles (as an improvement on the ZrO_2 probe, see above) [571].

to make D_O^δ sufficiently high[38] Only the interfacial equilibrium and, in the interior, the electronic equilibrium is established, i.e $\nabla\widetilde{\mu}_{e^-} = \mathbf{0}$, but $\nabla\widetilde{\mu}_{O^{2-}} \neq \mathbf{0} \neq \nabla\mu_O$, and a D_O^δ value as small as possible is desired[38] (see Table 7.2). Here contact with an oxydizing (reducing) gas such as O_2 (or CO) results in a trapping of electrons from the space charge zone in the adsorbate layer (or an injection into the space charge zone[39]) [313]. SnO_2 is a prototype material for this sensing mechanism (cf. Section 5.8). The conductivity behaviour in terms of the sensitivity and the selectivity is determined by the surface chemistry. As the diffusion step necessary in the bulk sensor is suppressed[38], the sensor can function at lower temperatures, but also leads to the disadvantage of lower selectivity. This can be partly compensated, by including temperature as a parameter, or by using whole arrays of oxide semiconductors, whose response pattern after calibration serves as a fingerprint in a quantitative sense. Figure 7.14 shows an example. Drift effects are expected to occur[38] when

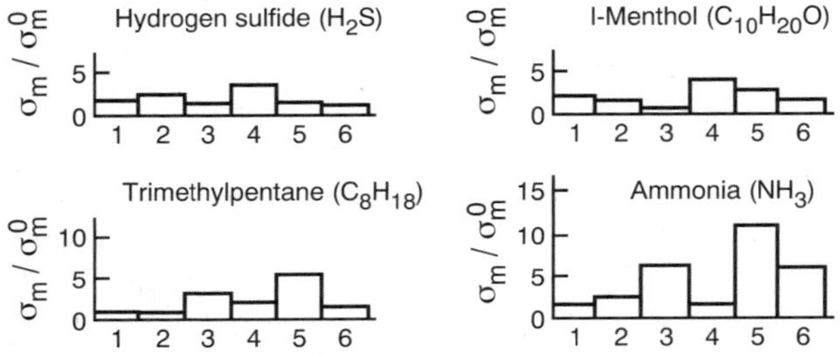

Fig. 7.14: Sensor array for selective gas sensing using boundary layer conductivity effects (Taguchi sensors). According to Ref. [541].

D_O^δ is not negligible [446]. In complete contrast to the bulk conductivity sensors, redoxactive impurities may now be of considerable advantage (see Section 6.6.1). The third type of sensor, in which a pure ion conductor is used, is the EMF sensor (potentiometric sensor), as represented by the λ–probe, dealt with in detail in Section 7.2. Again, only local equilibrium is set up there (see Table 7.2), thus, in the oxide $\nabla\mu_O \neq \mathbf{0}$, but now conversely $\nabla\widetilde{\mu}_{O^{2-}} = \mathbf{0} \neq \nabla\widetilde{\mu}_{e^-}$ (see previous section). Here there are problems if the electrons have an appreciable conductivity which may result in a reduced voltage signal (see Eq. (7.36)) and under certain conditions in drift effects (see Eq. (7.39)). The sensitivity of such cells can be very high, the temperature sensitivity is generally low ($E \propto T$ if the reference side is T–independent see Eq. (7.10)) and the selectivity is often considerable. Even though response is possible to parasitic processes on account of additional exchange equilibria[40] the process involving the highest exchange current density is generally dominant (cf. footnotes 20, 59). Selectivity can be increased by ampèrometric procedures in which signifi-

[38]Recent experiments on SnO_2 [437] show that the picture is more complicated: For small particles, the portion of the interfacial reaction succeeding the sensor action is rate determining, and not the diffusion step. (Then the statement $\nabla\widetilde{\mu}_{e^-} = 0$ may be violated.)

[39]Under usual conditions the effect of reducing gases on SnO_2 is an annihilation of oxydizing surface groups and thus a decrease of electron trapping.

[40]Thus, concentration cells with Nasicon can act as moisture sensors, and ones with window glass as oxygen sensors (cf. page 409).

cant currents are drawn. This makes processes with large polarization resistances less important. Generally, qualitative and quantitative detection is enabled by the kinetic behaviour in such procedures. Usually the concentration is determined by diffusion–limiting currents[41], as is the case in polarography [220].

All three types of sensor can be transformed for the purpose of detecting acid–base active gases such as H_2O, NH_3, CO_2, etc. [548,572] which primarily (i.e. with increased selectivity) react with ions ($H_2O + H^+ \rightleftharpoons H_3O^+$; $NH_3 + H^+ \rightleftharpoons NH_4^+$; $CO_2 + O^{2-} \rightleftharpoons CO_3^{2-}$).

In the case of the potentiometric sensor this means the construction of a chain, whose cell reaction does not involve valence changes. This principle led to the CO_2 sensors presented in Section 7.2 with open reference electrodes and zero sensitivity to oxygen.

An acid–base bulk conductivity sensor implies the dissolution of "complex" gases and, hence, the diffusion of two types of ions or of the neutral species itself. In principle, hydrates or water–dissolving perovskites come into question for the detection of water (cf. Chapter 5) and ammoniates for the detection of NH_3. On account of the kinetic difficulties (cf., for instance, $D^{\delta}_{H_2O}$ in Section 6.5) here it is necessary to use very thin films or high temperatures. On the other hand, high temperatures favour desorption and, hence, reduce sensitivity.

These problems can largely be solved by constructing an analogue to a boundary layer conductivity sensor. If we bring an acid–base active gas into contact with a suitable ion conductor (e.g. a Brønsted acid–base active gas with a proton conductor), the processes of interaction can be detected via the variation in surface conductivity. We have already discussed an instructive example, namely the detection of NH_3 by means of AgCl representing an acid–base interaction in a generalized sense (see Section 5.8).

Finally it should be mentioned that sensor research is aiming at producing high–performance "artificial noses" (and "eyes" or "ears") capable of learning using pattern recognition techniques. In this spirit the reader may consider the very last illustration in this book (Fig. 7.67).

In some sense the actuators, which generate action on command (i.e. information) form the counterpart to sensors which collect information about the surroundings. Actuators which change the composition have been implicitly addressed by the text (cf. e.g. electrochemical pumps). Actuators in a narrower sense cause mechanical effects (e.g. transfer of electrical into mechanical energy) and rely especially on the phenomenon of electrostriction (cf. piezoelectricity). For details the reader is referred to Ref. [573]. Also in this context perovskites play a dominant role (in particular perovskites based on lead titanate zirconates), and defect chemistry, even though not decisive, is of considerable importance (cf. Section 2.2.7).

---

[41] Diffusion current $\propto (c_{(gas)} - c_{surface})/\Delta x$. For hydrodynamic reasons Δx can be approximated by a finite effective boundary thickness. At high voltages $c_{surface}$ approaches zero and the signal is proportional to $c_{(gas)}$ (see also page 484).

7.3.2 Measurement cells

Let us now turn to the fundamental determination of electrochemical rate constants and, in particular, of the transport parameters[42]. For this purpose we shall consider the general case of a mixed conducting sample. For clarity, we shall take silver sulfide as the central material — the "electrolyte" as it were. The ion conduction of this mixed conductor is carried out by silver ions. The various methodologies can be classified according to the type of electrode used. If the mobile ions under consideration are silver ions, metallic silver is the candidate of choice for the reversible electrode, that is an electrode that allows passage of both ions and electrons, for which we use the notation[43] $(Ag^+, e^-\rangle$. An inert metal only letting through electrons but blocking silver ions at the phase boundary would be Pt (graphite would also be suitable at lower temperatures). Generally, such an "electronic electrode" $((e^-\rangle)$ only permits electrons to penetrate and blocks the ions selectively at the phase boundary. The arrangement $Ag\,|\,\alpha\text{–}AgI$ acts as a selective blocker for electrons and, hence, as an "ionic electrode" $((Ag^+\rangle)$. (The phase α–AgI is an appreciably better ionic conductor but an appreciably worse electronic conductor than Ag_2S.) For the cation–anion conductor MX combination of these possibilities yields the following six types [574]:

$$\text{Cell R} = \quad \ominus\,(M^+, e^-\rangle\,|MX|\,\langle M^+, e^-)\,\oplus$$

$$\text{Cell E1} = \quad \ominus\,(M^+, e^-\rangle\,|MX|\,\langle e^-)\,\oplus$$

$$\text{Cell E2} = \quad \ominus\,(e^-\rangle\,|MX|\,\langle e^-)\,\oplus$$

$$\text{Cell I1} = \quad \ominus\,(M^+\rangle\,|MX|\,\langle M^+, e^-)\,\oplus$$

$$\text{Cell I2} = \quad \ominus\,(M^+\rangle\,|MX|\,\langle M^+)\,\oplus$$

$$\text{Cell T} = \quad \ominus\,(e^-\rangle\,|MX|\,\langle M^+)\,\oplus$$

Specifically, for Ag_2S this may be

$$\text{Cell R} = \quad \ominus\,Ag|\,Ag_2S\,|\,Ag\oplus$$

$$\text{Cell E1} = \quad \ominus\,Ag|\,Ag_2S\,|\,Pt\oplus$$

$$\text{Cell E2} = \quad \ominus\,Pt|\,Ag_2S\,|\,Pt\oplus$$

---

[42] The presentation partially follows Refs. [391,393,431].

[43] The pointed brackets point towards the specified phase boundary.

Cell I1 = $\ominus$ Ag |AgI| Ag_2S | Ag$\oplus$

Cell I2 = $\ominus$ Ag |AgI| Ag_2S |AgI| Ag$\oplus$

Cell T = $\ominus$ Pt | Ag_2S |AgI| Ag $\oplus$.

Let us consider, as the second case, an oxide M_2O which conducts oxygen ions and electrons. There are several variants compared to the previous case. In principle, the parent metal M could serve as a reversible electrode if it is in phase equilibrium with M_2O. The electrode must then be protected from atmospheric oxygen. It is simpler to employ porous gas electrodes, that is Pt, O_2. This has the advantage that the chemical component potential can be tuned very simply. Conversely, if we wish to block oxygen conduction, we must avoid any gas exchange. In principle, this can be achieved by using sealed, inert electrodes with negligible oxygen solubility (for simplicity we shall select graphite (C) as our example); the free surfaces, however, should be protected from the atmosphere — preferably with suitable glass films[44]. (As mentioned below, it is also possible to use a reversible cathode, e.g. porous platinum if the O_2 gas chamber is sealed and exhaustible with respect to oxygen, i.e. small enough. However, the time dependence is complex here.) At sufficiently high temperatures the assembly Pt, $O_2|ZrO_2(Y_2O_3)$ is a suitable reversible ionic electrode[45].

Overall, available analogues are

Cell R = $\oplus$ Pt, O_2 |M_2O| O_2, Pt$\ominus$

Cell E1 = $\oplus$ Pt, O_2 |M_2O| C$\ominus$

Cell E2 = $\oplus$ C |M_2O| C$\ominus$

Cell I1 = $\oplus$ Pt, O_2 |$ZrO_2(Y_2O_3)$| M_2O | O_2, Pt$\ominus$

Cell I2 = $\oplus$ Pt, O_2 |$ZrO_2(Y_2O_3)$| M_2O |$ZrO_2(Y_2O_3)$| O_2, Pt$\ominus$

Cell T = $\oplus$ C |M_2O| $ZrO_2(Y_2O_3)$ | O_2, Pt$\ominus$,

whose modes of action will be discussed in order. For simplicity we wish to assume that we impose a constant external current and follow the voltage response with time. If not otherwise stated, the voltage is measured at the current–carrying electrodes using a high resistance instrument. It is, however, frequently advisable to use a multipoint arrangement (see Section 7.3.7).

[44] Ion blockage succeeds most elegantly if the gas exchange at the interface is kinetically inhibited. So long as the electrodes do not catalyse gas exchange, the arrangement corresponds to an automatic ion blockage (cf. Section 5.6, as well as Fig. 7.30, page 453).

[45] In order to avoid the inverse effect, namely that the electron conductor blocks the ion conduction in YSZ, it must be ensured that $\sigma_{eon}(YSZ) \ll \sigma_{ion}(M_2O)$ [575].

7.3.3 Bulk and phase boundary effects

a) Heuristic interpretation — equivalent circuits

First let us consider cell R of the oxide cells (page 425):
In the stationary state ($t = \infty$) the voltage is determined by the total resistance[46] $R = U(t{=}\infty)/I$ if interfacial effects are neglected. Evaluation of the electrode area (a) and the distance between the electrodes (L) yields (Section 6.3)

$$\sigma = \sigma_{ion} + \sigma_{eon} = \Sigma_k \sigma_k = \frac{L}{a} R^{-1}. \tag{7.40}$$

The contribution of ion conduction causes mass displacement[47]. In the case of the Pt, O_2 electrodes this can be monitored by analysis of the amount of oxygen addition and removal necessary to maintain the initial partial pressure. If M_2O is a cation conductor, a positional displacement of the oxide occurs, which can be analysed, in principle: With the polarity given, M^+ migrates from left to right, the oxide shrinks on the left and grows on the right; hence, the oxide mass migrates from left to right. In the case of a silver ion conductor sandwiched between two Ag electrodes with the polarity given in cell R on page 424, a growth of the left hand silver electrode and shrinkage of the right hand one will occur. A pure electron conductor does not lead to any mass changes at all. Suitably arranged experiments, which are named after Hittorf and Tubandt, can give quantitative information concerning cationic, anionic and electronic transference numbers; however, properly done experimental analyses are rare [576]. Let us return again to the general aspects and discuss the behaviour of the cells under galvanostatic polarization conditions (i.e. at $t{=}0$ a constant current I is suddenly switched on and is maintained constant)[48].
At the start of our d.c. experiment, i.e. immediately after switching on the current, in addition to the pure conduction current I_R there is also a capacitive displacement current flow (I_C), which is responsible for the time dependence. For small stimuli it follows that ($I_C = (\partial/\partial t)\, Q_C$; Q_C=capacitor charge)[7]

$$I_C = C\dot{U}. \tag{7.41}$$

The specific (dielectric) permeability responsible for the capacitance[7,8,46]

$$\varepsilon \equiv \varepsilon_r \varepsilon_0 = \frac{L}{a} C, \tag{7.42}$$

---

[46] In what follows we shall limit ourselves to small excitations, so that we do not need to distinguish between integral and differential resistance and capacitance elements. In more general cases we have to use $\bar{R}$ or $\bar{C}$.

[47] These are readily calculated from Faraday's law $n_k = \frac{Q_k}{z_k F} = \frac{m_k}{M_k}$ for every charge carrier k, whereby $\dot{Q}_k = I_k = U/R_k$ (Q: charge, m: mass, M: mole mass).

[48] It will be shown in Section 7.3.6 that, for small excitations, the response to any signal (as here the step function for current) contains the complete information.

is also made up of ionic and electronic contributions. While the capacitance measures the ability to store charge at a given voltage and depends on the geometry of the electrode arrangement, the dielectric number ε is an intensive quantity and determined by the atomistic polarizabilities according to the Clausius–Mosotti equation[49]. Equations (7.41, 7.42) can be obtained straight from the Poisson equation (5.219): Let us assume that there is internal electroneutrality (ρ=0), so the electrical potential is linear. The electrical field then is $-\Delta\phi/\mathrm{L}$. Since, at the boundary, the field changes from the internal value to the value zero in the metal, and since this change corresponds to the charge stored (more precisely to $\mathrm{Q}/(\varepsilon \mathrm{a})$, see Section 5.8), Eqs. (7.41, 7.42) follow.

As the electrical current can be carried by conduction *or* displacement current, the circuit elements, electrical resistance and capacitance are connected in parallel in the equivalent circuit. Since now the voltages across the resistance ($\mathrm{U} = \mathrm{I_R R}$) and capacitance (integral of $\dot{\mathrm{U}}_\mathrm{C} = \mathrm{I_C/C}$) are equal, and the contributing currents add to give a constant total current ($\mathrm{I} = (\mathrm{I_R} + \mathrm{I_C})$), this yields the differential equation

$$\dot{\mathrm{U}} = \frac{\mathrm{I_C}}{\mathrm{C}} = (\mathrm{I} - \mathrm{I_R})\,\frac{1}{\mathrm{C}} = \frac{\mathrm{IR} - \mathrm{U}}{\mathrm{RC}}. \tag{7.43}$$

By substituting $\mathrm{V} \equiv \mathrm{U} - \mathrm{IR}$ (i.e. $\dot{\mathrm{V}} = \dot{\mathrm{U}}$) and using the abbreviation $\tau = \mathrm{RC}$ we obtain a homogeneous differential equation ($\dot{\mathrm{V}} + \mathrm{V}/\tau = 0$) with the solution $\mathrm{V} = \mathrm{V(t{=}0)} \exp -\mathrm{t}/\tau$. If we reverse the substitution and take account of the fact that immediately on switching on the current the capacitor does not provide any resistance to current flow, i.e. $\mathrm{U(t{=}0)} \equiv \mathrm{V(t{=}0)} + \mathrm{IR} = 0$, then we obtain a monotonic asymptotic voltage behaviour rising to the stationary value IR according to

$$\mathrm{U} = \mathrm{IR}\,(1 - \exp -\mathrm{t}/\tau) = \mathrm{U_C}. \tag{7.44}$$

For the partial currents it follows that

$$\begin{aligned} \mathrm{I_R} &= \mathrm{I}(1 - \exp -\mathrm{t}/\tau), \\ \mathrm{I_C} &= \mathrm{I} \exp -\mathrm{t}/\tau. \end{aligned} \tag{7.45}$$

Thus, initially the whole current is a displacement current. As time passes the conduction current increases until it is representing the total current when the stationary state is reached. The capacitor is then completely impermeable and $\mathrm{I_C} = 0$. The breakdown of the conduction current into electronic and ionic current (the resistances $\mathrm{R_{eon}}$ and $\mathrm{R_{ion}}$ are in parallel, i.e. $\sigma = \sigma_\mathrm{eon} + \sigma_\mathrm{ion}$ or $\mathrm{R}^{-1} = \mathrm{R}_\mathrm{eon}^{-1} + \mathrm{R}_\mathrm{ion}^{-1}$) leads to (t: transference number)

$$\begin{aligned} \mathrm{I_{eon}} &= \frac{\mathrm{R_{ion}}}{\mathrm{R_{eon}} + \mathrm{R_{ion}}} \mathrm{I_R} = \frac{\sigma_\mathrm{eon}}{\sigma} \mathrm{I_R} \equiv \mathrm{t_{eon} I_R} \\ \mathrm{I_{ion}} &= \frac{\mathrm{R_{eon}}}{\mathrm{R_{eon}} + \mathrm{R_{ion}}} \mathrm{I_R} = \frac{\sigma_\mathrm{ion}}{\sigma} \mathrm{I_R} \equiv \mathrm{t_{ion} I_R} \end{aligned} \tag{7.46}$$

[49] See textbooks of physics and physical chemistry, e.g. Ref. [577]; cf. also page 110.

since $U = I_{eon}R_{eon} = I_{ion}R_{ion} = (I_R - I_{eon})R_{ion}$. It is not possible to separate σ_{eon} and σ_{ion} in such conductivity experiments with reversible electrodes.
The behaviour on switching off the current can be treated in a similar manner[50].

The situation is more complex if the interfaces that are necessarily present, such as the contacts between the sample and the electrodes, effect significant contributions to the resistance and the capacitance. Such an interfacial resistance (accompanied by a parallel capacitance) may be caused by a slow ionic or electronic charge transfer[51], but also by a sluggish charge carrier transport in the adjacent space charge zone (cf. Section 5.8). Generally speaking, it is necessary to assign a resistance to every elementary step (cf. oxygen diffusion, adsorption, dissociation, ionization, charge transfer), as already discussed in Chapter 6, for the case of zero external field. Internal interfaces, generally the grain boundaries, also influence current transport, by inhibiting transfer through the interfacial core and/or through the space charge zones either side of the core. (Grain boundaries can also provide highly conducting pathways that short–circuit the bulk. In this latter case the grain boundary contributions are connected in parallel to the bulk resistance. These effects will be discussed later.) In addition, we shall, so to speak, concentrate on the phenomenology and leave it open at first whether the second R–C–circuit refers to electrode contact, or grain–grain contact, and also whether it refers to boundary core or space charge (see page 438). We then get an approximation represented by the equivalent circuit illustrated in Fig. 7.15, in which bulk impedance and boundary layer impe-

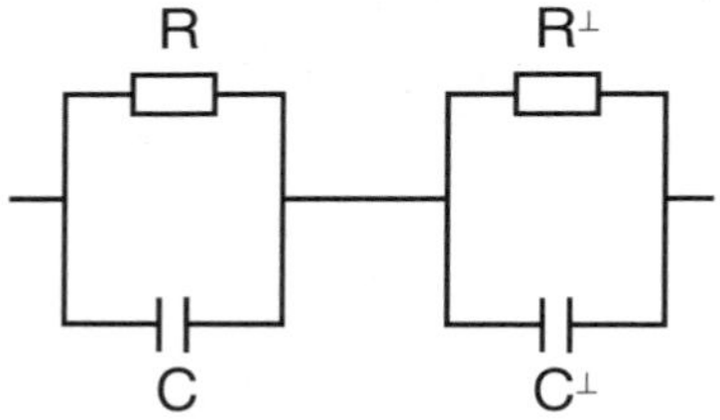

Fig. 7.15: Equivalent circuit for a sample with phase boundary impedance. The index $\perp$ shows that the charge transport takes place across the interface. It is assumed that the same charge carrier is responsible for R and $R^\perp$.

dance (comprising $R^\perp$ and $C^\perp$) are connected in series[52]. The interfacial capacitance takes account of the transient effects (change in charge as a result of variation of the electrical potential). For the purpose of clarity we shall denote the following bulk parameters by the subscript ∞. Since the voltage drops U_∞ and $U^\perp$ are additive, but the total current constant, the voltage response is readily obtained from two isomorphic partial problems. The result for the switching–on process is

$$U(t) = U^\perp(t) + U_\infty(t) = IR_\infty\left(1 - \exp -t/\tau_\infty\right) + IR^\perp\left(1 - \exp -t/\tau^\perp\right) \qquad (7.47)$$

[50] In a similar manner with I_P= polarization current ($I = I_{eon} + I_{ion} + I_C = 0$) we get $I_{eon} = t_{eon}I_P \exp -t/\tau$, $I_{ion} = t_{ion}I_P \exp -t/\tau$, $I_C = -I_P \exp -t/\tau$ and $U = I_P R \exp -t/\tau = U_C$.

[51] I.e. the electrodes are not ideally nonpolarizable.

[52] On account of the occurrence of ionic and electronic conduction phenomena, Fig. 7.15, in the primitive form given, applies precisely only if a single majority charge carrier determines both R and $R^\perp$.

with $\tau^\perp \equiv R^\perp C^\perp$ as relaxation time for the boundary process. These considerations refer to the linear range, that is the range of constant C and R values, and presuppose sufficiently small excitations.
The associated capacitance $C^\perp$ (core thickness typically 1 nm, space charge thickness typically a few nm up to 100nm (see page 438)) is now much greater than C_∞ (typical sample thickness 0.1mm–10mm). Thus, if $R^\perp$ and R_∞ are of a similar order of magnitude or even if $R^\perp > R_\infty$ (otherwise $R^\perp$ is not of interest in conductivity measurements) the corresponding time constants are very different:

$$\tau^\perp \gg \tau_\infty. \tag{7.48}$$

In polycrystalline materials there are many grain boundaries in series, i.e. the effective capacitance will then be much smaller than that of each individual contact, but still larger than that of the bulk. (The time constant remains invariant.) This ranking in the respective capacitances ($C_\infty \ll C^\perp_{\text{electrode}} \ll C^\perp_{\text{grain boundaries}}$) often makes an initial identification of the observed effect possible. Later we shall go into the possibilities for differentiating between electrode and grain boundary contacts by means of multipoint methods (Section 7.3.7). The reader may note at this stage that varying the sample thickness provides a further simple discrimination method, since the electrode contribution (as the bulk contribution) remains constant whilst the grain boundary contributions vary proportionally.
If the time constants of the bulk and boundary layer differ sufficiently, the phase boundary capacitor in Fig. 7.15 is still completely permeable at very short times $t \sim \tau_\infty \ll \tau^\perp$, so that the result is Eq. (7.44). At longer times ($t \sim \tau^\perp \gg \tau_\infty$) C_∞ has now become completely impermeable and the equivalent circuit reduces to R_∞ in series with the $R^\perp C^\perp$ parallel circuit. In other words

$$U\left(t \sim \tau^\perp \ll \tau_\infty\right) = IR_\infty + IR^\perp\left(1 - \exp -t/\tau^\perp\right). \tag{7.49}$$

remains. In this time resolution the bulk process appears simply as a jump ("IR drop") while the second charging process can be observed. After a very long period ($t \gg \tau^\perp \gg \tau_\infty$) all exponential functions become negligible, i.e. all capacitors are blocking and the final value is $I(R^\perp + R_\infty)$. It is, thus, possible to separate bulk and interfacial resistances using the initial and final values. Analysis of the associated time dependences yields the capacitances.
Figure 7.16 shows the results of an oscillographic measurement on PbO in a time range in which the interfacial process can be observed and the bulk process appears as a jump. The mechanistic interpretation of the bulk resistance was largely the subject of the previous chapters. The source of bulk capacitance has been discussed briefly above. We now wish to address interfacial layer parameters in more detail. First we neglect the influence of electrical capacitances by referring to the steady state and concentrate on the resistive parameters. Similarly, when we will discuss capacitive effects later on (Section 7.3.3) we will ignore Faradaic effects. The combination of both will, in a linear response approximation, be performed by a parallel

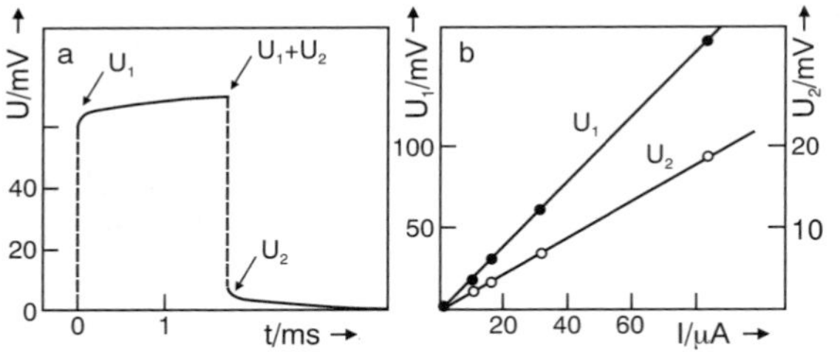

Fig. 7.16: Oscilloscopic observation of the voltage change during switching–on measurements at a PbO crystal ($U_1 = U_\infty, U_2 = U^\perp$) [578].

circuit. We also concentrate on the electrode interface, as grain boundary effects have been considered in detail in previous chapters (Sections 5.8 and 6.6).

b) Mechanistic interpretation — electrode kinetics

A mechanistic interpretation of the interfacial resistances and capacitances measured is as complex as the local electrochemical kinetics. In addition to the interfacial kinetics presented in Section 6.7, the voltage dependence enters the discussion explicitly. A systematic discussion cannot be presented here, for reasons of space[53], and only the basic principles of the treatment are considered highlighting important points. Since we refer to the steady state, we do not only neglect electrical capacitance but also (electro–) chemical storage phenomena (cf. Section 6.7). Let us select an ion conducting oxide (e.g. CaO–doped ZrO_2 or CeO_2) as our solid electrolyte and follow the presentation given in Ref. [579]. As we need, in addition to the gas phase and the oxide, an electrode phase, the electrochemical system under consideration is morphologically and mechanistically complex (cf. "three–phase contact"). In a simplified description we imagine this three–phase surface region as a homogeneous zone and assume all surface transport steps to the zone to be sufficiently fast. Similarly the lateral distribution of the incorporated oxygen is assumed to be fast. Then we can approximately ignore the surface inhomogeneities. We also neglect space charge effects due to the high carrier density[54].
The important elementary steps are assumed[55] to be dissociative oxygen sorption (on the electrode surface)

$$\text{Reaction S} = \qquad \frac{1}{2}O_2(g) + V_{ad} \underset{\overleftarrow{k}_s}{\overset{\overrightarrow{k}_s}{\rightleftharpoons}} O_{ad}, \tag{7.50}$$

and the introduction into the electrolyte phase. The assumption of neutral adsorbed oxygen atoms is artificial (real charge between zero and −2, cf. Section 6.7). Its

[53] Cf. Refs. [579,580,384].

[54] When doping is 10% the calculated Debye length is less than the nearest neighbour distance. However, the values may be larger because of nonidealities. In the Mott–Schottky case (see Section 5.8: constant dopant concentration, depletion of $V_O^{\cdot\cdot}$) the thickness of the zone increases with increasing space charge potential.

[55] This is of course an oversimplified model mechanism. The charge of oxygen species at the surface is certainly not zero (see [469]). See also Section 6.7.

incorporation into the solid electrolyte with simultaneous uptake of electrons from the electrode metal is taken to be the most difficult, and hence the rate–determining step:

$$\text{Reaction T} = \qquad O_{ad} + 2e' + V_O^{\cdot\cdot} \underset{\overleftarrow{k}_T}{\overset{\overrightarrow{k}_T}{\rightleftharpoons}} O_O^x + V_{ad}. \tag{7.51}$$

It is then possible to formulate an equilibrium reaction for the rapid sorption process $\left(K_s = \frac{\overrightarrow{k}_s}{\overleftarrow{k}_s}\right)$, from which it follows that the degree of coverage[56,57] is given by

$$\frac{\widehat{\Theta}}{1-\widehat{\Theta}} = K_s P^{1/2} \quad \text{i.e.} \quad \widehat{\Theta} = \frac{K_s P^{1/2}}{1+K_s P^{1/2}}. \tag{7.52}$$

The current density is proportional to the reaction rate, here the transfer rate. The application of the usual rate laws to Eq. (7.51) (cf. Section 6.1.2) yields

$$\begin{aligned} i &= |\overleftarrow{i}| - |\overrightarrow{i}| \\ &= zF(1-\widehat{\Theta})\exp\left(\overleftarrow{\alpha}\frac{zF\Delta\phi}{RT}\right)\overleftarrow{k}_T(T) - zF\widehat{\Theta}\exp\left(-\overrightarrow{\alpha}\frac{zF\Delta\phi}{RT}\right)\overrightarrow{k}_T(T). \end{aligned} \tag{7.53}$$

The electron concentration (constant in Pt) and the $V_O^{\cdot\cdot}$ concentration (constant in the oxide because of doping) are included in the k values as invariants. In Eq. (7.53) the rate constants $\widetilde{k}_T$ for the forward and back reaction have been separated into a chemical component k_T (see Section 6.1)

$$k_T = k_0 \exp -\frac{\Delta G_T^{\neq}}{RT} \tag{7.54}$$

and a field component. The latter component is given by $\exp -\frac{\overrightarrow{\alpha} zF\Delta\phi}{2RT}$ and $\exp +\frac{\overleftarrow{\alpha} zF\Delta\phi}{2RT}$, respectively for the forward and backward reactions (see Section 6.1). The corresponding free enthalpy change was already given in Fig. 6.2, page 277. Because of the approximately linear variation of the potential change $\Delta\phi$ along the reaction coordinate, the chemical activation threshold of the partial reaction is increased or decreased by half of the corresponding electrical energy effect ($\overrightarrow{\alpha} = \overleftarrow{\alpha} = 1/2$) depending on the direction, provided the maximum lies in the centre of the potential profile. Deviations from this, which are likely because of the asymmetry between educts and products, are taken into account in the electrochemical literature by symmetry factors $\overrightarrow{\alpha} = 1 - \overleftarrow{\alpha}$ that can deviate from 1/2 and lie between 0 and 1.

[56] Note that K_s includes the dissociation constant. The mass action constant of the fast surface redistribution has been set to unity. For smaller coverages it can also be incorporated into the rate constants of the transfer reaction without problem.

[57] This corresponds to the Langmuir adsorption (see Section 6.7).

The potential jump $\Delta\phi$ over the reaction coordinate (and approximately also over the distance Δx) can be separated into the contact equilibrium component also present in the state of zero bias[58] and the component due to the external voltage (transfer overvoltage η_T); so that finally we get

$$|\overleftarrow{i}| = |\overleftarrow{i}_0| \exp\left(\overleftarrow{\alpha}\frac{zF\eta_T}{RT}\right),$$
$$|\overrightarrow{i}| = |\overrightarrow{i}_0| \exp\left(-\overrightarrow{\alpha}\frac{zF\eta_T}{RT}\right). \tag{7.55}$$

The exchange current densities $\overleftarrow{i}_0, \overrightarrow{i}_0$ (cf. Section 6.7) contain the concentrations as well as the equilibrium potential terms. In addition, their absolute values must be equal since the partial current densities compensate each other for $\eta_T = 0$ (i=0). It follows, therefore,

$$|\overleftarrow{i}_0| = zF(1-\widehat{\Theta}) \exp\left(\overleftarrow{\alpha}\frac{zF\Delta\phi(i=0)}{RT}\right)\overleftarrow{k}_T =$$
$$|\overrightarrow{i}_0| = zF\widehat{\Theta} \exp\left(-\overrightarrow{\alpha}\frac{zF\Delta\phi(i=0)}{RT}\right)\overrightarrow{k}_T \equiv i_0. \tag{7.56}$$

Multiplication of the two expressions for $|\overleftarrow{i}_0|^{\overrightarrow{\alpha}}$ and $|\overrightarrow{i}_0|^{\overleftarrow{\alpha}}$ in Eq. (7.56) yields a symmetrical equation of the form

$$i_0 = zF\widehat{c}_{eff}k_{Teff} \tag{7.57}$$

that is independent of $\Delta\phi(i=0)$. The mean values $\widehat{c}_{eff}$, $k_{Teff}(\equiv \widehat{k}_{Teff})$ are defined as

$$\widehat{c}_{eff} \equiv \widehat{\Theta}^{\overleftarrow{\alpha}}(1-\widehat{\Theta})^{\overrightarrow{\alpha}} \simeq \sqrt{\widehat{\Theta}(1-\widehat{\Theta})},$$
$$k_{Teff} \equiv \overrightarrow{k}_T^{\overleftarrow{\alpha}}\overleftarrow{k}_T^{\overrightarrow{\alpha}} \simeq \sqrt{\overrightarrow{k}_T\overleftarrow{k}_T}. \tag{7.58}$$

The right hand side approximations in Eq. (7.58) which represent simple geometrical means, are obtained when $\overrightarrow{\alpha} \simeq \overleftarrow{\alpha} \simeq \frac{1}{2}$.

The reader may recognize the isomorphy of exchange current density (Eq. (7.57)) and specific conductivity $\sigma = zFc \cdot u$ in which the mobility is proportional to the hopping reaction constants (Chapter 6).

The total current thus takes on the well–known Butler–Volmer form[59] [552]

$$i/i_0 = I/I_0 = \exp\left(\overleftarrow{\alpha}\frac{zF\eta_T}{RT}\right) - \exp\left(-\overrightarrow{\alpha}\frac{zF\eta_T}{RT}\right) \tag{7.59}$$

---

58 In the semiconductor literature one speaks of the "built–in potential". If we use the notation employed in Section 6.7.3, then the exponential term in Eq. (7.55) results from the ratios $\widetilde{k}/\widehat{\widetilde{k}}$ for forward and back reactions.

59 Several potential–forming processes occur together in many cases [581]. Let E_1 be the potential at which the current disappears when process 1 takes place alone, that is $E_1 \equiv \Delta\phi(i_1=0)$, and analogously $E_2 \equiv \Delta\phi(i_2=0)$. Then, the total current is not zero at either E_1 or E_2 but at the "mixed potential" E_R. In this case the electrode is not in electrochemical equilibrium on

which we already encountered in Section 6.7 in a different notation (Eq. (6.161)). If the symmetry factor α deviates from 1/2, the current–voltage curve obtained is not symmetric about the origin[60]. Figure 7.17 shows this dependence as well as the

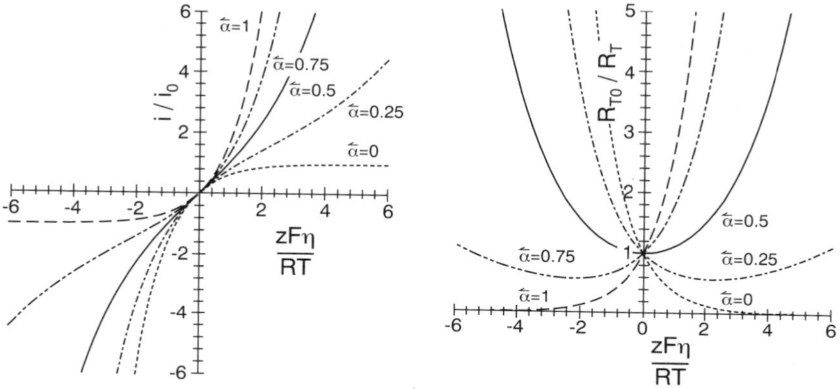

Fig. 7.17: Left: Current–voltage curve for various values of α. When $\overrightarrow{\alpha} = \overleftarrow{\alpha} = 1/2$ the sinh function (symmetrical about the origin) is obtained. Right: The differential transfer conductance (normalized with respect to $\frac{I_0 zF}{RT}$) for various values of α as a function of overvoltage. When $\overrightarrow{\alpha} = \overleftarrow{\alpha} = 1/2$ the result is a cosh function symmetrical about the ordinate axis.

differential conductance $R_T^{-1}(\eta) = \frac{\partial I}{\partial \eta}$. In the case of small overvoltages $\eta_T \ll \frac{2RT}{zF}$, an ohmic relationship results with the transfer resistance

$$R_T = \frac{RT}{zFI_0} = \frac{RT}{zFi_0 a} = \varrho_T/a. \tag{7.60}$$

The parameter ϱ_T is the surface–related transfer resistance and corresponds to the reciprocal of an area–related transfer conductance (see Eq. (7.57)).
A different representation makes use of the definition of effective permeabilities Λ^Q or effective rate constants $\bar{k}^Q$ (cf. general treatment in Section 6.7). Whereas the Λ^Q values, which are proportional to the exchange current densities, denote the ratio

account of the internal processes. If the current–voltage curves do not influence each other then $i(\Delta\phi) = i(\Delta\phi_1) + i(\Delta\phi_2)$ as a result of superposition of both (parallel connection of both processes, $\delta[e^-] = \delta[e^-]_1 + \delta[e^-]_2$). It is evident that the "most dynamic process" (i.e. the process with the least contact resistance, i.e. highest i_0) prevails. If $i_{01} \gg i_{02}$, the first curve in the plot i versus $\Delta\phi$ is much steeper and comes very close to the curve for the sum. This is of great relevance with respect to the selectivity of potentiometric sensors, and also for the treatment of local corrosion phenomena.

[60]The shape of the diode characteristic [219,252] (cf. p–n transition, Schottky contact, Section 5.8) follows for $\overrightarrow{\alpha} = 0$.

of flux density and the deviation of the local potential from the equilibrium value (overvoltage), the $\bar{k}^Q$ values denote the ratio of flux density and the deviation of the local concentration from its equilibrium value. The connection with i_0 is given by

$$\bar{k}^Q = \frac{i_0}{zF[O_O]}. \tag{7.61}$$

These definitions are particularly helpful in the case of complex processes. The parameter i_0 then refers to the rate determining step. At large overvoltages the current–voltage relationship described by Eq. (7.59) departs from linearity and takes on what is known as Tafel behaviour:
η is then linear with respect to the logarithm of the current. For if $|\eta_T| \gg \frac{2RT}{zF}$, then either the first or the second exponential function is negligible in Eq. (7.59). In both cases when $\vec{\alpha} = \overleftarrow{\alpha} = \frac{1}{2}$, it is found that

$$|I/I_0| = \exp\left|\frac{zF\eta_T}{2RT}\right|. \tag{7.62}$$

Thus, in the Tafel range the deformation of the activation threshold by the applied voltage causes a pronounced overproportional change in the reaction rate.
In the majority of cases (e.g. in fuel cells[61], see Section 7.4) the situation is more complex and different charge transfer processes have to be considered: The electron transfer to the adsorption site and transfer of partially ionized oxygen to the electrolyte phase, generally accompanied by further electron uptake; in addition, there can be lateral migration processes associated with electrical potential changes as well as transport processes through mixed conducting electrodes. A discussion of realistic situations in fuel cells is given in Refs. [582–584]. We will return to our simple model treatment which has been used by Wang and Nowick [579] to analyze the electrode kinetics of the Pt oxygen electrode on CaO–doped CeO_2. Let us look at their results:
Figure 7.18 shows that after subtraction of the bulk contribution the current–voltage situation at CeO_2 can be described approximately by the Butler–Volmer relationship. As far as the symmetry factors are concerned, it has to be taken into account that z=2 (cf. Fig. 7.18 and footnote 66). The exchange current densities obtained by extrapolation to small values of η_T will now be discussed as a function of the control variables temperature, oxygen partial pressure and doping content (as was done in detail for the bulk conductivity in Chapter 5).
On account of the exothermicity of the adsorption ($\Delta H_s^\circ < 0$) it follows that at high temperatures $K_s P^{1/2} \ll 1$, i.e. $\widehat{\Theta} \ll 1$; the same is naturally true at low partial pressures. Then the $\widehat{\Theta}(T, P)$ relationship (see Eq. (7.52)) simplifies to

$$\widehat{\Theta} = K_s P^{1/2}. \tag{7.63}$$

[61] Measurements of this kind been carried out in connection with the kinetics of fuel cell electrodes (Section 7.4.2).

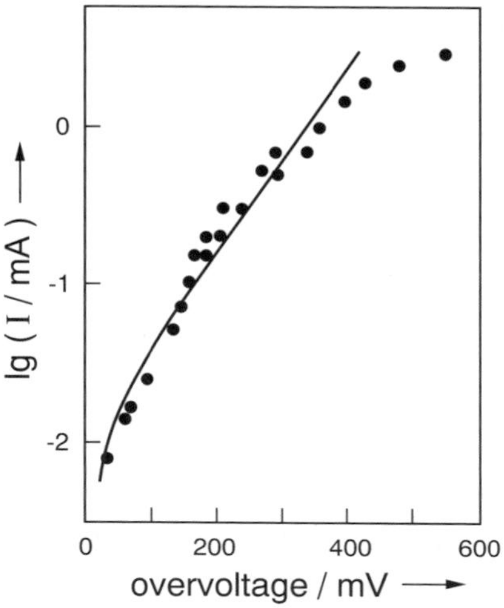

Fig. 7.18: Stationary current voltage characteristics for the symmetrical cell $Pt, O_2|ceria|Pt, O_2$ (ceria is CaO–doped) [579]. The transfer reaction determines the overall kinetics and makes for a markedly nonlinear behaviour. The fitting reveals $\vec{\alpha}z = \overleftarrow{\alpha}z = 1$. In the model z=2 and the symmetry factor 1/2. From Ref. [579].

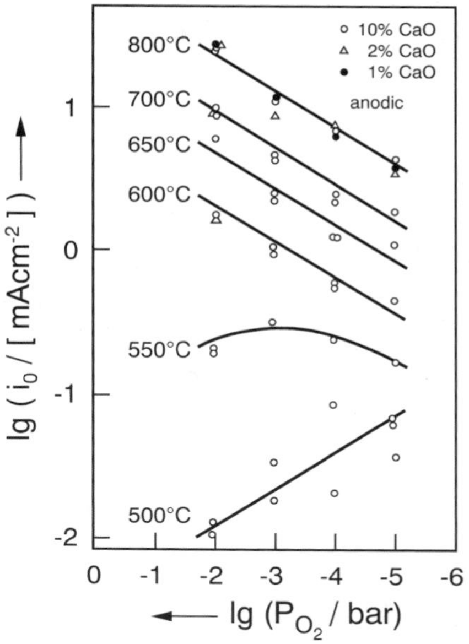

Fig. 7.19: Isotherms of the exchange current density (as a function of oxygen partial pressure) shown as a log–log plot, revealing that the slope $\partial \lg I_0 / \partial \lg P_{O_2}$ changes from +1/4 to -1/4 if the partial pressure increases and temperature decrease (see "transition isotherm" at 550°C). From Ref. [579].

Under these conditions (Eq. (7.57)) we obtain for the exchange current density

$$i_0 \propto \varrho_T^{-1} \propto K_s^{1/2} P^{1/4} k_{T_{eff}}. \tag{7.64}$$

The P and T dependences[62,63] of i_0 are then

$$\begin{aligned} \frac{\partial \ln i_0}{\partial \ln P} &= 1/4, \\ -\frac{\partial \ln i_0}{\partial 1/RT} &= \frac{1}{2}\Delta H_s^\circ + \Delta H_{T_{eff}}^{\neq} = \frac{1}{2}\Delta H_s^\circ + \frac{1}{2}\left(\overrightarrow{\Delta H}_T^{\neq} + \overleftarrow{\Delta H}_T^{\neq}\right). \end{aligned} \tag{7.65}$$

[62]The electrical potential drop $\widehat{\Delta\phi}$ is generally dependent on P_{O_2}. But this does not apply any longer for $k_{T_{eff}}$ (cf. Eqs. (7.56,7.57,7.58)).

[63]The enthalpy of dissociation is included in ΔH_S°. In the case of preceding or succeeding ionization (see footnote 66), these enthalpy contributions are formally contained in $\Delta H_T^{\neq}$. Under certain conditions the enthalpy of distribution (migration to the three–phase boundary) can also be important.

On the other hand in the region of high oxygen partial pressure or at low temperatures, $\widehat{\Theta}$ is approximately 1 (Eq. (7.52)) and so

$$1 - \widehat{\Theta} = K_s^{-1} P^{-1/2}, \tag{7.66}$$

that is

$$i_0 \propto K_s^{-1/2} P^{-1/4} k_{T_{eff}} \tag{7.67}$$

and hence[62,63]

$$\begin{aligned} \frac{\partial \ln i_0}{\partial \ln P} &= -1/4, \\ -\frac{\partial \ln i_0}{\partial 1/RT} &= -1/2 \Delta H_s^\circ + \frac{1}{2}\left(\Delta \vec{H}_T^{\neq} + \Delta \overleftarrow{H}_T^{\neq}\right). \end{aligned} \tag{7.68}$$

Figure 7.19 confirms impressively the predicted change in the slope of the $\log i_0 - \log P$ plot. The temperature dependence can be analyzed in the same manner (consider sum and difference of Eq. (7.65) and Eq. (7.68)). The dependence on doping is obtained from the $[V_O^{\cdot\cdot}]$ term as explicitly described in Chapter 5. See the original literature for details [579] (cf. also Ref. [384]).
The effective interfacial concentration can be approximately formulated in an analogous fashion to the bulk treatment[64]:

$$\widehat{c}_{eff} \propto P^{N'} \left(\Pi_r K_r^{\gamma_r'}\right). \tag{7.69}$$

The dependence of the exchange current density[65] follows from Eq. (7.57). As far as its temperature dependence is concerned, it is necessary to include $-\partial \ln k_{T_{eff}} / \partial(1/RT)$.
Butler–Volmer equations are not just obtained if the transfer reactions are rate–determining but also in more general cases[66]. A more complex current–voltage

[64]Note that the simple expression given by Eq. (7.69) which may be compared with Eq. (5.117) presumes neutral adsorption. If Θ refers to charged species which is realistically the case, then the results are more complex than in the bulk. The reason is the nonapplicability of the electroneutrality relationship.

[65]The analogy with σ is perfect, if we take into account that, in a better approximation, an expression of the form $c(c^{max} - c)$ must be included also in the case of the bulk. On account of the symmetry of the threshold $\Delta \vec{H}^{\neq} = \Delta \overleftarrow{H}^{\neq}$ for bulk transport.

[66]Parsons [585] showed that a series kinetics with one rate determining step can be approximately described by a Butler–Volmer equation, characterized by effective transfer coefficients $\vec{\beta}$ and $\overleftarrow{\beta}$ (in the text $\beta = \alpha z$), which are given by $\vec{\gamma}/\nu + r\vec{\alpha}$ and $\overleftarrow{\gamma}/\nu + r\overleftarrow{\alpha}$. The parameters r and $\vec{\gamma}$, $\overleftarrow{\gamma}$ are the numbers of the elementary charges transferred, in the rate–determining step and the steps before or after, respectively; ν counts how often the rate-determining step must be passed through for the whole reaction to take place once. The multistep kinetics analysis (see e.g. [219]) assumes rapid (pseudo 1st order) reactions before and after the rate–determining step and follows the principles of Section 6.7.2:
Let the reaction sequence be in a stationary state (and open on the electrode side!). The forward rate of the rate–determining step also involves (besides the electrochemical rate constants) concentrations, which can be associated with the educt via the preceding equilibria. In the same manner,

relationship is to be expected if neither adsorption nor transfer step alone determine the rate. In the stationary state we can set the rates of both reactions equal to each other (Eqs. (7.50), (7.51)) and after a lengthy[67] but elementary calculation we obtain

$$\mathrm{i}_0 = \frac{\mathrm{zF}\vec{\mathrm{k}}_\mathrm{T}\exp\left(-\vec{\alpha}\frac{\mathrm{zF}\Delta\phi(\mathrm{i}=0)}{\mathrm{RT}}\right)\vec{\mathrm{k}}_\mathrm{s}\mathrm{P}^{1/2}}{\vec{\mathrm{k}}_\mathrm{s}\mathrm{P}^{1/2} + \overleftarrow{\mathrm{k}}_\mathrm{s} + \vec{\mathrm{k}}_\mathrm{T}\exp\left(-\vec{\alpha}\frac{\mathrm{zF}\Delta\phi(\mathrm{i}=0)}{\mathrm{RT}}\right) + \overleftarrow{\mathrm{k}}_\mathrm{T}\exp\left(\overleftarrow{\alpha}\frac{\mathrm{zF}\Delta\phi(\mathrm{i}=0)}{\mathrm{RT}}\right)} \tag{7.70}$$

for the exchange current density. Neglect of the two transfer terms in the denominator yields Eq. (7.56) once again.
If conversely the adsorption rate is rate–determining then we can neglect $\vec{\mathrm{k}}_\mathrm{s}\mathrm{P}^{1/2}$ and $\overleftarrow{\mathrm{k}}_\mathrm{s}$ and find that

$$\mathrm{i}_0 = \frac{\mathrm{zF}\vec{\mathrm{k}}_\mathrm{s}\mathrm{P}^{1/2}}{1+\widetilde{\mathrm{K}}_\mathrm{T}^{-1}}. \tag{7.71}$$

We note that $\widetilde{\mathrm{K}}_\mathrm{T} = \mathrm{K}_\mathrm{T}\exp\left(-\frac{\mathrm{zF}\Delta\phi(\mathrm{i}=0)}{\mathrm{RT}}\right)$. On account of the mass action law it follows that $\mathrm{P}^{-1/2} = \mathrm{K}_\mathrm{s}\widetilde{\mathrm{K}}_\mathrm{T}$, whereby $\widetilde{\mathrm{K}}_\mathrm{T}$ is dependent on $\mathrm{P}_{\mathrm{O}_2}$ due to the field factor. From this and with the aid of Eq. (7.52) we obtain

$$\mathrm{i}_0 = \frac{\mathrm{zF}\overleftarrow{\mathrm{k}}_\mathrm{s}\vec{\mathrm{k}}_\mathrm{s}\mathrm{P}^{1/2}}{\overleftarrow{\mathrm{k}}_\mathrm{s}+\vec{\mathrm{k}}_\mathrm{s}\mathrm{P}^{1/2}} \simeq \mathrm{zF}\sqrt{\overleftarrow{\mathrm{k}}_\mathrm{s}\vec{\mathrm{k}}_\mathrm{s}\widehat{\Theta}(1-\widehat{\Theta})\mathrm{P}^{1/2}}, \tag{7.72}$$

that is, the exchange current density for the pure adsorption reaction. The expression on the right hand side shows that k_eff is determined via the sorption rate constants. In contrast to Eq. (7.58) c_eff still contains the factor $\mathrm{P}_{\mathrm{O}_2}^{1/2}$, which leads to a different $\mathrm{P}_{\mathrm{O}_2}$ dependence compared to the previous case. It should also be noted[68] that the expression obtained for the stationary current–voltage curve predicts asymptotic current behaviour (limiting current).
These are naturally only specific examples in the very complex field of electrode kinetics. (Analogous but usually less complex considerations have to be carried out for internal interfaces.) In view of the enormous technological importance (e.g. with fuel cells, see Section 7.4), thorough investigations of prototype systems are necessary. In practical systems, in particular in the case of gas electrodes, the inhomogeneity and three–dimensionality of the situation plays a decisive role for the kinetics as reflected in particular by the significance of bulk, surface and gas phase diffusion (see e.g. Ref. [584]).

---

the back reaction rate can be associated with the concentrations of the partners in the succeeding reactions and ultimately with the end product. The difference between the forward and back reaction rate describes the stationary current (cf. Eq. (6.120)). Rate and equilibrium constants include, in general, electrical field contributions, that are influenced by the applied voltage. This leads directly to the above effective transfer coefficients.

[67] We calculate Θ or $(1-\Theta)$ from the two rate equations (for Eqs. (7.50), (7.51)). We then obtain the total stationary rate by substituting Θ in Eq. (7.53). Its exchange rate is determined by Eq. (7.70).

[68] This follows from the considerations in footnote 66.

As discussed in Chapter 6, bulk transport only becomes nonlinear at very high overvoltages. Hence, nonlinearities that occur for overvoltages on the order of several RT/F point towards phase boundary limited, such as diffusion–, reaction– or transfer–limited processes. The first two frequently reveal themselves through the occurrence of limiting currents (see also Fig. 7.56 on page 484 and footnote 41 on page 423). This may permit a first statement concerning the mechanism.

If the boundary layer resistance $R^{\perp}$ is space charge–controlled, the case of the depletion layers is most interesting in this context. The analytical form of the space charge resistance close to equilibrium has been discussed in detail in Chapter 5. There we also discussed the P_{O_2} partial pressure dependence of the space charge resistance of $SrTiO_3$ grain boundaries. In the simplest case the contact resistance and the space charge resistance are connected in series with this serial connection being parallel to a series of corresponding capacitances. This highlights the fact that core charges and space charges are not independent of each other, and thus contact and space charge impedances are not simply decoupled. The more detailed literature [586,587] should be consulted for a more precise discussion. Below we wish to turn our attention more closely to the capacitances, which are important for the transient behaviour.

c) Mechanistic interpretation — interfacial capacitances

The (differential) electrical capacitance describes the variation of the stored charge with varying electrical potential difference[69]. At first we face the problem that, in principle, the Faradaic current ($R^{\perp}$ being finite) affects the charge distribution and, hence, the capacitance. This can frequently be ignored, to a first approximation, in the case of small currents, as we have already anticipated in the equivalent circuit (Fig. 7.15). The distribution considered for the purpose of calculating the capacitance is then the equilibrium distribution, and we can imagine the Faradaic current being set at zero. In fact most treatments in classical electrochemistry involve consideration of the ideally polarizable electrode[70] ($R^{\perp} \longrightarrow \infty$) (e.g. Hg in contact with an aqueous electrolyte).
Here too we shall first imagine the contact of a pure ion conductor (e.g. AgCl) with an electrode that does not permit transfer of the ions[71]; afterwards we will allow for more general cases simply by taking into account a finite transfer resistance. Since

[69] It has already been demonstrated in Section 5.4.4 that the excess charge of a polarized electrode is given by the negative change of the interfacial tension γ with the applied voltage U (Eq. (5.72)). As a result the differential capacitance can be obtained from the negative curvature of the γ(U) curve (electrocapillarity curve).

[70] It was shown in Ref. [588] that the results are also of importance for nonideally polarizable electrodes, but difficult to measure there.

[71] However, this is only unproblematic for the relevant short time period, since a stoichiometry polarization occurs because of the finite nature of the electronic resistance in AgCl (see Section 7.3.4).

the electrode phase can also adsorb cations or anions (cf. the adsorption of Ag^+ at the $RuO_2/AgCl$ contact, Section 5.8), a triple layer is approximately realized. We vary the electrode charge — more precisely the area–related charge density on the electrode side — by means of the applied voltage. (We denote this area–related quantity by Σ_E. The local coordinate on the electrode side at the contact is defined as s_E, that on the electrolyte side is defined as s.) The sum of Σ_E and the adsorbed charge at x = s (i.e. Σ_s) is compensated by the diffuse charge (Σ_{dif}). The latter possesses its local maximum value at x=0 (see Section 5.8). In classical electrochemistry we designate the two different layers at x=0 and x=s as the inner and outer Helmholtz layer. Irrespective of all similarities, the solid–solid contacts involve rearrangement and reorganization effects on a larger scale. Nonetheless we will ignore their impact in the following. The contact capacitance is then given by

$$\frac{C^{\perp}}{a} = \frac{\partial \Sigma_E}{\partial(\phi_E - \phi_\infty)} \tag{7.73}$$

(ϕ_E: potential of the electrode).

Let us first consider the case in which the solid electrolyte exhibits a high charge carrier density (α–AgI or highly doped AgCl). Then first the Debye length can be ignored (i.e. rigid triple layer with $\Sigma_{dif} \rightarrow \Sigma_0$) and second[72] $\phi_0 - \phi_\infty \simeq 0$. Since $\partial(\phi_E - \phi_\infty) \simeq \partial(\phi_E - \phi_0) = \partial(\phi_E - \phi_s) + \partial(\phi_s - \phi_0)$ and $\Sigma_E + \Sigma_s + \Sigma_0 = 0$ it follows that

$$\left(C^{\perp}/a\right)^{-1} = \frac{\partial(\phi_E - \phi_s)}{\partial \Sigma_E} + \frac{d\Sigma_0}{d\Sigma_E}\frac{\partial(\phi_s - \phi_0)}{\partial \Sigma_0} = (C_1/a)^{-1} + (C_2/a)^{-1}\left(1 + \frac{d\Sigma_s}{d\Sigma_E}\right). \tag{7.74}$$

Here C_1/a is the differential capacitance of the inner Helmholtz layer relative to the electrode ($\partial\Sigma_E/\partial(\phi_E - \phi_s)$), C_2/a the differential capacitance of the outer Helmholtz layer relative to the inner Helmholtz layer ($\partial\Sigma_0/\partial(\phi_0 - \phi_s)$). If chemisorption is marked, then Σ_s is only slightly dependent on Σ_E, and the triple layer behaves like two capacitors connected in series.

Next let us take into account a nonzero Debye length in the electrolyte, that is, let us permit diffuse charge distribution (between x=0 and x=∞), but neglect sorption effects for simplicity, i.e. $|\Sigma_E| = \left|\int_0^\infty \rho dx\right|$. Now we have to integrate the charge density between bulk and x=0 and to differentiate the result with respect to the potential at the interfacial position ($x = s_E$). (We have to extend the consideration analogously if a space charge also occurs in the electrode.) Because

$$\left(C^{\perp}/a\right)^{-1} = \frac{\partial(\phi_E - \phi_\infty)}{\partial \Sigma_E} = \frac{\partial(\phi_E - \phi_0)}{\partial \Sigma_E} + \frac{\partial(\phi_0 - \phi_\infty)}{\partial \Sigma_E} = (C_H/a)^{-1} + (C_{sc}/a)^{-1} \tag{7.75}$$

[72] Owing to the electrochemical potential being approximately constant, it holds that the $\Delta\phi$ scales with changes of logarithm of c and thus with $\Delta c/c$. If c is already high, this means that on comparable charge variation, the *relative* change in c and, hence, the absolute change in ϕ is small.

this can be expressed by connecting a Helmholtz capacitor (C_H) and space charge capacitor (C_{sc}) in series. Since we have not stored any charge between electrode and x=0, the ϕ profile is linear with the gradient[73] ($\phi'|_{x=0} =) \Sigma_E/\varepsilon$. On the other hand, the gradient is $(\phi_E - \phi_0)/|s_E|$. We, thus, obtain[74] (C_H/a) as $\varepsilon/|s_E|$.
When excitation is small, we expect a similar result of the form $\varepsilon/\mathcal{L}$ for the space charge contribution. It has been demonstrated that more generally for small signals the characteristic length $\mathcal{L}$ represents the centroid of charge perturbation [589]. For the area–related capacitance it then follows from Eq. (7.75) that $(C^{\perp}/a) = \varepsilon/(\mathcal{L} + |s_E|)$. Since, in a preliminary and approximate manner, we can identify $\mathcal{L}$ with the extent of the space charge zone, it can then be seen that (C_{sc}/a) will predominate in the above series connection for not too defective a solid.
Let us now inspect two simple cases more closely:
For small effects in the Gouy–Chapman situation (mobile majority charge carriers) we already know from Section 5.8 that the space charge potential is represented by an exponential function ($\phi = \phi_0 \exp(-x/\lambda)$). Integration of the space charge and, hence, of ϕ'' from the interior ($\phi' = 0$) to the phase boundary gives $\phi'|_{x=0}$ and, thus, $-\Sigma_E/\varepsilon$. The exponential character requires $\phi' = -\phi/\lambda$ yielding

$$(C_{sc}/a) = \frac{\varepsilon}{\lambda}, \tag{7.76}$$

as expected.
In the case of the Mott–Schottky boundary layer (majority charge carrier immobile, counterdefect depleted), we obtained a simple result for high depletion and could approximate the space charge profile by means of a rectangular function of width $\lambda^* \propto \lambda\sqrt{|\phi_0 - \phi_\infty|}$ (cf. Eq. (5.231) in Section 5.8)[75]. The total surface charge is approximately obtained by multiplication of λ^* with the constant doping concentration m resulting in[76]

$$C_{sc}/a = \frac{d(zFm\lambda^*)}{d\lambda^*}\frac{d\lambda^*}{d(\phi_0 - \phi_\infty)} = \sqrt{\frac{|z|Fm\varepsilon}{2|\phi_0 - \phi_\infty|}} = \frac{\varepsilon}{\lambda^*}. \tag{7.77a}$$

Evidently C_{sc} is, via λ^*, potential dependent. If $\widehat{\phi}_0 - \phi_\infty$ is the potential difference existing at equilibrium (i.e. without an external voltage), then $(\phi_0 - \phi_\infty) - (\widehat{\phi}_0 - \phi_\infty) = \phi_0 - \widehat{\phi}_0$ is the applied voltage (η). If the applied voltage takes the value $\eta^* \equiv -(\widehat{\phi}_0 - \phi_\infty)$, the depletion effect disappears; in the case of semiconductors

[73]After transition into the metal ϕ' jumps from the value ϕ'_0 to the value zero. The size of the jump is determined by Σ_E/ε.

[74]The normal even though questionable assumption is to set the dielectric constant of the Helmholtz layer the same as that of the volume dielectric constant.

[75]Such considerations are important for semiconductor electrodes in electrochemistry.

[76]In Eq. (7.77a) it is assumed that the charge density is determined by the doping concentration (see Section 5.8). When the influence of depleted charge carriers is taken into account, it is a better approximation to add the term $-\frac{RT}{|z|F}$ to the denominator in the root in Eq. (7.77a) [252]. This yields Eq. (7.77b).

this potential of zero charge[77] is also known as the flat band potential, since band bending approaches zero. According to this definition, we replace $\phi_0 - \phi_\infty$ with $\eta - \eta^*$ or $U - U^*$ (see Eq. (7.5)) and rewrite Eq. (7.77a) in the form[76]

$$\left(\frac{1}{C_{sc}/a}\right)^2 = \left|\frac{2}{zF\varepsilon m}\left(U - U^* - \frac{RT}{F}\right)\right|. \tag{7.77b}$$

The doping level and the flat band potential can be obtained from the Mott–Schottky plot [590], $(1/C_{sc})^2$ against U.

The fact that no potential dependence occurred in the first case ("Gouy–Chapman case"), is due to the assumption of small effects. The result for large effects is also easily obtained for the Gouy–Chapman case: Because $\Sigma_E \propto \sqrt{c_0}$ (see Eq. (5.251), Section 5.8.4), it follows that

$$\frac{C_{sc}}{a} = \frac{\varepsilon}{2\lambda}\zeta_{\pm 0}^{1/2} = \frac{\varepsilon}{2\lambda}\exp \mp \frac{|z|\,F\,(\phi_0 - \phi_\infty)}{2RT} = \frac{\varepsilon}{2\lambda}\exp \mp \frac{|z|F\,(U - U^*)}{2RT} \tag{7.78}$$

reflecting a marked voltage dependence[78].

Let us tackle a more general solution. (Yet, as before, we will neglect the influence of the Faradaic current on the charge distribution.) We do not have to start out from the concentration profile, for it is sufficient to calculate the surface charge density. This merely requires knowledge of $(\phi_0 - \phi_\infty)'$ which is given[79] by $\left[-\frac{2}{\varepsilon}\int_0^{(\phi_0-\phi_\infty)} \rho\,d\,(\phi - \phi_\infty)\right]^{1/2}$. In the intrinsic Gouy–Chapman case $\rho = -\,|z|\,Fc_\infty \sinh\left(|z|\,F\,(\phi - \phi_\infty)\,/RT\right)$ (cf. Eq. (5.215) Section 5.8.1) resulting in $C_{sc} \propto \cosh\left(|z|F\,(\phi_0 - \phi_\infty)\,/2RT\right)$. Considering this in detail, it follows that[80]

$$C_{sc}/a = \frac{\varepsilon}{\lambda}\cosh\frac{|z|\,F\,(\phi_0 - \phi_\infty)}{2RT} = \frac{\varepsilon}{\lambda}\cosh\frac{|z|F\,(U - U^*)}{2RT}. \tag{7.79}$$

For large values of $\phi_0 - \phi_\infty = U - U^*$ one of the two exponential functions dominates the hyperbolic cosine and Eq. (7.78) applies, while Eq. (7.76) is once again found as the approximation for small potential differences ($\cosh(...) \longrightarrow 1$). In addition ε/λ corresponds to the minimum value of the function obtained in Eq. (7.79) at $U = U^*$. This is shown in Fig. 7.20 and may be compared with the corresponding dependence of the differential resistance in Fig. 7.17. For slight variation of the

[77] For the electrochemical zero charge potential (point of zero charge), which can be obtained from the electrocapillarity curve, Σ_E disappears (cf. page 148, cf. footnote 69).

[78] It is interesting that an extrapolation (beyond the range of validity of Eq. (7.78)) to $\phi_0 - \phi_\infty = 0$ yields the capacitance $\varepsilon/(2\lambda)$. This is in agreement with the results from Section 5.8 that for large effects the term (2λ) plays the role of an effective thickness. The true space charge capacitance for small $\phi_0 - \phi_\infty$ is of course given by ε/λ (see Eq. (7.76) or Eq. (7.79)).

[79] Since $\frac{d}{dF}\left(F'^2\right) = \frac{d}{dx}\left(F'^2\right)\frac{dx}{dF} = \frac{d}{dx}\left(F'^2\right)\frac{1}{F'} = 2F''$ it follows that F'^2 is given by the integral of $2F''dF$. If we identify F with ϕ, the integrand is proportional to ρ/ε (Poisson equation).

[80] Note when working out the integral that $\sqrt{2\cosh x - 2} = 2\sinh(x/2)$.

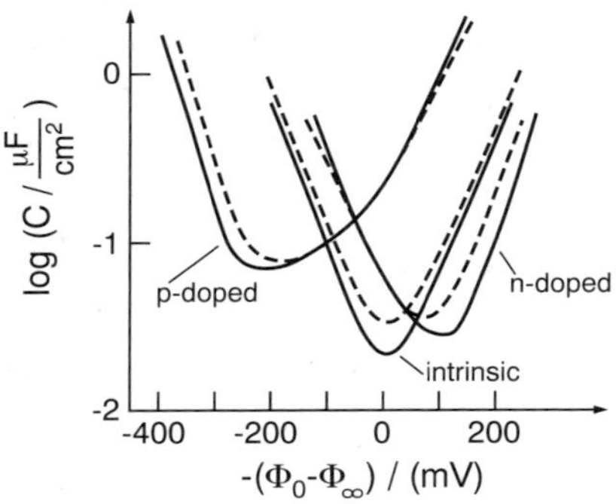

Fig. 7.20: Theoretical space charge capacitance according to Bohnenkamp and Engell as a function of the voltage drop over the boundary layer $-(\phi_0 - \phi_\infty)$ for intrinsic and doped semiconductors (donor concentration $N_D/V = 3.5 \cdot 10^{14} cm^{-3}$, acceptor concentration $N_A/V = 7.5 \cdot 10^{15} cm^{-3}$) at room temperature (dashed line: 45°C)[81]. According to Ref. [591].

electrical potential in the immediate neighbourhood of the minimum, the absolute value of the surface charge is increased proportionally; if we move within the diagram to higher absolute deviations to the left or right of the minimum, then $|\Sigma_E|$ changes overproportionally. The capacitances for a strongly positively or negatively doped material with immobile doping can be calculated in an analogous manner[81].
Let us consider the second case under more general conditions. Here in a Brouwer situation the negative charge carrier is directly obtained from the doping concentration m, while the positive mobile defects are given by K/m, if K represents the relevant mass action constant. The latter is in the minority when $U = U^*$, (i.e. $\phi_0 = \phi_\infty$). The voltage must be raised far over U^*, in order to make C_{sc} increase[81], while, in the second case, the minimum is displaced to the left (see Fig. 7.20). Such a variation in capacitance has been confirmed for the contact of a semiconductor (as electrode) with a liquid electrolyte [552,591]. Deviations from the theoretical curves are ascribed to surface states [592], saturation and finite size effects [593] and, at large voltages, to current flow no longer being negligible.
Figure 7.21 shows the potential dependence of the boundary layer capacitance of α–AgI in contact with a graphite electrode. Here the Debye length is vanishingly small (very high charge carrier concentration) and a negligible potential dependence is to be expected ($C/a \simeq \varepsilon/|s|$, Eq. (7.75)). When $\varepsilon \sim 8$ and $C^{\perp}/a \sim 2.5\mu F cm^{-2}$ the value of $|s|$ (see C_H on page 440) is obtained as ca. 30Å, which is unrealistically high. If a false estimate of the contact area or the dielectric constant can be ruled out, this points towards a more complicated situation, e.g. involving adsorption capacities (surface states). Such complications, together with the limited sites available [595] (Fermi–Dirac like distribution, see Chapter 5) are presumably responsible for the

[81] The analogous calculation yields [591]

$$C_{sc}/a = \sqrt{z^2F^2\frac{2\varepsilon m}{RT}}\frac{\exp\frac{|z|F(\phi_0-\phi_\infty)}{RT} - 1 + \left(\frac{K}{m^2}\right)2\sinh\frac{|z|F(\phi_0-\phi_\infty)}{RT}}{2\left(\exp\frac{|z|F(\phi_0-\phi_\infty)}{RT} - 1 - \frac{|z|F(\phi_0-\phi_\infty)}{RT} + \frac{K}{m^2}2\left(\cosh\left(\frac{|z|F(\phi_0-\phi_\infty)}{RT}\right)\right) - 1\right)^{1/2}}.$$

When there are high doping levels $(K/m^2) \to 0$, i.e.[positive defect] $\to 0$) and a sufficiently negative potential (depletion of the negative majority defect) the result is Eq. (7.77) again.

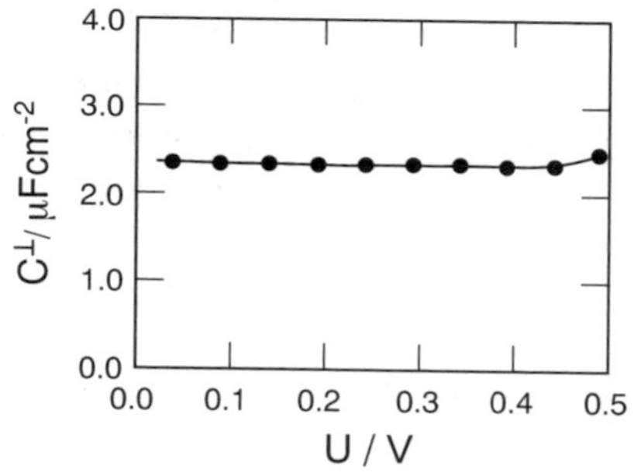

Fig. 7.21: Voltage dependence of the boundary layer capacitance for C|AgI at 175°C. From Ref. [594].

fact that the variations shown in Fig. 7.20 have not yet been clearly observed for solid ion conductors, even with considerable λ values[82].

Let us briefly consider the temperature dependence of C_{sc} for the simple Gouy–Chapman and Mott–Schottky cases. In the first case at small space charge potentials the temperature dependence follows from $\lambda(T)$ and, hence, from $c_\infty^{1/2}(T)$. In pure materials C_{sc} can thus increase considerably with temperature, while in highly doped materials C_{sc} will remain almost constant over a large temperature range. In the case of large effects the temperature dependence of the interfacial concentration, as discussed in Chapter 5, also becomes important.

In the second case it was found that $C_{sc} \propto 1/\lambda^*$ when the potential difference is large. Figure 7.22 shows the grain boundary capacitance of a $SrTiO_3$ polycrystal for

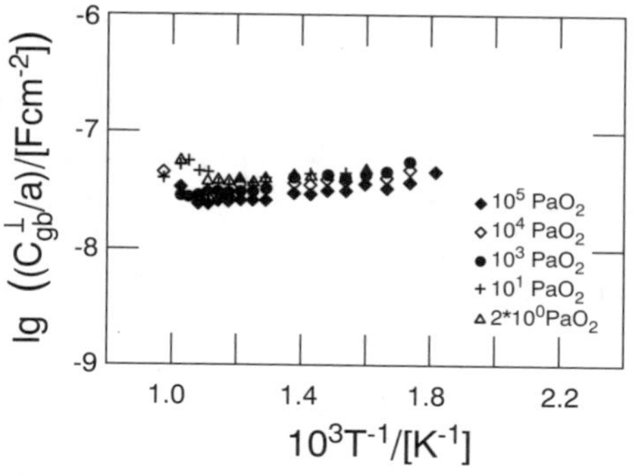

Fig. 7.22: Grain boundary capacitance of a Fe–doped $SrTiO_3$ polycrystal ($m_{Fe} = 6.5 \times 10^{19} cm^{-3}$), normalized to the electrode surface and measured at various oxygen partial pressures as a function of reciprocal temperature. The space charge potentials are significant. The bias used for the measurement is negligible [307]. Typical space charge potentials vary between 300 and 800mV.

which similar considerations are valid (double Schottky contact, cf. Section 5.8.5e). Since the dopant concentration is independent of temperature, and since the space charge potential does not significantly depend on T, λ^* and hence $C^\perp$ are almost temperature independent too. The reader may work out the dependence of component potential from analogous considerations.

Finally, we shall briefly note the exploitation of double–layer capacitances for en-

[82]When $\partial\Sigma$ is split up into space charge and adsorption contributions with approximately equal $\partial\phi$ (which may be a very rough approximation), the adsorption capacitor is obtained as a parallel element to C_{sc} [596].

ergy conversion. Owing to the low boundary layer thickness double–layer capacitors exhibit relatively high capacitances (typically $10\mu F/cm^2$). On account of the low portion of active zones that they possess (i.e. interfaces), they are not a substitute for batteries regarding long–term energy storage (cf. Section 7.4). On account of the very short time constants, however, fairly large quantities of energy can be stored and released very rapidly, e.g. within a few seconds (see also Fig. 7.55 on page 481). Specific capacitances of the order of magnitude of 100 F/g have been achieved in the case of carbon materials with large internal surface areas (e.g. aerogels). Such systems are of use as complementary storage elements to batteries (such as in automobiles, see Section 7.4) [597]. If the surface charge storage does not correspond to a purely dielectric process (polarization by charging current, generally formation of an adsorption layer), but a Faradaic interfacial reaction (e.g. surface redox reaction at RuO_2 electrodes by proton discharge [598]), this partly conforms to the principle of a solid state battery[83] (There, however, storage occurs in the bulk of the material, i.e. absorption instead of charge adsorption.) For more details on these supercapacitors and the increasing interest in this topic the reader is referred to Ref. [600].

7.3.4 Stoichiometry polarization

a) Measurement with selectively blocking electrodes

Although we used ideally polarizable electrodes to derive the relationships for the boundary layer capacitance, the whole of the previous section was nevertheless basically devoted to nearly reversible electrodes, i.e. electrodes which exchange ions and electrons but exhibit a certain contact resistance. Selectively blocking electrodes, which we now wish to discuss (i.e. we are considering cells of type E1, E2, I1 and I2 in the list given on page 424f), are electrodes which let through ions and block electrons or vice versa; in other words: They ideally possess an infinitely high contact resistance for one carrier type and zero–resistance for the countercarriers. Owing to the finite nature of the ionic or electronic bulk conductivity, stoichiometry polarization occurs in such cells. Initially, after the current has been turned on, both ions and electrons flow in the bulk according to their equilibrium conductivities. In the course of polarization at the given polarity, the blocked charge carrier gradually ceases to be available for transport as it is not supplied at the blocking electrode interface, and — if a second blocking electrode is used — cannot cross the other [324,325]. (Here it is important that the decomposition voltage is not exceeded.) It will be shown in the discussion that follows that in the case of small voltages the polarization behaviour of cells with one and two blocking electrodes only varies

[83] If the interfacial density is allowed to increase to any desired level, this principle approaches that of a solid state battery (Section 7.4). The conceptual transition between the two principles is found in nanocrystalline materials with grain sizes of the order of magnitude of the Debye length [599].

with respect to a factor of 4 in the time constant[84]. In addition, there is symmetry between ions and electrons, so that we can concentrate on a single cell.
In all cases stoichiometry gradients are built up during polarization. In the stationary state the gradient in the chemical potential of the blocked charge carrier compensates the electrical potential gradient. Thus, for example, in cell E1 or E2 in the stationary state $\mathbf{i}_{\text{ion}} = \mathbf{0}$, i.e. $\nabla\widetilde{\mu}_{\text{ion}} = \nabla\mu_{\text{ion}} + z_{\text{ion}}F\nabla\phi = \mathbf{0}$. Here we wish to ignore double–layer effects initially[85]. The stationary voltage is determined by the electrochemical potential gradient of the nonblocked charge carrier, whose conductivity can now be extracted.
Before we go into the thermodynamics let us estimate the effects using the heuristic equivalent circuit in Fig. 7.23, an approximation that we shall later find is suprisingly

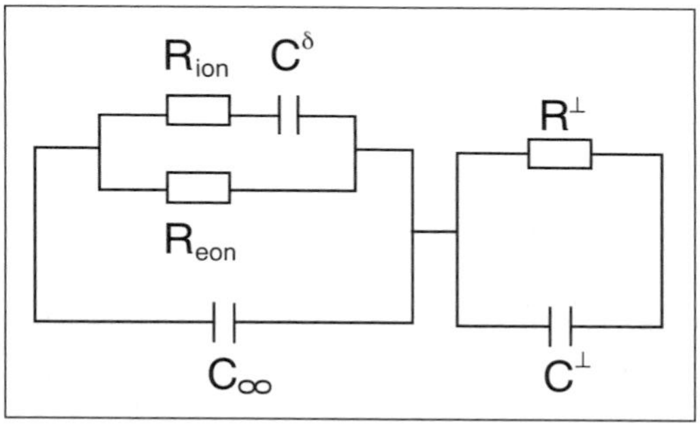

Fig. 7.23: Equivalent circuit for ion blockage. In the case of electron blockage the capacitor (C^δ) is in series with R_{eon}. Contact impedances in the cell are approximately taken into account by means of the $R^\perp$–$C^\perp$ parallel circuit[85] [431].

good. In Section 7.2 we already encountered the bulk part of it in discussing the EMF of the oxygen concentration cell with a mixed conductor (Fig. 7.9b). As we already know, the ionic and electronic contributions to the resistance are connected in parallel. The selective blockage is "effected" in the case assumed by putting a capacitance in series with R_{ion} [431]. The stoichiometry effect we are discussing differs markedly in its behaviour with time from a pure boundary effect, affects the whole bulk and is determined by the relatively slow ambipolar bulk diffusion. Hence C^δ is normally much larger than a typical double–layer capacitance.
At this point it is useful to lump together all the pure interfacial effects that take place at much smaller times ($\tau^\perp$) under a $R^\perp$–$C^\perp$ circuit in series, as done in Fig. 7.23. The contributions that this covers may include internal grain boundary impedances[86] or electrode effects of nonblocking electrodes. In this way also transfer impedances of the nonblocked species at the blocking electrode can be approximately taken account of (Section 7.3.4e).

[84] The time constant is proportional to the square of the thickness.

[85] In principle, specific transfer impedances must be considered for electronic and ionic carriers (see however Section 7.3.4e).

[86] As shown in Ref. [601] the situation is complicated by the fact that also grain boundaries (owing to the different transference numbers there compared to the bulk) can induce stoichiometry polarization. This is ignored here.

If we denote the time constant of the stoichiometry polarization by τ^δ (cf. C^δ), the relation

$$\tau^\delta \gg \tau^\perp \gg \tau_\infty. \tag{7.80}$$

is usually fulfilled.
Hence at comparatively short times $t \ll \tau^\delta$ the C^δ capacitor (in Fig. 7.23) is permeable and the problem reduces to what has already been discussed. For times of the order of τ^δ, the capacitors characterized by C_∞ and $C^\perp$ are no longer permeable and our equivalent circuit is reduced to the key element (C^δ in series with R_{ion} both parallel to R_{eon}) in series with $R^\perp$. We shall now consider the corresponding transient behaviour:
The voltage across R_{eon}, namely $I_{eon}R_{eon}$, is equal to the sum of $I_{ion}R_{ion}$ and the voltage U_C across C^δ. Since $I_{ion} = I - I_{eon}$ and $\dot{U}_C = I_{ion}/C^\delta$ the differential equation

$$\dot{I}_{ion} + I_{ion}/\tau_d = 0, \tag{7.81}$$

is obtained, whereby the constant τ^δ is introduced via

$$\tau^\delta = (R_{eon} + R_{ion})\, C^\delta \equiv R^\delta C^\delta. \tag{7.82}$$

Since the total current immediately upon switching on the circuit is determined by the pure parallel connection of R_{eon} and R_{ion} (the C^δ element being completely permeable), we find

$$I_{ion} = \frac{R_{eon}}{R_{eon} + R_{ion}} I \exp(-t/\tau^\delta) \tag{7.83}$$

as the solution. The electron current is therefore:

$$I_{eon} = I\left(1 - \frac{R_{eon}}{R_{eon} + R_{ion}} \exp -t/\tau^\delta\right). \tag{7.84}$$

Figures 7.24 and 7.25 emphasize the time evolution. Immediately after switching on the circuit, the current is partitioned as for the case of reversible electrodes. As the duration of polarization increases, I_{ion} diminishes and falls to zero in the stationary state in which the total current is completely carried by the electrons. The voltage across the capacitor increases ($\dot{U}_C = I_{ion}/C^\delta, U_C(t{=}0) = 0$) from zero to the stationary value IR_{eon} according to

$$U_C = IR_{eon}\left(1 - \exp -t/\tau^\delta\right). \tag{7.85}$$

The same stationary value IR_{eon} is obtained for the total voltage since

$$\begin{aligned} U = I_{eon}R_{eon} &= IR_{eon}\left(1 - \frac{R_{eon}}{R_{eon} + R_{ion}} \exp -t/\tau^\delta\right) \\ &= IR + IR_{eon}\frac{R_{eon}}{R_{eon} + R_{ion}}\left(1 - \exp -t/\tau^\delta\right). \end{aligned} \tag{7.86}$$

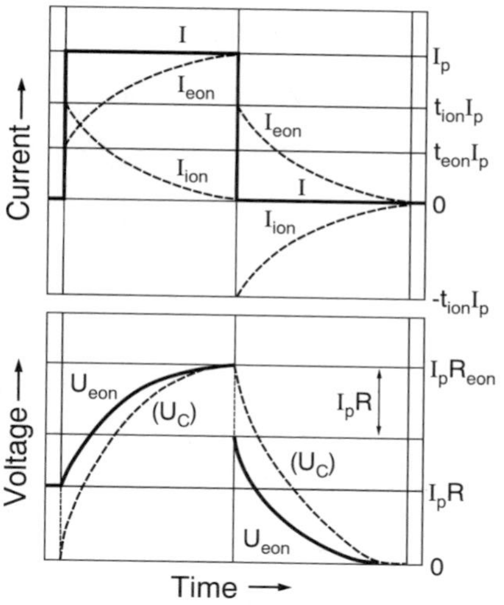

Fig. 7.24: The behaviour of voltage with time under galvanostatic load (switching–on and switching–off behaviour) of a cell blocking the ions. Contact resistances are neglected. Note that, to better approximation, the resistances have to be taken as dependent on concentration and, hence, on time and position (see below) [431].

When the interfacial resistance $R^{\perp}$ is taken into account, the stationary value becomes $R_{eon}+R^{\perp}$. Immediately after switching on the circuit (i.e. at t=0) U/I reduces to $\frac{R_{ion}R_{eon}}{R_{eon}+R_{ion}}$, that is to R_{∞}, or more accurately to $R_{\infty}+R^{\perp}$.

As mentioned at the start, the initial and final values permit — if $R^{\perp}$ has been corrected for — the discrimination between ionic and electronic conduction contributions. Since in the stationary state $\Delta\widetilde{\mu}_{ion}=0$ and so $U \propto \Delta\widetilde{\mu}_{eon} \propto \Delta\mu_{O_2}$, the voltage applied to U_C is the Nernst voltage, which would be measured in the absence of current in pure solid electrolytes[87] for the same chemical potential difference. On switching off[88] the circuit the analogous relationships

$$\begin{aligned} U_C &= I_pR_{eon}\exp -t/\tau^{\delta} \\ U &= I_p\frac{R_{eon}^2}{R_{eon}+R_{ion}}\exp -t/\tau^{\delta} \\ I_{eon} &= I_p\frac{R_{eon}}{R_{eon}+R_{ion}}\exp -t/\tau^{\delta} = -I_{ion} \end{aligned} \tag{7.87}$$

apply with I_p being the current that was flowing during the polarization phase. It is trivial but not unimportant to remark that again on switching off the current the initial voltage drop is given by IR_{∞} — or more accurately by $I(R_{\infty}+R^{\perp})$ (see Fig. 7.24).

Figure 7.25 illustrates the variation of voltage with time for a polarization with a constant current including a phase boundary impedance connected in parallel.

[87] Interestingly, this requires that a nonzero current is flowing and, hence, the drift term $IR = IR_{eon}R_{ion}/(R_{eon}+R_{ion})$ is included in the voltage. If we subtract this from IR_{eon} we obtain $IR_{eon}^2/(R_{eon}+R_{ion}) = IR_{eon}\cdot t_{ion}$. That is precisely the value expected from Eq. (7.36) (Nernst–voltage multiplied by the ionic transference number). It is also the value of the voltage immediately after switching off the circuit (Eq. (7.87)), provided the stoichiometric boundary values are the same.

[88] This discharge process actually belongs in Section 7.4. However, we limited ourselves to galvanic elements there.

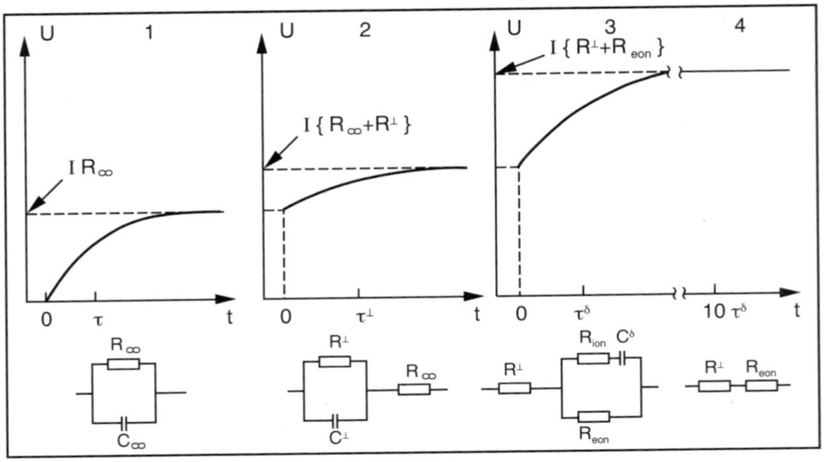

Fig. 7.25: The voltage development on polarization of an ion–blocking cell corresponding to Fig. 7.23 with a constant current for $\tau_\infty \ll \tau^\perp \ll \tau^\delta$ and the respective approximations of the equivalent circuit. For a time period of the order of τ_∞ immediately after switching on the current (1), $C^\perp$ and C^δ elements are "permeable" and it is only necessary to take account of the bulk process (R_∞ comprises the parallel connection of R_{ion} and R_{eon}). The C_∞ element is impermeable during the time period $\tau^\perp$ (2). The bulk process appears as a purely ohmic contribution (IR drop) and the phase boundary process becomes apparent. For times of the order of τ^δ (3) the phase boundary process is complete ($C^\perp$ elements impermeable). At this time resolution the initial jump becomes $I(R_\infty + R^\perp)$. The bulk polarization is stationary when $t \gg \tau^\delta$ (4) and all capacitive elements are blocked: $U(t{=}\infty) = R^\perp + R_{eon}$ [431].

The following treatment of the problem, using the methods of irreversible thermodynamics [324,602], is intended to test the validity of and to elaborate the heuristic model just discussed [431] as well as to establish a link with the underlying mechanism in particular as far as the parameter C^δ (and hence τ^δ) is concerned. Here we can basically rely on the treatment of chemical diffusion given in Chapter 6 (Section 6.5). Unlike there, however, ionic and electronic current contributions do not cancel each other out, but their sum gives the external current density. If no internal valence change or association occur, then

$$i = i_{O^{2-}} + i_{e^-}. \tag{7.88}$$

If we proceed analogously to Chapter 6 but allow for a constant nonzero current density, we obtain instead of Eqs. (6.51, 6.52)

$$i_{e^-} = -j_{e^-}F = \frac{\sigma_{e^-}}{F}\frac{\partial\widetilde{\mu}_{e^-}}{\partial x} = \frac{\sigma_{e^-}}{\sigma}i - \frac{\sigma_{O^{2-}}\sigma_{e^-}}{4F\sigma}\frac{\partial\mu_{O_2}}{\partial x} = \frac{\sigma_{e^-}}{\sigma}i + FD^\delta\frac{\partial c_{e^-}}{\partial x}, \tag{7.89}$$

$$i_{O^{2-}} = -j_{O^{2-}}2F = \frac{\sigma_{O^{2-}}}{\sigma}i + 2FD^\delta\frac{\partial c_{O^{2-}}}{\partial x}. \tag{7.90}$$

We note that the stoichiometry terms in Eqs. (7.89, 7.90) are equal and opposite for reasons of electroneutrality[89].
D^δ is given by Eq. (6.53) and is proportional to $\frac{\sigma_{e^-}\sigma_{O^{2-}}}{\sigma}\frac{d\mu_O}{dc_{O^{2-}}}$. Even though the partial current density now contains drift terms ($(\sigma_{e^-}/\sigma)i$, $(\sigma_{O^{2-}}/\sigma)i$) the increase in concentration with time is nevertheless given in the form of a "second Fick's law". We just have to assume small signal behaviour (small currents) for which the transport parameters remain locally constant. However, the boundary conditions differ from those discussed in Chapter 6 for the chemical relaxation measurement. At the contacts to the blocking electrodes the contribution of the partial current density that is blocked is zero. In the case of cell E1 the ion current disappears at the right phase boundary (x=L) and, hence, the concentration gradient according to Eq. (7.90) is constant:

$$\left.\frac{\partial c_{O^{2-}}}{\partial x}\right|_{x=L} = -\frac{\sigma_{O^{2-}}}{\sigma}\frac{i}{2FD^\delta}, \tag{7.91}$$

while, at the phase boundary of the reversible electrode (x=0), the concentration is maintained constant by the gas phase:

$$c_{O^{2-}}(x=0) = c_0. \tag{7.92}$$

Using the initial condition $c(x,t=0)=c_0$ (homogeneous starting condition), the solution is

$$c(x,t) = c_0 - \frac{\sigma_{O^{2-}}}{\sigma}\frac{iL}{2FD^\delta}\left\{\frac{x}{L} + \Theta\left(\frac{t}{\tau^\delta},\frac{x}{L}\right)\right\} \tag{7.93}$$

with the abbreviation

$$\Theta\left(\frac{t}{\tau^\delta},\frac{x}{L}\right) \equiv -\frac{8}{\pi^2}\sum_0^\infty(-1)^m(2m+1)^{-2}\exp\left[-(2m+1)^2\frac{t}{\tau^\delta}\right]\sin\left[(2m+1)\frac{\pi x}{2L}\right] \tag{7.94a}$$

and

$$\tau^\delta = 4L^2/(\pi^2 D^\delta). \tag{7.94b}$$

In a similar manner to the chemical relaxation case (see Section 6.7) the time–dependence reduces to a $\sqrt{t}$ law in the short–term approximation and, for longer periods of time, to an exponential law. When $t \to \infty$ (in practice $t \gtrsim 5\tau^\delta$), that is in the stationary state, the Θ function disappears and a linear concentration profile is the result, as it should be on account of $\partial c/\partial t = D^\delta \partial c^2/\partial x^2 = 0$ (Fig. 7.26). The behaviour on switching off the current is found analogously, if the stationary profile is taken as the initial condition. The reader is referred to Refs. [324,431] for an exhaustive discussion.

---

[89] $\partial c_O = \partial c_{O^{2-}} = -\partial c_{e^-}/2$.

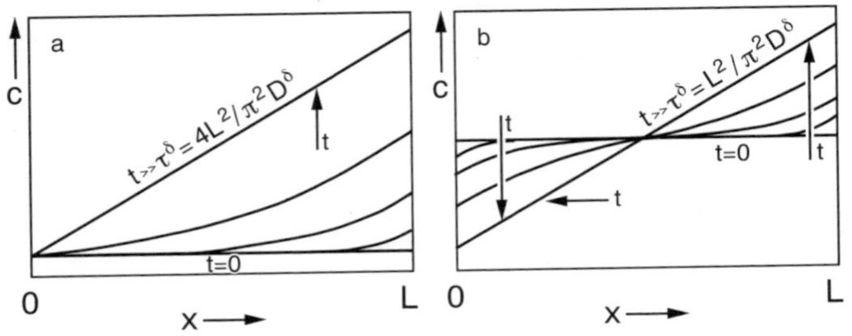

Fig. 7.26: Concentration profile during polarization experiments for cells containing one (a) and two (b) selective blocking electrodes [555].

Our measurement signal, however, is a voltage, not a concentration. Therefore let us consider Eq. (7.4) to establish the link between the two. According to Eqs. (7.1, 7.89) it follows that

$$\frac{\partial \widetilde{\mu}_{e^-}}{\partial x} = \frac{iF}{\sigma} - \frac{\sigma_{O^{2-}}}{4\sigma}\frac{\partial \mu_{O_2}}{\partial x}. \tag{7.95}$$

For i=0 this is identical to a result we have already met in EMF measurement. The voltage drop over the sample MX is now:

$$U_{MX} = \frac{1}{F}\int_0^L \frac{iF}{\sigma}dx - \frac{1}{F}\int_0^L \frac{\sigma_{O^{2-}}}{\sigma}\left(\frac{d\mu_{O_2}}{dx}\right)dx. \tag{7.96a}$$

$$\simeq \frac{iL}{\sigma} - \frac{1}{4F}\int_{\mu_{O_2}(0)}^{\mu_{O_2}(L)} \frac{\sigma_{O^{2-}}}{\sigma}d\mu_{O_2} \tag{7.96b}$$

$$\simeq \frac{iL}{\sigma} - \frac{1}{4F}\int_{c_{O^{2-}}(0)}^{c_{O^{2-}}(L)} \frac{\sigma_{O^{2-}}}{\sigma}\frac{d\mu_{O^{2-}}}{dc_{O^{2-}}}dc_{O^{2-}}. \tag{7.96c}$$

This important relationship[90] relates the voltage to drift and stoichiometry contributions and includes Ohm's law and the generalized Nernst equation (Eq. (7.36)). The integrand in Eq. (7.96c) is basically identical to D^δ/σ_{e^-}, so that, within the limits of our approximation, there is a linear relationship between the measurement signal and the concentration boundary values:

$$U_{MX} = \frac{iL}{\sigma} - \frac{2FD^\delta}{\sigma_{e^-}}\left(c_{O^{2-}}(L) - c_{O^{2-}}(0)\right). \tag{7.97}$$

For long times Eq. (7.93) yields the simple equation (cf. Section 6.5)

$$\begin{aligned} U_{MX} &= \frac{iL}{\sigma} + \frac{\sigma_{O^{2-}}}{\sigma}\frac{iL}{\sigma_{e^-}}\left(1 - \frac{8}{\pi^2}\exp - \frac{t}{\tau^\delta}\right) \\ &= IR + IR_{eon}\frac{R_{eon}}{R_{eon}+R_{ion}}\left(1 - \frac{8}{\pi^2}\exp - \frac{t}{\tau_d}\right). \end{aligned} \tag{7.98}$$

[90] It will be generalized in Subsection 7.3.4c.

We note that apart from the slight modification by the factor $\frac{8}{\pi^2} \simeq 1$, the heuristic approximation (Eq. (7.86)) is actually correct for long times. Eq. (7.94b) provides an atomistic interpretation of the time constants through D^δ and L. The comparison with the heuristic result (Eq. (7.82)) and with Eq. (6.54) ($D^\delta \propto \sigma_O^\delta / c_O^\delta$) leads[91] now to the interpretation of our diffusion–related capacitance [391,392] as[92]

$$C^\delta = \alpha V \left(\frac{d\mu_O}{dc_{O^{2-}}} \right)^{-1} \propto (c_0^\delta)^{-1}. \tag{7.99}$$

V is the volume enclosed by the electrodes, the proportionality factor α is $16F^2/\pi^2$ in the case of cells E1, I1 and $4F^2/\pi^2$ for cells E2 and I2. Thus, C^δ is apart from a proportionality factor identical to the chemical capacitance introduced in Section 6.7.4; it is inversely proportional to c_O^δ ($\propto D_O^\delta/\sigma_O^\delta$, cf. Eq. (6.54)), and gives the change in concentration as the chemical potential is changed, and, hence, is strongly dependent on the phase width. It measures the differential storage capacity of oxygen in the oxide[92,93] (cf. Section 6.7.4).
Our simple equivalent circuit in Fig. 7.23 breaks down at short times, at which according to Section 6.5 a $\sqrt{t/\tau^\delta}$ law is obtained[94]. Figure 7.27 shows a more

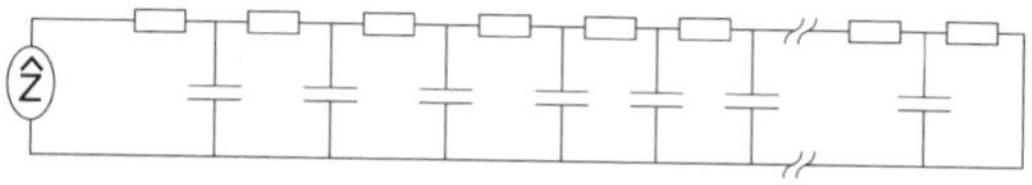

Fig. 7.27: Equivalent circuit of a Warburg impedance with differential local resistances and capacitances [603]; cf. also the more general circuit in Fig. 7.33.

appropriate equivalent circuit that describes stoichiometry polarization by taking into account an infinite number of differential elements. It will be shown in the discussion of impedance spectroscopy (Section 7.3) that the associated complex resistances can be regarded as Warburg impedances. Analogous considerations apply to the depolarization.
In the case of cell E2 the stoichiometry polarization is faster by a factor of 4 because two blocking electrodes are used. The boundary condition Eq. (7.91) now applies for both electrode contacts. The concentration function for the polarization is

$$c(x,t) = c_O - \frac{\sigma_{O^{2-}}}{\sigma} \frac{iL}{2FD^\delta} \left\{ \frac{x}{L} - \frac{1}{2} + \Xi\left(\frac{t}{\tau^\delta}, \frac{x}{L} \right) \right\} \tag{7.100a}$$

[91] Note that $\sigma_O^\delta = \sigma_{eon}\sigma_{ion}/\sigma \propto (R_{eon} + R_{ion})^{-1} = (R^\delta)^{-1}$.

[92] Cf. Section 4.2. With respect to the fact that the chemical capacitance (C^δ) can be treated as an electrical capacitance, see Ref. [392].

[93] A vivid exemplification of a chemical capacitance in a generalized sense is the following: It is often observed that the bodies of different persons react differently on the same input of food. While some stay slim others easily gain weight: They markedly differ in the chemical capacitance.

[94] This follows because $\sin[(2m+1)\pi/2] = (-1)^m$ and because of the identity given in footnote 90 on page 314.

with

$$\Xi = \left(\frac{t}{\tau^\delta}, \frac{x}{L}\right) = \frac{4}{\pi^2}\sum_0^\infty (2m+1)^{-2} \exp\left[-(2m+1)^2 \frac{t}{\tau^\delta}\right] \cos\left[(2m+1)\frac{\pi x}{L}\right], \tag{7.100b}$$

whereby

$$\tau^\delta = L^2/\left(\pi^2 D^\delta\right). \tag{7.100c}$$

On account of Eq. (7.100c) C^δ is also lower by the same factor of 4. The profiles now are symmetrical about the centre of the sample (see Fig. 7.26). Attributing the electronic conductivity measured in the stationary state to the exact stoichiometry represents an important difference between cell E1 and cell E2. This will be dealt with in the next section.

The relationships in cells I1 and I2 are analogous, apart from the fact that an additional ion conductor must be included. The voltage drop over the sample MX is now given by the difference between the electrochemical potentials of the ions[95]

$$U_{MX} = -\frac{1}{2F}\left(\widetilde{\mu}_{O^{2-}}(L) - \widetilde{\mu}_{O^{2-}}(0)\right) = -\frac{1}{2F}\int_O^L \left(\partial\widetilde{\mu}_{O^{2-}}/\partial x\right) dx. \tag{7.101}$$

It is of course necessary to take into account the voltage drops in the ion conductors implemented in series; however, these additional bulk or interfacial contributions do not affect the long–term behaviour of the cell (i.e. the diffusion). Thus, all of the relationships derived for U_{MX}, apply also here provided we exchange the suffixes

[95]For proof let us consider cell I2 (on page 425) and assume transfer equilibrium for the non-blocked charge carrier. The contact on the metal–side (metal, O_2/ion conductor) is indicated with 1 or (on the other side of the cell) with 8, the contact on the ion conductor side with 2 and 7 respectively. The ion conductor/sample contacts are denoted by 3 and 6 on the ion conductor side, and on the sample–side by 4 and 5. We then have

$$\begin{aligned} FU = \widetilde{\mu}_{e^-}^{(1)} - \widetilde{\mu}_{e^-}^{(8)} &= \frac{1}{2}\widetilde{\mu}_{O^{2-}}^{(2)} - \frac{1}{2}\widetilde{\mu}_{O^{2-}}^{(7)} - \left(\frac{1}{4}\mu_{O_2}^{(1)} - \frac{1}{4}\mu_{O_2}^{(8)}\right) \\ &= \frac{1}{2}\mu_{O^{2-}}^{(2)} - \frac{1}{2}\mu_{O^{2-}}^{(7)} - \frac{1}{4}\left(\mu_{O_2}^{(1)} - \mu_{O_2}^{(8)}\right) - \left(F\phi^{(2)} - F\phi^{(7)}\right). \end{aligned}$$

Because of $\mu_{O^{2-}} \simeq$ const in the ion conductor $\mu_{O^{2-}}^{(2)} - \mu_{O^{2-}}^{(7)} = \mu_{O^{2-}}^{(3)} - \mu_{O^{2-}}^{(6)} = \widetilde{\mu}_{O^{2-}}^{(3)} - \widetilde{\mu}_{O^{2-}}^{(6)} + F\phi^{(3)} - F\phi^{(6)} = \widetilde{\mu}_{O^{2-}}^{(4)} - \widetilde{\mu}_{O^{2-}}^{(5)} + F\phi^{(3)} - F\phi^{(6)}$. Since $\phi^{(3)} - \phi^{(2)} \equiv \Delta\phi_I$ and $\phi^{(7)} - \phi^{(6)} \equiv \Delta\phi_{II}$ it follows at the end if the partial pressures on both sides are the same that $U = \left\{\frac{1}{2F}\widetilde{\mu}_{O^{2-}}^{(4)} - \frac{1}{2F}\widetilde{\mu}_{O^{2-}}^{(5)}\right\} + F\left(\Delta\phi_I + \Delta\phi_{II}\right)$. The term in the braces is $U_{O^{2-}}$ and refers exclusively to the sample, while $\Delta\phi_I + \Delta\phi_{II}$ refers to the potential drops over the two ion conductor samples. If contact resistances also occur at other phase boundaries the drop in electrochemical potential does not equal zero; in the stationary state it is given by the product of current and contact resistance. Since, on the time scale of the stoichiometry polarization, electrical bulk processes and charge transfer processes actually behave in a quasi stationary manner, we may write that $U = \{...\} + \Sigma_i IR_i$, whereby i refers to all parts outside the sample.

"eon" and "ion". Specifically in the stationary state we measure σ_{ion} (constant phase boundary resistances and ZrO_2 resistance must be separated) and in the transient range D^δ.

Figures 7.28 to 7.30 show literature examples of the oxides $YBa_2Cu_3O_{6+x}$, PbO and $SrTiO_3$. Figures 7.28 and 7.29 refer to the blocking of electrons, Figure 7.30 to an ion blockage. It is of interest that no explicitly blocking electrodes were used in the

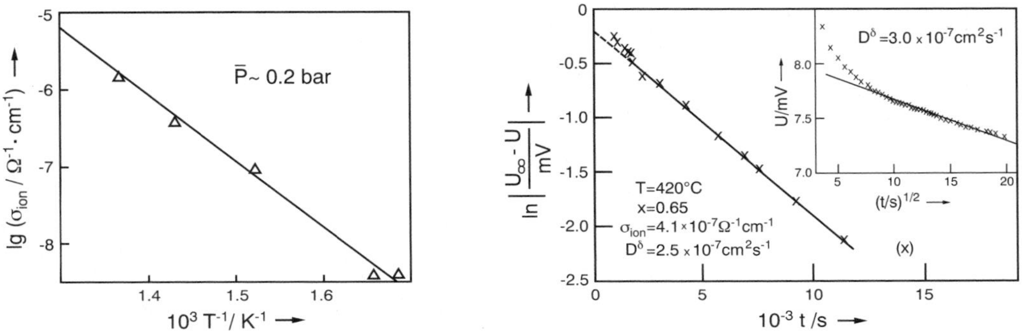

Fig. 7.28: Blocking the electrons permits the determination of the ionic conductivity (stationary behaviour) and of D^δ (transient behaviour) of the high–temperature superconductor material $YBa_2Cu_3O_{6+x}$. The comparatively high values are responsible for the rapid equilibration with the oxygen atmosphere during conditioning. The determination of the ionic conductivities, which is many orders of magnitude less than that of σ_{eon}, and the good agreement of the D^δ values obtained from long–term and short–term measurements (see Insert) are noteworthy. Sealing of the free surfaces by a suitable glass is important [604].

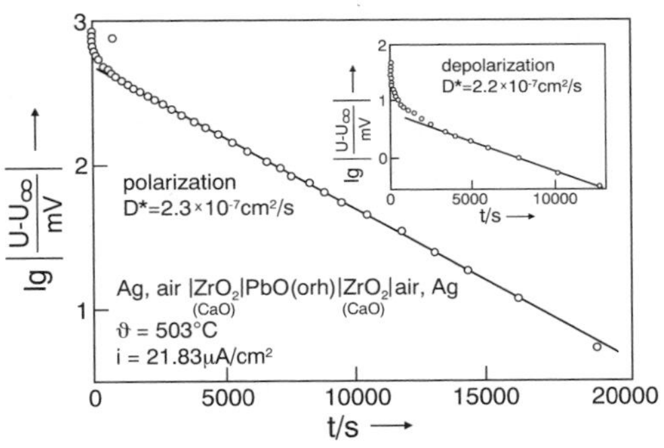

Fig. 7.29: Transient electron blockage in orthorhombic (yellow) PbO. The figure shows the good agreement between the switching–on and switching–off process. The noncontacted side surfaces are covered with glass [117].

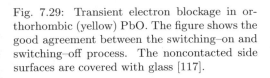

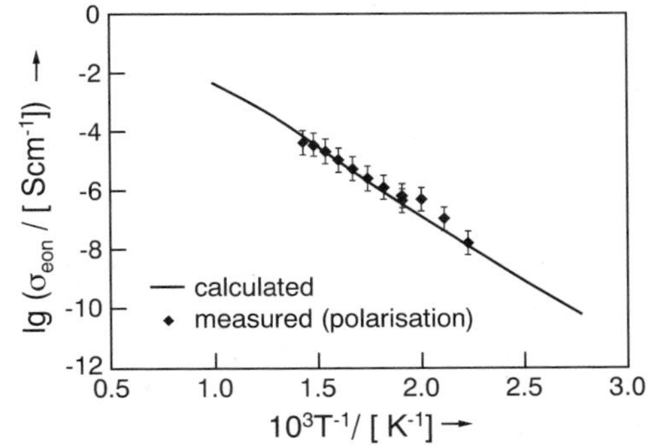

Fig. 7.30: In the case of $SrTiO_3$ it is possible to determine σ_{eon} and D^δ by electrochemical ion blockage, without using special electrodes and without encapsulation, simply on the basis of inhibited exchange kinetics. The electron conductivity determined is plotted as a function of temperature (cf. Chapter 5 for the calculation) [193].

last case. The polarization of the ions succeeded simply because the experiment was carried out in a parameter range where the exchange reaction with the gas phase is adequately kinetically inhibited (cf. Sections 5.6 and 6.8). Hence, the electrodes used must not act as catalysts in this respect.

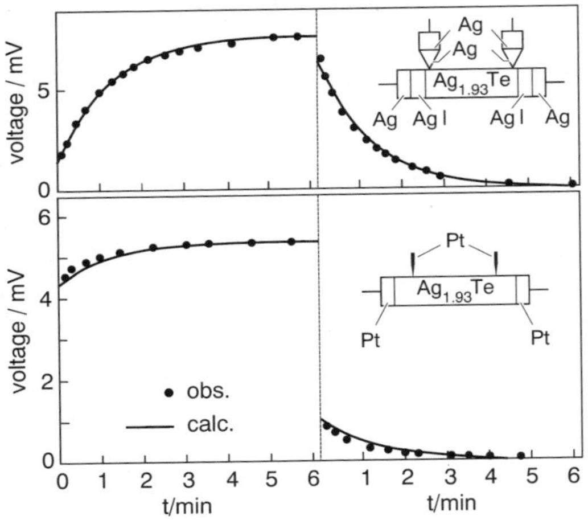

Fig. 7.31: Galvanostatic polarization of $Ag_{1.93}Te$ with ionic (top) and electronic electrodes (bottom). The fact that the voltage is measured separately via additional probes, only affects the analysis slightly (cf. four–point measurement: Section 7.3.7). From Ref. [324].

Figure 7.31 reproduces Yokota's classical polarization and depolarization curves for "Ag_2Te", in which, on the one hand, ionic electrodes (above), and, on the other hand, electronic electrodes (below) have been employed (corresponding to cells E2, I2). The independent results for diffusion coefficients and partial conductivities are consistent [324].

Like the conductivity, D^δ is obtained more precisely as a value averaged over the stoichiometry range covered by a single measurement. If this is small with respect to the total variation covered in the measurement series, then the construction of histograms makes D^δ and $\sigma_{eon,ion}$ accessible as functions of stoichiometry and, hence, of P_{O_2}. A more subtle discussion will now follow.

b) Stationary polarization state: Wagner–Hebb analysis and a simple correction procedure

The changes of conductivity with stoichiometry have been neglected in the treatments until now[96]. In other words: For appreciable gradients the conductivity value of the nonblocked species derived from currents and voltages in the stationary state is to be regarded as a mean value. For further analysis [555] let us first consider cells with blocking electrodes on each side (cells E2, I2) and consider the stationary state. As shown in Fig. 7.26b the profiles are symmetrical about the centre of the sample to a first approximation, leaving the midpoint (central layer) invariant. In addition, the concentration profile is linear in the stationary state corresponding to D^δ being approximately constant (cf. here Sections 6.4 and 6.5). Nonetheless, the profile can

[96] Cf. [15] with respect to higher order corrections.

now be utilized for a more precise, subsequent assignment of the measured averaged conductivity[96] to a defined stoichiometry. The result is not identical with the initial value, but it is[97]:

$$\langle \sigma_k \rangle = \frac{\sigma_k(0) - \sigma_k(L)}{\ln \sigma_k(0) - \ln \sigma_k(L)}; \tag{7.102}$$

where the parameters $\sigma_k(0)$ and $\sigma_k(L)$ are the boundary values. If $\sigma_k(0) \propto P^{N_k}(0)$ and $\sigma_k(L) \propto P^{N_k}(L)$ the measured conductivity value can be assigned to a partial pressure P, which is calculated from

$$\langle P^{N_k} \rangle = \frac{P^{N_k}(0) - P^{N_k}(L)}{\ln P^{N_k}(0) - \ln P^{N_k}(L)}. \tag{7.103}$$

When N_k is small this mean value is not very different from the geometric mean. The boundary values are readily determined from the initial value P(t=0) and the steady state voltage using the "Nernst equation" (see page 446)[98]. The above equation is also useful if conductivity measurements are carried out in the stationary state of a chemical polarization and can be used to correct the first order approximation in the case of chains including a reversible and a blocking electrode [555].
In order to construct the complete dependence on stoichiometry it is desirable to equilibrate the samples at various P_{O_2} before polarization. At very low polarization, subsequent correction of the mean is unnecessary.

There is not much to be gained by varying the polarization current or polarization voltage in the case of cells E2, I2. However, there is a great deal to be gained from such a procedure in the case of asymmetrical polarization cells (cells E1, I1), since here the stoichiometry is maintained constant on one side and the stoichiometry range covered is extended on the other side by increasing the polarization voltage (cf. Fig. 7.26a). In this case we speak of a Wagner–Hebb analysis [325,574]. Let us take ion blockage as an example (cell E1).
There the current in the stationary state is a pure electron current

$$i = -\frac{\sigma_{e^-}}{F}\frac{\partial \widetilde{\mu}_{e^-}}{\partial x} \tag{7.104a}$$

or after integration

$$i = -\frac{1}{L}\int_{\widetilde{\mu}_{e^-}(0)}^{\widetilde{\mu}_{e^-}(L)} \frac{\sigma_{e^-}}{F} d\widetilde{\mu}_{e^-} = \frac{1}{4L}\int_{\mu_{O_2}(0)}^{\mu_{O_2}(L)} \frac{\sigma_{e^-}}{F} d\mu_{O_2}. \tag{7.104b}$$

---

[97] According to the treatment in Section 5.8 the integral $\int_0^L \sigma_k^{-1} dx$ has to be evaluated in order to obtain $L/\langle \sigma_k \rangle$. For $\sigma_k = (\sigma_k(L) - \sigma_k(0))(x/L) + \sigma_k(0)$ Eq. (7.102) is obtained. The dielectric effects of the sample are unimportant in the stationary state. Cf. also Ref. [605].

[98] When the concentration profile is linear, the P^{N_k} profile is also linear, and $P^{N_k}(t=0) = \frac{P^{N_k}(L)+P^{N_k}(0)}{2}$; using $|U(t=\infty)| = \frac{RT}{4F}\left|\ln \frac{P(L)}{P(0)}\right|$ we obtain P(L) and P(0).

The expression on the right hand side is obtained because $\nabla\widetilde{\mu}_e = -\frac{1}{4}\nabla\mu_{O_2}$ since $\nabla\widetilde{\mu}_{O^{2-}} = \mathbf{0}$. This also implies that the voltage is of the Nernst type in spite of the current flow and in spite of electronic conduction. This perhaps surprising finding was explained in the previous section (see page 446). The electronic conductivity is precisely arrived at differentiating the current with respect to the oxygen potential at the contact of the blocking electrode[99] (x = L). Since $\partial\mu_{O_2}(L) = \partial(\mu_{O_2}(L) - \mu_{O_2}(0)) = -4\partial(\widetilde{\mu}_{e^-}(L) - \widetilde{\mu}_{e^-}(0)) = -4F\partial U$ it follows that

$$\partial i/\partial U = \sigma_{e^-}/L \tag{7.105}$$

and, hence, σ_{e^-} is obtained from the slope of the current–voltage curve. Equation (7.104b) can be integrated if the dependence on partial pressure is known; this is usually a power law of the form[100] $\sigma_{h^\cdot} \propto P^N \propto \sigma_{e'}^{-1}$. Since $\mu_{O_2} = \text{const} + RT\ln P_{O_2}$, it follows for the integrand that

$$\frac{\sigma_{h^\cdot}}{\sigma_{h^\cdot}(0)} = \exp\frac{N(\mu_{O_2} - \mu_{O_2}(0))}{RT} = \frac{\sigma_{e'}(0)}{\sigma_{e'}}, \tag{7.106}$$

if 0 refers to the value fixed on the reversible side. Integration (Eq. (7.104b)) yields $i = i_{eon} = \text{fct}(\mu_{O_2} - \mu_{O_2}(0))$ or $i = i_{eon} = \text{fct}(U)$. The result is

$$i_{eon} = \frac{RT}{4NFL}\left\{\sigma_{e'}(0)\left[\exp\left(+\frac{4NFU}{RT}\right) - 1\right] + \sigma_{h^\cdot}(0)\left[1 - \exp\left(-\frac{4NFU}{RT}\right)\right]\right\} \tag{7.107}$$

and simplifies for N=1/4 to the often cited Wagner–Hebb relation [325].
As it should be the case, there is an ohmic range for small voltages, viz.

$$i|_{U\longrightarrow 0} = \frac{U}{L}(\sigma_{e'}(0) + \sigma_{h^\cdot}(0)) = \frac{U}{L}\sigma_{eon}(0). \tag{7.108}$$

When the voltage is increased, there is a overproportional contribution from the conduction electrons, since the partial pressure of oxygen on the variable side is decreased (corresponding to an increase of the charge carrier concentration). At large voltages a Tafel region is created:

$$i|_{U\longrightarrow\infty} = \frac{RT}{4NFL}\sigma_{e'}(0)\exp\left(+\frac{4NFU}{RT}\right). \tag{7.109}$$

---

[99] If F is the antiderivative of f, that is $\Theta \equiv \int_{x_A}^{x} f(\xi)d\xi = F(x) - F(x_A)$, and if x in contrast to x_A is a variable integration limit, then $d\Theta/dx$ is equal to $dF(x)/dx = f(\xi = x)$ and hence determined via differentiation with respect to the variable limit. Note that $\mu_{O_2}(x=0) = \text{const}$.

[100] Obviously it is not necessary that $(\partial/\partial x)\mu_{O^{2-}} = 0$, i.e. $(\partial/\partial x)\phi = 0$, as sometimes falsely stated in the literature. This field–free situation for which Eq. (7.107) was originally derived, corresponds, for oxides, to a power law of N = 1/4. In the case of a power law with N≠1/4 the relevant chemical and electrical potential gradients are nonzero but proportional to each other [606,431,607]. If a power law is not fulfilled, the situation is more complex [607].

If $\sigma_{h^\cdot}(0) \gg \sigma_{e'}(0)$, then a plateau is obtained in the intermediate range; this is characterized by the holes as decisive charge carriers being reduced in their concentration

$$\frac{RT}{4NFL}\sigma_{h^\cdot}(0)\left[1-\exp\left(-\frac{4NF|U|}{RT}\right)\right] \longrightarrow \text{const} \propto \sigma_{h^\cdot}(0). \tag{7.110}$$

Figure 7.32 shows measurements of Y–doped ZrO_2 and ThO_2, which permit determination of the partial electronic conductivities of these ion conductors.

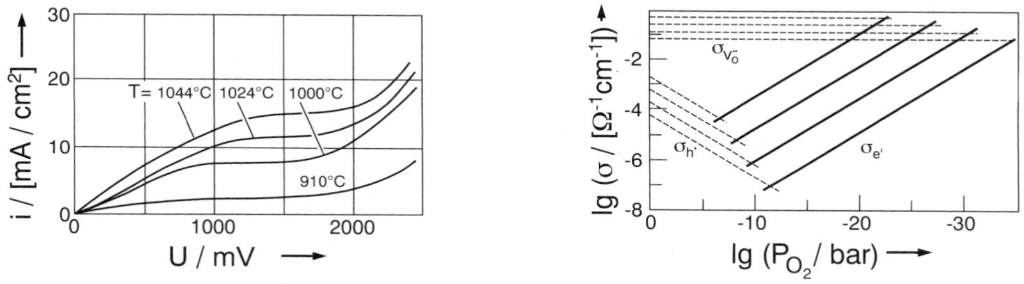

Fig. 7.32: Wagner–Hebb analysis for determination of the n and p conductivity in Y–doped O^{2-} conducting fluorites with the aid of the cell $\ominus N_2, Pt|YSZ$ or $YST|air, Pt\oplus$ (left: current voltage curve for $Th_{0.9}Y_{0.1}O_{1.95}$ (YST), right: electronic and ionic partial conductivities for $Zr_{0.9}Y_{0.1}O_{1.95}$ (YSZ)). From Ref. [608].

c) Effects of various valence states on the analysis

The assumptions may lose their validity in a material such as $YBa_2Cu_3O_{6+x}$, in which oxygen defects can possess different degrees of ionization and can all possibly be mobile [432]. Here we will treat the changes brought about by the occurrence of different charge states briefly and in a semiquantitative manner[101]. Let us consider the $ZrO_2/YBa_2Cu_3O_{6+x}$ contact and assume that O^{2-}, O^- and O^0 are mobile in $YBa_2Cu_3O_{6+x}$. However, the phase boundaries are only permeable to O^{2-}. Since on account of the internal equilibrium O^- and O^0 can dissociate into O^{2-} and $h^\cdot$, the stationary total current is not carried by $i_{O^{2-}}$ or by $i_{ion} = i_{O^{2-}} + i_{O^-}$ but by $i_{\{O\}} \equiv i_{O^{2-}} + 2i_{O^-} - 2Fj_{O^0} = i_{ion} + i_{O^-} - 2Fj_{O^0}$.

In the case of ion blockage the stationary current is $i_{\{e\}} \equiv i_{e^-} - i_{O^-} + 2Fj_{O^0}$. This corresponds to the fluxes of the corresponding "conservative ensembles", as introduced in Section 6.6.1. The basic current and voltage equations are now given in

[101] The exact treatment is given in Refs. [393,431,432].

generalized form as

$$
\begin{aligned}
\mathrm{i}_{\{O\}} &= \frac{\sigma_{\{O\}}}{\sigma}\mathrm{i} + 2\mathrm{F}\mathrm{D}^{\delta}\frac{\partial \mathrm{c}_{\{O\}}}{\partial \mathrm{x}} = \\
\mathrm{i}_{\{e\}} &= \frac{\sigma_{\{e\}}}{\sigma}\mathrm{i} + \mathrm{F}\mathrm{D}^{\delta}\frac{\partial \mathrm{c}_{\{e\}}}{\partial \mathrm{x}} \\
\mathrm{U}_{\mathrm{MX}} &= \begin{cases} \dfrac{\mathrm{iL}}{\sigma} - \dfrac{1}{4\mathrm{F}}\displaystyle\int \frac{\sigma_{\{O\}}}{\sigma}\mathrm{d}\mu_{\mathrm{O}_2} & \text{(cells E1, E2)} \\ \dfrac{\mathrm{iL}}{\sigma} + \dfrac{1}{4\mathrm{F}}\displaystyle\int \frac{\sigma_{\{e\}}}{\sigma}\mathrm{d}\mu_{\mathrm{O}_2} & \text{(cells I1, I2)} \end{cases} .
\end{aligned}
\tag{7.111}
$$

The parameters $c_{\{O\}}$, $c_{\{e\}}$ represent the ensemble concentrations defined in Section 6.6. Note that $\delta c_{\{O\}} = \delta c_O = -\delta c_{\{e\}}/2$. The conductivities of $\sigma_{\{O\}}$ and $\sigma_{\{e\}}$ are defined by $\sigma_{O^{2-}} + 2\sigma_{O^-}$ and $\sigma_{e^-} - \sigma_{O^-}$. In this case D^{δ} is determined by Eq. (6.89) (Chapter 6). Assuming constant transport coefficients the stationary state finally takes the form[101]

$$
\mathrm{U}_{\mathrm{MX}} = \begin{cases} \mathrm{iL}\dfrac{1 - 2\mathrm{Bh}_{\{e\}}}{\sigma_{\{e\}}} & \text{(cells E1, E2)} \\ \mathrm{iL}\dfrac{1 - 2\mathrm{Bh}_{\{O\}}}{\sigma_{\{O\}}} & \text{(cells I1, I2)} \end{cases}
\tag{7.112}
$$

with the abbreviations $B = \left(1 + 2h_{\{O\}} + 2h_{\{e\}}\right)^{-1}$, $h_{\{O\}} = \left(\sigma_{O^-} + 2s\right)/\sigma_{\{O\}}$ and $h_{\{e\}} = \left(\sigma_{O^-} + 2s\right)/\sigma_{\{e\}}$. We find the analogous result on generalizing the Wagner–Hebb analysis in differential form. In particular, it is found that in the extreme case of very large contributions by neutral oxygen ($s \gg \sigma_{O^-}, \sigma_{O^{2-}}$) there is no polarization at all. With a counterflux of neutral oxygen the ionic carriers can flow unimpeded as an effective internal electron flux.

The analysis of transients differs in the short–term behaviour, since — even concentration and transport parameters remain constant — the parameters h and B are involved (see above). In the case of the long–term approximation, however, D^{δ} is obtained from the slope as before. Nonetheless — as discussed in Chapter 6 — the interpretation of this parameter changes.

d) Interference of double–layer and stoichiometry effects: Electrical versus chemical capacitance

It has already been pointed out in Chapter 6 in the discussion of chemical diffusion that at boundaries chemical diffusion and space charge effects cannot, in general, be separated from each other. The same naturally applies here. However, the unified treatment of stoichiometry and space charge polarization can, under general conditions, only be carried out numerically [586]. An approximate analytical solution [587], in frequency space, which is valid to a good approximation for a large parameter window, is given in Section 7.3.6. It corresponds to the equivalent circuit

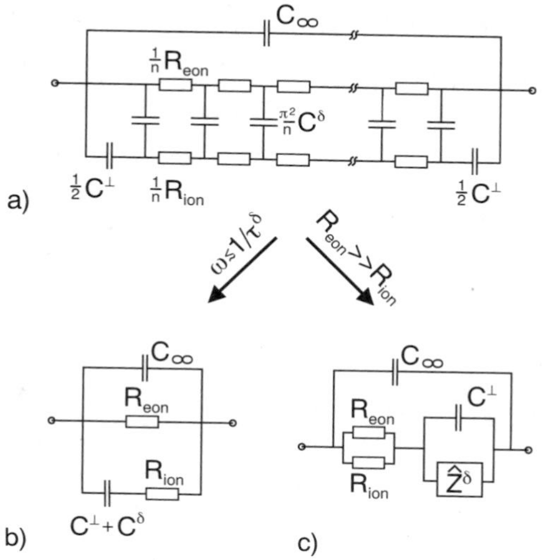

Fig. 7.33: Approximate equivalent circuit for cell E2, taking into account double–layers and the stoichiometry effect of the ion blockage. Approximation b) results for low frequencies and corresponds to the treatment above (Section 7.3.4). Approximation c) is a useful approximation for ion conductors (Randles' circuit [609, 610]). The factor π^2 occurring in (a) is not totally precise. In the case of impedance spectroscopy the factor 12, which is also not precise, is given instead (see Section 7.3.6). According to Ref. [610].

shown in Fig. 7.33. The contact resistance of the nonblocking species is neglected. For long times (small frequencies) a circuit analogous to Fig. 7.23 (apart from additional boundary layer impedances) is a good approximation; however the serial capacitance includes the double–layer and stoichiometry effects. The ratio of electrical (that is $C^\perp_{ion}$) to chemical (C^δ) capacitance decides which phenomenon dominates. We should note that when the double–layer capacitance is dominant (for long times or small frequencies) a polarization analogous to stoichiometry polarization is set up, but it remains uninfluenced by D^δ [610]. We shall return to this in Section 7.3.6. At the same time we shall comment on complications involving ultrashort behaviour or high frequency behaviour.

e) Nonideal selectively blocking electrodes

The relationships become complex if transfer hindrances also have to be considered for the nonblocked species, or if the blockage is not ideal. In the first case it turns out that in spite of the fact that we have to distinguish between different carriers and hence individual transfer mechanisms, the implementation of an additional parallel $R^\perp$–$C^\perp$ circuit in series (as already done in Fig. 7.23) is a reasonable first correction[102]. The steady state resistance is correspondingly increased. In the second case not only this steady state value is decreased but also the form of the polarization curve is affected [610]. We come back to this in Section 7.3.6.

[102] This in fact depends on the time resolution (d.c.) or on the frequency range (a.c.), see Refs. [611,612].

f) Bulk polarization because of internal boundaries

Like selectively blocking electrodes grain boundaries can also exhibit a filter–effect. This is due to changes of the transference number at grain boundaries. Hence the grain boundary is not only electrically active due to its own resistance and capacitance, also a bulk polarization (as discussed above) can be induced as was measured by impedance spectroscopy and d.c. experiments for $SrTiO_3$ [613,307, 614].

7.3.5 Coulometric titration

Last of all we shall discuss cell T (page 424f), which is made up of the sample MX and purely ionic and purely electronic electrodes. At the polarity given, a pure ion current flows into the sample via the ionic electrodes. On contact with the electronic electrode at the latest the ion is discharged by the electron current and, hence, the stoichiometry changes as a result of current flow (coulometric titration). In contrast to the cases discussed above a stationary state is not set up.
Experimentally we titrate the neutral component into or out of the sample in this manner for a certain time Δt, then switch off the current, await homogenization and repeat the procedure. Measurement of the voltage in the currentless homogeneous state (i.e. EMF E) allows determination of the stoichiometry variation and the associated activity. The time behaviour of the voltage during relaxation yields — diffusion controlled kinetics assumed — the chemical diffusion coefficients if the boundary conditions (and naturally the initial condition) are known [574,615].
Here let us consider the homogeneous state and initially just the extreme cases. For simplicity we wish to assume that only one oxide, namely "M_2O", exists in the phase diagram under the conditions of observation. At maximum removal of oxygen from $M_2O_{1+\delta}(\delta \lesseqgtr 0)$, i.e. minimum δ ($M_2O_{1+\delta_{min}}$) there is phase equilibrium with M. In other words we find ourselves at the reduced end of the homogeneity range of the phase, i.e. the chemical potential of the metal in "M_2O" is the same as that of the pure metal. In consequence the EMF is equal to that which would occur in equilibrium with the parent metal M (cf. Section 7.2):

$$\begin{aligned} \mathrm{E}\{\mathrm{Pt}|\mathrm{M_2O_{1+\delta_{min}}}|\mathrm{ZrO_2}|\mathrm{O_2,Pt}\} &= \mathrm{E}\{\mathrm{M}|\text{“}\mathrm{M_2O}\text{”}|\mathrm{ZrO_2}|\mathrm{O_2,Pt}\} \\ &= \tfrac{1}{2}\Delta_f \mathrm{G}^\circ_{\mathrm{M_2O}} - \tfrac{1}{4}\mathrm{RT}\ln \mathrm{P_{O_2}}. \end{aligned} \qquad (7.113)$$

As reduction continues the cell voltage no longer changes, since stoichiometry changes no longer occur. All that happens is that more M is produced in the two–phase region at the expense of "M_2O".
Conversely, at maximum O content ($M_2O_{1+\delta_{max}}$) there is phase equilibrium with the external partial pressure of oxygen. Further oxydation would correspond to the evolution of oxygen. The result is that the cell voltage falls to zero if no hydrostatic pressure differences occur and P_{O_2} is also 1 bar on the other side:

$$\mathrm{E}\{\mathrm{Pt}|\mathrm{M_2O_{1+\delta_{max}}}|\mathrm{ZrO_2}|\mathrm{O_2,Pt}\} \qquad (7.114)$$

$$= E\{O_2, Pt|\text{"}M_2O\text{"}|ZrO_2|Pt, O_2\} = 0.$$

According to the treatment in Chapter 5 (see Eq. (5.120), page 169) the deviation (δ) from the Dalton composition is proportional to $\sinh\left(\mu_{O_2} - \mu_{O_2(i)}\right)$, whereby the subscript i indicates the intrinsic point ($\delta = 0$). Since the cell voltage is given by $\Delta\mu_{O_2}$, it follows that $\delta \propto x_{(i)} \sinh\left(-\left(E - E_{(i)}\right)\right)$ with $x_{(i)}$ as intrinsic mole fraction. Figure 7.34 shows that for the example of Sr–doped La_2CuO_4 there is a distinct

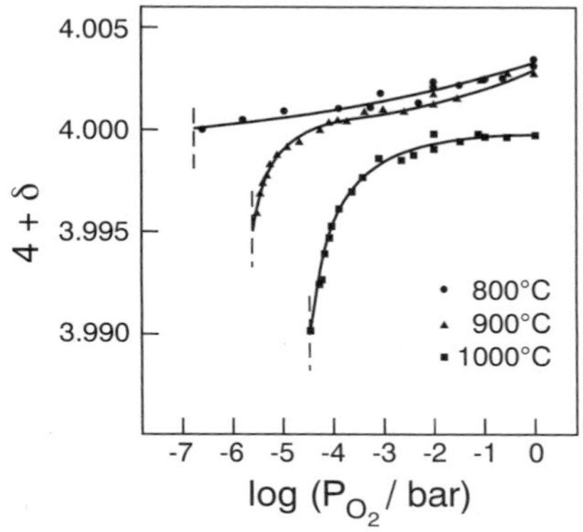

Fig. 7.34: Dependence of oxygen content in $La_{1.95}Sr_{0.05}CuO_{4+\delta}$ on $\log P_{O_2}$ at 800 ... 1000°C. The broken lines represent the decomposition partial pressures. From Ref. [616].

transition from oxygen deficiency ($\delta < 0$) to oxygen excess ($\delta > 0$) effected by coulometric titration, corresponding to the defect chemistry discussed in Sections 5.5 and 5.6.

In the case of cation conductors there are generally no sealing problems, and coulometric titration has been used intensively in such cases[103]. The evaluation is analogous. Figure 7.35 shows a typical titration curve and thus stoichiometry

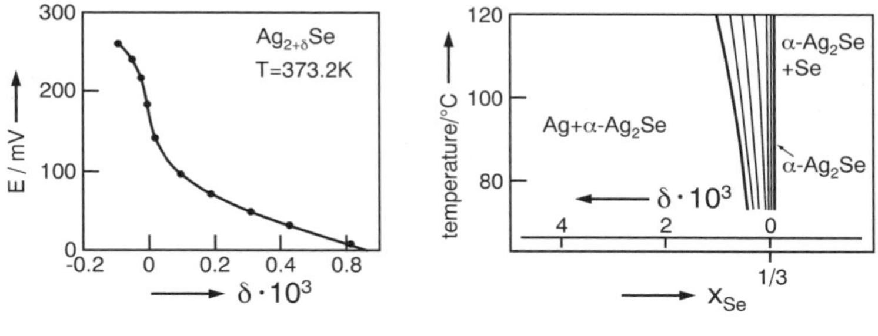

Fig. 7.35: Coulometric titration of α-Ag_2Se (left) and phase diagram (right). The β–phase is stable above ca. 130°C (cf. also footnote 104). According to Ref. [617].

variation in α–Ag_2Se[104], in which it is probably electronic defects that are in the majority. The intrinsic point ($E_{(i)}$, $\delta = 0$) is obtained (as also in Fig. 7.34) from the

[103]Conversely, in the case of oxides, it is frequently simpler and more elegant to carry out the "titration" nonelectrochemically, i.e. simply by setting and tuning the external partial pressure. If this is controlled by means of a λ probe, the construction is very similar to cell T; however, oxide and ZrO_2 are decoupled galvanically. The measurement of δ can be carried out by measuring the oxygen added or removed.

[104]The use of the symbols α and β is not uniform for silver chalcogenides.

point of inflection. These methods not only make it possible to investigate the phase diagram in a very sensitive manner (see, for example, Fig. 7.35) but also make it possible to draw detailed conclusions concerning defect concentrations, Fermi levels, band gap, effective masses and partial enthalpies and entropies of electrons and holes [617]. Because of the accuracy of the charge measurement the sensitivity is extremely high.

The analysis of the kinetics is similar to the description given above. It is necessary to take account of the boundary conditions, which are: the ion current disappears on one side and the electron current, on the other. The solution[105] permits determination of D^δ. It should be noted again that no stationary state is set up in contrast to cells R, E, I and that, after switching off, the EMF takes up a different value to before the current was switched on. Since the thermodynamic factor $\left(\frac{dE}{d\ln\delta} \propto \frac{d\mu}{d\ln c}\right)$ and the difference in the chemical capacitance $(dc/d\mu)$ can be obtained directly from the $E(\delta)$ curve, knowledge of D^δ also leads to the ambipolar conductivity σ^δ and, hence, to the conductivity of the less conducting species. Figure 7.36 shows the

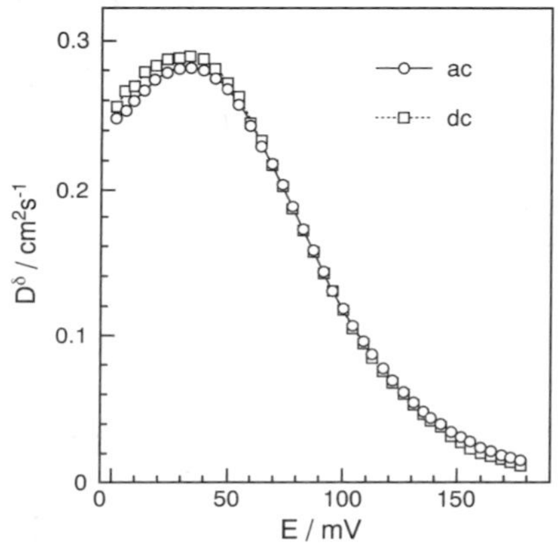

Fig. 7.36: The chemical diffusion coefficient of Ag in α–$Ag_{2+\delta}Te$ as a function of EMF of the cell Ag|AgI|Ag_2Te|Pt at 200 °C. The maximum in D^δ and, hence, the maximum in the thermodynamic factor $(d\mu_{Ag}/dc_{Ag})$ corresponds to the stoichiometric point (cf. Eq. (6.86), page 325). Note the enormous absolute values of D^δ, which make for a rapid propagation of the chemical signal "silver activity" in Ag_2Te. The D^δ–values can even exceed the diffusion coefficients of gas phase constituents. From Ref. [619].

chemical diffusion coefficient of silver in Ag_2Te as a function of EMF and, thus, of the precise stoichiometry. A combination of the measurements discussed earlier with a coulometric titration also makes it possible to follow the composition dependence of the partial conductivities in an elegant manner.

7.3.6 Impedance spectroscopy

Impedance spectroscopy, in which the voltage (current) response to an alternating current (voltage) is measured as a function of frequency, is an extremely important electrochemical method for determining kinetic parameters. The underlying concept is not fundamentally different from the previous considerations. Rather, in the case of small signals the information content is the same as in switching–on or switching–off measurements (response to step–like signals). It is only that the

[105] The general solution is given in Ref. [618].

inconvenient measurement of time dependence is now replaced by a simple measurement of amplitude and phase shift of the sinusoidal signals. Correspondingly, instead of investigating the current–voltage–time relationship once again using sinusoidal currents and appropriate boundary and initial conditions, we can employ a basic theorem of cybernetics, which is valid for small signals [620], and permits us to use the previous results. Put qualitatively it states that the response (output) to any excitation (input, e.g. of periodic nature) can be obtained from the response to a specific excitation (e.g. for the step function discussed above). This is rapidly demonstrated: For small signals we can presuppose that the system responds linearly to small excitations (cf. Section 6.1). This means that the response to a linear combination of stimuli, e.g. of step functions, is equal to the linear combination of the individual responses[106]. Since it is also possible to approximate every sine function by step functions[107] and vice versa, the corresponding response can be built up analogously. Figure 7.37 describes a transformation that is certainly not linear and belongs to the discussion in Section 6.10.

Fig. 7.37: The linear response theory does not apply to this input–output situation. From Ref. [621].

Under general conditions it is helpful to select a basic input function, which can easily be used to construct the stimulus under consideration. The response to this basic input will be termed transfer function. If we select the delta function[108], then

[106] A simple example may serve for illustration: The stimuli are currents, the responses voltages, and a simple Ohm's law is obeyed. If $U\{I\}$ is the voltage for current $I = I_1 + I_2$, then it follows that $U\{I_1 + I_2\} = R(I_1 + I_2) = RI_1 + RI_2 = U\{I_1\} + U\{I_2\}$. As R is assumed to be constant, we are concerned with the special case of time invariant small signal behaviour.

[107] Evidently every function can be approximated by rectangular functions, if the width of the rectangles ($\Delta\tau$) is only made small enough. A single rectangular function with the onset at $n\Delta\tau$ (i.e. zero to the left of $n\Delta\tau$, identical to the function value ($f_n(n\Delta\tau)$) between $n\Delta\tau$ and $(n+1)\Delta\tau$ and again zero on the other side of $(n+1)\Delta\tau$) can be written as the difference between two step functions, $H(t - n\Delta\tau) - H(t - n\Delta\tau - \Delta\tau)$, so that overall we get as approximation $\Sigma_n f_n(n\Delta\tau)(H(t - n\Delta\tau) - H(t - n\Delta\tau - \Delta\tau))$.

[108] The delta function $\delta(x)$ is actually a distribution, that disappears for all $x \neq 0$ and approaches infinity at $x = 0$, but encloses the area 1. The delta function can, for example, be thought of as being made up of a rectangular function whose width approaches zero while its height approaches infinity, such that the product of height and width remains constant ($= 1$). Accordingly it follows that

$$\int f(x)\delta(x - b)dx = f(b),$$

we get the following simple relationship[109] for the response (Rsp stands for response) to any desired excitation:

$$\mathrm{Rsp}\{\text{arbitrary function}\} = \mathrm{Rsp}\{\text{delta function}\} * \{\text{arbitrary function}\}. \quad (7.115)$$

According to Eq. (7.115), the desired response is obtained by convolution of the basic response (Rsp{delta function}) with the respective excitation function. The delta function is now just the derivative of the step function of interest to us, so that instead of Rsp{delta function} it is also possible to use the time derivative of the response to the step function, that we already know ($\partial/\partial t\,\mathrm{Rsp}\{\text{step function}\}$)[110]. Since the Laplace transformation[111] ($\mathcal{L}$) converts the convolution into a multiplication it is more concise to write

$$\mathcal{L}[\mathrm{Rsp}\{\text{arbitrary function}\}] = \mathcal{L}[\mathrm{Rsp}\,\{\text{delta function}\}] \cdot \mathcal{L}[\{\text{arbitrary function}\}] \quad (7.116)$$

or

$$\mathcal{L}[\mathrm{Rsp}\{\text{arbitrary function}\}] = \widehat{p}\mathcal{L}[\mathrm{Rsp}\,\{\text{step function}\}] \cdot \mathcal{L}[\{\text{arbitrary function}\}]. \quad (7.117)$$

In the last equation we exploited the fact that the Laplace transformation of a derivative means to multiply the transform by the function $\widehat{p} = j\omega$ (more generally:

---

if b lies within the limits of integration. This is evident since the integral can be split up into 3 contributions, one from the lower limit to $b-\varepsilon$, one from $b-\varepsilon$ to $b+\varepsilon$ and a third from $b+\varepsilon$ to the upper limit. Only the middle one differs from zero. If ε approaches zero then

$$f(b)\int \delta(x-b)dx = f(b).$$

Evidently $\delta(x) = \delta(-x)$ is also the derivative of the step function H(x).

[109] The convolution integral is defined as $a * b \equiv \int_0^t a(\tau)b(t-\tau)d\tau$. Since it is also possible to write $\int_t^0 a(t-\tau)b(\tau)d(t-\tau) = b * a$, there is commutativity. Usually the convolution is defined via the integral from $-\infty$ to $+\infty$. Evidently both definitions are identical if the functions vanish for negative arguments.
If $h \equiv \mathrm{Rsp}\{H\}$ is the response to the step function H, then it follows according to the previous footnote 107, after exploiting linearity that $\mathrm{Rsp}\{\Sigma_n f_n(H(t-n\Delta\tau)-H(t-n\Delta\tau-\Delta\tau))\} = \Sigma_n f_n([h(t-n\Delta\tau) - h(t-n\Delta\tau-\Delta\tau)]/\Delta\tau)\Delta\tau$. The limit $\Delta\tau \longrightarrow 0$ provides the convolution integral with $a \equiv f, b \equiv dh/dt$. Equation (7.115) follows from this, but with $\dot{h}$ instead of $\mathrm{Rsp}\{\delta\}$. That both are identical is proved in footnote 110, but this also follows immediately from the definition of the delta function $\int_{-\infty}^{+\infty} \dot{h}(t-\tau)\delta(\tau)d\tau = \int_{-\infty}^{+\infty} \dot{h}(\tau)\delta(t-\tau)d\tau$; this integral is identical to $\mathrm{Rsp}\{\delta\}$ and according to footnote 108 also equal to $\dot{h}(t)$. See Ref. [620] for a more detailed treatment.

[110] The fact that the delta function δ and the step function H are related via time derivation, also applies to the responses $\mathrm{Rsp}\{\delta\}$ and $\mathrm{Rsp}\{H\}$. This results from the following relationships:
$\delta = \frac{\partial}{\partial t}H$; $\mathrm{Rsp}\{H\} = \mathrm{Rsp}\,\{\delta\} * H$ according to Eq. (7.115);
$\frac{\partial}{\partial t}\mathrm{Rsp}\{H\} = \mathrm{Rsp}\,\{\delta\} * \frac{\partial}{\partial t}H = \mathrm{Rsp}\,\{\delta\} * \delta = \mathrm{Rsp}\{\delta\}$ according to Eq. (7.115).
In addition the fact is exploited that $\frac{\partial}{\partial t}(a * b) = \frac{\partial a}{\partial t} * b = a * \frac{\partial b}{\partial t}$.

[111] $\mathcal{L}\,[f(t)] \equiv \int_0^\infty f(t)e^{-\widehat{p}t}dt, \widehat{p} = s + j\omega$ (here $s = 0$), cf. mathematics textbooks, e.g. [341,622].

$s + j\omega$). The required response is then obtained by back transformation. In the case of sinusoidal functions the Laplace transform is directly relevant. In terms of an electrical input–output situation, it can be identified with the complex impedance $\widehat{Z}$, that is the complex a.c. resistance[112]:

$$\begin{aligned}\widehat{Z} &= \mathcal{L}\left[\text{Rsp}\left\{\text{delta function}\right\}\right] = \mathcal{L}\left[\partial/\partial t\text{Rsp}\left\{\text{step function}\right\}\right] \\ &= \widehat{p}\mathcal{L}\left[\text{Rsp}\left\{\text{step function}\right\}\right].\end{aligned} \tag{7.118}$$

According to this discussion, the following procedure is advisable:
We first solve the kinetic problem for a constant current, which is switched on at t=0 (i.e. step function). The subsequent Laplace transformation yields, in a straightforward way, the desired complex impedance[113].
If the problem can be mapped by means of an equivalent circuit, then the problem is solved by application of Kirchhoff's rules: $\widehat{Z} = \Sigma_i \widehat{Z}_i$ for series connection ($\widehat{Z}_i$: complex impedance of circuit element i), $\widehat{Z}^{-1} = \Sigma_i \widehat{Z}_i{}^{-1}$ for parallel connection, whereby $\widehat{Z} = \widehat{Z}_R = R$ for a resistor and $\widehat{Z} = \widehat{Z}_C = (j\omega C)^{-1}$ for a capacitor[114].
In the case of the simple $R_\infty C_\infty$ parallel circuit, which describes the bulk behaviour, it follows that

$$\begin{aligned}\widehat{Z}_\infty &= \frac{\widehat{Z}_{R_\infty}\widehat{Z}_{C_\infty}}{\widehat{Z}_{R_\infty} + \widehat{Z}_{C_\infty}} = \left(\frac{1}{R_\infty} + j\omega C_\infty\right)^{-1} \\ &= \frac{R_\infty}{1+\omega^2\tau_\infty^2} + j\left(-\frac{R_\infty \omega \tau_\infty}{1+\omega^2\tau_\infty^2}\right) = \text{Re}\widehat{Z}_\infty + j\text{Im}\widehat{Z}_\infty.\end{aligned} \tag{7.119}$$

The same result is obtained if Eq. (7.118) is applied to the exponential solution Eq. (7.44) (see also below). It is usual to represent the complex impedance by plotting

---

[112]If the current is sinusoidal, e.g. $I = I_0 \cos\omega t$, then the stationary (i.e. forced) voltage response is also sinusoidal with the same frequency but shifted in phase, i.e. $U = U_0 \cos(\omega t + \varphi)$ (cf. the particular solution of the corresponding *linear* differential equation). On account of Euler's equation $\exp j\omega t = \cos\omega t + j \sin\omega t$, it is also possible to write $\cosh j\omega t$ instead of $\cos\omega t$. This reformulation of the observable I is not advantageous. Rather, in the a.c. calculus one defines complex parameters, e.g. $\widehat{I} = I_0 \exp j\omega t$, which themselves have no real significance, but can be conveniently treated and can be used to obtain the real parameters (amplitude, phase) elegantly. Evidently $\text{Re}\widehat{I} = I = I_0 \cos\omega t$, $|\widehat{I}| = \sqrt{(\text{Re}\widehat{I})^2 + (\text{Im}\widehat{I})^2} = \sqrt{\widehat{I}\widehat{I}^*} = I_0$ and $\arctan(\text{Im}\widehat{I}/\text{Re}\widehat{I}) = \omega t$. Analogous relationships apply to $\widehat{U}$, whereby ωt is replaced by $(\omega t + \varphi)$. The quotient $\widehat{U}/\widehat{I} = \widehat{Z}$ is the complex impedance with $\widehat{Z} = (U_0/I_0)\exp j\varphi$. In particular $|\widehat{Z}| = U_0/I_0$ and $\varphi = \arctan(\text{Im}\widehat{Z}/\text{Re}\widehat{Z})$.

[113]In general (see Eq. (7.116)) $\widehat{Z}$ is the transfer function between the Laplace transforms of excitation and response: In the case of functions of type $e^{\widehat{p}t}$ the transfer function is also directly the quotient of the forced part of the output signal and the input signal [620], as we can readily convince ourselves using a capacitance as an example. In general this follows from the Laplace transform for $\cos(\omega t + \varphi)$. For $\mathcal{L}\{\cos(\omega t + \varphi)\}/\mathcal{L}\{\cos\omega t\}$ [623] we get $\cos\varphi - \frac{\omega}{\widehat{p}}\sin\varphi$; this yields for $\widehat{p} = j\omega$ and because of the linearity of $\mathcal{L}$ the identity $\mathcal{L}\{U\}/\mathcal{L}\{I\} = \widehat{U}/\widehat{I}$.

[114]Compare the transformation of the relationships in Section 7.3.3a.

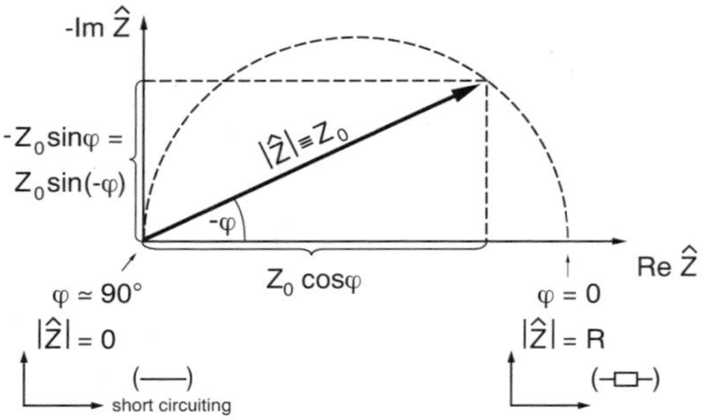

Fig. 7.38: Concerning impedance[112] of a parallel R–C circuit ($Z_0 \equiv |\widehat{Z}|$).

the negative imaginary part ($-\mathrm{Im}\widehat{Z}$) against the real part ($\mathrm{Re}\widehat{Z}$) (Fig. 7.38). As can be confirmed by substitution it follows that

$$\left(\mathrm{Re}\widehat{Z}_\infty - \frac{R_\infty}{2}\right)^2 - \left(\mathrm{Im}\widehat{Z}_\infty\right)^2 = \left(\frac{R_\infty}{2}\right)^2 . \tag{7.120}$$

This is the equation of a semicircle centred at $\mathrm{Re}\widehat{Z} = R_\infty/2$ in the first quadrant. The diameter (that is the difference between the two real values for which $\mathrm{Im}\widehat{Z}_\infty$ vanishes) is given by the d.c. resistance R_∞. The frequency at the maximum (that is at $\mathrm{Re}\widehat{Z} = R_\infty/2$) is $\omega_{\max} = \tau_\infty^{-1} = (R_\infty C_\infty)^{-1}$. It is simplest to obtain C_∞ from this. Naturally, it is also possible to obtain R_∞ and C_∞ from the frequency distribution by a fitting procedure.

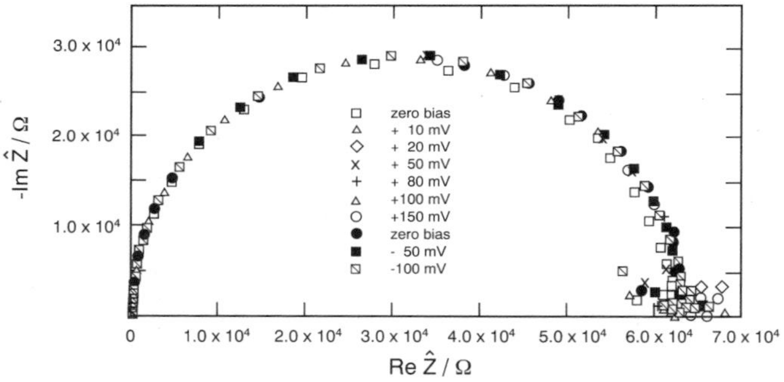

Fig. 7.39: Impedance measurements on the cell Ag|AgCl|Ag at 83°C [268].

Figure 7.39 shows impedance measurements on a single AgCl crystal (Ag electrodes). A bias dependence (i.e. dependence on an superimposed direct voltage) does not occur for the small amplitudes applied, as expected for linear systems (R_∞, C_∞ constant).

If phase boundary effects[115] are to be taken into account, then a further parallel RC circuit has to be added in series, and the impedance increases to $\widehat{Z}$ according to

$$\widehat{Z} = \widehat{Z}_\infty + \widehat{Z}^\perp . \tag{7.121}$$

[115] In order not to complicate the situation too much, bulk polarization effects caused by grain boundaries are neglected here. Cf. Section 7.3.4f.

If the relaxation times are sufficiently different[116], and, hence, the ω_{max} values as well, we obtain two joint semicircles one after the other: At very high frequencies $\widehat{Z}(\omega \sim \tau_\infty^{-1} \gg \tau^{\perp -1}) = \widehat{Z}_\infty$ (since the boundary capacitor is completely permeable), for small frequencies $\widehat{Z}(\omega \sim \tau^{\perp -1} \ll \tau_\infty^{-1}) = R_\infty + \widehat{Z}^\perp$ (the bulk capacitor blocks). The d.c. resistance, i.e. $\widehat{Z}(\omega \ll \tau^{\perp -1} \ll \tau_\infty^{-1})$, is, as expected, $R_\infty + R^\perp$ (both bulk and boundary layer capacitors block). Evaluation of diameters and maximum frequencies yields the four parameters R_∞, $C^\perp$, C_∞, $R^\perp$. Figure 7.40 shows that

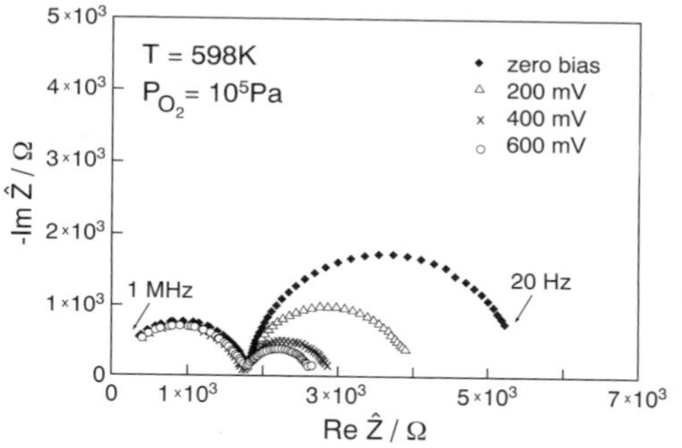

Fig. 7.40: Impedance measurements on $O_2, Pt|SrTiO_3|SrTiO_3|Pt, O_2$ as function of the superimposed d.c. bias. Electrodes are parallel to the bicrystal boundary ($\Sigma 5$ tilt grain boundary, iron content: $2\times10^{18}cm^{-3}$). Both bulk and boundary resistances are predominantly electronic resistances [307].

bulk and grain boundary contribution of a $SrTiO_3$ bicrystal can be obtained from the impedance spectrum. In contrast to the bulk, now a distinct bias–dependence, attributable to space charge effects (cf. Sections 5.8 and 7.3.3) is observed.

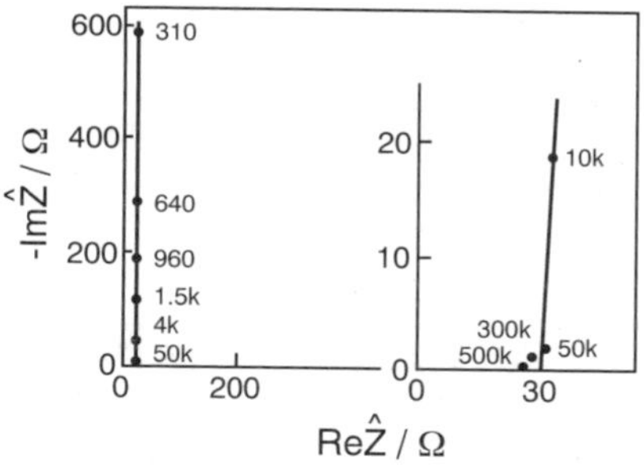

Fig. 7.41: Impedance spectrum of β–$Al_2O_3(Na_2O)$ single crystals (Pt electrodes). According to Ref. [624].

Figure 7.41 displays an impedance spectrum for Na^+–conducting "Na β–alumina", it is characterized by negligible bulk capacitance (in the frequency range observed) and virtually infinitely high contact resistance at the Pt electrodes[117] for Na^+ conductors.

Figure 7.42 refers to the formation of a resistive contact layer: The impedance signals of a thin LiCl film on a Li electrode can be conveniently used to follow the growth kinetics.

[116] If R_∞ and $R^\perp$ are of the same order of magnitude, the relaxation time τ_∞ is small with respect to $\tau^\perp$ because $C^\perp \gg C_\infty$.

[117] Adsorption or space charge effects lead to deviations from pure stoichiometry polarization, which expresses itself in the 45° angle (see Fig. 7.46).

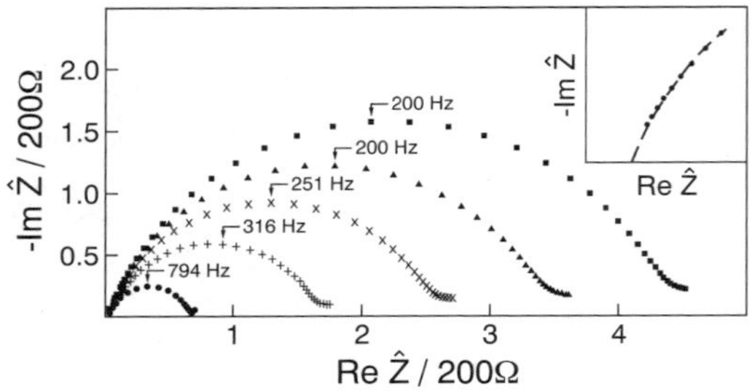

Fig. 7.42: Recording the impedance makes it possible to follow the growth of the LiCl film with time, on contact of a Li electrode with thionyl chloride (cf. also $Li/SOCl_2$ cell, Section 7.4). (The inset is an enlargement of the high frequency region.) According to Ref. [625].

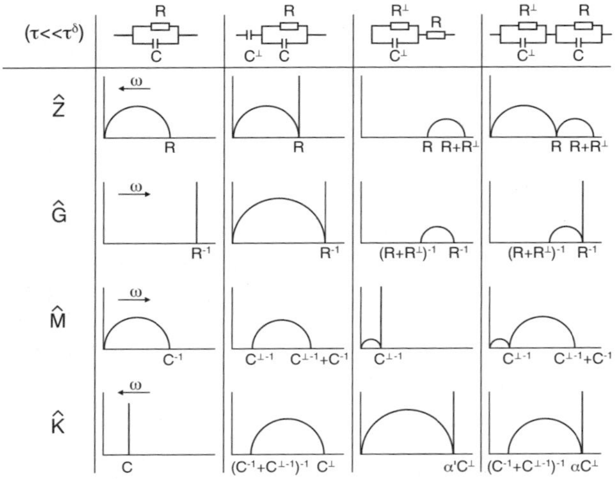

Fig. 7.43: Spectra of various system functions for simple equivalent circuits. ($\alpha' = R^{\perp 2}/(R + R^{\perp})^2$, $\alpha = (R^{\perp 2} + R^2C/C^{\perp})/(R + R^{\perp})^2$) [431].

Figure 7.43 shows the plots of other system functions, the complex admittance $\widehat{G} \equiv \widehat{Z}^{-1}$, the complex dielectric modulus $\widehat{M} \equiv j\omega\widehat{Z}$ and the complex capacitance $\widehat{K} \equiv \widehat{M}^{-1}$. In particular we can see that when $\widehat{M}$ is used the roles of R and C^{-1} are exchanged compared to the $\widehat{Z}$ plot, so this plot is particularly suitable for analysis of capacitances, although the information content is, in principle[118], the same for all system functions. The significance of the admittance plot consists in the fact that, for a parallel connection of R and C (and sufficiently high frequencies), the real part is identical with the sample conductance.

In the case of homogeneous samples it is useful to define complex specific parameters, such as the specific complex conductivity $\widehat{\sigma}$, which is obtained from $\widehat{G}$, in a manner analogous to obtaining σ from G. When Faradaic and dielectric effects are connected in parallel, then $\widehat{\sigma} = \sigma + j\omega\varepsilon$.

An exponential behaviour was obtained (see Section 7.3.4a) in the time domain for the switching–on measurements for long times in the case of selective ion or electron blocking electrodes (corresponding to pure stoichiometry polarization[115]); accordingly the diffusion impedance $\widehat{Z}^{\delta}$ (Warburg impedance) for very small frequencies

[118] I.e. when the resolution is as large as desired.

expresses itself in a semicircle in the $\widehat{Z}$ plane. However, only the right hand side of the semicircle is realistic. In the frequency range of the left hand part there is a straight line with gradient 1. This is the immediate result of the Laplace transformation of the $\sqrt{t}$ functionality[119] (see Eq. (7.118)) which describes the short time behaviour in the time domain:

$$\widehat{Z}^{\delta}(\omega > 1/\tau^{\delta}) \propto (1 - j)/\sqrt{\omega\tau^{\delta}} = \frac{1}{\sqrt{j\omega\tau^{\delta}/2}}. \quad (7.122)$$

In the case of cell E2 in the absence of defects with variable charges the proportionality factor is $\sqrt{2}\pi^{-1}R_{eon}^2/(R_{eon} + R_{ion})$. The Warburg impedance can itself be built up — as already shown in Fig. 7.27 — from an infinite series of R–C components [603], corresponding to the differential character of the problem. An experimental example is given in Fig. 7.44.

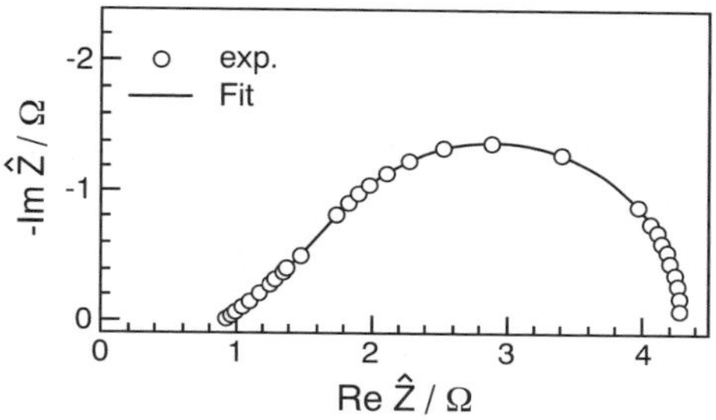

Fig. 7.44: The impedance spectrum of the cell Pt|Ag_2Te|Pt reveals the stoichiometry polarization (200°C, Dalton composition set up by coulometric titration). The transition from a straight 45° line to a semicircle, before the maximum frequency, is characteristic. According to Ref. [619].

The total impedance spectrum[120], which is the counterpart to Fig. 7.25 in the frequency domain, is shown schematically in Fig. 7.45. The diffusion behaviour described is exhibited at very small frequencies $\omega \ll 1/\tau^{\perp} \ll 1/\tau_{\infty}$ (bulk and boundary layer capacitors blocking). The l.h.s. intercept at the abscissa is at $Re\widehat{Z} = R_{\infty} + R^{\perp}$. The maximum frequency of the diffusion impedance[121] yields τ^{δ} and, hence, C^{δ}, if the partial resistance is known ($\omega_{max}^{-1} = \tau^{\delta} = R^{\delta}C^{\delta}$). The partial resistances are obtained from the limit $\omega \to 0$ ($\widehat{Z} = R^{\perp} + R_{eon,ion}$). Unlike the high values observed for the Ag_2Te–example (Fig. 7.44) the chemical diffusion coefficient and, hence, $1/\tau^{\delta}$ is usually relatively small. The maximum frequency then lies at very small values and recording the impedance spectrum in this range takes so long that it is easier

[119] If U describes the voltage response to the given current step function then, according to Eq. (7.118), $\widehat{Z}$ is determined from $\mathcal{L}\{U\}/I$. Initially (corresponding to relatively high frequencies) the voltage change obeys a $t^{-1/2}$ law, for which $\mathcal{L}\{t^{-1/2}\} = \sqrt{\pi/(j\omega)}$, while in the long–time range of interest (low frequencies) $\mathcal{L}\{\exp(-t/\tau)\} = (j\omega + 1/\tau)^{-1}$ is relevant. Note that $(1-j)/\sqrt{2} = \sqrt{-j} = 1/\sqrt{j} = \sqrt{2}/(1+j)$ as obvious after application of the third binomial formula.

[120] Note that stoichiometry polarizations can also be induced by grain boundaries [601] (cf. also footnote 86 on page 445).

[121] The junction between the Warburg straight line and semicircular behaviour near the maximum frequency is "bulged", as shown in the precise calculation (cf. also Fig. 7.44).

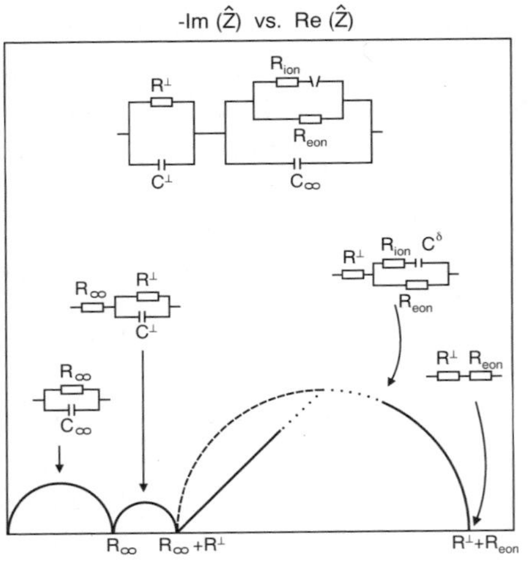

Fig. 7.45: Impedance spectrum and approximate equivalent circuit of a cell with blocking electrodes. (The galvanostatic response was given in Fig. 7.25.) The simplifications apply to the particular frequency ranges. The discussion is analogous to that for Fig. 7.25: Short/long times correspond to high/low frequencies. The approximations are not valid in the region of the broken line[121] and do not correspond to the accurate calculation (cf. also Fig. 7.44) [431].

to carry out a d.c. experiment. The resolution of $\sigma_{\mathrm{ion}}, \sigma_{\mathrm{eon}}$ and $\mathrm{R}^{\perp}$ is usually best achieved with a combination of DC measurements and AC measurements.
The theoretical formulation of the problem and also the consideration of the equivalent circuit show the analogy between chemical polarization (setting up a chemical gradient by applying different oxygen partial pressures to cell R) (see also Fig. 7.2) and the electrochemical polarization. The analogy becomes even more complete if the chemical relaxation is verified by means of a periodically alternating partial pressure. In this manner it is possible to speak of "chemical impedances", from whose amplitude and phase angle we can obtain analogous information in a simple manner[122].
While Fig. 7.45 is largely representative for cells of type R, E, I, in the case of the titration cell T, the low frequency part ends in a vertical line, corresponding to the fact that a stationary state is not set up in the d.c. galvanostatic experiment. This is also obtained on Laplace transformation of the associated solution [618].

At this point we must turn to a difficulty and apparent inconsistency: If we start with the consideration of the electrochemical polarization of a mixed conductor (according to Fig. 7.45) and allow σ_{eon} to approach zero, then we would expect an infinitely long Warburg increase for ion conductors when using electronic electrodes. On the other hand such measurements frequently exhibit a steep rise (as in Fig. 7.41), a "spike" behaviour that would be expected for the exponential solution (onset of a semicircle with almost infinite radius) directly according to Fig. 7.23 for single charge carrier systems ($\mathrm{R}_{\mathrm{eon}} = \infty$ in Fig. 7.23) in the case of ideally polarizable electrodes. In other words: When we neglect the electrons right from the beginning we obtain a qualitatively different result. The deeper reason for this apparent contradiction lies in the atomistic intermingling of volume polarization and space charge polarization

[122]See, for example, Ref. [626].

(and adsorption polarization, etc.) [610] which is not taken account of in Fig. 7.45. When the superposition of space charge and volume polarization is taken into account, then the approximate circuit already given in Fig. 7.33 applies, to which the impedance[123]

$$\widehat{Z}(\omega) = R_\infty + \frac{(R_{eon} - R_\infty)\tanh\sqrt{j\omega\pi^2 R^\delta C^\delta/4}}{\sqrt{j\omega\pi^2 R^\delta C^\delta/4} + j\omega R^\delta C^\perp \tanh\sqrt{j\omega\pi^2 R^\delta C^\delta/4}} \tag{7.123}$$

is assigned. Note that C^δ and $C^\perp$ refer to the whole sample.

When the frequencies are very small ($\omega < \left[R^\delta\left(C^\perp + \frac{\pi^2}{12}C^\delta\right)\right]^{-1} \simeq \left[R^\delta\left(C^\perp + C^\delta\right)\right]^{-1} \equiv 1/\tau'$) Eq. (7.123) simplifies to

$$\widehat{Z} = R_\infty + \frac{R_{eon} - R_\infty}{1 + i\omega\tau'} \tag{7.124}$$

and at comparatively high frequencies ($\omega \gtrsim 1/\tau'$) to

$$\widehat{Z} = R_\infty + \frac{R_{eon} - R_\infty}{\sqrt{j\omega\pi^2 R^\delta C^\delta/4} + j\omega R^\delta C^\perp}. \tag{7.125}$$

Both equations reveal the combination of double–layer and stoichiometry effects. If the electrical boundary capacitance ($C^\perp$) is large compared with the chemical capacitance (C^δ), then the time constant (τ') is unaffected by D^δ. Equation (7.125) demonstrates that a Warburg increase ($\sqrt{j\omega}$ term) only follows in fact for $C^\delta \gg C^\perp$, while if $C^\perp \gg C^\delta$ a behaviour is obtained that would be expected for a single charge carrier ($j\omega$ term). This transition is determined by the magnitude of the defect concentration[124] (C^δ generally increases more rapidly with the defect concentration than $C^\perp$). Figure 7.46 confirms this on the basis of numerical calculations. The

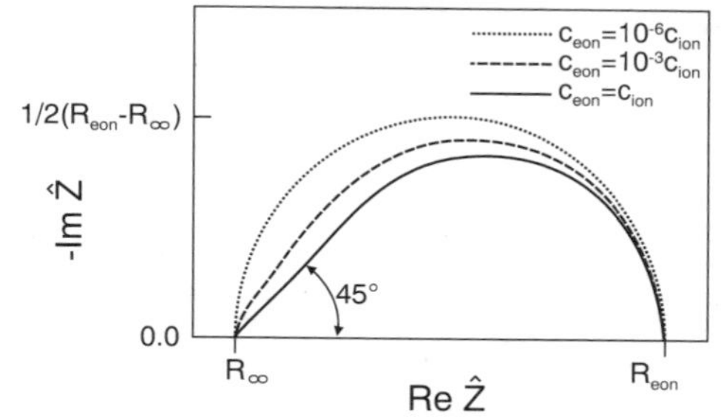

Fig. 7.46: Numerically calculated impedance for ion blockage. We can recognize the transition from Warburg to a pure semicircular behaviour as the defect concentration changes. The impedances are so normalized that the points of highest frequency coincide. The different curves correspond to various electronic defect concentrations ($u_{eon} = 10u_{ion}, L = 10^4\lambda, z_{ion} = 1 = -z_{eon}$). From Ref. [610].

[123] In the analysis of the frequency range C^δ, as already deduced in Eq. (7.99) from the switching–on behaviour, is found to be proportional to $\left(c^\delta\right)^{-1}$. However, π^2 is to be replaced by 12. This slight discrepancy is a consequence of the approximations made and occurs as a consequence of the different weighting of time/frequency domains. Insofar (cf. τ') $\pi^2/12$ can be set equal to unity in this connection. As before R^δ is $R_{ion} + R_{eon}$.

[124] Cf. Eqs. (5.222,7.78,7.79,7.99).

concepts of chemical resistance and chemical capacitance (cf. Section 6.7.4) and their analogies to the electrical parameters can be carried a very long way. Thus, it is possible to construct generalized equivalent circuits in which chemical potential gradients constitute the relevant driving forces and replace or complement electrical potential gradients. The reader is referred to Refs. [587,610] for details.

Concerning the impedance spectroscopy of cells with selectively blocking electrodes, it should be added (cf. also Subsection 7.3.4e) that additional hindrances with respect to the nonblocked carrier approximately result in an additional semicircle (cf. $R^{\perp}$–$C^{\perp}$ combination in Fig. 7.23). The real part for $\omega \to 0$ then follows as $R_{eon} + R^{\perp}$ in the case of an ion blockage. Conversely if the blocking is not perfect, the stoichiometry response in the impedance spectrum is deformed ($\mathrm{Re}\widehat{Z}(\omega = 0) < R_{eon}$ in the case of ion blockage) [587,610].

Generally, deviations from the idealized considerations lead to deformation of the ideal semicircles[125]. This is also valid for the high frequency bulk response for which the depression observed can be described in many cases by moving the centre of the semicircle below the abscissa [603]. Then circuit elements appear that cannot be characterized with R and C, but by elements with frequency–independent phase angles [627]. Proportionalities to ω^{α} with nontrivial α appear in the individual impedances. There are various reasons for this: Such frequency dependences appear, for instance, in cases of pronounced disorder, or of marked interaction between charge carriers [410,627,628], but can also be the result of complex geometry. We have already discussed such aspects in Chapter 6. Figure 6.35 (page 320) reproduced the frequency–dependent conductivity of $RbAg_4I_5$, expressed by a nonideal high frequency behaviour. The impedance of the tree electrode in Fig. 6.83 (page 394) showed nonideal behaviour, too. For such fractal geometries it is possible to relate α to the fractal dimension (see Section 6.10) [629]. The impedance spectra also adopt a flattened behaviour when individual semicircles overlap, which can be the case for inhomogeneous and heterogeneous samples (distribution of relaxation times)[125].

---

[125] Even if the dielectric constant is invariant, it is not possible to connect a series sequence of parallel R–C components into a single parallel R–C circuit. Let us consider, for simplicity, a series connection of two homogeneous but different sample parts, that are of the same size, then $\widehat{Z}(\omega) = \frac{R_1}{1+j\omega R_1 C} + \frac{R_2}{1+j\omega R_2 C}$. If the difference between τ_1 and τ_2 is not very great, a flattened semicircle is produced with the real axis intercept at $\omega = 0$, which now corresponds to the sum of the resistive components. However, if the measurement direction is perpendicular to the inhomogeneity coordinate, an ideal effective parallel–R–C circuit with a single relaxation time is formed whereby $R_{eff}^{-1} = R_1^{-1} + R_2^{-1}$ for all ω. This must be taken into account on measuring inhomogeneities (boundary layer problem, see Section 5.8), e.g. during a diffusion experiment (via chemical relaxation, see Section 6.5).

7.3.7 Inhomogeneities and heterogeneities: Many–point measurements and point electrodes

In this section we shall abandon the assumption of idealized homogenous samples between two electrodes, and take a serious look at inhomogeneities and heterogeneities (e.g. microstructural effects).
First we wish to acquaint ourselves with a simple method, which can be used to distinguish between contact impedances of internal interfaces and electrode interfaces; it is the method of many–point measurement (see also Fig. 7.31). In a four–point measurement we apply a constant (or sinusoidal) current I to the sample via electrodes applied to the end surfaces, but now we measure the voltage drop U with a high resistance voltmeter between two probes applied to the sample at a distance of Δx from each other. As before, the local current density is determined by the external current to be $i = I/a$ (a: contact area of external electrodes). The difference in the electrochemical potential of the electrons (or ions for "ionic measuring electrodes") at the probes is now proportional to the current density and to Δx. (Note that there are only marginal gradients in the relevant electrochemical potentials perpendicular to the current flow.) In the case of pure conductivity experiments, the specific conductivity becomes

$$\sigma = \frac{\Delta x}{a}\frac{I}{U}. \tag{7.126}$$

Now, it is only the impedances occuring between the measurement probes that are determined, including, for example, effects of grain boundaries between the probes, but not electrode effects. The impedances at the measurement probes do not play a significant role, since the current flowing through these is exceedingly small. In the same sense, a three–point arrangement makes it possible to study the behaviour of a single electrode[126].

In many cases, normal two–point impedance spectroscopy is frequently sufficient to distinguish between electrode and grain boundary effects: As already mentioned, in the case of polycrystalline samples the grain boundary capacitance is usually much lower than that of the electrode contact, on account of having many grain boundaries in series. Moreover, in some cases electrode impedances can be minimized by a careful electrode preparation. Figure 7.47 shows a two–point experiment on an AgBr bicrystal using Ag–electrodes. The low frequency response is solely due to the internal interface.

So far we have referred to blocking grain boundaries. Grain boundaries can also constitute high conduction pathways. In such cases the deconvolution by two– or multi–point impedance spectroscopy is not possible in this way, since their contribution is

[126] By suitable selection of the sensing probes it is possible to obtain evidence concerning various relevant potentials (cf. Fig. 7.31, cf. [630]).

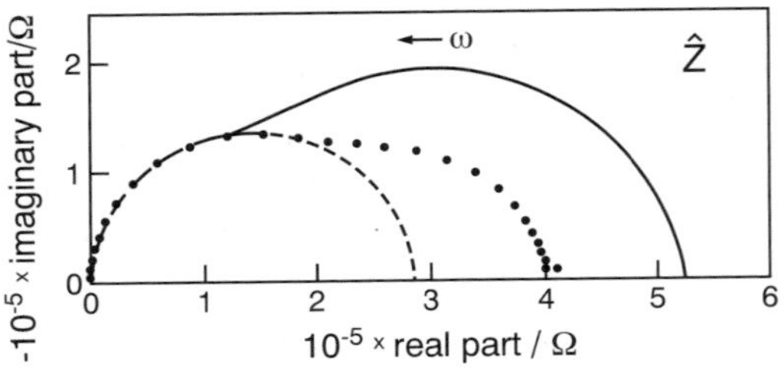

Fig. 7.47: Impedance spectra of an AgBr bicrystal [256]. Current direction is normal to the boundary.

parallel to the bulk contribution[127]. Moreover, on account of the inhomogeneity and, hence, anisotropy of the grain boundary (space charge zone plus core), blocking and highly conducting effects can even occur in the same material at the same time (see Chapter 5). Let us assume that the core exhibits an elevated resistance, while the space charge zone is highly conducting. The first effect is experienced on passage of a charge carrier from grain to grain ("perpendicular effect", $\widehat{Z}^{\perp}$), while "parallel" grain boundaries ($\widehat{Z}^{\parallel}$) can be used as bypasses[128]. The "brick layer model"[129] (Fig. 7.48) offers a first approximation. The equivalent circuit in Fig. 7.49 shows that

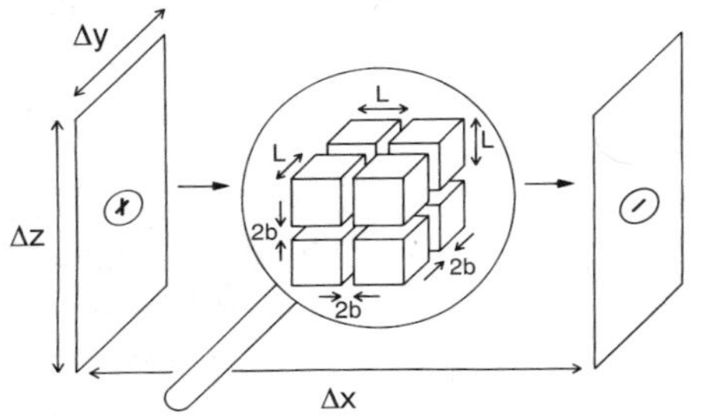

Fig. 7.48: "Brick layer model" as a simple structure model.

the parallel contributions short–circuit the bulk. (Whether the highly conducting boundary zones only short–circuit the grain interior, as is assumed here, or the whole grain impedance, may strongly depend on the behaviour of the intersections of the grain interfaces.) The resistance parallel to C_∞ ($\frac{R^{\parallel}R_\infty}{R^{\parallel}+R_\infty}$ instead of R_∞) is altered accordingly. Capacitance contributions parallel to the grain boundaries are very small (area small, thickness large). Conversely, the capacitance contributions, as a result of grain boundaries that are to be crossed perpendicularly, are very large (area large, thickness small). $R^{\perp}$ appears parallel to $C^{\perp}$ in a separate semicircle, as discussed above. (Compare here with the bicrystal impedance measurements in Fig. 7.47 and Fig. 7.40.)

127 Adequate discrimination is possible by changing grain size or by applying micro–electrodes (see Fig. 7.53).

128 Inversion space charge zones can cause such an anisotropy effect, without the core being affected (cf. Section 5.8).

129 Finite element calculations reveal that, in many realistic cases, the brick layer model [631] remains a good approximation in spite of nonideal microstructures [632].

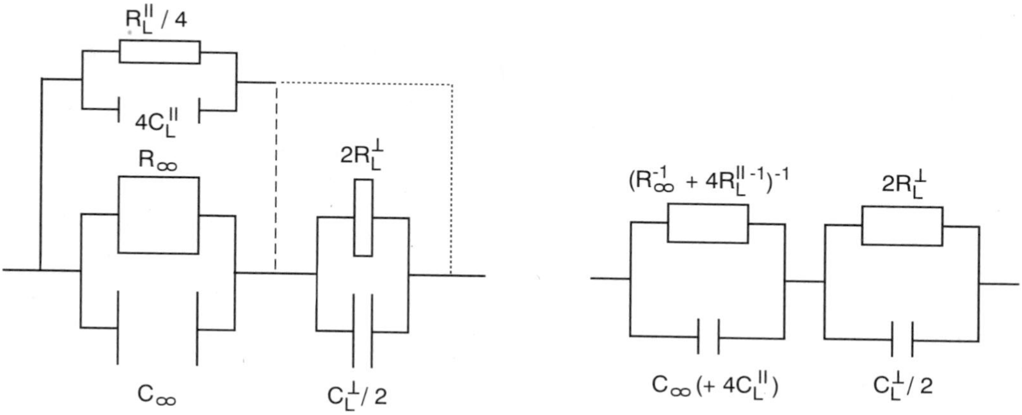

Fig. 7.49: The equivalent circuit of the "brick layer model". The validity is dependent on the local behaviour of the intersections of the grain boundaries. The approximation used in what follows refers to the broken connecting line. Then the highly conducting grain boundary paths (e.g. space charge zones) are blocked by the perpendicular boundary (e.g. core region). The factors 4 and 1/2 refer to the number of grain boundary contributions per grain (4 parallel and 2 serial half grain boundaries) according to Ref. [256].

By calculating the impedance of the brick layer polycrystal using Kirchhoff's rules and rewriting it as an effective complex specific conductivity[130] ($\widehat{\sigma}_m \equiv \widehat{Z}^{-1}L/a$) we obtain the approximation (cf. Sections 5.8, 6.6):

$$\widehat{\sigma}_m = \frac{\widehat{\sigma}_\infty \widehat{\sigma}_L^{\perp} + \beta_L^{\|} \varphi_L \widehat{\sigma}_L^{\|} \widehat{\sigma}_L^{\perp}}{\widehat{\sigma}_L^{\perp} + \beta_L^{\perp} \varphi_L \widehat{\sigma}_\infty}, \tag{7.127}$$

$\widehat{\sigma}^{\|}$ and $\widehat{\sigma}^{\perp}$ can be broken down further into core (co) and space charge (sc) contributions[131] [256]. The parameter φ_L describes the volume fraction of the grain boundaries (core and space charge zone), $\beta_L^{\|}$ and $\beta_L^{\perp}$ describe the proportion of this which is relevant for conduction in the direction of measurement, ideally: $\beta_L^{\perp} = 1/3$, $\beta_L^{\|} = 2/3$.

The effective space charge conductivities are obtained as calculated in Section 5.8. If $\tau_L \gg \tau_\infty$, Eq. (7.127) approximately results in at least two semicircles[132] in the impedance plot. For the high frequency semicircle $\widehat{\sigma}_m = \widehat{\sigma}_\infty + \frac{2}{3}\varphi_L \widehat{\sigma}_L{}^{\|}$; the boundary

[130] The presentation follows Ref. [256]. For inhomogeneous samples it is sensible to define complex effective specific values ($\widehat{\sigma}_m$) that are derived from the total impedance in the macroscopic measurements. As for the real values, it follows in the case of different homogeneous regions (i) in parallel that $\widehat{\sigma}_m = \Sigma_i \varphi \widehat{\sigma}_i$ whereby φ_i is the volume fraction. Analogously for series connection $\widehat{\sigma}_m^{-1} = \Sigma_i \varphi_i \widehat{\sigma}_i^{-1}$. If the inhomogeneity is continuous, integration is to be carried out over the local coordinates. Splitting up into $\widehat{\sigma}_m = \sigma_m + j\omega\varepsilon_m$ is only exact for the parallel case.

[131] Dislocations as one–dimensional defects primarily appear via parallel effects.

[132] Whether, for example, space charge and core effects can be separated depends on the specific situation. Usually this is not possible in a straightforward way, since core and space charges cannot be independently varied (cf. Section 5.8).

capacitance ($C_L^\perp$) is permeable for these frequencies. In addition to the bulk values the values of the parallel grain boundaries are included in this semicircle. The ⊥–contributions appear as a low frequency semicircle which is described by the approximation

$$\widehat{\sigma}_m^{-1} = \left[\sigma_\infty + \beta_L^{\|}\varphi_L\sigma_L^{\|}\right]^{-1} + \beta_L^\perp\varphi_L\widehat{\sigma}_L^{\perp^{-1}}. \tag{7.128}$$

In the d.c. limit $\widehat{\sigma}_m^{-1} = (\sigma_\infty + \beta_L^{\|}\varphi_L\sigma_L^{\|})^{-1} + \beta_L^\perp\varphi_L\sigma_L^{\perp^{-1}}$ corresponding to the real axis intercept in the impedance spectrum for $\omega \to 0$. In Eq. (7.128) the real term in square brackets describes the axis intercept where both semicircles join (blocking capacitor).

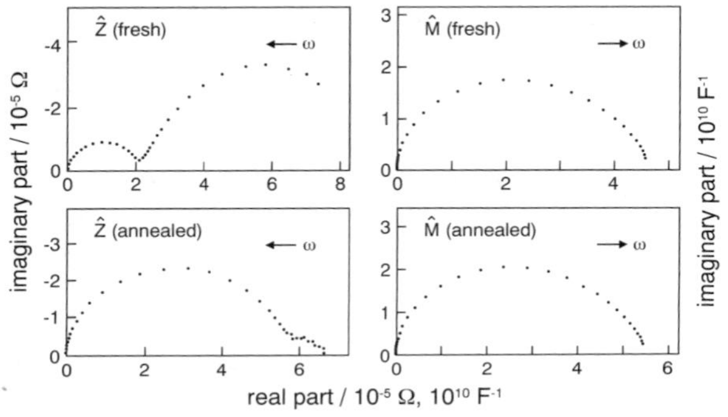

Fig. 7.50: $\widehat{Z}$ and $\widehat{M}$ spectra of an AgCl polycrystalline sample before and after being annealed [633]. Cf. also Fig. 7.43.

Measurements on silver halides (Figs. 7.47, 7.50) confirm this picture. In the case of the bicrystal (Fig. 7.47) with perpendicular grain boundaries the grain boundary effect manifests itself in a second semicircle which becomes, as expected, smaller upon intermediate annealing, while the bulk semicircle naturally remains the same. In the polycrystal (Fig. 7.50) we can also recognize two semicircles. During the annealing process the low frequency semicircle also shrinks; here, however, the high frequency semicircle grows. This is because the "bulk semicircle" now also contains the "favourable" parallel boundary contributions that are reduced on annealing. The analysis of the associated capacitances in Fig. 7.50 and, hence, of the effective thicknesses is consistent with this picture[133,134,135].

In the case of the bicrystal in Fig. 7.47, however, the effective thickness calculated from the low frequency semicircle is very much greater than expected. The fact that the activation energy is almost equal to that of the bulk also points to another phenomenon. This is a frequently encountered but frequently overlooked complication

[133] It is assumed that ε(grain boundary) = ε(bulk).

[134] The heterogeneous electrolytes described in Chapter 5 exhibit a very similar behaviour. Fine Al_2O_3 particles in the grain boundaries of AgCl effect an enormous increase in the parallel conductivity, whilst they reduce transport from grain to grain by constriction effects.

[135] Here it must be taken into account that the effective grain boundary thickness generally also involves the thickness of the space charge zone, since the appropriate impedance usually cannot be broken down into spatially separated RC elements.

which can be described as a current–constriction effect and is the result of nonideal contacting [634]. It occurs when there are lateral inhomogeneities or heterogeneities, e.g. when the crystal grains (or here two large single crystals) are not ideally sintered together, if pores or second phases are included[134], or at electrode boundary surfaces when the electrode contact is not homogeneous. Such lateral inhomogeneities can also, but generally with smaller effects occur in the local contact equilibrium (cf. the island model at the grain boundary in Section 5.4).
The insulating sites result in the direct current channel becoming narrower and only extending to normal size after some distance. These inhomogeneities formally introduce further capacitive elements and, hence, a new effective relaxation time. Although the effect is associated with nontrivial frequency–dependent potential distribution, numerical calculations [632] reveal that there are two semicircles formed in the general case (Figs. 7.51, 7.52). Insulating sites do not play any role in the high–frequency range since they are permeable dielectrically: The diameter of this semicircle yields the ideal bulk resistance ($\propto 1/\sigma$). The d.c. resistance (i.e. the sum

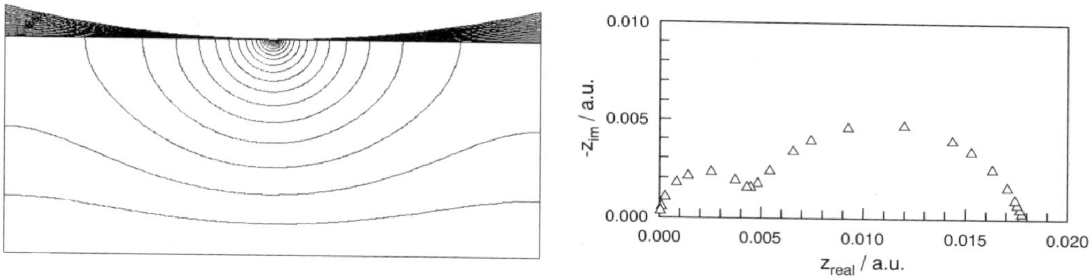

Fig. 7.51: The following can act as a prototype experiment with respect to current constriction phenomena: A metal sheet makes (ideal) contact with a single–crystalline solid electrolyte at only one point (left). Nonetheless it creates two semicircles (right). At high frequencies the very small "air gap" is dielectrically permeable. The diameter of the high frequency semicircle gives the resistance obtained if we had perfect contact over the entire area ($\propto 1/\sigma$). At low frequencies, the current constriction is predominant. (Compare here the electrical potential lines included in the left figure.) The real axis intercept at low frequencies is $R = \frac{\alpha}{\sigma}$ (see Eq. (7.129)), hence the diameter of the second semicircle is $\frac{\alpha}{\sigma} - \frac{\beta}{\sigma} \propto \frac{1}{\sigma}$ whereby α and β are purely geometric factors. For this reason both semicircle diameters are subject to the same thermal activation [632].

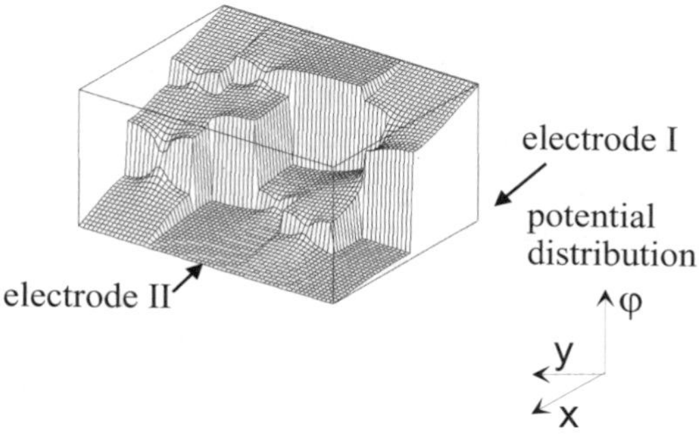

Fig. 7.52: Distribution of the electrical potential in a ^{2}D polycrystal at zero frequency [635].

of the two semicircle diameters) and, hence, the diameter of the second semicircle too, is in the ideal case also proportional to $1/\sigma$ as discussed below (cf. Eq. (7.129)). From this it follows that the corresponding activation energy is the same as that of the bulk[136].
Concerning the treatment of inhomogeneous systems by effective medium and percolation theory — which are especially helpful in the case of random distributions — the reader is referred to the literature [636]. Some remarks with respect to percolation theory have already been made in Section 6.10.3 (cf. also the experimental example in Fig. 5.98, page 252).

Current–carrying point electrodes can also be used specifically for measuring. It is evident that in a cell in which the thick sample under investigation is contacted over its entire surface on one side, but with only point contact (of radius b) at the other, the current lines must narrow down to the point electrode. The total resistance is determined by the immediate neighbourhood of the point contact, which acts as a bottleneck[137] and is, thus, independent of sample size L (if $b \ll L$). It follows, as already indicated, that [637]

$$R = \frac{\mathrm{const}}{b\sigma}. \tag{7.129}$$

The constant is 1/2 for a circular flat applied electrode (cf. Fig. 7.53 top right) and $1/\pi$ for an embedded hemispherical electrode (cf. Fig. 7.53 top left). It is obvious that point electrodes are very useful for studying inhomogeneities. Figure 7.53 indicates four experimental applications: The investigation of surface conductivity, which is scarcely amenable to investigation in any other way in the case of highly conducting boundary layers[138], the spatially resolved measurement of conductivity in a heterogeneous material, the investigation of a frozen–in diffusion profile and the investigation of the electrode kinetics of fuel cell cathodes.
In many cases, particularly for thin samples, the complex geometry prevents the use of simple measurement principles. Suitable techniques have been worked out for such cases. Table 7.3 gives a brief overview of the methods according to van der Pauw and Valdes. The reader is also referred to the specialist literature [642].

---

[136]Provided that morphology does not change. The dependences on component potential and doping have to be discussed analogously.

[137]A simplified treatment shows this [637]. Let the electrode considered be a sphere half–embedded in the sample and having a radius of b. The extended "normal" counterelectrode is separated by $L \gg b$. On account of the potential distribution (see Fig. 7.50) it can be assumed that the regions relevant for the conduction extend concentrically with increasing distance r from the contact point. It then follows for the resistance of an infinitesimal section at distance r, that $dR = \frac{2dr}{4\pi r^2 \sigma(r)}$. The integration is carried out between b and L. The result is $R = \frac{1}{2\pi\sigma b} - \frac{1}{2\pi\sigma L} \simeq \frac{1}{2\pi b}\frac{1}{\sigma}$. On account of the convergence of the integral for large L neither the exact sample thickness nor the exact geometry of the extended counterelectrode is important (see also Ref. [637]).

[138]Another but less precise method of measurement of the surface conductivity is the use of suitable multiple–electrode systems. If we lay a metallic ring round a circular electrode, that lies at the same potential as the flat extended counterelectrode or the circular electrode, it is possible to separate the surface current. Cf. here [641].

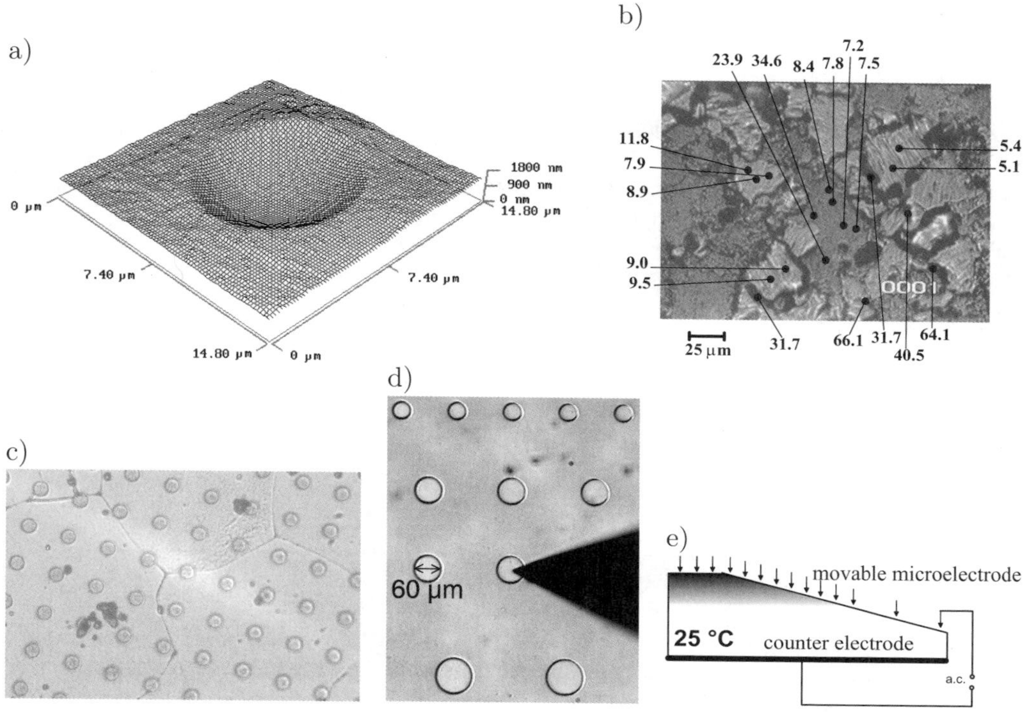

Fig. 7.53: Application of microcontact impedance technique for the measurement of local transport and reaction rates. The illustration top left (a) is an AFM image of an imprint of a microelectrode in AgCl, with which highly conducting surface films can be measured quite accurately (cf. Fig. 5.102) by varying the tip size. This technique has been used in Fig. b to detect enhanced grain boundary conductivities in the case of AgCl (see page 249) (figures in units of nS/cm) [638]. The application of contact points also allows the measurement of local conductivities and impedances. Part c refers to the measurement of depressed conductivities in the case of $SrTiO_3$ (see page 263) [639]. Part d refers to the study of electrode kinetics of oxygen reduction at $LaMnO_3$ on ZrO_2 (see page 484) [640]. Part e refers to $CdCl_2$ diffusion into AgCl: Scanning of a quenched instationary profile permits determination of the chemical diffusion coefficient [635].

Figure 7.54 shows how we can obtain local resolution by frequency variation, namely by means of constructing a "higher–dimensional" electrochemical chain (analogous to the construction of a field effect transistor[139]) [643]:

$$\text{Cell 2D} = \frac{\dfrac{\mathrm{Ag|AgCl}}{\mathrm{SiO_2}}}{\mathrm{Si}}$$

During measurement of the impedance between Ag and Si the depth of penetration of the signal into the AgCl under investigation is a function of frequency. This

[139] In the field effect transistor one is interested in the conductivity parallel to the $AgCl/SiO_2$ interface as a function of a field applied normal to it.

method proved helpful for measuring Ag|AgCl interfaces and also for localization of inhomogeneities.

Enormous advances have been made and are still being made in the development of local potential and conductivity measurements by the availability of scanning probe techniques (cf. e.g. "scanning impedance spectroscopy" [644]). Here, too, the reader is referred to the literature (cf. Fig. 5.112, page 264). Scanning electron microscopy can also be used for local detection of electrical properties, as briefly touched upon in Section 6.10 (cf. also Fig. 6.79, page 391) [645].

Table 7.3: Brief overview of the methods named after van der Pauw and Valdes [646,647].

van der Pauw method for plane parallel sheets of thickness d	any form 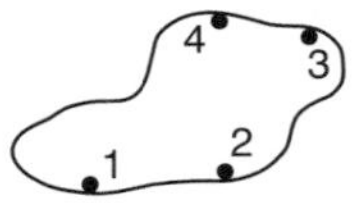	$\exp(-\pi d \sigma \frac{U_{43}}{I_{12}}) + \exp(-\pi d \sigma \frac{U_{14}}{I_{23}}) = 1$
	mirror symmetry 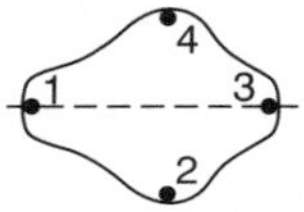	$\frac{U_{43}}{I_{12}} = \frac{U_{14}}{I_{23}} = \frac{\ln 2}{\pi d \sigma}$
Valdes method* for sheets of thickness d	linear array of 4 probes 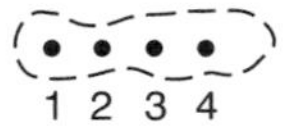	$\frac{U_{32}}{I_{14}} = \frac{\ln 2}{\pi d \sigma}$ (distance (1,4) << size)

* If the distance between the probes is not great compared with d, other correction factors have to be used [642].

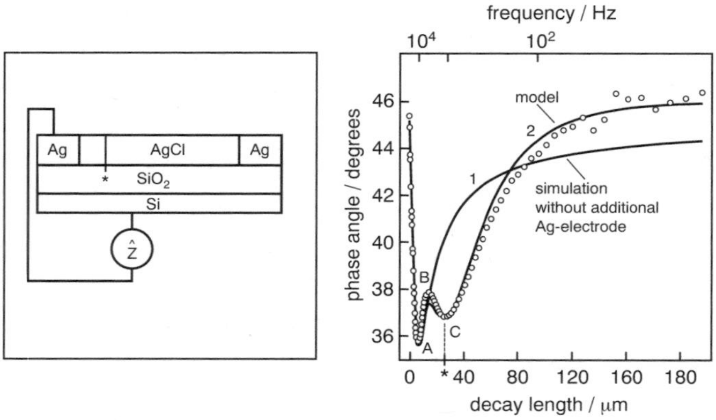

Fig. 7.54: The "penetration impedance": At high frequencies the SiO_2 is dielectrically short–circuited and the AgCl "overlooked". The lower the frequency, the more the signal penetrates the AgCl (left). The graph on the right shows an appreciable trough in the phase angle of the impedance, when the depth of penetration reaches the metallic inhomogeneity included at the position indicated by an asterisk [643].

7.4 Cells generating current

7.4.1 General

In the previous section, we analysed polarization cells, that is, cells to which we have applied a current and across which a chemical potential difference may have been generated. Here we will consider cells in which the difference in chemical potential inherently exists and a spontaneous external current flows when the cells are short–circuited[140]. In contrast to the "equilibrium cells" treated in Section 7.2.1 we are now interested in extracting current and power from galvanic elements, i.e. in the conversion of chemical to electrical energy. In what follows we shall primarily address technologically relevant systems.
However, to begin with we shall describe a method of separating ion and electron conduction that belongs in this chapter [648]. Let us consider cell R (page 424) in the stationary chemically polarized state, i.e. we assume that we have set up local equilibrium with the two oxygen partial pressures P_1 and P_2. Short–circuiting the two electrodes the external electron current immediately equalizes the electrochemical potentials of the electrons ($\mathbf{i} = \mathbf{i}_{\rm eon} \propto \sigma_{\rm eon}\nabla\widetilde{\mu}_{\rm eon}$) on account of the immensely high conductivity of the external connection. For this reason the driving force for an internal electronic current equals zero and the short–circuit current flowing internally must be purely ionic in nature. Hence, the internal current and the external one, being of the same size give the ionic conductivity if we know the partial pressures ($\sigma_{\rm ion} \propto {\rm i}/\Delta\widetilde{\mu}_{\rm ion} \propto {\rm i}/\Delta\mu_{\rm O_2}$) and, in combination with a measurement of the total conductivity, we also know the electronic contribution.

But now to the electrical applications.
We make a meaningful distinction between (i) primary systems, in which the cells are only discharged; i.e. chemical energy is transformed into electrical energy, the

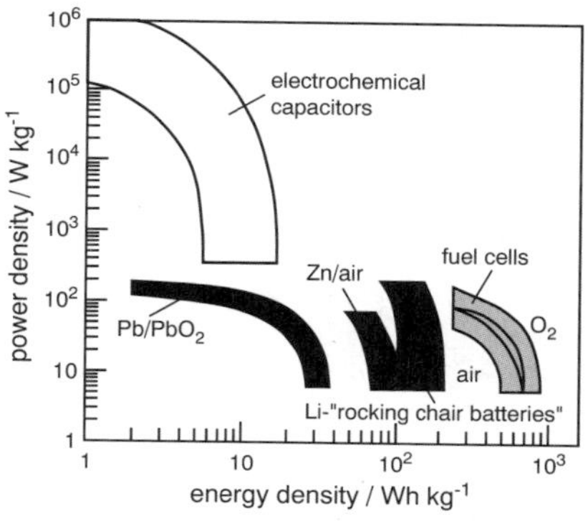

Fig. 7.55: For electrochemical applications the question of how much and how rapidly energy per unit mass can be "called up" is extremely significant. The figure gives information in this respect for the lead accumulator, zinc–air battery, lithium rocking chair battery and for modern fuel cells, and compares them with electrochemical capacitors (see also page 444). From Ref. [649].

[140]Strictly speaking, the depolarization process discussed in the previous section also belongs in this chapter. Supercapacitors, in particular, and their relationship to battery research (see Fig. 7.55), were discussed on page 444.

cell is then "flat"[141], (ii) secondary elements, which are returned by the reversal of the discharge process to the original conditions and (iii) fuel cells, in which the active components are provided (and the product is removed) continually ("metabolistic cells").

The interest in such battery systems is considerable [650–652], for they represent a means of locally supplying electrical energy, i.e. without the need for external powerlines; think of mobile household appliances, telephones, clocks, lap–tops and electrical traction.

The storage of electrical energy in secondary elements is also interesting for meeting peak power demands. Figure 7.55 gives an overview of mass specific energy and power densities as parameters for some systems, which play an important part in our treatment.

Let us start with fuel cells.

7.4.2 Fuel cells

In principle, the advantage of the direct transformation of chemical to electrical energy, which constitutes a large part of the fascination[142] of galvanic elements, is the avoidance of thermal processes and, hence, the higher theoretical efficiency [653]. Let us discuss this using the H_2/O_2 fuel cell as our example and consider the following set–up:

$$\text{Cell a} = \qquad H_2 \mid O_2 \tag{7.131a}$$

$$\text{Cell b} = \qquad H_2 \mid (O^{2-}, e^-) \mid O_2 \tag{7.131b}$$

$$\text{Cell c} = \qquad H_2 \mid (O^{2-}) \mid O_2. \tag{7.131c}$$

If we bring H_2 and O_2 into direct contact (if necessary over a catalyst) water is produced (cell a) and the stored chemical energy will be released in the form of the enthalpy of reaction $\Delta_r H$. The same applies in cell b, where H_2 and O_2 come into contact by diffusion through a mixed conductor (O^{2-}, e^-). The enthalpy of reaction can then be converted into mechanical energy by means of a heat engine, and subsequently into electrical energy by means of a dynamo. Even if we assume unrealistically that the latter does not involve losses, this procedure cannot transform more than a fraction w_C ($\equiv$ Carnot efficiency) of the chemical energy into electrical energy (reversible limit)[143]

$$w = \left| \frac{\text{electrical energy}}{\Delta_r H} \right| \leq w_C = \frac{T_2 - T_1}{T_2}, \tag{7.132}$$

[141] Re–use entails chemical recycling.

[142] A sociological (meta–) remark: The time dependence of the number of publications and patents issued concerning fuel cells is characterized by oscillations, which are at least partly attributable to nonlinear, autocatalytic frustration and enthusiasm effects (cf. Section 6.10).

[143] Please note that, unlike in the other parts of Chapter 7, here we follow our initial nomenclature: $\Delta_r G$, etc. are not referred to 1 mol, rather $\Delta_r G = n\Delta_r G_m$, etc.

whereby T_2 and T_1 are the temperatures between which the Carnot process is carried out. If T_2=583K (boiling point of water at 100bar) and T_1=313K, then $w_C = 46\%$; real values are appreciably less. However, if H_2 and O_2 are separated as in cell (c) by a pure oxygen ion conductor and the electron transport now takes place via the external leads, the chemical energy is directly transformed into electrical energy. The maximum possible efficiency (Eq. (7.12)) is given by w_g, for which ($\Delta_r G$ and $\Delta_r H$ are negative in the systems of interest):

$$w \leq w_g = \frac{\Delta_r G}{\Delta_r H} = 1 - \frac{T\Delta_r S}{\Delta_r H} = 1 + \frac{T\Delta_r S}{|\Delta_r H|}. \qquad (7.133)$$

This means that, in the case of reactions with positive reaction entropy, w_g can even be greater than 100%. This does not contravene the First Law, since the surroundings are cooled in such circumstances[144]. As translation contributions dominate in the entropy of reaction, this is the case if the number of gas molecules increases during reaction ($2C+O_2 \rightleftharpoons 2CO$). Table 7.4 (page 495) lists the w_g values of some fuel reactions.
Since we are interested in taking current, which unavoidably causes losses, overpotentials occur ($\eta = \Sigma_i \eta_i = \Sigma_i I\bar{R}_i(I)$)[145] and w_g is not reached in practice (see Section 7.1). Important contributions are voltage losses due to the electrolyte resistance, which is current–independent to a good approximation, the electrode overpotential, which, in the case of a transfer overpotential, increases underproportionally with increasing current (see Section 7.3.3), and the diffusion overpotential which may be caused by inhibition of gas transport. The latter increases asymptotically with the current and, thus, has a current–limiting effect. It reaches its limiting value at complete depletion[146] (Fig. 7.56). Such a behaviour is produced because the transport coefficient scarcely depends on the potential (Section 6.1), while, on the other hand, the driving force for diffusion cannot be increased to arbitrary values (cf. here footnote 41 on page 423). Table 7.5 (page 496) provides an overview of the electrode reactions of the most important types of fuel cells.
High temperature fuel cells and, in general, those with solid oxide electrolytes (SOFC = solid oxide fuel cells [652]), are able to run on not only H_2 but also other fuels, such as hydrocarbons (or even CO), directly at very high temperatures without noticeable electrode effects. At very high temperatures the electrolyte resistance contributes

[144] This does not contravene the Second Law, since it is not possible to construct a periodic machine, that carries out work merely by cooling.

[145] Here we define "voltage efficiency" $w_U = U/E$(theoretical). If nonideality can be attributed to the overvoltage alone, then this is $1 - \Sigma_i|\eta_i/E|$. In addition, it is necessary to take account of a "current efficiency" or better faradaic efficiency $w_F = It/nzF$; further operation losses result in additional efficiency–factors (cf., for example, "fuel usage" $n_{eff} < n$). If we combine the latter in an operating efficiency w_{op} the useable work is evidently $w_{op1} \cdot |UIt| = w_{op1} w_F |zFnU| = w_{op1} w_F w_U |zFnE| = w_{op1} w_F w_U w_g |n\Delta_r H_m| = w_{op1} w_F w_U w_g w_{op2} |\Delta_r H| = w_{op} w_F w_U w_g |\Delta_r H| = w_{total} |\Delta_r H|$. See also footnote 143.

[146] Limiting currents can also occur in reaction– (e.g. adsorption–) controlled kinetics (cf. Section 7.3). Cf. also footnote 41 on page 423.

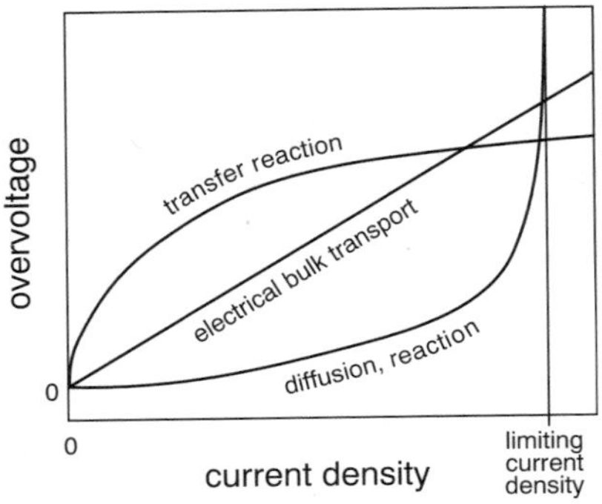

Fig. 7.56: Important types of overvoltage and their dependence on current density, schematic [654].

significantly to the overall losses[147]. The electrolyte of choice is still Y–doped ZrO_2. Scandium–doped ZrO_2 does exhibit a higher conductivity, but it is more expensive. Gd_2O_3–doped CeO_2 also has a higher conductivity, but this is counteracted by the appreciable electron conduction occurring on the anode side which brings about additional losses[148,149]. Very recently alkaline–earth doped $LaGaO_3$ perovskite has come to the fore[150].

For reasons of costs it is naturally desirable to reduce the operating temperature. However, this does not just raise the electrolyte resistance[151] but also causes marked electrode effects, although, as long as hydrogen is used, not so much on the anode side[152]. There metal–ceramic mixtures ("Cermet", (Ni-ZrO_2)) are used, in order to reduce the contact resistance. The behaviour of the kinetically complex cathode side is sketched schematically in Fig. 7.57. The detailed kinetics (see also Section 7.3) has not been completely elucidated for any configuration and naturally depends on the precise conditions of operation. Worthwhile information has been obtained by considering the dependence of the polarization resistance on geometric parameters (circumference, area, height of electrode spots). The results highlight the importance of lateral inhomogeneity and dimensional complexity as reflected in the respective contributions of bulk, surface and gas phase diffusion [584] (cf. Section 7.3.7 and Fig. 7.53d)[153].

The cathode material currently in favour is $LaMnO_3$ (Sr–doped). While in this material the three–phase contacts[153] [658,640] are considered to be the sites of electro-

[147] In addition, the SOFC cell is not sensitive to CO_2, so that air can be used.

[148] However, that does not categorically prevent application. Ref. [655] gives a calculation of the losses involved.

[149] By using layers of different electrolytes in series, it is also possible to use materials that would be unstable if subject to the total oxygen gradient [656].

[150] Major problems (at high temperatures) concern the chemical stability. Note the enormous requirements for a suitable electrolyte in terms of sufficient chemical, electrical, thermal and mechanical stabilities.

[151] That can be partially corrected using thinner membranes.

[152] It is not surprising, that the kinetics of H_2 oxydation are significantly less problematical than those of O_2 (or even of N_2) reduction.

[153] Current constriction phenomena at three–phase contacts must be taken into account; the neglect of these can lead to misinterpretations of electrode kinetics [657].

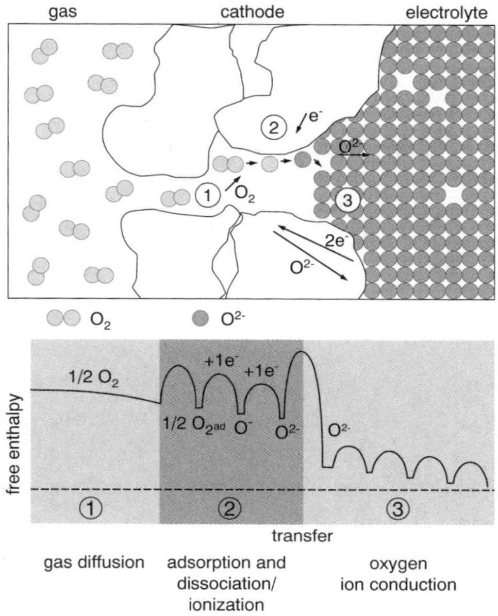

Fig. 7.57: A schematic diagram of cathode events for SOFC fuel cells. The processes actually taking place are usually much more complicated (cf. Sections 6.7 and 7.3.3), the same applies to the free enthalpy profile [654]. (Note that concerning gas diffusion (1) the configuration effect is included in the free enthalpy profile.) For simplicity and in contrast to reality, it is assumed that the oxygen is completely ionized when it enters the electrolyte.

chemical reaction, and the surface diffusion of absorbed oxygen to these constitutes an essential step, in the case of appropriately doped $LaCoO_3$ the permeability (σ^{δ}) for oxygen is so high that oxygen transport is considered to predominantly take place through the electrode mass[154]. However, the cobaltate is less stable in contact with ZrO_2, so that, at high working temperatures, buffer layers are necessary (CeO_2, for instance) [659].

An elegant solution would be the use of a single chemical substance which combines electrode and electrolyte functions as a result of suitable doping, and given the appropriate external chemical potentials. CeO_2 and the perovskite systems are relevant candidates for such monolithic cells. Thus, for example, as already mentioned Sr–doped $RGaO_3$ (R = La, Nd, etc.) is an excellent ion conductor while Co–doping brings about high electron conduction [660]. The pyrochlores, in particular (see Fig. 5.43, page 175), have been studied in this context [661]. The brownmillerites [556, 662] (see Fig. 7.10a) are also interesting material systems in this respect. On the other hand, the many criteria (mechanical, chemical, electrical, thermal stability; ecological and specific and general economic constraints) that must be fulfilled by a fuel cell, probably cannot be met by a single basic structure, in addition interdiffusion of dopant ions may constitute a problem.

Connecting individual cells together to form large stacks is not without complications (e.g. sealing). While Siemens has preferred for some years a planar structure,

[154]Recent experiments amend these statements, in that they show that also for the manganate the degree to which oxygen is incorporated at the three phase boundary depends on geometry (and bias) (see Fig. 7.53) [584].

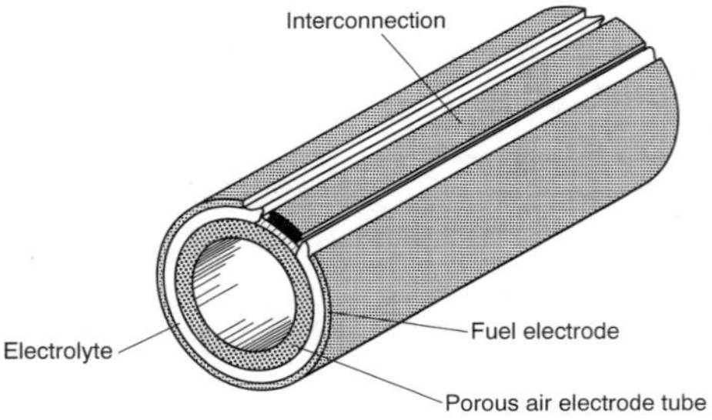

Fig. 7.58: The tubular stack–design of SOFC cells (a segment of a tube is shown) ("Westinghouse design"). The modules of stacked tubes do not require any high temperature seals. From Ref. [663].

Westinghouse favours a tubular one (Fig. 7.58)[155]. In the latter case the porous cathode ceramic acts as ceramic support. This is sealed with a ZrO_2 layer[156] and the anode material is then applied. Nowadays[157] modified planar configurations are en vogue again. A special problem at these high temperatures and extreme chemical conditions is the selection of a suitable material of high electronic conductivity as interconnect. The current favourite is doped lanthanum chromate ($LaCrO_3$).
The proton conductors, which can also naturally be exploited in H_2/O_2 fuel cells, constitute another important group. The fact that water forms on the air side, i.e. can be removed cheaply, as well as the higher conductivities constitute fundamental advantages. The "water–containing" perovskites, as discussed in Chapter 5, are possible candidates for high–temperature applications. Owing to their high conductivities a decrease of the operation temperature appears to be possible for the ceramic fuel cells which could lead to a substantial reduction of the materials problems (e.g. use of steel as interconnect). A serious disadvantage is the CO_2 sensitivity of most compounds. The most suitable candidate currently seems to be Y–doped $BaZrO_3$ [202,664].
High–temperature fuel cells are promising for localized power stations and for coupling with gas turbines. Fuel cells working in the medium– or low–temperature range are more suitable for small, nonstationary applications, such as desired in the automobile field (see Tables 7.6, page 496). The polymer fuel cells are the candidates of choice (see Fig. 7.59). The polymer electrolyte[158] generally used is Nafion [666, 667]. This consists of a perfluorinated hydrocarbon backbone with a high density of ether side chains exhibiting sulfonic acid end groups. The protons of the acid groups are dissociated away by the water present in the inner cavities in the swollen state. In a certain sense we can regard Nafion as an "acid sponge" (cf. Section 5.8.5). The proton transport here is similar to that in a fluid. Since the counterions do not

[155]The problems of including very recent information in a monograph are revealed by the fact that the Westinghouse division has been bought by Siemens during the writing. This confirms F. Dyson's statement: When a book is written at time t_0 with the intention of covering material up to $t_0 - \Delta t$, it is outdated by $t_0 + \Delta t$.
[156]The EVD method discussed for this purpose is an elegant but expensive preparation technique. Cf. here Section 6.9 [495].
[157]That is, at the time the book was written.
[158]Cf. also Section 5.8.5.

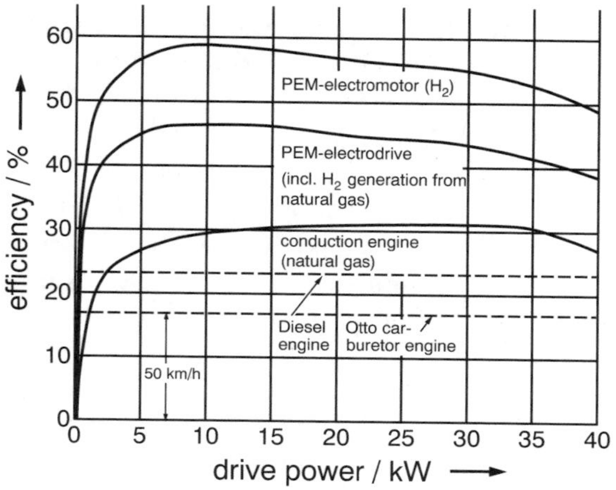

Fig. 7.59: Comparison of efficiencies of electrical traction using PEM cells with conventional internal combustion engines. Module blocks with power outputs of 50kW are state of the art for submarines. From Ref. [665].

migrate, Nafion belongs to the group of ion–exchange membrane electrolytes. Ion exchange fuel cells were used in the American Gemini program and have proved themselves particularly in submarines. The major disadvantage of Nafion is the price and, not independent of this, the inherent problems of the fluorine chemistry involved in manufacture, together with the permeability for water and for fuels such as methanol. In addition Nafion must always be kept moist. Currently work is being intensively carried out on modifications and alternatives, particularly those based on polyether–ketones (see e.g. [668,669]). A promising route consists in achieving proton transport via heterocycles with proton donor–acceptor functions built in the organic matrix [670]. Methanol–air fuel cells are appropriate candidates for electrotraction, in which methanol is used instead of gasoline (Fig. 7.59). At present a preceding reforming step is necessary, but the possibility of direct electrochemical methanol conversion does not appear hopeless.
With few exceptions, such as Na_2CO_3, K_2CO_3 molten electrolytes, liquid electrolytes are only of relevance for low–temperature systems. Classical systems are based on phosphoric acid or potassium hydroxide (see Tables 7.5 and 7.6, page 496) [671].

7.4.3 Batteries

a) Primary systems

In the case of primary batteries it is most important to consider the intended purpose. Miniaturizable low power batteries of great reliability find important applications in measurement devices (e.g. clocks), circuits and heart pacemakers. One system which refers to the last mentioned, medical application and which is now classic, is the Li-I_2 battery. Here the iodine is introduced as an electronically conducting charge transfer complex (e.g. as polyvinylpyridine) [672]. This does reduce the chemical potential of iodine and, hence, the cell voltage but guarantees a sufficient electronic conductivity and improves the handling properties. The key is that

the LiI film, which forms on contact, serves as a self–healing electrolyte: If cracks occur, LiI is reformed on contact. LiI:Al_2O_3, as a better conducting heterogeneous electrolyte (cf. Section 5.8.3), was also used in this context.
Li batteries with Li as negative electrode are of special interest for battery systems providing higher power per unit mass on account of their low weight and high electropositivity, which in general leads to a high cell voltage[159]. The Li–$SOCl_2$ cell [673], which is able to deliver high power levels, was in vogue for a long period. Here the active, liquid cathode material ($SOCl_2$) reacts with Li to give SO_2, sulfur and LiCl. On initial contact LiCl (see Fig. 7.42, page 468), as Li^+–conducting solid electrolyte, is produced in a purely chemical manner. At later stages LiCl is deposited electrochemically at the porous "current collector" graphite at the cathode and limits the power production. The Li-MnO_2 primary cell finds use currently (see e.g. [649–651]). Since, in principle, MnO_2 can, as many other transition metal oxide, also accommodate lithium reversibly, this combination will be discussed in the context of the Li secondary batteries again.
Zinc plays an important role as a cheap anode material. In the Leclanché cell, put simply, Zn is oxydized to ZnO by MnO_2. The Zn air battery[160] is also a primary battery but exhibits similarities with a fuel cell. Zn systems using HgO, Ag_2O_2 as cathodes are all excellent primary systems that have long been in use [650,651,674]. Table 7.7 (page 497) gives an overview (cf. also Tables 7.11, 7.12 on page 499).
Solid electrolyte batteries using silver systems have played a prominent role historically. The superionic conductor[161] α–AgI is a suitable solid electrolyte for an Ag-I_2 battery, but is only stable above 146°C (see Section 6.2.1). On the other hand, $RbAg_4I_5$ is an excellent conductor even at room temperature, but unstable with respect to I_2[162]. For this reason, $(R_4N^+I^-)$–I_2 adducts are used; the resulting reduction of the iodine activity and hence the reduction of the already modest cell voltage are the attendant compromises. In all these cases, the formation of AgI is disadvantageous [672].
This brings us to two important points: On the one hand the electrolyte must be stable with respect to the active components, that is, possess a suitable redox window[163] (Table 7.8 on page 497 shows this for Li^+ solid electrolytes); on the

---

[159] In addition, Li, like Na, is ecologically unobjectionable.

[160] The negative electrode is metallic zinc, and the positive electrode is a porous air electrode, usually carbon-based. The ZnO formed is dissolved in the circulating alkali electrolyte and washed away. During charging the ZnO–containing solution is washed back again. This "reversibility" may be compared to that of a secondary cell.

[161] Cf. Sections 6.2.1 and 5.7.2.

[162] Decomposition to $RbI_3 + 4AgI$.

[163] The electrolyte must naturally be stable towards both anode and cathode. This also implies that the width of the redox window (thermostatic) must not exceed the decomposition voltage (i.e. $|zF\Delta_f G|$). Sometimes the band gap of the electrolyte is used as the relevant measure [675]. However, this is an extremely rough upper estimate, because it only refers to electronic effects. Note that the free energy of decomposition of NaCl into Na and Cl_2 as separate phases is very much lower than the formation of Na^0 and Cl^0 within NaCl (band–band transition, cf. Sections 2.2 and 5.3).

other hand, ideal cathodes should not only discharge ions but also accommodate the product. If the element is reversibly stored (i.e. it can also be easily released) then such electrodes are suitable for secondary cells, as is the case for TiS_2 as an intercalation electrode for lithium batteries [676].

b) Secondary systems

Figure 7.60 left shows an example of a Li-TiS_2 microbattery with glass electrolyte. The capacities are very small but adequate for some low power applications, and

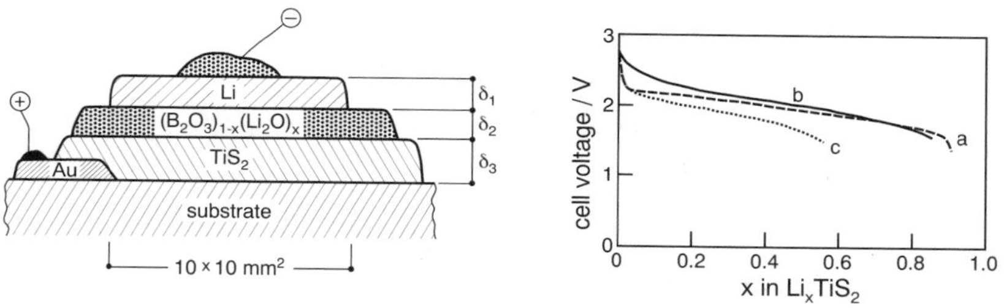

Fig. 7.60: Left: Li/TiS_2 microbattery with Li^+ conducting glass electrolyte. Right: Discharge curves for Li/TiS_2 cells with amorphous TiS_2 film (b: Current density = $3\mu Acm^{-2}$, c: $16\mu Acm^{-2}$) and crystalline TiS_2 (a: $10mAcm^{-2}$) using liquid electrolyte. From Ref. [650].

the discharge curves are adequately flat. Because the glass film is rather thin, its moderate ionic conductivity is sufficient for operation.
In principle, chemical energy can be removed from the cell

$$Li|Li^+\text{–electrolyte}|Li\text{–intercalation compound}$$

until the chemical potential of the lithium in the intercalation electrode (positive electrode, i.e. cathode during discharge) becomes equal to that of the pure lithium (negative electrode, i.e. anode during discharge). Although the Li–activity of the electrode must fall continuously with discharge, nevertheless, the discharge curves can very closely approach ideal behaviour (step function) — as shown in Fig. 7.60 right[164]. The intercalation of the Li into the layer structure of the TiS_2 also guarantees a good reversibility, so that cells of this type play a role as typical secondary

[164] In some cases, as the intercalation of Li in the spinel $LiMn_2O_4$ (with the formation of nominal $Li_{1+x}Mn_2O_4$), the stationary discharge curve is necessarily horizontal on account of the formation of a two–phase region (see Ref. [677]) (cf. Sections 4.3 and 7.3.5). The discharge curve (U against Q) of an ideal capacitor (including supercapacitors, see page 444) is linear and, corresponding to the integral, the energy stored at equal maximal charge, is only half as large as in the case of the ideal battery (step function) [675] (cf. Fig. 7.55).

cells. It is obviously important that the electrodes have a high D^{δ}_{Li} value at the operating temperature[165] [676,679].
Hence, in principle, all stable compounds with adequate stoichiometry widths, with high electronic conductivities and sufficient Li^+ mobilities, in which lithium has a low enough chemical potential, are suitable for use as positive electrodes in secondary cells. In general, materials based on layered transition metal compounds such as the $LiMO_2$ phases of the α–$NaFeO_2$ type (M= Ni,Co,V), or spinel phases such $LiMn_2O_4$ are relevant candidates [680].
Figure 7.61 shows the charge and discharge curves of such high–performance cells with nonaqueous liquid electrolytes. (Figure 7.62 provides information concerning

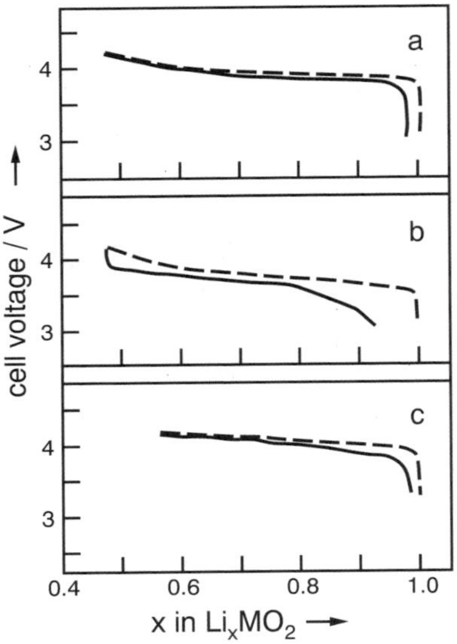

Fig. 7.61: Charging and discharging curves of high performance cells with Li as anode, Li_xCoO_2(a), Li_xNiO_2(b), $Li_xMn_2O_4$(c) as cathodes. The crystal structure of the latter is of the spinel type. From Ref. [650].

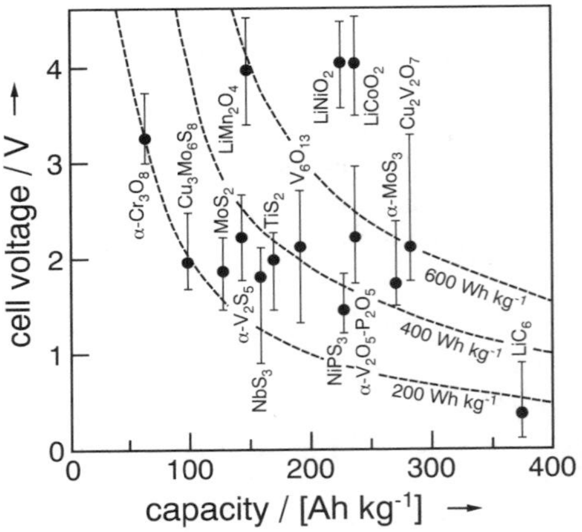

Fig. 7.62: Cell voltage and capacity of cathode materials for secondary Li intercalation cells. From Ref. [650].

the cell voltage, capacity and energy density.) Such discharge curves, when they are recorded in a quasi–stationary manner, correspond to the coulometric titration curves described in Section 7.3.5. In many cases these discharge curves exhibit plateaux corresponding to the formation of more or less ordered intermediate phases. The cell voltage of Li–free CoO_2 with respect to Li is ca. 5.1V; the operable cell voltage is ca. 4V [681]. Partial substitution of the Co by Al allows this to be

[165] In specific cases [678], such redox reactions are accompanied by reversible colour changes. We speak of "electrochromism". If the colour change corresponds to the darkening of a transparent sample then such a material is suitable for controlling light transparency ("electrochromic window").

increased [682]. Partial substitution with Fe or Mn even allows the operating voltage to be raised to more than 5V. In these compounds separate charge changes of Co (Co^{3+} to Co^{4+}) and of Mn (Mn^{3+}/Mn^{4+}) or Fe (Fe^{3+}/Fe^{4+}) can be distinguished in the discharge curves [683,684]. The use of nanostructured electrodes (and hence tiny diffusion lengths) renders possible the use of active components with low conductivities[166] [685,687].
Inherent safety and corrosion problems can largely be overcome when the negative electrode is also an intercalation electrode. Carbon is particularly relevant in practice: Loss of Li activity, and hence of cell voltage is low (ca. 0.1V) and the cycling properties are much improved. In such "rocking chair" batteries (see Fig. 7.55) Li is rocked backwards and forwards between two intercalation electrodes[167], e.g. Li_xC and $Li_xMO_2 (M = Co, Ni, Mn)$,

$$Li_{x_1}C|Li\text{–electrolyte}|Li_{x_2}MO_2.$$

The most advanced at the moment is the Sony cell ($Li_{x_1}C - Li_{x_2}CoO_2$). Figure 7.63 shows the intercalation process for Li in graphite. While the layers in graphite (cf.

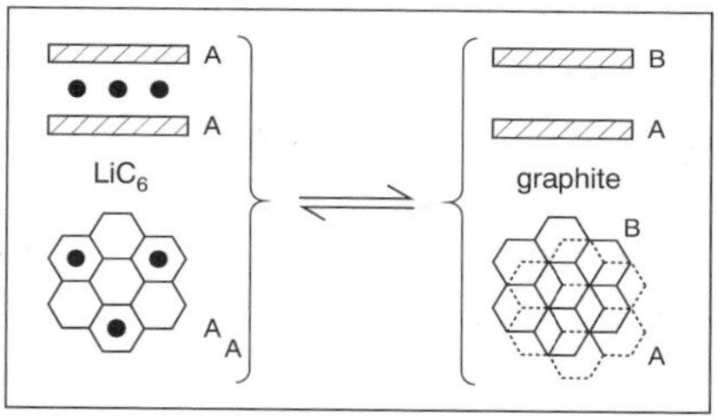

Fig. 7.63: On intercalation of lithium in graphite there is a displacement of the layer stacking sequence (see text).

Section 2.2 and Fig. 2.22) exhibit an AB stacking sequence[168], in LiC_6 the Li separates layers that are arranged exactly on top of each other (AA). The layer distance increases thereby by about 10%. LiC_6 corresponds approximately to the maximum composition on charging the battery. Compounds richer in Li (e.g. LiC_2) are chemically accessible. This field has been additionally stimulated by the discovery of fullerenes (see Section 2.2.3).
If nanocrystalline oxides are used, the lithium can also be reversibly stored via a heterogeneous reaction. Ref. [688] demonstrated the usefulness of CoO as a negative electrode component. Upon discharge Li_2O and Co are formed. Because of the tiny

[166] Li_xFePO_4–carbon composites are adequate examples. If the carbon network is fine enough even moderately thin phosphate particles can be fully intercalated [685]. Other experiments refer to the possibility of drastically increasing the electronic conductivity of this compound by doping [686].

[167] See, for example, Refs. [419,649–651,683]. The term "lithium ion batteries" for batteries with polar Li–compounds as negative electrodes is unfortunate.

[168] Graphite with the stacking sequence ABC also exists. The energy differences between the different sequences are only small, and many types of stacking faults are realized accordingly. Other highly disordered graphitic and nongraphitic structures are also of interest on account of their low price.

diffusion length (approximately 1nm) the reaction is reversible in spite of the low chemical diffusivities (cf. also Refs. [689–691]). The reader is referred to Ref. [272] for more details on the impact of nano–crystallinity on thermodynamic and kinetic properties of Li–batteries.

Because of the good contact and mechanical flexibility, liquid electrolytes are generally used[169] as ion conductors (e.g. $LiPF_6$ in ethylene carbonate / dimethyl carbonate). These are sometimes immobilized by polymers such as PMMA (polymethyl methacrylate) [692] to improve the mechanical properties. The conductivity of pure polymeric electrolytes like (Li^+–containing) PEO (see Fig. 6.12, page 290) is too low for many applications and can be improved by inorganic fillers[170] (cf. Section 5.8.5a) [368].

The reader can find helpful tables on pages 498 and 499. Table 7.9 provides information concerning important operating data while Table 7.10 gives an overview of the historical development of important secondary elements.

High performance batteries are of enormous importance for electrotraction (see here Tables 7.10, 7.12, 7.11, Fig. 7.55). The classical battery type is the lead–acid accumulator[171]. The large molar mass of lead is reflected in a very negative manner in the energy content per unit weight and the operating range of automobiles driven by it (Tables 7.12, 7.11). If one wished to drive a car of sensible size and performance with this alone one would require a battery weight of 1t. Ni–Cd accumulators are not much better in this respect[172].

For this reason there is considerable interest in high performance Li– or Na–based accumulators (cf., here Table 7.11). The Na-S cells [693]

$$\mathrm{Na}|\beta''\text{–}\mathrm{Al_2O_3}\,(\mathrm{Na_2O})\,|\mathrm{S}$$

were regarded as being very promising for a long while (Fig. 7.64). Here the electrolyte is a β''–Al_2O_3–Na_2O ceramic ("β-alumina", cf. Section 6.2). The use of a solid electrolyte makes it possible to use liquid electrodes (sodium and sulfur), also ensuring good contact. The operating temperature is ca. 300°C. The sulfide formed dissolves in the sulfur with the formation of polysulfide, thus, solving the storage problem:

$$2\mathrm{Na} + \mathrm{xS} \rightleftharpoons \mathrm{Na_2S_x}.$$

The voltage of the cell under open circuit conditions is shown in Fig. 7.65 as a function of the Na content of the cathode. With good thermal insulation the operating temperature can be maintained by the waste heat. In spite of the good performance

[169] Protic solvents are naturally unsuitable (see also footnote 163). But the aprotic liquid electrolytes used are typically thermodynamically unstable, too. The formation of a thin passivizing layer is highly important here [649].

[170] The "soggy sand electrolytes" described in Ref. [269] provide an interesting materials class that should be useful for Li–batteries. Cf. also footnote 188 on page 246.

[171] $Pb + PbO_2 + 2H_2SO_4 \rightleftharpoons 2PbSO_4 + 2H_2O$. The electrolyte is H_2SO_4.

[172] $Cd + 2NiO(OH) + 2H_2O \rightleftharpoons Cd(OH)_2 + 2Ni(OH)_2$. Sodium hydroxide solution is used as an electrolyte.

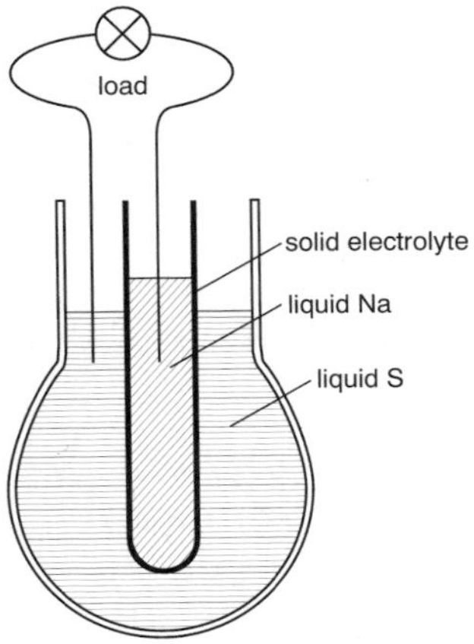

Fig. 7.64: Schematic diagram of a Na/S cell.

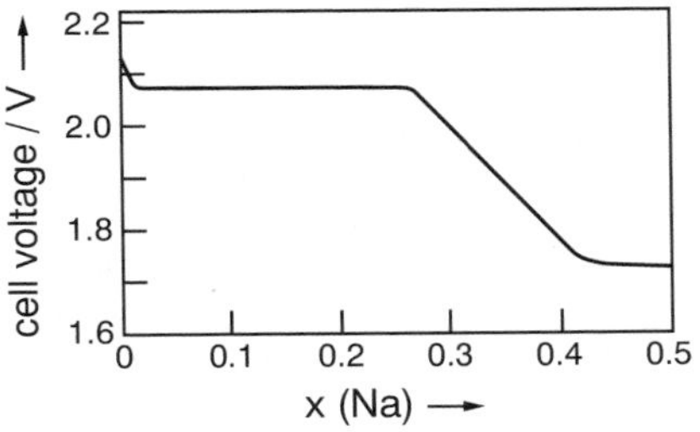

Fig. 7.65: Open circuit voltage of the Na/S cell as a function of the Na content of the cathode. From Ref. [650].

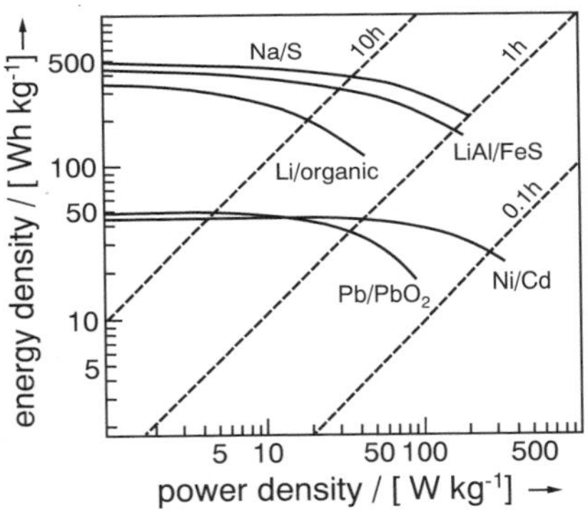

Fig. 7.66: Energy density as a function of the power density for various battery systems. The energy density of gasoline is $3 \times 10^4 \mathrm{Whkg}^{-1}$ [694]. From Ref. [650].

data (see Fig. 7.66), the high cell voltage of 3V and the mass related energy content being 10 times better that the lead–acid accumulator, investigations with respect to electrotraction have been mostly abandoned. The dangers of crack formation and resulting catastrophic local chemical reaction are the main reason.

The zebra cell is a related cell that possesses much less damage potential in this respect [695]. The solid electrolyte is the same as in the Na–S battery, and the operating temperature is similar. However, the cell reaction comprises the reaction of Na with $NiCl_2$ to Ni and NaCl, which is contained in a $NaAlCl_4$ melt (open circuit voltage per cell $\simeq$ 2.6V). If the electrolyte ceramic breaks, Na reacts with $NaAlCl_4$ to yield Al. The short circuit that occurs ensures that the cell chain maintains its function even if 5% of the cell has been destroyed in this manner.

Other candidates for electrotraction purposes are (besides fuel cells) naturally Li accumulators (see above), Ni metal hydride systems[173] and metal air systems like the zinc–air batteries already described[160] (Fig. 7.55). The latter are not proper secondary systems, but nonetheless relevant for fleet operators (e.g. taxi enterprises).

c) Outlook

The history of the application of electrochemical systems is clearly not just that of the development of concepts but also of the development of materials and their integration into an appropriate system.
Finally, and more as a stimulus than for the purpose of contemplation, we consider a sketch of an autonomous system (Fig. 7.67) that combines many of the functions

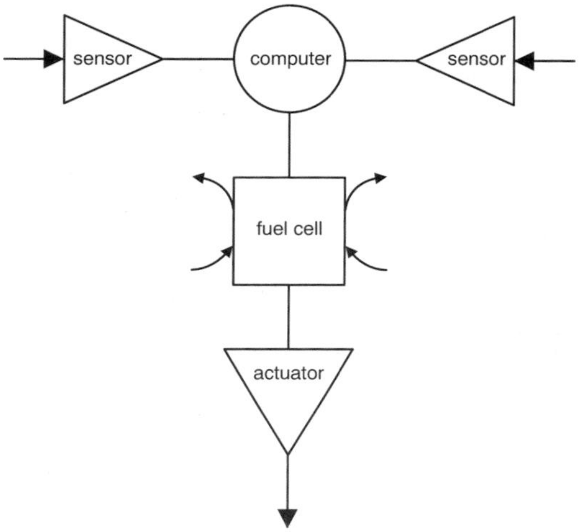

Fig. 7.67: Autonomous system of functional materials [696].

described in this book. It is already reality in its basic outline (robotics) if we refer to macroscopic components but is a project of the future in terms of nano–integration (see also Table 6.7, page 393).
The sensor signal produced by an excitation from the environment is processed by the computer "brain". Actuators influence the environment correspondingly, and bring about autonomous movement. The necessary "metabolism" is made possible by fuel cells. In all cases the core materials of the "organs" are functional materials, that carry out most of their functions as a result of defects, and the behaviour of these is determined by the physical chemistry of materials. It is these interactions between the "chemical environment" and the charge carriers of the solid in most of the organs ("internal chemical life") that endows solid state electrochemistry with a central role in the development and understanding of function and stability.

---

[173] They rely on the reduction of NiO(OH) by hydrogen. The latter is stored in solid phases such as $LaNi_5H_x$. Thus, if we ignore the anode, the cell is similar to the Ni–Cd accumulator (cf. footnote 172).

7.4.4 Tables

Table 7.4: Thermodynamic data and efficiencies for important cell reactions. According to Ref. [674].

cell reaction	temp. (°C)	$\Delta_r H^\circ$ ($Jmol^{-1}$)	$\Delta_r S^\circ$ ($Jmol^{-1}K^{-1}$)	$\Delta_r G^\circ$ ($Jmol^{-1}$)	-U (V)	dU/dT ($mV^{-1}K^{-1}$)	$w_g^\circ = \frac{\Delta_r G^\circ}{\Delta_r H^\circ}$
$H_2+\frac{1}{2}O_2 \rightarrow H_2O_{(l)}$	25	-285800	-162.40	-237400	1.23	+0.840	0.83
	60	-285050	-159.00	-231830	1.20	+0.820	0.81
	100	-283300	-155.00	-220370	1.17	+0.800	0.78
$H_2+\frac{1}{2}O_2 \rightarrow H_2O_{(g)}$	25	-241830	-44.40	-228580	1.18	+0.230	0.945
	60	-242180	-45.60	-226990	1.18	+0.230	0.94
	100	-242580	-46.60	-225160	1.17	+0.240	0.93
	500	-246180	-55.10	-203530	1.05	+0.280	0.83
$NH_{3(g)}+\frac{3}{4}O_2 \rightarrow \frac{1}{2}N_2+\frac{3}{2}H_2O_{(l)}$	25	-382510	-145.50	-339120	1.17	+0.500	0.89
$NH_{3(g)}+\frac{3}{4}O_2 \rightarrow \frac{1}{2}N_2+\frac{3}{2}H_2O_{(g)}$	25				1.13		
$N_2H_{4(l)}+O_2 \rightarrow N_2+2H_2O_{(l)}$	25	-621100	+5.10	-622600	1.61	-0.010	1.00
$2Na+H_2O_{(l)}+\frac{1}{2}O_2 \rightarrow 2NaOH_{(aq)}$	25	-653210	-174.90	-601050	3.11	+0.900	0.92
$C_{Gr}+\frac{1}{2}O_2 \rightarrow CO$	25	-110500	+89.10	-137080	0.71	-0.460	1.24
	500	110800	+89.90	-180300	0.93	-0.460	1.63
$C_{Gr}+O_2 \rightarrow CO_2$	25	-393500	+2.87	-394350	1.02	-0.007	1.00
$CO+\frac{1}{2}O_2 \rightarrow CO_2$	25	-283000	-86.20	-257300	1.33	+0.440	0.91
$CH_3OH_{(l)}+\frac{3}{2}O_2 \rightarrow CO_2+2H_2O_{(l)}$	25	-726260	-76.50	-703700	1.21	+0.130	0.97
$CH_4+2O_2 \rightarrow CO_2+2H_2O_{(g)}$	25	-802400	-6.00	-800600	1.04	+0.007	1.00
	60	-802060	-4.90	-800420	1.04	+0.006	1.00
	100	-801700	-3.90	-800200	1.04	+0.005	1.00
	500	-800300	-1.70	-798900	1.03	+0.002	1.00
$CH_4+2O_2 \rightarrow CO_2+2H_2O_{(l)}$	25	-890200	-242.60	-817900	1.06	+0.310	0.92
$C_2H_4+3O_2 \rightarrow 2CO_2+2H_2O_{(l)}$	25	-1306320	-62.10	-1287810	1.11	+0.050	0.99
$C_2H_4+3O_2 \rightarrow 2CO_2+2H_2O_{(g)}$	25				1.09		
$C_3H_8+5O_2 \rightarrow 3CO_2+4H_2O_{(l)}$	25	-2218900	-374.00	-2107440	1.09	+0.190	0.95
$C_3H_8+5O_2 \rightarrow 3CO_2+4H_2O_{(g)}$	25	-2044000	+108.00	-2076380	1.07	-0.050	1.02
$n\text{-}C_4H_{10}+6\frac{1}{2}O_2 \rightarrow 4CO_2+5H_2O_{(l)}$	25	-2878270	-438.00	-2747930	1.09	+0.170	0.955

Table 7.5: Overall electrode reactions for important fuel cell types [654].

fuel cell types	anode reaction	cathode reaction
phosphoric acid (PAFC) and polymer membrane fuel cell (PEMFC)	$H_2 \longrightarrow \mathbf{2H^+} + 2e^-$	$1/2O_2 + \mathbf{2H^+} + 2e^- \longrightarrow H_2O$
alkali fuel cell (AFC)	$H_2 + \mathbf{2OH^-} \longrightarrow 2H_2O + 2e^-$	$1/2O_2 + H_2O + 2e^- \longrightarrow \mathbf{2OH^-}$
molten carbonate fuel cell (MCFC)	$H_2 + \mathbf{CO_3^{2-}} \longrightarrow H_2O + CO_2 + 2e^-$ $CO + \mathbf{CO_3^{2-}} \longrightarrow 2CO_2 + 2e^-$	$1/2O_2 + CO_2 + 2e^- \longrightarrow \mathbf{CO_3^{2-}}$
solid oxide fuel cell (SOFC)	$H_2 + \mathbf{O^{2-}} \longrightarrow H_2O + 2e^-$ $CO + \mathbf{O^{2-}} \longrightarrow CO_2 + 2e^-$ $CH_4 + \mathbf{4O^{2-}} \longrightarrow 2H_2O + CO_2 + 8e^-$	$1/2O_2 + 2e^- \longrightarrow \mathbf{O^{2-}}$

Table 7.6: Overview of important fuel cell types [671].

name and abbreviation	temperature in degree centigrade	electrolyte	fuel	effiency in percent	envisaged application
alkali fuel cell (AFC)	80 to 90	potash lye	hydrogen	50 to 65	transport, space flight, shipping
polymer electrolyte membrane fuel cell (PEMFC)	80 to 90	polymer membrane (Nafion)	hydrogen, reformed methanol or methane	50 to 60	transport, electro car, space flight, shipping
phosphoric acid fuel cell (PAFC)	200	phosphoric acid	hydrogen, reformed methane	35 to 45	1 upto 100 MW power plants, 5 to 500 kW heating power station
molten carbonate fuel cell (MCFC)	650	calcium carbonate	hydrogen, methane	45 to 60	1 upto 100 MW power plants, 5 to 500 kW heating power station
solid oxide fuel cell (SOFC)	850 to 1000	zirconia	hydrogen, methane	50 to 60	1 upto 100 MW power plants, 5 to 500 kW heating power station

Table 7.7: Important zinc–based primary elements. The price per kWh increases monotonically within the table (from top to bottom) and is about an order of magnitude greater for the Ag–Zn cell than for the Leclanché cell (date 1996) [697].

cell	reaction	working voltage (V)	energy density (Wh/kg)	(Wh/l)
Leclanché	$Zn + 2MnO_2 \xrightarrow{(NH_4Cl)} ZnO + Mn_2O_3$	1.25	50 - 60	100 - 120
Leclanché alkal.	$Zn + 2MnO_2 \xrightarrow{(KOH)} ZnO + Mn_2O_3$	1.1 - 1.2	80	210
mercury oxide cell	$Zn + HgO \xrightarrow{(KOH)} ZnO + Hg$	1.1 - 1.3	80 - 100	270 - 370
silver zinc cell	$Zn + AgO \xrightarrow{(KOH)} ZnO + Ag$	1.5	120 - 190	370

Table 7.8: Stability windows of some Li^+ conductors. If $U_{min} > 0$ this means that the electrolyte is not stable with respect to Li. From Ref. [650].

Stability windows of some lithium ion conductors

Material	Temperature (°C)	Voltage vs. Li U_{min}	U_{max}
LiI	25	0	2.79
Li_2O	150	0	2.84
LiCl	25	0	3.98
Li_3N	25	0	0.44
$LiAlCl_4$	25	1.68	4.36
$Li_9N_2Cl_3$	100	0	2.50
Li_6NBr_3	176	0	1.30
$Li_{9.1}N_{2.7}I$	316	0	0.90
$LiNO_3$	150	2.50	4.20
Li_4SiO_4	415	0.14	3.06
$Li_2Si_2O_5$	415	1.31	3.31
Li_8ZrO_6	325	0	2.65
Li_2ZrO_3	325	0.35	3.06

Table 7.9: Characteristics of some Li secondary cells. From Ref. [650].

Anode	Electrolyte	Cathode	Current density ($mA\ cm^{-2}$)	Cycles	Producer
WO_2	$LiClO_4$-PC	TiS_2	0.1	68	Rome Univ.
MoO_2	$LiAsF_6$-PC	$LiCoO_2$	0.1	-	AT&T
WO_2	$LiAsF_6$-PC	$LiCoO_2$	0.1	-	AT&T
Li_xC_6	$LiClO_4$-PC	$LiMn_2O_4$	0.8	25	AT&T
$Li_9Mo_6Se_6$	$LiAsF_6$-THF	Mo_6Se_6	1.2	10	Bell Comm.
TiS_2	$LiAsF_6$-AN	$LiCoO_2$	0.5	500	U.S. Army
Coke	$LiN(CF_3SO_2)_2$	$LiNiO_2$	1.0	1000	Moly Energy
Li_xC_6	$LiPF_6$-PC-DEC	$LiCoO_2$	0.5	100	Sony

Table 7.10: Secondary cells of importance in historical development. From Ref. [650].

secondary cells

date	type	cell
1860	lead acid cell	$PbO_2/H_2SO_4/Pb$
1900	Edison's cell	NiOOH/KOH/Fe
	Ni-Cd cell	NiOOH/KOH/Cd
1965	Na-S cell	Na/β-Al_2O_3/S
1970	Zn-Cl_2 cell	Zn/elect./Cl_2
1980-90	Li-liquid electrolyte	Li/PC-Li_2ClO_4/MX_2
	-polymer cells	Li/PEO-$LiClO_4$/TiS_2
	-glass cells	Li/Li^+-glass/TiS_2
1991	Li-microbatteries	Li/Li^+-glass/TiS_2
1992	rocking chair cells	$LiMn_2O_4$/elect./carbon
		$LiCoO_2$/elect./carbon
		$LiNiO_2$/elect./carbon

Table 7.11: Cell voltage and cycle numbers of older accumulator systems. From Ref. [674].

system	cell voltage (V)	energy density theoret. (Wh/kg)	energy density pract. (Wh/kg)	number of cycles
Pb cell	2.0	161	10 - 35	500 - 1500
Ni-Fe cell	1.6	250	25 - 35	up to 3000
Ni-Cd cell	1.35	210	25 - 30	1000 - 2000
Ag-Zn cell	1.6	220 (1.stage)		
		350 (2.stage)	70 - 130	50
Ag-Cd cell	1.4	156 (1.stage)	30 - 45 (1.stage)	ca. 200
		245 (2.stage)	45 - 60 (2.stage)	
Zn-air cell	ca. 1.45	1450	90 - 180	

For comparison: Modern lithium rocking chair cells work at a cell voltage of 4V over 1000 cycles with energy densities exceeding 100 Wh/kg [698].

Table 7.12: Expected ranges of electromobiles equipped with accumulators (estimated values). From Ref. [674].

system	energy density (Wh/kg)	range (km)
Pb cell	10- 35	<50
Ag-Zn cell	70-130	80-160
Zn-air cell	90-180	130-250
Na-S cell	ca. 300	>100
Li-Cl_2 cell	ca. 300	>200
gasoline engine (120 kWh per tank filling)	375	400-560

8 Bibliography

[1] J. Frenkel, Z. Physik **53** (1926) 652.

[2] C. Wagner, W. Schottky, Z. Phys. Chem. **B11** (1930) 163; C. Wagner, Z. Phys. Chem. **B32** (1936) 447.

[3] F. A. Kröger, *Chemistry of Imperfect Crystals*, North–Holland, Amsterdam, 1964.

[4] K. Hauffe, *Reaktionen in und an festen Stoffen*. The figure is reprinted from this reference. Copyright 1955, with permission from Springer–Verlag, Berlin.

[5] H. Rickert, *Electrochemistry of Solids*, Springer–Verlag, Berlin, 1982.

[6] H. Schmalzried, A. Navrotsky, *Festkörperthermodynamik*, VCH, Weinheim, 1975.

[7] H. Schmalzried, *Solid State Reactions*, VCH, Weinheim, 1981.

[8] H. Schmalzried, *Chemical Kinetics of Solids*, VCH, Weinheim, 1995.

[9] A. R. Allnatt, A. B. Lidiard, *Atomic Transport in Solids*, Cambridge University Press, Cambridge, 1993; A. B. Lidiard, in: *Handbuch der Physik*, S. Flügge (ed.), Vol. 20, p. 246, Springer–Verlag, Berlin, 1957.

[10] J. Corish, P. W. M. Jacobs, Surf. Def. Prop. Solids **2** (1973) 160; ibid. **6** (1976) 219.

[11] F. Aguillo-Lopez, C. R. A. Catlow, P. D. Townsend, *Point Defects in Materials*, Academic Press, New York, 1988.

[12] P. Kofstad, *Nonstoichiometry, Diffusion and Electrical Conductivity in Binary Oxides*, John Wiley & Sons, Ltd., New York, 1972.

[13] Solid State Ionics, International Journal on Diffusion and Reactions, Elsevier, Amsterdam.

[14] J. Maier, Angew. Chem. Int. Ed. Engl. **32** (3) (1993) 313; Angew. Chem. **105** (1993) 333.

[15] J. Maier, Chemistry — A European Journal **7** (22) (2001) 4762.

[16] L. Pauling, Chem. Rev. **5** (1928) 173.

[17] J. E. Lennard–Jones, Trans. Farad. Soc. **25** (1929) 668.

[18] F. Hund, Z. Physik **51** (1928) 759; R. S. Mulliken, Phys. Rev. **32** (1928) 186.

[19] L. Pauling, E. B. Wilson, *Introduction to Quantum Mechanics*, McGraw–Hill, London, 1935.

[20] W. Kutzelnigg, *Einführung in die theoretische Chemie*, VCH, Weinheim, 1993.

[21] C. A. Coulson, *Valence*, Oxford Univ. Press, London, 1961.

[22] L. Zülicke, *Quantenchemie*, Vol. 1–2, Hüthig–Verlag, Heidelberg–Berlin, 1978.

[23] L. D. Landau, E. M. Lifshitz, *Quantum Mechanics (Course of Theoretical Physics)*, Vol. III, Pergamon Press, Oxford, 1965.

[24] K. Ruedenberg, Rev. Mod. Phys. **34** (1962) 326.

[25] L. Pauling, *The nature of the chemical bond and the structure of molecules and crystals*, Cornell Univ. Press, Ithaca, NY, 1967.

[26] W. A. Harrison, *Electronic Structure and the Properties of Solids*, W. H. Freeman, San Francisco, 1980.

[27] H. Chirgwin, C. A. Coulson, Proc. Royal Soc. **A20** (1950) 196.

[28] H. Hellmann, *Einführung in die Quantenmechanik*, Deuticke, Leipzig–Wien, 1937.

[29] R. P. Feynman, Phys. Rev. **56** (1939) 340.

[30] N. Figgis, *Introduction to Crystal Fields*, Interscience Publishers, New York, 1966; C. J. Ballhausen, *Introduction to Ligand Field Theory*, McGraw–Hill, New York, 1966; L. G. Burns, *Mineralogical Applications of Crystal Field Theory*, Cambridge University Press, Cambridge, 1970.

[31] A. A. Levin, *Solid State Quantum Chemistry*, McGraw–Hill, New York, 1977.

[32] J. K. Burdett, *Chemical Bonding in Solids*, Oxford University Press, New York–Oxford, 1995.

[33] G. A. L. Mie, Ann. Phys. **11** (1903) 1936.

[34] P. M. Morse, Phys. Rev. **34** (1929) 57.

[35] J. E. Lennard–Jones, in: *Statistical Mechanistics*, R. H. Fowler (ed.), Cambridge University Press, London, 1936.

[36] M. P. Allen, D. J. Tildesley, *Computer Simulations of Liquids*, Clarendon Press, Oxford, 1987.

[37] M. P. Allen, D. J. Tildesley, *Computer Simulation in Chemical Physics*, Kluwer, Dordrecht, 1993; M. Parrinello, Solid State Commun. **102** (1997) 107.

[38] R. Car, M. Parrinello, Phys. Rev. Lett. **55** (1985) 2471.

[39] T. P. Martin, in: *Festkörperprobleme*, P. Grosse (ed.), Vol. XXIV, p. 1, Vieweg, Braunschweig, 1984; Phys. Reps. **95** (1983) 167, Elsevier, Amsterdam.

[40] H. G. Fritsche, Phys. Stat. Sol. (b) **154** (1989) 603.

[41] R. L. Kronig, W. G. Penney, Proc. Royal Soc. (London) **A130** (1931) 499.

[42] Ch. Kittel, *Introduction to Solid State Physics*, John Wiley & Sons, Ltd., New York, 1986.

[43] G. M. Barrow, *Physikalische Chemie*, Vieweg, Braunschweig, 1974.

[44] S. Roth, *One Dimensional Metals*, VCH, Weinheim, 1995.

[45] R. Hoffmann, *Solids and Surfaces*, VCH, Weinheim, 1988.

[46] J. K. Burdett, Progr. Solid State Chem. **15** (1984) 173.

[47] N. F. Mott, *Metal–Insulator Transitions*, Taylor & Francis, London, 1974.

[48] R. J. Borg, G. J. Dienes: *The Physical Chemistry of Solids.* The figure is reprinted from this reference. Copyright 1992, with permission from Elsevier.

[49] P. Y. Yu, M. Cardona, *Fundamentals of Semiconductors*, Springer–Verlag, Berlin, 1996.

[50] E. F. Schubert, *Doping in III–V–Semiconductors*, Cambridge University Press, Cambridge, 1993.

[51] M. L. Cohen, J. Chelikowsky, *Electronic Structure and Optical Properties of Semiconductors*, Springer Ser. Solid State Sci., Vol. 75, Springer–Verlag, Berlin, 1989.

[52] P. Hohenberg, W. Kohn, Phys. Rev. **136** (1964) B864; W. Kohn, L. J. Sham, Phys. Rev. **140** (1965) A1133.

[53] D. Wolf, Phys. Rev. Lett. **68** (1992) 3315; Solid State Ionics **75** (1995) 3.

[54] H. G. von Schnering, private communication.

[55] A. Weiss, H. Witte, *Kristallstruktur und chemische Bindung*, VCH, Weinheim, 1983.

[56] E. A. Moelwyn–Hughes, *Physical Chemistry*, Pergamon Press, Oxford, 1961.

[57] Data compiled according to [10] and references therein.

[58] P. A. Cox, *The Electronic Structure and Chemistry of Solids*, Oxford Science Publications, Oxford, 1987.

[59] R. Waser, in: *Keramik*, H. Schaumburg (ed.), B. G. Teubner, Stuttgart, 1994.

[60] M. S. Dresselhaus, G. Dresselhaus, P. C. Eklund, *Science of Fullerenes and Carbon Nanotubes.* The figure is reprinted from this reference. Copyright 1996, with permission from Elsevier.

[61] E. Schönherr, K. Matsumoto, M. Freiberg, Fullerene Sci. Technol. (1999), 7(3), 455.

[62] A. F. Hollemann, E. Wiberg, *Lehrbuch der Anorganischen Chemie*, de Gruyter, Berlin, 1995.

[63] According to F. Haber, as described in [64].

[64] K. Meyer, *Physikalisch–Chemische Kristallographie*, VEB–Verlag, Leipzig, 1968.

[65] A. Simon, Structure and Bonding **36** (1979) 81; Angew. Chem. **100** (1988) 163.

[66] H. G. von Schnering, Angew. Chem. **93** (1981) 44; H. G. von Schnering, W. Hönle, Chem. Rev. **88** (1988) 243.

[67] U. Müller, *Inorganic Structural Chemistry*, John Wiley & Sons, Ltd., Chichester, 1993.

[68] J. C. Schön, M. Jansen, Angew. Chem. Int. Ed. Engl. **35** (1996) 1286.

[69] A. F. Wells, *Structural Inorganic Chemistry*, Clarendon Press, Oxford, 1975.

[70] W. G. Addison, *Structural Principles in Inorganic Compounds.* The figure is reprinted from this reference. Copyright 1961, with permission from Pearson Education, UK.

[71] H. Jaffe, W. R. Cook, B. Jaffe, *Piezoelectric Ceramics*, Academic Press, London, 1971.

[72] IFF–Ferienkurs, *Elektrokeramische Materialien*, Forschungszentrum Jülich, 1995.

[73] D. Seyferth, G. Mignani, J. Mater. Sci. Lett. **7** (1988) 487; H. P. Baldus, O. Wagner, M. Jansen, in: Mat. Res. Soc. Symp. Proc., Vol. 271, p. 821, MRS, Pittsburgh (PA), 1992; R. Riedel, G. Passing, H. S. Schönfelder, R. J. Brook, Nature **355** (1992) 714; J. Bill, F. Aldinger, Adv. Mater. **7** (1995) 775.

[74] R. J. Haug, K. von Klitzing, FED–Journal **6** (1995) 4; K. Eberl, P. M. Petroff (eds.), *Low Dimensional Structures Prepared by Epitaxial Growth or Growth on Patterned Substrates*, Proc. NATO Advanced Res. Workshop, Ringberg, Kluwer, Dordrecht, 1995; P. C. Klipstein, R. A. Stradling (eds.), *Growth and Characterisation of Semiconductors*, Hilger, Bristol, 1990; K. Ploog, in: *Semiconductor Interfaces: Formation and Properties*, G. LeLay, J. Denien, N. Boccara (eds.), Springer–Verlag, Berlin, 1987.

[75] D. A. Bonnell (ed.), *Scanning Tunneling Microscopy and Spectroscopy*, VCH, Weinheim, 1993; R. Wiesendanger, *Scanning Probe Microscopy and Spectroscopy: Methods and Applications*, Cambridge University Press, Cambridge, 1994.

[76] J. M. Lehn, *Supramolekulare Chemie*, VCH, Weinheim, 1995.

[77] A. Einstein, Ann. Physik **22** (1906) 800; **34** (1911) 170.

[78] L. D. Landau, E. M. Lifshitz, *Mechanics (Course of Theoretical Physics)*, Vol. I, Pergamon Press, Oxford, 1969.

[79] P. Debye, Ann. Physik **39** (1912) 789.

[80] K.-H. Hellwege, *Einführung in die Festkörperphysik*, Springer–Verlag, Berlin, 1976.

[81] See e.g. H. Schilling, *Festkörperphysik*, Verlag Harri Deutsch, Thun, 1977.

[82] G. Burns, *Solid State Physics*, Academic Press, New York, 1985.

[83] W. H. Liehn, N. E. Phillips, Proc. 7th Int. Conf. Low Temp. Phys., University of Toronto Press, Toronto, 1961.

[84] F. Seitz, *The Modern Theory of Solids*, McGraw–Hill, New York, 1990.
[85] D. R. Lide (ed.), *Handbook of Chemistry and Physics*, CRC Press, Boca Raton, 1999.
[86] According to H. P. L. G. Lindemann, c.f. [64,87].
[87] A. Swalin, *Thermodynamics of Solids*, John Wiley & Sons, Ltd., New York, 1972.
[88] E. Grüneisen, Ann. Physik **26** (1908) 393.
[89] C. H. P. Lupis, *Chemical Thermodynamics of Materials*, North–Holland, Amsterdam, 1983.
[90] This illustrative comparison has been given by H. Schmalzried.
[91] L. D. Landau, E. M. Lifshitz, *Theory of Elasticity (Course of Theoretical Physics)*, Vol. VII, Pergamon Press, Oxford, 1986.
[92] A. Sanfeld, in: *Physical Chemistry, An Advanced Treatise*, Vol. I, *Thermodynamics*, H. Eyring, D. Henderson, W. Jost (eds.), p. 245, Academic Press, New York, 1971.
[93] R. Haase, in: *Physical Chemistry, An Advanced Treatise*, Vol. I, *Thermodynamics*, H. Eyring, D. Henderson, W. Jost (eds.), Academic Press, New York, 1971.
[94] P. Glansdorff, I. Prigogine, *Thermodynamic Theory of Structure, Stability and Fluctuations*, John Wiley & Sons, Ltd., New York, 1971.
[95] J. W. Gibbs, Collected Works, Yale University Press, New Haven, 1948.
[96] A. Sanfeld, in: *Physical Chemistry, An Advanced Treatise*, Vol. I, *Thermodynamics*, H. Eyring, D. Henderson, W. Jost (eds.), p. 99, 217, Academic Press, New York, 1971.
[97] J. W. Cahn, Acta Met. **7** (1959) 18; J. E. Hilliard, *Phase Transformations*, Ch. 12, Am. Soc. Met., Metals Park, 1970.
[98] I. Barin, *Thermodynamical Data of Pure Substances*, VCH, Weinheim, 1989.
[99] Ch. Kittel, H. Krömer, *Thermal Physics*, W. H. Freeman, San Francisco, 1980.
[100] J. Maier, U. Warhus and E. Gmelin, Solid State Ionics **18/19** (1986) 969.
[101] A. A. Gribb, J. F. Banfield, Am. Mineral. **82** (1997) 717.
[102] J. Maier, Phys. Chem. Chem. Phys. **5** (11) (2003) 2164.
[103] J. J. van Laar, Z. Phys. Chem. **53** (1908) 216; **64** (1908) 257.
[104] J. W. Cahn, Trans. Met. Soc. AIME **242** (1968) 166.
[105] E. A. Guggenheim, *Mixtures*, Oxford University Press, Oxford, 1952.
[106] E. Heifets, E. A. Kotomin, J. Maier, Surf. Sci. **462** (2000) 19.
[107] C. Herring, J. Appl. Phys. **21** (1950) 437; J. C. M. Li, R. A. Oriani, L. S. Darken, Z. Phys. Chem. N. F. **49** (1960) 271; K. Weiss, Z. Phys. Chem. N. F. **91** (1974) 77.
[108] M. Lannoo, J. Bourgoin, *Point Defects in Semiconductors I*, Springer–Verlag, Berlin, 1981.
[109] M. Born, Z. Physik **1** (1920) 45.
[110] W. Jost, J. Chem. Phys. **1** (1933) 466.
[111] N. F. Mott, M. J. Littleton, Trans. Faraday Soc. **34** (1938) 485.
[112] C. Chaillout, S. W. Cheong, Z. Fisk, M. S. Lehmann, M. Marezio, B. Morosin, J. E. Schirber: The crystal structure of superconducting $La_2CuO_{4.032}$ by neutron diffraction; Physica C **158** (1989) 183. The figure is reprinted from this reference. Copyright 1989, with permission from Elsevier.
[113] See e.g. H. J. Wollenberger, in: *Physical Metallurgy*, Part II, R. W. Cahn, P. Haasen (eds.), North–Holland, Amsterdam, 1983.

[114] N. H. March, M. P. Tosi, J. Phys. Chem. Solids **46** (1985) 757.

[115] N. Hainovsky and J. Maier, Phys. Rev. B **51** (22) (1995) 15789.

[116] L. Heyne, U. M. Beekmans, A. de Beer, J. Electrochem. Soc. **119** (1972) 77.

[117] J. Maier and G. Schwitzgebel, Mater. Res. Bull. **18** (1983) 601.

[118] W. Hayes, A. M. Stoneham, *Defect and Defect Processes in Nonmetallic Solids*, John Wiley & Sons, Ltd., New York, 1985.

[119] W. Münch, unpublished.

[120] K. D. Kreuer, W. Münch, U. Traub, J. Maier, Ber. Bunsenges. Phys. Chem. **102** (1998) 552, VCH, Weinheim.

[121] P. Murugaraj, J. Maier, A. Rabenau, Solid State Commun. **71** (1989) 167; D. G. Hinks, B. Dabrowski, K. Zhang, C. U. Segre, J. D. Jorgensen, L. Soderholm, M. A. Beno, Proc. MRS, p. 9, Boston, 1988.

[122] W. Schottky, *Halbleiterprobleme I*, W. Schottky (ed.), Vieweg, Braunschweig, 1954.

[123] R. O. Simmons, R. W. Baluffi, Phys. Rev. **117** (1960) 52.

[124] N. B. Hannay, *Solid–state Chemistry*, Prentice–Hall, Englewood Cliffs, 1967.

[125] S. Wißmann, V. v. Wurmb, F. J. Litterst, R. Dieckmann and K. D. Becker, J. Phys. Chem. Solids **59** (3) (1998) 321.

[126] See e.g. P. Ehrhart, in: *Elektrokeramische Materialien*, 26. IFF–Ferienkurs, Forschungszentrum Jülich, 1995.

[127] T. Bieger, J. Maier and R. Waser, Ber. Bunsenges. Phys. Chem. **97** (9) (1993) 1098.

[128] O. Madelung, *Grundlagen der Halbleiterphysik*. The figure is reprinted from this reference. Copyright 1970, with permission from Springer–Verlag, Berlin.

[129] O. Madelung, *Introduction to Solid State Theory*, Springer–Verlag, Berlin, 1978.

[130] J. Maier, Electrochemistry **68** (6) (2000) 395.

[131] P. Haasen, *Physikalische Metallkunde*. The figure is reprinted from this reference. Copyright 1984, with permission from Springer–Verlag, Berlin.

[132] J. Bohm, *Realstruktur von Kristallen*, E. Schweizerbart'sche Verlagsbuchhandlung, Stuttgart, 1995.

[133] W. D. Kingery, H. K. Bowen, D. R. Uhlmann, *An Introduction to Ceramics*, John Wiley & Sons, Ltd., New York, 1976.

[134] C. Herring, J. Appl. Phys. **21** (1950) 437; R. L. Coble, J. Appl. Phys. **34** (1963) 679.

[135] Courtesy of Oliver Kienzle, Max–Planck–Institut für Metallforschung, Stuttgart, 1999.

[136] U. T. Read, W. Shockley, Phys. Rev. **78** (1950) 275.

[137] R. Becker, Ann. Phys. **32** (1938) 128.

[138] G. Hasson, C. Goux: Interfacial energies of tilt boundaries in aluminium. Experimental and theoretical determination; Scripta Materialia, **5** (1971) 889. The figure is reprinted from this reference. Copyright 1971, with permission from Elsevier.

[139] O. Kienzle, F. Ernst, J. Am. Ceram. Soc. **80** (1997) 1639, The American Ceramic Society, Westerville (OH).

[140] I. Denk, Ph.D. Thesis, University of Stuttgart, 1995; I. Denk, F. Noll and J. Maier, J. Am. Ceram. Soc. **80** (2) (1997) 279, The American Ceramic Society, Westerville (OH).

[141] R. H. French, R. M. Cannon, L. K. DeNoyer, Y.–M. Chiang, Solid State Ionics **75** (1995) 13.

[142] D. R. Clarke, J. Am. Ceram. Soc. **70** (1987) 15.

[143] G. Gottstein. L. S. Shindlerman, *Grain Boundary Migration in Metals*, CRC Press, New York, 1999.

[144] D. Wolf, in: *Materials Interfaces. Atomistic–level Structure and Properties*, D. Wolf and S. Yip (eds.), Chapman & Hall, London, 1992; A. P. Sutton, R. W. Balluffi, *Interfaces in Crystalline Materials*, Clarendon Press, Oxford, 1995.

[145] G. Binnig and H. Rohrer, Helv. Phys. Acta **55** (1982) 726.

[146] See e.g. M. Elbaum, S. G. Lipson, J. G. Dash, J. Crystal Growth **129** (1993) 491.

[147] G. Ertl, Adv. Catalysis **37** (1990) 213; G. Ertl, in: *Handbook of Heterogeneous Catalysis*, G. Ertl, H. Knözinger, J. Weitkamp (eds.), Vol. 3, p. 1032, VCH, Weinheim, 1997.

[148] R. Defay, I. Prigogine, A. Bellemans, H. Everett, *Surface Tension and Adsorption*, John Wiley & Sons, Ltd., New York, 1960.

[149] A. I. Rusanov, *Phasengleichgewichte und Grenzflächenerscheinungen*, Akademie–Verlag, Berlin, 1978.

[150] G. Bakker, *Handbuch der Experimentalphysik*, Vol. 6, *Kapillarität und Oberflächenspannung*, Harms, Leipzig, 1928.

[151] R. Parsons, in: *Modern Aspects of Electrochemistry*, J. O'M. Bockris (ed.), Vol. 1, p. 103, Butterworth, London, 1954.

[152] J. O'M. Bockris, in: *Physical Chemistry, An Advanced Treatise IXA*, H. Eyring, D. Henderson, W. Jost (eds.), Academic Press, New York, 1970.

[153] G. Kortüm, *Treatise on Electrochemistry*, Elsevier, Amsterdam, 1965.

[154] G. Wulff, Z. Krist. **34** (1901) 449.

[155] L. D. Landau, E. M. Lifshitz, *Statistical Physics (Course of Theoretical Physics)*, Vol. V, Pergamon Press, Oxford, 1968.

[156] R. Kaischew, Bull. Acad. Sci. Phys. **1** (1950) 100; Bull. Acad. Sci. Phys. **2** (1951) 191.

[157] K. Tsuruta, A. Omeltchenko, R. K. Kalia, P. Vashishta, Europhys. Lett. **33** (1996) 441, EDP Sciences, Les Ulis, France.

[158] R. E. Johnson, J. Phys. Chem. **63** (1959) 1655.

[159] C. Herring, in: *Physics of Powder Metallurgy*, W. E. Kingston (ed.), McGraw–Hill, New York, 1951.

[160] C. S. Smith, *Metal Interfaces*, American Society of Metallurgists, Cleveland, 1952.

[161] H. E. Exner, Z. Metallk. **46** (1973) 273.

[162] H. E. Exner and E. Arzt, in: *Physical Metallurgy*, R. W. Cahn, P. Haasen (eds.), North–Holland, Amsterdam, 1983.

[163] K. Tsuruta, A. Omeltchenko, R.K. Kalia, P. Vashishta, in: Mat. Res. Soc. Symp. Proc., Vol. 408, p. 181, MRS, Pittburgh (PA), 1996.

[164] J.-R. Lee, Y.-M. Chiang: Bi segregation at ZnO grain boundaries in equilibrium with Bi_2O_3–ZnO liquid; Solid State Ionics, **75** (1995) 79. The figure is reprinted from this reference. Copyright 1995, with permission from Elsevier.

[165] C. H. P. Lupis, *Chemical Thermodynamics of Materials*, North–Holland, Amsterdam, 1983.

[166] W. Schottky, *Halbleiterprobleme IV*, W. Schottky (ed.), P. 235, Vieweg, Braunschweig, 1958.

[167] F. A. Kröger, H. J. Vink, in: *Solid State Physics*, F. Seitz, D. Turnbull (eds.), Vol. 3, p. 307, Academic Press, New York, 1956.

[168] J. Maier: Defect chemistry and ion transport in nanostructured materials. Part II. Aspects of nano-ionics; Solid State Ionics **157** (2003) 327. The figure is reprinted from this reference. Copyright 2003, with permission from Elsevier.

[169] G. Brouwer, Philips Res. Repts. **9** (1954) 366.

[170] K. R. Popper, *The Logic of Scientific Discovery*, Hutchinson, London, 1968.

[171] C. G. Fonstad and R. H. Rediker, J. Appl. Phys. **42** (1971) 2911.

[172] J. Maier, W. Göpel, J. Solid State Chem. **72** (1988) 293.

[173] H. H. von Baumbach, C. Wagner, Z. Phys. Chem. **B22** (1993) 199.

[174] H. Dünwald, C. Wagner, Z. Phys. Chem. **B22** (1933) 212; J. Gundermann, C. Wagner, Z. Phys. Chem. **B37** (1937) 155.

[175] H. H. von Baumbach, C. Wagner, Z. Phys. Chem. **B24** (1934) 59.

[176] J. Maier, in: *Recent Trends in Superionic Solids and Solid Electrolytes*, S. Chandra, A. Laskar (eds.), p. 137, Academic Press, New York, 1989.

[177] M. Spaeth, K. D. Kreuer, C. Cramer, J. Maier, J. Solid State Chem. **148** (1999) 169.

[178] H. H. von Baumbach, H. Dünwald, C. Wagner, Z. Phys. Chem. **B22** (1933) 226.

[179] M. T. Tsai, E. J. Opila, H. L. Tuller, in: Mat. Res. Soc. Proc., Vol. 169, p. 65, MRS, Pittsburgh (PA), 1989.

[180] J. Maier and G. Schwitzgebel, Mater. Res. Bull. **17** (1982) 1061.

[181] P. K. Moon, H. L. Tuller, Sensors and Actuators B **1** (1990) 199, Elsevier, Amsterdam.

[182] J. Mizusaki, K. Fueki, Solid State Ionics **6** (1982) 55.

[183] E. Koch. C. Wagner, Z. Phys. Chem. **135** (1950) 197.

[184] J. Corish, P. W. M. Jacobs, J. Phys. Chem. Solids **33** (1972) 1799.

[185] J. Teltow, Z. Phys. Chem. **195** (1950) 213; Ann. Phys. **5** (1950) 63.

[186] J. G. Bednorz, K. A. Müller, Z. Physik B **64** (1986) 189.

[187] J. Maier and G. Pfundtner, Adv. Mater. **30** (6) (1991) 292.

[188] H. Tannenberger, H. Schachner, P. Kovacs, Revue EPE (Journées Internationales d'Etudes des Piles à Combustibles, Brussels, 1965), **III** (1) (1966) 19.

[189] J. Daniels, K. H. Härdtl, D. Hennings, R. Wernicke, Philips Res. Repts. **31** (1978) 489.

[190] G. M. Choi and H. L. Tuller, J. Am. Ceram. Soc. **71** (4) (1988) 201.

[191] S. Steinsvik, T. Norby, P. Kofstad, in: *Electroceramics IV*, R. Waser, S. Hoffmann, D. Bonnenberg, Ch. Hoffmann (eds.), Vol. II, p. 691, Augustinus Buchhandlung, Aachen, 1994.

[192] T. He, K.-D. Kreuer, Yu. M. Baikov and J. Maier, Solid State Ionics **95** (1997) 301.

[193] I. Denk, W. Münch and J. Maier, J. Am. Ceram. Soc. **78** (12) (1995) 3265.

[194] K. Sasaki and J. Maier, J. Appl. Phys. **86** (10) (1999) 5422; 5434.

[195] F. A. Kröger, H. J. Vink, J. van den Boomgaard, Z. Phys. Chem. **203** (1954) 1; D. M. Smyth, M. P. Harmer, P. Peng, J. Am. Ceram. Soc. **72** (1989) 2276; R. Waser, J. Am. Ceram. Soc. **74** (1991) 1934.

[196] S. Stotz and C. Wagner, Ber. Bunsenges. Phys. Chem. **70** (1966) 781.
[197] R. Waser, Ber. Bunsenges. Phys. Chem. **90** (1986) 1223.
[198] H. Iwahara, T. Esaka, H. Uchida, N. Maeda, Solid State Ionics **314** (1981) 259.
[199] P. Murugaraj, K. D. Kreuer, T. He, T. Schober, J. Maier, Solid State Ionics **98** (1997) 1.
[200] D. G. Thomas, I. J. Lander, J. Chem. Phys. **25** (1956) 1136.
[201] Y. Larring, T. Norby, Solid State Ionics **77** (1995) 147; A. S. Nowick, Solid State Ionics **77** (1995) 137.
[202] K. D. Kreuer, Chem. Mater. **8** (1996) 610.
[203] N. Bjerrum, Kgl. Danske Videnskab. Selskab. Mat. Fys. Medd. **7** (1926) 3.
[204] A. B. Lidiard, in: *Handbuch der Physik*, S. Flügge (ed.), Vol. 20, p. 246, Springer–Verlag, Berlin, 1957.
[205] P. Debye, E. Hückel, Phys. Z. **24** (1923) 185; 305.
[206] J. Maier, Solid State Ionics **143** (2001) 17.
[207] R. Merkle and J. Maier, Phys. Chem. Chem. Phys. **5** (11) (2003) 2297.
[208] E. A. Kotomin, A. I. Popov, Nucl. Inst. and Meth. in Phys. Res. B **141** (1998) 1.
[209] M. Quilitz and J. Maier, J. Supercond. **9** (1) (1996) 121.
[210] W. Preis, M. Holzinger and W. Sitte, Monatshefte für Chemie 132 (2001) 499.
[211] J. J. Markham, *F–Centres in Alkali Halides*, Academic Press, New York, 1966.
[212] D. Knödler, P. Pendzig, W. Dieterich, Solid State Ionics **86-88** (1996) 29.
[213] J. Brynestad, H. Flood, Z. Elektrochem. **62** (1958) 953; F. Koch, J. B. Cohen, Acta Cryst. B **25** (1969) 275.
[214] B. T. M. Willis, Nature **197** (1963) 755.
[215] A. Magnéli, Arkiv kemi **1** (1949) 213; 512.
[216] Data based on R. R. Merrit and B. G. Hyde, Phil. Trans. Roy. Soc. **274A** (1973) 1245.
[217] S. N. Ruddlesden, P. Popper, Acta Cryst. **10** (1957) 538.
[218] R. J. Cava, in: *Processing and Properties of High–T_C–Superconductors*, S. Jin (ed.), p. 13, World Scientific, Singapore, 1993.
[219] J. O'M. Bockris and A. K. V. Reddy, *Modern Electrochemistry*, Plenum Press, New York, 1970.
[220] A. J. Bard, L. R. Faulkner, *Electrochemical Methods*, John Wiley & Sons, Ltd., New York, 1980.
[221] J. S. Newman, *Electrochemical Systems*, Prentice–Hall, Englewood Cliffs, 1991.
[222] A. Münster, *Statistical Thermodynamics*, Springer–Verlag, Berlin, 1974.
[223] H. Ted Davis, in: *Physical Chemistry. An Advanced Treatise*, Vol. II, *Statistical Mechanics*, H. Eyring, D. Henderson, W. Jost (eds.), Academic Press, New York, 1967.
[224] J. G. Kirkwood, J. Chem. Phys. **14** (1946) 180.
[225] J. Mayer, J. Chem. Phys. **18** (1950) 1426.
[226] A. R. Allnatt, M. H. Cohen, J. Chem. Phys. **40** (1964) 1860, 1871; R. A. Sevenich, K. L. Kliewer, J. Chem. Phys. **48** (1964) 3045; A. R. Allnatt, E. Loftus, L. A. Rowley, Crystal Lattice Defects **3** (1972) 77.
[227] S. Ling, Solid State Ionics **70/71** (1994) 686.

[228] F. Zimmer, P. Ballone, J. Maier, and M. Parrinello, Ber. Bunsenges. Phys. Chem. **101** (9) (1997) 1333; F. Zimmer, P. Ballone, J. Maier, M. Parrinello, J. Chem. Phys. **112** (14) (2000) 6416; F. Zimmer, P. Ballone, M. Parrinello, J. Maier, Solid State Ionics **127** (2000) 277.
[229] H. P. D. Lanyon, R. A. Tuft, *Bandgap Narrowing in Heavily Doped Silicon*, p. 316, IEEE Tech. Dig., Int. Electron. Device Meet., 1978.
[230] J. G. Ghosh, J. Chem. Soc. **113** (1918) 707.
[231] H. S. Frank, P. T. Thompson, Ber. Bunsenges. Phys. Chem. **67** (1963) 836.
[232] R. Huberman, Phys. Rev. Lett. **32** (1974) 100.
[233] H. Schmalzried, Z. Phys. Chem. **22** (1959) 199; Ber. Bunsenges. Phys. Chem. **84** (1980) 120.
[234] A. Bunde, Z. Physik B **36** (1980) 251.
[235] J. E. Lennard-Jones and A. F. Devonshire, Proc. R. Soc. London A **169** (1946) 170; **169** (1946) 317; **169** (1946) 464.
[236] J. Maier and W. Münch, Z. Anorg. Chem. **626** (2000) 264.
[237] R. Kirchheim, Progr. Mater. Sci. **32** (1988) 261.
[238] M. H. Anderson, J. R. Ensher, M. R. Matthews, C. E. Wiemann and E. A. Cornell, Science **269** (1995) 198.
[239] K. B. Davis, M.-O. Mewes, M. R. Andrews, N. J. van Druten, D. S. Durfee, D. M. Kurn and W. Ketterle, Phys. Rev. Lett. **75** (1995) 3969.
[240] J. E. Mayer, M. G. Mayer, *Statistical Mechanics*, John Wiley & Sons, Ltd., New York, 1940; S. Ling, Solid State Ionics **70/71** (1994) 686.
[241] J. Jamnik and J. Maier, Solid State Ionics **94** (1997) 189.
[242] D. Gillespie, W. Nonner, D. Henderson and R. S. Eisenberg, Phys. Chem. Chem. Phys. **4** (2002) 4763.
[243] B. Katz, *Nerve, Muscle and Synapse*, McGraw–Hill, New York, 1960.
[244] J. Maier, Prog. Solid State Chem. **23** (3) (1995) 171.
[245] E. K. H. Salje, *Phase Transitions in Ferroelastic and Co-elastic Crystals*, Cambridge University Press, Cambridge, 1990.
[246] R. C. Baetzold, Y. T. Tan, P. W. Tasker, Surf. Sci. **195** (1988) 579.
[247] G. Gouy, J. Physique **9** (1910) 457; D. L. Chapman, Phil. Mag. **25** (1913) 475.
[248] J. Maier, J. Electrochem. Soc. **134** (1987) 1524.
[249] J. Maier, J. Phys. Chem. Solids **46** (1985) 309.
[250] G. Farlow, A. Blose, Sr. J. Feldott, B. Lounsberry, L. Slifkin, Radiation Effects **75** (1983) 1, Gordon and Breach Publishers, Lausanne.
[251] R. de Souza, unpublished.
[252] S. M. Sze, *Semiconductor Devices*, John Wiley & Sons, Ltd., New York, 1985.
[253] H. K. Henisch, *Semiconductor Contacts*, Clarendon Press, Oxford, 1984.
[254] S. Kim, J. Fleig and J. Maier, Phys. Chem. Chem. Phys. **5** (11) (2003) 2268.
[255] J. Maier, Ber. Bunsenges. Phys. Chem. **93** (1989) 1468; 1474.
[256] J. Maier, Ber. Bunsenges. Phys. Chem. **90** (1986) 26.
[257] J. Jamnik, J. Maier, S. Pejovnik, Solid State Ionics **75** (1995) 51.
[258] R. B. Poeppel, J. M. Blakely, Surf. Sci. **15** (1969) 507.
[259] K. L. Kliewer, J. S. Koehler, Phys. Rev. A **140** (1965) 1226.
[260] C. C. Liang, J. Electrochem. Soc. **120** (1973) 1298.

[261] K. Shahi, J. B. Wagner. Appl. Phys. Lett. **37** (1980) 757.

[262] J.-S. Lee, S. Adams, J. Maier, J. Electrochem. Soc. **147** (6) (2000) 2407.

[263] J. Maier, J. Eur. Ceram. Soc. **19** (6-7) (1999) 675, Elsevier, Amsterdam.

[264] U. Riedel, J. Maier and R. Brook, J. Eur. Ceram. Soc. **9** (3) (1992) 205.

[265] A. Bunde, W. Dieterich, E. Roman, Solid State Ionics **18/19** (1986) 147; E. Roman, A. Bunde, W. Dieterich, Phys. Rev. B **34** (1986) 331; P. Knauth, G. Albinet, J. M. Debierre, Ber. Bunsenges. Phys. Chem. **98** (7) (1998) 945; J. C. Wang, N. J. Dudney, Solid State Ionics **18/19** (1986) 112.

[266] J. Electroceramics **5** (2) (2000), Special Volume on Composite Materials (J. Fleig, J. Maier, eds.); J. Fleig and J. Maier, J. Am. Ceram. Soc., **82** (1999) 3485.

[267] Y. Saito and J. Maier, J. Electrochem. Soc. **142** (9) (1995) 3078.

[268] U. Lauer and J. Maier, J. Electrochem. Soc. **139** (5) (1992) 1472.

[269] A. J. Bhattacharyya and J. Maier, Adv. Mater. , 2004, in press.

[270] F. Croce, G. B. Appetecchi, L. Persi and B. Scrosati, Nature **394** (1998) 456.

[271] W. Wieczorek, Z. Florjancyk, J. R. Stevens, Electrochim. Acta **40** (1995) 2251.

[272] J. Jamnik and J. Maier, Phys. Chem. Chem. Phys. **5** (23) (2003) 5215.

[273] W. Petuskey, Solid State Ionics **21** (1986) 117.

[274] Y.-M. Chiang, E. B. Lavik, I. Kosacki, H. L. Tuller, J. Y. Ying, J. Electroceramics **1** (1997) 7.

[275] S. Kim, R. Merkle, J. Maier, Surf. Sci. **549** (3) (2004) 196.

[276] J. Maier, Ber. Bunsenges. Phys. Chem. **89** (1985) 355.

[277] U. Lauer and J. Maier, Ber. Bunsenges. Phys. Chem. **96** (1992) 111.

[278] A. Chandra and J. Maier, Solid State Ionics **148** (2002) 153.

[279] P. Manoravi and K. Shahi, Solid State Ionics **58** (1992) 243.

[280] P. Manoravi and K. Shahi, Journal of Physics and Chemistry of Solids **52** (1991) 527.

[281] A. Bunde, S. Havlin: Percolation, in: *Fractals and Disordered Systems*, A. Bunde, S. Havlin (eds.). The figure is reprinted from this reference. Copyright 1996, with permission from Springer–Verlag, Berlin. A. Bunde, S. Havlin, *Fractals in Science*, Springer–Verlag, Berlin, 1994.

[282] F. Granzer, J. Imag. Sci. **33** (1989) 207.

[283] N. Sata, K. Eberman, K. Eberl and J. Maier, Nature **408** (2000) 946.

[284] J. Maier, U. Lauer, Ber. Bunsenges. Phys. Chem. **94** (1990) 973.

[285] P. Lauque, M. Bendahan, J. L. Seguin, M. Pasquinelli, P. Knauth, Solid State Ionics **136** (2000) 603; P. Lauque, J.-M. Laugier, C. Jacolin, M. Bendahan, C. Lemire, P. Knauth, Sensors and Actuators **B87** (2002) 431.

[286] M. Holzinger, J. Fleig, J. Maier and W. Sitte, Ber. Bunsenges. Phys. Chem. **99** (11) (1995) 1427.

[287] J. Maier, Phys. Stat. Sol. (a) **112** (1989) 115.

[288] J. Maier, Solid State Ionics **23** (1987) 59.

[289] J. Maier, Z. Phys. Chem. **217** (2003) 415.

[290] B. Wassermann, T. P. Martin, J. Maier, Solid State Ionics **28-30** (1988) 1514.

[291] N. Starbov, J. Inf. Rec. Mater. **13** (1985) 307.

[292] E. Schreck, K. Länger, K. Dransfeld, Z. Physik B **62** (1986) 33.

[293] W. Puin, S. Rodewald, R. Ramlau, P. Heitjans, J. Maier, Solid State Ionics, **131** (1, 2) (2000) 159; W. Puin and P. Heitjans, Nano–Struct. Materials **6** (1995) 885.
[294] S. Kim, J. Maier, J. Electrochem. Soc. **149** (10) (2002) J73.
[295] A. Tschöpe, Solid State Ionics **139** (2001) 267.
[296] A. Tschöpe, E. Sommer and R. Birringer, Solid State Ionics **139** (2001) 255.
[297] K. D. Kreuer, in: *Solid State Ionics: Science & Technology*, B. V. R. Chowdari, K. Lal, S. A. Agnihotry, N. Khare, S. S. Sekhon, P. C. Srivastava, S. Chandra (eds.), p. 263, World Scientific Publishing Co., Singapore, 1998.
[298] I. Lubomirsky, J. Fleig, J. Maier, J. Appl. Phys. **92** (2002) 6819.
[299] J. Maier, Solid State Ionics **154-155** (2002) 291.
[300] R. Lipowsky, *Phasenübergänge an Oberflächen* (IFF–Ferienkurs), p. 9.1, Forschungszentrum Jülich GmbH, 1993; Springer Tracts in Mod. Phys., Vol. 127.
[301] A. I. Baranov, V. V. Sinitsyn, E. G. Ponyatovskii, L. A. Shuvalov, JETP Lett. **44** (1986) 237.
[302] N. F. Uvarov, E. F. Hairetdinov, A. I. Rykov, Yu. T. Pavlyukhin, Solid State Ionics **96** (1997) 233.
[303] B. L. Davies, L. R. Johnson, Crystal Lattice Defects **5** (1974) 235.
[304] M. Vossen, F. Forstmann, A. Krämer, Solid State Ionics **94** (1997) 1.
[305] G. E. Pike, Phys. Rev. B **30** (1984) 795; G. E. Pike, C. H. Seager, J. Appl. Phys. **50** (1979) 3414.
[306] M. Vollmann, R. Waser, J. Am. Ceram. Soc. **77** (1994) 235.
[307] I. Denk, J. Claus and J. Maier, J. Electrochem. Soc. **144** (10) (1997) 3526.
[308] X. Guo, J. Fleig, J. Maier, J. Electrochem. Soc. **148** (2001) 121.
[309] M. Vollmann, R. Hagenbeck, R. Waser, J. Am. Ceram. Soc. **80** (1997) 2301.
[310] D. A. Bonnell, J. Am. Ceram. Soc. **81** (1998) 3049, The American Ceramic Society, Westerville (OH).
[311] W. Heywang, Solid State Electronics **3** (1961) 51.
[312] G. H. Jonker, Solid State Electronics **7** (1964) 895.
[313] T. Seiyama, A. Kato, K. Fujiishi, M. Nagatani, Anal. Chem. **34** (1962) 1502; **38** (1966) 1069; N. Taguchi, Jpn. Patent 45-38200, 1962.
[314] C. Wagner, J. Phys. Chem. Solids **33** (1972) 1051.
[315] M. P. Setter, J. B. Wagner, Solid State Ionics **28/30** (1988) 1579.
[316] R. A. De Souza, J. Fleig, J. Maier, O. Kienzle, Z. Zhang, W. Sigle, and R. Rühle, J. Am. Ceram. Soc. **86** (6) (2003) 922; R. A. de Souza et al., in preparation.
[317] E. L. Brus, J. Chem. Phys. **80** (1984) 4403.
[318] A. Fojtik, H. Weller, U. Koch, A. Henglein, Ber. Bunsenges. Phys. Chem. **88** (1984) 969.
[319] J. Lüning, Ph.D. Thesis, University of Köln, 1998; J. Lüning, J. Rockenberger, S. Eisebitt, J.-E. Rubensson, A. Karl, A. Kornowski, H. Weller, W. Eberhardt, Solid State Commun. **112** (1999) 5.
[320] D. M. Kolb, R. Ullmann, J. C. Ziegler, Electrochim. Acta **43** (1998) 2751.
[321] H. P. Strunk, Lecture at 9th Europhys. Conf. on Defects in Insulating Materials (EURODIM 2002), Wroclaw, Radiation Eff. Def. Sol.
[322] J. Schoonman, Solid State Ionics, **135** (2000) 5.
[323] H. L. Tuller, Solid State Ionics **131** (2000) 143.

[324] I. Yokota, J. Phys. Soc. Japan **16** (1961) 2213.

[325] C. Wagner, Z. Elektrochem. **60** (1956) 4; M. H. Hebb, J. Chem. Phys. **20** (1952) 185.

[326] S. R. de Groot, *Thermodynamics of Irreversible Processes*, North–Holland, Amsterdam 1960.

[327] A. Höpfner, *Irreversible Thermodynamik für Chemiker*, de Gruyter, Berlin, 1972.

[328] R. Landauer, J. Stat. Phys. **13** (1975) 1.

[329] J. Maier, in preparation.

[330] L. Onsager, Phys. Rev. **37** (1931) 405; **38** (1931) 2265.

[331] I. Prigogine, *Introduction to Thermodynamics of Irreversible Thermodynamics*, Interscience Publishing, New York, 1967.

[332] A. Sanfeld, in: *Physical Chemistry. An Advanced Treatise*, H. Eyring, D. Henderson, W. Jost (eds.), Vol. I, p. 246, Academic Press, New York, 1967.

[333] R. Haase, *Thermodynamik der irreversiblen Prozesse*, Steinkopff, Darmstadt, 1963.

[334] J. Maier, Angew. Chem. Int. Ed. Engl. **32** (4) (1993) 528.

[335] W. Nernst, Z. Phys. Chem. **2** (1888) 613.

[336] K. J. Laidler, *Chemical Kinetics*, McGraw–Hill, New York, 1973.

[337] H. Eyring, J. Chem. Phys. **3** (1935) 3; M. G. Evans, M. Polanyi, Trans. Faraday Soc. **33** (1937) 448.

[338] G. H. Vineyard, J. Phys. Chem. Solids **3** (1957) 121.

[339] M. Eigen: Selforganization of Matter and the Evolution of Biological Macromolecules, Naturwiss. **58** (1971) 465. The figure is reprinted from this reference. Copyright 1971, with permission from Springer–Verlag, Berlin.

[340] H. Haken, *Synergetik*. The figure is reprinted from this reference. Copyright 1990, with permission from Springer–Verlag, Berlin.

[341] H. Margenau, G. M. Murphy, *The Mathematics of Physics and Chemistry*, Van Nostrand, Princeton, NJ, 1962.

[342] J. Tafel, Z. Phys. Chem. **50** (1905) 641.

[343] see e.g. Ref. [7].

[344] P. Müller, Phys. Stat. Sol. **13** (1965) 775.

[345] S. Chandra, *Superionic Solids*, North–Holland, Amsterdam, 1981; S. Chandra, A. S. Laskar (eds.), *Superionic Solids and Solid Electrolytes*, Academic Press, New York, 1989.

[346] O. Yamamoto, Y. Takedo, R. Kanno, Kagaku **38** (1983) 387, Kagaku-Dojin Publishing Company, Kyoto.

[347] L. W. Strock, Z. Phys. Chem. **B25** (1934) 441; R. Cava, F. Reidinger, B. J. Wuensch, Solid State Commun. **24** (1977) 411.

[348] Data compiled by K. Sasaki according to: T. Kudo and K. Fueki, *Solid State Ionics*, Kodansha, 1990, and references therein; A. Rabenau, Solid State Ionics **6** (1982) 277; K. D. Kreuer, Chem. Mater. **8** (1996) 610 and references therein; K. D. Kreuer, Th. Dippel, J. Maier, in: Proc. Electrochem. Soc. **PV 95-23** (1995) 241;K. Schmidt-Rohr, J. Clauss, B. Blümich and H. W. Spiess, Magn. Reson. in Chem. **28** (1990) 3; K. D. Kreuer, Solid State Ionics **97** (1997) 1; H. Iwahara, T. Esaka, H. Uchida and N. Maeda, Solid State Ionics **3/4** (1981) 539; B. C. H. Steele, Solid State Ionics **75** (1995) 157; T. Takahashi, H. Iwahara and T. Esaka, J. Electrochem. Soc. **124** (1977)

1563; T. Ishihara, H. Furutani, H. Nishiguchi and Y. Takita, Electr. Soc. Proc. **97-24** (1998) 834; T. Ishihara, H. Matsuda and Y. Takita, J. Am. Ceram. Soc. **116** (1994) 3801; J. B. Goodenough, J. E. Ruiz-Diaz and Y. S. Zhen, Solid State Ionics **44** (1990) 21; I. Kontoulis, Ch. P. Ftikos and B. C. H. Steele, Mater. Sci. Eng. **B22** (1994) 313; H. L. Tuller, Solid State Ionics **94** (1997) 63; J. T. Kummer, Progr. Solid St. Chem. **7** (1972) 141; G. Farrington, B. Dunn, Solid State Ionics **7** (1982) 287.

[349] B. Ma, J. P. Hodges, J. D. Jorgensen, D. J. Miller, J. W. Richardson Jr, U. Balachandran, J. Solid State Chem. **141** (1998) 576.

[350] W. Meyer, H. Neldel, Z. Techn. Phys. **12** (1937) 588; A. S. Nowick, W. K. Lee, H. Jain, Solid State Ionics **28/30** (1988) 89; K. L. Ngai, Solid State Ionics **105** (1998) 231.

[351] W. Jost, Z. Phys. Chem. A **169** (1934) 129.

[352] K. Funke, in: *Superionic Solids and Solid Electrolytes*, A. S. Laskar, S. Chandra (eds.), p. 569, Academic Press, New York, 1989.

[353] J. N. Bradley, P. D. Green, Trans. Faraday Soc. **62** (1966) 2069.

[354] B. B. Owens, J. E. Oxley, A. F. Sammells, in: *Solid Electrolytes*, S. Geller (ed.), p. 67, Springer–Verlag, Berlin 1977.

[355] C. A. Beevers, M. A. R. Ross, Z. Krist. **95** (1937) 59; J. T. Kummer, Progr. Solid State Chem. **7** (1972) 141.

[356] J. H. Kennedy: The β–Aluminas; in: *Solid Electrolytes*, S. Geller (ed.), Vol. 21, p. 105. The figure is reprinted from this reference. Copyright 1977, with permission from Springer–Verlag, Berlin.

[357] T. Takahashi, O. Yamamoto, S. Yamada, S. Hayashi, J. Electrochem. Soc. **126** (1979) 1654.

[358] G. Farrington, B. Dunn, Solid State Ionics **7** (1982) 287.

[359] M. Meyer, P. Maass, A. Bunde, J. Chem. Phys. **109** (1998) 109.

[360] A. Rabenau, Solid State Ionics **6** (1982) 277.

[361] K. D. Kreuer, H. Kohler and J. Maier, in: *High Conductivity Ionic Conductors: Recent Trends and Applications*, T. Takahashi (ed.), World Scientific, Singapore (1989) 242.

[362] H. Kohler, H. Schulz: Nasicon Solid Electrolytes – Part I: The Na^+-diffusion path and its relation to the structure, Mater. Res. Bull. **20** (1985) 1461. The figure is reprinted from this reference. Copyright 1985, with permission from Elsevier.

[363] J. B. Goodenough, H. Y.-P. Hong, J. A. Kafalas, Mater. Res. Bull. **11** (1976) 203.

[364] J. P. Boilot, P. Colomban, G. Collin, Solid State Ionics **28-30** (1988) 403; H. Kohler, H. Schultz, O. Melnikov, Mater. Res. Bull. **18** (1983) 589.

[365] J. Maier and U. Warhus, J. Chem. Thermodynamics **18** (1986) 309.

[366] A. Lundén, A. Bengtzelius, R. Kaber, L. Nilsson, K. Schroeder, R. Törneberg, Solid State Ionics **9/10** (1983) 89; L. Nilssen, J. O. Thomas, B. C. Tofield, J. Phys. **C13** (1980) 6441; D. Wilmer, R. D. Banhatti, J. Fitter, K. Funke, M. Jansen, G. Korus, R. E. Lechner, Physica B **241/243** (1998) 338.

[367] C. A. Vincent, *Chemistry in Britain*, April, 1981, p. 391.

[368] B. Scrosati, in: *Lithium Ion Batteries*, M. Wakihara, O. Yamamoto (eds.), VCH, Weinheim, 1998.

[369] P. Maass, M. Meyer, A. Bunde, W. Dieterich, Phys. Rev. Lett. **77** (1996) 1528; P. Pendzig, W. Dieterich, A. Nitzan, J. Non-Cryst. Solids **235-237** (1998) 748.

[370] M. B. Armand, Ann. Rev. Mat. Sci. **6** (1986) 245; M. A. Ratner, D. F. Shriver, Mater. Res. Bull. **14** (1989) 39.

[371] M. Tuckermann, L. Laasonen, M. Sprik, M. Parrinello, J. Phys.: Condens. Matter. **6** (Suppl. 23A) (1994) 99.

[372] P. Schuster, G. Zundel, C. Sandorfy (eds.), *The Hydrogen Bond*, North–Holland, Amsterdam, 1976.

[373] R. Hempelmann, Ch. Karmonik, Th. Matzke, M. Capadonia, U. Stimming, T. Springer, M. Adams, Solid State Ionics **77** (1995) 152.

[374] K. D. Kreuer, Solid State Ionics **136-137** (2000) 149; Chem. Phys. Chem. **3** (2002) 771.

[375] E. S. Lewis, in: *Proton Transfer Reactions*, E. Caldin, V. Gold (eds.), p. 317, Chapman & Hall, New York, 1975; M. Cappadonia, H. T. von der Heyden, U. Stimming, Solid State Ionics **94** (1997) 9.

[376] H. J. von Daal: Polar optical-mode scattering of electrons in SnO_2, Solid State Commun. **6** (1968) 5. The figure is reprinted from this reference. Copyright 1968, with permission from Elsevier.

[377] H. L. Tuller, A. S. Nowick, J. Phys. Chem. Solids **38** (1977) 859; H. L. Tuller, Solid State Ionics **94** (1997) 63.

[378] R. E. Peierls, *Quantum Theory of Solids*, Clarendon Press, Oxford, 1955.

[379] cf. Ref. [44].

[380] M. Tinkham, *Introduction to Superconductivity*, McGraw–Hill, New York, 1975; W. Buckel, *Supraleitung*, VCH, Weinheim, 1990.

[381] R. J. Cava, R. B. van Dover, B. Batlogg, E. A. Rietman, Phys. Rev. Lett. **58** (1987) 408, The American Physical Society, College Park (MD).

[382] J. C. Philips: *Physics of High Temperature Superconductors.* The figure is reprinted from this reference. Copyright 1989, with permission from Elsevier.

[383] D. Christen, J. Narayam, L. Schneemeyer (eds.), Mat. Res. Soc. Symp. Proc., Vol. 169, Materials Research Society, Pittsburgh (PA), 1990; T. Ruf, Physik in unserer Zeit **29** (1998) 160.

[384] J. Maier: On the correlation of macroscopic and microscopic rate constants in solid state chemistry; Solid State Ionics **112** (1998) 197. The figure is reprinted from this reference. Copyright 1998, with permission from Elsevier.

[385] R. Moos, K. H. Härdtl, J. Am. Ceram. Soc. **80** (1997) 2549.

[386] J. R. Manning, *Diffusion Kinetics for Atoms in Crystals*, Van Nostrand, Princeton, 1968; A. D. Le Claire, A. B. Lidiard, Phil. Mag. **1** (1956) 1.

[387] R. Dieckmann, H. Schmalzried, Ber. Bunsenges. Phys. Chem. **81** (1977) 344, VCH, Weinheim.

[388] E. J. Opila, H. L. Tuller, B. J. Wuensch and J. Maier, in: Mat. Res. Soc. Symp. Proc. **209** (1991) 795, MRS, Pittsburgh (PA).

[389] C. Wagner, Progr. Solid State Chem. **10** (1975) 3.

[390] M. Martin, H. Schmalzried, Solid State Ionics **20** (1986) 75; H. I. Yoo, H. Schmalzried, M. Martin, J. Janek, Z. Phys. Chem. NF **168** (1990) 129.

[391] J. Maier, Solid State Phenomena **39-40** (1994) 35.

[392] J. Jamnik and J. Maier, J. Electrochem. Soc. **145** (5) (1998) 1762.

[393] J. Maier, J. Am. Ceram. Soc. **76** (5) (1993) 1212; 1218; 1223; 1228.

[394] L. Heyne in: *Solid State Electrolytes*, S. Geller (ed.), 169 pp., Springer–Verlag, Berlin, 1977.

[395] H. S. Carslaw, J. C. Jäger, *Conduction of Heat in Solids*, Clarendon Press, Oxford, 1959; with regard to the transformation to diffusion processes see J. Crank, *Mathematics of Diffusion*, Clarendon Press, Oxford, 1975.

[396] W. Jost, *Diffusion in Solids, Liquids and Gases*, Academic Press, New York, 1960.

[397] J.-H. Lee, M. Martin and H. I. Yoo, Korean J. Ceram. **4** (1998) 90, The Korean Ceramic Society, Seoul.

[398] A. J. Millis, Nature **392** (1998) 147.

[399] R. A. De Souza, J. A. Kilner: Oxygen transport in $La_{1-x}Sr_xMn_{1-y}Co_yO_{3\pm\delta}$ perovskites Part I; Solid State Ionics **106** (1998) 175. The figure is reprinted from this reference. Copyright 1998, with permission from Elsevier.

[400] G. Pfundtner, Ph.D. Thesis, University of Tübingen, 1993.

[401] Data compiled by K. Sasaki, according to: K. D. Becker, H. Schmalzried and V. von Wurmb, Solid State Ionics **11** (1983) 213; I. Rom and W. Sitte, Solid State Ionics **70/71** (1994) 147; J. Mizusaki, K. Fueki, Solid State Ionics **6** (1982) 85; K. Sasaki, M. Haseidl and J. Maier, Proc. EUROSOLID 4, A. Negro and L. Montanaro (eds.), Politecnico die Torino, Turin, 1997, p. 123; R. I. Merino, N. Nicoloso, J. Maier, in: *Ceramic Oxygen Ion Conductors and their Technological Applications* (British Ceram. Proc.), B. C. H. Steele (ed.), The University Press Cambridge (1996), p. 43; M. H. R. Kankhorst and H. J. M. Bouwmeester, J. Electrochem. Soc. **144** (1997) 1261; F. Millot and P. de Mierry, J. Phys. Chem. Solids **46** (1985) 797; A. Belzner, T. M. Gür, R. A. Huggins, Solid State Ionics **40/41** (1990) 535.

[402] R. I. Merino, N. Nicoloso, J. Maier, in: *Ceramic Oxygen Ion Conductors and Their Technological Applications* (British Ceram. Proc.), B. C. H. Steele (ed.), The University Press Cambridge (1996), p. 43.

[403] M. Quilitz, G. Pfundtner, J. Maier, in: *Hochleistungskeramiken — Herstellung, Aufbau, Eigenschaften*, G. Petzow, J. Tobolski and R. Telle (eds.), VCH, Weinheim, 1996, p. 314.

[404] K. D. Becker, F. Rau, Ber. Bunsenges. Phys. Chem. **91** (1987) 1279.

[405] K. Sasaki and J. Maier, Phys. Chem. Chem. Phys. **2** (2000) 3055.

[406] I. Denk, U. Traub, F. Noll and J. Maier, Ber. Bunsenges. Phys. Chem. **99** (6) (1995) 798; M. Leonhardt, Ph.D. Thesis, University of Stuttgart, 1999.

[407] H. G. Zachmann, *Mathematik für Chemiker*, p. 360, VCH, Weinheim, 1974.

[408] D. L. Bleke, Defect and Diffusion Forum **129-130** (1996) 9.

[409] J. W. Cahn, Acta Met. **9** (1961) 795.

[410] K. Funke: Jump relaxation in solid electrolytes; Prog. Solid State Chem. **22** (1993) 111. The figure is reprinted from this reference. Copyright 1993, with permission from Elsevier.

[411] L. Onsager, Phys. Z. **27** (1926) 388; **28** (1927) 277; P. Debye, H. Falkenhagen, Phys. Z. **29** (1928) 121, 401.

[412] K. Funke: Ion transport in fast ion conductors - spectra and models; Solid State Ionics **94** (1997) 27. The figure is reprinted from this reference. Copyright 1997, with permission from Elsevier.

[413] K. Funke, B. Roling, M. Lange, Solid State Ionics **105** (1998) 195.
[414] K. Funke, I. Riess, Z. Phys. Chem. NF **140** (1984) 217, R. Oldenbourg Verlag, Munich.
[415] A. K. Jonscher, Nature **267** (1977) 673; K. L. Ngai, Comments Solid State Phys. **9** (1979) 127; W. K. Lee, J. F. Liu, A. S. Nowick, Phys. Rev. Lett. **67** (1991) 1559.
[416] P. Maass, J. Petersen, A. Bunde, W. Dieterich, H. E. Roman, Phys. Rev. Lett. **66** (1991) 52; J. Petersen, W. Dieterich, Phil. Mag. B **65** (1992) 231; B. Rinn, W. Dieterich, P. Maass, Phil. Mag. B **77** (1998) 1283; P. Maass, M. Meyer, A. Bunde, Phys. Rev. B **51** (1995) 8164.
[417] K. E. Wapenaar, J. L. van Koesfeld, J. Schoonman, Solid State Ionics **2** (1981) 145.
[418] W. Dieterich, P. Maass, Chemical Physics **284** (2002) 439.
[419] J. A. Bruce, M. D. Ingram, Solid State Ionics **9/10** (1983) 717.
[420] G. V. Chandrashekar, L. M. Foster, Solid State Commun. **27** (1978) 269.
[421] P. K. Davies, G. I. Pfeiffer, S. Canfield, Solid State Ionics **18/19** (1996) 704.
[422] M. Tatsumisago, Y. Akamatsu, T. Minami, Mater. Sci. Forum **32-33** (1088) 617.
[423] H. Hrugchka, E. E. Lissel, M. Jansen, Solid State Ionics **28-30** (1988) 159.
[424] J. A. Bruce, R. A. Howie, M. D. Ingram: Mechanism of mixed cation effects in β-alumina; Solid State Ionics **18/19** (1986) 1129. The figure is reprinted from this reference. Copyright 1986, with permission from Elsevier.
[425] M. Meyer, V. Jaenisch, P. Maass, A. Bunde, Phys. Rev. Lett. **76** (1996) 2338.
[426] A. Bunde, Solid State Ionics **105** (1981) 1; A. Bunde, M. D. Ingram, P. Maass, J. Non-Cryst. Solids **172/174** (1994) 1222.
[427] P. Pendzig, W. Dieterich, Solid State Ionics **105** (1998) 209.
[428] E. O. Kirkendall, Trans. AIME **147** (1942) 104; A. D. Smigelskas, E. O. Kirkendall, Trans. AIME Techn. Publ. 2071 (1946).
[429] G. Kutsche, H. Schmalzried, Solid State Ionics **43** (1990) 43.
[430] T. Pfeiffer, K. Winters, Phil. Mag. **A61** (1990) 685.
[431] J. Maier, Z. Phys. Chem. NF **140** (1984) 191.
[432] J. Maier, G. Schwitzgebel, Phys. Stat. Sol. (b) **113** (1982) 535.
[433] This representation of the conservative ensembles goes back to M. H. R. Lankhorst, H. J. M. Bouwmeester, H. Verweij, in: *Electroceramics IV*, R. Waser, S. Hoffmann, D. Bonnenberg, Ch. Hoffmann (eds.), Vol. II, 697 pp., Augustinus Buchhandlung, Aachen, 1994.
[434] M. Widerøe, W. Münch, Y. Larring, T. Norby, Solid State Ionics **154/155** (2002) 669.
[435] J. Claus, I. Denk, M. Leonhardt and J. Maier, Ber. Bunsenges. Phys. Chem. **101** (9) (1997) 1386; H. J. Schlüter, M. Barsoum and J. Maier, Solid State Ionics **101-103** (1997) 509; K. Sasaki and J. Maier, Phys. Chem. Chem. Phys. **2** (2000) 3055.
[436] J. Maier, in: *Ionic and Mixed Conducting Ceramics*, T. A. Ramanarayanan, W. L. Worrell, H. L. Tuller (eds.), Vol. 94-12, The Electrochem. Soc., Pennington (NJ) (1994), p. 542; J. Maier and W. Münch, J. Chem. Soc., Faraday Trans. **92** (12) (1996) 2143.
[437] B. Kamp, R. Merkle, J. Maier, Sensors and Actuators B **77** (2001) 534; J. Jamnik, B. Kamp, R. Merkle and J. Maier, Solid State Ionics **150** (2002) 157.
[438] M. Spaeth, K.-D. Kreuer, Th. Dippel and J. Maier, Solid State Ionics **97** (1997) 291.

[439] H. Schmalzried, J. Janek, Ber. Bunsenges. Phys. Chem. **102** (1998) 127.

[440] J. C. Fischer, J. Appl. Phys. **22** (1951) 74.

[441] R. T. Whipple, Phil. Mag. **45** (1954) 1225; A. D. LeClaire, Brit. J. Appl. Phys. **14** (1963) 351; Y.-Ch. Chung and B. J. Wuensch, J. Appl. Phys. **79** (1996) 8323; O. Preis, W. Sitte, J. Appl. Phys. **79** (1996) 2986.

[442] J. Jamnik and J. Maier, Ber. Bunsenges. Phys. Chem. **101** (1) (1997) 23; J. Jamnik and J. Maier, J. Phys. Chem. Solids **59** (9) (1998) 1555.

[443] J. Jamnik, in: *Solid State Ionics: Science & Technology*, B. V. R. Chowdari, K. Lal, S. A. Agnihotry, N. Khare, S. S. Sekhon, P. C. Srivastava, S. Chandra (eds.), p. 13, World Scientific Publishing Co., Singapore, 1998.

[444] J. Maier: Interfaces; in: Oxygen Ion and Mixed Conductors and Their Technological Applications (H. L. Tuller, J. Schoonman and I. Riess, (eds.)), NATO SCIENCE SERIES: E Applied Sciences **368** (2000) 75. The figure is reprinted from this reference. Copyright 2000, with permission from Kluwer Academic Publishers.

[445] R. Meyer, R. Waser, J. Eur. Ceram. Soc. **21** (2001) 1743.

[446] J. Jamnik, B. Kamp, R. Merkle, J. Maier, Solid State Ionics **150** (2002) 157.

[447] M. Leonhardt, J. Jamnik and J. Maier: In situ Monitoring and Quantitative Analysis of Oxygen Diffusion Through Schottky-Barriers in $SrTiO_3$ Bicrystals; Electrochemical and Solid-State Letters **2** (7) (1999) 333. The figure is reprinted from this reference. Copyright 1999, with permission from The Electrochemical Society, Inc.

[448] J. Jamnik, J. Maier, in: *High Temperature Electrochemistry: Ceramics and Metals* (F. W. Poulsen, N. Bonanos, S. Linderoth, M. Mogensen and B. Zachau-Christiansen (eds.)), Risø National Laboratory, Roskilde, Denmark, 1996, p. 287.

[449] J. Jamnik and J. Maier, Solid State Ionics **119** (1999) 191.

[450] S. Brunauer, *The Adsorption of Gases and Vapors*, Oxford University Press, London, 1944; L. J. Slutsley, G. D. Halsey, in: *Physical Chemistry, An Advanced Treatise*, H. Eyring, D. Henderson, W. Jost (eds.), Academic Press, New York, 1967, p. 479.

[451] I. Langmuir, J. Am. Chem. Soc. **38** (1917) 2221; **40** (1918) 136.

[452] J. Maier: Interaction of oxygen with oxides: How to interpret measured effective rate constants?; Solid State Ionics **135** (1-4) (2000) 575. The figure is reprinted from this reference. Copyright 2000, with permission from Elsevier.

[453] M. Bodenstein, Z. phys. Chem. **85** (1913) 329.

[454] M. Leonhardt, Ph.D. Thesis, University of Stuttgart, 1999.

[455] E. Opila, H. L. Tuller, B. J. Wuensch, J. Maier, J. Am. Ceram. Soc. **76** (9) (1993) 2363, The American Ceramic Society, Westerville (OH).

[456] R. Merkle, R. A. De Souza, J. Maier, Angew. Chemie Int. Ed. Engl. **40** (11) (2001) 2126.

[457] B. C. H. Steele: Interfacial reactions associated with ceramic ion transport membranes; Solid State Ionics **75** (1995) 157. The figure is reprinted from this reference. Copyright 1995, with permission from Elsevier.

[458] R. A. De Souza, J. A. Kilner: Oxygen transport in $La_{1-x}Sr_xMn_{1-y}Co_yO_{3\perp\delta}$ perovskites Part II; Solid State Ionics **126** (1999) 153. The figure is reprinted from this reference. Copyright 1999, with permission from Elsevier.

[459] I. C. Fullarton, J.-P. Jacobs, H. E. van Benthem, J. A. Kilner, H. H. Brongersma, P. J. Scanlon, B. C. H. Steele, Ionics **1** (1995) 51.

[460] S. J. Benson, R. J. Chater, J. A. Kilner, in: *Ionic and Mixed Conducting Ceramics*, T. A. Ramanarayanan, W. L. Worrell, H. L. Tuller, M. Mogensen, A. C. Khandkar (eds.), Vol. PV 97-24, p. 596, The Electrochemical Society, Pennington (NJ), 1997.

[461] R. A. De Souza, J. A. Kilner, C. Jeynes, Solid State Ionics **97** (1997) 409.

[462] E. Ruiz-Trejo, J. A. Kilner, in: *High Temperature Electrochemistry: Ceramics and Metals*, F. W. Poulsen, N. Bonanos, S. Linderoth, M. Mogensen and B. Zachau-Christiansen (eds.), p. 411, Risø National Laboratory, Roskilde, Dänemark, 1996.

[463] T. Ishihara, J. A. Kilner, M. Honda, Y. Takita, J. Am. Chem. Soc. **119** (1997) 2747.

[464] J. D. Sirman, Ph.D. Thesis, University of London (1998).

[465] J. D. Sirman, J. A. Kilner, in: *High Temperature Electrochemistry: Ceramics and Metals*, F. W. Poulsen, N. Bonanos, S. Linderoth, M. Mogensen and B. Zachau-Christiansen (eds.), p. 417, Risø National Laboratory, Roskilde, Denmark, 1996.

[466] P. S. Manning, J. D. Sirman, R. A. De Souza, J. A. Kilner, Solid State Ionics **100** (1997) 1.

[467] E. Ruiz-Trejo, J. D. Sirman, Yu. M. Baikov, J. A. Kilner, Solid State Ionics **113-115** (1998) 565.

[468] J. Maier, in: *Solid State Ionics V*, G.-A. Nazri, C. Julien, A. Rougier (eds.), MRS, Pittsburgh, 1999; Mat. Res. Soc. Proc. **548** (1999) 415; Solid State Ionics **135** (1-4) (2000) 575.

[469] R. Merkle and J. Maier, Phys. Chem. Chem. Phys., **4** (17) (2002) 4140.

[470] K. Sasaki and J. Maier, Solid State Ionics **161** (2003) 145.

[471] M. Leonhardt, R. A. De Souza, J. Claus and J. Maier, J. Electrochem. Soc. **149** (2) (2002) J19.

[472] J. Jamnik and J. Maier, Phys. Chem. Chem. Phys. **3** (9) (2001) 1668; J. Maier, J. Jamnik, M. Leonhardt, Solid State Ionics **129** (1-4) (2000) 25.

[473] R. Schlögl, in: *Handbook of Heterogeneous Catalysis*, G. Ertl, H. Knözinger, J. Weitkamp (eds.), Vol. 4, p. 1697, VCH, Weinheim, 1997; G. Ertl, J. Vac. Sci. Technol. A **1** (1983) 1247.

[474] E. G. Schlosser, *Heterogene Katalyse*, Chemische Taschenbücher, VCH, Weinheim, 1972.

[475] M. I. Temkin, J. Physik. Chem. (Auss.) **31** (1957) 3.

[476] G. C. Vayenas, S. Bebelis, S. Neophytides, J. Phys. Chem. **92** (1988) 5085.

[477] C. G. Vayenas, S. Bebelis, C. Pliangos, S. Brosda, D. Tsiplakides, T. S. Stein (eds.), *Electrochemical Activation of Catalysis: Promotion, Electrochemical Promotion, and Metal-Support Interactions*, Plenum Publishing Corporation (2002).

[478] J. Maier, P. Murugaraj, Solid State Ionics **40/41** (1990) 1017.

[479] G. Simkovich, C. Wagner: The role of ionic defects in the catalytic activity of ionic crystals; J. Catal. **1** (1962) 521. The figure is reprinted from this reference. Copyright 1962, with permission from Elsevier.

[480] F. Corà, C. R. A. Catlow, D. W. Lewis, J. Mol. Cat. A (Chem) **166** (1) (2001) 123.

[481] Yu. Zhukovskii, E. A. Kotomin, P. W. M. Jacobs, A. M. Stoneham, Phys. Rev. Lett. **84** (2000) 1256.

[482] G. Tammann, Z. Allg. Anorg. Chem. **111** (1920) 78; C. Wagner, J. Electrochem. Soc. **103** (1956) 571.

[483] N. B. Pilling, R. G. Bedworth, J. Inst. Metals **19** (1923) 529.

[484] C. Wagner, Z. Phys. Chem. **B21** (1933) 25; Corr. Sci. **9** (1969) 91.
[485] C. Gensch, K. Hauffe, Z. Phys. Chem. **196** (1951) 427.
[486] L. S. Richardson, N. J. Grant, J. Metals **6** (1954) 69.
[487] J. Paidassi, Trans AIME, J. Metals **4** (1952) 536.
[488] C. Wagner, in: *Handbuch Metallphysik*, S. Flügge (ed.), Vol. I, Leipzig, 1940; K. Hauffe, W. Schottky, in: *Halbleiterprobleme V* W. Schottky (ed.), p. 259, Vieweg, Braunschweig, 1960.
[489] M. Vollmer, *Kinetik der Phasenbildung*, Steinkopff, Dresden, 1939; M. Kahlweit, *Grenzflächenerscheinungen*, Steinkopff, Darmstadt, 1981.
[490] R. Becker and W. Döring, Ann. Phys. Leipzig **119** (1926) 277; E. Budevski, G. Staikov, W. J. Lorenz, *Electrochemical Phase Formation and Growth*, VCH, Weinheim, 1996.
[491] A. T. Fromhold, *Theory of Metal Oxidation*, North–Holland, Amsterdam, 1980.
[492] D. D. Macdonald, J. Electrochem. Soc. **139** (1992) 3434.
[493] M. Avrami, J. Chem. Phys. **7** (1939) 1103; J. Chem. Phys. **8** (1940) 212; J. Chem. Phys. **9** (1941) 177.
[494] W. A. Johnson, R. F. Mehl, Trans. AIME **135** (1939) 416.
[495] J. Schoonman, in: Oxygen Ion and Mixed Conductors and Their Technological Applications (H. L. Tuller, J. Schoonman and I. Riess, (eds.)), NATO SCIENCE SERIES: E Applied Sciences **368** (2000) 295.
[496] A. M. Ginstling and B. I. Brounshtein, J. Appl. Chem. (USSR) **23** (1950) 1249.
[497] R. E. Carter, J. Chem. Phys. **34** (1961) 2010.
[498] M. Jansen, Angew. Chem. Int. Ed. Engl. **41** (2002) 3746.
[499] J. A. Pask and L. K. Templeton, in: *Kinetics of High–Temperature Processes*, W. D. Kingery (ed.), Technology Press of Massachusetts Institute of Technology, Cambridge (MA), 1959.
[500] C. Wagner, J. Electrochem. Soc. **103** (1956) 571; W. N. Mullins, R. F. Sekerka, J. Appl. Phys. **35** (1964) 444; R. T. Delves, in: *Crystal Growth*, B. R. Pamplin (ed.), p. 40, Pergamon Press, Oxford, 1975.
[501] M. Backhaus–Ricoult, H. Schmalzried, Ber. Bunsenges. Phys. Chem. **89** (1985) 1323.
[502] M. Martin, P. Tigelmann, S. Schimschal-Thölke, G. Schulz, Solid State Ionics **75** (1995) 219.
[503] W. Hahn, *The Stability of Motion*, Springer–Verlag, Berlin, 1967; L. S. Pontryagin, *Ordinary Differential Equations*, Addison–Wesley, Reading, 1962.
[504] H. R. Oswald, J. R. Günter, in: *1976 Crystal Growth and Materials*, E. Kaldis, H. J. Sheel (eds.), p. 416, North–Holland, Amsterdam, 1977.
[505] W. Laqua, H. Schmalzried, in: *High Temperature Corrosion*, R. A. Rapp (ed.), 115 pp., NACE, Houston, 1983; H. Schmalzried, W. Laqua, Oxid. Metals **15** (1981) 339.
[506] M. Martin, Ceram. Trans. **24** (1991) 91; M. Martin, Proc. Electrochem. Soc. **99-19**, p. 308, Pennington (NJ), 1999; O. Teller and M. Martin, Solid State Ionics **101-103** (1997) 475; O. Teller and M. Martin, Ber. Bunsenges. Phys. Chem. **101** (1997) 1377.
[507] R. W. Cahn, P. Haasen (eds.), *Physical Metallurgy*, North–Holland, Amsterdam, 1983.
[508] J. Meixner, Z. Naturforschung **4A** (1949) 594; G. Nicolis, I. Prigogine, *Self-Organisation in Non-Equilibrium Systems*, John Wiley & Sons, Ltd., New York, 1971.

[509] W. Ebeling, *Strukturbildung bei irreversiblen Prozessen*, B. G. Teubner, Leipzig, 1976.

[510] A. Sanfeld, in: *Physical Chemistry, An Advanced Treatise*, Vol. I, *Thermodynamics*, H. Eyring, D. Henderson, W. Jost (eds.), p. 234, Academic Press, New York, 1971.

[511] J. Keizer, *Statistical Thermodynamics of Nonequilibrium Processes*, Springer, New York, 1987.

[512] P. van Rysselberghe, *Thermodynamics of Irreversible Processes*, Hermann, Paris, 1963.

[513] E. Schöll, *Phase Transitions in Semiconductors*, Springer–Verlag, Berlin, 1987.

[514] F. J. Dyson, J. Mol. Evol. **18** (1982) 344.

[515] M. Eigen, P. Schuster, *The Hypercycle*, Springer–Verlag, Heidelberg, 1979.

[516] H. Poincaré, *Les Méthodes Nouvelles de la Mécanique Céleste*, Gauthier–Villars, Paris, 1892; E. N. Lorenz, J. Atmos. Sci. **20** (1963) 130; H. G. Schuster, *Deterministic Chaos*, VCH, Weinheim, 1984.

[517] M. Eiswirth et al., in: *Jahrbuch der MPG*, Verlag Vandenhoeck & Ruprecht, Göttingen, 1991.

[518] A. E. McCombs, A. G. Milnes, Int. J. Electron. **32** (1972) 361; S. H. Koenig, R. D. Brown, W. Schillinger, Phys. Rev. **128** (1962) 1668; M. E. Cohen, P. T. Landsberg, Phys. Rev. **154** (1972) 683.

[519] S. W. Teitsworth, R. M. Westervelt, E. E. Haller: Nonlinear Oscillations and Chaos in Electrical Breakdown in Ge; Phys. Rev. Lett. **51** (1983) 825 , The figure is reprinted from this reference. Copyright 1983, with permission from The American Physical Society.

[520] K. M. Mayer, R. Gross, J. Parisi, J. Peinke, R. P. Huebener: Spatially resolved observation of current filament dynamics in semiconductors; Solid State Commun. **63** (1987) 55. The figure is reprinted from this reference. Copyright 1987, with permission from Elsevier.

[521] C. Wagner, J. Colloid Sci. **5** (1950) 85.

[522] W. Wang and A. V. Virkar, Proc. Joint Int. Meeting ECS/ISE, San Francisco, Sept 2-7, 2001, Abstract No. 1574; E. D. Wachsman, T. L. Clites, J. Electrochem. Soc. **149** (2002) A242; S. Martin, W. Martienssen, Z. Physik B –Cond. Matter **68** (1987) 299 and references therein; N. I. Ionescu, M. Caldararu, D. Sprinceana, Rev. Roum. Chimie **43** (1998) 71.

[523] J. Janek, S. Majoni, Ber. Bunsenges. Phys. Chem. **99** (1995) 14, VCH, Weinheim.

[524] H. W. Roesky, K. Möckel, *Chemische Kabinettsstückchen*, VCH, Weinheim, 1994.

[525] H. Meinhard, A. Gierer, J. Cell. Sci. **15** (1974) 312.

[526] T. Ihle, H. Müller–Krumbhaar, Phys. Rev. E **49** (1994) 2972.

[527] T. A. Witten, L. M. Sander, Phys. Rev. Lett. **47** (1981) 1400.

[528] J. Nittmann, H. E. Stanley, Nature **321** (1986) 663.

[529] H. E. Stanley: Fractals and Multifractals: The Interplay of Physics and Geometry, in: A. Bunde, S. Havlin, *Fractals and Disordered Systems*. The figure is reprinted from this reference. Copyright 1996, with permission from Springer–Verlag, Berlin.

[530] G. Daccord, L. Lenormand, Nature **325** (1987) 41.

[531] S. Busch, U. Schwarz, R. Kniep, Chem. Mater. **13** (10) (2001) 3260. The figure is reprinted from this reference. Copyright 2001, with permission from American Chemical Society.

[532] B. B. Mandelbrot, *Die fraktale Geometrie der Natur*, Birkhäuser Verlag, Basel, 1987; *The Fractal Geometry of Nature*, W. H. Freeman, New York, 1983.
[533] D. Avnir (ed.), *The Fractal Approach to Heterogeneous Chemistry*, John Wiley & Sons, Ltd., New York, 1989.
[534] B. Sapoval, Solid State Ionics **75** (1995) 269; B. Sapoval, J. N. Chazalviel, J. Peyrierre, Solid State Ionics **28/30** (1988) 1441; A. Le Méhauté, G. Crépy, Solid State Ionics **9/10** (1983) 17.
[535] H.-D. Wiemhöfer, Habilitation Thesis, University of Tübingen, 1991.
[536] H. L. Tuller, J. Phys. Chem. Solids **55** (1991) 1393.
[537] S. z. B. M. Henzler, W. Göpel, *Oberflächenphysik des Festkörpers*, B. G. Teubner, Stuttgart, 1991; W. Göpel, C. Ziegler, *Grundlagen, Mikroskopie und Spektroskopie*, B. G. Teubner, Stuttgart, 1994; W. Göpel, C. Ziegler, *Struktur der Materie: Grundlagen, Mikroskopie und Spektroskopie*, B. G. Teubner, Stuttgart, 1994.
[538] H. Dietz, W. Haecker, H. Jahnke, in: *Advances in Electrochemistry and Electrochemical Engineering*, H. Gerischer, C. W. Tobias (ed.), John Wiley & Sons, Ltd., New York, 1977.
[539] W. A. Fritsche, D. Jahnke, *Metallurgische Elektrochemie*, Springer–Verlag, Berlin, 1975.
[540] T. Yajama, H. Kazeoka, T. Yogo, H. Iwahara, Solid State Ionics **47** (1991) 271.
[541] W. Göpel, J. Hesse, J. N. Zemel (eds.), *Sensors, A Comprehensive Study*, VCH, Weinheim, 1987.
[542] J. Janata, *Principles of Chemical Sensors*, Plenum Press, New York, 1989.
[543] P. Henderson, Z. Phys. Chem. **59** (1907) 118.
[544] J. Maier, in: *Science and Technology of Fast Ion Conductors*, H. L. Tuller, M. Balkanski (eds.), p. 299, Plenum Press, New York, 1989.
[545] H.-H. Möbius, P. Shuk, W. Zastrow, Fresenius J. Anal. Chem. **349** (1996) 684.
[546] N. Miura, S. Yao, Y. Shimizu and N. Yamazoe, J. Electrochem. Soc. **139** (1992) 1384.
[547] J. Maier, M. Holzinger, W. Sitte, Solid State Ionics **74** (1994) 5.
[548] J. Maier, Solid State Ionics **62** (1,2) (1993) 105.
[549] M. Holzinger, J. Maier, W. Sitte, Solid State Ionics **94** (1997) 217.
[550] J. J. Egan, J. Phys. Chem. **68** (1964) 978; R. J. Heus, J. J. Egan, Z. Phys. Chem. **49** (1966) 38.
[551] C. Wagner, Z. Elektrochem. **60** (1956) 4.
[552] See textbooks on electrochemistry [153,219–221,580].
[553] H. Schmalzried, J. Chem. Phys. **33** (1960) 940; S. Mitoff, ibid. **36** (1962) 1383.
[554] H. Schmalzried, Z. Phys. Chem. NF **38** (1963) 87, R. Oldenbourg Verlag, München.
[555] J. Maier, J. Phys. Chem. Solids **46** (1985) 197.
[556] J. B. Goodenough, J. E. Ruiz-Diaz, Y. S. Zhen: Oxide-ion conduction in $Ba_2In_2O_5$ and $Ba_3In_2MO_8$ (M=Ce,Hf, or Zr); Solid State Ionics **44** (1990) 21. The figure is reprinted from this reference. Copyright 1990, with permission from Elsevier.
[557] T. Norby, P. Kofstad, Solid State Ionics **20** (1986) 164.
[558] T. Norby, O. Dyrlie, P. Kofstad, Solid State Ionics **53** (1992) 446.
[559] S. Yuan, U. Pal, K. C. Chou, in: *Ionic and Mixed Conducting Ceramics*, T. A. Ramanarayanan, W. L. Worrell, H. L. Tuller (eds.), Vol. 94-12, p. 46, The Electrochemical Society, Pennington (NJ), 1994.

[560] J. E. ten Elshof, H. J. M. Bouwmeester, H. Verweij, Solid State Ionics **81** (1995) 97; B. Ma, U. Balachandran, J.-H. Park, C. U. Segre, J. Electrochem. Soc. **143** (1996) 1736; S. Kim, Y. L. Yang, A. J. Jacobson, B. Abeles, Solid State Ionics **106** (1998) 189.

[561] M. Zhou, H. Deng, B. Abeles, Solid State Ionics **93** (1997) 133.

[562] H. Iwahara, Chem. Solid State Mater. **2** (1992) 122, Cambridge University Press, Cambridge (UK).

[563] H. Iwahara: Technological challenges in the application of proton conducting ceramics; Solid State Ionics **77** (1995) 289. The figure is reprinted from this reference. Copyright 1995, with permission from Elsevier.

[564] G. Marnellos and M. Stoukides, Science **282** (1998) 98.

[565] H. Gerischer, J. Electroanal. Chem. **58** (1975) 263; H. Tributsch, J. Electrochem. Soc. **125** (1978) 1087; C. Gutierrez, P. Salvador, J. B. Goodenough, J. Electroanal. Chem. **134** (1982) 325.

[566] M. Graetzel, Cattech **3** (1999) 3.

[567] B. O'Reagan, M. Grätzel, Nature **353** (1991) 737.

[568] M. Grätzel, Nature **414** (2001) 338.

[569] H. Kaiser, Z. Anal. Chem. **260** (1972) 252.

[570] J. Maier, in: 'Oxygen Ion and Mixed Conductors and their Technological Applications' (H. L. Tuller and J. Schoonman and I. Riess, (eds.)), NATO SCIENCE SERIES: E Applied Sciences, Kluwer Academic Publishers, **368** (2000) 399; Sensors and Actuators B **65** (2000) 199.

[571] Ch. Tragut, K. H. Härdtl, Sensors and Actuators B **4** (1991) 425; J. Gerblinger, H. Meixner, Sensors and Actuators B **4** (1991) 99.

[572] J. Kircher, M. Alouani, M. Garriga, P. Murugaraj, J. Maier, C. Thomson, M. Cardona, O. K. Andersen and O. Jepsen, Phys. Rev. B **40** (1989) 7368.

[573] K. Uchino, *Piezoelectric Actuators and Ultrasonic Motors*, Kluwer Academic Publishers (1997); B. Jaffe, W. R. Cook and H. Jaffe, *Piezoelectric Ceramics*, Academic Press (1971).

[574] C. Wagner, in: Proc. 7th Meeting Int. Comm. on Electrochem. Thermodynamics and Kinetics, Butterworth, London, 1957.

[575] I. Riess, Mater. Sci. Eng. B **12** (1992) 351.

[576] C. Tubandt, F. Lorenz, Z. Phys. Chem. **87** (1913) 543.

[577] S. z. B. H. Fröhlich, *Theory of Dielectrics: Dielectric Constant and Dielectric Loss*, Oxford University Press, Oxford, 1958.

[578] J. Maier, Ph.D. Thesis, University of Saarland, Saarbrücken, 1982.

[579] D. Y. Wang, A. S. Nowick: Cathodic and Anodic Polarization Phenomena at Platinum Electrodes with Doped CeO_2 as Electrolyte; J. Electrochem. Soc. **126** (1979) 1155. The figure is reprinted from this reference. Copyright 1979, with permission from The Electrochemical Society, Inc.

[580] K. J. Vetter, *Electrochemical Kinetics*, Academic Press, New York, 1967,

[581] C. Wagner, W. Traud, Z. Elektrochem. **44** (1938) 391; E. Lange, Z. Elektrochem. **55** (1951) 76.

[582] M. Mogensen, S. Skaarup, Solid State Ionics **86/88** (1996) 1151; J. Mizusaki, H. Tagawa, K. Tsuneyoshi, A. Sawata, J. Electrochem. Soc. **138** (1991) 1867; F. H. van Heuveln, H. J. M. Bouwmeester, J. Electrochem. Soc. **144** (1997) 134; R. Jiminez, T. Kloidt, M. Kleitz, J. Electrochem. Soc. **144** (1997) 5823; S. D. Adler, J. A. Lane, B. C. H. Steele, J. Electrochem. Soc. **144** (1997) 1884; H. Hu, M. Liu, J. Electrochem. Soc. **144** (1997) 3561.

[583] I. Riess, M. Gödickemeier, L. J. Gauckler, Solid State Ionics **90** (1990) 9; B. A. Boukamp, I. C. Vinke, J. J. De Vries, A. J. Burggraaf, Solid State Ionics **32/33** (1998) 918.

[584] V. Brichzin, J. Fleig, H.-U. Habermeier and J. Maier, Electrochem. Solid State Let. **3** (9) (2000) 403.

[585] R. Parsons, Trans. Faraday Soc. **47** (1951) 1332.

[586] J. R. Macdonald, in: *Impedance Spectroscopy*, (J. R. Macdonald, Ed.), John Wiley & Sons 1987, p. 114; J. R. Macdonald, J. Chem. Phys. **58** (1973) 4982.

[587] J. Jamnik and J. Maier, Phys. Chem. Chem. Phys. **3** (9) (2001) 1668.

[588] D. C. Grahame, Chem. Rev. **41** (1947) 441.

[589] J. Jamnik, Appl. Phys. **A55** (1992) 518.

[590] H. Gerischer, in: *Physical Chemistry, An Advanced Treatise IXA*, H. Eyring, D. Henderson, W. Jost (eds.), Academic Press, New York, 1970; W. Schottky, Z. Physik **113** (1939) 367; **118** (1942) 359; N. F. Mott, Proc. Royal Soc. (London) **A 171** (1939) 27.

[591] K. Bohnenkamp, H.-J. Engell, Z. Elektrochem. **61** (1957) 1184, VCH, Weinheim.

[592] R. D. Armstrong and B. R. Horrocks, Solid State Ionics **94** (1997) 181.

[593] J. Jamnik, H.-U. Habermeier and J. Maier, Physica B **204** (1995) 57.

[594] R. D. Armstrong, R. Mason: Double layer capacity measurements involving solid electrolytes; J. Electroanal. Chem. **41** (1973) 231. The figure is reprinted from this reference. Copyright 1973, with permission from Elsevier.

[595] R. D. Armstrong and B. R. Horrocks, Solid State Ionics **94** (1997) 181.

[596] J. R. Macdonald, D. R. Franceschetti and A. P. Lehnen, J. Chem. Phys. **73** (1980) 5272.

[597] J. C. Lassègues, J. Grondin, T. Becker, L. Servant, M. Hernandez, Solid State Ionics **77** (1995) 311; J. Fricke, Plenarvortrag, Bunsenhauptversammlung, '98, Münster. E. T. Eisenmann, in: Proc. Electrochem. Soc., Vol. 95–29, p. 255, Pennington (NJ), 1995.

[598] A. Yamamada, J. B. Goodenough, J. Electrochem. Soc. **145** (1998) 737.

[599] J. Maier, Solid State Ionics **148** (3, 4) (2002) 367.

[600] B. E. Conway, *Electrochemical Supercapacitors: Scientific Fundamentals and Technological Applications*, Kluwer Academic Publishers, New York.

[601] J. Jamnik, X. Guo and J. Maier, Appl. Phys. Lett. **82** (17) (2003) 2820.

[602] G. J. Dudley and B. C. H. Steele, J. Solid State Chem. **10** (1980) 233.

[603] J. R. Macdonald (ed.), *Impedance Spectroscopy*, John Wiley & Sons, Ltd., New York, 1987; J. R. Macdonald, J. Schoonman, A. Lehnen, J. Electroanal. Chem. **131** (1982) 77.

[604] J. Maier, P. Murugaraj, G. Pfundtner, W. Sitte, Ber. Bunsenges. Phys. Chem. **93** (1989) 1350.

[605] J. Fleig and J. Maier, in preparation.

[606] J. Mizusaki and K. Fueki, Revue de Chimie Minerale **17** (4) (1980) 356.

[607] X. Guo and J. Maier, Solid State Ionics **130** (2000) 267.

[608] L. D. Burke, H. Rickert, R. Steiner, Z. Phys. Chem. NF **74** (1971) 146.

[609] J. E. B. Randles, Discuss. Faraday Soc. **1** (1947) 11.

[610] J. Jamnik, J. Maier and S. Pejovnik: A Powerful Electrical network Model for the Impedance of Mixed Conductors; Electrochim. Acta, **44** (1999) 4139. The figure is reprinted from this reference. Copyright 1989, with permission from Elsevier.

[611] J. Jamnik and J. Maier, J. Electrochem. Soc. **146** (11) (1999) 4183.

[612] J. Jamnik, Solid State Ionics **157** (2002) 19.

[613] X. Guo, J. Jamnik, J. Maier, to be published.

[614] R. Waser, T. Baiatu, and K.-H. Härdtl, J. Am. Ceram. Soc. **73** (1990) 1645.

[615] N. Valverde, Z. Phys. Chem. NF **74** (1971) 146; W. Piekarcyak, W. Weppner, A. Rabenau, Z. Naturforschung **34a** (1979) 430; W. Weppner, R. A. Huggins, J. Electrochem. Soc. **124** (1977) 10.

[616] J. Mizusaki: Defect Chemistry of $La_{2-x}Sr_xCu_{4-\delta}$: Oxygen Nonstoichiometry and Thermodynamic Stability; J. Solid State Chem. **131** (1997) 150. The figure is reprinted from this reference. Copyright 1997, with permission from Elsevier.

[617] U. von Oehsen, H. Schmalzried, Ber. Bunsenges. Phys. Chem. **85** (1981) 7, VCH, Weinheim.

[618] W. Preis, W. Sitte, J. Chem. Soc., Faraday Trans. **92** (1996) 1197.

[619] R. Andreaus, W. Sitte: Ionic Transport Properties of Mixed Conductors: Application of AC and DC Methods to Silver Telluride; J. Electrochem. Soc. **144** (1997) 1040. The figure is reprinted from this reference. Copyright 1997, with permission from The Electrochemical Society, Inc.

[620] N. Wiener, *Cybernetics or Control and Communication in the Animal and the Machine*, John Wiley & Sons, Ltd., New York, 1948; G. J. Murphy, *Basic Automatic Control Theory*, van Nostrand, Princeton (NJ), 1957; K. Göldner, *Mathematische Grundlagen der Systemanalyse*, Verlag Harri Deutsch, Thun, 1981; W. Hahn, F. L. Bauer, *Physikalische und elektrotechnische Grundlagen für Information*, Springer–Verlag, Berlin, 1975.

[621] Picture postcard TM 10 Putput, ©Tom, Berlin.

[622] D. Widder, *Laplace Transforms*, Princeton University Press, Princeton, 1941.

[623] I. N. Bronstein, K. A. Semendjajew, G. Musiol, H. Mühlig, *Handbook of mathematics*, Springer-Verlag, Berlin, 1997.

[624] R. D. Armstrong: Impedance Diagrams for Solid Electrolyte cells; in: *Electrode Processes in Solid State Ionics*, M. Kleitz and J. Dupuy (eds.). The figure is reprinted from this reference. Copyright 1975, with permission from Kluwer Academic Publishers.

[625] M. Gaberšček, J. Jamnik and S. Pejovnik: A.C. impedance studies of the anodic passivating layer in lithium-$SOCl_2$ batteries; J. Power Sources **25** (1989) 123. The figure is reprinted from this reference. Copyright 1989, with permission from Elsevier.

[626] C. Tragut, Ph.D. Thesis, University of Karlsruhe, 1992.

[627] A. K. Jonscher, Nature **267** (1977) 673; K. L. Ngai, A. K. Jonscher, Nature **277** (1979) 185.

[628] K. L. Ngai, U. Strom, Phys. Rev. B **38** (1988) 10350.
[629] A. Le Mehauté, J. Stat. Phys. **36** (1984) 665.
[630] T. Große, H. Schmalzried, Z. Phys. Chem. NF **172** (1991) 197; W. Preis, W. Sitte, Solid State Ionics **76** (1995) 5.
[631] M. J. Verkerk, B. J. Middlehuis, A. J. Burggraaf, Solid State Ionics **6** (1982) 159.
[632] J. Fleig and J. Maier, Electrochim. Acta **41** (7/8) (1996) 1003.
[633] J. Maier, S. Prill and B. Reichert, Solid State Ionics **28-30** (1988) 1465.
[634] M. Kleitz, H. Bernard, E. Fernandez and E. Schouler, Adv. Ceram. Sci. Tech. Zirconia **3** (1981) 310; M. Meyer, H. Rickert, and U. Schwaitzer, Solid State Ionics **9, 10** (1983) 689; J. E. Bauerle, J. Phys. Chem. Solids **30** (1969) 2657; H. Rickert, H.-D. Wiemhöfer, Ber. Bunsenges. Phys. Chem. **87** (1983) 236.
[635] J. Fleig, J. Maier, Solid State Ionics **86-88** (1996) 1351.
[636] R. Landauer, in: *Electrical Transport and Optical Properties of Inhomogeneous Media*, J. C. Garland, D. B. Tanner (eds.), AIP Conf. Proc. 40, New York, 1987, p. 2; D. Stauffer and A. Aharony, *Percolation and Conduction*, VCH, Weinheim, 1973, p. 574; D. S. McLachlan, M. Blaskiewicz, R. E. Newnham, J. Am. Ceram. Soc. **73** (1990) 2187; A. Bunde, Solid State Ionics **75** (1995) 147.
[637] R. Holm, *Electrical Contacts Handbook*, Springer–Verlag, Berlin, 1958; J. Newman, J. Electrochem. Soc. **113** (1966) 501.
[638] J. Fleig, J. Maier, Solid State Ionics **85** (1996) 9.
[639] J. Fleig, S. Rodewald, J. Maier, J. Appl. Phys. **87** (5) (2000) 2372.
[640] V. Brichzin, J. Fleig, H. U. Habermeier, G. Cristiani, J. Maier, Solid State Ionics **152-153** (2002) 499.
[641] W. G. Amey, F. Hamburger, Proc. Am. Soc. Testing Mater. **49** (1949) 1079.
[642] A. Uhlir, Bell. Syst. Tech. J. **34** (1955) 105.
[643] J. Jamnik, H.-U. Habermeier and J. Maier, Physica B **204** (1995) 57.
[644] S. V. Kalinin and D. A. Bonnell, J. Appl. Phys. **91** (2) (2002) 832.
[645] M. B. H. Breere, D. N. Jamieson, P. J. C. King, *Materials Analysis with a Nuclear Microprobe*, John Wiley & Sons, Ltd., New York, 1995.
[646] L. J. van der Pauw, Philips Res. Repts. **13** (1958) 1; I. Riess, D. S. Tannhauser, Solid State Ionics **7** (1982) 307.
[647] L. B. Valdes, Proc. I.R.E. **42** (1954) 420.
[648] I. Riess, Solid State Ionics **44** (1991) 207; I. Riess, S. Kramer, H. L. Tuller, in: *Solid State Ionics*, M. Balkanski, T. Takahashi, H. L. Tuller (eds.), 429 pp., Elsevier, Amsterdam, 1992.
[649] M. Winter, J. O. Besenhard, M. E. Spahr, P. Novák, Adv. Mater. **10** (1998) 725, VCH, Weinheim.
[650] Ch. Julien, G.-A. Nazri, *Solid State Batteries: Materials, Design and Optimization*, Kluwer Academic Publishers, New York, 1994.
[651] M. Wakihara. O. Yamamoto (eds.), *Lithium Ion Batteries*, VCH, Weinheim, 1998.
[652] B. C. H. Steele, Phil. Trans. R. Soc. London A **354** (1996) 1695; T. Kanada and H. Yokokawa, Key Engineering Materials **125-126** (1997) 187; S. C. Singhal, in: *Solid Oxide Fuel Cells V*, U. Stimming, S. C. Singhal, H. Tagawa, W. Lehnert, (eds.), Vol. PV 97-40, 37 pp., The Electrochemical Society, Pennington (NJ), 1997.
[653] W. Ostwald, Z. Elektrochem. **1** (1894/95) 122.

[654] K.-D. Kreuer, J. Maier, Spektrum der Wissenschaft **7** (1995) 92.
[655] I. Riess, M. Gödickemeier, L. J. Gauckler, Solid State Ionics **90** (1996) 91.
[656] E. D. Wachsman, Solid State Ionics **152-153** (2002) 657.
[657] J. Fleig and J. Maier, J. Electrochem. Soc. Letters **144** (1997) L302.
[658] T. Horita, K. Yamaji, N. Sakai, H. Yokokawa, T. Kawada, T. Kato, Solid State Ionics **127** (2000) 55.
[659] C. C. Chen, M. M. Nasrallah, H. U. Anderson, in: *Solid Oxide Fuel Cells*, S. C. Singhal, H. Iwahara (eds.), Vol. PV 93-4, p. 598, The Electrochemical Society, Pennington (NJ), 1993.
[660] T. Ishihara, H. Matsuda, Y. Takita, J. Am. Chem. Soc. **116** (1994) 3801; T. Ishihara, H. Furutani, H. Nishiguchi,Y. Takita, in: *Ionic and Mixed Conducting Ceramics III*, T. A. Ramanarayanan, W. L. Worrell, H. L. Tuller, M. Mogensen, A. C. Khandkar (eds.), Vol. PV 97-24, p. 834, The Electrochemical Society, Pennington (NJ), 1997.
[661] H. L. Tuller, in: Proc. 17th Risø Int. Symp. Materials Science: High Temperature Electrochemistry: Ceramics and Metals, F. W. Poulsen, N. Bonanos, S. Linderoth, M. Mogensen, B. Zachau-Christiansen (eds.), p. 139, Risø National Laboratory, Roskilde, Denmark, 1996.
[662] S. B. Adler, J. A. Reimer, J. Baltisberger, U. Werner, J. Am. Ceram. Soc. **116** (1994) 675; G. B. Zhang, D. M. Smyth, Solid State Ionics **82** (1996) 161.
[663] S. C. Singhal: Recent Progress in Tubular Solid Oxide Fuel Cell Technology; in: *Solid Oxide Fuel Cells V*, U. Stimming, S. C. Singhal, H. Tagawa, W. Lehnert, (eds.), Vol. PV 97-40, 37 pp. The figure is reprinted from this reference. Copyright 1997, with permission from The Electrochemical Society, Inc.
[664] H. Iwahara, T. Yajima, T. Hibino, H. Ushida, J. Electrochem. Soc. **140** (1993) 1687.
[665] W. Drenckhahn, H. Vollmar, Siemens–Zeitschrift **5** (1995) 31, Siemens AG, Munich.
[666] A. Eisenberg, H. L. Yeager (eds.), *Perfluorinated Ionomer Membranes*, The American Chemical Society, Washington (DC), 1982.
[667] G. G. Scherer, H. P. Brack, F. N. Buchi, B. Gupta, O. Haas, M. Rota, in: Proc. 11th World Hydrogen Energy Conf., T. N. Veziroglu (eds.), Vol. 2, p. 1727, International Association for Hydrogen Energy, Coral Gables (FL), 1996; K.-D. Kreuer, Th. Dippel, J. Maier, in: Proc. Electrochem. Soc. **PV 95-23** (1995) 241.
[668] M. Rehahn, A. D. Schlüter, G. Wegner, Makromol. Chem. **191** (1990) 1991.
[669] K. D. Kreuer, Solid State Ionics **97** (1997) 1.
[670] K. D. Kreuer, A. Fuchs, M. Ise, M. Spaeth, J. Maier, Electrochim. Acta **43** (10-11) (1998) 1281; K. D. Kreuer, J. Membrane Sci. **185** (2001) 29; M. Schuster, W. H. Meyer, G. Wegner, H. G. Herz, M. Ise, K. D. Kreuer, and J. Maier, Solid State Ionics **145** (2001) 85; H. G. Herz, K. D. Kreuer, J. Maier, G. Scharfenberger, M. F. H. Schuster and W. H. Meyer, Electrochim. Acta **48** (2003) 2165.
[671] Data from W. Gajewski, Spektrum der Wissenschaft **7** (1995) 88 and K. Sasaki, personal communication.
[672] B. B. Owens, J. E. Oxley, A. F. Sammells, in: *Solid Electrolytes*, S. Geller (ed.), p. 67, Springer–Verlag, Berlin, 1977.
[673] A. J. Hills and N. A. Hampson, J. Power Sources **24** (1988) 253.
[674] F. von Sturm, *Elektrochemische Stromerzeugung*, VCH, Weinheim, 1969.
[675] J. B. Goodenough, NATO ASI Series B **217** (1990) 157.

[676] P. G. Dickens, M. S. Whittingham, Quart. Rev. **22** (1968) 30; M. S. Whittingham, R. A. Huggins, J. Chem. Phys. **54** (1971) 414; M. S. Whittingham, J. Electrochem. Soc. **125** (1976) 315; M. S. Whittingham, Prog. Solid State Chem. **12** (1978) 41.
[677] J. B. Goodenough, in: *Lithium Ion Batteries*, M. Wakihara, O. Yamamoto (eds.), p. 1, VCH, Weinheim, 1998.
[678] S. z. B. P. M. S. Mouk, J. A. Duffy, M. D. Ingram, Electrochim. Acta **38** (1993) 2759.
[679] P. Hagenmüller, Prog. Solid State Chem. **5** (1971) 71.
[680] K. Mizushima, P. C. Jones, J. B. Goodenough, J. Electrochem. Soc. **132** (1985) 783.
[681] K. Mizushima, P. C. Jones, P. J. Wiseman, J. B. Goodenough, Mater. Res. Bull. **17** (1980) 785.
[682] E. Ceder, Y.–M. Chiang, D. R. Sadoway, M. K. Aydinol, Y.–J. Jang, B. Huang, Nature **392** (1998) 694.
[683] G. G. Amatucci, J. M. Tarascon, L. C. Klein, J. Electrochem. Soc. **143** (1996) 1114; T. Ohzuku, A. Ueda, M. Nagayama, J. Electrochem. Soc. **140** (1993) 1862; T. Ohzuku et al., J. Electrochem. Soc. **137** (1990) 769.
[684] H. Kawai, N. Nagata, H. Tsukamoto, A. R. West, J. Mater. Chem. **8** (1998) 837.
[685] N. Ravet, J. B. Goodenough, S. Besner, M. Simoneau, P. Hovington and M. Armand, in: Proceedings of the 196th ECS Meeting, Honolulu **99-2** (1999) Abstr. 127.
[686] S. Y. Chung, J. T. Bloking, Y. M. Chiang, Nature Materials **2** (2002) 123.
[687] N. Ravet, Y. Chouinard, J. F. Magnan, S. Besner, M. Gauthier, M. Armand, Journal of Power Sources **97-98** (2001) 503.
[688] P. Poizot, S. Laruelle, S. Grugeon, L. Dupont and J.-M. Tarascon, Nature **407** (2000) 496.
[689] H. Li, G. Richter, J. Maier, Adv. Mater. **15** (9) (2003) 736.
[690] P. Balaya, H. Li, L. Kienle and J. Maier, Adv. Funct. Mater. **13** (8) (2003) 621.
[691] L. F. Nazar et al., International Journal of Inorganic Materials **3**, 191-200 (2001).
[692] M. Armand, Solid State Ionics **69** (1994) 309; G. Feullade, P. Perche, J. Appl. Electrochem. **5** (1975) 63.
[693] J. H. Kennedy, in: *Solid Electrolytes*, S. Geller (ed.), p. 105, Springer–Verlag, Berlin, 1977.
[694] R. Selim, P. Bro, J. Electrochem. Soc. **121** (1974) 1457.
[695] A. van Zyl, Solid State Ionics **86/88** (1996) 883.
[696] J. Maier, in: Proc. Werkstoffwoche '96: *Werkstoff-Verfahrenstechnik*, G. Ziegler, H. Cherdron, W. Hermel, J. Hirsch, H. Kolaska (eds.), 3 pp., DGM Informationsgesellschaft mbH, 1997.
[697] J. Maier, *Einführung in die Physikalische Festkörperchemie*, Lecture Notes, Tübingen-Stuttgart, 1990.
[698] B. Scrosati, Nature **373** (1995) 557.

Index